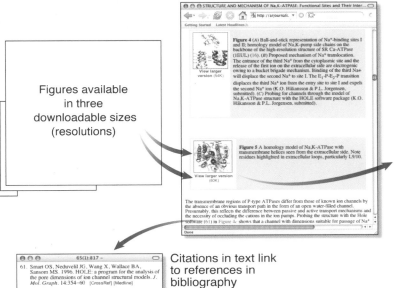

Figures available in three downloadable sizes (resolutions)

C0-ARF-888

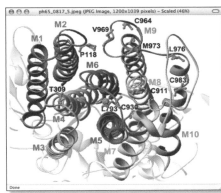

Citations in text link to references in bibliography

References in Annual Reviews article bibliography link out to sources of cited articles online

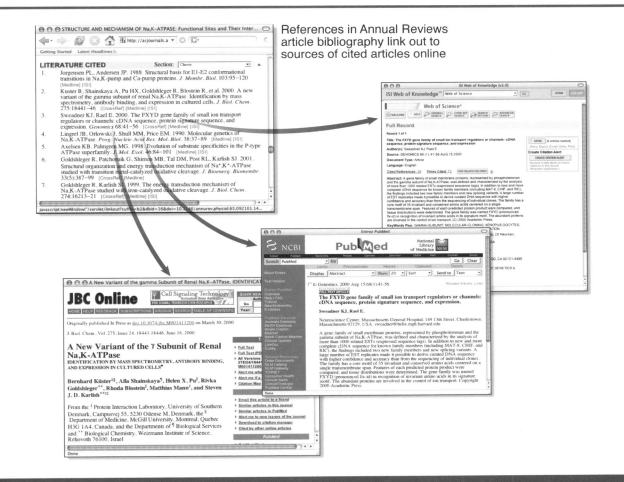

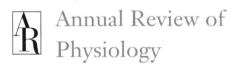

Annual Review of
Physiology

CABRINI COLLEGE LIBRARY
610 KING OF PRUSSIA ROAD
RADNOR, PA 19087

QP
1
.A535
v. 71
2009

#1481511

Editorial Committee (2009)

James M. Anderson, University of North Carolina, Chapel Hill
Richard C. Boucher, Jr., University of North Carolina, Chapel Hill
David E. Clapham, Harvard Medical School
Martin E. Feder, University of Chicago
Gerhard H. Giebisch, Yale University School of Medicine
Holly A. Ingraham, University of California, San Francisco
David Julius, University of California, San Francisco
Roger Nicoll, University of California, San Francisco
Jeffrey Robbins, Children's Hospital Research Foundation

Responsible for the Organization of Volume 71
(Editorial Committee, 2007)

James M. Anderson
Richard C. Boucher, Jr.
David E. Clapham
Jeffrey M. Drazen
Martin E. Feder
Gerhard H. Giebisch
Holly A. Ingraham
David Julius
Roger Nicoll
Jeffrey Robbins

Production Editor: Shirley S. Park
Managing Editor: Jennifer L. Jongsma
Bibliographic Quality Control: Mary A. Glass
Electronic Content Coordinator: Suzanne K. Moses
Illustration Editor: Glenda Lee Mahoney

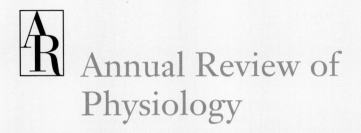

Annual Review of Physiology

Volume 71, 2009

David Julius, *Editor*
University of California, San Francisco

David E. Clapham, *Associate Editor*
Harvard Medical School

www.annualreviews.org • science@annualreviews.org • 650-493-4400

Annual Reviews
4139 El Camino Way • P.O. Box 10139 • Palo Alto, California 94303-0139

Annual Reviews
Palo Alto, California, USA

COPYRIGHT © 2009 BY ANNUAL REVIEWS, PALO ALTO, CALIFORNIA, USA. ALL RIGHTS RESERVED. The appearance of the code at the bottom of the first page of an article in this serial indicates the copyright owner's consent that copies of the article may be made for personal or internal use, or for the personal or internal use of specific clients. This consent is given on the condition that the copier pay the stated per-copy fee of $20.00 per article through the Copyright Clearance Center, Inc. (222 Rosewood Drive, Danvers, MA 01923) for copying beyond that permitted by Section 107 or 108 of the U.S. Copyright Law. The per-copy fee of $20.00 per article also applies to the copying, under the stated conditions, of articles published in any *Annual Review* serial before January 1, 1978. Individual readers, and nonprofit libraries acting for them, are permitted to make a single copy of an article without charge for use in research or teaching. This consent does not extend to other kinds of copying, such as copying for general distribution, for advertising or promotional purposes, for creating new collective works, or for resale. For such uses, written permission is required. Write to Permissions Dept., Annual Reviews, 4139 El Camino Way, P.O. Box 10139, Palo Alto, CA 94303-0139 USA.

International Standard Serial Number: 0066-4278
International Standard Book Number: 978-0-8243-0371-6
Library of Congress Catalog Card Number: 39-15404

All Annual Reviews and publication titles are registered trademarks of Annual Reviews.

⊗ The paper used in this publication meets the minimum requirements of American National Standards for Information Sciences—Permanence of Paper for Printed Library Materials, ANSI Z39.48-1992. The paper contains 20% postconsumer recycled content.

Annual Reviews and the Editors of its publications assume no responsibility for the statements expressed by the contributors to this *Annual Review*.

TYPESET BY APTARA
PRINTED AND BOUND BY SHERIDAN BOOKS, INC., CHELSEA, MICHIGAN

Preface

Welcome to Volume 71 of the *Annual Review of Physiology*, which features a rather large selection of topically diverse yet complementary reviews representing a wide range of issues and questions pertaining to the biochemical, cellular, organismal, and behavioral aspects of physiology—from unicellular eukaryotes on up!

Each year we feature a Special Topic section, which is meant to highlight an area of physiology that resonates with our readership from both a basic science and a clinical perspective. This year we focus on asthma, a problem of growing proportions that likely reflects a complex interplay between genetics and the environment. These articles review and discuss asthma from a variety of perspectives that touch on the involvement of inflammation and immunity, including the interplay between airway smooth muscle hypertrophy and the immune response. We are extremely grateful to Jeffrey Drazen, Professor of Environmental Health, Distinguished Parker B. Francis Professor of Medicine at Harvard Medical School, and Editor-in-Chief of *New England Journal of Medicine*, for organizing this very topical and fascinating section. The Special Topic section is nicely complemented by articles in the Respiratory Physiology section that review the involvement of histone deacetylase (HDAC2) or ion channels (amiloride-sensitive sodium or calcium-activated chloride channels) in lung function under normal and pathological conditions.

More by happenstance than design, this volume also features a wonderful collection of chemosensory articles that discuss current models and controversies concerning the molecules and neural pathways that govern the detection of, perception of, and behavioral response to olfactory and pheromonal cues. These contributions give a comprehensive and comparative overview of olfactory systems at both molecular and integrative levels as gleaned from genetic, electrophysiological, and anatomical studies of invertebrate and vertebrate organisms. The differences between and similarities of these chemosensory systems highlight specific signal detection, transduction, and coding mechanisms that have evolved to suit a particular niche or behavioral trait.

This volume also contains an unusually high preponderance of articles that have as their focus ion channels. Despite what some may think, the "channelophiles" on the editorial board did not arrange this intentionally—again, the result is a consequence more of happenstance than of design. Nonetheless, by whatever mechanism, this year's volume features a very fine collection of reviews that describe the workings and/or roles of cationic and anionic channels in the regulation of processes ranging from the cell cycle to synaptic plasticity to plant cell signaling and membrane transport.

Finally, there are other articles that do not fall into one of the above categories, but that nevertheless cover exciting and topical areas, including (but not limited to) the identification of stem cells that contribute to continual renewal of the intestinal organ system, transcriptional mechanisms that regulate mitochondrial biogenesis, and the role of calcium-sensing receptors in modulating fluid secretion and absorption by the gut.

This year we have taken a brief hiatus from our tradition of commissioning a Perspectives article, in which a scientist of note provides a personal account of the experiences and circumstances that shaped his or her life, career, and discoveries. Next year, however, we shall revive the Perspectives theme with a very exciting and unique contribution that involves a wonderful collaboration between a guest interviewer and a legendary figure in the world of developmental neurobiology. Stay tuned for this wonderful multimedia experience!

David Julius, Editor, for the *Annual Review of Physiology* Editorial Committee

Contents

Annual Review of
Physiology

Volume 71, 2009

SPECIAL TOPIC, ASTHMA, *Jeffrey M. Drazen, Special Topic Editor*

Indexes

Errata

An online log of corrections to *Annual Review of Physiology* articles may be found at
http://physiol.annualreviews.org/errata.shtml

Other Reviews of Interest to Physiologists

From the *Annual Review of Pathology: Mechanisms of Disease*, Volume 3 (2008)

Airway Smooth Muscle in Asthma
Marc B. Hershenson, Melanie Brown, Blanca Camoretti-Mercado, and Julian Solway

Molecular Pathobiology of Gastrointestinal Stromal Sarcomas
Christopher L. Corless and Michael C. Heinrich

From the *Annual Review of Pharmacology and Toxicology*, Volume 49 (2009)

The COXIB Experience: A Look in the Rearview Mirror
Lawrence J. Marnett

The TRPC Class of Ion Channels: A Critical Review of Their Roles in Slow,
Sustained Increases of Intracellular Ca^{2+} Concentrations
Lutz Birnbaumer

From the *Annual Review of Plant Biology*, Volume 59 (2008)

Plant Aquaporins: Membrane Channels with Multiple Integrated Functions
Christophe Maurel, Lionel Verdoucq, Doan-Trung Luu, and Véronique Santoni

Annual Reviews is a nonprofit scientific publisher established to promote the advancement of the sciences. Beginning in 1932 with the *Annual Review of Biochemistry*, the Company has pursued as its principal function the publication of high-quality, reasonably priced *Annual Review* volumes. The volumes are organized by Editors and Editorial Committees who invite qualified authors to contribute critical articles reviewing significant developments within each major discipline. The Editor-in-Chief invites those interested in serving as future Editorial Committee members to communicate directly with him. Annual Reviews is administered by a Board of Directors, whose members serve without compensation.

2009 Board of Directors, Annual Reviews

Richard N. Zare, *Chairman of Annual Reviews, Marguerite Blake Wilbur Professor of Chemistry, Stanford University*

John I. Brauman, *J.G. Jackson–C.J. Wood Professor of Chemistry, Stanford University*

Peter F. Carpenter, *Founder, Mission and Values Institute, Atherton, California*

Karen S. Cook, *Chair of Department of Sociology and Ray Lyman Wilbur Professor of Sociology, Stanford University*

Sandra M. Faber, *Professor of Astronomy and Astronomer at Lick Observatory, University of California at Santa Cruz*

Susan T. Fiske, *Professor of Psychology, Princeton University*

Eugene Garfield, *Publisher, The Scientist*

Samuel Gubins, *President and Editor-in-Chief, Annual Reviews*

Steven E. Hyman, *Provost, Harvard University*

Sharon R. Long, *Professor of Biological Sciences, Stanford University*

J. Boyce Nute, *Palo Alto, California*

Michael E. Peskin, *Professor of Theoretical Physics, Stanford Linear Accelerator Center*

Harriet A. Zuckerman, *Vice President, The Andrew W. Mellon Foundation*

Management of Annual Reviews

Samuel Gubins, President and Editor-in-Chief
Richard L. Burke, Director for Production
Paul J. Calvi Jr., Director of Information Technology
Steven J. Castro, Chief Financial Officer and Director of Marketing & Sales
Jeanne M. Kunz, Human Resources Manager and Secretary to the Board

Annual Reviews of

Analytical Chemistry
Anthropology
Astronomy and Astrophysics
Biochemistry
Biomedical Engineering
Biophysics
Cell and Developmental Biology
Clinical Psychology
Earth and Planetary Sciences
Ecology, Evolution, and
   Systematics
Economics
Entomology
Environment and Resources
Financial Economics

Fluid Mechanics
Genetics
Genomics and Human Genetics
Immunology
Law and Social Science
Marine Science
Materials Research
Medicine
Microbiology
Neuroscience
Nuclear and Particle Science
Nutrition
Pathology: Mechanisms
   of Disease
Pharmacology and Toxicology

Physical Chemistry
Physiology
Phytopathology
Plant Biology
Political Science
Psychology
Public Health
Resource Economics
Sociology

SPECIAL PUBLICATIONS
Excitement and Fascination of
   Science, Vols. 1, 2, 3, and 4

Sex-Based Cardiac Physiology

Elizabeth D. Luczak and Leslie A. Leinwand

Department of Molecular, Cellular, and Developmental Biology,
University of Colorado, Boulder, Colorado 80309;
email: elizabeth.head@colorado.edu, leslie.leinwand@colorado.edu

Annu. Rev. Physiol. 2009. 71:1–18

First published online as a Review in Advance on
October 2, 2008

The *Annual Review of Physiology* is online at
physiol.annualreviews.org

This article's doi:
10.1146/annurev.physiol.010908.163156

Copyright © 2009 by Annual Reviews.
All rights reserved

0066-4278/09/0315-0001$20.00

Key Words

estrogen, phytoestrogen, testosterone, gender

Abstract

Biological sex plays an important role in normal cardiac physiology
as well as in the heart's response to cardiac disease. Women generally have better cardiac function and survival than do men in the face
of cardiac disease; however, this sex difference is lost when comparing postmenopausal women with age-matched men. Animal models of
cardiac disease mirror what is seen in humans. Sex steroid hormones
contribute significantly to sex-based differences in cardiac disease outcomes. Estrogen is generally considered to be cardioprotective, whereas
testosterone is thought to be detrimental to heart function. Environmental estrogen-like molecules, such as phytoestrogens, can also affect
cardiac physiology in both a positive and a negative manner.

INTRODUCTION

Traditionally, cardiovascular studies in both humans and animals have been carried out in males. However, it is becoming increasingly clear that biological sex plays a significant role in the function of the healthy heart, as well as in the outcome of cardiovascular disease. Females tend to have improved cardiac function compared with males in a variety of settings; however, this sex difference is largely lost when comparing postmenopausal women with age-matched men. This has led to the hypothesis that estrogen is cardioprotective, and provides a potential mechanism for sex-specific differences in the heart.

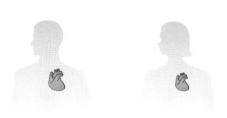

	Men	Premenopausal women	Postmenopausal women
Cardiac myocyte size	++	+	+
Systolic function	+	+	+
Diastolic function	+	++	+
Estrogen	+	10+	+
Testosterone	10+	+	+
ERα	+	+	+
ERβ	++	+	+

Figure 1

Summary of sex differences in the heart. Cardiac size and function differ between men and women. Sex hormone levels are largely different between men and women, but both estrogen and testosterone are found in both sexes. Additionally, androgen and estrogen receptors are found in the hearts of men and women at comparable levels. + denotes relative size/function/levels.

HUMAN HEARTS

Normal Structure and Function

Before puberty no significant differences in heart size or body size are seen between males and females. This suggests that males and females are born with the same number of cardiac myocytes and that these cells are the same size in males and females (1). Postnatal growth of the heart occurs primarily by an increase in cardiac myocyte size, or hypertrophy, and it is widely accepted that the adult heart is a postmitotic tissue because myocytes lose their ability to proliferate shortly after birth (2). After puberty, heart mass is 15–30% larger in men than in women. This growth is proportional to body size. Given that males and females have equivalent numbers of myocytes at infancy, male myocytes must undergo a greater degree of hypertrophy than do female myocytes. This difference in myocyte hypertrophy is proportionally symmetrical throughout the heart and results in no sex difference in relative wall thickness of the heart (1).

Aging leads to an increase in both septal and wall thickness in both males and females. However, left ventricular diameter increases with aging only in males (3) and results in a loss of myocardial mass in men, but not in women. The loss of myocardium in older men is attributed to a reduced number of myocytes. Between the ages of 17 and 89, men lose on average 1 gram per year of cardiac mass, which is equivalent to approximately 64 million cardiac myocytes. Myocyte loss also leads to a compensatory myocyte hypertrophy in men to maintain adequate heart mass. In women, myocyte number and size are preserved with aging (4).

Considering heart function, young women have better diastolic function compared with men. However, both men and women have decreasing diastolic function with aging (3). Additionally, men have decreased systolic function with aging, whereas women do not (**Figure 1**).

Studies of exercise performance and cardiac adaptation in healthy men and women have described varying results. Increases in cardiac output during exercise occur but are not different between men and women (5); however, different mechanisms to increase cardiac output have been seen (6). These results suggest that the mechanisms for cardiac adaptation to exercise are inherently different in male and female hearts.

Disease

Clinical studies have revealed sex differences in cardiovascular disease outcomes. In general, premenopausal women fare better than men in most cardiovascular-related diseases. For example, premenopausal women have a better prognosis than do men in response to hypertension, aortic stenosis, and hypertrophic cardiomyopathy (HCM). Cardiac function is preserved in women, whereas men experience poor cardiac contractility owing to chamber dilation and wall thinning (7, 8). Women with congestive heart failure have also been shown to survive better than men in some studies (9, 10). Another study, however, reports no difference in congestive heart failure prognosis for men and women (11). Women also have a better long-term prognosis following an ischemic event, even though their immediate death rate is elevated compared with men (12, 13).

Sex-related differences in cardiovascular disease also appear to be affected by age. The National Center for Health Statistics reports that more men than women die from cardiovascular disease between ages 45 and 64, but after age 65 the death rate due to cardiovascular disease is greater in women than in men. Additionally, this study reports a greater prevalence of congestive heart failure in men than in women ages 65–74 and no difference between men and women after age 74 (14).

The discrepancies in the outcomes of clinical studies may be attributed to the fact that women have been underrepresented in clinical trials (15, 16). This underrepresentation may be because women can be difficult to enroll in clinical studies. Women often do not meet enrollment criteria because they display pre-served systolic function, whereas men display systolic dysfunction (17). Additionally, many studies have age-based exclusion criteria and will not enroll older individuals (18). Women generally do not develop cardiovascular disease until an older age and are thus excluded from these studies.

Women, however, do not always fare better than men with respect to cardiovascular disease. In cases of idiopathic dilated cardiomyopathy, women have a poorer prognosis than do men (19). Additionally, women are more sensitive than men to alcohol-induced cardiac disease (20). In these cases, estrogen may be detrimental in disease progression or may not contribute at all.

SEX DIFFERENCES IN ANIMAL MODELS

Traditionally, researchers study only one sex or do not distinguish between the sexes in studies involving laboratory animals. However, it is becoming increasingly clear that the sex differences that are seen in humans extend to animals as well. Animal models provide a valuable tool for studying the mechanisms of sexual dimorphisms in healthy as well as diseased hearts.

Exercise

Exercise elicits a sexually dimorphic response in animals. Treadmill training results in cardiac hypertrophy in both male and female rats. However, only the hearts of male rats develop improved performance (21). In contrast, swim training in rats results in an increased hypertrophic response in females compared with males (22, 23). The hypertrophy in females is associated with increased tension development of the heart (24).

Mice also display a sexually dimorphic response to exercise. Female mice have an increased capacity to exercise in both voluntary wheel and treadmill paradigms. Female mice run more on a cage wheel than males independently of strain and age. Moreover, females perform better in a treadmill-based endurance test and a stress test, which is an indicator

HCM: hypertrophic cardiomyopathy

TAC: thoracic aortic constriction

MI: myocardial infarction

SHHF: spontaneous hypertensive heart failure

MyHC: myosin heavy chain

cTnT: cardiac troponin T

TNF-α: tumor necrosis factor-α

of increased cardiovascular performance. Cage wheel running induces significant cardiac hypertrophy in both male and female mice, but females have a greater percent increase in heart weight compared with males (25).

Disease

As seen in studies of human cardiovascular disease, in animal models of pathological hypertrophy females generally fare better than males. For example, in a pressure overload model in rats, males develop dilation, diastolic dysfunction, and elevated wall stress 20 weeks after thoracic aortic constriction (TAC). Females develop elevated systolic pressure but do not progress to heart failure (8). A similar outcome is seen in models of myocardial infarction (MI) in both rats and mice. Progression to heart failure is attenuated in female compared with male rats following MI (26, 27). In a similar model in mice, male hearts rupture more readily, have poorer left ventricular function, and have enhanced dilation and hypertrophy compared with females (28).

Females also fare better than males in hypertensive rat models. Salt-sensitive Dahl rats develop hypertension and significant hypertrophy when fed a high-salt diet. Following MI, both male and female Dahl rats develop significant hypertrophy compared with controls. However, male rats undergo dilation, whereas female rats experience no chamber dilation and have elevated contractile function compared with males (29, 30). The spontaneously hypertensive heart failure (SHHF) rat is another well-studied model in which rats develop hypertension spontaneously and are either homo- or heterozygous for an obesity gene. Both male and female SHHF rats develop heart failure independently of the obesity gene (31, 32). Progression to heart failure is accelerated in males compared with females in lean rats, with males displaying signs of heart failure by 16 months versus 22 months in females (31, 33). Likewise, obese male rats develop heart failure before their female counterparts, although the onset of failure is earlier in both sexes (34).

Sex differences have also been documented in mouse models of genetic heart disease such as HCM. Transgenic mice with a mutant myosin heavy chain (MyHC) gene display sex differences (35–37). Both sexes display significant ventricular hypertrophy, but males have more pronounced histological disease features and altered electrophysiological parameters (36, 38). Additionally, males develop progressive ventricular dilation and contractile dysfunction, whereas females maintain a hypertrophic state and have preserved contractile function (35, 37).

Mutations in cardiac troponin T (cTnT) contributing to HCM also display sex-dependent phenotypes. An R92Q missense mutation results in smaller ventricles in males, but not in females (39, 40). Another cTnT mutation that results in a carboxy-terminal truncation displays no sex difference at baseline. Both males and females with the truncated cTnT mutation have smaller hearts as well as diastolic and systolic dysfunction (41). However, treatment of the truncated cTnT mutant mice with angiotensin II causes hypertrophy in females, but not males. Treatment of both cTnT mutants, truncated and R92Q, with isoproterenol and phenylephrine simultaneously leads to sudden cardiac death in all the males, whereas all the females survive (39). These data suggest an underlying sex-based difference in the heart's response to adrenergic stimulation.

Signaling Pathways

Mechanisms for how sex-based differences occur have not been well described. Transgenic mouse models with altered signal transduction pathways can begin to aid in identifying important pathways. Sex-based differences have been described in several transgenic mouse models. However, there is potential for more differences to be identified because many of the phenotypes of these mice are not described in both sexes.

TNF-α. Circulating tumor necrosis factor-α (TNF-α), a proinflammatory cytokine with pleiotropic biological effects, is elevated in patients with congestive heart failure (42), and its levels may be correlated with the severity of the disease (43). Human myocardium produces TNF-α (44) and expresses TNF-α receptor (45), suggesting a role for TNF-α in the heart.

Mice overexpressing TNF-α in the heart display ventricular hypertrophy, dilation, and fibrosis and have increased expression of atrial natriuretic factor (ANF) (46). Females have left ventricular thickening without dilation, whereas males progress to dilation. Both sexes have equivalent expression of TNF-α and other proinflammatory cytokines; however, the sex difference appears to stem from increased TNF receptor mRNA expression in males only (47).

Adrenergic signaling. Chronic stimulation of neurohormonal pathways elicits pathological remodeling of the heart (48). Sex differences in adrenergic receptor signaling, as well as downstream effectors, have been observed in mice (49–52). These studies indicate that males have a reduced ability to contend with increased adrenergic drive.

Cardiac-specific β_2-adrenergic receptor overexpression results in decreased survival in males compared with females, with 13% and 56% survival rates, respectively, at 15 months of age. Additionally, males develop left ventricular dilation and reduced contractile function, symptoms of heart failure, increased myocyte size, and fibrosis. All these features are less marked in female transgenic mice (49).

In mice doubly null for $\alpha_{1A/C}$- and α_{1B}-adrenergic receptors, males have small hearts owing to smaller myocyte size, whereas females have normal heart size. In addition, males also have decreased exercise capacity and increased mortality following pressure overload (50), suggesting a reduced adaptive response to cardiac stress in the male $\alpha_{1A/C}$- and α_{1B}-adrenergic receptor double-null mice.

Transgenic mice with a fourfold overexpression of phospholamban (51) display a sex difference in the onset of cardiac dysfunction and heart failure. Males develop ventricular hypertrophy and show increased mortality at 15 months of age, whereas this phenotype in females is delayed until 22 months of age. Yet males and females express similar levels of phospholamban and sarcoplasmic reticulum Ca^{2+}-ATPase 2 (SERCA2) (52). Mice overexpressing the superinhibitory $Val^{49} \rightarrow Gly$ mutant of phospholamban display a similar sexually dimorphic response. Males show depressed contractile parameters, left ventricular hypertrophy and dilation, fetal gene expression, fibrosis, and death by six months of age. Females display hypertrophy but have normal function up to 12 months of age (53).

PPARα. Cellular lipid homeostasis is critical for normal heart function because the heart is a highly aerobic tissue that requires fatty acids for fuel. Peroxisome proliferator–activated receptor α (PPARα), a member of the nuclear receptor superfamily of transcription factors that targets genes involved in fatty-acid metabolism, is important for maintaining lipid homeostasis in the heart. Mice lacking PPARα have also revealed a sex difference in the heart's response to altered lipid metabolism. Inhibition of cellular fatty-acid flux in PPARα null mice caused cardiac lipid accumulation, leading to death in 100% of males and only 25% of females (54). Additionally, overexpression of lipoprotein lipase, an enzyme that liberates free fatty acids from lipoproteins, in PPARα null mice also shows a sex-based difference. This genetic combination results in premature death in males with 55% mortality by 4 months of age, and the remaining males die by 11 months. Females, however, survive beyond 12 months of age (55).

Relaxin. Relaxin is an insulin-like peptide hormone that was originally described to be an important factor in tissue remodeling during female reproduction (56). More recently, relaxin has been shown to act on the cardiovascular system in both males and females. Both relaxin mRNA and the relaxin receptor are expressed in the heart (57). Functionally, relaxin increases the rate and force of cardiac contractions (58). Mice null for the relaxin gene display a sexually dimorphic phenotype. Male

ANF: atrial natriuretic factor

SERCA2: sarcoplasmic reticulum Ca^{2+}-ATPase 2

PPARα: peroxisome proliferator–activated receptor α

GSK-3β: glycogen synthase kinase-3β

ER: estrogen receptor

ERE: estrogen-responsive element

PLC: phospholipase C

IGFR: insulin-like growth factor receptor

relaxin null mice have impaired diastolic filling and increased atrial weights owing to increased fibrosis of the ventricle. Age-matched females display none of these characteristics (59).

GSK-3β. Glycogen synthase kinase-3β (GSK-3β) acts as an antihypertrophic signaling molecule in response to pathological stimuli. Transgenic mice expressing constitutively active GSK-3β have reduced cardiac growth in response to several pathological stimuli including β-adrenergic stimulation, pressure overload (60), and an α-MyHC mutation (61). Mice doubly transgenic for constitutively active GSK-3β and an α-MyHC mutation display a sexually dimorphic phenotype. Males have increased mortality, with approximately 75% mortality by 18 months, whereas female mice all have equivalent survival independent of genotype (61). Doubly transgenic males and females have equivalent cardiac function; however, in this model (61) females appear to have an increased capacity to cope with the lack of hypertrophic growth.

SEX HORMONES

Sex steroid hormones have been hypothesized to contribute to sex-based differences in the heart. Women have a lower incidence of cardiac disease and better prognosis when faced with disease than do men. Additionally, the disappearance of sex differences with aging is correlated with the loss of estrogen in postmenopausal women. This has led to the hypothesis that estrogen is cardioprotective and that testosterone may be detrimental to cardiac function.

Both estrogen receptors (ERs) and androgen receptors are expressed in both male and female hearts (62, 63), supporting a role for estrogen and testosterone in cardiac physiology (**Figure 1**). Additionally, the removal of sex hormones by gonadectomy in rodents has provided further evidence for their role in cardiac physiology. Gonadectomy in both rats and mice results in decreased heart weight compared with sham-operated controls (35, 64). Moreover, a

study of gonadectomized rats showed decreased cardiac function and a shift in MyHC isoform expression to the β isoform, which is indicative of a pathological state (64). Another study showed that gonadectomy plus supplementation with a sex-inappropriate hormone caused the hearts of α-MyHC mutant mice to be most similar in size to the opposite-sex sham-operated animals (35). These data suggest that sex hormones are a major determinant of heart size.

Estrogen

The classical, or genomic, view of estrogen action occurs via nuclear ERs. Two ERs, ERα and ERβ, are expressed from separate genes and have distinct tissue distribution. 17β-Estradiol, the major circulating estrogen, binds equally to both receptors in the nucleus. The estrogen-ER complex acts as a transcription factor by directly binding a specific DNA sequence, or estrogen-responsive element (ERE). Alternatively, estrogen can act on transcription via indirect tethering with other transcriptional activators (65).

ERs can also function through nongenomic mechanisms involving the activation of protein kinase cascades in the cell. This occurs via ERs that are located in or adjacent to the plasma membrane or through non-ER plasma membrane–associated estrogen-binding proteins. Nongenomic ER signaling can induce signaling pathways involved in cardiac hypertrophy, including phospholipase C (PLC), G protein, and MAP kinase pathways, as well as Ca^{2+} flux and inositol triphosphate (IP_3) generation. Additionally, ERs can utilize other growth factor receptors, such as insulin-like growth factor receptor (IGFR), to enhance the activation of downstream targets. This type of response occurs quickly, within minutes of estrogen stimulation of cells (65) (**Figure 2**).

Expression of ERs in the heart certainly implies a function for estrogen in the heart. ERα expression levels are equivalent in the hearts of men and women, whereas ERβ expression levels are higher in the hearts of men than in those of women (63). Both ERα and

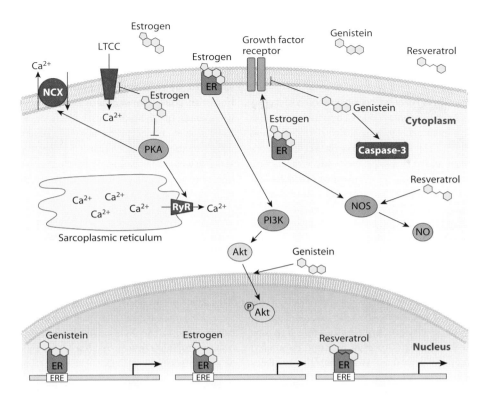

Figure 2

Mechanisms of estrogen and phytoestrogen signaling in the heart. Estrogen can act via the estrogen receptor (ER) and directly activate the transcription of genes containing an estrogen-responsive element (ERE) in the nucleus of cardiac myocytes. Additionally, estrogen influences other signaling pathways in the heart including Ca^{2+} flux, growth factor receptor signaling, and nitric oxide (NO) production. Phytoestrogens can also activate transcription similarly to estrogen via the ER. In addition, phytoestrogens can have effects similar to estrogen in other pathways such as Akt activation and nitric oxide synthase (NOS) production. However, phytoestrogens can have opposing effects to estrogen such as the inhibition of growth factor receptor signaling and the induction of caspases. Other abbreviations used: LTCC, L-type calcium channel; NCX, sodium-calcium exchange; PI3K, phosphoinositol 3-kinase; PKA, protein kinase A; RyR, ryanodine receptor.

ERβ are upregulated in both males and females during aortic stenosis (63). In addition, ERα mRNA is upregulated in end-stage dilated cardiomyopathy in both males and females (66). Moreover, both men and women produce estrogen. Men have approximately 10–20-fold-lower circulating estrogen levels than do premenopausal women at the highest point of their cycle (67). Postmenopausal women have approximately the same estrogen levels as do men.

The generation of ER null mice has provided additional evidence that ER activity is required in both sexes, most notably with regard to fertility. Both male and female ERα null mice are completely sterile. ERα null females have immature uterine development and improper pituitary function (68), whereas ERα null males lose sperm owing to progressive testicular deterioration (69). Fertility is also affected in ERβ null females, which have suboptimal pregnancies, but male fertility is not affected (69). These studies provide evidence for functional differences between the two ER receptors. Additionally, the role of estrogen in males is unclear, but

iNOS: inducible nitric oxide synthase

eNOS: endothelial nitric oxide synthase

NRVM: neonatal rat ventricular myocyte

that ER null males are phenotypically affected provides evidence for a functional role for estrogen in males.

Studies in animal models have also shown that ERs and estrogen are responsible for some of the cardioprotection observed in females. Cardioprotection against ischemia/reperfusion injury in female mice requires both ERα and ERβ (70, 71). ERβ mediates the sex difference in response to pressure overload: Females fare better than males (72). Furthermore, ERβ attenuates the transition to heart failure seen in males following MI (73).

The loss of estrogen during menopause is associated with increased risk for cardiovascular disease. Animal models of ovariectomy mimic the loss of estrogen in women, providing a model to study the changes that occur postmenopausally. In a rat model of chronic volume overload, postpubertally ovariectomized females developed significantly more left ventricular hypertrophy and dilation compared with intact controls. Moreover, the ovariectomized females displayed symptoms of heart failure, including pulmonary edema and decreased left ventricular function, comparable to those reported for males (74). In pressure overload models, ovariectomized females attained significantly more cardiac hypertrophy than did intact or estrogen-supplemented females (75, 76). In addition, ovariectomized rats showed deteriorated heart function and increased mortality following treatment with a β-adrenergic agonist (75).

Hormone Replacement Therapy

The strong association of increased mortality and morbidity from cardiovascular disease in postmenopausal women suggests a protective role for estrogen in the heart (9, 10, 77). Many observational studies of postmenopausal women receiving hormone replacement therapy showed an approximately 40% reduction in cardiovascular events over the course of the study (78). The cardioprotective role of estrogen, however, was challenged by the findings of the Women's Health Initiative study, as well

as a recent study published in the *British Medical Journal* (79, 80). In these studies, hormone-replacement therapy increased cardiovascular events in postmenopausal women (79, 80). One caveat to the results of these studies is that the hormone replacement was not initiated until the women in the studies were many years past menopause. The average age of women enrolled in the Women's Health Initiative study was 63, whereas women involved in one of the observational studies with a positive outcome were on average 51 years of age (81).

Mechanism of Estrogen Action in the Heart

The mechanism of cardioprotection via estrogen still remains unclear. However, researchers have described several roles for estrogen in the heart that include both genomic and nongenomic actions. It is likely that all these mechanisms contribute to the cardioprotective effects of estrogen.

Nitric oxide. Nitric oxide (NO) production plays an important role in regulating cardiac muscle function. Estrogen stimulates expression of inducible nitric oxide synthase (iNOS) and endothelial nitric oxide synthase (eNOS) in neonatal rat ventricular myocytes (NRVMs) and adult rat myocytes. This stimulation of expression can be blocked by the ER antagonist ICI 182,780. Additionally, iNOS and eNOS levels are reduced in ovariectomized female rats (82). Estrogen stimulates expression of eNOS mRNA in endothelial cells but does not directly bind promoter DNA. Instead, it increases activity of the Sp1 transcription factor, which is essential for eNOS transcription (83).

Calcium handling. Ca^{2+} is essential for relaying the electrical stimulus for contraction to the myofilament and directly activates contraction of the heart muscle. Alteration of Ca^{2+} handling in the heart alters its contractile properties. Defects in Ca^{2+} handling are correlated with heart failure. ER null cardiac myocytes have increased expression of the L-type calcium channel (LTCC), the inward Ca^{2+} channel in the plasma membrane, as well as inward Ca^{2+} current through the channel. ER null mice also have a prolonged QT interval, which may lead to increased risk of arrhythmia and cardiovascular disease (84). Ovariectomy also increased expression of LTCC in rat heart, which was normalized by estrogen supplementation (85). Estrogen treatment of guinea pig cardiac myocytes reduces inward Ca^{2+} current and intracellular Ca^{2+} concentration. This suggests that estrogen inhibits Ca^{2+} channels in the plasma membrane (86).

In addition to modulating inward Ca^{2+} current, ovariectomy in rats also increased Ca^{2+} flux in the heart via the ryanodine receptor (RyR), the Ca^{2+} leak channel in the sarcoplasmic reticular membrane, and the Na^+/Ca^{2+} exchanger (NCX), a Ca^{2+} extruder in the plasma membrane. Both of these changes are reversed by estrogen replacement or by protein kinase A (PKA) inhibition. Ovariectomy also increased PKA expression in the rat heart, and the hyperactivity of RyR and NCX can be attributed to the upregulation of PKA (87).

Cell survival. Cell death may contribute to the progression to decompensated cardiac hypertrophy, dilation, and ultimately heart failure. Survival of cardiac myocytes in the face of injury or disease is critical for maintaining adequate function of the heart. Estrogen promotes cell survival. Estrogen supplementation in ovariectomized mice reduces cardiac myocyte apoptosis following MI compared with ovariectomized controls. Estrogen treatment of isolated NRVMs also reduces apoptosis following daunorubicin treatment (88).

Estrogen's ability to promote cell survival is dependent on phosphoinositol 3-kinase (PI3K) and Akt. Akt is activated in vivo as well as in NRVMs following treatment with estrogen (88). Active, or phosphorylated, Akt induces cell survival of cardiomyocytes when it is localized to the nucleus (89). Young women have higher levels of nucleus-localized phospho-Akt than do age-matched men or postmenopausal women. This is also true in mouse hearts, where both nuclear localization of phospho-Akt and Akt activity are elevated in females compared with males. Additionally, cultured NRVMs treated with estrogen also have nucleus-localized phospho-Akt (90).

Testosterone

Like estrogen, testosterone action occurs via nuclear receptors. Upon testosterone binding, the androgen receptor binds to specific DNA sequences and can modulate transcription. Androgen receptor expression occurs in the hearts of both males and females in multiple species, including humans (62). Moreover, testosterone is produced by both men and women. Adult men have approximately tenfold-higher levels of circulating testosterone than do females (67). This suggests that testosterone functions in the hearts of both males and females.

Testosterone is far less studied than estrogen in terms of its effect on cardiac physiology. Nevertheless, studies suggest that testosterone has a negative impact on the heart. Treatment of NRVMs with testosterone elicits a hypertrophic response including expression of the pathological marker ANF (62). A study conducted using a mouse model of MI showed that testosterone and estrogen have opposing effects on cardiac remodeling and function (91). In gonadectomized animals, testosterone supplementation resulted in decreased cardiac function and increased myocyte cross-sectional area in both males and females, whereas estrogen had the opposite effect. Additionally, ovaries produce significant amounts of testosterone postmenopausally (92). This provides a second mechanism for the increased risk of cardiovascular disease in postmenopausal women.

Studies of athletes using anabolic steroids have provided further evidence for the detrimental effects of testosterone on cardiac physiology. Anabolic steroids are chemically modified derivatives of testosterone that are less susceptible to metabolic inactivation. Weightlifters using anabolic steroids have increased cardiac hypertrophy (93), as well as

LTCC: L-type calcium channel

RyR: ryanodine receptor

NCX: sodium-calcium exchanger

PKA: protein kinase A

PI3K: phosphoinositol 3-kinase

impaired diastolic function (94). Moreover, anabolic steroid use has been associated with an increased risk for MI and sudden cardiac death (95).

Both male and female mice null for aromatase, the enzyme that converts testosterone to estrogen, have nearly undetectable levels of estrogen and increased levels of testosterone. The phenotypes of these mice differ from those of the ERα null mice, suggesting that increased testosterone is also contributing to their phenotype (96). Although no cardiac-specific phenotype has been described, aromatase null mice have reduced blood pressure, reduced heart rate, and impaired baroreceptor reflex (97). These cardiac-related phenotypes, and the known effects of testosterone on the heart, suggest that these mice also display cardiac-specific phenotypes that are yet to be identified.

DIETARY PHYTOESTROGENS

Recently, diet has also been identified as a source of supplementary estrogen. Many plant-based foods contain highly bioactive compounds that can mimic endogenous hormones to activate signaling cascades in many tissues, including the heart (**Figure 2**). Plant estrogens, or phytoestrogens, have both beneficial and detrimental effects in humans and animals. The mechanism of action is not clear in all cases because phytoestrogens can also function independently of the estrogen receptor pathways.

Soy

Estrogenic compounds that mimic the actions of endogenous estrogen are abundant in plant-based food sources, especially soy-based foods. Soy phytoestrogens fall into the class of isoflavones, which are structurally similar to 17β-estradiol (**Figure 2**). The most abundant isoflavones found in soy, genistein, and daidzein and its metabolite, equol, bind to both ERs but have a higher affinity for ERβ (98). The relative potency of phytoestrogens in activating transcription of estrogen-responsive genes

is 1000- to 10,000-fold less compared with 17β-estradiol (99).

Genistein possesses estrogen-like activity in a variety of cell types and tissues. It activates an ERE-luciferase reporter construct in ER-positive MCF-7 breast cancer cells and NIH-3T3 fibroblasts transfected with either ERα or ERβ (100). Additionally, genistein injections activate an ERE-eGFP reporter in aromatase null mice (101). Genistein treatment also promotes nuclear localization of phospho-Akt in NRVMs. Nucleus-localized phospho-Akt is also enhanced in neonatal mice exposed to genistein (90).

Genistein elicits ER-independent effects in addition to activating ER activity. Genistein is widely used as an inhibitor of tyrosine kinases in the laboratory setting. It inhibits the activation of Akt, a prohypertrophic signaling molecule (102). Genistein also induces apoptosis through caspase-3 activation (103) and a reduction in mitochondrial membrane potential (104). One study demonstrated that an ER antagonist only partially blocks the negative effects of genistein on thymus size (105). This finding provides evidence that the ER-independent effects of genistein may be physiologically relevant in contributing to disease.

Gene expression profiling of adipose tissue from mice treated with genistein or estrogen revealed largely nonoverlapping expression profiles (106). However, another study identified genistein as having a gene expression signature highly similar to that of estrogen in the ER-positive MCF-7 cell line (107). These studies demonstrate that genistein has the ability to act like estrogen but also possesses unique functions that distinguish it from estrogen.

The perception that soy-based foods benefit health has, in part, increased consumption of these foods. Some studies have suggested that soy phytoestrogens have a beneficial impact on blood lipid profiles, osteoporosis, and breast cancer and lessen menopausal symptoms. Recently, however, the American Heart Association reversed its endorsement of soy-containing foods. This scientific advisory found no significant effect of soy proteins or phytoestrogens on serum lipids, blood pressure, or menopausal symptoms (108). A recent study contradicts these results, suggesting that phytoestrogens may have a beneficial effect on heart function. In this study of a Japanese cohort, high intake of dietary phytoestrogens was associated with decreased risk of MI in women that was more pronounced in postmenopausal women, with no change in risk for men (109).

Detrimental effects of high levels of dietary phytoestrogens have been documented in humans and animals, especially with regard to fertility. For example, male rats have decreased sperm production (110), and female mice produce fewer offspring (111) when exposed to high levels of phytoestrogens, similar to those found in soy-based rodent chow. Sheep infertility syndrome was linked to clover that had very high levels of phytoestrogens (112), whereas quail have evolved to eat high-phytoestrogen-containing plants in times of drought to lower reproduction (113). Additionally, cheetahs in captivity were infertile until soy was reduced in the diet (114).

Recently, our group found that the traditional soy-based chow worsens the phenotype in male mice with an α-MyHC mutation (35). We found that male mice no longer progress to dilation and heart failure when raised on a diet derived from milk proteins. This difference between diets is not seen in female mice.

Of interest to the scientific community is that most laboratory rodents are maintained on a soy-based chow that contains high levels of phytoestrogens. Serum levels of phytoestrogens in mice eating soy-based commercial chow have been reported to reach 8.5 μM (115), which can exceed the endogenous estrogen concentration by 30,000- to 60,000-fold (116).

Most human diets do not result in phytoestrogen levels as high as those in laboratory animals, even when humans eat a soy-rich diet. For example, a study of Japanese men who eat a diet rich in soy-based foods reached serum levels of phytoestrogens of only 0.16–0.89 μM (117). However, there are a few circumstances in which high levels of dietary phytoestrogens lead to circulating levels similar to those seen in laboratory rodents. Consumption of a single soy-rich meal can result in peak phytoestrogen levels of up to 4.4 μM (118, 119). Additionally, women consuming three meals per day containing soy milk or people who consume a 100 mg supplement of soy phytoestrogens daily can have serum levels of phytoestrogens of up to 4 μM (120, 121). Of greatest concern are the approximately 25% of infants in the United States fed soy-based formula. Serum concentrations of phytoestrogens can reach up to 4.4 μM, which is 200 times that of infants fed cow or breast milk and 10 times higher than that of adults eating a soy-rich diet (122).

Resveratrol

Resveratrol is a naturally occurring polyphenol found in a variety of plants, including grapes, blueberries, and peanuts. Recently, resveratrol has gained attention owing to its ability to extend the life spans of various organisms such as yeast (123); *Caenorhabditis elegans*; *Drosophila*; and *Nothobranchius furzeri* (124), a short-lived species of fish. Resveratrol slows or prevents the progression of cancer, cardiovascular disease, and ischemic injury (125).

Many biological activities have been attributed to resveratrol, including the modulation of lipid metabolism, anti-inflammatory activity, vasorelaxation, free-radical scavenging, and estrogenic activity (126). Several of these activities are associated with improvement in cardiovascular function or are important in the maintenance of function following a cardiac insult.

Resveratrol possesses estrogen-like activities. Unlike genistein, resveratrol binds equally to both ERs, but with 7000-fold-lower affinity compared with estrogen (127). It activates an ERE-luciferase reporter in CHO-K1 cells transfected with either ERα or ERβ (128). One study described an increase in NOS induction with resveratrol treatment in aortic-banded rats (129). Another study showed that the consumption of resveratrol results in an upregulation of iNOS and eNOS in rats following ischemic injury (130). Additionally, treatment of guinea pig ventricular myocytes with resveratrol led to enhanced myofilament Ca^{2+} sensitivity resulting from decreased Ca^{2+}-transient amplitude and altered cell shortening (131).

Treatment with resveratrol improves functional outcomes following cardiac insult in rodent models. Consumption of resveratrol reduced infarct size and cardiac myocyte apoptosis following ischemic injury in rats (130). Resveratrol treatment improved recovery of ventricular function and reduced infarct size following ischemia reperfusion injury in Langendorff-perfused rat hearts (132). Additionally, treatment following aortic banding in rats prevented left ventricular hypertrophy and impaired cardiac function (129).

CONCLUSION

It is becoming increasingly clear that the hearts of men and women are not equivalent. Several studies have shown that sex steroid hormones play an important role in the risk and outcome of cardiac disease. Hormone replacement therapy for postmenopausal women was thought to increase risk for cardiovascular disease, but a more detail analysis of the data has revealed that this may not be the case. A more carefully designed study of hormone replacement in women immediately following menopause may yield different results. The benefits of phytoestrogen intake with regard to cardiac function and disease have recently been popularized. Although phytoestrogens have the potential to mimic the effects of estrogen, they are not pure agonists. Phytoestrogens like genistein and resveratrol possess additional functional activities that may be detrimental to cardiac physiology; therefore, caution should be used when one is considering dietary supplementation with these molecules.

SUMMARY POINTS

1. Premenopausal women generally have improved cardiac disease outcomes compared with men; however, this sex difference is lost when comparing postmenopausal women with age-matched men.

2. Sex-based differences have been identified in animal models of exercise and cardiac disease, as well as in transgenic mice with altered signal transduction.

3. Sex steroid hormones likely play an important role in cardiac disease outcomes. Estrogen and testosterone have opposing effects on the heart, with estrogen being protective and testosterone detrimental to heart function.

4. Environmental influences, such as dietary phytoestrogens, can both positively and negatively influence cardiac function.

DISCLOSURE STATEMENT

The authors are not aware of any biases that might be perceived as affecting the objectivity of this review.

LITERATURE CITED

1. de Simone G, Devereux RB, Daniels SR, Meyer RA. 1995. Gender differences in left ventricular growth. *Hypertension* 26:979–83
2. Zak R. 1974. Development and proliferative capacity of cardiac muscle cells. *Circ. Res.* 35(Suppl. II):17–26
3. Grandi AM, Venco A, Barzizza F, Scalise F, Pantaleo P, Finardi G. 1992. Influence of age and sex on left ventricular anatomy and function in normals. *Cardiology* 81:8–13

4. Olivetti G, Giordano G, Corradi D, Melissari M, Lagrasta C, et al. 1995. Gender differences and aging: effects on the human heart. *J. Am. Coll. Cardiol.* 26:1068–79

5. Sullivan MJ, Cobb FR, Higginbotham MB. 1991. Stroke volume increases by similar mechanisms during upright exercise in normal men and women. *Am. J. Cardiol.* 67:1405–12

6. Higginbotham MB, Morris KG, Coleman RE, Cobb FR. 1984. Sex-related differences in the normal cardiac response to upright exercise. *Circulation* 70:357–66

7. De Maria R, Gavazzi A, Recalcati F, Baroldi G, De Vita C, Camerini F. 1993. Comparison of clinical findings in idiopathic dilated cardiomyopathy in women versus men. The Italian Multicenter Cardiomyopathy Study Group (SPIC). *Am. J. Cardiol.* 72:580–85

8. Douglas PS, Katz SE, Weinberg EO, Chen MH, Bishop SP, Lorell BH. 1998. Hypertrophic remodeling: gender differences in the early response to left ventricular pressure overload. *J. Am. Coll. Cardiol.* 32:1118–25

9. Deswal A, Bozkurt B. 2006. Comparison of morbidity in women versus men with heart failure and preserved ejection fraction. *Am. J. Cardiol.* 97:1228–31

10. Schocken DD, Arrieta MI, Leaverton PE, Ross EA. 1992. Prevalence and mortality rate of congestive heart failure in the United States. *J. Am. Coll. Cardiol.* 20:301–6

11. Hofman A, Grobbee DE, de Jong PT, Van Den Ouweland FA. 1991. Determinants of disease and disability in the elderly: the Rotterdam Elderly Study. *Eur. J. Epidemiol.* 7:403–22

12. Martin CA, Thompson PL, Armstrong BK, Hobbs MS, de Klerk N. 1983. Long-term prognosis after recovery from myocardial infarction: a nine year follow-up of the Perth Coronary Register. *Circulation* 68:961–69

13. Vaccarino V, Parsons L, Every NR, Barron HV, Krumholz HM. 1999. Sex-based differences in early mortality after myocardial infarction. National Registry of Myocardial Infarction 2 Participants. *N. Engl. J. Med.* 341:217–25

14. Ni H. 2003. Prevalence of self-reported heart failure among US adults: results from the 1999 National Health Interview Survey. *Am. Heart J.* 146:121–28

15. Lindenfeld J, Krause-Steinrauf H, Salerno J. 1997. Where are all the women with heart failure? *J. Am. Coll. Cardiol.* 30:1417–19

16. Wenger NK. 1992. Exclusion of the elderly and women from coronary trials. Is their quality of care compromised? *JAMA* 268:1460–61

17. Hayward CS, Kalnins WV, Kelly RP. 2001. Gender-related differences in left ventricular chamber function. *Cardiovasc. Res.* 49:340–50

18. Gurwitz JH, Col NF, Avorn J. 1992. The exclusion of the elderly and women from clinical trials in acute myocardial infarction. *JAMA* 268:1417–22

19. Mohan SB, Parker M, Wehbi M, Douglass P. 2002. Idiopathic dilated cardiomyopathy: a common but mystifying cause of heart failure. *Clevel. Clin. J. Med.* 69:481–87

20. Fernandez-Sola J, Nicolas-Arfelis JM. 2002. Gender differences in alcoholic cardiomyopathy. *J. Gend. Specif. Med.* 5:41–47

21. Schaible TF, Penpargkul S, Scheuer J. 1981. Cardiac responses to exercise training in male and female rats. *J. Appl. Physiol.* 50:112–17

22. Schaible TF, Scheuer J. 1979. Effects of physical training by running or swimming on ventricular performance of rat hearts. *J. Appl. Physiol.* 46:854–60

23. Schaible TF, Scheuer J. 1981. Cardiac function in hypertrophied hearts from chronically exercised female rats. *J. Appl. Physiol.* 50:1140–45

24. Mole PA. 1978. Increased contractile potential of papillary muscles from exercise-trained rat hearts. *Am. J. Physiol.* 234:H421–25

25. Konhilas JP, Maass AH, Luckey SW, Stauffer BL, Olson EN, Leinwand LA. 2004. Sex modifies exercise and cardiac adaptation in mice. *Am. J. Physiol. Heart Circ. Physiol.* 287:H2768–76

26. Giuberti K, Pereira RB, Bianchi PR, Paigel AS, Vassallo DV, Stefanon I. 2007. Influence of ovariectomy in the right ventricular contractility in heart failure rats. *Arch. Med. Res.* 38:170–75

27. Smith PJ, Ornatsky O, Stewart DJ, Picard P, Dawood F, et al. 2000. Effects of estrogen replacement on infarct size, cardiac remodeling, and the endothelin system after myocardial infarction in ovariectomized rats. *Circulation* 102:2983–89

28. Cavasin MA, Tao Z, Menon S, Yang XP. 2004. Gender differences in cardiac function during early remodeling after acute myocardial infarction in mice. *Life Sci.* 75:2181–92

29. Jain M, Liao R, Podesser BK, Ngoy S, Apstein CS, Eberli FR. 2002. Influence of gender on the response to hemodynamic overload after myocardial infarction. *Am. J. Physiol. Heart Circ. Physiol.* 283:H2544–50

30. Podesser BK, Jain M, Ngoy S, Apstein CS, Eberli FR. 2007. Unveiling gender differences in demand ischemia: a study in a rat model of genetic hypertension. *Eur. J. Cardiothorac. Surg.* 31:298–304

31. Gerdes AM, Onodera T, Wang X, McCune SA. 1996. Myocyte remodeling during the progression to failure in rats with hypertension. *Hypertension* 28:609–14

32. McCune SA, Park S, Radin MJ, Jurin RR. 1996. Renal and heart function in the SHHF/Mcc-cp rat. In *Mechanisms of Heart Failure*, ed. PK Singal, IMC Dixon, RE Beamish, NS Dhalla, MA Norwell, pp. 91–106. Dordrecht, Neth.: Kluwer Acad.

33. Tamura T, Said S, Gerdes AM. 1999. Gender-related differences in myocyte remodeling in progression to heart failure. *Hypertension* 33:676–80

34. McCune SA, Baker PB, Stills HFJ. 1990. SHHF/Mcc-cp rat: model of obesity, noninsulin-dependent diabetes, and congestive heart failure. *ILAR News* 32:23–27

35. Stauffer BL, Konhilas JP, Luczak ED, Leinwand LA. 2006. Soy diet worsens heart disease in mice. *J. Clin. Investig.* 116:209–16

36. Geisterfer-Lowrance AA, Kass S, Tanigawa G, Vosberg HP, McKenna W, et al. 1990. A molecular basis for familial hypertrophic cardiomyopathy: a β cardiac myosin heavy chain gene missense mutation. *Cell* 62:999–1006

37. Vikstrom KL, Factor SM, Leinwand LA. 1996. Mice expressing mutant myosin heavy chains are a model for familial hypertrophic cardiomyopathy. *Mol. Med.* 2:556–67

38. Berul CI, Christe ME, Aronovitz MJ, Seidman CE, Seidman JG, Mendelsohn ME. 1997. Electrophysiological abnormalities and arrhythmias in α MHC mutant familial hypertrophic cardiomyopathy mice. *J. Clin. Investig.* 99:570–76

39. Maass AH, Ikeda K, Oberdorf-Maass S, Maier SK, Leinwand LA. 2004. Hypertrophy, fibrosis, and sudden cardiac death in response to pathological stimuli in mice with mutations in cardiac troponin T. *Circulation* 110:2102–9

40. Tardiff JC, Hewett TE, Palmer BM, Olsson C, Factor SM, et al. 1999. Cardiac troponin T mutations result in allele-specific phenotypes in a mouse model for hypertrophic cardiomyopathy. *J. Clin. Investig.* 104:469–81

41. Tardiff JC, Factor SM, Tompkins BD, Hewett TE, Palmer BM, et al. 1998. A truncated cardiac troponin T molecule in transgenic mice suggests multiple cellular mechanisms for familial hypertrophic cardiomyopathy. *J. Clin. Investig.* 101:2800–11

42. Levine B, Kalman J, Mayer L, Fillit HM, Packer M. 1990. Elevated circulating levels of tumor necrosis factor in severe chronic heart failure. *N. Engl. J. Med.* 323:236–41

43. Ferrari R, Bachetti T, Confortini R, Opasich C, Febo O, et al. 1995. Tumor necrosis factor soluble receptors in patients with various degrees of congestive heart failure. *Circulation* 92:1479–86

44. Doyama K, Fujiwara H, Fukumoto M, Tanaka M, Fujiwara Y, et al. 1996. Tumour necrosis factor is expressed in cardiac tissues of patients with heart failure. *Int. J. Cardiol.* 54:217–25

45. Torre-Amione G, Kapadia S, Lee J, Durand JB, Bies RD, et al. 1996. Tumor necrosis factor-α and tumor necrosis factor receptors in the failing human heart. *Circulation* 93:704–11

46. Kubota T, McTiernan CF, Frye CS, Slawson SE, Lemster BH, et al. 1997. Dilated cardiomyopathy in transgenic mice with cardiac-specific overexpression of tumor necrosis factor-α. *Circ. Res.* 81:627–35

47. Kadokami T, McTiernan CF, Kubota T, Frye CS, Feldman AM. 2000. Sex-related survival differences in murine cardiomyopathy are associated with differences in TNF-receptor expression. *J. Clin. Investig.* 106:589–97

48. Rockman HA, Koch WJ, Lefkowitz RJ. 2002. Seven-transmembrane-spanning receptors and heart function. *Nature* 415:206–12

49. Gao XM, Agrotis A, Autelitano DJ, Percy E, Woodcock EA, et al. 2003. Sex hormones and cardiomyopathic phenotype induced by cardiac β2-adrenergic receptor overexpression. *Endocrinology* 144:4097–105

50. O'Connell TD, Ishizaka S, Nakamura A, Swigart PM, Rodrigo MC, et al. 2003. The $\alpha_{1A/C}$- and α_{1B}-adrenergic receptors are required for physiological cardiac hypertrophy in the double-knockout mouse. *J. Clin. Investig.* 111:1783–91

51. Dash R, Kadambi V, Schmidt AG, Tepe NM, Biniakiewicz D, et al. 2001. Interactions between phospholamban and β-adrenergic drive may lead to cardiomyopathy and early mortality. *Circulation* 103:889–96

52. Dash R, Schmidt AG, Pathak A, Gerst MJ, Biniakiewicz D, et al. 2003. Differential regulation of p38 mitogen-activated protein kinase mediates gender-dependent catecholamine-induced hypertrophy. *Cardiovasc. Res.* 57:704–14

53. Haghighi K, Schmidt AG, Hoit BD, Brittsan AG, Yatani A, et al. 2001. Superinhibition of sarcoplasmic reticulum function by phospholamban induces cardiac contractile failure. *J. Biol. Chem.* 276:24145–52

54. Djouadi F, Weinheimer CJ, Saffitz JE, Pitchford C, Bastin J, et al. 1998. A gender-related defect in lipid metabolism and glucose homeostasis in peroxisome proliferator-activated receptor α-deficient mice. *J. Clin. Investig.* 102:1083–91

55. Nohammer C, Brunner F, Wolkart G, Staber PB, Steyrer E, et al. 2003. Myocardial dysfunction and male mortality in peroxisome proliferator-activated receptor α knockout mice overexpressing lipoprotein lipase in muscle. *Lab. Investig.* 83:259–69

56. Bryant-Greenwood GD, Schwabe C. 1994. Human relaxins: chemistry and biology. *Endocr. Rev.* 15:5–26

57. Osheroff PL, Ho WH. 1993. Expression of relaxin mRNA and relaxin receptors in postnatal and adult rat brains and hearts. Localization and developmental patterns. *J. Biol. Chem.* 268:15193–99

58. Kakouris H, Eddie LW, Summers RJ. 1992. Cardiac effects of relaxin in rats. *Lancet* 339:1076–78

59. Du XJ, Samuel CS, Gao XM, Zhao L, Parry LJ, Tregear GW. 2003. Increased myocardial collagen and ventricular diastolic dysfunction in relaxin deficient mice: a gender-specific phenotype. *Cardiovasc. Res.* 57:395–404

60. Antos CL, McKinsey TA, Frey N, Kutschke W, McAnally J, et al. 2002. Activated glycogen synthase-3β suppresses cardiac hypertrophy in vivo. *Proc. Natl. Acad. Sci. USA* 99:907–12

61. Luckey SW, Mansoori J, Fair K, Antos CL, Olson EN, Leinwand LA. 2007. Blocking cardiac growth in hypertrophic cardiomyopathy induces cardiac dysfunction and decreased survival only in males. *Am. J. Physiol. Heart Circ. Physiol.* 292:H838–45

62. Marsh JD, Lehmann MH, Ritchie RH, Gwathmey JK, Green GE, Schiebinger RJ. 1998. Androgen receptors mediate hypertrophy in cardiac myocytes. *Circulation* 98:256–61

63. Nordmeyer J, Eder S, Mahmoodzadeh S, Martus P, Fielitz J, et al. 2004. Upregulation of myocardial estrogen receptors in human aortic stenosis. *Circulation* 110:3270–75

64. Schaible TF, Malhotra A, Ciambrone G, Scheuer J. 1984. The effects of gonadectomy on left ventricular function and cardiac contractile proteins in male and female rats. *Circ. Res.* 54:38–49

65. Levin ER. 2001. Cell localization, physiology, and nongenomic actions of estrogen receptors. *J. Appl. Physiol.* 91:1860–67

66. Mahmoodzadeh S, Eder S, Nordmeyer J, Ehler E, Huber O, et al. 2006. Estrogen receptor α upregulation and redistribution in human heart failure. *FASEB J.* 20:926–34

67. Greenspan FS, Gardner DG. 2004. *Basic & Clinical Endocrinology.* New York: Lange Medical Books/McGraw-Hill. 976 pp.

68. Hewitt SC, Harrell JC, Korach KS. 2005. Lessons in estrogen biology from knockout and transgenic animals. *Annu. Rev. Physiol.* 67:285–308

69. Couse JF, Korach KS. 1999. Estrogen receptor null mice: What have we learned and where will they lead us? *Endocr. Rev.* 20:358–417

70. Gabel SA, Walker VR, London RE, Steenbergen C, Korach KS, Murphy E. 2005. Estrogen receptor β mediates gender differences in ischemia/reperfusion injury. *J. Mol. Cell Cardiol.* 38:289–97

71. Zhai P, Eurell TE, Cooke PS, Lubahn DB, Gross DR. 2000. Myocardial ischemia-reperfusion injury in estrogen receptor-α knockout and wild-type mice. *Am. J. Physiol. Heart Circ. Physiol.* 278:H1640–47

72. Skavdahl M, Steenbergen C, Clark J, Myers P, Demianenko T, et al. 2005. Estrogen receptor-β mediates male-female differences in the development of pressure overload hypertrophy. *Am. J. Physiol. Heart Circ. Physiol.* 288:H469–76

73. Pelzer T, Loza PA, Hu K, Bayer B, Dienesch C, et al. 2005. Increased mortality and aggravation of heart failure in estrogen receptor-β knockout mice after myocardial infarction. *Circulation* 111:1492–98

74. Brower GL, Gardner JD, Janicki JS. 2003. Gender mediated cardiac protection from adverse ventricular remodeling is abolished by ovariectomy. *Mol. Cell Biochem.* 251:89–95

75. Bhuiyan MS, Shioda N, Fukunaga K. 2007. Ovariectomy augments pressure overload-induced hypertrophy associated with changes in Akt and nitric oxide synthase signaling pathways in female rats. *Am. J. Physiol. Endocrinol. Metab.* 293:E1606–14

76. van Eickels M, Grohe C, Cleutjens JP, Janssen BJ, Wellens HJ, Doevendans PA. 2001. 17β-Estradiol attenuates the development of pressure-overload hypertrophy. *Circulation* 104:1419–23

77. Ho KK, Anderson KM, Kannel WB, Grossman W, Levy D. 1993. Survival after the onset of congestive heart failure in Framingham Heart Study subjects. *Circulation* 88:107–15

78. Harman SM. 2006. Estrogen replacement in menopausal women: recent and current prospective studies, the WHI and the KEEPS. *Gend. Med.* 3:254–69

79. Rossouw JE, Anderson GL, Prentice RL, LaCroix AZ, Kooperberg C, et al. 2002. Risks and benefits of estrogen plus progestin in healthy postmenopausal women: principal results from the Women's Health Initiative randomized controlled trial. *JAMA* 288:321–33

80. Vickers MR, Maclennan AH, Lawton B, Ford D, Martin J, et al. 2007. Main morbidities recorded in the women's international study of long duration oestrogen after menopause (WISDOM): a randomised controlled trial of hormone replacement therapy in postmenopausal women. *Br. Med. J.* 335:239

81. Grodstein F, Stampfer MJ, Manson JE, Colditz GA, Willett WC, et al. 1996. Postmenopausal estrogen and progestin use and the risk of cardiovascular disease. *N. Engl. J. Med.* 335:453–61

82. Nuedling S, Kahlert S, Loebbert K, Doevendans PA, Meyer R, et al. 1999. 17β-Estradiol stimulates expression of endothelial and inducible NO synthase in rat myocardium in-vitro and in-vivo. *Cardiovasc. Res.* 43:666–74

83. Kleinert H, Wallerath T, Euchenhofer C, Ihrig-Biedert I, Li H, Forstermann U. 1998. Estrogens increase transcription of the human endothelial NO synthase gene: analysis of the transcription factors involved. *Hypertension* 31:582–88

84. Johnson BD, Zheng W, Korach KS, Scheuer T, Catterall WA, Rubanyi GM. 1997. Increased expression of the cardiac L-type calcium channel in estrogen receptor-deficient mice. *J. Gen. Physiol.* 110:135–40

85. Chu SH, Goldspink P, Kowalski J, Beck J, Schwertz DW. 2006. Effect of estrogen on calcium-handling proteins, β-adrenergic receptors, and function in rat heart. *Life Sci.* 79:1257–67

86. Jiang C, Poole-Wilson PA, Sarrel PM, Mochizuki S, Collins P, MacLeod KT. 1992. Effect of 17β-oestradiol on contraction, Ca^{2+} current and intracellular free Ca^{2+} in guinea-pig isolated cardiac myocytes. *Br. J. Pharmacol.* 106:739–45

87. Kravtsov GM, Kam KW, Liu J, Wu S, Wong TM. 2007. Altered Ca^{2+} handling by ryanodine receptor and Na^{+}-Ca^{2+} exchange in the heart from ovariectomized rats: role of protein kinase A. *Am. J. Physiol. Cell Physiol.* 292:C1625–35

88. Patten RD, Pourati I, Aronovitz MJ, Baur J, Celestin F, et al. 2004. 17β-Estradiol reduces cardiomyocyte apoptosis in vivo and in vitro via activation of phospho-inositide-3 kinase/Akt signaling. *Circ. Res.* 95:692–99

89. Shiraishi I, Melendez J, Ahn Y, Skavdahl M, Murphy E, et al. 2004. Nuclear targeting of Akt enhances kinase activity and survival of cardiomyocytes. *Circ. Res.* 94:884–91

90. Camper-Kirby D, Welch S, Walker A, Shiraishi I, Setchell KDR, et al. 2001. Myocardial Akt activation and gender: increased nuclear activity in females versus males. *Circ. Res.* 88:1020–27

91. Cavasin MA, Sankey SS, Yu AL, Menon S, Yang XP. 2003. Estrogen and testosterone have opposing effects on chronic cardiac remodeling and function in mice with myocardial infarction. *Am. J. Physiol. Heart Circ. Physiol.* 284:H1560–69

92. Sluijmer AV, Heineman MJ, De Jong FH, Evers JL. 1995. Endocrine activity of the postmenopausal ovary: the effects of pituitary down-regulation and oophorectomy. *J. Clin. Endocrinol. Metab.* 80:2163–67

93. Sachtleben TR, Berg KE, Elias BA, Cheatham JP, Felix GL, Hofschire PJ. 1993. The effects of anabolic steroids on myocardial structure and cardiovascular fitness. *Med. Sci. Sports Exerc.* 25:1240–45

94. Urhausen A, Holpes R, Kindermann W. 1989. One- and two-dimensional echocardiography in body-builders using anabolic steroids. *Eur. J. Appl. Physiol. Occup. Physiol.* 58:633–40

95. Sullivan ML, Martinez CM, Gennis P, Gallagher EJ. 1998. The cardiac toxicity of anabolic steroids. *Prog. Cardiovasc. Dis.* 41:1–15

96. Fisher CR, Graves KH, Parlow AF, Simpson ER. 1998. Characterization of mice deficient in aromatase (ArKO) because of targeted disruption of the *cyp19* gene. *Proc. Natl. Acad. Sci. USA* 95:6965–70

97. Head GA, Obeyesekere VR, Jones ME, Simpson ER, Krozowski ZS. 2004. Aromatase-deficient (ArKO) mice have reduced blood pressure and baroreflex sensitivity. *Endocrinology* 145:4286–91

98. Setchell KD, Brown NM, Lydeking-Olsen E. 2002. The clinical importance of the metabolite equol: a clue to the effectiveness of soy and its isoflavones. *J. Nutr.* 132:3577–84

99. Safford B, Dickens A, Halleron N, Briggs D, Carthew P, Baker V. 2003. A model to estimate the oestrogen receptor mediated effects from exposure to soy isoflavones in food. *Regul. Toxicol. Pharmacol.* 38:196–209

100. Fujimoto N, Honda H, Kitamura S. 2004. Effects of environmental estrogenic chemicals on AP1 mediated transcription with estrogen receptors α and β. *J. Steroid Biochem. Mol. Biol.* 88:53–59

101. Toda K, Hayashi Y, Okada T, Morohashi K, Saibara T. 2005. Expression of the estrogen-inducible EGFP gene in aromatase-null mice reveals differential tissue responses to estrogenic compounds. *Mol. Cell. Endocrinol.* 229:119–26

102. Matsui T, Li L, Wu JC, Cook SA, Nagoshi T, et al. 2002. Phenotypic spectrum caused by transgenic overexpression of activated Akt in the heart. *J. Biol. Chem.* 277:22896–901

103. Kumi-Diaka J, Sanderson NA, Hall A. 2000. The mediating role of caspase-3 protease in the intracellular mechanism of genistein-induced apoptosis in human prostatic carcinoma cell lines, DU145 and LNCaP. *Biol. Cell* 92:595–604

104. Yoon HS, Moon SC, Kim ND, Park BS, Jeong MH, Yoo YH. 2000. Genistein induces apoptosis of RPE-J cells by opening mitochondrial PTP. *Biochem. Biophys. Res. Commun.* 276:151–56

105. Yellayi S, Naaz A, Szewczykowski MA, Sato T, Woods JA, et al. 2002. The phytoestrogen genistein induces thymic and immune changes: a human health concern? *Proc. Natl. Acad. Sci. USA* 99:7616–21

106. Penza M, Montani C, Romani A, Vignolini P, Pampaloni B, et al. 2006. Genistein affects adipose tissue deposition in a dose-dependent and gender-specific manner. *Endocrinology* 147:5740–51

107. Lamb J, Crawford ED, Peck D, Modell JW, Blat IC, et al. 2006. The Connectivity Map: using gene-expression signatures to connect small molecules, genes, and disease. *Science* 313:1929–35

108. Sacks FM, Lichtenstein A, Van Horn L, Harris W, Kris-Etherton P, Winston M. 2006. Soy protein, isoflavones, and cardiovascular health: an American Heart Association Science Advisory for professionals from the Nutrition Committee. *Circulation* 113:1034–44

109. Kokubo Y, Iso H, Ishihara J, Okada K, Inoue M, Tsugane S. 2007. Association of dietary intake of soy, beans, and isoflavones with risk of cerebral and myocardial infarctions in Japanese populations: the Japan Public Health Center-based (JPHC) study cohort I. *Circulation* 116:2553–62

110. Glover A, Assinder SJ. 2006. Acute exposure of adult male rats to dietary phytoestrogens reduces fecundity and alters epididymal steroid hormone receptor expression. *J. Endocrinol.* 189:565–73

111. Jefferson WN, Padilla-Banks E, Newbold RR. 2005. Adverse effects on female development and reproduction in CD-1 mice following neonatal exposure to the phytoestrogen genistein at environmentally relevant doses. *Biol. Reprod.* 73:798–806

112. Adams NR. 1995. Detection of the effects of phytoestrogens on sheep and cattle. *J. Anim. Sci.* 73:1509–15

113. Leopold AS, Erwin M, Oh J, Browning B. 1976. Phytoestrogens: adverse effects on reproduction in California quail. *Science* 191:98–100

114. Setchell KD, Gosselin SJ, Welsh MB, Johnston JO, Balistreri WF, et al. 1987. Dietary estrogens—a probable cause of infertility and liver disease in captive cheetahs. *Gastroenterology* 93:225–33

115. Brown NM, Setchell KD. 2001. Animal models impacted by phytoestrogens in commercial chow: implications for pathways influenced by hormones. *Lab. Invest.* 81:735–47

116. Thigpen JE, Setchell KD, Ahlmark KB, Locklear J, Spahr T, et al. 1999. Phytoestrogen content of purified, open- and closed-formula laboratory animal diets. *Lab. Anim. Sci.* 49:530–36

117. Adlercreutz H, Markkanen H, Watanabe S. 1993. Plasma concentrations of phyto-oestrogens in Japanese men. *Lancet* 342:1209–10

118. King RA, Bursill DB. 1998. Plasma and urinary kinetics of the isoflavones daidzein and genistein after a single soy meal in humans. *Am. J. Clin. Nutr.* 67:867–72

119. Watanabe S, Yamaguchi M, Sobue T, Takahashi T, Miura T, et al. 1998. Pharmacokinetics of soybean isoflavones in plasma, urine and feces of men after ingestion of 60 g baked soybean powder (kinako). *J. Nutr.* 128:1710–15

120. Xu X, Harris KS, Wang HJ, Murphy PA, Hendrich S. 1995. Bioavailability of soybean isoflavones depends upon gut microflora in women. *J. Nutr.* 125:2307–15

121. Djuric Z, Chen G, Doerge DR, Heilbrun LK, Kucuk O. 2001. Effect of soy isoflavone supplementation on markers of oxidative stress in men and women. *Cancer Lett.* 172:1–6

122. Setchell KD, Zimmer-Nechemias L, Cai J, Heubi JE. 1997. Exposure of infants to phyto-oestrogens from soy-based infant formula. *Lancet* 350:23–27

123. Howitz KT, Bitterman KJ, Cohen HY, Lamming DW, Lavu S, et al. 2003. Small molecule activators of sirtuins extend *Saccharomyces cerevisiae* lifespan. *Nature* 425:191–96

124. Valenzano DR, Terzibasi E, Genade T, Cattaneo A, Domenici L, Cellerino A. 2006. Resveratrol prolongs lifespan and retards the onset of age-related markers in a short-lived vertebrate. *Curr. Biol.* 16:296–300

125. Baur JA, Sinclair DA. 2006. Therapeutic potential of resveratrol: the in vivo evidence. *Nat. Rev. Drug Discov.* 5:493–506

126. Fremont L. 2000. Biological effects of resveratrol. *Life Sci.* 66:663–73

127. Bowers JL, Tyulmenkov VV, Jernigan SC, Klinge CM. 2000. Resveratrol acts as a mixed agonist/antagonist for estrogen receptors α and β. *Endocrinology* 141:3657–67

128. Klinge CM, Risinger KE, Watts MB, Beck V, Eder R, Jungbauer A. 2003. Estrogenic activity in white and red wine extracts. *J. Agric. Food Chem.* 51:1850–57

129. Juric D, Wojciechowski P, Das DK, Netticadan T. 2007. Prevention of concentric hypertrophy and diastolic impairment in aortic-banded rats treated with resveratrol. *Am. J. Physiol. Heart Circ. Physiol.* 292:H2138–43

130. Das S, Alagappan VK, Bagchi D, Sharma HS, Maulik N, Das DK. 2005. Coordinated induction of iNOS-VEGF-KDR-eNOS after resveratrol consumption: a potential mechanism for resveratrol preconditioning of the heart. *Vascul. Pharmacol.* 42:281–89

131. Liew R, Stagg MA, MacLeod KT, Collins P. 2005. The red wine polyphenol, resveratrol, exerts acute direct actions on guinea-pig ventricular myocytes. *Eur. J. Pharmacol.* 519:1–8

132. Ray PS, Maulik G, Cordis GA, Bertelli AA, Bertelli A, Das DK. 1999. The red wine antioxidant resveratrol protects isolated rat hearts from ischemia reperfusion injury. *Free Radic. Biol. Med.* 27:160–69

Convergent Evolution of Alternative Splices at Domain Boundaries of the BK Channel

Anthony A. Fodor[1] and Richard W. Aldrich[2]

[1]Bioinformatics Research Center, Cameron Applied Research Center, University of North Carolina, Charlotte, North Carolina 28223; email: anthony.fodor@gmail.com

[2]Section of Neurobiology, University of Texas, Austin, Texas 78712

Annu. Rev. Physiol. 2009. 71:19–36

First published online as a Review in Advance on August 11, 2008

The *Annual Review of Physiology* is online at physiol.annualreviews.org

This article's doi:
10.1146/annurev.physiol.010908.163124

Copyright © 2009 by Annual Reviews.
All rights reserved

0066-4278/09/0315-0019$20.00

Key Words

calcium-activated potassium channel, bioinformatics, constitutive exons, comparative genomics

Abstract

Alternative splicing is a widespread mechanism for generating transcript diversity in higher eukaryotic genomes. The alternative splices of the large-conductance calcium-activated potassium (BK) channel have been the subject of a good deal of experimental functional characterization in the Arthropoda, Chordata, and Nematoda phyla. In this review, we examine a list of splices of the BK channel by manual curation of Unigene clusters mapped to mouse, human, chicken, *Drosophila*, and *Caenorhabditis elegans* genomes. We find that BK alternative splices do not appear to be conserved across phyla. Despite this lack of conservation, splices occur in both vertebrates and invertebrates at identical regions of the channel at experimentally established domain boundaries. The fact that, across phyla, unique splices occur at experimentally established domain boundaries suggests a prominent role for the convergent evolution of alternative splices that produce functional changes via changes in interdomain communication.

INTRODUCTION

The large-conductance calcium- and voltage-activated potassium (BK) channel is a widely expressed protein that has been implicated in a variety of phenotypes, including vascular tone (1), vasoregulation (2), urinary bladder function (3), circadian rhythms (4), and susceptibility to hearing damage (5). BK channels in different tissues have very different electrophysiological properties (2, 3, 6–8). Physiological tuning of BK channels in native tissues has been explained by a variety of interacting mechanisms including modulation by different β subunits (9–11), posttranslational modification (12, 13), and alternative splicing of the main α subunit. The α subunit of the BK channel is a well-conserved protein consisting of at least three distinct domains (14). Numerous labs have described alternatively spliced isoforms of the BK channel (7, 15–31). Because alternative splicing changes the properties of the BK channel and the BK channel modulates cellular excitability, alternative splicing of BK channels may control a number of phenotypes, including auditory turning (20), neuronal excitability (18), and secretion of epinephrine (22).

When the first alternative splices of the BK channel were described in the early 1990s, alternative splicing was thought to be a relatively rare phenomenon. Since then, new technologies have revealed splicing to be far more prevalent than imagined in previous decades. The genomic era, and the associated drop in the cost of sequencing, has given rise to huge sequence databases from which multiple splices of the same genes can be mined. A shotgun technique that gained popularity in the 1990s was the use of expressed sequence tags (ESTs) (32). In this technique, RNA from a tissue of interest is reverse transcribed to cDNA, which is used to create bacterial libraries. Clones from the libraries are picked at random, and a single sequence read is created for each picked clone. Compared with full-length sequences, ESTs produce only gene fragments, but with greatly reduced experimental effort.

Initial analyses of EST and full-length RNA databases suggested that between 35% and 60% of all genes had at least one alternatively spliced form (33–36). However, a significant fraction of these splice sequences appear to contain errors (37, 38). Many splices in coding regions introduce stop codons into the middle of proteins. Furthermore, splices that introduce premature stop codons tend to occur only once in sequence databases and tend to be from regions of the genome that are poorly conserved between mouse and human (37). In contrast, splices that are multiples of three DNA nucleotides and hence do not produce frame-shift mutations are much more likely to be conserved between human and mouse and are more likely to be represented by more than one entry in a sequence database (37). A very interesting question is whether splicing "errors" represent experimental noise, biological noise, or alternatively some unknown function such as gene downregulation (39). Many alternative splices are found only in EST databases, and many ESTs are derived from cancer tissues that prominently feature nonfunctional, aberrant transcription (37). This observation argues for the hypothesis that a significant number of splices represent errors in the splicing machinery that do not serve a biological function.

Microarray technology has further expanded the universe of possible splices and given more urgency to the question of to what extent splicing represents biological or experimental noise or biological function. High-density microarray technology is capable of capturing transcription across large swaths of genomes. For example, by hybridization of cDNA or RNA to a tiling array, it is possible to perform a large-scale experiment in which, with some algorithmic manipulation, one can detect nearly every transcribed region of a genome in a more or less unbiased manner (40–42). These experiments have discovered a rate of alternative splicing far higher than was expected on the basis of the results of analyzing EST and cDNA databases (40, 41, 43). On the basis of

BK channel: calcium- and voltage-activated potassium channel

Alternative splicing: the combination of exons in different ways to produce multiple transcripts from the same region of the genome

Expressed sequence tag (EST): a partial sequence generated by shotgun sequencing of cDNA copies of RNA

these results, it appears that alternative splicing is rampant in mammalian genomes. Nearly every gene is spliced, and there appears to be a universe of splices not captured by current sequence databases. The question remains: Does this newfound extent of splicing serve a biological purpose, or does it just reflect that we now have very sensitive techniques that can measure nonsense aberrant splices? Does rampant mammalian alternative splicing reflect quickly degraded errors in the splicing machinery or real biological signals with unknown function?

These new findings regarding alternative splicing represent both a challenge and an opportunity to the large literature on BK channel splices. On the one hand, if it turns out that the splicing machinery makes mistakes that are not biologically important, perhaps some reported BK splices may not be physiologically relevant. If the splicing machinery makes frequent mistakes, the isolation of a novel BK splice via a PCR experiment is not necessarily meaningful. Likewise, if the global splicing machinery responds differently to different signals, experiments that find different splices of the BK channel in different tissues do not necessarily imply a functional role for these different splices. On the other hand, the BK channel is an ion channel, and we of course have a method for determining protein function with the patch clamp. Because the BK channel has been the target of so many studies of alternative splicing, an unusually large number of full-length RNAs describing alternative splices of the BK channel have been deposited in GenBank. Moreover, the functional differences caused by a number of BK alternative splices have been characterized at the protein level with the patch-clamp technique (7, 17, 18, 22, 25, 26, 29). These experimental data give us a tool that may allow us to discriminate true splices from aberrant splicing in a manner that does not depend only on computational methods of validating splice prediction.

A limitation in understanding the patterns of alternative splicing in BK channels is bioinformatic in nature. Although a large number of labs have sequenced alternative splices of the BK channel, each lab tends to use its own nomenclature (see **Table 1**). This makes comparing splices, especially across species for which the sequences may not be identical, very difficult. In addition, large-scale sequencing projects, such as EST projects, have identified possible alternative splices that did not initially come to the attention of researchers interested in the BK channel. In this review, we perform a manual curation of sequence databases and combine the results with a literature search. With this achieved, we argue for a hypothesis of the evolution of the BK channel in which alternative splicing tunes the function of the channel by altering the sequence at domain boundaries.

blast: the most commonly used algorithm for finding a query sequence in a sequence database

RESULTS

Genomic Sequences Can Be Used to Anchor Transcripts

Given a set of transcripts of the same gene from different species, it is not a trivial problem to detect all the alternative splices. If genome sequences are available, a natural approach is to map the transcripts back to genomic sequence. This allows for alternative patterns of introns and exons to be visualized and provides a framework for comparing species across a wide phylogenetic space. The development of algorithms that map transcripts to each other and to genomes is a very active area of research (44, 45). Algorithms such as Sim4 are designed to map a transcript back to a genome of the same species (46). These programs take advantage of canonical splice donor and acceptor sites, which are dinucleotide signals that bound nearly all exons. These kinds of programs are very useful for providing the best picture of the exact intron/exon boundary of a given transcript. They are not designed, however, to map transcripts to distantly related genomes. Because our goal was to compare BK channel transcripts across multiple phyla, we chose to map with the venerable program blast (47). Although blast does not take into account splice donor and acceptor sites and therefore will sometimes not correctly give the exact exon

Table 1 Evidence supporting splices[a]

Vertebrate splices		
Splice	**GenBank evidence**	**Other names and functional characterizations**
A	MR: U09383	
	ME: BU611487, BC065068	
	RR: U55995, AF083341, U93052	
	HR: U11717, AB113382, AB113575, AY040849, U13913, NM_002247	
	HE: (CD614365, CD614364, CD614362)	
B	RR: NM_031828, U40603, AF135265	R: X0_60AA (30)
C	ME: (BB642344, BY742482)	
D	ME: (BB642344, BY742482)	R: X2 (30) (without the stop codon)
		C: site 5 (19) (without the stop codon)
		M: site 2—exon 12 (31)
E	HE: AL705754	
F	ME: (BB831822, BB833079, BB836985, BY378603)	
G	MR: U09383	H: site1_4aa (18)
	ME: BC065068	M: site C1-e17 (7)
	RR: AF083341, NM_031828, U40603, AF135265	C: site3_4 (20), site 3 (19); T: SS1_4 (21)
	TR: AF036627, AF036626	M: site 3: exon 19—type III deletion (31)
H	HE: AK128392, BG216619, BQ446412, AW305218	
I	MR: L16912	H: site2_3aa (18); decreased open probability
	ME: (BB282848, BB280770)	M: site C2-e20 (7); calcium and voltage sensitivity similar to wild type; increased single-channel conductance (7)
	TR: AF036628.1	C: site4_3 (20)
		T: SS2_3 (21)
		M: site 4: exon 22 (31)
J	MR: AF156674, NM_010610	R: X4_61aa (30)
	ME: BC065068, BF581457	M: siteC2-e21 (7); left shift in IV relationship; slower deactivation (7)
	HR: AB113382	T: SS2_61 (21)
	HE: (BF111069, BE219872)	R: Strex-1 (22); currents activated at more negative voltages; speeds activation, slows deactivation (22)
	TR: AF036627.1	
K	HE: (BG220672, BG198117, BG189025, BG185868), CB047110	
L	HR: BX648925	H: gBK (29); when compared with splice N, slowed activation, produced greater currents at low Ca^{2+} concentrations (29)
M	HR: U02632	H: site2_29aa (18)
N	HE: BQ017892	
O	HR: AB113575	
P	MR: U09383	H: site4_27aa (18)
	ME: CD802839	C: site6_28 (20), site 5 (19)
	HE: AI002717	R: X6_27aa (30); rSlo27 (25); small increase in speed of activation (25)

(Continued)

Table 1 (*Continued*)

Vertebrate splices		
Splice	GenBank evidence	Other names and functional characterizations
	RE: AF135265	
Q	RR: AY330290	
R	MR: L16912	
Drosophila splices		
T		A2 (15)
U	DR: BT004876	A3 (15); decreased single-channel conductance, caused faster activation times, decreased open probability (17)
V	DR: BT004876	C2 (15)
W	DR: AI402739	E1 (15); decreased open probability (17)
X	DR: M96840, NM_079762	G5 (15)
Y	DR: BT004876	
Z	DR: BT004876	I1 (15)
Caenorhabditis elegans splices		
AA	WR: AV195353, AF431893, NM_171559	Alternative exon 9 (26)
BB	WR: AF431893, NM_171559	Alternative exon 11b (26)
CC	WR: AF431891, NM_171936	Lacking exon 13 (26); no effect in patch clamp

[a]GenBank IDs in parentheses are from the same clone library and are counted only once in the evidence tally. The sequences and positions of each splice are given in **Figures 3–5**. The species and evidence codes are as given in **Figure 6**. Reference numbers are in parentheses.

boundaries, it has tremendous flexibility. In particular, we used it to perform two different kinds of searches. Within a phylum, we used blastn, which directly compares nucleotide transcripts with a nucleotide genome. Across phyla, however, such a direct approach yielded few hits (data not shown). Owing to the degenerate nature of the nucleotide code, however, protein sequences are often conserved when the underlying nucleotide sequences are not. Across phyla, therefore, we use a search called tblastx, in which the nucleotide genome and the nucleotide transcript are both searched by translating all possible reading frames.

Searches involving translating the query and database sequences are potentially much more sensitive than plain nucleotide searches, but they are also much more dependent on the parameters of the algorithm used. We were able to successfully map vertebrate transcripts to invertebrate genomes (and vice versa), using tblastx from blast version 2.2.8 (see below for why we think we were successful in this endeavor!). More recent versions of blast, however, largely failed to map BK sequences across the phyla with tblastx (data not shown). This again shows the difficulty of mapping sequences across a wide phylogenetic space and suggests that it may be difficult to automate the kinds of manual annotation analyses we performed here.

Constitutive Exons Are Conserved Between Vertebrates and Invertebrates, Whereas Alternatively Spliced Exons Are Unique to Each Phylum

Our goal was to catalog alternative splices of the BK channel across a diversity of species. To view the results of mapping BK transcripts to genomic sequence with blast, we constructed a Java-based custom BK channel genome browser that aligns the mouse, human, chicken, *Drosophila*, and *Caenorhabditis elegans* Unigene transcripts to the genomes from each of these species (**http://www.afodor.net**; requires a Java-enabled browser). **Figure 1** shows a screen shot from this custom browser showing

blastn: compares a nucleotide sequence with a nucleotide database

tblastx: compares two nucleotide sequences by translating both sequences to all six reading frames; is more sensitive, but slower, than blastn

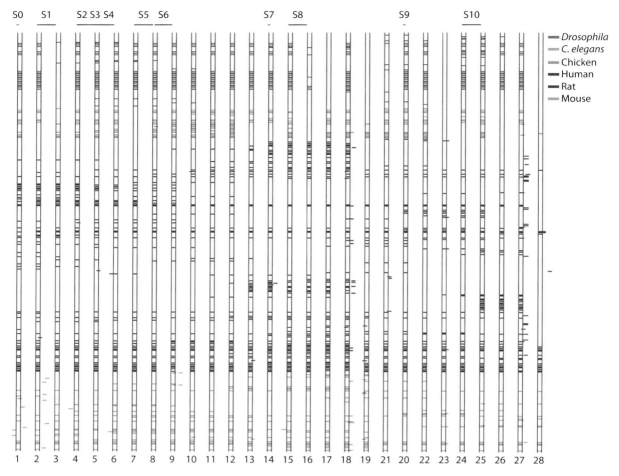

Figure 1

Unigene clusters mapped via blastn (mouse, rat, cow, human) or tblastx (chicken, *Caenorhabditis elegans*, *Drosophila*) to the mouse genome. Each row represents a distinct GenBank sequence. Only regions of homology that had a blast score of e-score cutoff (e) < = 0.001 are shown. Constitutive exons 1–28 are defined as exons that are always present in each mouse transcript or transcript fragment. (Mouse transcript AF465244, the bottom-most transcript, is made up only of these constitutive exons.) Introns have been constrained so that the size between each constitutive exon is held constant. An interactive version of this figure can be seen at **http://www.afodor.net** (Java-enabled Web browser required).

Constitutive exon:
an exon that is present in all known sequences of a given gene

BK mRNA and EST transcripts from a number of species aligned via blast to the mouse genome. Each row in **Figure 1** represents a distinct RNA or EST sequence mapped to the mouse genome. We define an exon as constitutive if it is seen in every transcript or transcript fragment from a given species. By this standard, we define 28 constitutive exons for the mouse channel, labeled as 1–28 in **Figure 1**. Using the thin gray vertical lines, we can compare the 28 reference exons we observed in mouse with transcripts from other species. By inspection of **Figure 1** we note the following patterns of splices in the BK channel: (*a*) The constitutive exons are largely conserved across phyla, (*b*) there are hot spots of alternative splicing (such as between exons 18 and 19) in which many different alternative splices are observed, and (*c*) the alternative splices seen in vertebrates are not conserved to *Drosophila* and *C. elegans*.

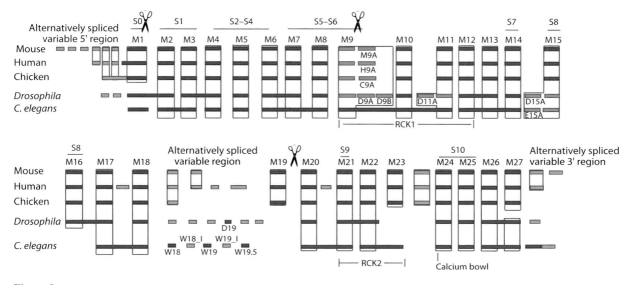

Figure 2

Conservation pattern of constitutive (*blue*), tandem duplicate (*green*), and inserted (*orange*) exons. We label the constitutive exons from mouse M1–M27. (Exon 28 contains a stop codon early in its sequence and is not well conserved across species.) S0–S10 annotations indicate hydrophobic regions of the channel (10). Gray boxes indicate a conserved exon with amino acid sequence identity > = 30% for coding regions or a blast score of $e < = 0.001$ for noncoding regions (see Methods, below; sequences are given in **Supplemental Table 1** online). Scissor icons indicate known domain boundaries where either separate transcripts could be coinjected to produce functional channels [S0 + rest of the channel (10); S0–S8 + S9–S10 (54)] or a truncated S0–S6 channel by itself formed functional channels (53). Splices that are only supported by ESTs from one experiment are excluded from this figure, as are insertion splices that are too small to reliably map between species (e.g., splice G from **Figure 3**).

To summarize the data from our custom BK genome browser (**http://www.afodor.net**), we present in **Figure 2** an abstract view of the constitutive and alternatively spliced exons for five species. Each row in **Figure 2** represents all exons from all transcripts of a given species mapped to the genome of that species. Nearly all the constitutive exons in mouse can be mapped at this level of sequence identity from vertebrates to *Drosophila* and *C. elegans*. This supports our first observation: Constitutive exons are well conserved across phyla. This also gives support to our assertion that our use of blast version 2.2.8 was successful, even though some later versions of blast fail to align many of the conserved constitutive exons across phlya with tblastx.

The translated sequences of transcripts that contain alternatives to the constitutive exons are shown for vertebrates (**Figure 3**), *Drosophila* (**Figure 4**), and *C. elegans* (**Figure 5**). The re-sults in **Figures 2–5** are from our database search, with rather strict criteria for inclusion. As outlined in the introduction, recent literature in alternative splicing suggests that at least some alternative splicing events represent noise with no biological function. In examining the results of our blast searches, we therefore need to establish some criteria as to which splices we believe are not aberrant. We therefore refer to a splice as well supported if it is part of a full-length mRNA sequence, if it is derived from a region that is conserved in the genome of another species, or if it has been observed in ESTs from multiple experiments. We refer to a splice as poorly supported if it has only been seen in a single EST or PCR experiment. To the right of each protein sequence in **Figures 3–5**, we have placed a three-character code demonstrating the level of support for each splice. The key for these three-letter codes is shown in **Figure 6**. The first letter represents the species, the

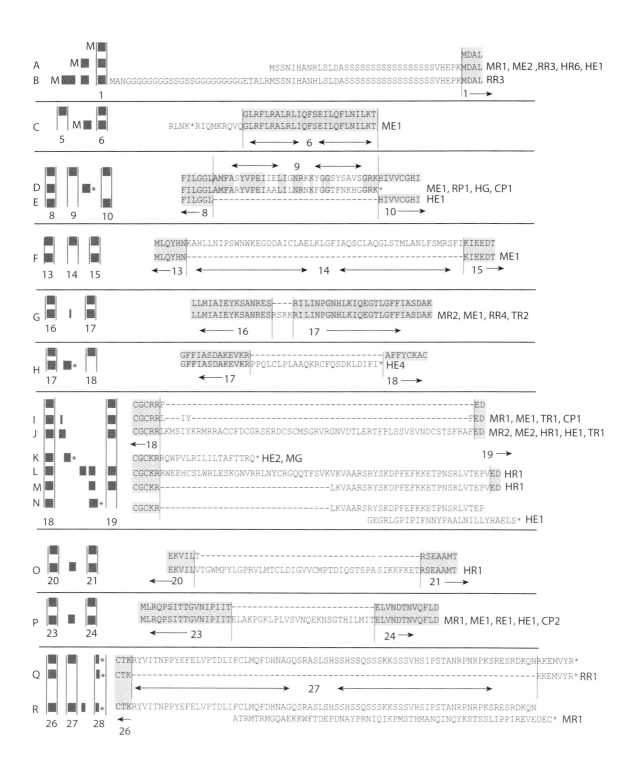

A   M   MSSNIHANHLSLDASSSSSSSSSSSSSSSSSSSSSVHEPKMDAL   MR1, ME2 ,RR3, HR6, HE1

B   M   MANGGGGGGGGSSGSSGGGGGGGGGGGETALRMSSNIHANHLSLDASSSSSSSSSSSSSSSSSSSSSVHEPKMDAL   RR3

1

C   M   RLNK*RIQMKRQVQGLRFLRALRLIQFSEILQFLNILKT   ME1

5   6   6

D   9   FILGGLAMFASYVPEIIELIGNRKKYGGSYSAVSGRKHIVVCGHI   ME1, RP1, HG, CP1

E   FILGGLAMFARYVPEIAALILNRNKFGGTFNKHGGRK* / FILGGL--------------------------HIVVCGHI   HE1

8   9   10   8   10

F   MLQYHNKAHLLNIPSWNWKEGDDAICLAELKLGFIAQSCLAQGLSTMLANLFSMRSFIKIEEDT / MLQYHN--KIEEDT   ME1

13   14   15   13   14   15

G   LLMIAIEYKSANRES----RILINPGNHLKIQEGTLGFFIASDAK / LLMIAIEYKSANRESRSRKRILINPGNHLKIQEGTLGFFIASDAK   MR2, ME1, RR4, TR2

16   17   16   17

H   GFFIASDAKEVKR-------------------------AFFYCKAC / GFFIASDAKEVKRPPQLCLPLAAQKRCFQSDKLDIFI*   HE4

17   18   17   18

I   CGCRRP-------------------------------------ED / CGCRRL---IY-------------------------------FED   MR1, ME1, TR1, CP1

J   CGCRRLKMSIYKRMRRACCFDCGRSERDCSCMSGRVRGNVDTLERTFPLSSVSVNDCSTSFRAFED   MR2, ME2, HR1, HE1, TR1

K   CGCKRRQWPVLRILILTAFTTRQ*   HE2, MG

L   CGCKRRWEEHCSLWRLESKGNVRRLNYCRGQQTFSVKVKVAARSRYSKDPFEFKKETPNSRLVTEPVED   HR1

M   CGCKR-------------------------LKVAARSRYSKDPFEFKKETPNSRLVTEPVED   HR1

N   CGCKR-------------------------LKVAARSRYSKDPFEFKKETPNSRLVTEP / GEGRLGPIPIFNNYPAALNILLYHAELS*   HE1

18   19   18   19

O   EKVILT-----------------------------------RSEAAMT / EKVILVTGWMPYLGPRVLMTCLDIGVVCMPTDIQSTSPASIKKFKETRSEAAMT   HR1

20   21   20   21

P   MLRQPSITTGVNIPIIT--------------------------ELVNDTNVQFLD / MLRQPSITTGVNIPIITELAKPGKLPLVSVNQEKNSGTHILMITELVNDTNVQFLD   MR1, ME1, RE1, HE1, CP2

23   24   23   24

Q   CTKRYVITNPPYEFELVPTDLIFCLMQFDHNAGQSRASLSHSSHSSQSSSKKSSSVHSIPSTANRPNRPKSRESRDKQNRKEMVYR*

CTK---RKEMVYR*   RR1

27

R   CTKRYVITNPPYEFELVPTDLIFCLMQFDHNAGQSRASLSHSSHSSQSSSKKSSSVHSIPSTANRPNRPKSRESRDKQN / ATRMTRMGQAEKKWFTDEPDNAYPRNIQIKPMSTHMANQINQYKSTSSLIPPIREVEDEC*   MR1

26   27   28   26

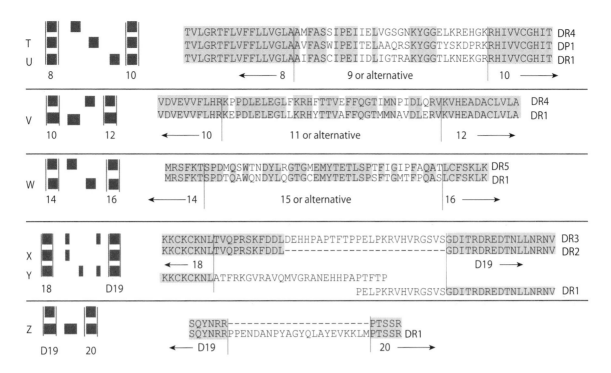

Figure 4

Drosophila alternative splices.

second letter represents the type of evidence, and the last digit is the number of times that evidence has been observed. For example, splice H, with an evidence code HE4 in **Figure 3**, has been observed in four distinct EST experiments. The GenBank IDs and literature references referred to by these codes are provided in **Table 1**. As a consequence of our criteria for inclusion, and of only examining a subset of species, the data in **Figures 2–5** are not exhaustive summaries of splices that have been reported in all species in the literature (see Methods, below, for more details).

We observe two kinds of nonconstitutive, alternatively spliced BK exons that by the above criteria we consider well supported. Insertion exons are inserted in between constitutive exons. They produce in some transcripts an additional sequence. We see that, in contrast to constitutive exons, alternatively spliced insertion exons are not conserved between vertebrates and invertebrates. We also observe for BK channels a number of tandem duplicate exons (**Figure 2**). In these cases, an exon has undergone a duplication event. Although in principle both duplicated exons could be included

Insertion exons: are inserted in between constitutive exons in some sequences

Tandem duplicate exon: an exon that has undergone a duplication event. Only one form of the exon is usually observed in each sequence

Figure 3

Vertebrate alternative splices. Each splice is given an arbitrary single character name as shown in the left-most column. An M in the left column indicates a transcript that has a possible start Met. An asterisk in the left column indicates a transcript with a stop codon in frame. Exon numbers are as in **Figure 2**. To the right of the protein sequence of each splice, the number of times each splice has been observed is given with the evidence codes displayed in **Figure 6**. GenBank accession numbers and references for the lines of evidence are given in **Table 1**.

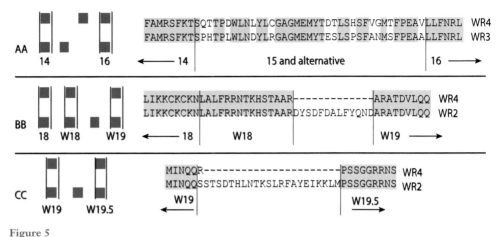

Figure 5

Caenorhabditis elegans alternative splices.

in a BK transcript, across species we always observed only one of the duplicate exons in any transcript. As has been previously reported (48), there is evidence that tandem duplicate exons arose independently in BK for vertebrates and invertebrates. **Figure 7** shows alignments of the tandem duplicates across species for exons 9, 11, and 15. The dendrograms in the right panel of **Figure 7** show that for exon 9 the vertebrate

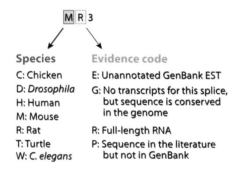

Figure 6

Key for the evidence codes used in **Figures 3–5**. The first character indicates species. The second character indicates the type of evidence. The final character indicates the number of times the evidence has been seen. We count multiple ESTs from the same experiment and tissue as only one line of support. In the example given, the code MR3 would indicate that a splice has been seen in three distinct full-length mouse RNAs.

forms of the duplicate exons are more closely related to one another than to the invertebrate forms. This suggests that the duplication event for exon 9 arose after the separation of vertebrates from invertebrates. Likewise, the dendrogram for exon 15 suggests that the duplication of exon 15 took place after the separation of *C. elegans* from *Drosophila*. These phylogenetic analyses provide strong evidence for the hypothesis that the exon duplication events occurred independently within each phylum (see also Reference 48).

We conclude that alternative splicing events observed for the BK channel are unique to each phylum. It is, of course, impossible to prove that sequences with little in common do not in fact share some ancient common ancestor. When examining the distinct insertion exons that are in similar regions of the channel sequence in different phyla, we therefore do not know if these distinct alternative splices arose independently or if, under much less selection pressure than were constitutive regions, alternatively spliced exons simply evolved rapidly to reach their distinct sequences. In either case, we arrive at a picture of channel function in which rigidly conserved constitutive exons appear to be required across phyla for normal channel function whereas more poorly conserved insertion exons are occasionally used in some tissues

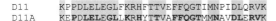

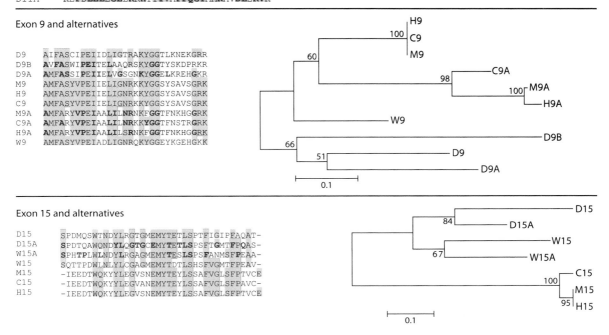

Figure 7

Tandem duplicates of BK channels. Phylogenetic trees were drawn with the program MEGA by the use of the Neighbor-Joining algorithm with the default parameters (60). Scale bars indicate fraction identity. Silent mutations are shown in red for D11A (relative to D11, $p < 10^{-6}$), D9A and D9B [relative to D9, $p = 0.02$ (D9A), $p = 0.12$ (D9B)], M9A relative to M9 ($p = 0.014$), C9A relative to C9 ($p = 0.001$), H9A relative to H9 ($p = 0.04$), D15A relative to D15 ($p < 0.001$), and W15A relative to W15 ($p = 0.03$). p-values indicate evaluation of the null hypothesis that the prevalence of the silent mutations is what would be expected by chance substitution; see Methods, below.

to tune channel function. We are on firmer phylogenetic ground with the duplication events at exons 9 and 15, where, because the duplicated exons still have significant sequence identity to one another, we can trace the phylogenetic relationship between the duplicated pieces (**Figure 7**).

Rare Vertebrate Duplicate Exons Show Less Conservation than More Common Exons but Are Still Under Protein Selection Pressure

Within vertebrates for exon 9, we can discriminate between forms of the duplicate exons that appear often in their respective species (H9,

M9, C9 for human, mouse, and chicken, respectively) and the forms that rarely appear (H9A, M9A, and C9A). Of the ~50 transcripts in the human, mouse, and chicken Unigene clusters that cover the exon 9 region in **Figure 1**, only two ESTs from the same experiment (BY742482, BB642344) feature the M9A, H9A, or C9A variant. Moreover, both of these mouse transcripts feature a stop codon at the end of exon M9A (**Figure 3**; splice D). Every other available transcript that met our criteria for inclusion in this study (see Methods, below) from vertebrates features H9, C9, or M9 rather than H9A, C9A, or M9A [although exon C9A and a rat exon equivalent to M9A have been described in the literature (19, 30); see

Table 1]. We see in the dendrogram for exon 9 in **Figure 7** that the more frequently observed vertebrate forms of exon 9 (H9, M9, C9) are perfectly conserved among vertebrates. There is more variation, however, among the more poorly expressed alternative forms of the exons (H9A, M9A, and C9A). This is consistent with the idea that there is less selection pressure on rare, poorly expressed forms of the channel sequence than on more frequently observed exons.

How can we be sure that the exon duplication event represented by exon 9 in vertebrates represents a real protein variant and not genetic drift based on an aberrant splicing event? In *Drosophila*, one of the alternatives to exon 9 has been functionally characterized (**Table 1**, splice U), but this does not guarantee that the vertebrate forms, which as we saw above appear to have arisen independently, are under protein selection pressure. It has been argued that the presence of premature stop codons is a hallmark of nonfunctional splicing (37, 49). The EST mouse transcripts that contain H9A contain stop codons (**Figure 3**; splice D). Given the lack of full-length transcripts containing H9A, M9A, or C9A, is there other evidence that we can bring to bear?

To test the hypothesis that the alternative forms of the tandem duplicate exons are under protein selection pressure, we looked in the nucleotide sequences for the prevalence of silent mutations. For each of the alternative forms of the exons in **Figure 7**, we looked for silent mutations that occurred when compared with the main form. So, for example, we compared exon D9A and D9B from *Drosophila* with D9. Likewise, we compared H9A from human with H9, C9A from chicken with C9, and so forth. By applying a simple statistical test based on the binomial distribution (see Methods, below) we find that for all the vertebrate alternatives to exon 9, there is a statistically significant ($p < 0.05$) number of silent mutations relative to the more commonly expressed form. This observation strongly suggests that, despite their rarity, the alternative forms of the exons are indeed under protein selection pressure.

Splices Are Associated with Experimentally Established Domain Boundaries

How does the domain structure of the BK channel relate to the patterns of alternative splicing that we observe across species? It is well established that the S5–S6 region of the channel forms the ion-conducting pore region (50). An informatics approach that has been used to discover other domains within BK channels looks for regions of homology between the BK sequence and domains of known structure (14). It has been hypothesized that the post-S6 region of BK contains two regulator of calcium conductance (RCK) domains (51, 52). The locations of these putative domain boundaries, based on published alignments (51, 52), are shown in **Figure 2**. The second putative RCK domain is substantially smaller than the first; alignments suggest that RCK2 is missing four β strands and three α helices found in RCK1 and in the crystal structure of MthK (52). Because putative RCK2 is so much shorter than RCK1, it is much more difficult to detect significant homology between MthK and RCK2 than between MthK and RCK1 (14). The sequence similarity between BK and the proteins for which these domains have been solved by crystal structure is quite weak, well below the 30% sequence identity level. It is difficult at these low levels of sequence similarity to precisely define domain boundaries, so the RCK1 and RCK2 annotations in **Figure 2** may only mark approximately where these putative domains begin and end.

For the BK channel we do not have to rely only on bioinformatic methods to determine domain boundaries because functional studies have directly shown experimental demarcations of several domain boundaries within the channel. For example, a mouse BK transcript truncated immediately after the S6 region produced functional channels in an oocyte exogenous expression system, albeit at very low levels of expression (53). This truncation site (**Figure 2**) is near a hot spot of alternative splicing, with vertebrates and *Drosophila* exhibiting alternative forms of exon 9 (**Figures 2**, **3**, and **7**). Another experimentally determined domain boundary is at the S8 region of the channel. A transcript truncated at S8, when coinjected into oocytes with a transcript consisting of the S9–S10 region, produced functional channels (54). This truncation site after S8 (**Figure 2**) corresponds to another hot spot of channel splicing activity with multiple unique insertion splices occurring across species between exons 18 and 19 (**Figure 2**). Finally, the beginning and end of the channel are by definition domain boundaries and are also sites where alternative splicing frequently occurs.

There are a number of possible explanations for the correlation of splicing activity with experimentally established domain boundaries in the channel. Functional channels may tolerate changes in sequence at domain boundaries much more easily than changes in the middle of a domain. If this were the case, there would be more opportunity for the introduction of successful splices at these domain boundaries, and across evolutionary space, we would expect to see splices arise independently. Alternatively, the presence of the stop codons in some splices after exon 9 (**Figure 3**; splice D), and in several well-supported splices that occur between S8 and S9 (**Figure 3**; splices H and K), suggests that transcription of the different domains of the channel may occur independently. That is, some forms of the channel may be assembled in vivo in a manner similar to the coinjection experiments in which the S0–S8 and S9–S10 domains were coinjected in oocytes (54). Alternative splicing is often associated with the 5′- and 3′-terminal regions of genes (36). This presumably allows for modulation of regions of sequences that control gene expression and protein trafficking. If the S0–S8 and S9–S10 regions are indeed transcribed independently, phylum-specific trafficking signals may account for some of the observed splice variations in the S8 region.

DISCUSSION

A recent genome-wide study of alternative splicing concluded that alternatively spliced exons are more poorly conserved than are constitutive exons (55). This observation is consistent with our results that constitutive, but not alternatively spliced, BK exons are conserved across phyla. Our results suggest a number of possible future experimental explorations of BK channel function. We hypothesize that insertion alternative splices between exons 18 and 19 are common and poorly conserved because the channel can tolerate insertions and variation in this interdomain region. If one did an experiment in which random sequences were inserted throughout the channel sequence and screened for functional channels, we would predict a disproportionate number of successful insertions in the regions between exons M18 and M19. As novel BK sequences are found by new sequencing efforts, we predict that additional insertion splices will be discovered in the M18–M19 region and that many of these splices will produce functional channels. Splices in this region that have been functionally characterized change the biophysical properties of the channel modulating the relationship between open probability and voltage (**Table 1**) and changing the susceptibility of the channel to regulation by phosphorylation (7, 56). The diversity of sequences seen in this region across evolutionary space therefore apparently provides a mechanism to support functional tuning of the channel. This modulation in turn may be important in setting the excitability of a variety of tissue types.

Another region of the channel at which multiple splices occur is in exon M9 at the end of the S5–S6 pore region. The alternative vertebrate forms of exon 9 are seen in only a few ESTs, some of which feature a stop codon. In the absence of other evidence, this may suggest that these splices are aberrant (37, 38). We can tell by the prevalence of silent mutations (**Figure 7**) in the vertebrate forms of exon 9, however, that the alternative forms of the exon are likely under protein selection pressure. Moreover, the *Drosophila* version of this exon (without the stop codon) has been functionally characterized at the protein level (**Table 1**, splice U). We have, therefore, a high level of confidence that the splices at the S6 domain boundary are valid. As might be expected for an alteration to the channel in such an important region, the functionally characterized alternative form of exon 9 from *Drosophila* has a profound impact on channel function, decreasing the single-channel conductance and open probability while causing faster activation times (15; **Table 1**). The alternative forms of exon 9 have apparently arisen independently in *Drosophila* and vertebrates (**Figure 7**; 48). If this region were the target of true convergent evolution, we would predict that an experiment in which the alternative *Drosophila* forms of exon 9 were introduced into a vertebrate sequence would yield a similar phenotype to that of a channel in which the alternative vertebrate form of exon 9 were introduced into a *Drosophila* sequence.

Previous genome-wide studies of alternative splicing (57, 58) have suggested that splices often disrupt or remove domains. By contrast, in the BK channel we see convergent evolution of alternative splicing at experimentally established domain boundaries. Genome-wide studies of alternative splicing may not have emphasized our observed pattern of splicing because their results, on average, are derived from "average" genes that are not as extensively spliced or as long as the BK channel. As the cost of sequencing continues to drop and more experimental data are integrated with the sequence databases, it will be interesting to see if alternative splicing at domain boundaries of multidomain proteins becomes a more prominent theme.

METHODS

Splice Discovery

For all searches we used NCBI-blast version 2.2.8 for the Windows XP platform. Genomic regions of the mouse, human, chicken, *C. elegans*, and *Drosophila* BK α subunit were downloaded from the University of California at Santa Cruz genome Web site (**http://genome.ucsc.edu/**). With an e-score cutoff (*e*) < 0.001, we used blastn (for identical species or to map within mammals) or tblastx to map Unigene clusters for BK to the genomic regions of human, mouse, chicken, *Drosophila*, and *C. elegans*. Blast results were visualized with a custom BK genome browser (**http://www.afodor.net**; code available upon request). To control for genomic contamination, we only considered transcripts that mapped to at least three distinct exons. **Figure 2** shows only well-supported splices, by which we mean splices that are supported by multiple lines of evidence or are part of a full-length transcript. By these criteria, we excluded from **Figure 2** splices that were only seen in an EST from a single experiment. In **Figures 3–6**, we display only splices that have consequences for protein sequence. We therefore exclude 5′ and 3′ splices that occur before the first possible start Met or after a stop codon.

To be included in our survey, detected BK sequences had to be included in the Unigene clusters on the date that we downloaded them from NCBI (November 2006). In addition, our requirement for a well-supported Unigene transcript is conservative and excludes some published splices. For example, our survey does not include S0 splices, which do not form functional channels, a proposed dominant-negative regulator of BK channel expression (27, 59). Our requirement that splices be described in Unigene transcripts that map to at least three distinct genomic regions also excluded some splices that had been described in the literature at the protein sequence level from chicken (19, 20), rat (30), and *Drosophila* (15, 17). Likewise, because we did not consider all species, we excluded some well-supported splices that occurred only in turtle (21, 23). Finally, our Unigene search did not pick up two additional insertion splices between exons 18 and 19 that had been functionally characterized (7). One of these, e22, is poorly conserved between mouse and human (7), whereas the other, e23, introduces a stop codon and produces nonfunctional channels (7).

Criteria for a Conserved Exon

For each exon in **Figure 2**, we looked for matching hits in genomic regions of each species shown in **Figure 2**, using blastn ($e < 0.1$) for the noncoding splices (before exon M1 or after exon M27) and tblastx ($e < 0.01$) for the coding regions (between exons M1 and M27). At these blast cutoff levels, we found no hits for the insertion splices across phyla. We cannot, of course, guarantee that a more sensitive search would not have found nontrivial similarities in cases in which we report no conservation. There may also be as-yet-undescribed splices that are conserved between vertebrates and invertebrates. However, given that, for example, the matching constitutive exons from **Figure 2** map from mouse to *Drosophila* with tblastx e-scores ranging from $e < 10^{-5}$ to $e < 10^{-62}$, and our blast searches at $e < 0.01$ failed for alternatively spliced insertion exons, we feel that our conclusion that the constitutively and alternatively spliced insertion exons are under very different degrees of conservation pressure is highly reasonable.

Supplemental Table 1 shows alignments for the exons in coding regions shown in **Figure 2** that have at least a 30% amino acid sequence identity to every other exon in the group (to access supplemental material, follow the Supplemental Material link from the Annual Reviews home page at **http://www.annualreviews.org**). Because some of the exon fragments were quite short, we here define sequence identity in a conservative way as the number of identical residues divided by the number of ungapped residues in the target or query sequence that had the most number of ungapped residues in the alignment. That is, if M is the number of matching residues, Q is the number of ungapped residues from the query sequence in the alignment, and T is the number of ungapped residues in the target sequence from the alignment, we define percent identity as

$$\frac{M}{\max(Q, T)}.$$

Alignments were created with ClustalW with the default parameters.

Calculation of *p*-Values for Silent Mutations

If one makes a random change to a codon, changing any one of the three nucleotides to another base, the odds that this change will produce a silent mutation are ~0.24. If one blasts via blastn with a cutoff of $e < 0.01$ the genomic region of the human BK channel to the genomic region of the mouse BK channel, one will obtain 343 high-scoring pairs. If one appends these 343 high-scoring pairs together and looks at all the nucleotide changes, the odds of seeing a silent mutation change are 0.25, similar to the 0.24 value obtained by changing bases at random. Using the more conservative value of 0.25, if, for a given reading frame, one sees N codon changes at the DNA level between two aligned sequences and n of these codon changes are silent mutations, the probability that one would see that many silent mutations by chance is given by

$$\sum_{n}^{N} \left(\frac{N!}{n!(N-n)!} \right) \cdot 25^n \cdot 75^{(N-n)}.$$

SUMMARY POINTS

1. The BK channel is extensively spliced.

2. New technology has shown alternative splicing to be far more prevalent in mammalian genomes than previously thought.

3. We do not know how many of the recently discovered alternative splices are important biologically.

4. Because different splices of ion channel proteins can be functionally characterized at the protein level with the patch clamp, ion channels may be a useful assay for discriminating important and unimportant splices.

5. Constitutive exons of the BK channels are conserved across phlya.

6. Alternatively spliced exons are not conserved across phyla.

7. Across phyla, certain regions of the BK channel are prone to splicing events, suggesting convergent evolution.

DISCLOSURE STATEMENT

The authors are not aware of any biases that might be perceived as affecting the objectivity of this review.

ACKNOWLEDGMENTS

We thank Weiyan Li, Andrea Meredith, Jon Sack, and Christina Wilkens for reading an earlier version of the manuscript. This work was supported by the Mathers Foundation.

LITERATURE CITED

1. Nelson MT, Cheng H, Rubart M, Santana LF, Bonev AD, et al. 1995. Relaxation of arterial smooth muscle by calcium sparks. *Science* 270:633–37

2. Brenner R, Perez GJ, Bonev AD, Eckman DM, Kosek JC, et al. 2000. Vasoregulation by the $\beta 1$ subunit of the calcium-activated potassium channel. *Nature* 407:870–76

3. Meredith AL, Thorneloe KS, Werner ME, Nelson MT, Aldrich RW. 2004. Overactive bladder and incontinence in the absence of the BK large conductance Ca^{2+}-activated K^{+} channel. *J. Biol. Chem.* 279:36746–52

4. Meredith AL, Wiler SW, Miller BH, Takahashi JS, Fodor AA, et al. 2006. BK calcium-activated potassium channels regulate circadian behavioral rhythms and pacemaker output. *Nat. Neurosci.* 9:1041–49

5. Pyott SJ, Meredith AL, Fodor AA, Vazquez AE, Yamoah EN, Aldrich RW. 2007. Cochlear function in mice lacking the BK channel α, $\beta 1$, or $\beta 4$ subunits. *J. Biol. Chem.* 282:3312–24

6. Werner ME, Zvara P, Meredith AL, Aldrich RW, Nelson MT. 2005. Erectile dysfunction in mice lacking the large-conductance calcium-activated potassium (BK) channel. *J. Physiol.* 567:545–56

7. Chen L, Tian L, MacDonald SH, McClafferty H, Hammond MS, et al. 2005. Functionally diverse complement of large conductance calcium- and voltage-activated potassium channel (BK) α-subunits generated from a single site of splicing. *J. Biol. Chem.* 280:33599–609

8. Ricci AJ, Gray-Keller M, Fettiplace R. 2000. Tonotopic variations of calcium signalling in turtle auditory hair cells. *J. Physiol.* 524(Pt. 2):423–36

9. Dworetzky SI, Boissard CG, Lum-Ragan JT, McKay MC, Post-Munson DJ, et al. 1996. Phenotypic alteration of a human BK (hSlo) channel by hSloβ subunit coexpression: changes in blocker sensitivity, activation/relaxation and inactivation kinetics, and protein kinase A modulation. *J. Neurosci.* 16:4543–50

10. Wallner M, Meera P, Toro L. 1996. Determinant for β-subunit regulation in high-conductance voltage-activated and Ca^{2+}-sensitive K^{+} channels: an additional transmembrane region at the N terminus. *Proc. Natl. Acad. Sci. USA* 93:14922–27

11. Brenner R, Jegla TJ, Wickenden A, Liu Y, Aldrich RW. 2000. Cloning and functional characterization of novel large conductance calcium-activated potassium channel β subunits, hKCNMB3 and hKCNMB4. *J. Biol. Chem.* 275:6453–61

12. Chung SK, Reinhart PH, Martin BL, Brautigan D, Levitan IB. 1991. Protein kinase activity closely associated with a reconstituted calcium-activated potassium channel. *Science* 253:560–62

13. Reinhart PH, Chung S, Martin BL, Brautigan DL, Levitan IB. 1991. Modulation of calcium-activated potassium channels from rat brain by protein kinase A and phosphatase 2A. *J. Neurosci.* 11:1627–35

14. Fodor AA, Aldrich RW. 2006. Statistical limits to the identification of ion channel domains by sequence similarity. *J. Gen. Physiol.* 127:755–66

15. Adelman JP, Shen KZ, Kavanaugh MP, Warren RA, Wu YN, et al. 1992. Calcium-activated potassium channels expressed from cloned complementary DNAs. *Neuron* 9:209–16

16. Butler A, Tsunoda S, McCobb DP, Wei A, Salkoff L. 1993. mSlo, a complex mouse gene encoding "maxi" calcium-activated potassium channels. *Science* 261:221–24

17. Lagrutta A, Shen KZ, North RA, Adelman JP. 1994. Functional differences among alternatively spliced variants of Slowpoke, a *Drosophila* calcium-activated potassium channel. *J. Biol. Chem.* 269:20347–51

18. Tseng-Crank J, Foster CD, Krause JD, Mertz R, Godinot N, et al. 1994. Cloning, expression, and distribution of functionally distinct Ca^{2+}-activated K^+ channel isoforms from human brain. *Neuron* 13:1315–30

19. Navaratnam DS, Bell TJ, Tu TD, Cohen EL, Oberholtzer JC. 1997. Differential distribution of Ca^{2+}-activated K^+ channel splice variants among hair cells along the tonotopic axis of the chick cochlea. *Neuron* 19:1077–85

20. Rosenblatt KP, Sun ZP, Heller S, Hudspeth AJ. 1997. Distribution of Ca^{2+}-activated K^+ channel isoforms along the tonotopic gradient of the chicken's cochlea. *Neuron* 19:1061–75

21. Jones EM, Laus C, Fettiplace R. 1998. Identification of Ca^{2+}-activated K^+ channel splice variants and their distribution in the turtle cochlea. *Proc. R. Soc. London Ser. B* 265:685–92

22. Xie J, McCobb DP. 1998. Control of alternative splicing of potassium channels by stress hormones. *Science* 280:443–46

23. Jones EM, Gray-Keller M, Fettiplace R. 1999. The role of Ca^{2+}-activated K^+ channel spliced variants in the tonotopic organization of the turtle cochlea. *J. Physiol.* 518(Pt. 3):653–65

24. Jones EM, Gray-Keller M, Art JJ, Fettiplace R. 1999. The functional role of alternative splicing of Ca^{2+}-activated K^+ channels in auditory hair cells. *Ann. N. Y. Acad. Sci.* 868:379–85

25. Ha TS, Jeong SY, Cho SW, Jeon H, Roh GS, et al. 2000. Functional characteristics of two BKCa channel variants differentially expressed in rat brain tissues. *Eur. J. Biochem.* 267:910–18

26. Wang ZW, Saifee O, Nonet ML, Salkoff L. 2001. SLO-1 potassium channels control quantal content of neurotransmitter release at the *C. elegans* neuromuscular junction. *Neuron* 32:867–81

27. Zarei MM, Zhu N, Alioua A, Eghbali M, Stefani E, Toro L. 2001. A novel MaxiK splice variant exhibits dominant-negative properties for surface expression. *J. Biol. Chem.* 276:16232–39

28. Erxleben C, Everhart AL, Romeo C, Florance H, Bauer MB, et al. 2002. Interacting effects of N-terminal variation and strex exon splicing on *slo* potassium channel regulation by calcium, phosphorylation, and oxidation. *J. Biol. Chem.* 277:27045–52

29. Liu X, Chang Y, Reinhart PH, Sontheimer H. 2002. Cloning and characterization of glioma BK, a novel BK channel isoform highly expressed in human glioma cells. *J. Neurosci.* 22:1840–49

30. Langer P, Grunder S, Rusch A. 2003. Expression of Ca^{2+}-activated BK channel mRNA and its splice variants in the rat cochlea. *J. Comp. Neurol.* 455:198–209

31. Beisel KW, Rocha-Sanchez SM, Ziegenbein SJ, Morris KA, Kai C, et al. 2007. Diversity of Ca^{2+}-activated K^+ channel transcripts in inner ear hair cells. *Gene* 386:11–23

32. Adams MD, Kelley JM, Gocayne JD, Dubnick M, Polymeropoulos MH, et al. 1991. Complementary DNA sequencing: expressed sequence tags and human genome project. *Science* 252:1651–56

33. Mironov AA, Fickett JW, Gelfand MS. 1999. Frequent alternative splicing of human genes. *Genome Res.* 9:1288–93

34. Lander ES, Linton LM, Birren B, Nusbaum C, Zody MC, et al. 2001. Initial sequencing and analysis of the human genome. *Nature* 409:860–921

35. Eyras E, Caccamo M, Curwen V, Clamp M. 2004. ESTGenes: alternative splicing from ESTs in Ensembl. *Genome Res.* 14:976–87

36. Zavolan M, van Nimwegen E, Gaasterland T. 2002. Splice variation in mouse full-length cDNAs identified by mapping to the mouse genome. *Genome Res.* 12:1377–85

37. Sorek R, Shamir R, Ast G. 2004. How prevalent is functional alternative splicing in the human genome? *Trends Genet.* 20:68–71

38. Green RE, Lewis BP, Hillman RT, Blanchette M, Lareau LF, et al. 2003. Widespread predicted nonsense-mediated mRNA decay of alternatively-spliced transcripts of human normal and disease genes. *Bioinformatics* 19(Suppl. 1):i118–21

39. Lareau LF, Green RE, Bhatnagar RS, Brenner SE. 2004. The evolving roles of alternative splicing. *Curr. Opin. Struct. Biol.* 14:273–82

40. Kampa D, Cheng J, Kapranov P, Yamanaka M, Brubaker S, et al. 2004. Novel RNAs identified from an in-depth analysis of the transcriptome of human chromosomes 21 and 22. *Genome Res.* 14:331–42

41. Kapranov P, Cawley SE, Drenkow J, Bekiranov S, Strausberg RL, et al. 2002. Large-scale transcriptional activity in chromosomes 21 and 22. *Science* 296:916–19

42. Bertone P, Stolc V, Royce TE, Rozowsky JS, Urban AE, et al. 2004. Global identification of human transcribed sequences with genome tiling arrays. *Science* 306:2242–46

43. Johnson JM, Castle J, Garrett-Engele P, Kan Z, Loerch PM, et al. 2003. Genome-wide survey of human alternative premRNA splicing with exon junction microarrays. *Science* 302:2141–44

44. Bonizzoni P, Rizzi R, Pesole G. 2005. ASPIC: a novel method to predict the exon-intron structure of a gene that is optimally compatible to a set of transcript sequences. *BMC Bioinformatics* 6:244

45. Bonizzoni P, Rizzi R, Pesole G. 2006. Computational methods for alternative splicing prediction. *Brief Funct. Genomic Proteomic* 5:46–51

46. Florea L, Hartzell G, Zhang Z, Rubin GM, Miller W. 1998. A computer program for aligning a cDNA sequence with a genomic DNA sequence. *Genome Res.* 8:967–74

47. Altschul SF, Madden TL, Schaffer AA, Zhang J, Zhang Z, et al. 1997. Gapped BLAST and PSI-BLAST: a new generation of protein database search programs. *Nucleic Acids Res.* 25:3389–402

48. Copley RR. 2004. Evolutionary convergence of alternative splicing in ion channels. *Trends Genet.* 20:171–76

49. Lewis BP, Green RE, Brenner SE. 2003. Evidence for the widespread coupling of alternative splicing and nonsense-mediated mRNA decay in humans. *Proc. Natl. Acad. Sci. USA* 100:189–92

50. Hille B. 2001. *Ion Channels of Excitable Membranes*. Sunderland, MA: Sinauer

51. Jiang Y, Pico A, Cadene M, Chait BT, MacKinnon R. 2001. Structure of the RCK domain from the *E. coli* K^+ channel and demonstration of its presence in the human BK channel. *Neuron* 29:593–601

52. Kim HJ, Lim HH, Rho SH, Eom SH, Park CS. 2006. Hydrophobic interface between two regulators of K^+ conductance domains critical for calcium-dependent activation of large conductance Ca^{2+}-activated K^+ channels. *J. Biol. Chem.* 281:38573–81

53. Piskorowski R, Aldrich RW. 2002. Calcium activation of BK(Ca) potassium channels lacking the calcium bowl and RCK domains. *Nature* 420:499–502

54. Wei A, Solaro C, Lingle C, Salkoff L. 1994. Calcium sensitivity of BK-type KCa channels determined by a separable domain. *Neuron* 13:671–81

55. Modrek B, Lee CJ. 2003. Alternative splicing in the human, mouse and rat genomes is associated with an increased frequency of exon creation and/or loss. *Nat. Genet.* 34:177–80

56. Tian L, Duncan RR, Hammond MS, Coghill LS, Wen H, et al. 2001. Alternative splicing switches potassium channel sensitivity to protein phosphorylation. *J. Biol. Chem.* 276:7717–20

57. Liu S, Altman RB. 2003. Large scale study of protein domain distribution in the context of alternative splicing. *Nucleic Acids Res.* 31:4828–35

58. Resch A, Xing Y, Modrek B, Gorlick M, Riley R, Lee C. 2004. Assessing the impact of alternative splicing on domain interactions in the human proteome. *J. Proteome Res.* 3:76–83

59. Zarei MM, Eghbali M, Alioua A, Song M, Knaus HG, et al. 2004. An endoplasmic reticulum trafficking signal prevents surface expression of a voltage- and Ca^{2+}-activated K^+ channel splice variant. *Proc. Natl. Acad. Sci. USA* 101:10072–77

60. Kumar S, Tamura K, Jakobsen IB, Nei M. 2001. MEGA2: molecular evolutionary genetics analysis software. *Bioinformatics* 17:1244–45

Mechanisms of Muscle Degeneration, Regeneration, and Repair in the Muscular Dystrophies

Gregory Q. Wallace[1] and Elizabeth M. McNally[1,2]

[1]Department of Medicine and [2]Department of Human Genetics, The University of Chicago, Chicago, Illinois 60637; email: emcnally@uchicago.edu

Annu. Rev. Physiol. 2009. 71:37–57

First published online as a Review in Advance on September 22, 2008

The *Annual Review of Physiology* is online at physiol.annualreviews.org

This article's doi: 10.1146/annurev.physiol.010908.163216

Copyright © 2009 by Annual Reviews. All rights reserved

0066-4278/09/0315-0037$20.00

Key Words

dystrophin, sarcoglycan, dysferlin, lamin A/C, nitric oxide synthase

Abstract

To withstand the rigors of contraction, muscle fibers have specialized protein complexes that buffer against mechanical stress and a multifaceted repair system that is rapidly activated after injury. Genetic studies first identified the mechanosensory signaling network that connects the structural elements of muscle and, more recently, have identified repair elements of muscle. Defects in the genes encoding the components of these systems lead to muscular dystrophy, a family of genetic disorders characterized by progressive muscle wasting. Although the age of onset, affected muscles, and severity vary considerably, all muscular dystrophies are characterized by muscle necrosis that overtakes the regenerative capacity of muscle. The resulting replacement of muscle by fatty and fibrous tissue leaves muscle increasingly weak and nonfunctional. This review discusses the cellular mechanisms that are primarily and secondarily disrupted in muscular dystrophy, focusing on membrane degeneration, muscle regeneration, and the repair of muscle.

Dysferlin: a protein at the membrane of muscle that mediates membrane repair; mutations in dysferlin lead to limb girdle muscular dystrophy

Satellite cells: stem cells of muscle that are defined by their position between the basement membrane and the plasma membrane of muscle and that are capable of muscle regeneration

Duchenne/Becker muscular dystrophy (DMD and BMD): the severe and milder forms of muscular dystrophy caused by dystrophin gene mutations

Dystrophin glycoprotein complex (DGC): the collection of transmembrane and cytoplasmic proteins that copurify with dystrophin

Sarcoglycan: a subcomplex within the dystrophin glycoprotein complex; mutations in the sarcoglycan genes cause limb girdle muscular dystrophy

Dystrobrevin: a cytoplasmic component of the dystrophin glycoprotein complex that interacts with syntrophins

Syntrophin: a cytoplasmic component of the dystrophin complex that binds to nitric oxide synthase

OVERVIEW

Muscular dystrophy, a phenotype produced by many different genetic mutations, is characterized by progressive muscle weakness and degeneration with cycles of muscle necrosis and regeneration as the pathophysiological hallmarks of the disease. Muscle damage incited by membrane disruption triggers an increase in intracellular calcium. This increase in intracellular calcium has both negative and positive consequences; it activates proteolysis that can exacerbate muscle damage and also stimulates a muscle membrane repair system containing the protein dysferlin. The dysferlin repair system is important for myofibers because it participates in vesicle-mediated patching to repair focal membrane damage. Extensive muscle damage additionally activates satellite cells, resident muscle stem cells located between the sarcolemma and basement membrane. Normally quiescent, satellite cells divide in response to injury and use the basal lamina as a scaffold to repair or regenerate necrotic fibers. As muscle disease advances, muscle repair cannot adequately compensate for damage, and muscle is gradually replaced by fibrofatty tissue. It is unclear whether fibrofatty infiltration represents defective regeneration or an alternative scar pathway. With either possibility, the loss of myofibers is a primary effect. Different forms of muscular dystrophy target specific muscle groups. The most severe dystrophies are characterized by widespread muscle weakness that can lead to loss of ambulation, cardiopulmonary failure, and death.

The first identified muscular dystrophy gene was the dystrophin gene on the X chromosome, which encodes a structural protein. Muscle lacking dystrophin is especially susceptible to contraction-induced muscle degeneration. More recently, muscular dystrophy–associated mutations have been found in genes encoding proteins critical for muscle repair and regeneration. The axis of structural integrity, muscle repair, and muscle regeneration must be intact for normal muscle function.

MUSCLE DEGENERATION

Muscular Dystrophies Arising from Defects in the Connection to the Extracellular Matrix

Duchenne muscular dystrophy (DMD) is the most common and best understood muscular dystrophy, with the genetic defect identified two decades ago. Boys with DMD usually manifest with motor difficulties by six years of age. Muscle weakness progresses, leaving patients wheelchair-bound by their teens, with death occurring in their twenties owing to respiratory and cardiovascular failure. The gene defect for DMD was mapped to an X chromosome gene that encodes the 427-kDa intracellular protein dystrophin. The milder Becker muscular dystrophy (BMD) is also caused by mutations in dystrophin, although BMD patients often have an internally truncated and partially functional dystrophin protein.

Dystrophin is part of a multimeric protein complex, the dystrophin glycoprotein complex (DGC) (**Figure 1**). Other protein members of the DGC include the sarcoglycans, sarcospan, dystroglycans, dystrobrevins, and syntrophins (1, 2). The amino-terminal and spectrin-repeat domains of dystrophin bind filamentous actin in the muscle cytoskeleton (3). Through its cysteine-rich domain, dystrophin binds the transmembrane protein β-dystroglycan, which, in turn, is noncovalently linked to α-dystroglycan to form the dystroglycan subcomplex. To complete the link from the cytoskeleton to the extracellular matrix, the highly glycosylated α-dystroglycan binds laminin in the basal lamina (4). The dystrobrevins and syntrophins are cytoplasmic proteins anchored to dystrophin's carboxyterminal domain. The dystrobrevins and syntrophins can directly interact with each other, and the syntrophins also bind nitric oxide synthase (NOS). This cytoplasmic aspect of the DGC regulates cell signaling, although the extent of this signaling is not fully understood (5).

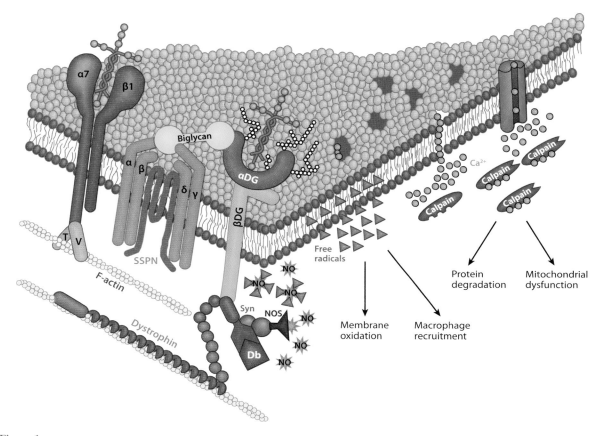

Figure 1

Mechanisms for membrane degeneration in dystrophic muscle. The α7β1 integrin linkage system and dystrophin glycoprotein complex (DGC) are distinct, multimeric protein complexes that link the extracellular matrix component to the actin cytoskeleton in muscle cells. (*Left*) A model of these complexes in normal muscle. The α7β1 integrin dimer binds laminin (*blue cruciforms*) extracellularly and associates intracellularly with actin-binding proteins, including vinculin (V) and talin (T). The DGC consists of the sarcoglycan subcomplex of α, β, γ, δ sarcoglycans and sarcospan (SSPN); this subcomplex associates with the matrix through biglycan. The α-dystroglycan subunit is heavily glycosylated, and these carbohydrate moieties are necessary for laminin binding. The β-dystroglycan subunit associates directly with dystrophin, and dystrophin binds directly to filamentous actin. The carboxy-terminal domain associates with dystrobrevins (Db) and syntrophins (Syn) that, in turn, bind nitric oxide synthase (NOS). Absence of components of either complex can cause muscular dystrophies characterized by sarcolemmal disruptions and muscle degeneration (*right*). In the absence of proper membrane-matrix attachment, tears may simply develop as a result of the mechanical stress. An increase in free radicals (*green triangles*) is thought to be caused by the displacement of NOS from the plasma membrane. The high levels of free radicals in dystrophic muscle are thought to contribute to muscle degeneration via the oxidation of muscle membranes and recruitment macrophages. Calcium-sensitive pathways also contribute to muscle degeneration in muscular dystrophy. Calcium (*pink spheres*) may enter dystrophic muscle through membrane lesions or through calcium channels. Calcium dysregulation may also lead to abnormal mitochondrial function as well as the activation of the calcium-dependent protease calpain to degrade muscle membrane proteins.

A distinct subcomplex within the DGC can be biochemically isolated and consists of five transmembrane proteins, including α-, β-, γ-, and δ-sarcoglycan as well as the small tetraspanin protein sarcospan. Loss-of-function mutations in the genes encoding α-, β-, γ-, and δ-sarcoglycans cause limb girdle muscular dystrophy (LGMD) types 2D, 2E, 2C, and 2F, respectively (reviewed in References 6 and 7). Mutations in the sarcoglycan genes phenocopy what is seen in dystrophin mutations. The sarcoglycanopathies and

NOS: nitric oxide synthase

Limb girdle
muscular dystrophy
(LGMD): a similar
phenotype to
Duchenne and Becker
muscular dystrophy
that arises from
mutations in many
different genes

Biglycan: a small
extracellular matrix
protein that binds to
the sarcoglycans

Congenital muscular
dystrophy (CMD): a
severe muscular
dystrophy present at
birth

FKRP: fukutin-
related protein

dystrophinopathies share the common phenotype of progressive, focal muscle degeneration that usually involves cardiac muscle. The similarity of phenotypes is likely explained by the secondary destabilization of the entire DGC that results from deficiency of only one member. Additionally, the sarcoglycan complex itself participates in protein interactions between the membrane and the matrix. In their extracellular portions, α-sarcoglycan and γ-sarcoglycan each bind the small proteoglycan biglycan (8). Biglycan also plays an essential role organizing the intracellular components of the DGC, including the syntrophins and dystrobrevins, mostly likely by way of its interaction with α-dystroglycan and the sarcoglycans (9). Like the entire DGC, the sarcoglycan complex likely has both structural and signaling roles (10).

Mutations in genes that encode extracellular proteins such as laminin subunits or posttranslational modifiers of α-dystroglycan can cause congenital muscular dystrophy (CMD). Congenital muscular dystrophy 1A, also known as MDC1A, is an autosomal recessive CMD caused by mutations in laminin-$\alpha2$ (11) and modeled by the *dy* mutant mouse. Although no human mutations have been described in the dystroglycan gene, the enzymes involved in posttranslational modification of α-dystroglycan have been linked to muscular dystrophy. Dystroglycan undergoes *O*-linked glycosylation, and abnormal glycosylation is thought to weaken the attachment of dystroglycan to the extracellular matrix. To bind laminin strongly, dystroglycan must be extensively glycosylated (12). Defects in genes encoding POMT1/2, POMGnT1, fukutin, and the fukutin-related protein (FKRP) result in Walker-Warberg syndrome, Muscle-Eye-Brain disease, CMD1C, CMD1D, and Fukuyama CMD (13). FKRP gene mutations associate with a spectrum of disease that also includes milder disease in the form of LGMD (14).

Like the DGC, the $\alpha7\beta1$ integrin complex links laminin to actin (**Figure 1**). The extracellular domain of the integrin binds directly to laminin. Intracellular associations with proteins such as vinculin and talin are thought to mediate binding to the actin cytoskeleton. Mutations in the $\alpha7$ integrin chain cause CMD (15). Recent experiments suggest that the $\alpha7\beta1$ integrin complex and DGC have compensatory abilities. Increased expression of $\alpha7\beta1$ integrin (16, 17) and integrin-binding proteins (18) was seen in muscle lacking DGC components. Mice lacking the DGC and integrin have a much more severe phenotype than do mice lacking only one of the linkage systems (17, 19, 20). Transgenic overexpression of $\alpha7$ integrin prevents muscle damage in the absence of the DGC (21), again suggesting that the membrane-matrix connection can be augmented through other means.

Muscle Degeneration from Membrane Fragility

Disrupting the cytoskeleton-membrane-matrix connection renders the muscle membrane fragile. Membrane tears induce a series of cellular responses that may be protective or pathological. The presence of elevated muscle enzymes in serum of dystrophic patients has commonly been interpreted as evidence of sarcolemmal fragility in dystrophic muscle. Indeed, early electron microscopy analysis of DMD muscle biopsies revealed numerous membrane lesions (22), explaining the leakage of muscle proteins, increased uptake of membrane-impermeable dyes, and presence of nonmuscle proteins inside of dystrophic muscle (23, 24). Myofibers lacking dystrophin display abnormal osmotic properties and cortical stiffness (25, 26).

The DGC and $\alpha7\beta1$ integrin complexes are localized at the sarcolemma in a discrete lattice of costameres. By linking the cytoskeleton to the extracellular matrix at costameres, these complexes can organize the sarcolemma and transmit contractile forces laterally to the basal lamina (3, 27). Costameres are disorganized in the muscle of humans and mice lacking dystrophin and laminin, and it is thought that this disorganization increases susceptibility to contraction-induced muscle injury (3, 27, 28).

The membrane fragility hypothesis has been advanced by the demonstration of elevated muscle damage with mechanical stress of dystrophin-deficient muscle using eccentric contractions. Eccentric contraction refers to muscle contraction of a maximally lengthened muscle, and this arrangement renders significant stress on the muscle membrane. Subjecting *mdx* muscle to repeated eccentric contraction in the presence of membrane-impermeable dyes demonstrated a marked increase in membrane damage compared with wild-type muscle (29, 30). Similarly, dystrophin-deficient cardiac muscle also has enhanced damage when exposed to adrenergic stress (31). The observations of costamere disruption and increased membrane permeability that is exacerbated by muscle use strongly support the hypothesis that the connection between actin and laminin is vital for the structural integrity of muscle.

Muscle Degeneration from Disrupted Calcium Homeostasis

Disruption of myofibers by membrane instability is associated with abnormal intracellular calcium and dysregulated calcium-responsive pathways (**Figure 2**). Elevated calcium in muscle fibers from DMD patients and *mdx* mice has been demonstrated (32–34), but the means by which calcium accumulates in dystrophic fibers remains unknown. Early studies speculated that extracellular calcium simply enters dystrophic fibers through microlesions in the sarcolemma (35). Recordings from dystrophic muscle suggest that nonselective cation calcium-permeable channels are activated by stretch (36–40). Although several TRP channels have been proposed as stretch-activated channels, no consensus has emerged for the gene(s) responsible. The calcium-permeable, growth factor–regulated channel (also known as TRPV2) is elevated in the hearts of *mdx* and sarcoglycan mutant models, and transgenic overexpression of this channel leads to cardiomyopathy (40), suggesting that this channel may be pathologically important.

Regardless of the mechanism of elevated calcium, persistently high calcium levels are thought to subvert the muscle's ability to maintain physiologic calcium levels and then to activate calcium-dependent proteases, including the calpains. Calpain activity is increased in *mdx* mice, and calpain can degrade many muscle membrane proteins, contributing to myonecrosis (39, 41). Increased calcium uptake may also cause abnormal mitochondrial function in dystrophic myotubes as an alternative pathway by which increased calcium leads to cell death (33, 42). In support of the mitochondrial effect, inhibition of the mitochondrial transition pore, through either genetic ablation of the cyclophilin gene or pharmacologic inhibition of this isomerase, can improve the phenotype in muscular dystrophy (43).

Increased intracellular calcium represents an attractive target for therapy to treat muscle degeneration. Introducing the stretch-activated calcium channel blocker streptomycin into the drinking water of *mdx* mice decreased muscle damage (38). Degeneration is also ameliorated by injecting *mdx* mice with protease inhibitors (44) or with transgenic overexpression of the calpain inhibitor calpastatin (45). There is strong evidence for a role of calcium dysregulation in the pathophysiology of muscular dystrophy, and restoring calcium homeostasis or regulating calcium-dependent proteases may be an effective strategy to suppress muscle degeneration.

Muscle Degeneration from Free Radicals

The finding that dystrophin-deficient myotubes are susceptible to oxidative damage (46) launched investigation into the role of free radicals in muscle degeneration (47). Neuronal nitric oxide synthase (nNOS) directly binds the syntrophins in the DGC, and in the absence of dystrophin, nNOS is displaced from the plasma membrane into the cytoplasm (5). NOS produces nitric oxide (NO), a molecule that reacts with free radicals and regulates vascular tone. In *mdx* muscle, a reduction of

Calpain: a calcium-activated protease; mutations in muscle-specific calpain subunits lead to limb girdle muscular dystrophy

Myotubes: multinucleated cells that arise from the fusion of myoblasts to each other or the fusion of myoblasts to myotubes

NO: nitric oxide

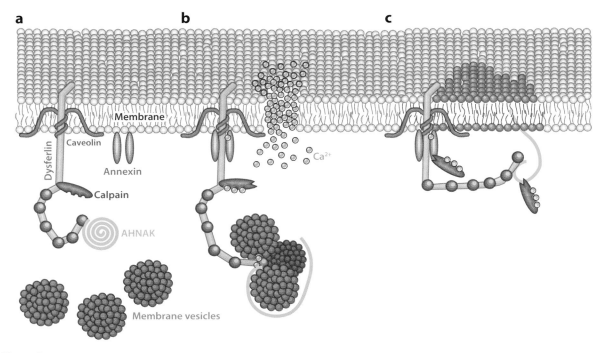

Figure 2

A model for dysferlin-mediated membrane repair. (*a*) Dysferlin (*green*) is localized at the sarcolemma as part of a membrane repair complex. Caveolin-3 (*purple*), a muscle-specific caveolin, interacts with dysferlin. Dysferlin also interacts with AHNAK (*yellow*) and calpain (*dark blue*), a calcium-activated protease. (*b*) Tears in the sarcolemma result in an influx of calcium (*pink spheres*), which activates and alters the binding properties of proteins in the membrane repair complex. Annexins (*light blue*) bind dysferlin and phospholipids with higher affinity in the presence of calcium, the C2A domain of dysferlin binds phospholipids in a calcium-dependent manner, and calpains are activated. These interactions are thought to encourage the recruitment of internal vesicle structures (*red*). (*c*) Within seconds of activation, membrane lesions are resealed, calcium concentrations are normalized, and the repair complex is deactivated. The deactivation of the complex may be mediated in part by calpain-dependent cleavage of annexins and AHNAK. Mutations in the genes encoding proteins of this sarcolemmal repair complex cause LGMD2B and Miyoshi myopathy (dysferlin), LGMD1C (caveolin), and LGMD2A (calpain-3).

membrane-associated nNOS results in a concomitant decrease in NO (48) and may lead to free radical–mediated damage. Forced expression of an nNOS transgene in *mdx* mice prevented muscle degeneration (48). Likewise, nNOS is reduced in mice lacking α-dystrobrevin, which also have muscular dystrophy (49).

Interestingly, mice lacking nNOS have normal muscle architecture (50), leading to the proposal of a two-hit hypothesis of free radical–mediated muscle degeneration (51). Hypotheses regarding the role of NO in muscular dystrophy pathology are complicated by its dual function in promoting oxidative reactions after reacting with superoxides as well as in acting as an antioxidant (52). According to this hypothesis, absence of NO alone is insufficient for extensive oxidative muscle damage unless mechanical stress is applied to dystrophic muscle. This hypothesis is supported by recent studies that demonstrated that the combined effects of stretch-mediated mechanical stress and ischemia/reperfusion on sarcolemma damage were greater than the sum of either alone (53).

Markers of oxidative stress have been extensively investigated in *mdx* mice and DMD patients. Byproducts of free radical–mediated lipid damage are increased in DMD muscle (54). Presumably in an attempt to guard against oxidative damage, muscle fibers from DMD patients have elevated levels of the antioxidant molecules vitamin E and coenzyme Q (55) and antioxidant enzymes superoxide dismutase-1, catalase, and glutathione peroxidase (54, 56, 57). Increasing antioxidants using green tea extract (58) or feeding *mdx* mice a low-iron diet (59) resulted in decreased muscle degeneration, but similar benefits were not demonstrated in DMD patients (60, 61).

The mechanism of free radical–mediated myonecrosis may be more complicated than simple membrane damage and may involve inflammatory cells. Although membrane lipid peroxidation is increased in prenecrotic *mdx* muscle fibers (56), free radicals can also activate signaling pathways that recruit inflammatory cells (62). Mechanical stress of diaphragm muscles from *mdx* mice activated the NF-κB pathway and increased the downstream proinflammatory cytokines TNFα and IL-1β in a free radical–dependent manner (63). Release of proinflammatory cytokines may, in turn, recruit cytolytic macrophages to stressed muscle, resulting in necrosis. Experiments in *mdx* mice support the hypothesis of macrophage-mediated muscle damage. Transgenic expression of nNOS reduced degeneration and decreased macrophage accumulation in *mdx* muscle (48). Similar benefits were seen following macrophage depletion (48), block of TNFα activity (64, 65), treatment with immunosuppressants (66), or administration of anti-inflammatory molecules (65, 67, 68).

MUSCLE MEMBRANE REPAIR

Dysferlinopathies

Even in normal muscle, contraction-induced sarcolemmal injury is a commonplace event. Large syncytial cells such as myofibers have an efficient sarcolemmal repair system to mediate response to local damage (69). Small tears within membranes normally are sealed within seconds. The protein product of the LGMD 2B locus, dysferlin, is a major mediator of membrane resealing in muscle. Dysferlin is a 230-kDa, membrane-anchored protein that is abundantly expressed in skeletal and cardiac muscle (70). A role for dysferlin in membrane fusion was initially suggested by its homology to the *Caenorhabditis elegans* protein FER1, which is essential for vesicle fusion to the sperm plasma membrane (71). Dysferlin, like FER1, contains multiple C2 domains. C2 domains are found in a number of membrane-associated proteins, including those involved in membrane fusion events, such as the synaptotagmins. The first C2 domain of dysferlin binds to negatively charged phospholipids in a calcium-sensitive manner (72).

Loss-of-function mutations in the dysferlin gene cause LGMD2B (73) and the milder allelic Miyoshi myopathy (74). Like other LGMDs, dysferlin mutations result in proximal muscle weakness. In contrast, Miyoshi myopathy preferentially affects the gastrocnemius muscle, with later involvement of distal musculature. Surprisingly, identical dysferlin mutations cause both forms of dysferlinopathy, suggesting that modifiers mediate substantial aspects of the disease pathology (75). Dysferlin-deficient muscle from patients and mouse models shows membrane discontinuities, submembranous vesicle aggregates, papillary projections, and thickening of the basal lamina (69, 76, 77). Additional roles for dysferlin in muscle differentiation and regeneration are suggested by the delayed fusion and differentiation observed in cultured dysferlin-null myoblasts (78). Distinct from the DGC and its role in maintaining muscle integrity, dysferlin mutant muscle has defective membrane repair. This defective repair leaves dysferlin-null muscle unable to cope with normal cell injury, resulting in progressive myonecrosis.

Dysferlin-Mediated Membrane Repair

Although the means by which dysferlin regulates sarcolemma repair are largely unknown, a patch model of dysferlin-mediated membrane resealing has emerged as a leading mechanism (**Figure 2**). Ex vivo myofiber microinjury assays in the presence of a membrane-impermeable

Myoblasts: with activation, satellite cells can become myoblasts, precursor cells of muscle

Caveolin-3: a muscle-specific form of caveolin that participates in membrane trafficking

fluorescent dye (69) or injuring cells containing a fluorescent calcium indicator (79) unequivocally demonstrated a role for dysferlin in membrane repair. Transient changes in fluorescence kinetics were quantified as indicators of the time frame of membrane resealing and showed that recovery after injury occurs on the order of seconds in normal myofibers but is disrupted in fibers from dysferlin-null mice.

Moreover, dysferlin repair is calcium sensitive. Myofibers injured in the presence of the calcium chelator EGTA did not effectively repair their damaged membranes (69). These data are consistent with in vitro studies showing that the first C2 domain (C2A) of dysferlin binds negatively charged phospholipids only in the presence of biologically relevant calcium concentrations (72). Further supporting the idea that dysferlin mediates membrane repair through calcium-sensitive lipid binding, introduction of the LGMD2B-associated mutation V67D abolished phospholipid binding of C2A (72). Whether the remaining C2 domains bind lipids and/or participate in protein binding remains to be demonstrated.

Dysferlin-Interacting Proteins

Dysferlin associates with caveolin-3. Caveolin-3 is a muscle-specific structural component of caveolar membranes that also associates with the DGC (80). Mutations in caveolin cause LGMD1C (81) and result in mislocalization of dysferlin (82). Dysferlin and caveolin associate in normal muscle (82). Because dysferlin accumulates in the Golgi apparatus of caveolin-null cells, caveolin is hypothesized to transport dysferlin to the sarcolemma (83). Recently, caveolin has been implicated in maintaining normal levels of dysferlin at the muscle membrane by preventing its endocytosis as well (84).

Patients with LGMD2A have mutations in the gene encoding the muscle-specific protease calpain-3 (85). Calpain-3 is a calcium-dependent protease that coimmunoprecipitates with dysferlin in skeletal muscle lysates (86). Levels of calpain-3 are reduced in dysferlinopathy patients (87), and dysferlin is de-

creased in LGMD2A patients. Proteolytic targets of calpain-3 include annexins and AHNAK (Hebrew for "giant") (88), both proteins with established associations to dysferlin (79, 89).

Annexins are calcium-dependent, phospholipid-binding proteins that also participate in vesicle aggregation. Annexin A1 and A2 are members of this protein family that associate with dysferlin in a calcium-dependent manner at the sarcolemma (79). Cleavage of annexin A1 by calpain enhances its calcium-sensitive affinity for phospholipids (90). Both annexin A1 and A2 are abnormally expressed in patients with LGMD2B, Miyoshi myopathy, LGMD1C, and DMD (91). The critical role of annexins in membrane repair has been demonstrated in vitro because antibody blockade was shown to inhibit membrane resealing (92).

AHNAK (also known as desmoyokin) is a ubiquitous protein that is highly expressed in skeletal muscle. AHNAK can be found in the nucleus or associated with the plasma membrane (93). Recently, AHNAK was found to have a calcium-independent interaction with the C2A domain of dysferlin, and loss of dysferlin leads to reduction of AHNAK (89). Cleavage by calpain abolishes the interaction between AHNAK and dysferlin (88). Dysferlin and AHNAK both redistribute to the cytoplasm during muscle regeneration (89). Although the exact function of AHNAK is unknown, its presence in vesicles proposed to repair membranes (e.g., enlargeosomes) has led to the speculation that AHNAK may play an important role in recruiting these internal membrane structures to the sarcolemma to seal membrane tears (88).

The abundance of calcium-dependent proteins has suggested a model for sarcolemmal repair in which a calcium bolus enters damaged muscle through sarcolemmal tears. The calcium influx and activated proteases initiate the membrane repair system by strengthening interactions between repair complex proteins and phospholipids on internal vesicles and sections of the sarcolemma adjacent to the membrane lesion. As a result, vesicles accumulate beneath the damaged membrane and eventually fuse to repair the damage in a process similar to synaptic vesicle fusion (94). Membrane patches contain dysferlin and lysosomal-associated membrane protein-1 (LAMP-1) but lack other sarcolemmal markers, possibly suggesting that repair patches originate via exocytosis of internal vesicles rather than existing muscle membrane (79). Once the tears are repaired, calcium influx is abated, and the repair system is deactivated.

Myoferlin was identified on the basis of its structural relationship to dysferlin (95). Like dysferlin, myoferlin is a 220-kDa protein with at least six C2 domains in its large cytoplasmic domain. The gene structure is conserved between myoferlin and dysferlin. The first C2 domain of myoferlin also binds negatively charged phospholipids in the presence of calcium (72). In contrast to dysferlin, myoferlin is detected at higher levels in myoblasts. Myoferlin is specifically enriched at the membrane interface of fusing myoblasts and myotubes and then is decreased upon muscle differentiation and in mature myofibers (96). Myoferlin expression increases after muscle injury, and myoferlin-null muscle fibers are small in vitro and in vivo, demonstrating a role for myoferlin in muscle formation that requires myoblast fusion (96). Myoferlin is increased in *mdx* muscle, and this increase may reflect enhanced regeneration (95). It is unclear whether myoferlin can substitute for dysferlin because myoferlin is not increased in patients with dysferlinopathy (97). Myoferlin and dysferlin may have distinct functions.

REGENERATION DEFECTS IN MUSCULAR DYSTROPHY

The dysferlin repair system is activated within myotubes as a first intracellular response to muscle damage. More extensive damage is accompanied by the activation of satellite cells, the stem cells in muscle that reside between the sarcolemma and basal lamina. Muscle injury is thought to stimulate release of mitogens, including hepatocyte growth factor (HGF), that awaken quiescent satellite cells from G_0 ar-

rest (98). Upon activation, satellite cells are attracted by chemotaxis to the site of injury, where they produce precursor cells. Additional proliferation by muscle precursor cells is followed by their withdrawal from the cell cycle, after which they undergo a regenerative process that largely recapitulates the cellular and molecular events of myogenesis. Muscle-specific transcription factors, including MyoD and myogenin, mark the differentiation of muscle precursor cells into myoblasts. The myoblasts can repair injured myofibers by fusing to fibers at the sites of injury. Severely injured muscle fibers are often completely phagocytosed by immune cells, whereafter satellite cells generate entirely new, small-caliber fibers by fusing to each other. After the initial wave of fusion, hypertrophy of regenerated fibers occurs by fusion of additional myoblasts. To reestablish their cytoarchitecture, regenerated fibers transiently revert to a pattern of gene expression that resembles embryonic development, including expression of embryonic forms of myosin heavy chain. Thus, within days of injury, a normally functioning satellite cell repair system is able to regenerate damaged muscle completely (**Figure 3**).

In addition to muscle precursor cells, asymmetric division of satellite cells also maintains the satellite cell pool. Differential activation of the Delta-Notch pathway is thought to be a mechanism by which satellite cells accomplish this. Asymmetric distribution of the Notch antagonist Numb to muscle precursor cells activates a number of myogenic genes, committing them to the myoblast lineage (99, 100). In contrast, the daughter satellite cell reestablishes its quiescence and remains as a reserve cell for future muscle regeneration. Although satellite cells are the primary source of muscle regenerative cells (101), cells from nonmuscle sources, including the bone marrow, may minimally contribute to regeneration.

The ability of the satellite cell pool to be reactivated many times underlies the persistent regenerative capacity of this compartment. Nevertheless, the progressive replacement of muscle with fibrofatty tissue in dystrophic

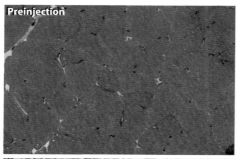

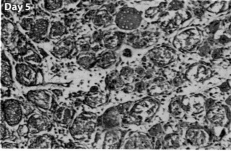

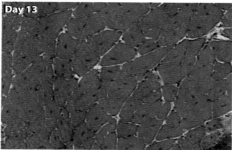

Figure 3

Muscle regeneration after cardiotoxin injury. (*Top panel*) Normal muscle fibers are closely apposed with peripherally located nuclei. (*Middle panel*) Injection of cardiotoxin, a component of cobra venom, leads to widespread muscle necrosis after injection. (*Bottom panel*) By day 10–13 postinjection, satellite cells have been activated to restore muscle architecture in normal mice. Newly regenerated muscle fibers are marked by their centrally positioned nuclei. Over time, these nuclei will move to their normal position at the periphery of the myofiber.

patients highlights the inability of the satellite cell system to compensate for dystrophic damage. The apparent exhaustion of regenerative capacity with time in dystrophic muscle is not fully understood, but several hypotheses have been proposed (**Figure 4**). Replicative aging

may occur, leading to the senescence of satellite cells. Alternatively, there may be a progressive inability of muscle precursor cells to differentiate properly. Finally, the microenvironment of degenerating muscle may be unfavorable for regeneration.

Myoblasts isolated from DMD patients demonstrate decreased proliferation potential in vitro (102, 103). A failure in the proliferative ability of satellite cells would likely manifest itself in vivo as a decrease in the absolute number of cells in dystrophic muscle. Fewer satellite cells or subsets of satellite cells were detected in some studies of dystrophic muscle (103, 104), but other reports describe normal or increased levels of satellite cells (105–107).

One compelling argument for reduced satellite cell proliferation is a mechanism involving replicative aging. According to this hypothesis, the rapid turnover of muscle fibers from cycles of degeneration and regeneration results in the continuous activation and proliferation of satellite cells. Consequently, telomeres of dystrophic satellite cells become rapidly eroded, leading to premature senescence. Short telomeres have been observed in DMD (108–110) and LGMD2C (108) patients as well as *mdx* mice (111). Another study did not detect significant telomere shortening (112), suggesting that satellite cell exhaustion from replicative aging may not be the sole source of regenerative defects in dystrophic patients.

It is also possible that dystrophic satellite cells do not effectively differentiate or fuse during regeneration. Two early studies (113, 114) described delayed myoblast fusion and abnormal differentiation when comparing muscle cultures from DMD patients with normal controls, despite normal doubling times. One study suggested that TGF-β1 may, in part, be a causative factor (115). Normal myoblast cultures demonstrated aberrant differentiation when exposed to conditioned media from DMD myoblasts, which could be partially reversed by TGF-β1-blocking antibodies. In this model, TGF-β1 released by DMD myoblasts exerts negative effects on regeneration via an autocrine pathway. This work was further

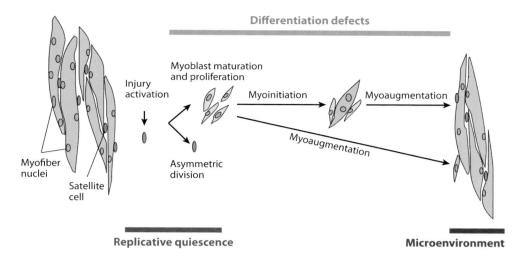

Figure 4

Muscle regeneration defects in muscular dystrophies. In normal muscle, satellite cells are mitotically activated by injury or degeneration. Satellite cells can differentiate into activated myoblasts in response to injury. Satellite cells can also replenish the satellite cell pool. Upon activation, myoblasts undergo additional mitoses, generating a population of fusogenic cells. Daughter myoblasts can fuse to multinucleated myofibers (myoaugmentation) or can fuse with each other to form new myofibers (myoinitiation). Newly formed myofibers can also undergo growth through the fusion of additional myoblasts (myoaugmentation). In dystrophic muscle, regenerative failure may occur as a result of replicative quiescence, in which satellite cells ineffectively generate muscle precursor cells. Alternatively, satellite cells may generate sufficient numbers of muscle precursor cells, but the cells may fail to differentiate adequately. In another possibility, the microenvironment of dystrophic muscle may not be permissive for continued regeneration. In this last scenario, satellite cells are activated and divide correctly, but environmental factors of the dystrophic tissue prevent myofiber replacement.

supported in vivo using anti-TGFβ antibodies in the *mdx* mouse or using the angiotensin receptor inhibitor to bolster muscle regeneration (116).

Decreased NOS in dystrophic muscle may also negatively affect regeneration by altering satellite cell proliferation and differentiation. NOS activity was increased in rat muscle following stretch assays, releasing NO and stimulating HGF, a growth factor that promotes satellite cell proliferation (117). Pretreating *mdx* mice with NOS inhibitors disrupted muscle repair (118). Detailed studies on chick myoblasts also demonstrated that an increase in NOS is necessary for proper fusion and that incubation with NO promoted fusion (119). These data suggest that abnormalities in the NOS pathway may affect not only membrane integrity but regeneration as well.

Muscle degeneration and regeneration are complex processes involving multiple cell types. In addition to satellite cells, myoblasts, and myofibers, immune cells scavenge necrotic tissue, adipose cells replace dystrophic muscle, and fibroblasts deposit a fibrous meshwork of connective tissue. These nonmuscle cells also influence satellite cell function. Fibroblasts from DMD patients secrete high levels of IGF-binding proteins in vitro (120), which may sequester IGF and interfere with IGF signaling, a critical pathway for efficient muscle regeneration (121). A twofold increase in muscle IGF-I increased regeneration in transgenic *mdx* mice (122). When myoblasts were cultured in the same dish with macrophages, myoblast proliferation was enhanced and differentiation was delayed, suggesting a role for macrophages in abnormal muscle regeneration (123). Thus,

EDMD: Emery-
Dreifuss muscular
dystrophy

Emerin: a membrane
protein of the nuclear
membrane

Lamin A/C: the type
V intermediate
filament proteins
found under the inner
nuclear membrane

cell-intrinsic and cell-extrinsic factors are both essential for ongoing regeneration in muscular dystrophy.

The importance of a suitable environment for efficient muscle regeneration has been investigated in aged muscle, which also exhibits ineffective regeneration (124). Strong evidence points to an age-related impairment of the Delta-Notch pathway as a key factor in regeneration defects. Levels of Notch were normal, but the Notch ligand, Delta, was not upregulated following injury in aged muscle (99). Regenerative potential was restored with a Notch-activating antibody (99). Interestingly, an aged mouse parabiotically joined to a young mouse demonstrated Delta upregulation following injury along with improved regeneration (125). This finding suggests that serum proteins also impact muscle regeneration and that failures in paracrine, environmental, and cellular factors all contribute to the complex phenotype of regenerative failure in muscular dystrophy. Because of their remarkable ability to continually regenerate muscle, satellite cells are currently being explored for their ability to supplement the myogenic cell pool of dystrophic muscle (126).

THE NUCLEAR MEMBRANE AND MUSCLE DISEASE

Emery Dreifuss muscular dystrophy (EDMD) is the most thoroughly studied of the muscular dystrophies caused by mutations in nuclear proteins. Emery & Dreifuss first established EDMD as an X-linked dystrophy clinically distinct from DMD in the 1960s. Genetic linkage analysis mapped EDMD to emerin on the X chromosome (127). Autosomal forms of EDMD were identified and mapped to the gene encoding lamins A and C (*LMNA*) on human chromosome 1 (128). Mutations in the lamin gene have been also been linked to LGMD1B (129, 130) and dilated cardiomyopathy 1A (DCM1A) (131).

Lamin and emerin directly interact at the inner nuclear membrane (132) (**Figure 5**). Emerin is an integral protein at the nuclear en-

velope, whereas lamins polymerize to form a protein meshwork known as the nuclear lamina under the inner nuclear membrane (133). Lamin and emerin, together with other nuclear proteins, are thought to maintain nuclear morphology, regulate chromatin organization and gene transcription, link the cytoskeleton to the nuclear skeleton, and serve as a scaffold for other nuclear proteins involved in gene regulation and DNA replication. Emerin and lamin proteins are expressed in all tissues but have tissue-specific phenotypes when mutated. The reason why muscle tissue is particularly sensitive to emerin and lamin mutations is unknown, but two predominant hypotheses have been proposed and supported by experimental evidence.

First, the nuclear fragility model predicts that nuclei in EDMD patients may be abnormally susceptible to damage from mechanical forces during contraction. Nuclei are joined to the actin cytoskeleton via the LINC complex, which links the nucleus to the cytoplasm (133, 134). This network includes the SUN and nesprin proteins and creates a protein link from the intermediate filament network under the inner nuclear membrane across the periplasmic lumen and ultimately to the actin cytoskeleton (**Figure 5**). Through this network, contractile stress can pass from the sarcolemma through the cytoskeleton to the nucleus. In EDMD patients, this force may be enough to disrupt nuclear structure and lead to the subsequent death of dystrophic muscle fibers. Nuclei of EDMD patients show nuclear blebbing and aberrant lamin localization in many cells (135–138). Studies on cultured emerin-null and lamin-null fibroblasts also show abnormal nuclear morphology in support of the nuclear fragility model (139, 140). When lamin-null and wild-type fibroblasts were subjected to mechanical strain, less force was required to rupture lamin-null fibroblasts than to rupture wild-type cells (140, 141), a phenomenon not duplicated in emerin-null fibroblasts (139). Finally, defects in nuclear architecture were seen in cardiomyoctyes of lamin-null mice (142).

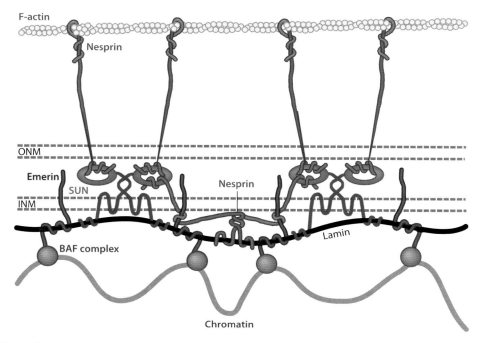

F-actin

Nesprin

ONM

Emerin

SUN

Nesprin

INM

Lamin

BAF complex

Chromatin

Figure 5

The nuclear membrane complex in skeletal muscle. Skeletal muscle nuclei contain a protein complex that links filamentous actin in the cytoskeleton to the nucleoplasm. The large forms of nesprin (*blue*) have an actin-binding site at their amino terminus. These large forms are thought to reside on the outer nuclear membrane (ONM), where they interact with the SUN proteins (*green*). The SUN proteins can also interact with short forms of nesprin (also shown in *blue*) on the inner nuclear membrane (INM). This complex also directly interacts with lamins A and C (*black*), which form the meshwork under the INM. Emerin (*brown*) also bind lamins and interacts with chromatin through its association with BAF (*red*). Patients with Emery Dreifuss muscular dystrophy (EDMD) have mutations in the genes encoding emerin, lamin A/C, or nesprin and are characterized by nuclear defects and abnormal transcription.

An alternate, although not necessarily mutually exclusive, explanation of EDMD pathology is the gene expression hypothesis, which proposes that specific chromatin sites are altered when their interactions with emerin-lamin complexes are disrupted. In this model, expression of critical muscle genes is consequently altered, resulting in downstream defects in apoptosis, mechanotransduction, and muscle differentiation. Mechanotransduction and antiapoptotic gene expression were decreased in lamin-null or emerin-null fibroblasts (139). Microarray analyses of regenerating muscle from emerin-null mice (143) and EDMD patients (144) found differential expression of genes involved in cell cycle and differentiation.

Among the most interesting misregulated pathways identified were those regulated by retinoblastoma protein 1 (Rb1) and MyoD, both important for muscle differentiation. This finding is consistent with the impaired muscle differentiation in emerin-null mice (143) and myoblasts from mice lacking lamin or emerin (145). Myoblasts expressing emerin and lamins with EDMD-causing mutations also demonstrated abnormal differentiation coupled with abnormal levels of Rb1 and MyoD (146–148). Restoring MyoD expression in lamin-null cells using a retroviral vector enhanced their regenerative potential to near-wild-type levels (145). Proliferation and differentiation defects have been described in satellite cells from mice lacking MyoD (149–152), and *mdx* mice lacking

MyoD have severe muscular dystrophy, likely owing to compromised regenerative capacity (151). Thus, decreased MyoD expression has a strong influence on satellite cell function and represents one interesting explanation of EDMD pathology.

CONCLUSIONS

Since the first historical accounts of muscular dystrophies in the 1830s by Sir Charles Bell, substantial efforts across science disciplines have painted a complex picture of the mutations, molecular events, cell biology, and physiological abnormalities that accompany these diseases. To date, more than 30 muscular dystrophies and more than 100 muscle diseases have been identified. Critical examination of initial hypotheses that muscular dystrophies are caused primarily by structural defects in muscle has revealed that muscle disease can result from defects in a variety of processes essential for muscle function. Nevertheless, shared secondary defects in muscular dystrophies are attractive targets to develop broadly effective treatment strategies. Although there is currently no cure for muscular dystrophy, a better understanding of muscle disease processes over the past several years has launched substantial efforts toward managing the progression of muscle disease. Promising avenues in the areas of gene replacement, stem cell transplantation, pharmaceuticals, and upregulation of endogenous compensatory proteins have all been explored. Many of these studies have yielded exciting results that have spawned clinical trials that are in their infancy. As clinical trials are expanded and investigation into mechanisms of muscle disease advances, there is hope that effective therapies are on the horizon.

SUMMARY POINTS

1. The dystrophin glycoprotein complex mediates membrane stability in muscle.

2. The dystrophin glycoprotein complex has both mechanical/structural roles and signaling roles to maintain muscle membrane integrity.

3. Defects in the muscle membrane lead to increased intracellular calcium.

4. Dysferlin is a membrane-associated protein that mediates muscle membrane repair.

5. The nuclear membrane of muscle is important for regulating the positioning of nuclei as well as nuclear function, including gene replication, transcription, and nuclear transport.

6. Regeneration in muscle is ongoing in muscular dystrophy and is mediated by the stem cell of muscle, the satellite cell.

FUTURE ISSUES

1. Identifying modifiers of muscular dystrophy may lead to novel pathways important for therapy.

2. Membrane stabilizers are being developed to help treat muscular dystrophy.

3. Increased intracellular calcium is a target for therapeutic intervention in muscular dystrophy.

4. The role of nuclear membrane protein genes in the pathogenesis of muscular dystrophy may include both mechanical and nonmechanical roles.

5. Regeneration in muscle is a major target of future research because muscle is highly regenerative.

DISCLOSURE STATEMENT

The authors are not aware of any biases that might be perceived as affecting the objectivity of this review.

ACKNOWLEDGMENTS

E.M.M. is supported by NIH HL61322, NIH ND47726, the Muscular Dystrophy Association, and the Doris Duke Charitable Foundation. G.Q.W. is supported by T32HL007381, F32AR054700, and the Muscular Dystrophy Association.

LITERATURE CITED

1. Ervasti JM, Ohlendieck K, Kahl SD, Gaver MG, Campbell KP. 1990. Deficiency of a glycoprotein component of the dystrophin complex in dystrophic muscle. *Nature* 345:315–19
2. Yoshida M, Ozawa E. 1990. Glycoprotein complex anchoring dystrophin to sarcolemma. *J. Biochem.* 108:748–52
3. Rybakova IN, Patel JR, Ervasti JM. 2000. The dystrophin complex forms a mechanically strong link between the sarcolemma and costameric actin. *J. Cell Biol.* 150:1209–14
4. Ibraghimov-Beskrovnaya O, Ervasti JM, Leveille CJ, Slaughter CA, Sernett SW, Campbell KP. 1992. Primary structure of dystrophin-associated glycoproteins linking dystrophin to the extracellular matrix. *Nature* 355:696–702
5. Brenman JE, Chao DS, Gee SH, McGee AW, Craven SE, et al. 1996. Interaction of nitric oxide synthase with the postsynaptic density protein PSD-95 and α1-syntrophin mediated by PDZ domains. *Cell* 84:757–67
6. Heydemann A, Doherty KR, McNally EM. 2007. Genetic modifiers of muscular dystrophy: implications for therapy. *Biochim. Biophys. Acta* 1772:216–28
7. Wagner KR. 2002. Genetic diseases of muscle. *Neurol. Clin.* 20:645–78
8. Rafii MS, Hagiwara H, Mercado ML, Seo NS, Xu T, et al. 2006. Biglycan binds to α- and γ-sarcoglycan and regulates their expression during development. *J. Cell Physiol.* 209:439–47
9. Mercado ML, Amenta AR, Hagiwara H, Rafii MS, Lechner BE, et al. 2006. Biglycan regulates the expression and sarcolemmal localization of dystrobrevin, syntrophin, and nNOS. *FASEB J.* 20:1724–26
10. Barton ER. 2006. Impact of sarcoglycan complex on mechanical signal transduction in murine skeletal muscle. *Am. J. Physiol. Cell Physiol.* 290:C411–19
11. Helbling-Leclerc A, Zhang X, Topaloglu H, Cruaud C, Tesson F, et al. 1995. Mutations in the laminin α2-chain gene (*LAMA2*) cause merosin-deficient congenital muscular dystrophy. *Nat. Genet.* 11:216–18
12. Michele DE, Barresi R, Kanagawa M, Saito F, Cohn RD, et al. 2002. Post-translational disruption of dystroglycan-ligand interactions in congenital muscular dystrophies. *Nature* 418:417–22
13. Martin PT. 2007. Congenital muscular dystrophies involving the *O*-mannose pathway. *Curr. Mol. Med.* 7:417–25
14. Brockington M, Yuva Y, Prandini P, Brown SC, Torelli S, et al. 2001. Mutations in the fukutin-related protein gene (*FKRP*) identify limb girdle muscular dystrophy 2I as a milder allelic variant of congenital muscular dystrophy MDC1C. *Hum. Mol. Genet.* 10:2851–59
15. Hayashi YK, Chou FL, Engvall E, Ogawa M, Matsuda C, et al. 1998. Mutations in the integrin α7 gene cause congenital myopathy. *Nat. Genet.* 19:94–97

16. Hodges BL, Hayashi YK, Nonaka I, Wang W, Arahata K, Kaufman SJ. 1997. Altered expression of the α7β1 integrin in human and murine muscular dystrophies. *J. Cell Sci.* 110(Pt. 22):2873–81

17. Allikian MJ, Hack AA, Mewborn S, Mayer U, McNally EM. 2004. Genetic compensation for sarcoglycan loss by integrin α7β1 in muscle. *J. Cell Sci.* 117:3821–30

18. Thompson TG, Chan YM, Hack AA, Brosius M, Rajala M, et al. 2000. Filamin 2 (FLN2): a muscle-specific sarcoglycan interacting protein. *J. Cell Biol.* 148:115–26

19. Guo C, Willem M, Werner A, Raivich G, Emerson M, et al. 2006. Absence of α7 integrin in dystrophin-deficient mice causes a myopathy similar to Duchenne muscular dystrophy. *Hum. Mol. Genet.* 15:989–98

20. Rooney JE, Welser JV, Dechert MA, Flintoff-Dye NL, Kaufman SJ, Burkin DJ. 2006. Severe muscular dystrophy in mice that lack dystrophin and α7 integrin. *J. Cell Sci.* 119:2185–95

21. Burkin DJ, Wallace GQ, Nicol KJ, Kaufman DJ, Kaufman SJ. 2001. Enhanced expression of the α7β1 integrin reduces muscular dystrophy and restores viability in dystrophic mice. *J. Cell Biol.* 152:1207–18

22. Mokri B, Engel AG. 1975. Duchenne dystrophy: electron microscopic findings pointing to a basic or early abnormality in the plasma membrane of the muscle fiber. *Neurology* 25:1111–20

23. Straub V, Donahue KM, Allamand V, Davisson RL, Kim YR, Campbell KP. 2000. Contrast agent-enhanced magnetic resonance imaging of skeletal muscle damage in animal models of muscular dystrophy. *Magn. Reson. Med.* 44:655–59

24. Pestronk A, Parhad IM, Drachman DB, Price DL. 1982. Membrane myopathy: morphological similarities to Duchenne muscular dystrophy. *Muscle Nerve* 5:209–14

25. Menke A, Jockusch H. 1991. Decreased osmotic stability of dystrophin-less muscle cells from the *mdx* mouse. *Nature* 349:69–71

26. Pasternak C, Wong S, Elson EL. 1995. Mechanical function of dystrophin in muscle cells. *J. Cell Biol.* 128:355–61

27. Williams MW, Bloch RJ. 1999. Extensive but coordinated reorganization of the membrane skeleton in myofibers of dystrophic (*mdx*) mice. *J. Cell Biol.* 144:1259–70

28. Minetti C, Tanji K, Bonilla E. 1992. Immunologic study of vinculin in Duchenne muscular dystrophy. *Neurology* 42:1751–54

29. Cox GA, Cole NM, Matsumura K, Phelps SF, Hauschka SD, et al. 1993. Overexpression of dystrophin in transgenic *mdx* mice eliminates dystrophic symptoms without toxicity. *Nature* 364:725–29

30. Petrof BJ, Shrager JB, Stedman HH, Kelly AM, Sweeney HL. 1993. Dystrophin protects the sarcolemma from stresses developed during muscle contraction. *Proc. Natl. Acad. Sci. USA* 90:3710–14

31. Danialou G, Comtois AS, Dudley R, Karpati G, Vincent G, et al. 2001. Dystrophin-deficient cardiomyocytes are abnormally vulnerable to mechanical stress-induced contractile failure and injury. *FASEB J.* 15:1655–57

32. De Backer F, Vandebrouck C, Gailly P, Gillis JM. 2002. Long-term study of Ca^{2+} homeostasis and of survival in collagenase-isolated muscle fibres from normal and *mdx* mice. *J. Physiol.* 542:855–65

33. Robert V, Massimino ML, Tosello V, Marsault R, Cantini M, et al. 2001. Alteration in calcium handling at the subcellular level in *mdx* myotubes. *J. Biol. Chem.* 276:4647–51

34. Turner PR, Fong PY, Denetclaw WF, Steinhardt RA. 1991. Increased calcium influx in dystrophic muscle. *J. Cell Biol.* 115:1701–12

35. Bodensteiner JB, Engel AG. 1978. Intracellular calcium accumulation in Duchenne dystrophy and other myopathies: a study of 567,000 muscle fibers in 114 biopsies. *Neurology* 28:439–46

36. Franco A Jr, Lansman JB. 1990. Calcium entry through stretch-inactivated ion channels in *mdx* myotubes. *Nature* 344:670–73

37. Vandebrouck C, Martin D, Colson-Van Schoor M, Debaix H, Gailly P. 2002. Involvement of TRPC in the abnormal calcium influx observed in dystrophic (*mdx*) mouse skeletal muscle fibers. *J. Cell Biol.* 158:1089–96

38. Yeung EW, Whitehead NP, Suchyna TM, Gottlieb PA, Sachs F, Allen DG. 2005. Effects of stretch-activated channel blockers on $[Ca^{2+}]_i$ and muscle damage in the *mdx* mouse. *J. Physiol.* 562:367–80

39. Alderton JM, Steinhardt RA. 2000. How calcium influx through calcium leak channels is responsible for the elevated levels of calcium-dependent proteolysis in dystrophic myotubes. *Trends Cardiovasc. Med.* 10:268–72

40. Iwata Y, Katanosaka Y, Arai Y, Komamura K, Miyatake K, Shigekawa M. 2003. A novel mechanism of myocyte degeneration involving the Ca^{2+}-permeable growth factor-regulated channel. *J. Cell Biol.* 161:957–67

41. Goll DE, Thompson VF, Li H, Wei W, Cong J. 2003. The calpain system. *Physiol. Rev.* 83:731–801

42. Vandebrouck A, Ducret T, Basset O, Sebille S, Raymond G, et al. 2006. Regulation of store-operated calcium entries and mitochondrial uptake by minidystrophin expression in cultured myotubes. *FASEB J.* 20:136–38

43. Millay DP, Sargent MA, Osinska H, Baines CP, Barton ER, et al. 2008. Genetic and pharmacologic inhibition of mitochondrial-dependent necrosis attenuates muscular dystrophy. *Nat. Med.* 14:442–47

44. Bonuccelli G, Sotgia F, Schubert W, Park DS, Frank PG, et al. 2003. Proteasome inhibitor (MG-132) treatment of *mdx* mice rescues the expression and membrane localization of dystrophin and dystrophin-associated proteins. *Am. J. Pathol.* 163:1663–75

45. Spencer MJ, Mellgren RL. 2002. Overexpression of a calpastatin transgene in *mdx* muscle reduces dystrophic pathology. *Hum. Mol. Genet.* 11:2645–55

46. Rando TA, Disatnik MH, Yu Y, Franco A. 1998. Muscle cells from *mdx* mice have an increased susceptibility to oxidative stress. *Neuromuscular Disord.* 8:14–21

47. Tidball JG. 2005. Inflammatory processes in muscle injury and repair. *Am. J. Physiol. Regul. Integr. Comp. Physiol.* 288:R345–53

48. Wehling M, Spencer MJ, Tidball JG. 2001. A nitric oxide synthase transgene ameliorates muscular dystrophy in *mdx* mice. *J. Cell Biol.* 155:123–31

49. Grady RM, Grange RW, Lau KS, Maimone MM, Nichol MC, et al. 1999. Role for α-dystrobrevin in the pathogenesis of dystrophin-dependent muscular dystrophies. *Nat. Cell Biol.* 1:215–20

50. Crosbie RH, Straub V, Yun HY, Lee JC, Rafael JA, et al. 1998. *mdx* muscle pathology is independent of nNOS perturbation. *Hum. Mol. Genet.* 7:823–29

51. Rando TA. 2001. Role of nitric oxide in the pathogenesis of muscular dystrophies: a "two hit" hypothesis of the cause of muscle necrosis. *Microsc. Res. Tech.* 55:223–35

52. Miles AM, Bohle DS, Glassbrenner PA, Hanser B, Wink DA, Grisham MB. 1996. Modulation of superoxide-dependent oxidation and hydroxylation reactions by nitric oxide. *J. Biol. Chem.* 271:40–47

53. Dudley RW, Danialou G, Govindaraju K, Lands L, Eidelman DE, Petrof BJ. 2006. Sarcolemmal damage in dystrophin deficiency is modulated by synergistic interactions between mechanical and oxidative/nitrosative stresses. *Am. J. Pathol.* 168:1276–87

54. Kar NC, Pearson CM. 1979. Catalase, superoxide dismutase, glutathione reductase and thiobarbituric acid-reactive products in normal and dystrophic human muscle. *Clin. Chim. Acta* 94:277–80

55. Touboul D, Brunelle A, Halgand F, De La Porte S, Laprevote O. 2005. Lipid imaging by gold cluster time-of-flight secondary ion mass spectrometry: application to Duchenne muscular dystrophy. *J. Lipid. Res.* 46:1388–95

56. Disatnik MH, Dhawan J, Yu Y, Beal MF, Whirl MM, et al. 1998. Evidence of oxidative stress in *mdx* mouse muscle: studies of the prenecrotic state. *J. Neurol. Sci.* 161:77–84

57. Ragusa RJ, Chow CK, Porter JD. 1997. Oxidative stress as a potential pathogenic mechanism in an animal model of Duchenne muscular dystrophy. *Neuromuscular Disord.* 7:379–86

58. Buetler TM, Renard M, Offord EA, Schneider H, Ruegg UT. 2002. Green tea extract decreases muscle necrosis in *mdx* mice and protects against reactive oxygen species. *Am. J. Clin. Nutr.* 75:749–53

59. Bornman L, Rossouw H, Gericke GS, Polla BS. 1998. Effects of iron deprivation on the pathology and stress protein expression in murine X-linked muscular dystrophy. *Biochem. Pharmacol.* 56:751–57

60. Backman E, Nylander E, Johansson I, Henriksson KG, Tagesson C. 1988. Selenium and vitamin E treatment of Duchenne muscular dystrophy: no effect on muscle function. *Acta Neurol. Scand.* 78:429–35

61. Fenichel GM, Brooke MH, Griggs RC, Mendell JR, Miller JP, et al. 1988. Clinical investigation in Duchenne muscular dystrophy: penicillamine and vitamin E. *Muscle Nerve* 11:1164–68

62. Messina S, Altavilla D, Aguennouz M, Seminara P, Minutoli L, et al. 2006. Lipid peroxidation inhibition blunts nuclear factor-κB activation, reduces skeletal muscle degeneration, and enhances muscle function in *mdx* mice. *Am. J. Pathol.* 168:918–26

63. Kumar A, Boriek AM. 2003. Mechanical stress activates the nuclear factor-κB pathway in skeletal muscle fibers: a possible role in Duchenne muscular dystrophy. *FASEB J.* 17:386–96

64. Gosselin LE, Martinez DA. 2004. Impact of TNF-α blockade on TGF-β1 and type I collagen mRNA expression in dystrophic muscle. *Muscle Nerve* 30:244–46

65. Grounds MD, Torrisi J. 2004. Anti-TNFα (Remicade) therapy protects dystrophic skeletal muscle from necrosis. *FASEB J.* 18:676–82

66. De Luca A, Nico B, Liantonio A, Didonna MP, Fraysse B, et al. 2005. A multidisciplinary evaluation of the effectiveness of cyclosporine A in dystrophic *mdx* mice. *Am. J. Pathol.* 166:477–89

67. Messina S, Bitto A, Aguennouz M, Minutoli L, Monici MC, et al. 2006. Nuclear factor κ-B blockade reduces skeletal muscle degeneration and enhances muscle function in *Mdx* mice. *Exp. Neurol.* 198:234–41

68. Porter JD, Khanna S, Kaminski HJ, Rao JS, Merriam AP, et al. 2002. A chronic inflammatory response dominates the skeletal muscle molecular signature in dystrophin-deficient *mdx* mice. *Hum. Mol. Genet.* 11:263–72

69. Bansal D, Miyake K, Vogel SS, Groh S, Chen CC, et al. 2003. Defective membrane repair in dysferlin-deficient muscular dystrophy. *Nature* 423:168–72

70. Anderson LV, Davison K, Moss JA, Young C, Cullen MJ, et al. 1999. Dysferlin is a plasma membrane protein and is expressed early in human development. *Hum. Mol. Genet.* 8:855–61

71. Achanzar WE, Ward S. 1997. A nematode gene required for sperm vesicle fusion. *J. Cell Sci.* 110(Pt. 9):1073–81

72. Davis DB, Doherty KR, Delmonte AJ, McNally EM. 2002. Calcium-sensitive phospholipid binding properties of normal and mutant ferlin C2 domains. *J. Biol. Chem.* 277:22883–88

73. Bashir R, Britton S, Strachan T, Keers S, Vafiadaki E, et al. 1998. A gene related to *Caenorhabditis elegans* spermatogenesis factor *fer-1* is mutated in limb-girdle muscular dystrophy type 2B. *Nat. Genet.* 20:37–42

74. Liu J, Aoki M, Illa I, Wu C, Fardeau M, et al. 1998. Dysferlin, a novel skeletal muscle gene, is mutated in Miyoshi myopathy and limb girdle muscular dystrophy. *Nat. Genet.* 20:31–36

75. Weiler T, Bashir R, Anderson LV, Davison K, Moss JA, et al. 1999. Identical mutation in patients with limb girdle muscular dystrophy type 2B or Miyoshi myopathy suggests a role for modifier gene(s). *Hum. Mol. Genet.* 8:871–77

76. Piccolo F, Moore SA, Ford GC, Campbell KP. 2000. Intracellular accumulation and reduced sarcolemmal expression of dysferlin in limb-girdle muscular dystrophies. *Ann. Neurol.* 48:902–12

77. Selcen D, Stilling G, Engel AG. 2001. The earliest pathologic alterations in dysferlinopathy. *Neurology* 56:1472–81

78. de Luna N, Gallardo E, Soriano M, Dominguez-Perles R, de la Torre C, et al. 2006. Absence of dysferlin alters myogenin expression and delays human muscle differentiation "in vitro". *J. Biol. Chem.* 281:17092–98

79. Lennon NJ, Kho A, Bacskai BJ, Perlmutter SL, Hyman BT, Brown RH Jr. 2003. Dysferlin interacts with annexins A1 and A2 and mediates sarcolemmal wound-healing. *J. Biol. Chem.* 278:50466–73

80. Parton RG. 1996. Caveolae and caveolins. *Curr. Opin. Cell Biol.* 8:542–48

81. Minetti C, Sotgia F, Bruno C, Scartezzini P, Broda P, et al. 1998. Mutations in the caveolin-3 gene cause autosomal dominant limb-girdle muscular dystrophy. *Nat. Genet.* 18:365–68

82. Matsuda C, Hayashi YK, Ogawa M, Aoki M, Murayama K, et al. 2001. The sarcolemmal proteins dysferlin and caveolin-3 interact in skeletal muscle. *Hum. Mol. Genet.* 10:1761–66

83. Hernandez-Deviez DJ, Martin S, Laval SH, Lo HP, Cooper ST, et al. 2006. Aberrant dysferlin trafficking in cells lacking caveolin or expressing dystrophy mutants of caveolin-3. *Hum. Mol. Genet.* 15:129–42

84. Hernandez-Deviez DJ, Howes MT, Laval SH, Bushby K, Hancock JF, Parton RG. 2008. Caveolin regulates endocytosis of the muscle repair protein, dysferlin. *J. Biol. Chem.* 283:6476–88

85. Richard I, Broux O, Allamand V, Fougerousse F, Chiannilkulchai N, et al. 1995. Mutations in the proteolytic enzyme calpain 3 cause limb-girdle muscular dystrophy type 2A. *Cell* 81:27–40

86. Huang Y, Verheesen P, Roussis A, Frankhuizen W, Ginjaar I, et al. 2005. Protein studies in dysferlinopathy patients using llama-derived antibody fragments selected by phage display. *Eur. J. Hum. Genet.* 13:721–30

87. Anderson LV, Harrison RM, Pogue R, Vafiadaki E, Pollitt C, et al. 2000. Secondary reduction in calpain 3 expression in patients with limb girdle muscular dystrophy type 2B and Miyoshi myopathy (primary dysferlinopathies). *Neuromuscular Disord.* 10:553–59

88. Huang Y, de Morree A, van Remoortere A, Bushby K, Frants RR, et al. 2008. Calpain 3 is a modulator of the dysferlin protein complex in skeletal muscle. *Hum. Mol. Genet.* 7:1855–66

89. Huang Y, Laval SH, van Remoortere A, Baudier J, Benaud C, et al. 2007. AHNAK, a novel component of the dysferlin protein complex, redistributes to the cytoplasm with dysferlin during skeletal muscle regeneration. *FASEB J.* 21:732–42

90. Ando Y, Imamura S, Hong YM, Owada MK, Kakunaga T, Kannagi R. 1989. Enhancement of calcium sensitivity of lipocortin I in phospholipid binding induced by limited proteolysis and phosphorylation at the amino terminus as analyzed by phospholipid affinity column chromatography. *J. Biol. Chem.* 264:6948–55

91. Cagliani R, Magri F, Toscano A, Merlini L, Fortunato F, et al. 2005. Mutation finding in patients with dysferlin deficiency and role of the dysferlin interacting proteins annexin A1 and A2 in muscular dystrophies. *Hum. Mutat.* 26:283

92. McNeil AK, Rescher U, Gerke V, McNeil PL. 2006. Requirement for annexin A1 in plasma membrane repair. *J. Biol. Chem.* 281:35202–7

93. Sussman J, Stokoe D, Ossina N, Shtivelman E. 2001. Protein kinase B phosphorylates AHNAK and regulates its subcellular localization. *J. Cell Biol.* 154:1019–30

94. Detrait ER, Yoo S, Eddleman CS, Fukuda M, Bittner GD, Fishman HM. 2000. Plasmalemmal repair of severed neurites of PC12 cells requires Ca^{2+} and synaptotagmin. *J. Neurosci. Res.* 62:566–73

95. Davis DB, Delmonte AJ, Ly CT, McNally EM. 2000. Myoferlin, a candidate gene and potential modifier of muscular dystrophy. *Hum. Mol. Genet.* 9:217–26

96. Doherty KR, Cave A, Davis DB, Delmonte AJ, Posey A, et al. 2005. Normal myoblast fusion requires myoferlin. *Development* 132:5565–75

97. Vainzof M, Anderson LV, McNally EM, Davis DB, Faulkner G, et al. 2001. Dysferlin protein analysis in limb-girdle muscular dystrophies. *J. Mol. Neurosci.* 17:71–80

98. Tatsumi R, Anderson JE, Nevoret CJ, Halevy O, Allen RE. 1998. HGF/SF is present in normal adult skeletal muscle and is capable of activating satellite cells. *Dev. Biol.* 194:114–28

99. Conboy IM, Conboy MJ, Smythe GM, Rando TA. 2003. Notch-mediated restoration of regenerative potential to aged muscle. *Science* 302:1575–77

100. Conboy IM, Rando TA. 2002. The regulation of Notch signaling controls satellite cell activation and cell fate determination in postnatal myogenesis. *Dev. Cell* 3:397–409

101. Collins CA, Olsen I, Zammit PS, Heslop L, Petrie A, et al. 2005. Stem cell function, self-renewal, and behavioral heterogeneity of cells from the adult muscle satellite cell niche. *Cell* 122:289–301

102. Blau HM, Webster C, Pavlath GK. 1983. Defective myoblasts identified in Duchenne muscular dystrophy. *Proc. Natl. Acad. Sci. USA* 80:4856–60

103. Webster C, Blau HM. 1990. Accelerated age-related decline in replicative life-span of Duchenne muscular dystrophy myoblasts: implications for cell and gene therapy. *Somat. Cell Mol. Genet.* 16:557–65

104. Heslop L, Morgan JE, Partridge TA. 2000. Evidence for a myogenic stem cell that is exhausted in dystrophic muscle. *J. Cell Sci.* 113(Pt. 12):2299–308

105. Wakayama Y, Schotland DL, Bonilla E, Orecchio E. 1979. Quantitative ultrastructural study of muscle satellite cells in Duchenne dystrophy. *Neurology* 29:401–7

106. Ishimoto S, Goto I, Ohta M, Kuroiwa Y. 1983. A quantitative study of the muscle satellite cells in various neuromuscular disorders. *J. Neurol. Sci.* 62:303–14

107. Matecki S, Guibinga GH, Petrof BJ. 2004. Regenerative capacity of the dystrophic (*mdx*) diaphragm after induced injury. *Am. J. Physiol. Regul. Integr. Comp. Physiol.* 287:R961–68

108. Decary S, Hamida CB, Mouly V, Barbet JP, Hentati F, Butler-Browne GS. 2000. Shorter telomeres in dystrophic muscle consistent with extensive regeneration in young children. *Neuromuscular Disord.* 10:113–20

109. Decary S, Mouly V, Hamida CB, Sautet A, Barbet JP, Butler-Browne GS. 1997. Replicative potential and telomere length in human skeletal muscle: implications for satellite cell-mediated gene therapy. *Hum. Gene Ther.* 8:1429–38

110. Renault V, Piron-Hamelin G, Forestier C, DiDonna S, Decary S, et al. 2000. Skeletal muscle regeneration and the mitotic clock. *Exp. Gerontol.* 35:711–19

111. Lund TC, Grange RW, Lowe DA. 2007. Telomere shortening in diaphragm and tibialis anterior muscles of aged *mdx* mice. *Muscle Nerve* 36:387–90

112. Oexle K, Zwirner A, Freudenberg K, Kohlschutter A, Speer A. 1997. Examination of telomere lengths in muscle tissue casts doubt on replicative aging as cause of progression in Duchenne muscular dystrophy. *Pediatr. Res.* 42:226–31

113. Delaporte C, Dehaupas M, Fardeau M. 1984. Comparison between the growth pattern of cell cultures from normal and Duchenne dystrophy muscle. *J. Neurol. Sci.* 64:149–60

114. Jasmin G, Tautu C, Vanasse M, Brochu P, Simoneau R. 1984. Impaired muscle differentiation in explant cultures of Duchenne muscular dystrophy. *Lab. Investig.* 50:197–207

115. Melone MA, Peluso G, Petillo O, Galderisi U, Cotrufo R. 1999. Defective growth in vitro of Duchenne Muscular Dystrophy myoblasts: the molecular and biochemical basis. *J. Cell Biochem.* 76:118–32

116. Cohn RD, van Erp C, Habashi JP, Soleimani AA, Klein EC, et al. 2007. Angiotensin II type 1 receptor blockade attenuates TGF-β-induced failure of muscle regeneration in multiple myopathic states. *Nat. Med.* 13:204–210

117. Tatsumi R, Hattori A, Ikeuchi Y, Anderson JE, Allen RE. 2002. Release of hepatocyte growth factor from mechanically stretched skeletal muscle satellite cells and role of pH and nitric oxide. *Mol. Biol. Cell* 13:2909–2918

118. Anderson JE. 2000. A role for nitric oxide in muscle repair: nitric oxide-mediated activation of muscle satellite cells. *Mol. Biol. Cell* 11:1859–74

119. Lee KH, Baek MY, Moon KY, Song WK, Chung CH, et al. 1994. Nitric oxide as a messenger molecule for myoblast fusion. *J. Biol. Chem.* 269:14371–74

120. Melone MA, Peluso G, Galderisi U, Petillo O, Cotrufo R. 2000. Increased expression of IGF-binding protein-5 in Duchenne muscular dystrophy (DMD) fibroblasts correlates with the fibroblast-induced downregulation of DMD myoblast growth: an in vitro analysis. *J. Cell Physiol.* 185:143–53

121. Philippou A, Halapas A, Maridaki M, Koutsilieris M. 2007. Type I insulin-like growth factor receptor signaling in skeletal muscle regeneration and hypertrophy. *J. Musculoskelet. Neuronal Interact.* 7:208–18

122. Barton ER, Morris L, Musaro A, Rosenthal N, Sweeney HL. 2002. Muscle-specific expression of insulin-like growth factor I counters muscle decline in *mdx* mice. *J. Cell Biol.* 157:137–48

123. Merly F, Lescaudron L, Rouaud T, Crossin F, Gardahaut MF. 1999. Macrophages enhance muscle satellite cell proliferation and delay their differentiation. *Muscle Nerve* 22:724–32

124. Grounds MD. 1998. Age-associated changes in the response of skeletal muscle cells to exercise and regeneration. *Ann. N. Y. Acad. Sci.* 854:78–91

125. Conboy IM, Conboy MJ, Wagers AJ, Girma ER, Weissman IL, Rando TA. 2005. Rejuvenation of aged progenitor cells by exposure to a young systemic environment. *Nature* 433:760–64

126. Wallace GQ, Lapidos KA, Kenik JS, McNally EM. 2008. Long-term stem survival of transplanted cells in immunocompetent mice with muscular dystrophy. *Am. J. Pathol.* 173:792–802

127. Bione S, Maestrini E, Rivella S, Mancini M, Regis S, et al. 1994. Identification of a novel X-linked gene responsible for Emery-Dreifuss muscular dystrophy. *Nat. Genet.* 8:323–27

128. Bonne G, Di Barletta MR, Varnous S, Becane HM, Hammouda EH, et al. 1999. Mutations in the gene encoding lamin A/C cause autosomal dominant Emery-Dreifuss muscular dystrophy. *Nat. Genet.* 21:285–88

129. Muchir A, Bonne G, Van Der Kooi AJ, van Meegen M, Baas F, et al. 2000. Identification of mutations in the gene encoding lamins A/C in autosomal dominant limb girdle muscular dystrophy with atrioventricular conduction disturbances (LGMD1B). *Hum. Mol. Genet.* 9:1453–59

130. Todorova A, Halliger-Keller B, Walter MC, Dabauvalle MC, Lochmuller H, Muller CR. 2003. A synonymous codon change in the *LMNA* gene alters mRNA splicing and causes limb girdle muscular dystrophy type 1B. *J. Med. Genet.* 40:e115

131. Fatkin D, MacRae C, Sasaki T, Wolff MR, Porcu M, et al. 1999. Missense mutations in the rod domain of the lamin A/C gene as causes of dilated cardiomyopathy and conduction-system disease. *N. Engl. J. Med.* 341:1715–24

132. Manilal S, Nguyen TM, Sewry CA, Morris GE. 1996. The Emery-Dreifuss muscular dystrophy protein, emerin, is a nuclear membrane protein. *Hum. Mol. Genet.* 5:801–8

133. Goldman RD, Goldman AE, Shumaker DK. 2005. Nuclear lamins: building blocks of nuclear structure and function. *Novartis Found. Symp.* 264:3–16

134. Crisp M, Liu Q, Roux K, Rattner JB, Shanahan C, et al. 2006. Coupling of the nucleus and cytoplasm: role of the LINC complex. *J. Cell Biol.* 172:41–53

135. Capanni C, Cenni V, Mattioli E, Sabatelli P, Ognibene A, et al. 2003. Failure of lamin A/C to functionally assemble in R482L mutated familial partial lipodystrophy fibroblasts: altered intermolecular interaction with emerin and implications for gene transcription. *Exp. Cell Res.* 291:122–34

136. Favreau C, Dubosclard E, Ostlund C, Vigouroux C, Capeau J, et al. 2003. Expression of lamin A mutated in the carboxyl-terminal tail generates an aberrant nuclear phenotype similar to that observed in cells from patients with Dunnigan-type partial lipodystrophy and Emery-Dreifuss muscular dystrophy. *Exp. Cell Res.* 282:14–23

137. Muchir A, Medioni J, Laluc M, Massart C, Arimura T, et al. 2004. Nuclear envelope alterations in fibroblasts from patients with muscular dystrophy, cardiomyopathy, and partial lipodystrophy carrying lamin A/C gene mutations. *Muscle Nerve* 30:444–50

138. Vigouroux C, Auclair M, Dubosclard E, Pouchelet M, Capeau J, et al. 2001. Nuclear envelope disorganization in fibroblasts from lipodystrophic patients with heterozygous R482Q/W mutations in the lamin A/C gene. *J. Cell Sci.* 114:4459–68

139. Lammerding J, Hsiao J, Schulze PC, Kozlov S, Stewart CL, Lee RT. 2005. Abnormal nuclear shape and impaired mechanotransduction in emerin-deficient cells. *J. Cell Biol.* 170:781–91

140. Lammerding J, Schulze PC, Takahashi T, Kozlov S, Sullivan T, et al. 2004. Lamin A/C deficiency causes defective nuclear mechanics and mechanotransduction. *J. Clin. Investig.* 113:370–78

141. Broers JL, Peeters EA, Kuijpers HJ, Endert J, Bouten CV, et al. 2004. Decreased mechanical stiffness in LMNA$^{-/-}$ cells is caused by defective nucleo-cytoskeletal integrity: implications for the development of laminopathies. *Hum. Mol. Genet.* 13:2567–80

142. Nikolova V, Leimena C, McMahon AC, Tan JC, Chandar S, et al. 2004. Defects in nuclear structure and function promote dilated cardiomyopathy in lamin A/C-deficient mice. *J. Clin. Investig.* 113:357–69

143. Melcon G, Kozlov S, Cutler DA, Sullivan T, Hernandez L, et al. 2006. Loss of emerin at the nuclear envelope disrupts the Rb1/E2F and MyoD pathways during muscle regeneration. *Hum. Mol. Genet.* 15:637–51

144. Bakay M, Wang Z, Melcon G, Schiltz L, Xuan J, et al. 2006. Nuclear envelope dystrophies show a transcriptional fingerprint suggesting disruption of Rb-MyoD pathways in muscle regeneration. *Brain* 129:996–1013

145. Frock RL, Kudlow BA, Evans AM, Jameson SA, Hauschka SD, Kennedy BK. 2006. Lamin A/C and emerin are critical for skeletal muscle satellite cell differentiation. *Genes Dev.* 20:486–500

146. Favreau C, Higuet D, Courvalin JC, Buendia B. 2004. Expression of a mutant lamin A that causes Emery-Dreifuss muscular dystrophy inhibits in vitro differentiation of C2C12 myoblasts. *Mol. Cell. Biol.* 24:1481–92

147. Markiewicz E, Ledran M, Hutchison CJ. 2005. Remodelling of the nuclear lamina and nucleoskeleton is required for skeletal muscle differentiation in vitro. *J. Cell Sci.* 118:409–20

148. Parnaik VK, Manju K. 2006. Laminopathies: multiple disorders arising from defects in nuclear architecture. *J. Biosci.* 31:405–21

149. Sabourin LA, Girgis-Gabardo A, Seale P, Asakura A, Rudnicki MA. 1999. Reduced differentiation potential of primary MyoD$^{-/-}$ myogenic cells derived from adult skeletal muscle. *J. Cell Biol.* 144:631–43

150. Yablonka-Reuveni Z, Rudnicki MA, Rivera AJ, Primig M, Anderson JE, Natanson P. 1999. The transition from proliferation to differentiation is delayed in satellite cells from mice lacking MyoD. *Dev. Biol.* 210:440–55

151. Megeney LA, Kablar B, Garrett K, Anderson JE, Rudnicki MA. 1996. MyoD is required for myogenic stem cell function in adult skeletal muscle. *Genes Dev.* 10:1173–83

152. Cornelison DD, Olwin BB, Rudnicki MA, Wold BJ. 2000. MyoD$^{-/-}$ satellite cells in single-fiber culture are differentiation defective and MRF4 deficient. *Dev. Biol.* 224:122–37

Plant Ion Channels: Gene Families, Physiology, and Functional Genomics Analyses

John M. Ward,[1] Pascal Mäser,[2] and Julian I. Schroeder[3]

[1] Department of Plant Biology, University of Minnesota, St. Paul, Minnesota 55108; email: jward@umn.edu

[2] Institute of Cell Biology, University of Bern, CH-3012 Bern, Switzerland

[3] Division of Biological Sciences, Cell and Developmental Biology Section, University of California, San Diego, La Jolla, California 92093; email: julian@biomail.ucsd.edu

Annu. Rev. Physiol. 2009. 71:59–82

First published online as a Review in Advance on October 8, 2008

The *Annual Review of Physiology* is online at physiol.annualreviews.org

This article's doi: 10.1146/annurev.physiol.010908.163204

Copyright © 2009 by Annual Reviews. All rights reserved

0066-4278/09/0315-0059$20.00

Key Words

guard cells, stomatal regulation, *Arabidopsis*, comparative genomics

Abstract

Distinct potassium, anion, and calcium channels in the plasma membrane and vacuolar membrane of plant cells have been identified and characterized by patch clamping. Primarily owing to advances in *Arabidopsis* genetics and genomics, and yeast functional complementation, many of the corresponding genes have been identified. Recent advances in our understanding of ion channel genes that mediate signal transduction and ion transport are discussed here. Some plant ion channels, for example, ALMT and SLAC anion channel subunits, are unique. The majority of plant ion channel families exhibit homology to animal genes; such families include both hyperpolarization- and depolarization-activated Shaker-type potassium channels, CLC chloride transporters/channels, cyclic nucleotide–gated channels, and ionotropic glutamate receptor homologs. These plant ion channels offer unique opportunities to analyze the structural mechanisms and functions of ion channels. Here we review gene families of selected plant ion channel classes and discuss unique structure-function aspects and their physiological roles in plant cell signaling and transport.

INTRODUCTION

Ion channels have long been known to function in action potential generation in animal and plant algal cells (1–3). In plants the depolarizing phase of action potentials is, however, not mediated by Na^+ channels, owing to the lack of a suitable Na^+ gradient and to the fact that Na^+ ions are toxic to most plant cells. Rather, the depolarizing phase of action potentials is mediated by Cl^- efflux channels and by Ca^{2+} channels (3–6). Recent research has led to initial characterizations of Cl^-/anion channel–encoding genes with unique structures (7–11).

In addition to their classical roles in action potential mediation, several classes of plant ion channels function in long-term net uptake or release of ions from cells or organelles (12–23). In this respect, several classes of plant ion channels do not show the typical rapid inactivation of animal ion channels. For example, inward- and outward-rectifying K^+ channels in guard cells can remain activated for >30 min (24). This property allows these ion channels to function in long-term ion transport, thus mediating plant movements that are osmotically driven by so-called motor cells, which can show large changes in their turgor (25, 26).

Plant ion channels provide potent models to dissect general properties of ion channels that are presently of great interest and intensely debated. For example, the mechanisms allowing voltage sensor domains to activate and deactivate ion channels can be analyzed in plant ion channels because relatively closely related K^+ channels exist with opposite voltage dependencies (27–32). Recent studies have shown that different members of the CLC family of membrane proteins can encode for either Cl^- channels or endosomal Cl^-/H^+ exchangers (33–36). However, the physiological relevance of this Cl^-/H^+ exchange mechanism remains incompletely understood. Recent research has directly analyzed and demonstrated anion/H^+ exchange mediated by an *Arabidopsis* CLC in the intact large vacuolar organelles of plant cells and has illustrated the physiological necessity of this type of transporter for concentrating nitrate reserves at 50-fold-higher concentrations inside vacuoles than in the cytoplasm (37).

A few studies have characterized depolarization-activated Ca^{2+}-permeable channels in plant cells (38–40), and this type of ion channel has only recently been recorded in *Arabidopsis* (189). These Ca^{2+} channels may have specific functions, including the rapid responses of Venus flytrap (*Dionaea muscipula*) leaves (5). Most plasma membrane Ca^{2+} channels that have been recorded in plants show increased activity upon hyperpolarization (41–44). Interestingly, the genes that encode these Ca^{2+} permeable channels have not yet been characterized. Candidate genes in *Arabidopsis* may be the family of 20 cyclic nucleotide–gated (CNG) channel homologs (45) and 20 genes encoding homologs to animal glutamate receptor channels (46). However, sequenced genomes and EST databases indicate that higher plants may not include TRP channels (47) or ORAI homologs (48) within their genetic repertoires.

Analyses of cells with specific functions have facilitated the characterization of plant ion channel functions and regulation. Guard cells have been developed as a model system for illuminating plant ion channel function and signaling (**Figure 1**). Pairs of guard cells in the epidermis of leaves form stomatal pores (**Figure 1a,b**) that enable plants to take in CO_2 from the atmosphere for carbon fixation. However, stomatal pores also provide the major conduits for water loss of plants, via transpiration to the atmosphere. Stomatal pores can close (**Figure 1a**) and open (**Figure 1b**), thus regulating the exchange of CO_2 and water with the atmosphere. Physiological stimuli such as light and low CO_2 concentrations regulate the osmotic opening of stomatal pores by increasing the K^+, anion, and organic solute concentrations of guard cells (**Figure 1b;c**, right). Stomatal closing, in contrast, is mediated by the release of K^+ and anions from guard cells, resulting in guard cell turgor reduction (**Figure 1a;c**, left). Stomatal closing reduces water loss in plants and is triggered by drought conditions, elevated CO_2, and darkness. The adaptation of patch clamping and Ca^{2+}-imaging methods to the relatively small guard cells of *Arabidopsis* (49, 50) opened up the possibility of combining powerful molecular genetic/genomic analyses in *Arabidopsis* with single-cell ion channel regulation and cell signaling analyses. Analyses of ion channel regulation in *Arabidopsis* have greatly increased our understanding of ion channel functions and signaling mechanisms in plants. Several reviews are available for background information on upstream signaling mechanisms that regulate plant ion channel activity (25, 26, 51, 52), and recently reported examples of ion channel signaling are reviewed here.

GUARD CELLS AS A MODEL SYSTEM TO STUDY PHYSIOLOGICAL FUNCTIONS OF PLANT ION CHANNELS

Guard cells regulate the aperture of stomatal pores in response to many physiological

Figure 1

Model for ion channel and transporter classes and their demonstrated or predicted roles during stomatal movements. (*a*) A closed stomate in a leaf from *Vicia faba* (broad bean). (*b*) An open stomate from *Vicia faba*. Two guard cells surround the stomatal pore and regulate the aperture of the central pore. (*c*) Model for regulation and activity of guard cell ion channels and transporters. (*Left*) (*1*) Signals that induce stomatal closing include the hormone abscisic acid (ABA), high $[CO_2]$, high extracellular $[Ca^{2+}]$, and high ozone $[O_3]$. Many of the receptors involved remain to be identified (see Reference 26). Stomatal closing requires net cellular efflux of solutes, in particular K^+, Cl^-, and malate. ABA induces reactive oxygen species (ROS) production (*2*), which activates Ca^{2+}-permeable I_{Ca} channels (*3*). Cytosolic Ca^{2+} concentration ($[Ca^{2+}]_{cyt}$) is a central regulator of transport mechanisms in guard cells and activates slow/sustained (S-type) anion channels (*4*) and vacuolar SV (TPC) channels (*7*) and VK (TPK) channels (*8*). Cl^- and malate efflux through S-type and rapid (R-type) anion channels (*4*) and possibly CLC Cl^-/H^+ antiporters (*6*) causes depolarization and drives K^+ efflux through outward-rectifying K^+ channels (*5*). At the vacuole membrane, SV (TPC1) channels (*7*) are Ca^{2+}-activated and Ca^{2+}-permeable voltage-dependent channels. VK (TPK) channels (*8*) are Ca^{2+} activated, are highly selective for K^+, and are proposed to allow K^+ release from the vacuole during stomatal closing. (*Right*) Mechanisms that function in guard cell ion uptake and stomatal opening. Blue light (*10*) activates phototropins (receptor/kinase), leading to stomatal opening. Signaling results in 14-3-3 binding to plasma membrane proton pumps (*11*), leading to hyperpolarization and acidification of the extracellular space. Hyperpolarization activates inward-rectifying K^+ channels (*12*). At the vacuolar membrane, proton pumps (*13*) acidify the vacuole lumen and drive K^+/H^+ antiporters (*14*). Cl^- and malate may accumulate in the vacuole through anion channel and anion antiporters (*15*).

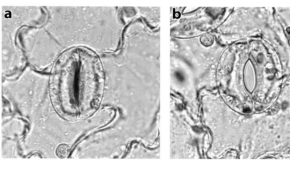

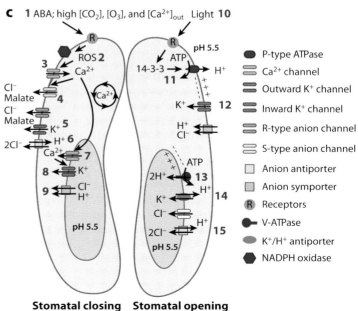

Stomatal closing   Stomatal opening

Legend:
- P-type ATPase
- Ca^{2+} channel
- Outward K^+ channel
- Inward K^+ channel
- R-type anion channel
- S-type anion channel
- Anion antiporter
- Anion symporter
- R Receptors
- V-ATPase
- K^+/H^+ antiporter
- NADPH oxidase

stimuli such as light, CO_2 concentration, and hormones (26, 52). Additional features have made guard cells an attractive model system for studies of ion channels: These cells are located in the epidermis and can be easily isolated, they are not connected to adjacent cells by plasmodesmata, and their cellular movements are rapid and driven by osmotic changes. Early work showed that cellular ion flux mediates guard cell osmotic changes (53–55). The inward and outward-rectifying K^+ channels (14, 24) and anion channels (15, 16) were characterized and proposed to mediate stomatal movements (**Figure 1c**). The development of the plant *Arabidopsis thaliana* as a genetic model organism opened the possibility of combining electrophysiology, cell biology, molecular genetics, and functional genomics to study ion channel function and signal transduction in plants. This has allowed several approaches to relate specific ion channel genes to physiological functions in guard cells. Insertional mutants have been used to identify physiological functions for specific channel genes (10, 21–23, 56). Ion channel activities have been analyzed in *Arabidopsis* signal transduction mutants to understand how mutations in upstream abscisic acid (ABA) signal transduction components such as protein phosphatases impair ABA regulation of anion channels that control stomatal responses (49). The ability to apply the transgenic cameleon Ca^{2+} indicator to analyze *Arabidopsis* signaling mutants has allowed further resolution of the roles of cytosolic Ca^{2+} in the control of stomatal aperture (50, 57, 58).

During stomatal opening, guard cell volume increases by approximately 40% (59, 60). This increase occurs by activation of plasma membrane H^+ pumps, K^+ uptake through inward-rectifying K^+ channels, organic acid (mainly malate) production, and the uptake of inorganic anions (Cl^-), presumably through proton-coupled uptake transporters (**Figure 1c**, right). This lowers the water potential in guard cells and drives osmotic water uptake into guard cells. Most of the volume change in guard cells is accounted for by increased vacuolar volume, with multiple smaller vacuoles fusing to form

a large central vacuole (60). K^+ uptake into vacuoles during stomatal opening is likely due to H^+/K^+ antiporter activity (61), whereas anion uptake into vacuoles may occur through a combination of low-affinity anion channels (9, 62, 63) and an active H^+/anion exchange mechanism (37).

During stomatal closing, guard cell volume decreases owing to net cellular efflux of solutes. Depolarization due to plasma membrane anion channel activity was proposed to drive K^+ efflux through outward-rectifying K^+ channels (14–16). The resulting cellular export of K, Cl^-, and organic ions such as malate drives water efflux (**Figure 1a;c**, left). At the vacuole membrane, Ca^{2+}-activated vacuolar K^+ (VK) channels are required for K^+ release to the cytoplasm (17, 22, 64).

In the following sections, we present an overview of plant ion channel classes, how they relate in structure and activity to channels in other systems, and how they function in plants.

PLANT CHLORIDE/ANION CHANNELS AND TRANSPORTERS

Cl^- channels in algae mediate a major component of the depolarizing phase of action potentials, consistent with a downhill Cl^-/anion gradient from the cytoplasm to the cell wall space of plants cells that is typically opposite to the gradient found in animal cells (3, 6, 65, 66). In guard cells, roots, and other cell types, anion channels may provide major mechanisms for net anion efflux and transport across the plasma membrane (7, 15, 16, 67) (**Figure 1**). In addition, many responses in plants, including plant pathogen responses, involve membrane potential depolarization accompanied by Cl^- efflux, and anion channel blockers inhibit pathogen-induced signaling, suggesting that anion channels may play roles in these responses as well (68). Indeed, anion channel–mediated depolarization may contribute to the propagation of signals because most plant cells are cytoplasmically connected to one another via plasmodesmata.

Aluminum Tolerance Mediated by ALMT Anion Channels/Transporters

In addition to their proposed functions as major regulators of stomatal closing as discussed above, plant anion channels have been of great interest in understanding how plants can acquire tolerance to aluminum stress. Aluminum is the third most abundant element in the earth's lithosphere (crust). In acidic soils, which cause solubilization of Al^{3+}, Al^{3+} is highly toxic to plant growth (69). Research on different plant species has demonstrated that plant cultivars that show tolerance to Al^{3+} are able to secrete Al^{3+}-chelating organic anions, such as malate and citrate at the root surface, in response to Al^{3+} exposure (70–72). Electrophysiological studies have shown that Al^{3+} activates anion channel currents in cortical cells in the root apex (71, 73).

Major genetic Al^{3+} tolerance loci have been cloned and mapped in several plants, and several of these loci encode members of a family of membrane proteins known as Al^{3+}-activated malate transporters (ALMTs) (**Table 1**) (7, 8, 74). ALMTs have 6–8 predicted membrane domains. Interestingly, expression of individual ALMTs in *Xenopus* oocytes is sufficient to mediate Al^{3+}-induced malate currents, suggesting that the ALMT transporters are direct sensors or receptors of Al^{3+} stress (7, 8, 75). Reversal potential shifts in *Xenopus* oocytes show Nernstian behavior correlating with the malate gradient (8), leading to the suggestion that ALMTs encode anion channels.

A recent study has shown that the *Arabidopsis* AtALMT9 protein is targeted to the vacuolar membrane of plant cells and mediates low-affinity malate uptake into plant vacuoles (9). Vacuolar (V-type) proton pumps acidify the lumen of vacuoles and provide the driving force for low- and high-affinity vacuolar malate uptake. Plant vacuoles have malate uptake

Table 1 Number of predicted transporter genes in the genomes of *Arabidopsis*, poplar, rice, and *Chlamydomonas*[a]

	A. thaliana	*P. trichocarpa*	*O. sativa*	*C. reinhardtii*
Predicted proteome size	31,711	58,036	26,841	14,598
Cation transporters				
KcsA type	1	0	0	0
Shaker type	9	11	6	0
TPK/KCO	5	10	4/3	0
Unclassified 6TM1P	0	0	0	9
CNGC	20	12	10	0
TPC	1	1	1	0
HKT	1	2	6	0
KUP/HAK	13	28	25	3
Anion channels				
CLC	7	8	5	5
SLAC	5	4	7	1
ALMT	13	22	8	0
Calcium channels				
VDCC	0	0	0	9
Ins-3P-R	0	0	0	1
GluR (plant type)	20	61	13	0

[a]The predicted proteomes were screened with HMMer profiles (188) made from homology-reduced sets of representative proteins of each family. See text for abbreviations of family names. The results are affected by differences in sequence coverage and quality of gene predictions between species.

channels (62, 63, 76). Patch clamp analysis of intact vacuoles has shown that this vacuolar malate current is absent in *ALMT9* knockout mutant plants (9). Further structure-function analyses of ALMTs should shed light into the mechanisms of Al^{3+} activation, anion transport, and selectivity.

CLC Chloride Channels and Proton/Anion Exchangers

In animal cells several classes of Cl^- channels are known. These include the CLC family (77), GABA receptor and glycine receptor channels (78), volume-regulated anion channels (VRACs), and bestrophins (79). In plants no bestrophin homologs or direct GABA/glycine receptor homologs have been reported. Small domains with homology to these ion channels have been reported within the 20-member glutamate receptor homolog family (80), although no evidence has been reported that these function as anion channels.

In the case of the membrane protein CLC genes, a family of seven homologs is present in the *Arabidopsis* genome (**Table 1**). Several plant CLC proteins are targeted to intracellular organelles (37, 81–83). The first electrophysiologically characterized plant CLC protein was AtCLCa, which is targeted to plant vacuoles (37). Interestingly, AtCLCa mediates proton/anion (nitrate) exchanger activity in the vacuolar membrane, rather than anion channel activity (37). Previous studies of the bacterial ClC-e1 and the endosomal CLC-4 and CLC-5 membrane proteins showed that they encode proton/Cl^- exchangers (33–35). However, the physiological relevance of proton/Cl^- exchange activity compared with Cl^- channel activity remained unclear in bacteria and animal cells. For example, both Cl^- channels and proton/Cl^- exchangers would be predicted to mediate Cl^- uptake into endosomes. Research on plant vacuoles showed that AtCLCa is energetically suited for the physiological accumulation of nitrate into vacuoles at lumenal concentrations that are 50-fold higher than typical cytosolic nitrate concentrations (37). Passive nitrate channels would not allow such physiological nitrate accumulation in plant vacuoles that may generate membrane potentials of –30 mV (on the cytoplasmic membrane side relative to the vacuole lumen).

Analyses of bacterial CLC proton/Cl^- exchanger ClC-ec1 have shown that two glutamate residues (E148 and E203) are important for proton/Cl^- exchange activity vis-à-vis Cl^- channel activity of other CLC proteins (36). Interestingly, four of seven *Arabidopsis* CLC proteins have glutamate residues at both corresponding positions (e.g., E203 and E270 of AtCLC-a) (37, 84), indicating that only three *Arabidopsis* CLCs may encode anion channels, whereas most plant CLC transporters may encode proton/anion exchangers.

Expression of a proton/anion exchanger CLC in the plasma membrane of plant cells may be relevant for anion efflux and membrane depolarization (84) because the extracellular space of plant cells has an acidic pH (e.g., pH 5.5 to 6). For example, motor cells that rapidly reduce their turgor via ion release (**Figure 1*a***) could build up high cell wall anion concentrations (25, 84), which would limit the ability of passive anion channels to continue driving turgor reduction (**Figure 1*c***, left). However, proton/anion exchangers would energetically favor continued anion efflux across the plasma membrane and could continue to mediate anion efflux and depolarization, which is required for K^+ efflux via K^+ channels from motor cells. Further research is needed to determine whether a plant CLC protein is targeted to the plasma membrane and what type of biophysical anion transport mechanism such transporters utilize.

Slow Anion Channels Require the SLAC1 Membrane Protein

Research on guard cells has led to the model that anion channels are critical mechanisms for controlling stomatal closing and anion efflux from plant cells (15, 16). However, apart from genes encoding the ALMT malate channels/transporters, anion channel–encoding genes have remained unknown in plants. Recent genetic screens for ozone sensitivity of leaves and for CO_2 insensitivity of leaf temperature have led to isolation of a gene, named *SLAC1*, that encodes a plasma membrane protein in guard cells (10, 11). Interestingly, *slac1* mutant alleles show insensitivity to CO_2-, ABA-, and ozone-induced stomatal closing (10, 11). This ozone insensitivity leads to an increased entry of ozone into leaves through open stomata (**Figure 1**) and thus to enhanced ozone damage in *slac1* mutants (10).

Patch clamp analyses of two *slac1* mutant alleles showed that slow/sustained (S-type) anion channel currents were greatly impaired, whereas the rapid (R-type) anion channels and the plasma membrane Ca^{2+}-permeable I_{Ca} channels were intact in guard cells (10). Thus, these data provide strong genetic evidence for the model that S-type anion channels provide one of the critical mechanisms for signal-induced stomatal closing (15). These analyses do not exclude a role for R-type anion channels in stomatal closing because, for example, light-dark transitions, ABA, and transitions to low humidity showed a slowed or a partial stomatal closing response in *slac1* mutant alleles that may be mediated by the remaining R-type anion channels or other anion efflux conductances (10). As discussed above in reference to CLCs, there may be a third type of anion efflux channel/transporter that also contributes to the control of stomatal closing. Furthermore, analyses of *slac1* mutants show that R-type and S-type anion channels do not require identical membrane proteins (10), although these findings do not exclude the possibility that other proteins are shared among these two types of plant anion channels (85).

SLAC1 is part of a gene family with five members in the *Arabidopsis* genome (**Table 1**). SLAC1 encodes a membrane protein that has nine to ten predicted transmembrane domains with large N- and C-terminal domains. SLAC proteins show a weak homology in their predicted transmembrane domains to a fungal (86) and a bacterial malate transporter in the TDT family. Both the S-type and the R-type anion channels allow Cl^- and malate efflux from guard cells (16, 87), which correlates with malate and Cl^- efflux from guard cells during stomatal closing (88, 89). Thus, the homology of SLAC1 to malate transporters and the disruption of S-type anion currents in *slac1* mutants indicate that SLAC1 is a member of a novel plant anion/Cl^- channel family. Further research on SLAC proteins should determine their regulation mechanisms and functions in various plant tissues.

PLANT Ca^{2+} CHANNELS

Several distinct Ca^{2+} channel activities have been characterized in the plasma membrane of plant cells. However, to date, the corresponding genes have not been identified. Land plants encode one type of voltage-dependent Ca^{2+} channel homolog (TPC1) that has been well characterized and is expressed in the vacuolar membrane (see below). Candidate genes that possibly encode plasma membrane Ca^{2+} channels are a family of 20 glutamate receptor homologs (GluR) (46) and a family of 20 CNG channel homologs (CNGC) (45, 90). Glutamate receptor homologs are required for glutamate-induced depolarization and intracellular $[Ca^{2+}]$ elevation in roots (91, 92). Further voltage clamp recordings of these channels will be valuable for comparisons to known root Ca^{2+} channels (42, 93). For more information on plant homologs of ionotropic glutamate receptors, see Lacombe et al. (46).

Several studies have identified hyperpolarization-activated Ca^{2+} channels (41–44). ABA enhances this I_{Ca} activity (44) by shifting the activation voltage to more positive potentials (43). ABA regulation of I_{Ca} can occur through reactive oxygen species (ROS) production (44, 94), which induces increases in cytosolic Ca^{2+} concentration ($[Ca^{2+}]_{cyt}$). ABA induction of ROS and ABA activation of I_{Ca} channels are impaired in guard cells in which the plasma membrane NADPH oxidase *AtrBOHD* and *AtrBOHF* genes are disrupted, providing genetic evidence for a role of ROS in I_{Ca} regulation by ABA (94). Furthermore, I_{Ca} is regulated by Ca^{2+}-dependent protein kinases and protein phosphorylation events (95, 96).

Hyperpolarization-activated Ca^{2+}-permeable channels have also been analyzed in root cells (42, 97). Differences in ROS sensitivity of these channels in mature versus elongating root cells indicates that more than one class of hyperpolarization-activated channel may exist in plants (93). A study in the root hair development mutant *rhd2*, which has very short root hair cells, identified the *rhd2* mutation in the transmembrane NADPH oxidase AtRBOHC (98). This study further showed a defect in *rhd2* in establishing a root hair tip–focused $[Ca^{2+}]_{cyt}$ gradient. Application of hydrogen peroxide to cells reestablished Ca^{2+} influx and Ca^{2+} channel activation in root hair cells (98; see Note Added in Proof following the Literature Cited).

Depolarization-activated Ca^{2+} channels have been characterized in plant cells (38–40, 99) but appear to be less ubiquitous than hyperpolarization-activated channels. The relative scarcity of depolarization-activated Ca^{2+} channels may be because optimal conditions for analyzing these channels have not yet been developed or other plant types and cell systems may need to be analyzed for this purpose. Studies of action potentials in Venus flytrap suggest that depolarization-activated Ca^{2+} channels may function in this electrically propagated response (5, 100). Plants also have stretch-activated Ca^{2+} channels (101), for which the genes and regulation mechanisms remain less studied to date. Mechanical stimulation of *Arabidopsis* roots causes a rapid rise in $[Ca^{2+}]_{cyt}$ (102). This system may be well suited for advancing models of functions, regulation, and genes mediating mechanosensitive $[Ca^{2+}]_{cyt}$ elevations. A recent study identified *Arabidopsis* Mca1, a potential component of mechanosensitive Ca^{2+} channels (103; see Note Added in Proof).

COMPARATIVE GENOMICS OF HIGHER PLANT ION CHANNELS WITH *CHLAMYDOMONAS*

The unicellular green alga *Chlamydomonas reinhardtii* has been a model organism primarily for the study of photosynthesis and chloroplast biogenesis and its ability to grow heterotrophically in the dark, allowing for isolation of photosynthetic mutations that would be lethal in vascular plants. Since the completion of the *Chlamydomonas* genome project (104), *C. reinhardtii* has been useful for comparative genomics with other members of the kingdom Viridiplantae. These genomic analyses allow extrapolation to the common ancestor of all green plants, a photosynthetic organism that lived more than a billion years ago before the divergence of the chlorophytes (green algae) and the streptophytes (land plants). The *C. reinhardtii* genome was found to encode a number of ion channels that had been thought to be specific to animals (105; **Table 1**): an inositol-1,4,5-trisphosphate-gated Ca^{2+} channel, nine predicted 4×6 transmembrane domain–containing Ca^{2+} channel α subunits, and seven 6TM1P proteins with similarity to hyperpolarization-activated cyclic nucleotide–gated (HCN)-type K^+ channels. The physiological function of these ion channels in *C. reinhardtii* remains to be elucidated. On the bases of these findings and the lack of known close homologs in land plant genomes (**Table 1**), *Chlamydomonas* can be considered a "green animal" (104). The presence of such animal-type ion channel genes in a green alga suggests that they had also been part of the ion channel repertoire of the ancestral viridiplant and were subsequently eliminated in the streptophyte lineage (105). Likewise, Na^+ and K^+ ion transporter/channels such as HKT/Trk or the KUP/HAKs (discussed below), which occur not only in plants but also in fungi and bacteria, may have been present in an ancestor of animals as well and were lost in the course of metazoan evolution. Thus, parsimonious comparison of transporter genes indicates that the common ancestor of plants and animals had a larger repertoire of ion channel classes than did extant members of either kingdom. Presumably such an organism was heterotrophic and unicellular.

POTASSIUM CHANNELS

Cation Channels with Pore Loops

We refer the reader to reviews on the physiology, regulation, and structure-function analyses of plant K^+ channels for background information (106, 107, 108). Here we review some of the major and new insights into K^+ channel structure and function and also recent advances from functional genetic analyses in *Arabidopsis*. Loop structures, in contrast to α-helical transmembrane domains, allow channels to engage moieties of the polypeptide backbone in K^+ ion binding (109). Thus, the K^+ ions inside the pore of the K^+ channel interact with the carbonyl oxygens of the filter triad glycine-tyrosine-glycine (GYG) located at the end of the channel's pore loop (P-loop) (110). Together, the carbonyl groups of the four filters of a tetrameric channel coordinate two K^+ ions separated by one molecule of H_2O in a similar way to the hydration sphere in water, explaining at once the near diffusion limit V_{max} of K^+ channels and the strong discrimination against the smaller cation Na^+ (110). The GYG triad is characteristic for K^+ channels (110); less selective cation channels/transporters also possess P-loops, but with a less stringent filter motif (111–113). Nevertheless, the P-loops are evolutionarily conserved to a degree that renders them readily detectable in silico (**Figure 2**). To identify all types of P-loop-containing cation channels from plants, we scanned the predicted proteomes of *A. thaliana*, *Populus trichocarpa*, *Oryza sativa*, and *C. reinhardtii* with a hidden Markov model–based profile for P-loops. The results are shown in **Figure 2**. Ubiquitous throughout all kingdoms of life are the tetrameric cation channel with cytosolic N and C termini and, per subunit, six transmembrane domains and one P-loop between TM5 and TM6. This topology, 6TM1P, is shared between voltage-gated K^+ (Kv) channels, CNG channels, and HCN channels, suggesting a common evolutionary origin of these families. The smallest K^+ channel subunits, represented by KcsA from *Streptomyces lividans* (110), consist of just one

P-loop between two transmembrane domains. Other classes of cation channels carry two P-loops per subunit and presumably function as dimers. In plants these are the KCO/TPK (two-pore) K^+ channels with four transmembrane domains; the P-loops are located between TM1 and TM2 and between TM3 and TM4. The schematic topologies are depicted in **Figure 2**.

Shaker-Type K^+ Channels

Named after the *Drosophila* shaker mutant, Shaker-type K^+ channels are the archetypal voltage-gated K^+ channels. Crystal structures have been resolved for the archaebacterial KvAP and the mammalian channel Kv1.2 (110). Shaker channels are activated by depolarization of the membrane potential. The voltage sensitivity is conferred by TM4, which contains a series of positively charged amino acids. Movement inside the membrane of this voltage sensor in response to depolarization opens the channel (114–117). The first depolarization-activated K^+ channel described from plants was SKOR, the stelar K^+ outward rectifier from *A. thaliana* (29). SKOR is expressed in the vascular cylinder of the root, where it is thought to mediate the release of K^+ into the xylem (conducting tissue) for root-to-leaf transport. This model is in accordance with the lowered K^+ content in xylem sap of T-DNA-disrupted *skor* null mutant *Arabidopsis* (29). SKOR mainly allows unidirectional K^+ efflux from cells because the activation potential of plant outward-rectifying K^+ channels shifts to more positive voltages upon an increase in the extracellular K^+ concentration (24, 118, 119). A related channel from *Arabidopsis*, GORK (guard cell outward-rectifying K^+ channel) (120), is the major outward-rectifying K^+ channel in guard cells and contributes to stomatal closure. Depolarization of guard cells by anion efflux activates GORK, and the resulting K^+ efflux further contributes to water loss and stomatal closure. Disruption of *GORK* by T-DNA insertion impaired stomatal closure (21). GORK is also expressed in root hairs (121). Although there are apparently no further

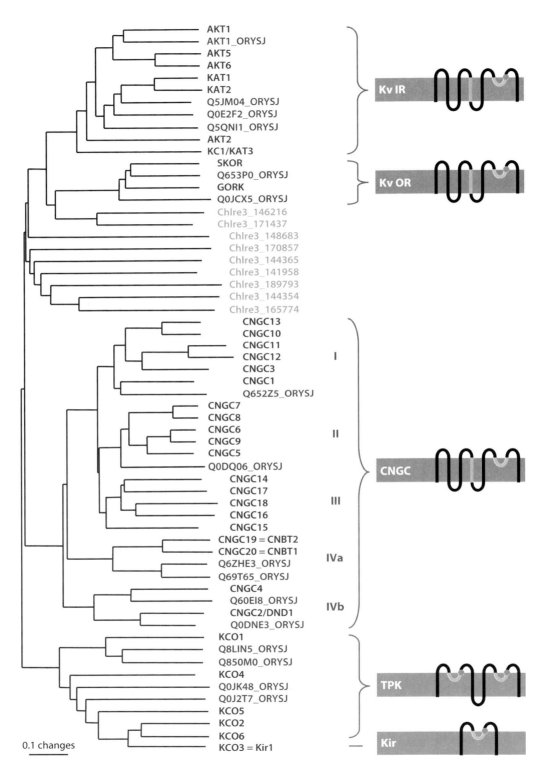

genes for outward-rectifying K$^+$ channels in the *Arabidopsis* genome, the situation is more complex: When coexpressed in *Saccharomyces cerevisiae* or *Xenopus laevis* oocytes, SKOR and GORK subunits are able to form heteromeric channels, and thus heteromeric channels may form in vivo (122).

Inward-Rectifying Shaker K$^+$ Channels

Plants possess members of the 6TM1P superfamily that carry all the hallmarks of Shaker channels, including the voltage sensor. These channels function as inward-rectifiers, activated by hyperpolarization of the membrane potential (14, 27, 123). The family of inward-rectifying Shaker channels comprises the first K$^+$ channels identified from plants, *Arabidopsis* AKT1 and KAT1. Both genes were identified by functional complementation in K$^+$ uptake–deficient yeast selected on limiting [K$^+$] (124, 125).

AKT1 is expressed in *Arabidopsis* roots. Homozygously disrupted *akt1* plants exhibit an impaired growth phenotype on micromolar extracellular [K$^+$], indicating that a channel can contribute to root K$^+$ accumulation from soil provided the membrane potential is large enough to enable concentrative uptake (18). The growth phenotype is apparent only in the presence of NH$_4^+$ (18). These data correlate with the finding that ammonium blocks alternative transporters for root K$^+$ uptake (126). AKT1 is regulated by the kinase CIPK23, which in turn is activated by calcineurin B–like Ca^{2+} sensors (127, 128). The *Arabidopsis* null mutant *cipk23* also exhibits an impaired growth phenotype at low K$^+$ concentrations in the presence of NH$_4^+$ (129). *Arabidopsis* guard cells express several of the inward-rectifying Shaker channels: AKT1, AKT2, KAT1, KAT2, and the regulatory subunit KC1 (19). These channels mediate hyperpolarization-induced K$^+$ influx, resulting in water influx and stomatal opening in response to signals such as sunlight (**Figure 1b;c**, right). Expression of dominant-negative KAT1 variants suppressed light-induced K$^+$ uptake in guard cells and reduced stomatal opening (20). In accordance with an important role of KAT1 in stomatal opening, KAT1 activity is regulated at multiple levels. Treatment with ABA reduced *KAT1* expression in guard cells and triggered endocytosis of the channel (130), whereas auxin enhanced *KAT1* expression in seedlings (131). When expressed in heterologous systems, KAT1 was activated by external acidification (132) or coexpression of 14-3-3 proteins (133) and was inhibited by coexpression of the regulatory subunit KC1 (134). Again, the properties of the inward-rectifying Shaker channels in planta may be multiplexed by the formation of heteromeric channels (135–138).

HCN Channels and CNG Channels

HCN and CNG cation channels have a 6TM1P topology, with positive charges in TM4 and a C-terminal cyclic nucleotide (cNMP) binding domain, and presumably function as tetramers. Both channel types are allosterically

---

Figure 2

Phylogenetic tree of P-loop–containing proteins from *Arabidopsis thaliana* (*dark blue*), rice (*Oryza sativa*; *red*), and *Chlamydomonas reinhardtii* (*light blue*). Where applicable, schematized membrane topologies are indicated: P-loop, *green*; transmembrane domain 4 (TM4) including positively charged amino acids, *orange*; K$^+$ selectivity filter residues, *magenta*. Cytosolic elements such as long C-terminal domains, cyclic nucleotide (cNMP) binding domains, or ankyrin repeats are not depicted. The photosynthetic alga *C. reinhardtii* has a unique set of ion channel genes (see Reference 105 and text). The P-loop–containing channels from *C. reinhardtii* all have a positively charged TM4, but only five (179857, 144365, 189793, 144354, and 165774) have a K$^+$ selectivity filter. Because these do not form a clade, and because the *C. reinhardtii* channels are poorly resolved in general, they could not be attributed to specific classes. Scale indicates the number of amino acid substitutions per site. Kv IR, inward-rectifying K$^+$ channel; Kv OR, outward-rectifying K$^+$ channel; CNGC, cyclic nucleotide–gated channel; TPK, two-pore K$^+$ channel.

activated by cyclic nucleotides, but only the HCN channels are voltage sensitive, activated by membrane hyperpolarization. In accordance with the less stringent filter sequence, HCN K^+ channels are also permeable to Na^+. CNG channels are even less selective and also conduct divalent cations. In mammals, HCN channels function as cardiac pacemakers, whereas CNG channels generate electrical signals in olfactory neurons and photoreceptors. Land plants do not appear to possess HCN channels. The photosynthetic green alga *C. reinhardtii*, however, encodes seven unclassified channels related to HCN channels (**Table 1**; **Figure 2**) that represent the predominant GTM1P channels in this organism.

The *Arabidopsis* genome encodes 20 CNGC genes (45). Forward and reverse genetic studies have identified functions of *Arabidopsis CNGC* genes. The *Arabidopsis dnd1* (defense, no death) mutant has a defect in the *CNGC2* gene (139). *dnd1* mutants exhibit a reduced localized programmed cell death response to pathogens, termed a hypersensitive response, but have elevated systemic acquired resistance. AtCNGC11 and -12 are important for resistance against an avirulent *Hyaloperonospora* pathogen (140). These results indicate roles for CNG channels in pathogen defense responses. Furthermore, a recent study showed that null mutations in the *Arabidopsis CNGC18* gene abolishes polarized growth of pollen tubes (141). cAMP but not cGMP activates a hypopolarization-dependent Ca^{2+} current in guard cells (142). Apart from these advances, the in vivo electrophysiological regulation properties of AtCNGCs remain less well understood and will require further analyses.

HKT TRANSPORTERS

Plant HKT (high-affinity K^+ transporter) proteins belong to the Trk/HKT superfamily of monovalent cation transporters, together with bacterial TrkH, KtrB, and fungal Trk proteins (111). Trk/HKT transporters carry eight transmembrane domains and four P-loop-like domains in one polypeptide chain and are therefore thought to function as monomers that resemble KcsA-type K^+ channel tetramers in structure (111, 143). The first gene cloned from plants was wheat *HKT1* (144), which, when expressed in *X. laevis* oocytes or in *S. cerevisiae*, functions as a K^+/Na^+ cotransporter/channel (145). The majority of the subsequently characterized HKTs from plants are permeable to Na^+ as well (146–148). HKT transporters in *Arabidopsis*, rice, and wheat protect plant leaves from overaccumulating Na^+ to toxic levels (149–151). In plant physiology, Na^+ has a dual role: Under K^+ starvation, Na^+ can replace the nutrient K^+ (152) in its nonspecific function as an osmolyte in the vacuole. However, in the cytosol, Na^+ competes with K^+-specific functions and is toxic to most plants. *Arabidopsis AtHKT1* is a single-copy gene that transports Na^+ when expressed in oocytes or yeast (146). Site-directed mutagenesis showed that a point mutation in the predicted selectivity filter contributes to determining K^+/Na^+ selectivity (112). Forward genetic (150, 153–155) as well as reverse genetic (149, 151, 156) approaches have identified AtHKT1 as a key component in protection of plants against salinity stress and in controlling shoot/root Na^+ distribution. *Arabidopsis athkt1* null mutants exhibit lower root Na^+ content and higher shoot Na^+ content compared with wild-type plants (149). This indicates that AtHKT1 functions to counteract Na^+ accumulation in the shoot, where photosynthetic tissue is particularly Na^+ sensitive. AtHKT1 and the rice ortholog *OsHKT8* (*OsHKT1;5*) are expressed in xylem parenchyma cells in leaves (150, 151). Both function in the retrieval of Na^+ from the xylem sap (150, 151, 156, 157), thus protecting leaves from overaccumulating Na^+. A recent review describes plant HKT proteins in more detail (158).

KUP/HAK/KT K^+ TRANSPORTERS

The KUP/HAK/KTs are a family of permeases that are selective for K^+ but do not have P-loops. They have a predicted topology of 10 to 14 predicted transmembrane domains

(159), but the K$^+$ transport mechanism (channel or other) remains unknown. The founding members of this family were cloned from bacteria (named KUP) (160) and from fungi named HAK (high-affinity K$^+$ uptake mechanism) (161). In plants, the first KUP/HAK/KT transporters were discovered independently by several laboratories from EST sequences and via yeast complementation (126, 162–164). Bacteria and fungi usually have only one or two *KUP/HAK/KT* genes (if any; *S. cerevisiae*, for instance, has none), whereas in plants the *KUP/HAK/KTs* form multigene families, with 13 members in *Arabidopsis*, 25 in rice, and 24 in poplar (90, 158, 159, 165). *C. reinhardtii* has three predicted *KUP/HAK/KT* genes (104). The expansion of the KUP/HAK/KT family in vascular plants is likely to reflect the speciation of individual KUP/HAK/KT paralogs and the importance of K$^+$ transport for many different physiological processes. The physiological functions of individual KUP/HAK/KT members are therefore challenging to assign by reverse genetics and require knowledge of the gene's expression pattern. The promoter of *A. thaliana AtHAK5* is active in the root vasculature and epidermis (166), and *AtHAK5* expression is upregulated at low [K$^+$] (166–168). K$^+$ starvation also induced expression of orthologs from rice and barley (165, 169). Genetic disruption of *AtHAK5* did not greatly affect plant growth. However, *athak5* null mutant plants contained less K$^+$ and displayed impaired Rb$^+$ uptake kinetics. In particular, the K$^+$-starvation inducible high-affinity component of root Rb$^+$/K$^+$ absorption (170) was diminished in *athak5* null mutant *Arabidopsis* (166). For two other KUP/HAK members, a physiological function could be attributed by forward genetics: The *Arabidopsis* mutant *Shy3-1* (short hypocotyl in the dark) results from a point mutation in *AtKUP2* (At2g30070) (171), and the mutant *Thr1* (tiny root hair) is caused by disruption of *AtKUP4* (At4g23640) (172), indicating a role for KUP/HAK transporters in cell expansion. A recent review discusses the KUP/HAKs in more detail (158).

VACUOLAR ION CHANNELS

TPK1 Twin-Pore K$^+$ Channels

VK (vacuolar K$^+$) channels were first identified in the guard cell vacuole membranes and are activated by elevated [Ca^{2+}]$_{cyt}$ in the range of 1 μM and highly selective for K$^+$ (17). VK channels were suggested to function as a pathway for organellar K$^+$ release into the cytosol in response to elevated [Ca^{2+}]$_{cyt}$ during stomatal closure (17). VK channels show a slight rectification that favors K$^+$ efflux from vacuoles, and VK channel activity is also stimulated by low cytosolic pH (17, 173). Expression of an *Arabidopsis* cDNA *TPK1/KCO1* (At5g55630) in yeast revealed vacuolar channel activity consistent with hallmark properties of VK channels (174). *Arabidopsis* mutants disrupted in *TPK1* lacked detectable VK channel activity and exhibited slower stomatal closing in response to ABA (22).

TPK1 is localized in vacuole membranes, and the *TPK1* gene is expressed at high levels in most tissues (22, 175). TPK1 is a member of the two-pore-domain K$^+$ channel family, first described in yeast (176; **Figure 2**). There are five members of this family in *Arabidopsis* and three in rice, and all share the same topology of four transmembrane spans and two pore domains (**Figure 2**; **Table 1**). TPK1 has two EF hand Ca^{2+} binding domains in the C terminus, which makes TPK1 distinct from animal TPK channels (177) and indicates that Ca^{2+} binds and activates the channel directly (17, 22).

The KcsA-Type Channel Kir

Consisting of just two transmembrane domains and one P-loop per subunit, the KcsA-type channels are the most rudimentary of the K$^+$ channels (110). The *Arabidopsis* genome encodes only one KcsA-type channel, Kir1, whereas rice and poplar apparently lack such channels. Kir1 was formerly named KCO3 because it belongs to the KCO family of TPK channels (**Figure 2**). Thus, *Kir1/KCO3* may be the direct descendant of a gene that, upon duplication, became the founding member of the two-pore family. Alternatively, *Kir1/KCO3* may be a truncated descendant of that family. The position of Kir1/KCO3 in the phylogenetic tree (**Figure 2**) does not support the first hypothesis, and the circumstance that other plants lack KcsA-type genes also points toward *Kir1/KCO3* being a descendant of the KCO/TPK family. The function of Kir1/KCO3 in *Arabidopsis* remains to be investigated.

TPC1 Voltage-Dependent Calcium Channel: Slow Vacuolar Channel

The slow vacuolar (SV) channel was shown to be Ca^{2+} activated in vacuole membranes from beet storage tissue (178) and has subsequently been found in vacuole membranes from many plant cell types. SV channels are activated by Ca^{2+} and positive voltages on the cytoplasmic membrane side and are large-conductance, nonselective cation channels. In guard cells SV channels were found to be Ca^{2+}-permeable (17), leading to the hypothesis that SV channels participate in Ca^{2+} signaling by releasing Ca^{2+} from the vacuole.

In 2005, Peiter et al. (56) identified the gene encoding SV channels in *Arabidopsis* as *TPC1* (At4g03560). *TPC1* is a unique gene in plants, the only homolog of voltage-dependent Ca^{2+} channels in the *Arabidopsis* genome. Similar to TPC channels in animals (179), TPC1 in plants has two pore domains rather than the four pore domains usually associated with voltage-dependent Ca^{2+} channels. In plants, *TPC1* is highly expressed in all tissues, consistent with the ubiquity of SV channel activity in plant vacuoles. The voltage dependence of SV channels limits vacuolar Ca^{2+} release, although recordings of deactivating (tail) currents show a clear Ca^{2+} permeation toward the cytoplasmic membrane side (17, 180). For physiological Ca^{2+} release (181), it has been proposed that the voltage dependence of SV channels would need to be shifted by modulators, as was recently found in *Arabidopsis* mesophyll vacuoles (182). An *Arabidopsis* TPC1 knockout mutant was defec-

tive in the inhibition of stomatal opening by external Ca^{2+} (56). This finding is consistent with a function for SV channels in guard cell Ca^{2+} signaling. However, ABA inhibited stomatal opening in the mutant similarly to wild type (56). The TPC1 mutant was, however, less sensitive to ABA inhibition of seed germination, indicating that SV channels participate in a subset of ABA signaling pathways in plants. Now that *TPC1* has been identified as the SV channel–encoding gene (56), the physiological functions of this ubiquitous channel can be analyzed in diverse tissue types and under diverse conditions.

Elevated extracellular Ca^{2+} causes cytoplasmic $[Ca^{2+}]$ elevations in plant guard cells (57, 183). The CAS1 Ca^{2+} receptor is required for this response, triggering $[Ca^{2+}]_{cyt}$ oscillations that then inhibit stomatal opening and induce stomatal closing (184, 185). CAS1 has a single putative transmembrane domain and binds Ca^{2+} ions with a low affinity (184). No CAS1 homologs have been identified in nonplant species. CAS1 has been localized to the plasma membrane (184) and to thylakoid membranes in chloroplasts (185). More work will be required to understand how extracellular Ca^{2+} causes CAS1-dependent intracellular $[Ca^{2+}]$ elevations (184, 185).

Fast Vacuolar Channels

The fast vacuolar (FV) channel is a low-conductance, nonselective cation channel that is inhibited by elevated $[Ca^{2+}]_{cyt}$. FV channels were first identified in vacuole membranes from beet storage tissue (178). FV channels were subsequently analyzed in guard cells (173). The genes encoding FV channels have not been identified.

In guard cell vacuoles, if K^+ accumulates relative to the cytoplasm during stomatal opening, then FV channels should be downregulated, allowing K^+/H^+ antiporters such as CHX20 (61) to generate a K^+ gradient. FV channels can be very active at resting $[Ca^{2+}]_{cyt}$. However, physiological concentrations of Mg^{2+} inhibit FV channels (186), and FV channels are voltage dependent, with lower open probabilities at membrane potentials close to 0 mV (178, 186). Elevated vacuolar K^+ causes an increase in the open probability of FV channels (187). FV channels may function as a pathway for K^+ flux from the vacuole into the cytoplasm during proposed Ca^{2+}-independent stomatal closing.

CONCLUSIONS

Detailed studies have revealed the activities of K^+, anion, and Ca^{2+} channels and transporters in plant cells. Recent advances, especially patch clamping of *Arabidopsis* cell types, the sequencing of the *Arabidopsis* genome, and the development of genomic approaches in *Arabidopsis*, have facilitated the identification of genes encoding specific ion channel and transporter activities. The availability of mutants with defects in signal transduction pathways combined with electrophysiological assays have allowed rapid progress in understanding physiological functions of specific ion channels, as exemplified for guard cell ion channels. Guard cells have been developed as model plant cells for the study of ion channels because they respond to many physiological stimuli, are easily observed and isolated, and undergo osmotically driven reversible cellular movements requiring ion channel activity. Notably, related ion channels are found in most other cell types and are important for many other physiological processes, including mineral nutrition, signal transduction, tropisms, cell elongation, the export of toxic ions and organic acids, and interactions with pathogens and symbiotic organisms.

The finding that most plant ion channels have homologs in animals has increased interest in plants as model systems. Analysis of differences in ion channel gating, kinetics, selectivity, and regulation is contributing to understanding the structure-function relation of ion channels in general. In addition, plant cells provide advantages for analysis of intracellular channels: The plant vacuole is large and can be directly patch clamped, whereas the equivalent endosomes or lysosomes in animal cells are not as readily accessible.

Understanding the genetic bases of ion channels raises the possibility of engineering new traits of importance for agriculture and the environment. For example, ALMT anion channels function in the export of organic acids that allow plants to survive toxic Al^{3+} concentrations that are a major global problem in acidic soils. ALMT activity or expression could be modified in agricultural crops to allow food production on marginal land. Another example is the recent identification of SLAC1. SLAC1 has a central role in regulating stomatal aperture, which controls CO_2 uptake and water loss. Water use efficiency and resistance to drought are increasingly important in agriculture. Atmospheric CO_2 concentrations are predicted to continue to increase. Modification of SLAC1 may provide a mechanism to limit water loss and improve drought resistance while maintaining high carbon fixation rates. Thus, many new and unexpected properties of plant ion channels of importance to humans and the environment are undoubtedly awaiting discovery.

DISCLOSURE STATEMENT

The authors are not aware of any biases that might be perceived as affecting the objectivity of this review.

ACKNOWLEDGMENTS

We thank colleagues for providing preprints of publications. We thank Glenn Wheeler and Colin Brownlee for sharing results prior to publication. We apologize to those authors whose work we did not cite owing to length and reference limitations. Research in the authors' laboratories was supported by grants from the DOE (FG03-94-ER20148), NIH (R01GM60396 P42 ES01 0337), and NSF (MCB 0417118) to J.I.S., from the NSF (IOS-0419695) to J.W. and J.I.S., and from the Swiss NSF to P.M.

LITERATURE CITED

1. Cole KS, Curtis HJ. 1938. Electrical impedance of *Nitella* during activity. *J. Gen. Physiol.* 22:37–64
2. Hodgkin AL, Huxley AF. 1952. The components of membrane conductance in the giant axon of Loligo. *J. Physiol.* 116:473–96
3. Gaffey CT, Mullins LJ. 1958. Ion fluxes during the action potential in *Chara*. *J. Physiol.* 144:505–24
4. Hope AB, Findlay GP. 1962. The action potential in Chara. *Plant Cell Physiol.* 5:377–79
5. Hodick D, Sievers A. 1988. The action potential of *Dionaea muscipula* Ellis. *Planta* 174:8–18
6. Kaneko T, Saito C, Shimmen T, Kikuyama M. 2005. Possible involvement of mechanosensitive Ca^{2+} channels of plasma membrane in mechanoperception in *Chara*. *Plant Cell Physiol.* 46:130–35
7. Sasaki T, Yamamoto Y, Ezaki B, Katsuhara M, Ahn SJ, et al. 2004. A wheat gene encoding an aluminum-activated malate transporter. *Plant J.* 37:645–53
8. Hoekenga OA, Maron LG, Pineros MA, Cancado GM, Shaff J, et al. 2006. AtALMT1, which encodes a malate transporter, is identified as one of several genes critical for aluminum tolerance in *Arabidopsis*. *Proc. Natl. Acad. Sci. USA* 103:9738–43
9. Kovermann P, Meyer S, Hortensteiner S, Picco C, Scholz-Starke J, et al. 2007. The *Arabidopsis* vacuolar malate channel is a member of the ALMT family. *Plant J.* 52:1169–80
10. Vahisalu T, Kollist H, Wang YF, Nishimura N, Chan WY, et al. 2008. SLAC1 is required for plant guard cell S-type anion channel function in stomatal signalling. *Nature* 452:487–91
11. Negi J, Matsuda O, Nagasawa T, Oba Y, Takahashi H, et al. 2008. CO_2 regulator SLAC1 and its homologues are essential for anion homeostasis in plant cells. *Nature* 452:483–86
12. Schroeder JI, Hedrich R, Fernandez JM. 1984. Potassium-selective single channels in guard cell protoplasts of *Vicia faba*. *Nature* 312:361–62
13. Moran N, Ehrenstein G, Iwasa K, Bare C, Mischke C. 1984. Ion channels in the plasmalemma of wheat protoplasts. *Science* 226:835-37
14. Schroeder JI, Raschke K, Neher E. 1987. Voltage dependence of K^+ channels in guard cell protoplasts. *Proc. Natl. Acad. Sci. USA* 84:4108–12
15. Schroeder JI, Hagiwara S. 1989. Cytosolic calcium regulates ion channels in the plasma membrane of *Vicia faba* guard cells. *Nature* 338:427–30
16. Keller BU, Hedrich R, Raschke K. 1989. Voltage-dependent anion channels in the plasma membrane of guard cells. *Nature* 341:250–53
17. Ward JM, Schroeder JI. 1994. Calcium-activated K^+ channels and calcium-induced calcium release by slow vacuolar ion channels in guard cell vacuoles implicated in the control of stomatal closure. *Plant Cell* 6:669–83
18. Hirsch RE, Lewis BD, Spalding EP, Sussman MR. 1998. A role for the AKT1 potassium channel in plant nutrition. *Science* 280:918–21
19. Szyroki A, Ivashikina N, Dietrich P, Roelfsema MR, Ache P, et al. 2001. KAT1 is not essential for stomatal opening. *Proc. Natl. Acad. Sci. USA* 98:2917–21
20. Kwak JM, Murata Y, Baizabal-Aguirre VM, Merrill J, Wang M, et al. 2001. Dominant negative guard cell K^+ channel mutants reduce inward-rectifying K^+ currents and light-induced stomatal opening in *Arabidopsis*. *Plant Physiol.* 127:473–85
21. Hosy E, Vavasseur A, Mouline K, Dreyer I, Gaymard F, et al. 2003. The *Arabidopsis* outward K^+ channel GORK is involved in regulation of stomatal movements and plant transpiration. *Proc. Natl. Acad. Sci. USA* 100:5549–54
22. Gobert A, Isayenkov S, Voelker C, Czempinski K, Maathuis FJ. 2007. The two-pore channel *TPK1* gene encodes the vacuolar K^+ conductance and plays a role in K^+ homeostasis. *Proc. Natl. Acad. Sci. USA* 104:10726–31
23. Lebaudy A, Vavasseur A, Hosy E, Dreyer I, Leonhardt N, et al. 2008. Plant adaptation to fluctuating environment and biomass production are strongly dependent on guard cell potassium channels. *Proc. Natl. Acad. Sci. USA* 105:5271–76

24. Schroeder JI. 1988. K$^+$ transport properties of K$^+$ channels in the plasma membrane of *Vicia faba* guard cells. *J. Gen. Physiol.* 92:667–83

25. Moran N. 2007. Osmoregulation of leaf motor cells. *FEBS Lett.* 581:2337–47

26. Pandey S, Zhang W, Assmann SM. 2007. Roles of ion channels and transporters in guard cell signal transduction. *FEBS Lett.* 581:2325–36

27. Schachtman DP, Schroeder JI, Lucas WJ, Anderson JA, Gaber RF. 1992. Expression of an inward-rectifying potassium channel by the *Arabidopsis* KAT1 cDNA. *Science* 258:1654–58

28. Cao Y, Crawford NM, Schroeder JI. 1995. Amino terminus and the first four membrane-spanning segments of the *Arabidopsis* K$^+$ channel KAT1 confer inward-rectification property of plant-animal chimeric channels. *J. Biol. Chem.* 270:17697–701

29. Gaymard F, Pilot G, Lacombe B, Bouchez D, Bruneau D, et al. 1998. Identification and disruption of a plant shaker-like outward channel involved in K$^+$ release into the xylem sap. *Cell* 94:647–55

30. Latorre R, Olcese R, Basso C, Gonzalez C, Muñoz F, et al. 2003. Molecular coupling between voltage sensor and pore opening in the *Arabidopsis* inward rectifier K$^+$ channel KAT1. *J. Gen. Physiol.* 122:459–69

31. Lai HC, Grabe M, Jan YN, Jan LY. 2005. The S4 voltage sensor packs against the pore domain in the KAT1 voltage-gated potassium channel. *Neuron* 47:395–406

32. Grabe M, Lai HC, Jain M, Jan YN, Jan LY. 2007. Structure prediction for the down state of a potassium channel voltage sensor. *Nature* 445:550–53

33. Accardi A, Miller C. 2004. Secondary active transport mediated by a prokaryotic homologue of ClC Cl$^-$ channels. *Nature* 427:803–7

34. Picollo A, Pusch M. 2005. Chloride/proton antiporter activity of mammalian CLC proteins ClC-4 and ClC-5. *Nature* 436:420–23

35. Scheel O, Zdebik AA, Lourdel S, Jentsch TJ. 2005. Voltage-dependent electrogenic chloride/proton exchange by endosomal CLC proteins. *Nature* 436:424–27

36. Miller C. 2006. ClC chloride channels viewed through a transporter lens. *Nature* 440:484–89

37. De Angeli A, Monachello D, Ephritikhine G, Frachisse JM, Thomine S, et al. 2006. The nitrate/proton antiporter AtCLCa mediates nitrate accumulation in plant vacuoles. *Nature* 442:939–42

38. Huang JW, Grunes DL, Kochian LV. 1994. Voltage-dependent Ca^{2+} influx into right-side-out plasma membrane vesicles isolated from wheat roots: characterization of a putative Ca^{2+} channel. *Proc. Natl. Acad. Sci. USA* 91:3473–77

39. Thuleau P, Ward JM, Ranjeva R, Schroeder JI. 1994. Voltage-dependent calcium-permeable channels in the plasma membrane of a higher plant cell. *EMBO J.* 13:2970–75

40. Marshall J, Corzo A, Leigh RA, Sanders D. 1994. Membrane potential-dependent calcium transport in right-side-out plasma membrane vesicles from *Zea mays* L. roots. *Plant J.* 5:683–94

41. Gelli A, Blumwald E. 1997. Hyperpolarization-activated Ca^{2+}-permeable channels in the plasma membrane of tomato cells. *J. Membr. Biol.* 155:35–45

42. Very AA, Davies JM. 2000. Hyperpolarization-activated calcium channels at the tip of *Arabidopsis* root hairs. *Proc. Natl. Acad. Sci. USA* 97:9801–6

43. Hamilton DW, Hills A, Kohler B, Blatt MR. 2000. Ca^{2+} channels at the plasma membrane of stomatal guard cells are activated by hyperpolarization and abscisic acid. *Proc. Natl. Acad. Sci. USA* 97:4967–72

44. Pei ZM, Murata Y, Benning G, Thomine S, Klusener B, et al. 2000. Calcium channels activated by hydrogen peroxide mediate abscisic acid signalling in guard cells. *Nature* 406:731–34

45. Kaplan B, Sherman T, Fromm H. 2007. Cyclic nucleotide-gated channels in plants. *FEBS Lett.* 581:2237–46

46. Lacombe B, Becker D, Hedrich R, DeSalle R, Hollmann M, et al. 2001. The identity of plant glutamate receptors. *Science* 292:1486–87

47. Clapham DE. 2003. TRP channels as cellular sensors. *Nature* 426:517–24

48. Feske S, Gwack Y, Prakriya M, Srikanth S, Puppel SH, et al. 2006. A mutation in Orai1 causes immune deficiency by abrogating CRAC channel function. *Nature* 441:179–85

49. Pei ZM, Kuchitsu K, Ward JM, Schwarz M, Schroeder JI. 1997. Differential abscisic acid regulation of guard cell slow anion channels in *Arabidopsis* wild-type and *abi1* and *abi2* mutants. *Plant Cell* 9:409–23

50. Allen GJ, Kwak JM, Chu SP, Llopis J, Tsien RY, et al. 1999. Cameleon calcium indicator reports cytoplasmic calcium dynamics in *Arabidopsis* guard cells. *Plant J.* 19:735–47

51. Blatt MR. 2000. Cellular signaling and volume control in stomatal movements in plants. *Annu. Rev. Cell Dev. Biol.* 16:221–41

52. Schroeder JI, Allen GJ, Hugouvieux V, Kwak JM, Waner D. 2001. Guard cell signal transduction. *Annu. Rev. Plant Physiol. Plant Mol. Biol.* 52:627–58

53. Humble GD, Raschke K. 1971. Stomatal opening quantitatively related to potassium transport: evidence from electron probe analysis. *Plant Physiol.* 48:447–53

54. Van Kirk CA, Raschke K. 1978. Presence of chloride reduces malate production in epidermis during stomatal opening. *Plant Physiol.* 61:361–64

55. MacRobbie EAC. 1981. Ion fluxes in 'isolated' guard cells of *Commelina communis* L. *J. Exp. Bot.* 32:545–62

56. Peiter E, Maathuis FJ, Mills LN, Knight H, Pelloux J, et al. 2005. The vacuolar Ca^{2+}-activated channel TPC1 regulates germination and stomatal movement. *Nature* 434:404–8

57. Allen GJ, Chu SP, Schumacher K, Shimazaki CT, Vafeados D, et al. 2000. Alteration of stimulus-specific guard cell calcium oscillations and stomatal closing in *Arabidopsis det3* mutant. *Science* 289:2338–42

58. Young JJ, Mehta S, Israelsson M, Godoski J, Grill E, Schroeder JI. 2006. CO_2 signaling in guard cells: calcium sensitivity response modulation, a Ca^{2+}-independent phase, and CO_2 insensitivity of the *gca2* mutant. *Proc. Natl. Acad. Sci. USA* 103:7506–11

59. Shope JC, DeWald DB, Mott KA. 2003. Changes in surface area of intact guard cells are correlated with membrane internalization. *Plant Physiol.* 133:1314–21

60. Gao XQ, Li CG, Wei PC, Zhang XY, Chen J, Wang XC. 2005. The dynamic changes of tonoplasts in guard cells are important for stomatal movement in *Vicia faba. Plant Physiol.* 139:1207–16

61. Padmanaban S, Chanroj S, Kwak JM, Li X, Ward JM, Sze H. 2007. Participation of endomembrane cation/H^+ exchanger AtCHX20 in osmoregulation of guard cells. *Plant Physiol.* 144:82–93

62. Pei ZM, Ward JM, Harper JF, Schroeder JI. 1996. A novel chloride channel in *Vicia faba* guard cell vacuoles activated by the serine/threonine kinase, CDPK. *EMBO J.* 15:6564–74

63. Hafke JB, Hafke Y, Smith JA, Luttge U, Thiel G. 2003. Vacuolar malate uptake is mediated by an anion-selective inward rectifier. *Plant J.* 35:116–28

64. MacRobbie EA. 1998. Signal transduction and ion channels in guard cells. *Philos. Trans. R. Soc. London Ser. B* 353:1475–88

65. Homann U, Thiel G. 1994. Cl^- and K^+ channel currents during the action potential in *Chara*. Simultaneous recording of membrane voltage and patch currents. *J. Membr. Biol.* 141:297–309

66. Shimmen T. 1997. Studies on mechano-perception in Characeae: effects of external Ca^{2+} and Cl^-. *Plant Cell Physiol.* 38:691–97

67. Kohler B, Wegner LH, Osipov V, Raschke K. 2002. Loading of nitrate into the xylem: Apoplastic nitrate controls the voltage dependence of X-QUAC, the main anion conductance in xylem-parenchyma cells of barley roots. *Plant J.* 30:133–42

68. Jabs T, Tschope M, Colling C, Hahlbrock K, Scheel D. 1997. Elicitor-stimulated ion fluxes and O_2^- from the oxidative burst are essential components in triggering defense gene activation and phytoalexin synthesis in parsley. *Proc. Natl. Acad. Sci. USA* 94:4800–5

69. Delhaize E, Ryan PR. 1995. Aluminum toxicity and tolerance in plants. *Plant Physiol.* 107:315–21

70. Delhaize E, Ryan PR, Randall PJ. 1993. Aluminum tolerance in wheat (*Triticum aestivum* L.). II. Aluminum-stimulated excretion of malic acid from root apices. *Plant Physiol.* 103:695–702

71. Pineros MA, Kochian LV. 2001. A patch-clamp study on the physiology of aluminum toxicity and aluminum tolerance in maize. Identification and characterization of Al^{3+}-induced anion channels. *Plant Physiol.* 125:292–305

72. Delhaize E, Gruber BD, Ryan PR. 2007. The roles of organic anion permeases in aluminium resistance and mineral nutrition. *FEBS Lett.* 581:2255–62

73. Kollmeier M, Dietrich P, Bauer CS, Horst WJ, Hedrich R. 2001. Aluminum activates a citrate-permeable anion channel in the aluminum-sensitive zone of the maize root apex. A comparison between an aluminum-sensitive and an aluminum-resistant cultivar. *Plant Physiol.* 126:397–410

74. Ligaba A, Katsuhara M, Ryan PR, Shibasaka M, Matsumoto H. 2006. The *BnALMT1* and *BnALMT2* genes from rape encode aluminum-activated malate transporters that enhance the aluminum resistance of plant cells. *Plant Physiol.* 142:1294–303

75. Pineros MA, Cancado GM, Maron LG, Lyi SM, Menossi M, Kochian LV. 2008. Not all ALMT1-type transporters mediate aluminum-activated organic acid responses: the case of ZmALMT1—an anion-selective transporter. *Plant J.* 53:352–67

76. Pantoja O, Smith JA. 2002. Sensitivity of the plant vacuolar malate channel to pH, Ca^{2+} and anion-channel blockers. *J. Membr. Biol.* 186:31–42

77. Jentsch TJ. 2008. CLC chloride channels and transporters: from genes to protein structure, pathology and physiology. *Crit. Rev. Biochem. Mol. Biol.* 43:3–36

78. Colquhoun D, Sivilotti LG. 2004. Function and structure in glycine receptors and some of their relatives. *Trends Neurosci.* 27:337–44

79. Hartzell HC, Qu Z, Yu K, Xiao Q, Chien LT. 2008. Molecular physiology of bestrophins: multifunctional membrane proteins linked to best disease and other retinopathies. *Physiol. Rev.* 88:639–72

80. Brenner ED, Martinez-Barboza N, Clark AP, Liang QS, Stevenson DW, Coruzzi GM. 2000. *Arabidopsis* mutants resistant to S(+)-β-methyl-α, β-diaminopropionic acid, a cycad-derived glutamate receptor agonist. *Plant Physiol.* 124:1615–24

81. Hechenberger M, Schwappach B, Fischer WN, Frommer WB, Jentsch TJ, Steinmeyer K. 1996. A family of putative chloride channels from *Arabidopsis* and functional complementation of a yeast strain with a CLC gene disruption. *J. Biol. Chem.* 271:33632–38

82. Teardo E, Frare E, Segalla A, De Marco V, Giacometti GM, Szabo I. 2005. Localization of a putative ClC chloride channel in spinach chloroplasts. *FEBS Lett.* 579:4991–96

83. von der Fecht-Bartenbach J, Bogner M, Krebs M, Stierhof YD, Schumacher K, Ludewig U. 2007. Function of the anion transporter AtCLC-d in the *trans*-Golgi network. *Plant J.* 50:466–74

84. Schroeder JI. 2006. Physiology: nitrate at the ion exchange. *Nature* 442:877–78

85. Linder B, Raschke K. 1992. A slow anion channel in guard cells, activating at large hyperpolarization, may be principal for stomatal closing. *FEBS Lett.* 313:27–30

86. Camarasa C, Bidard F, Bony M, Barre P, Dequin S. 2001. Characterization of *Schizosaccharomyces pombe* malate permease by expression in *Saccharomyces cerevisiae*. *Appl. Environ. Microbiol.* 67:4144–51

87. Schmidt C, Schroeder JI. 1994. Anion selectivity of slow anion channels in the plasma membrane of guard cells (large nitrate permeability). *Plant Physiol.* 106:383–91

88. Van Kirk CA, Raschke K. 1978. Release of malate from epidermal strips during stomatal closure. *Plant Physiol.* 61:474–75

89. MacRobbie EAC. 1981. Effects of ABA on 'isolated' guard cells of *Commelina communis* L. *J. Exp. Bot.* 32:563–72

90. Mäser P, Thomine S, Schroeder JI, Ward JM, Hirschi K, et al. 2001. Phylogenetic relationships within cation transporter families of *Arabidopsis*. *Plant Physiol.* 126:1646–67

91. Qi Z, Stephens NR, Spalding EP. 2006. Calcium entry mediated by GLR3.3, an *Arabidopsis* glutamate receptor with a broad agonist profile. *Plant Physiol.* 142:963–71

92. Stephens NR, Qi Z, Spalding EP. 2008. Glutamate receptor subtypes evidenced by differences in desensitization and dependence on the *GLR3.3* and *GLR3.4* genes. *Plant Physiol.* 146:529–38

93. Demidchik V, Shabala SN, Davies JM. 2007. Spatial variation in H_2O_2 response of *Arabidopsis thaliana* root epidermal Ca^{2+} flux and plasma membrane Ca^{2+} channels. *Plant J.* 49:377–86

94. Kwak JM, Mori IC, Pei ZM, Leonhardt N, Torres MA, et al. 2003. NADPH oxidase *AtrbohD* and *AtrbohF* genes function in ROS-dependent ABA signaling in *Arabidopsis*. *EMBO J.* 22:2623–33

95. Kohler B, Blatt MR. 2002. Protein phosphorylation activates the guard cell Ca^{2+} channel and is a prerequisite for gating by abscisic acid. *Plant J.* 32:185–94

96. Mori IC, Murata Y, Yang Y, Munemasa S, Wang YF, et al. 2006. CDPKs CPK6 and CPK3 function in ABA regulation of guard cell S-type anion- and Ca^{2+}-permeable channels and stomatal closure. *PLoS Biol.* 4:e327

97. Demidchik V, Bowen HC, Maathuis FJ, Shabala SN, Tester MA, et al. 2002. *Arabidopsis thaliana* root nonselective cation channels mediate calcium uptake and are involved in growth. *Plant J.* 32:799–808

98. Foreman J, Demidchik V, Bothwell JH, Mylona P, Miedema H, et al. 2003. Reactive oxygen species produced by NADPH oxidase regulate plant cell growth. *Nature* 422:442–46

99. Ranjeva R, Graziana A, Mazars C, Thuleau PP. 1992. Putative L-type calcium channels in plants: biochemical properties and subcellular localisation. In *Transport and Receptor Proteins of Plant Membranes*, ed. DT Cooke, DT Clarkson, pp. 145–53. New York: Plenum

100. Shimmen T. 2001. Involvement of receptor potentials and action potentials in mechano-perception in plants. *Aust. J. Plant Physiol.* 28:567–76

101. Cosgrove DJ, Hedrich R. 1991. Stretch-activated chloride, potassium, and calcium channels coexisting in plasma membranes of guard cells of *Vicia faba* L. *Planta* 186:143–53

102. Legue V, Blancaflor E, Wymer C, Perbal G, Fantin D, Gilroy S. 1997. Cytoplasmic free Ca^{2+} in *Arabidopsis* roots changes in response to touch but not gravity. *Plant Physiol.* 114:789–800

103. Nakagawa Y, Katagiri T, Shinozaki K, Qi Z, Tatsumi H, et al. 2007. *Arabidopsis* plasma membrane protein crucial for Ca^{2+} influx and touch sensing in roots. *Proc. Natl. Acad. Sci. USA* 104:3639–44

104. Merchant SS, Prochnik SE, Vallon O, Harris EH, Karpowicz SJ, et al. 2007. The *Chlamydomonas* genome reveals the evolution of key animal and plant functions. *Science* 318:245–50

105. Wheeler GL, Brownlee C. 2008. Ca^{2+} signalling in plants and green algae: changing channels. *Trends Plant Sci.* 13:506–14

106. Zimmermann S, Sentenac H. 1999. Plant ion channels: from molecular structures to physiological functions. *Curr. Opin. Plant Biol.* 2:477–82

107. Lebaudy A, Very AA, Sentenac H. 2007. K^+ channel activity in plants: genes, regulations and functions. *FEBS Lett.* 581:2357–66

108. Gambale F, Uozumi N. 2006. Properties of shaker-type potassium channels in higher plants. *J. Membr. Biol.* 210:1–19

109. MacKinnon R. 1995. Pore loops: an emerging theme in ion channel structure. *Neuron* 14:889–92

110. MacKinnon R. 2003. Potassium channels. *FEBS Lett.* 555:62–65

111. Durell SR, Guy HR. 1999. Structural models of the KtrB, TrkH, and Trk1,2 symporters based on the structure of the KcsA K^+ channel. *Biophys J.* 77:789–807

112. Mäser P, Hosoo Y, Goshima S, Horie T, Eckelman B, et al. 2002. Glycine residues in potassium channel-like selectivity filters determine potassium selectivity in four-loop-per-subunit HKT transporters from plants. *Proc. Natl. Acad. Sci. USA* 99:6428–33

113. Tholema N, Vor der Bruggen M, Mäser P, Nakamura T, Schroeder JI, et al. 2005. All four putative selectivity filter glycine residues in KtrB are essential for high affinity and selective K^+ uptake by the KtrAB system from *Vibrio alginolyticus*. *J. Biol. Chem.* 280:41146–54

114. Papazian DM, Timpe LC, Jan YN, Jan LY. 1991. Alteration of voltage-dependence of Shaker potassium channel by mutations in the S4 sequence. *Nature* 349:305–10

115. Cha A, Snyder GE, Selvin PR, Bezanilla F. 1999. Atomic scale movement of the voltage-sensing region in a potassium channel measured via spectroscopy. *Nature* 402:809–13

116. Chanda B, Asamoah OK, Blunck R, Roux B, Bezanilla F. 2005. Gating charge displacement in voltage-gated ion channels involves limited transmembrane movement. *Nature* 436:852–56

117. Long SB, Tao X, Campbell EB, MacKinnon R. 2007. Atomic structure of a voltage-dependent K^+ channel in a lipid membrane-like environment. *Nature* 450:376–82

118. Vergani P, Miosga T, Jarvis SM, Blatt MR. 1997. Extracellular K^+ and Ba^{2+} mediate voltage-dependent inactivation of the outward-rectifying K^+ channel encoded by the yeast gene *TOK1*. *FEBS Lett.* 405:337–44

119. Johansson I, Wulfetange K, Poree F, Michard E, Gajdanowicz P, et al. 2006. External K^+ modulates the activity of the *Arabidopsis* potassium channel SKOR via an unusual mechanism. *Plant J.* 46:269–81

120. Ache P, Becker D, Ivashikina N, Dietrich P, Roelfsema MR, Hedrich R. 2000. GORK, a delayed outward rectifier expressed in guard cells of *Arabidopsis thaliana*, is a K^+-selective, K^+-sensing ion channel. *FEBS Lett.* 486:93–98

121. Ivashikina N, Becker D, Ache P, Meyerhoff O, Felle HH, Hedrich R. 2001. K^+ channel profile and electrical properties of *Arabidopsis* root hairs. *FEBS Lett.* 508:463–69

122. Dreyer I, Poree F, Schneider A, Mittelstadt J, Bertl A, et al. 2004. Assembly of plant Shaker-like K_{out} channels requires two distinct sites of the channel α-subunit. *Biophys. J.* 87:858–72

123. Kourie J, Goldsmith MH. 1992. K channels are responsible for an inwardly rectifying current in the plasma membrane of mesophyll protoplasts of *Avena sativa*. *Plant Physiol.* 98:1087–97

124. Sentenac H, Bonneaud N, Minet M, Lacroute F, Salmon J-M, et al. 1992. Cloning and expression in yeast of a plant potassium ion transport system. *Science* 256:663–65

125. Anderson JA, Huprikar SS, Kochian LV, Lucas WJ, Gaber RF. 1992. Functional expression of a probable *Arabidopsis thaliana* potassium channel in *Saccharomyces cerevisiae*. *Proc. Natl. Acad. Sci. USA* 89:3736–40

126. Santa-Maria GE, Rubio F, Dubcovsky J, Rodriguez-Navarro A. 1997. The HAK1 gene of barley is a member of a large gene family and encodes a high-affinity potassium transporter. *Plant Cell* 9:2281–89

127. Xu J, Li HD, Chen LQ, Wang Y, Liu LL, et al. 2006. A protein kinase, interacting with two calcineurin B-like proteins, regulates K^+ transporter AKT1 in *Arabidopsis*. *Cell* 125:1347–60

128. Lee SC, Lan WZ, Kim BG, Li L, Cheong YH, et al. 2007. A protein phosphorylation/dephosphorylation network regulates a plant potassium channel. *Proc. Natl. Acad. Sci. USA* 104:15959–64

129. Cheong YH, Pandey GK, Grant JJ, Batistic O, Li L, et al. 2007. Two calcineurin B-like calcium sensors, interacting with protein kinase CIPK23, regulate leaf transpiration and root potassium uptake in *Arabidopsis*. *Plant J.* 52:223–39

130. Sutter JU, Sieben C, Hartel A, Eisenach C, Thiel G, Blatt MR. 2007. Abscisic acid triggers the endocytosis of the *Arabidopsis* KAT1 K^+ channel and its recycling to the plasma membrane. *Curr. Biol.* 17:1396–402

131. Philippar K, Ivashikina N, Ache P, Christian M, Luthen H, et al. 2004. Auxin activates *KAT1* and *KAT2*, two K^+-channel genes expressed in seedlings of *Arabidopsis thaliana*. *Plant J.* 37:815–27

132. Very AA, Gaymard F, Bosseux C, Sentenac H, Thibaud JB. 1995. Expression of a cloned plant K^+ channel in *Xenopus* oocytes: analysis of macroscopic currents. *Plant J.* 7:321–32

133. Sottocornola B, Visconti S, Orsi S, Gazzarrini S, Giacometti S, et al. 2006. The potassium channel KAT1 is activated by plant and animal 14-3-3 proteins. *J. Biol. Chem.* 281:35735–41

134. Duby G, Hosy E, Fizames C, Alcon C, Costa A, et al. 2008. AtKC1, a conditionally targeted Shaker-type subunit, regulates the activity of plant K^+ channels. *Plant J.* 53:115–23

135. Dreyer I, Antunes S, Hoshi T, Muller-Rober B, Palme K, et al. 1997. Plant K^+ channel α-subunits assemble indiscriminately. *Biophys. J.* 72:2143–50

136. Obrdlik P, El-Bakkoury M, Hamacher T, Cappellaro C, Vilarino C, et al. 2004. K^+ channel interactions detected by a genetic system optimized for systematic studies of membrane protein interactions. *Proc. Natl. Acad. Sci. USA* 101:12242–47

137. Pilot G, Gaymard F, Mouline K, Cherel I, Sentenac H. 2003. Regulated expression of *Arabidopsis* Shaker K^+ channel genes involved in K^+ uptake and distribution in the plant. *Plant Mol. Biol.* 51:773–87

138. Xicluna J, Lacombe B, Dreyer I, Alcon C, Jeanguenin L, et al. 2007. Increased functional diversity of plant K^+ channels by preferential heteromerization of the Shaker-like subunits AKT2 and KAT2. *J. Biol. Chem.* 282:486–94

139. Clough SJ, Fengler KA, Yu IC, Lippok B, Smith RK Jr, Bent AF. 2000. The *Arabidopsis dnd1* "defense, no death" gene encodes a mutated cyclic nucleotide-gated ion channel. *Proc. Natl. Acad. Sci. USA* 97:9323–28

140. Yoshioka K, Moeder W, Kang HG, Kachroo P, Masmoudi K, et al. 2006. The chimeric *Arabidopsis* CYCLIC NUCLEOTIDE-GATED ION CHANNEL11/12 activates multiple pathogen resistance responses. *Plant Cell* 18:747–63

141. Frietsch S, Wang YF, Sladek C, Poulsen LR, Romanowsky SM, et al. 2007. A cyclic nucleotide-gated channel is essential for polarized tip growth of pollen. *Proc. Natl. Acad. Sci. USA* 104:14531–36

142. Lemtiri-Chlieh F, Berkowitz GA. 2004. Cyclic adenosine monophosphate regulates calcium channels in the plasma membrane of *Arabidopsis* leaf guard and mesophyll cells. *J. Biol. Chem.* 279:35306–12

143. Kato Y, Sakaguchi M, Mori Y, Saito K, Nakamura T, et al. 2001. Evidence in support of a four transmembrane-pore-transmembrane topology model for the *Arabidopsis thaliana* Na^+/K^+ translocating AtHKT1 protein, a member of the superfamily of K^+ transporters. *Proc. Natl. Acad. Sci. USA* 98:6488–93

144. Schachtman DP, Schroeder JI. 1994. Structure and transport mechanism of a high-affinity potassium uptake transporter from higher plants. *Nature* 370:655–58

145. Rubio F, Gassmann W, Schroeder JI. 1995. Sodium-driven potassium uptake by the plant potassium transporter HKT1 and mutations conferring salt tolerance. *Science* 270:1660–63

146. Uozumi N, Kim EJ, Rubio F, Yamaguchi T, Muto S, et al. 2000. The *Arabidopsis HKT1* gene homolog mediates inward Na$^+$ currents in *Xenopus laevis* oocytes and Na$^+$ uptake in *Saccharomyces cerevisiae*. *Plant Physiol.* 122:1249–59

147. Fairbairn DJ, Liu W, Schachtman DP, Gomez-Gallego S, Day SR, Teasdale RD. 2000. Characterisation of two distinct HKT1-like potassium transporters from *Eucalyptus camaldulensis*. *Plant Mol. Biol.* 43:515–25

148. Horie T, Yoshida K, Nakayama H, Yamada K, Oiki S, Shinmyo A. 2001. Two types of HKT transporters with different properties of Na$^+$ and K$^+$ transport in *Oryza sativa*. *Plant J.* 27:129–38

149. Mäser P, Eckelman B, Vaidyanathan R, Horie T, Fairbairn DJ, et al. 2002. Altered shoot/root Na$^+$ distribution and bifurcating salt sensitivity in *Arabidopsis* by genetic disruption of the Na$^+$ transporter AtHKT1. *FEBS Lett.* 531:157–61

150. Ren ZH, Gao JP, Li LG, Cai XL, Huang W, et al. 2005. A rice quantitative trait locus for salt tolerance encodes a sodium transporter. *Nat. Genet.* 37:1141–46

151. Sunarpi, Horie T, Motoda J, Kubo M, Yang H, et al. 2005. Enhanced salt tolerance mediated by AtHKT1 transporter-induced Na unloading from xylem vessels to xylem parenchyma cells. *Plant J.* 44:928–38

152. Horie T, Costa A, Kim TH, Han MJ, Horie R, et al. 2007. Rice OsHKT2;1 transporter mediates large Na$^+$ influx component into K$^+$-starved roots for growth. *EMBO J.* 26:3003–14

153. Berthomieu P, Conejero G, Nublat A, Brackenbury WJ, Lambert C, et al. 2003. Functional analysis of AtHKT1 in *Arabidopsis* shows that Na$^+$ recirculation by the phloem is crucial for salt tolerance. *EMBO J.* 22:2004–14

154. Gong JM, Waner DA, Horie T, Li SL, Horie R, et al. 2004. Microarray-based rapid cloning of an ion accumulation deletion mutant in *Arabidopsis thaliana*. *Proc. Natl. Acad. Sci. USA* 101:15404–9

155. Rus A, Baxter I, Muthukumar B, Gustin J, Lahner B, et al. 2006. Natural variants of AtHKT1 enhance Na$^+$ accumulation in two wild populations of *Arabidopsis*. *PLoS Genet.* 2:e210

156. Horie T, Horie R, Chan WY, Leung HY, Schroeder JI. 2006. Calcium regulation of sodium hypersensitivities of *sos3* and *athkt1* mutants. *Plant Cell Physiol.* 47:622–33

157. Davenport RJ, Muñoz-Mayor A, Jha D, Essah PA, Rus A, Tester M. 2007. The Na$^+$ transporter AtHKT1;1 controls retrieval of Na$^+$ from the xylem in *Arabidopsis*. *Plant Cell Env.* 30:497–507

158. Gierth M, Mäser P. 2007. Potassium transporters in plants—involvement in K$^+$ acquisition, redistribution and homeostasis. *FEBS Lett.* 581:2348–56

159. Schwacke R, Schneider A, Van Der Graaff E, Fischer K, Catoni E, et al. 2003. ARAMEMNON, a novel database for *Arabidopsis* integral membrane proteins. *Plant Physiol.* 131:16–26

160. Schleyer M, Bakker EP. 1993. Nucleotide sequence and 3′-end deletion studies indicate that the K$^+$-uptake protein kup from *Escherichia coli* is composed of a hydrophobic core linked to a large and partially essential hydrophilic C terminus. *J. Bacteriol.* 175:6925–31

161. Banuelos MA, Klein RD, Alexander-Bowman SJ, Rodriguez-Navarro A. 1995. A potassium transporter of the yeast *Schwanniomyces occidentalis* homologous to the Kup system of *Escherichia coli* has a high concentrative capacity. *EMBO J.* 14:3021–27

162. Quintero FJ, Blatt MR. 1997. A new family of K$^+$ transporters from *Arabidopsis* that are conserved across phyla. *FEBS Lett.* 415:206–11

163. Fu HH, Luan S. 1998. AtKuP1: a dual-affinity K$^+$ transporter from *Arabidopsis*. *Plant Cell* 10:63–73

164. Kim EJ, Kwak JM, Uozumi N, Schroeder JI. 1998. *AtKUP1*: an *Arabidopsis* gene encoding high-affinity potassium transport activity. *Plant Cell* 10:51–62

165. Banuelos MA, Garciadeblas B, Cubero B, Rodriguez-Navarro A. 2002. Inventory and functional characterization of the HAK potassium transporters of rice. *Plant Physiol.* 130:784–95

166. Gierth M, Mäser P, Schroeder JI. 2005. The potassium transporter AtHAK5 functions in K$^+$ deprivation-induced high-affinity K$^+$ uptake and AKT1 K$^+$ channel contribution to K$^+$ uptake kinetics in *Arabidopsis* roots. *Plant Physiol.* 137:1105–14

167. Shin R, Schachtman DP. 2004. Hydrogen peroxide mediates plant root cell response to nutrient deprivation. *Proc. Natl. Acad. Sci. USA* 101:8827–32

168. Armengaud P, Breitling R, Amtmann A. 2004. The potassium-dependent transcriptome of *Arabidopsis* reveals a prominent role of jasmonic acid in nutrient signaling. *Plant Physiol.* 136:2556–76

169. Rubio F, Santa-María GE, Rodríguez-Navarro A. 2000. Cloning of *Arabidopsis* and barley cDNAs encoding HAK potassium transporters in root and shoot cells. *Physiol. Plant.* 109:34–43

170. Epstein E, Rains DW, Elzam OE. 1963. Resolution of dual mechanisms of potassium absorption by barley roots. *Proc. Natl. Acad. Sci. USA* 49:684–92

171. Elumalai RP, Nagpal P, Reed JW. 2002. A mutation in the *Arabidopsis KT2/KUP2* potassium transporter gene affects shoot cell expansion. *Plant Cell* 14:119–31

172. Rigas S, Debrosses G, Haralampidis K, Vicente-Agullo F, Feldmann KA, et al. 2001. TRH1 encodes a potassium transporter required for tip growth in *Arabidopsis* root hairs. *Plant Cell* 13:139–51

173. Allen GJ, Sanders D. 1996. Control of ionic currents in guard cell vacuoles by cytosolic and luminal calcium. *Plant J.* 10:1055–69

174. Bihler H, Eing C, Hebeisen S, Roller A, Czempinski K, Bertl A. 2005. TPK1 is a vacuolar ion channel different from the slow-vacuolar cation channel. *Plant Physiol.* 139:417–24

175. Czempinski K, Frachisse JM, Maurel C, Barbier-Brygoo H, Mueller-Roeber B. 2002. Vacuolar membrane localization of the *Arabidopsis* 'two-pore' K$^+$ channel KCO1. *Plant J.* 29:809–20

176. Ketchum KA, Joiner WJ, Sellers AJ, Kaczmarek LK, Goldstein SA. 1995. A new family of outwardly rectifying potassium channel proteins with two pore domains in tandem. *Nature* 376:690–95

177. Lesage F, Lazdunski M. 2000. Molecular and functional properties of two-pore-domain potassium channels. *Am. J. Physiol. Renal Physiol.* 279:F793–801

178. Hedrich R, Neher E. 1987. Cytoplasmic calcium regulates voltage dependent ion channels in plant vacuoles. *Nature* 329:833–35

179. Ishibashi K, Suzuki M, Imai M. 2000. Molecular cloning of a novel form (two-repeat) protein related to voltage-gated sodium and calcium channels. *Biochem. Biophys. Res. Commun.* 270:370–76

180. Ward JM, Pei ZM, Schroeder JI. 1995. Roles of ion channels in initiation of signal transduction in higher plants. *Plant Cell* 7:833–44

181. Bewell MA, Maathuis FJ, Allen GJ, Sanders D. 1999. Calcium-induced calcium release mediated by a voltage-activated cation channel in vacuolar vesicles from red beet. *FEBS Lett.* 458:41–44

182. Scholz-Starke J, Carpaneto A, Gambale F. 2006. On the interaction of neomycin with the slow vacuolar channel of *Arabidopsis thaliana*. *J. Gen. Physiol.* 127:329–40

183. McAinsh MR, Webb A, Taylor JE, Hetherington AM. 1995. Stimulus-induced oscillations in guard cell cytosolic free calcium. *Plant Cell* 7:1207–19

184. Han S, Tang R, Anderson LK, Woerner TE, Pei ZM. 2003. A cell surface receptor mediates extracellular Ca^{2+} sensing in guard cells. *Nature* 425:196–200

185. Nomura H, Komori T, Kobori M, Nakahira Y, Shiina T. 2008. Evidence for chloroplast control of external Ca^{2+}-induced cytosolic Ca^{2+} transients and stomatal closure. *Plant J.* 53:988–98

186. Pei ZM, Ward JM, Schroeder JI. 1999. Magnesium sensitizes slow vacuolar channels to physiological cytosolic calcium and inhibits fast vacuolar channels in fava bean guard cell vacuoles. *Plant Physiol.* 121:977–86

187. Pottosin II, Martinez-Estevez M. 2003. Regulation of the fast vacuolar channel by cytosolic and vacuolar potassium. *Biophys. J.* 84:977–86

188. Eddy SR. 1998. Profile hidden Markov models. *Bioinformatics* 14:755–63

189. Miedema H, Demidchik V, Very AA, Bothwell JH, Brownlee C, Davies JM. 2008. Two voltage-dependent calcium channels co-exist in the apical plasma membrane of *Arabidopsis thaliana* root hairs. *New Phytol.* 179:378–85

NOTE ADDED IN PROOF

Concerning Ca^{2+} influx and tip growth: Cytoplasmic $[Ca^{2+}]$ oscillates in root hair tips with the same period as oscillations in root hair growth (Monhausen et al. 2008), and the timing of Ca^{2+} influx indicates a function in limiting root hair growth.

Concerning mechanosensitive ion channels: MscS genes have also recently been characterized as required for plant stretch–activated channels (Haswell et al. 2008).

Monshausen GB, Messerli MA, Gilroy S. 2008. Imaging of the Yellow Cameleon 3.6 indicator reveals that elevations in cytosolic Ca^{2+} follow oscillating increases in growth in root hairs of *Arabidopsis*. *Plant Physiol.* 147:1690–98

Haswell ES, Peyronnet R, Barbier-Brygoo H, Meyerowitz EM, Frachisse JM. 2008. Two MscS homologs provide mechanosensitive channel activities in the *Arabidopsis* root. *Curr. Biol.* 18:730–34

Polycystins and Primary Cilia: Primers for Cell Cycle Progression

Jing Zhou

Renal Division, Department of Medicine, Brigham and Women's Hospital and Harvard Medical School, Boston, Massachusetts 02115; email: zhou@rics.bwh.harvard.edu

Annu. Rev. Physiol. 2009. 71:83–113

The *Annual Review of Physiology* is online at physiol.annualreviews.org

This article's doi:
10.1146/annurev.physiol.70.113006.100621

Copyright © 2009 by Annual Reviews.
All rights reserved

0066-4278/09/0315-0083$20.00

Key Words

cilia, cell proliferation, polycystic kidney disease, calcium signaling

Abstract

Polycystins are a family of eight-transmembrane proteins united by sequence homology. The name stems from the identification of mutations in genes encoding polycystin-1 and -2 in polycystic kidney diseases. This review discusses recent topics in polycystin research, with a focus on the role of polycystin-1 and polycystin-2 in primary cilia and the cell cycle. Polycystins appear to play key roles during development, but a major question is their function in mature organs. Their roles in primary cilia, shear stress sensation, alteration of intracellular calcium, and planar cell polarity are examined. The third-hit hypothesis of polycystic kidney disease is discussed.

INTRODUCTION

The term polycystin originally referred to the product of the major autosomal dominant polycystic kidney disease (PKD) gene, *PKD1*, which was identified by positional cloning. Seven more polycystins were identified later through homology cloning, and together the eight polycystins compose a novel protein family. Polycystin-1 (PC1) has a large N-terminal extracellular domain, eleven transmembrane domains, and a short intracellular C-terminal tail. PC1 is proposed to function as a G protein–coupled receptor (GPCR). Polycystin-2 (PC2, TRPP2), encoded by *PKD2*, has six transmembrane domains and is presumed to form a cation-selective ion channel permeable to Ca^{2+} (**Figure 1**). The polycystins are divided into two subfamilies on the ba-

sis of protein structure and putative function: (*a*) PC1 receptor–like molecules, including PC1, polycystin-REJ, polycystin-1L1 (PCL), polycystin-1L2, and polycystin-1L3, and (*b*) PC2 ion channel–like proteins, including PC2, polycystin-L (or polycystin-2L1), and polycystin-2L2 (**Table 1**). The resemblance of PC2 and PC2-like molecules in sequence and topology to transient receptor potential (TRP) channels has led to their inclusion as a subfamily of the TRP channel superfamily, TRPP (TRPP2 or PC2, TRPP3 or PC2-L1, and TRPP5 or PC2-L2). All polycystins share significant peptide sequence homologies in their last six transmembrane domains, and thus all polycystins may at some point be included in the TRPP channel subfamily as their functions are elucidated.

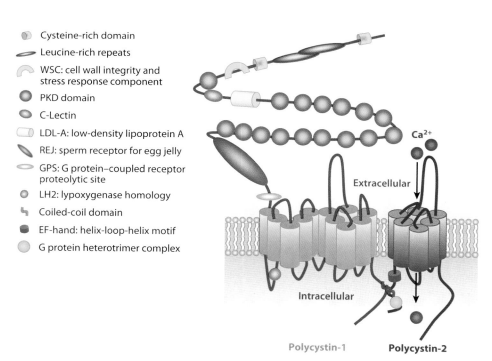

- Cysteine-rich domain
- Leucine-rich repeats
- WSC: cell wall integrity and stress response component
- PKD domain
- C-Lectin
- LDL-A: low-density lipoprotein A
- REJ: sperm receptor for egg jelly
- GPS: G protein–coupled receptor proteolytic site
- LH2: lypoxygenase homology
- Coiled-coil domain
- EF-hand: helix-loop-helix motif
- G protein heterotrimer complex

Extracellular

Ca^{2+}

Intracellular

Polycystin-1   Polycystin-2

Figure 1

Topology of polycystin-1 (PC1, or PKD1) and polycystin-2 (PC2, PKD2, or TRPP1). PC1 and PC2 family members all share significant peptide sequence and domain structure in their last six transmembrane domains. PC1 is proposed to function as a G protein–coupled receptor, and PC2 as a cation-selective ion channel permeable to Ca^{2+}. PC1 and PC2 physically interact via their respective C-terminal coiled-coil domains. The PC1/2 complex is presumably activated by stimuli such as shear stress from fluid flow across the surface of the cell.

Table 1 The polycystin protein family[a]

Protein	Gene	Chromosome locus	Expression	G protein binding/activation	Pore forming	Reference(s)
PC1-like						
Polycystin-1	*PKD1*	16p13.3	Widespread	+	−	43, 56, 126, 153
Polycystin-REJ	*PKDREJ*	22q13	Testis (controls the timing of fertilization)	n.d.	n.d.	154
Polycystin-1L1	*PKD1L1*	16p12–13	Relatively widespread, with higher levels in heart, testis	+	n.d.	155
Polycystin-1L2	*PKD1L2*	16q23	Relatively widespread, with higher levels in heart, testis	+	n.d.	49, 156
Polycystin-1L3	*PKD1L3*	16q22	Relatively widespread, but not in skeletal muscle	n.d.	n.d.	156
PC2-like						
Polycystin-2	*PKD2*	4q21–23	Widespread	n.d.	+	157, 158
Polycystin-L (Polycystin-2L1)	*PKDL* (*PKD2L1*)	10q24–25	Relatively widespread; found in taste buds, testis	n.d.	+	29, 159
Polycystin-2L2	*PKD2L2*	5q31	Heart, testis	n.d.	n.d.	160, 161

[a]n.d., not determined.

PC1 and PC2 are the best-known members of the polycystin family and are the focus of this review. Research on these two polycystins is largely facilitated by their involvement in autosomal dominant polycystic kidney disease (ADPKD) (code 173900 in Online Mendelian Inheritance of Man: **http://www.ncbi.nlm. nih.gov/omim/**). ADPKD is the most common monogenic disorder of the kidney, affecting all ethnic groups worldwide, with a frequency of 1:400 to 1:1000 (1–3). Approximately 600,000 people are affected by this disease in the United States. The disease is characterized by the progressive formation and enlargement of epithelial lined cysts in the kidney, typically resulting in chronic renal failure by late middle life. It is a multisystemic disorder with potentially serious extrarenal complications such as liver and pancreatic cysts, cardiac valve defects, colonic diverticulosis, abdominal wall hernias, and intracranial arterial aneurysms (4, 5). ADPKD is caused by a mutation in either of two genes, *PKD1* or *PKD2*, encoding PC1 and PC2, respectively. More than 85% of ADPKD patients have mutations in *PKD1*; the remaining patients have mutations in *PKD2* (6).

POLYCYSTIN LOCALIZATION

Polycystins Are Widely Distributed and Found on the Primary Cilia

All polycystins are expressed in the testes (see **Table 1**). PC1 and PC2 are expressed in a number of tissues and organs, including the ductal epithelial cells in the kidney, liver, pancreas, and breast; the smooth muscle and endothelial cells in the vasculature; and astrocytes in the brain (7–15). The subcellular localization of polycystins is complex and debated by researchers. PC1 and PC2 are localized by antibodies on the shaft and basal body of the primary cilia, a microtubule-based structure that projects at the apical membrane of a cell (see section entitled Primary Cilia, Mechanosensation, and Polycystic Kidney Disease, below, for more detail). Ciliary localization for PC1 and PC2 was mainly established in cultured cells, including inner medullary collecting ducts (IMCDs) (16, 17), Madin-Darby canine kidney (MDCK) cells, and mouse embryonic kidney (MEK) cells (17, 18). PC2 was also detected in the cilia of kidney tubules (19) and in a subset of cilia in the embryonic node (20). A short motif localized within the N-terminal domain of PC2 RVxP has been proposed to be responsible for ciliary targeting for PC2 (21). PCL may also reside in the cilia of postconfluent kidney epithelial cells derived from the inner medulla. It appears to colocalize with PC1 in centrosomes of subconfluent cells (22).

Nonciliary Localization of Polycystins

In addition to their ciliary localization, PC1 and PC2 are also found in other subcellular compartments and membrane domains. PC1 is found at apical membranes and the adherent and desmosomal junctions (7, 8, 23, 24). PC2 has been detected in the cytoplasm (25) and in the apical and basolateral membranes (12, 26, 27) in the kidney. In cultured cells, it has been found in the endoplasmic reticulum (ER) and plasma membrane (17, 26). Recently, PC1 and PC2 have been found in urinary exosomes, suggesting that polycystins are present in the internal vesicles of multivesicular bodies and are secreted into the urine (28). Although all these sites may contain polycystins, some cross-reactivity due to antibody specificity cannot be ruled out. PC1 and PC2 have not been found in tight junctions.

Polycystin 2-L1 (PC2-L1, or TRPP3) is expressed in multiple tissues, including retina, testis, liver, pancreas, heart, and spleen (29). It is found on the plasma membranes of *Xenopus* oocytes (30) and in the apical region of human and mouse kidney tubules (31). Recent evidence suggests that PC1-L3 and PC2-L1 are coexpressed in a subset of taste receptor cells in specific taste areas of the tongue (32). PC2-L2 (33) is detectable in the plasma membrane of spermatocytes and round spermatids as well as in the head and tail of elongating spermatids within seminiferous tubules in mouse testis (34).

Regulation of Polycystin-1 and Polycystin-2 Expression and Localization

PC1 is expressed early in development. *Pkd1* transcripts can be detected in mouse embryonic stem cells (W. Lu & J. Zhou, unpublished data). PC1 expression is high in developing tissues and low in adult tissues (7). In the mouse kidney, PC1 expression is found in nearly all tubular segments of the nephron. PC1 expression peaks at embryonic day (E)15; two weeks after birth, it falls to a low level that is maintained in adult life (8). Coordinated downregulation of PC1 (7, 8) and PC2 (12) in mature kidney is most prominent in the proximal tubules. In mature kidneys, PC2 expression levels also decrease in cortical and medullary collecting ducts, and PC2 is expressed at low levels in the thick ascending limbs of loops of Henle and distal convoluted tubules [where the expression of PC1 is also low (12)].

Expression studies of PC1 trafficking have been difficult owing to the large size of the ~14-kb transcript. PC2, only one-fourth of the size of PC1, has been more amenable to study. We previously suggested that the subcellular localization of PC2 is cell type specific and dynamic (35). PC2 protein and channel activities are detected on the cell plasma membrane (17) and in the ER (36) of renal tubular epithelia. PC2 localization to the plasma membrane is modulated by chemical chaperones, proteasome inhibitors, protein-protein interactions, and phosphorylation. PC2 translocates to the cell membrane in the presence of chemical chaperones and proteasome inhibitors and in the presence of PC1 in cultured cells (37). Its localization to the plasma membrane is in part modulated by phosphorylation at serine 76, probably by glycogen synthase kinase 3 (GSK3) (38). An acidic cluster phosphorylated by casein kinase 2 on serine 812 in the C-terminal cytoplasmic domain of human PC2 modulates the trafficking of PC2 among the ER, Golgi, and plasma membrane (39). This acidic cluster allows the binding of two adaptor proteins, phosphofurin acidic cluster sorting protein (PACS)1 and PACS2, that mediate the retrieval of PC2 to the trans-Golgi network and the ER, respectively. A phosphor-defective mutation at serine 812 abolishes the retrieval mechanism and allows PC2 to reach other subcellular regions such as the plasma membrane. A phosphor-mimicking mutation enhances the interaction of PC2 with PACS proteins and retains PC2 in the ER (39). The forward trafficking of PC2 from ER to Golgi is augmented by PIGEA, a coiled-coil-containing protein that links PC2 to GM130, a Golgi matrix protein (40). Nek8, a serine/threonine kinase, also modulates PC2 phosphorylation and trafficking to primary cilia. A mutation in the RCC (regulator of chromosome condensation) domain of Nek8 kinase facilitates the phosphorylation of PC2 and the expression and trafficking of PC1 and PC2 to the primary cilia in the kidney (41).

Interdependent Localization of Polycystin-1 and -2

PC1 and PC2 physically interact through their C-terminal cytoplasmic domains. However, whether and how they regulate each other's subcellular localization remain controversial. A coexpression study of PC1 and PC2 in Chinese hamster ovary cells suggested that PC1 assists targeting of PC2 to cell plasma membranes (42); transfection of PC2 alone results in its accumulation in the ER, whereas cotransfection of PC2 with PC1 results in cell surface expression of both proteins. This finding was supported by several studies of (a) heterologous expression of PC1 and PC2 in sympathetic neu-

rons (43) and (b) immunostaining of PC2 in kidney epithelial cells derived from homozygous Pkd1-targeted mice (18) or in cysts from human ADPKD patients with complete loss of ciliary PC1 (44, 45). However, in cyst-lining epithelial cells from human ADPKD kidneys, PC2 remained on cilia in a small percentage of cells, suggesting that an alternative pathway(s) may regulate PC2 trafficking to cilia (44). Nevertheless, PC2 is present on cilia in a line of mouse kidney tubular epithelial cells lacking PC1 (21). These findings are in contrast to the findings of Grimm et al. (46), who proposed that PC2 regulates the subcellular distribution of PC1. In the absence of PC2, PC1 is located on the plasma membrane, whereas cotransfection of both PC1 and PC2 results in ER localization of PC1 (46). It is noteworthy that PC1 and PC2 are not always colocalized. For example, only PC2 is found in cilia of the embryonic node (20). The discordant expression pattern of PC1 and PC2 in the embryonic node correlates well with the difference in body axis phenotype such that only Pkd2 knockout mice, but not Pkd1 knockout mice, develop left-right symmetry defects (47). Specific polycystins may pair up as a receptor-channel complex in a given cell type such that a PC1-like molecule may work together with PC2 in the node.

FUNCTIONS OF POLYCYSTINS

G Protein Signaling by Polycystins

The complex domain structure of PC1 suggests that PC1 acts as a cell surface receptor in both cell-cell and cell-matrix interactions. Heterologous expression of full-length PC1 in sympathetic neurons revealed that PC1 behaves in a manner consistent with GPCRs (43). When expressed alone, PC1 activates a G protein signaling pathway by direct binding and activation of heterotrimeric Gαi/o proteins (43, 48, 49) and subsequent modulation of voltage-gated Ca^{2+} channels and GIRK K^+ channels via the release of Gβγ subunits (43). PC1-L1 and PC1-L2 bind G proteins, but with a subunit preference differing from that of PC1 (49). GPCRs typically contain a seven-transmembrane-spanning segment; the 11-transmembrane-domain topology makes PC1 and PC1-like molecules atypical GPCRs. Although the last six transmembrane domains of PC1 share significant sequence homology with TRP channels and domains of Na^+ and Ca^{2+} channels, expression of PC1 alone does not yield measurable channel activity, suggesting that PC1 molecules are unable to form ion channels alone (42, 43). Similar to the sea urchin version of polycystin-REJ (50), some PC1 molecules are cleaved at the GPS domain (51). This cleavage occurs ubiquitously, but most of the cleaved extracellular domains of PC1 remain tethered at the cell surface, and only a small number are secreted. The coexistence of uncleaved and cleaved PC1 molecules suggests a physiological relevance; mice with a mutated cleavage site expressing only full-length PC1 develop late-onset PKD (52).

PC2 and PC1 interact in vitro via their C-terminal cytoplasmic tails (53, 54). Coexpression of PC2 with PC1 strongly represses the activation of heterotrimeric G proteins by PC1 (43). This finding has several implications. First, when the PC1-PC2 protein complex is normally silent, mutations leading to the distortion of the stoichiometry of the polycystin protein complex (e.g., resulting in overexpression of PC1 or underexpression of PC2) may trigger abnormal G protein signaling and lead to cyst formation in ADPKD (43). Second, if the polycystin protein complex is activated by physiological stimuli, a conformational change in the protein complex can release the inhibition of PC2 on PC1 and initiate G protein signaling. Whether PC1-mediated G protein signaling is physiological in vivo remains unknown at present. However, at least in cells coexpressing PC1 and PC2, stimulation of PC1 with an antibody (MR3) to the extracellular domain of PC1 in vitro simultaneously activated both G proteins and the PC2 channel (55).

The C-tail of PC1 activates the alpha subunit of all four families of heterotrimeric G proteins in HEK293 cells (56). However, these data are difficult to interpret because the isolated C-tail may mimic PC1 signaling or achieve a dominant negative effect on endogenous PC1 signaling or a gain-of-function effect through activating unrelated pathways. One study comparing the expression of the C-tail with that of the full-length construct found that the C-tail of PC1 could either augment or interrupt PC1 signaling, depending on the background level of full-length PC1 (57). Whether PC1 can activate G proteins in a physiologically relevant cell type is currently unknown. The physiological activator of PC1-mediated G protein signaling has not been identified. The negative regulators of G protein signaling proteins (RGS) terminate G protein signaling by accelerating the intrinsic GTPase activity of specific $G\alpha$ subunits. A member of the RGS family, RGS7, was reported to bind to the C terminus of PC1 (58). However, the functional activation of Gi/o proteins by full-length PC1 is independent of RGS protein, at least in a model expression system (43).

Although there is no direct evidence for PC1 activation of G proteins in kidney epithelial cells, increases in cellular cAMP levels have been observed in cyst-lining epithelial cells from human ADPKD patients as well as in cystic kidneys of animal models for PKD. Such animal models include γGT-Cre:$Pkd1^{flox/flox}$ mice (59), pck rat, and pcy and $Pkd2^{WS25/-}$ mice (60). cAMP stimulates DNA synthesis, cell proliferation, and fluid secretion in cyst-lining epithelial cells from ADPKD patients, but not in normal renal tubular epithelial cells (61, 62). cAMP-dependent fluid secretion is likely through the activation of cystic fibrosis transmembrane conductance regulator (CFTR), which is expressed in the apical membrane in cyst-lining epithelial cells. The critical role of cAMP in ADPKD is further demonstrated by the successful treatment of PKD in several animal models with drugs that reduce cAMP levels (60, 63, 64). Inhibition of CFTR activity with thiazolidinone and glycine hydrazide classes of small molecules also slowed cyst expansion in in vitro and in vivo models of PKD (65). For a recent review on cAMP and PKD, please see Reference 66.

Ca²⁺ SIGNALING BY POLYCYSTINS

The elucidation of the Ca^{2+}-modulated Ca^{2+}-permeable cation-selective channel property of PCL (30) facilitated an understanding of the function of PC2 (37, 36, 67). Like PC2-L1, PC2 is reportedly permeable to Ca^{2+}, Na^+, and K^+, and its channel activity is modulated by raising the intracellular Ca^{2+} concentration. The PC2 channel is found in both plasma and intracellular membranes. It is insensitive to ryanodine and inositol triphosphate (IP_3) (37). With its reported ER location, PC2 was proposed to be a member of a third class of Ca^{2+} release channels in addition to ryanodine receptors and IP_3 receptors (37). This hypothesis was supported by another study using lipid bilayers (36). PC2 may also contribute to the K^+ influx pathways in the ER that are coupled to Ca^{2+} release as part of a highly cooperative ion-exchange mechanism (37). Mutant PC2 channels appear to have partial function (68). Serine phosphorylation at position 812 appears to mediate the Ca^{2+} dependence of the PC2 channel (69). PC2 mutant proteins lacking the known C-terminal dimerization domain were still able to form oligomers and coimmunoprecipitate full-length PC2, suggesting the existence of a proximal dimerization domain (70). PC2 has also been claimed to form heteromeric channels with TRPC1 and TRPV4 (71, 72). Kottgen et al. (72) proposed that TRPP2 and TRPV4 form a mechanical and thermal sensor in cilia and suggested that TRPV4 is the mechanically sensitive component of the complex. TRPP2 and TRPV4 appear to colocalize in cilia, net current was increased when coexpressed in *Xenopus* oocytes, and flow-induced Ca^{2+} responses were decreased when TRPV4 was knocked down in renal endothelial cells. Because zebrafish lacking TRPV4 do not form renal cysts, these researchers interpret their findings as challenging the concept that defective ciliary flow sensing is fundamental to cystogenesis. However, the work did not establish that TRPP2 and TRPV4 form a heteromeric channel, because the pore properties of the presumed heteromer were not examined (72). For a recent review on PC2 channels, see Reference 73. PC2 reportedly interacts with a number of proteins (**Table 2**); some of the proteins modulate the activity of PC2 by affecting PC2 trafficking to the plasma membrane (see Polycystin Localization section, above), whereas other proteins modulate PC2 activation and gating. Polycystins and flow-induced Ca^{2+} signaling are discussed in the context of cilia in the following section.

PRIMARY CILIA, MECHANOSENSATION, AND POLYCYSTIC KIDNEY DISEASE

A primary role of PC2-mediated mechanosensation and Ca^{2+} signaling (18) is believed to be related to the primary cilia. Recent evidence also suggests that the PC2/TRPV4 heteromeric channel participates in thermosensation (72). Moreover, PC1-L1 and PC2-L1 (TRPP3) form a sour taste receptor and mediate chemosensation when heterologously expressed in HEK 293T cells (74). However, in vivo behavior analyses revealed that *PkdL* knockout mice retain normal sensation to sweet, sour, bitter, and salty tastes (L. Guo & J. Zhou, unpublished data).

Primary cilia have become a major focus for studies of PKD in the past five years. Primary cilia are thin hair-like structures that typically appear as a single projection from the apical surface of a cell. In contrast to the 9+2 arrangement in motile cilia, primary cilia have a 9+0 microtubule arrangement (**Figure 2a**). The lack of a central pair of microtubules structurally differentiates primary cilia from motile cilia, although some cilia with a 9+0 microtubule arrangement such as those in the node are also motile. Primary cilia have been documented for decades, but their function in the kidney has only recently become known.

The assembly, maintenance, and function of eukaryotic cilia or flagella require the movement of proteins along ciliary microtubules, a process called intraflagellar transport (IFT) (**Figure 2b**). Two types of microtubule-based motors, kinesin-II and dynein 1b, have

Table 2 Polycystin-2-binding proteins

Protein name[a]	Protein function	Effects on PC2
PC1 (43, 55)	GPCR	Channel gating
EGFR (108)	Growth factor receptor	Channel activation
TRPC1 (71)	Ion channel	Heteromultimeric channel
TRPV4 (72)	Ion channel	Heteromultimeric channel
PIGEA-12 (40)	Adaptor protein	ER-Golgi forward trafficking
PACS1, PACS2 (39)	Adaptor protein	Retrieval of PC2 to the trans-Golgi network and ER
CD2AP (162)	Scaffolding protein	Unclear
Hax1 (163)	Substrate of Src family tyrosine kinases, antiapoptotic	Links to cell cytoskeleton
α-Actinin (164)	Actin-bundling protein	Stimulates channel activity
Tropomysin-1 (165)	Actin-binding protein that regulates actin mechanics	Links to cell cytoskeleton
Triponin-1 (166)	Tropomysin-binding protein	Links to cell cytoskeleton
Pericentrin (167)	Centrosome protein	Links to cilia and centrosome
Casein kinase 2 (69, 168)	Serine/threonine kinase	Phosphorylation
ATPase p97 (169)	ER-associated protein degradation	Degradation
Glycogen synthase kinase (38)	Serine/threonine protein kinase	Phosphorylation, cell surface localization
Phospholipase C-γ2 (108)	Signaling	Suppresses PC2 channel activity
Id2 (101)	Transcription regulator	PC2 regulates cell cycle and transcription
mDia1 (152, 170)	Cytoskeletal organization, cytokinesis, and signal transduction	Gating
Elf2α (171)	Translation initiation factor, autophagy	PC2 inhibits cell proliferation
PERK (171)	ER-resident eIF2α kinase	PC2 promotes PERK-mediated eIF2α phosphorylation
HERP (169)	Ubiquitin-like protein	Degradation
FPC (109, 172, 173)	ARPKD protein, receptor	Modulates flow-induced Ca^{2+} response and expression levels, stimulates PC2 activity

[a]References are in parentheses.

been proposed to drive the anterograde and retrograde movement of IFT particles along the axoneme, respectively. The IFT particles consist of two subcomplexes, IFTA (linked to retrograde transport) and IFTB (linked to anterograde transport), which furnish multiple sites for binding the large number of axoneme, matrix, and membrane proteins that must be delivered to the distal tip of the cilium (75). The IFT particles move bidirectionally along the axoneme at rates on the order of a micrometer per second (76, 78). They carry their cargo from the site of synthesis in the cell body to the microtubule, ending at the cilium tip, where they un-load their cargo for the growth of the axoneme (77–82). For a review of primary cilia, please see References 83 and 84.

That disruptions in IFT components such as IFT88 (also called polaris) (85) or in a subunit of the kinesin-II motor protein that mediates anterograde ciliary IFT (86) cause loss of primary cilia and the PKD phenotype strongly suggests that the structural integrity of primary cilia is required for normal tubule structure. In human ADPKD kidneys or mouse kidneys with a targeted *Pkd1* mutation, however, the primary cilia of cyst-lining epithelial cells are neither absent nor shortened (**Figure 3**).

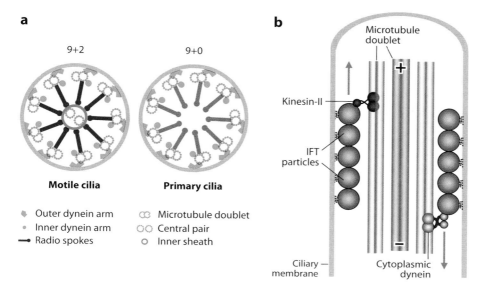

Figure 2

(*a*) Microtubule arrangement of cilia and (*b*) the intraflagellar transport (IFT) machinery. (*a*) Primary cilia (*right*) lack a central pair of microtubules, which distinguish themselves from the motile cilia (*left*). (*b*) During IFT, linear arrays of IFT particles are transported toward the plus (distal) ends of the cilia outer doublet microtubules by kinesin-II and toward the minus (proximal) ends of the microtubules by cytoplasmic dynein 1b (*gray*). The IFT particles, which comprise at least 17 different proteins, carry precursors necessary for the assembly of the ciliary axoneme and a number of signaling proteins. The IFT particles are linked to the ciliary membrane, indicating membrane proteins in the cargo. Modified from Reference 83.

Polycystins and Mechanosensation in Kidney Epithelial Cells

To test the functionality of renal cilia in PC1 mutants, a flow-Ca^{2+} imaging system was developed to test MEK epithelial cells (18) based on a previously described study in MDCK cells (87). Although the wild-type MEK cells responded to fluid flow at a rate that is similar to the physiological urine flow rate of 0.75 dyne cm^{-2}, *Pkd1* knockout cells failed to respond to a wide range of fluid flow rates including those higher or lower than the rate to which wild-type cells responded (which is 0.75 dyne cm^{-2}) (18). The flow-induced Ca^2 response likely involves PC2 because blocking antibodies to the large extracellular loop of PC2 abolished the Ca^{2+} response to flow (18). Given the idea that PC1 and PC2 suppress each other's

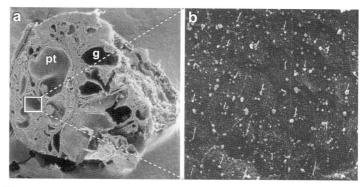

Figure 3

Scanning electron micrograph (SEM) of polycystic kidneys from *Pkd1*[del34/del34] mice at embryonic day (E)15.5. Panel *a* shows a low-power SEM of the cystic kidney. The area in the white box is magnified in panel *b*. Cysts, found mostly in the glomerulus and proximal tubules at this stage, develop at ~E17.5 in the collecting ducts. pt, proximal tubule cyst; g, glomerular cyst. Panel *b* shows the presence of primary cilia in cyst-lining epithelial cells.

activities when complexed (43), and the observation that stimulation of the extracellular domain of PC1 activates the PC2 channel (55), we proposed that PC1 and PC2 form a complex in the primary cilia of kidney epithelia, where they sense fluid flow and transduce this mechanical signal into a Ca^{2+} signal. In this model, the large extracellular domain of PC1 serves as an extracellular sensor for mechanical force such as urine flow, and subsequent conformational changes of PC1 activate the PC2 channel, resulting in Ca^{2+} entry. Recent studies using atomic force microscopy revealed that the PKD domains in PC1 exhibit remarkable mechanical strength similar to that of immunoglobulin domains in the giant muscle protein titin (88, 89), further supporting PC1 as a mechanosensor.

Primary Cilia and Polycystin-1 Also Sense Fluid Flow Shear Stress in Endothelial Cells

Recent work on endothelial cells supports the contention that PC1 functions in primary cilia as a mechanosensor (90). Like most other human cells, vascular endothelial cells develop primary cilia and express PC1 on this organelle. $Pkd1^{null/null}$ vascular endothelial cells lack PC1 and are unable to transduce extracellular shear stress into intracellular Ca^{2+} signaling. $Tg737^{orpk/orpk}$ vascular endothelial cells lacking polaris and cilia do not respond to flow with an increase in intracellular $[Ca^{2+}]$, although PC1 is present at the basal body, suggesting that the presence of both cilia and ciliary PC1 is required to sense fluid shear stress.

Under extracellular fluid–shear stress stimulation, vascular endothelial cells release nitric oxide (NO), an important vasodilator. The release of NO is dependent on an increase in intracellular $[Ca^{2+}]$ and thus the presence of primary cilia and PC1. By simultaneously recording cytosolic $[Ca^{2+}]$ and NO release in wild-type cells, Nauli et al. (90) showed that flow-induced shear stress triggered an intracellular Ca^{2+} response and released NO. Cells lacking either PC1 or polaris failed to release NO upon shear stress stimulation, although they are able to respond to other mechanical (e.g., touch) or pharmacological (e.g., ATP) stimuli. These data provide evidence that PC1 (required for cilia function) and polaris (required for cilia structure) are crucial mechanosensitive molecules in endothelial cells.

The Complexity of Flow-Induced Ca^{2+} Signaling: Entry and Amplification

In the original model (18), the polycystin mechanosensors were proposed to function through a sensory role of PC1 mediated by its large extracellular domain, followed by the activation of the PC2 channel. Studies in human kidney epithelial cells supported this model (44, 45). Later studies have shown that other Ca^{2+}-permeable channels that interact with PC2, such as TRPV4 (72) and TRPC1 (91, 92), also reside on the primary cilia, which raises the possibility that the flow-induced Ca^{2+} entry pathway in primary cilia may involve heteromeric channels such as PC2/TRPC1 and PC2/TRPV4. The PC2/TRPC1 channel is activated in response to G protein–coupled receptor activation (71). The channel is blocked by amiloride, although to a lesser degree than is the PC1/PC2 channel. Whether the PC2/TRPC1 channel is activated in response to flow stimulation is not known. Coimmunoprecipitation and fluorescence resonance energy transfer in cells transfected with both proteins support a model of direct interaction of TRPV4 with PC2 (72). Endogenous TRPV4 and PC2 colocalize on cilia in MDCK cells. That knockdown of endogenous TRPV4 by siRNA in MDCK cells abolished the shear stress–induced Ca^{2+} signal (72) suggests that TRPV4 is an essential component in ciliary flow sensation. As of now, however, there has been no electrophysiological evidence for a TRPV4/PC2 heteromeric channel in kidney epithelial cells.

The initial flow-induced Ca^{2+} entry signal on primary cilia has not yet been detected. A major difficulty is the ability to distinguish the Ca^{2+} signal in the tiny ciliary compartment from the Ca^{2+} dye background of the cell body. In MDCK cells, the initial fluid flow shear stress–dependent Ca^{2+} entry signal is believed to be amplified through IP_3 receptors. In mouse kidney epithelial cells, shear stress–induced Ca^{2+} is inhibited by ryanodine, but not by IP_3 receptor inhibitors, suggesting that ryanodine receptors participate in the Ca^{2+}-induced Ca^{2+} release amplification of this Ca^{2+} signal (18). This finding was also confirmed in human kidney epithelial cells (44, 93).

Using a double-chamber system, Praetorius and colleagues (94) recently observed that the application of pressure on MDCK cells activates a Ca^{2+} signal, presumably via apyrase-sensitive ATP release. The pressure-induced Ca^{2+} signal was not dependent on the cilium because MDCK cells without cilia remained responsive (94). At least in dissected mouse thick ascending limbs (95), P2Y receptors contribute to flow-induced Ca^{2+} elevation subsequent to nucleotide release. Flow-induced Ca^{2+} elevation was reduced to approximately 36% in tubules dissected from $P2Y_2$ knockout mice. Because in vitro perfusion of tubules is accompanied by an expansion in tubule lumen size, which may activate stretch-activated Ca^{2+} ion channels (96), it remains to be determined whether the P2Y receptor contributes to shear stress–induced Ca^{2+} signaling in tubular epithelial cells.

One possibility is that cilia activation results in nucleotide release, which in turn activates the P2Y receptors and P2Y receptor–dependent Ca^{2+} signals. Hovater et al. (97) have shown that loss of cilia impairs ATP secretion across the apical cell surface and ATP-dependent flow-induced Ca^{2+} signals. The authors used $Tg737^{orpk/orpk}$ mutant cells grown on permeable supports, either with or without rescue with a wild-type $Tg737$ expression construct, and tested the role of cilia in ATP release by harsh pipetting of medium directly onto the center of a monolayer of cells. These researchers found that such mechanical stimulation and hypotonic cell swelling trigger the release of a common ATP pool. Such pipette-induced flow also evoked an increase in cytosolic free Ca^{2+} concentration that was sensitive to extracellular Gd^{3+} and extracellular apyrase.

Most importantly, the authors found that mechanically stimulated Ca^{2+} signaling and stimulated ATP release are impaired in cilium-deficient monolayers (97), suggesting that ATP release is downstream from cilia activation. Although they are unable to respond to fluid flow shear stress with a Ca^{2+} signal (18), $Pkd1$ mutant endothelial (90) or epithelial (S.M. Nauli & J. Zhou, unpublished data) cells retain their Ca^{2+} response to mechanical stimuli such as touch of the cell membrane.

Selectivity of Flow Rates

The initial work on polycystin and mechanosensation employed MEK epithelial cells from $Pkd1^{null/null}$-targeted mice and their normal littermates (18). Follow-up studies using human kidney epithelial cells from ADPKD patients and normal individuals (44, 45) verified the requirement of PC1 for mechanosensation of shear stress in kidney tubular epithelial cells. Both studies (44, 45) found that confluent normal human kidney epithelial cells respond to flow with an increase in intracellular $[Ca^{2+}]$ that required extracellular Ca^{2+} influx and was sensitive to ryanodine (45). The kinetics of flow-induced Ca^{2+} signaling is similar. However, human cells require a higher flow rate than do mouse cells. Nauli et al. (44) reported 0.78 dyne cm^{-2} as an optimal shear stress for MEK epithelial cells and 1.2 dyne cm^{-2} as an optimal shear stress for human kidney epithelial cells; each evoked a similar Ca^{2+} response. Xu et al. (45) reproduced these data and found that human cells also respond to shear stress as high as 10 dyne cm^{-2}.

Although it is evident that the primary cilium in human kidney cells acts as a mechanosensory organelle, it is not clear that a higher set point for a response is necessary. The greater lumen size and/or rate of urine flow in the adult human kidney compared with the embryonic mouse kidney may be related to the higher shear stress needed to trigger a Ca^{2+} response in human cells (44). Such conservation in two-dimensional culture also suggests that the set point is not dependent solely on the tubule geometry. The collecting duct in the rat is thought to experience shear stress values in a range of 0.2–20 dyne cm^{-2}, depending on the rates of urine production (98). Thus, the higher shear stress values experienced by human cells are still within the minimum range of possible shear stress values in the collecting duct. Furthermore, the mechanosensitivity of kidney epithelia varies according to developmental stage (99). Could the expression levels of mechanosensitive proteins contribute to these differences? A higher level of polycystin expression has been reported in the developing kidney compared with the adult kidney (8, 100). Although there is a wide range of physiologically relevant flow rates, primary cilia selectively sense a narrow range of low shear stress, at least in mouse cells. This selectivity may be necessary for the control of the rate and orientation of cell proliferation and differentiation of tubular epithelia and, in turn, the maintenance of normal tubular lumen size and architecture. In disease states when ciliary signaling is substantially altered, the tubule lumen may enlarge, possibly owing to the loss of control of cell proliferation (101) and the guidance of planar cell polarity (PCP) such as oriented cell division.

Flow-Induced Subcellular Activity

It has long been known that fluid flow modulates K$^+$ secretion (102) and Na$^+$ absorption (103) in renal tubular epithelial cells through channel activation. Collective evidence has shown that shear stress activates the epithelial Na$^+$ channel by increasing channel open probability (104, 105). Frictional force of the renal tubular flow also modulates glucose transport (106) and magnesium levels (107). The downstream signaling pathways triggered by primary cilia–dependent fluid flow shear stress are not known. Among the proteins/lipids that are localized to the primary cilia, phosphatidylinositol-4,5-bisphosphate (PIP2) (108), TRPC1 (91), Nek8 (41), fibrocystin/polyductin (FPC) (109), and TRPV4 (72) have been found in the polycystin protein complex, suggesting that these proteins may participate or modulate flow sensing by polycystins.

Significance of Flow Shear Stress–Induced Ca^{2+} Signaling

Flow-induced Ca^{2+} signaling is believed to modulate a number of cellular functions that are abnormal in PKD, such as cell proliferation and cell differentiation, because this signal is lost in *Pkd1* mutant cells from the predisease stage to the early disease stage (18, 110). Because TRPV4 knockdown abolishes flow-induced Ca^{2+} signaling and TRPV4 knockout mice do not develop the polycystic kidney phenotype, Kottgen et al. (72) interpret their findings as questioning the importance of flow-mediated Ca^{2+} signaling in cystogenesis. Notably, FPC knockdown also largely reduces the flow-induced Ca^{2+} signal, and knockout of *Pkhd1* at many sites causes virtually no renal cyst phenotype. The role of FPC in cyst formation is unequivocal because mutations in *PKHD1* are responsible for the autosomal recessive form of PKD. Whether TRPV4 is a major component of flow-induced Ca^{2+} signaling remains to be determined. Flow-induced Ca^{2+} signaling requires multiple components such as the integrity of the cell cytoskeleton. Disruptions of actin filaments or intermediate filaments or alterations in microtubule stability all affect flow-induced Ca^{2+} signals (111).

Shear stress–induced Ca^{2+} increase, at least in epithelial cells, cannot be triggered within 30 min after a previous stimulation in human and mouse renal epithelial cells (44). This refractory period was slightly longer in experiments carried out below 37°C (45). Protein trafficking, channel inactivation/desensitization, cytoskeletal modifications, and cleavage and recycling of PC1 (112) in response to shear stress and the calcium spike all may contribute to the 30-min refractory period (44). However, the significance of this refractory period is presently unknown. Together with the dependence on flow rate, however, the 30-min refractory period may provide a high specificity for mechanosensing through the primary cilia. For example, if fluid flow in the renal tubules were continuous, the 30-min period would suggest that flow-induced Ca^{2+} is normally suppressed. Consequently, flow-induced Ca^{2+} signaling would provide a tonic signal for the maintenance of tubule architecture in the developed kidney. Whether and how this 30-min refractory period is modulated remain to be investigated.

Flow-Independent Activation of Polycystin-2 Channel

A common theme that links many channels, including TRP channels, is their modulation by phosphatidylinositol signal transduction pathways (113). Phosphatidylinositol is an important lipid, both as a key membrane constituent and as a participant in essential metabolic processes in all plants and animals. Because TRP channels are often activated by PLC and PIP2, Ma et al. (108) tested the activation of PC2 by PLC and phosphatidylinositol. They searched for components of the PLC activation pathway that have been reported to cause renal cystic disease. Because epidermal growth factor receptor (EGFR)-mutant mice develop dilated collecting tubules (114), Ma et al. tested a functional interaction between the EGFR and PC2. Activated EGFR binds PLC-γ1 and -γ2, which cleave PIP2 and produce IP$_3$ and diacylglycerol (DAG). Activated EGFR also activates phosphoinositide 3-kinase (PI3K) to phosphorylate PIP2 to phosphatidylinositol-3,4,5-triphosphate (PIP3). Using PC2 antibodies as a blocking agent, Ma et al. found that PC2 mediates EGF-induced currents in LLC-PK$_1$ cells. Moreover, recombinant PC2 coimmunoprecipitated PLC-γ2 and EGFR (108). Although it is not known whether endogenous PC2 also coimmunoprecipitates PLC-γ2, PIP2 colocalizes with PC2 in primary cilia of LLC-PK$_1$ cells, suggesting that the EGF-PC2 channel activation pathway is active on primary cilia.

Consequence of Fluid Flow Shear Stress: Inhibition of Polycystin-1 Cytoplasmic Tail Cleavage?

An interesting question arising from the ciliary polycystin mechanosensation hypothesis is how the initial calcium signal triggered by shear stress is mediated within a cell. Two recent studies have reported that the C-tail of PC1 is cleaved under static conditions; these studies proposed that fluid flow inhibits the cleavage of the PC1 C-tail (112, 115). Both groups showed that the cleaved C-tail is translocated to the nucleus, where it initiates nuclear signaling. Chauvet et al. (112) detected the cleaved C-tail in kidneys of wild-type animals after unilateral ureteral ligation and in mice without renal cilia because of conditional inactivation of the KIF3A subunit, suggesting that fluid flow normally inhibits the cleavage and that the cleavage occurs under pathological conditions. Low et al. (115) showed that the cleavage of the PC1 C-tail occurs under static cell culture conditions. The cleavage fragments identified by these two groups were of different sizes (34 kDa and 14 kDa, respectively), suggesting that there are two distinct cleavage sites in the C-tail. It appears that the 34-kDa nuclear PC1 C-tail can activate the AP1 pathway (112), whereas the 14-kDa fragment binds to STAT6 (signal transducer and activator of transcription protein 6) and its coactivator P100 and activates STAT6-dependent gene transcription (115). Ureteral ligation, the method used to study fluid flow by Chauvet et al. (112), may affect a number of factors other than flow and pressure and may initiate nonspecific injury-related events. In contrast, Low et al. (115) induced flow through orbital shaking, a condition with unknown effects on Ca^{2+} signaling. One of the major efforts under way is to identify the downstream signaling cascades that eventually lead to the abnormal cellular behaviors seen in PKD.

POLYCYSTINS AND CELL CYCLE CONTROL

Among the long-noted cellular abnormalities in PKD, increased cell proliferation has become the key feature of a polycystic kidney; cysts are "neoplasia in disguise" (116). Other abnormalities include cell differentiation, fluid secretion, extracellular matrix production, and apoptosis. These cellular features identified in polycystic kidneys have led to speculation on the roles of polycystins in the control of cell proliferation, differentiation, and death. Studies on polycystin function through overexpression of full-length PC1 in kidney cells have provided evidence that PC1 controls the cell cycle. Cells overexpressing PC1 arrest the cell cycle at the G0/G1 phase (117). Cells with *Pkd1* mutations displayed enhanced S phase entry (101). PC1-overexpressing cells are also resistant to apoptosis and induce tubulogenesis in the absence of hepatic growth factor (118). Three major pathways by which polycystins control the cell cycle through direct binding are discussed below.

The JAK-STAT Pathway

PC1 signaling pathways involve at least two STATs (STAT1 and STAT6). One study used full-length PC1-stable cell lines to show that PC1 induces STAT1 activation by direct association and activation of JAK2, which in turn induces p21 expression and modulates the cell cycle. However, the activation of JAK2 by PC1 requires PC2 because the R4227X truncation mutant of PC1 was able to bind but not activate JAK2, and full-length PC1 was unable to activate JAK2 in cells lacking PC2 (117). Another study showed that the expression of the cleaved C-tail of PC1 activated STAT6 by directly binding to P100 (115). In immune cells, STAT6 is activated by JAK1 or JAK3. Because PC1 does not bind to JAK1 and its interaction with JAK3 was not tested, the authors proposed that JAK3 may be involved in PC1-dependent activation of STAT6. Phosphorylation of STAT6 by PC1-activated JAK2 is not excluded. In this study, a pathological role of STAT6 was suggested. The authors examined the localization of STAT6 in human ADPKD kidneys and found increased expression of STAT6 in the nucleus of cyst-lining epithelial cells in ADPKD kidneys. These investigators proposed that, in normal renal tubular lumen with fluid flow and normal PC1 expression, STAT6 is sequestered in the cilia by PC1. In the absence of urine flow or PC1 or overexpression of mutant PC1, STAT6 translocates from the cilia to the nucleus to initiate STAT6-dependent transcription (115). Although several factors such as fluid flow, calcium influx, and cytokine stimulation have been speculated to facilitate the activation of the JAK-STAT pathway, the mechanism of STAT6 upregulation and activation in ADPKD remains unclear and requires further study.

The Id Pathway

The Id family (where Id refers to inhibitor of DNA binding or inhibitor of differentiation) comprises four relatively new transcriptional regulators in the superfamily of helix-loop-helix proteins. Id proteins inhibit differentiation of certain cell lineages and can stimulate proliferation. A recent study has shown that PC2 directly associates with Id2 and controls cell cycle progression (101). Membrane-anchored PC2 sequesters Id2 in the cytosol through direct binding and prevents Id2 nuclear translocation. Notably, Id2 appears to bind to phosphorylated PC2. Whereas overexpression of PC1 causes an increase in PC2 phosphorylation and PC2-Id2 interaction, mutations in polycystins disrupt this interaction and lead to increased Id2 nuclear accumulation in *Pkd1*-targeted mice and in human ADPKD patients with either *PKD1* or *PKD2* mutations (101). Nuclear Id2 antagonizes the function of E2A/E47-dependent growth-suppressive gene transcription and induces cell proliferation. Interestingly, inhibition of Id2 mRNA by RNA interference reduced the proliferative cell cycle profile of *Pkd1* knockout cells close to normal (101), suggesting that modulating Id2 expression levels or subcellular localization may be one way to lower the cell proliferation rate in PKD kidneys. PC2 phosphorylation is modulated by a number of factors including PC1 (101), GSK3 (38), and casein kinase 2 (69). Because Id2-PC2 interaction and cell surface expression (38) depend on the phosphorylation status of PC2, therapies directed at the modulation of PC2 phosphorylation may be effective for patients with PC1 mutations that affect PC2 phosphorylation.

The PC2-Id2 pathway links to several important signaling proteins that may play a role in the pathogenesis of PKD. These include β-catenin (119), E-cadherin (120), c-myc, and Rb (121), which may also contribute to the hyperproliferative phenotype in cyst-lining epithelia. In particular, β-catenin, which coimmunoprecipitates with PC1, can activate Id2 transcription, probably through c-myc, which is overexpressed in cystic kidneys. In addition, a role of Id2 in cell proliferation and differentiation has been linked to Smad4-dependent transforming growth factor-β (TGF-β) (120) and bone morphogenetic protein (BMP)-mediated pathways (122). The effective arrest of cystic disease by a cyclin-dependent kinase inhibitor (roscovitine) in two animal models of PKD (*cpk* and *jck*) (123) supports the notion that a dysregulated cell cycle is a proximal cause of cystogenesis.

The mTOR Pathway

The pathway involving the mammalian target-of-rapamycin (mTOR) complex has attracted great attention in the PKD field in the past two years. mTOR is a serine/threonine protein kinase, a central regulator of cell growth and metabolism in all eukaryotes (recently reviewed by Reference 124). mTOR is part of two distinct multiprotein complexes: mTORC1 and mTORC2. mTORC1 is sensitive to rapamycin and responsive to cellular energy status, nutrient availability, stress, and the presence of growth factors. mTORC1 signaling is a positive regulator of protein synthesis. mTORC2 is insensitive to rapamycin and regulates spatial aspects of cell growth (124). Both complexes are upregulated in a number of tumors. The role of mTOR in PKD was first tested by Wahl et al. (125), who used rapamycin (sirolimus) to treat PKD in a dominant rat model, the Han:SPRD rat model. These researchers found that cyst

volume density in the heterozygotes (cy/+) was reduced by 18% after a three-month treatment accompanied by a drop in S6 kinase activity (125).

A molecular link between mTOR and the polycystin signaling pathway came from a number of molecular studies on *TSC2* and *TSC1* genes encoding tuberin and hamartin, respectively. Mutations in *TSC2* and *TSC1* genes cause tuberous sclerosis, a genetic disease with renal cysts and tumors in a number of tissues. mTOR is regulated by the *TSC2* gene product tuberin through the small Ras-related GTPase Rheb (Ras homolog enriched in the brain). Rheb activates mTOR and is inactivated by the Rheb-GTPase-activating protein (RhebGAP), better known as the tuberin-hamartin protein complex. The tuberin encoding *TSC2* is evolutionarily linked to *PKD1*. *TSC2* and *PKD1* are localized tail to tail on chromosome 16p13.3, only approximately 60 bp apart (126). An earlier study using a tuberin-deficient renal cell line suggested that tuberin expression is required for the plasma membrane targeting of PC1 (127). Shillingford et al. (128) recently provided biochemical evidence that the transfected N- but not the C-terminal segment of the PC1 C-tail coimmunoprecipitated with endogenous tuberin in MDCK cells. Cotransfection of tuberin with the N-terminal fragment of PC1 shifted tuberin from its punctate cytoplasmic pattern to the Golgi apparatus, where the PC1 fragment is localized. Given the lack of coimmunoprecipitation data on endogenous proteins, these data suggest that PC1 and tuberin transiently interact and modulate each other's localization. Another interesting finding of this study is the upregulation of the phosphorylated, active form of mTOR and its downstream effector S6 kinase in cyst-lining epithelial cells from human ADPKD kidneys. This upregulation was also seen in cysts, but not in normal tubules in *Pkd1* mutant mice and two other mouse PKD models (*MAL* and *orpk* due to myelin and lymphocyte protein overexpression and mutation in a ciliary protein, Tg737, respectively). The inhibition of mTOR by rapamycin significantly alleviated the cystic phenotype in these animal models. Shillingford et al. (128) also performed a small retrospective study and found a reduction in renal cyst volume in ADPKD patients who underwent kidney transplantation and with treatment with rapamycin. Several clinical trials using rapamycin on ADPKD patients are ongoing because rapamycin is an FDA-approved immunosuppressant. The effectiveness of rapamycin was recently also shown in 16 ADPKD patients with polycystic liver disease (129), in whom rapamycin effectively inhibited the activation of S6 kinase, a readout of mTOR activation, and reduced the volume of polycystic livers.

DEBATES AGAINST A DIRECT ROLE OF POLYCYSTIN-1 IN CELL PROLIFERATION

Several recent studies of *Pkd1*-targeted mutants have raised questions about the role of polycystins in cell proliferation and the cell cycle (130–132). Takakura et al. (130) have shown that inactivation of both *Pkd1* alleles in one-week-old developing kidney resulted in rapid extensive cyst formation (**Figure 4**), similar to germline knockout of *Pkd1* (**Table 3**), which leads to extensive kidney cyst formation in utero (133, 134). Surprisingly, if *Pkd1* inactivation is induced in five-week-old mature kidneys, either there are no cysts or only focal cysts can be

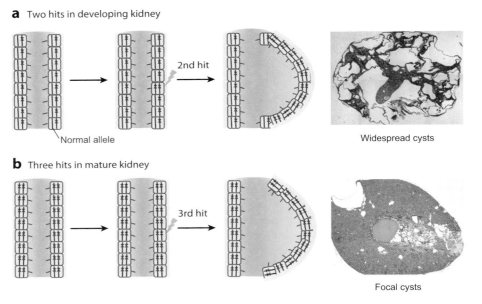

a Two hits in developing kidney

Normal allele

2nd hit

Widespread cysts

b Three hits in mature kidney

3rd hit

Focal cysts

Figure 4

Schematic presentation of the two-hit and three-hit hypotheses in the pathogenesis of autosomal dominant polycystic kidney disease. (*a*) In a developing kidney, a germline mutation (first hit) in one of the two copies of PKD1 in a given cell is not sufficient to alter the phenotype. A cyst develops when a somatic mutation knocks out the normal allele (the second hit) such that neither allele is functional and no polycystin function remains. The right-most panel shows widespread cysts in a mouse kidney with one germline *Pkd1*null mutation and that received a somatic second hit in *Pkd1* in collecting tubules during postnatal development (one week). (*b*) In mature kidney, however, loss of polycystin-1 due to mutations inactivating both alleles of PKD1 or by other means in a nephron segment is not sufficient to cause a cyst. A cyst forms upon a third hit, which could be a somatic mutation inactivating a growth suppressor or activating a growth activator or other cellular insults triggering cell proliferation or cell reorganization. The right-most panel shows focal cysts in a mouse kidney with one germline *Pkd1*null mutation and that received a somatic second hit in *Pkd1* in collecting tubules postdevelopment (at five weeks).

Table 3   Phenotypes of *Pkd1* and *Pkd2* knockout mice[a]

Strain[b]	Mutation	Allele	EL	Kidney and pancreatic cysts	Edema	Cardiovascular defects	Skeletal defects	Left-right defects	Heterozygous phenotype
*Pkd1*del34 (133, 174)	Exon 34 deletion	*Pkd1*tm1Jzh	+	+	+	−	+	−	Kidney, liver, pancreatic cysts
*Pkd1*null (134)	Exon 4 insertion	*Pkd1*tm1Jzh	+	+	+	Subcutaneous bleeding (<1%)	+	−	Kidney, liver, pancreatic cysts
*Pkd1*del43 (175)	Exon 43–45 deletion	*Pkd1*tm1Maa	+	+	+	Vascular leek	n.d.	n.d	n.d.
*Pkd1*del17 (176)	Exon 17–21 deletion	*Pkd1*tm1Rsa	+	+	+	+	+	−	Kidney, liver cysts
Pkd1− (177)	Exon 2–6 deletion	*Pkd1*tm1Shh	+	+	+	Double outlet right ventricle	n.d.	n.d.	n.d.
Pkd1− (178)	Single amino acid substitution	*Pkd1*m1Bei	+	+	+	n.d	n.d.	n.d.	Kidney, liver, pancreatic cysts
*Pkd1*nl (179)	Insertion of neo cassette in intron 1 causes 80% splicing defects	*Pkd1*nl/nl	−	+	−	Aorta aneurysms	n.d.	n.d.	No cysts
*Pkd1*flox (59)	Exon 2–6 deletion, γGT.Cre	*Pkd1*del2−6	−	Kidney +; pancreas −					
*Pkd1*cond (180)	Exon 2–4, MMTV.Cre		−	Kidney ±	−	−	−	−	n.d.
*Pkd2*null (141, 181)	Exon 1 disruption	*Pkd2*tm1Som	+	+	+	+	n.d.	+	n.d.
*Pkd2*ws25 (181)	Exon 1 duplication causing unstable allele	*Pkd2*tm2Som	+	+	n.d.	n.d	n.d.	+	K. cysts
Pkd2−LacZ (47)	Exon 1 deletion with LacZ promoter trap	*Pkd2*tm1Blum	+	+	n.d.	+	+	+[c]	n.d.

[a]EL, embryonic lethality with death occurring between embryonic day 13.5 and birth. All mutations are targeted mutations with the exception of one *N*-ethyl-nitrosourea-induced point mutation (*Pkd1*Bei). n.d., not described.

[b]References are in parentheses.

[c]Randomization, right pulmonary isomerism.

Table 4 *Pkd1* inducible knockouts

Inducible knockouts[a]	Mutation	Allele	Early induction[b]	Renal cysts	Onset	Late induction	Renal cysts	Onset
*Pkd1*flox (130)	Exon 2–6 deletion, *Mx1Cre*, pIpC inducible	*Pkd1*del2−6	One week	Widespread distal nephron segments	Six weeks later	Five weeks	Focal cysts	Six weeks
*Pkd1*cond (180)	Exon 2–11, tam-KspCad-CreERT2, tamoxifen inducible	*Pkd1*del2−11	P4	Massive cysts	One month later	One month	Microscopic cysts	Three months after
*Pkd1*cond (132)	Exon 2–4 deletion *Cre/Esr1*, tamoxifen inducible	*Pkd1*del2−4	<P12	Widespread, all nephron segments	Two weeks later	>P14 up to six weeks	Tubule dilation in almost all segments[c]	Six months

[a]References are in parentheses.
[b]P: postnatal days.
[c]Normal histology at three months.

detected weeks later. Once cysts form, there is an induction of local fibrosis and a clear regional progression of cyst formation, inflammation, and fibrosis (**Figure 4**). Severe polycystic kidneys develop nearly one year after inactivation of both *Pkd1* alleles. Two independent groups also found a slow onset of cystic phenotype in mice with different sets of *Pkd1* exons deleted in adulthood (131, 132) (**Table 4**). That induction of the same *Pkd1* mutations in younger mice produced severe PKD by all three groups rules out the possibility that the lack of phenotypes in mice with PC1 removal in adulthood is a result of specific mutations.

More than a decade ago, Geng et al. (8) dissected a series of embryonic and postnatal developing mouse kidneys and showed that PC1 expression level parallels kidney development: PC1 expression is high during late embryogenesis (in E15.5 kidneys) and remains high until two weeks after birth, when kidney maturation is complete. Only a low level of PC1 expression is maintained in adult kidneys (8). This developmentally regulated expression pattern led to the proposal that PC1 plays a role in tubule elongation and maturation during development and in the maintenance of the mature state during adult life (8). As described above, data from

cultured cells using kidney cells overexpressing full-length PC1 and PC2 have shown that both polycystins modulate cell cycle progression (101, 117). What factors account for the lack of obvious phenotype in adult kidneys lacking PC1 for weeks to months?

A major difference between one-week-old and five-week-old kidneys is that there is active ongoing postnatal renal development with rapid tubule formation, elongation, and maturation in the younger kidneys. Is the high cell proliferation rate during kidney development a major contributing factor to this difference? The cell proliferation rate in the five-week-old mouse kidney is ~50-fold lower if proliferation cell nuclear antigen (PCNA) expression levels are used as a measure (130). The higher rate of cell proliferation appears to persist until postnatal day 14 but drops at postnatal day 16, although the rate is still several-fold higher than in the adult kidney (132).

Increased cell proliferation has been widely observed in human ADPKD kidneys and in mouse PKD models (135–137), shown mostly by increased PCNA staining. Similarly, increased cell proliferation rates were seen in cyst-lining epithelial cells and normal tubules around cysts in mice with adult inactivation of

Pkd1 (130). However, in precystic adult kidneys after more than a week of *Pkd1* inactivation, no increase in PCNA expression was detected. In cystic kidneys in the adult *Pkd1* inactivation mouse model, immunolocalization with PCNA and Ki67 showed that neither tubular epithelia nor interstitial cells in areas distant from cysts had increased cell proliferation index, suggesting that *Pkd1* inactivation does not cause immediate cell proliferation (figure 7b in Reference 130).

In rodents, kidney development completes two weeks after birth, coinciding with downregulation of PC1 expression (8). In the postnatal kidney, a high cell proliferation rate accompanies the sustained activation of a number of developmental signaling pathways. Piontek et al. (132) identified postnatal days 12 to 14 as a critical window for the sensitivity of the kidney to *Pkd1* loss. *Pkd1* deletion induced at postnatal day 12 leads to rapid cyst formation, whereas deletion induced at postnatal day 14 is associated with the late onset of cysts. Notably, 14 days after birth is also the time point when cellular PC1 expression is significantly downregulated to a low level, which raises the possibility that there is a change in the role of PC1 in the kidney at this time point, probably from a more active role in tubule construction to a latent housekeeping role, which is maintained throughout adulthood.

Because the higher rate of cell proliferation appears to persist until postnatal day 14, yet inactivation of *Pkd1* at this time point results in a late-onset phenotype similar to adult loss of *Pkd1*, Piontek et al. (132) concluded that defective growth regulation is probably not the primary defect resulting from loss of *Pkd1* and that the relationship between cellular proliferation and cyst formation may be indirect. By microarray analyses, Piontek et al. found that ~827 genes are differentially expressed before (postnatal days 11 and 12) and after (postnatal days 14 and 15) the breakpoint. Many of these differentially expressed genes encode proteins with transporter and catalytic activity.

Three-Hit Hypothesis versus Two-Hit Hypothesis

In human ADPKD, cysts form focally in only ~5% of nephrons. A two-hit hypothesis has been proposed to explain the focal nature of cyst development (138): A germline mutation (first hit) in one of the two copies (alleles) of, for example, *PKD1* in a given cell is not sufficient to alter the phenotype and initiate cyst formation because the second allele is functioning normally. Cystogenesis occurs when a somatic mutation knocks out the normal allele (the second hit) such that neither allele is functional and no polycystin function remains. A cyst then forms through proliferation of the cell that has received two hits. In support of this hypothesis, somatic *PKD1* or *PKD2* mutations have been found in kidney (139) and liver (140) cysts from human ADPKD patients. The two-hit hypothesis is validated by conventional mouse knockout data: Loss of both copies of either the *Pkd1* or the *Pkd2* gene in the germline is sufficient to cause cyst formation in developing kidneys (133, 134, 141). Conditional knockout of *Pkd1* in postnatal developing kidney shown by somatic deletion of a *Pkd1* allele in mice carrying a germline mutation in *Pkd1* also supports the two-hit hypothesis (59, 130, 131). Therefore, this hypothesis remains valid in the context of renal development.

However, the two-hit hypothesis cannot explain the focal nature of cyst formation and regional progression of cyst development in mature organs of *Pkd1* inducible knockout mice. In these kidneys, every cell carries a germline-inherited *Pkd1* mutation, and almost all distal nephron segments positive for *Dolichos biflorus* agglutinin and Tamm-Horsfall protein have received a somatic hit (130). Therefore, we proposed that, in addition to the two hits to *Pkd1* alleles, a third hit (**Figure 4**) in a cell is necessary for rapid cyst formation (130). The loss of PC1 may predispose the tubular epithelial cells to a third hit, at which time the cell enters the cell cycle and cysts form. An alternative hypothesis is that PC1 loss has a tonic effect such that the mutant tubular epithelial cells are destined to develop cysts without any additional trigger. This model may explain the sudden onset of widespread tubule dilation in the *Esr1Cre Pkd1*cond/cond model (132) at six months after PC1 removal, but it is difficult to explain the focal renal disease in the *Mx1Cre Pkd1*flox/flox and *Mx1Cre Pkd1*flox/null model (130) (**Table 4**).

What Is the Third Hit?

It is still difficult to separate the potential requirement of a pulse of cell proliferation from activation of a developmental program as a trigger for cystogenesis in cells that received two hits in *PKD1*. Together with the developmentally regulated expression of PC1, however, it is tempting to speculate that PC1 is needed either when the renal developmental program is reactivated or when cell proliferation is increased during kidney regeneration in response to renal injury or other genetic or nongenetic insults. The resulting phenotype due to the loss of PC1 would depend on when the kidney received such a hit/stimulus. The group of cells receiving such a hit/insult would begin cystogenesis. The resultant cyst lesion(s) would initially be focal but would expand with accruement of hits/insults over time. The third-hit hypothesis can at least in part explain why somatic inactivation of *Pkd1* in adult mice causes no, or only a mild, cystic phenotype. These experimental mice, unlike humans, are kept in a nearly sterile environment with standardized food and are not exposed to any stress or toxins. One argument against the third-hit theory is that in human ADPKD patients, both kidneys enlarge in volume at a given rate per year (142). It is not known when the second or third hits are received in the human disease condition. However, there is a twofold increase in cell proliferation index in *Pkd1* heterozygous mice with adult inactivation of the normal allele (figure 7b in Reference 130) that better resembles the human condition than do mice with both *Pkd1* alleles inactivated in adulthood because all patients carry a germline mutation and receive a second hit later in life. The low rate of cell proliferation, probably coupled with defects in cell division, is not sufficient for cyst formation. Instead, it places the cells in a more vulnerable status prone to cyst formation upon a third hit. Once a cyst is formed, it may grow in size at a more steady rate. Moreover, the third hit, such as renal injuries that are required for the development of cysts in adult inactivation of PKD1, may either occur frequently in both kidneys as to achieve a steady rate, or occur in multiple forms at variable frequency so that their total impact is similar on both kidneys over a period of time (e.g., a year).

Does the slow onset of cystic phenotype caused by adult removal of PC1 conflict with the ciliary hypothesis for PKD? Adult loss of cilia through inducible knockout of two ciliogenesis genes, *Tg737* and *Kif3a* (143, 144), resulted in a late-onset polycystic kidney phenotype similar to the loss of *Pkd1*, suggesting that the requirement for ciliary signaling is similar to, if not the same as, that for PC1 signaling in mature kidneys.

PLANAR CELL POLARITY AND POLYCYSTIC KIDNEY DISEASE

Cell polarity commonly refers to polarization along the apical/basal axis of a cell, which is a universal feature of epithelial cells and is important for specialized epithelial functions. PCP refers to the polarization of a field of cells within the plane of a cell sheet. This form of epithelial polarity controls cell division orientation and is involved in a variety of developmental processes that determine cell differentiation fate and contribute to tissue and organ morphogenesis.

Most of the genes that are known to act in the mechanisms of PCP were identified through the studies of genetic mutants in *Drosophila melanogaster*. The core PCP genes in *Drosophila* include *Frizzled (Fz)*, *Flamingo (Fmi)*, *Strabismus (Stbm)/Van Gogh (Vang)*, *Prickle (Pk)*, *Dishevelled (Dsh)*, *Diego (Dgo)*, and the G protein Gαo subunit. Disruption of noncanonical Wnt signaling results in abnormalities in oriented cell division during zebrafish gastrulation, indicating that oriented cell division is a driving force for axis elongation (145).

Recent work by Fischer et al. (146) shows that lengthening of renal tubules is associated with the orientation of cells along the tubule axis, an intrinsic PCP process. Fischer et al. described a defect in the mitotic orientation in PCK rats and HNF1β knockout mice (models of autosomal recessive PKD) before overt tubule dilation and cyst formation, suggesting that defects in PCP are responsible for cyst development (**Figure 5**). Mitotic angle is defined as the difference in orientation between mitotic separating chromosomes during anaphase or telophase and the tubule axis. Fisher et al. (146) proposed that oriented cell division dictates the control of tubular lumen size during tubule elongation.

An attractive hypothesis is that the primary cilia of renal tubular epithelial cells sense the direction of primary urine flow and provide a cue for PCP and oriented cell division during tubule elongation through the function of polycystins. Evidence from amphibian motile cilia showed that tissue patterning coupled with fluid flow act in a positive feedback loop to direct the planar polarity of cilia in developing *Xenopus* larval skin (147). Cilia are required to produce flow and sense flow by responding with reorientation. Using developing skin explants exposed to a constant laminar flow in a flow chamber, the authors found that a shear stress of 0.5 dyne cm^{-2} provided a cue to cilia to generate a directional flow (147).

In the kidney, Igarashi and colleagues (144) have recently analyzed mice with kidney-specific knockout of Kif3a. These mice lack cilia in their kidney tubules, and in precystic kidneys at postnatal day 7 to day 10, the orientation of mitotic spindles was significantly different in mutant and control animals (144). In the Kif3a mutant kidneys, only 46% of the mitotic spindles were oriented within 20° of the axis of the collecting ducts. Twenty-four percent of the mitotic spindles were oriented at an angle greater than 40° from the axis of the tubule. This is in contrast to control kidneys, in which 91% of the mitotic spindles were oriented within 20° of the longitudinal axis of the collecting ducts. Therefore, primary cilia are required for the maintenance of PCP in the

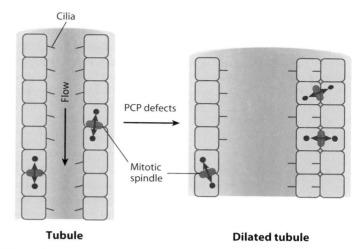

Figure 5

Schematic presentation of how planar cell polarity (PCP) defects such as abnormal oriented cell division may contribute to tubule dilation in polycystic kidney disease (PKD). In normal tubular epithelial cells, primary cilia may sense the direction of urine flow and provide a cue for oriented cell division that is parallel to the tubular axis and leads to tubule elongation. Disruption in primary cilia function or the flow sensing or responding pathway may result in abnormal oriented cell division and tubule dilation.

mammalian kidney, and the loss of cilia results in aberrant PCP prior to cyst formation.

A study on inversin, a protein involved in left-right asymmetry and kidney cyst formation, has provided some clues about how fluid flow shear stress may regulate the Wnt signaling pathway (148). Fluid flow over a kidney cell line increased inversin levels and slightly reduced β-catenin levels (~19%) (151). Inversin facilitates the degradation of cytosolic dishevelled, which is involved in canonical Wnt signaling, but not plasma membrane–bound dishevelled, which mediates noncanonical Wnt signaling. Furthermore, Simons et al. (148) showed that inversin is required for convergent extension movements in gastrulating *Xenopus laevis* embryos, a process mediated by the β-catenin-independent noncanonical Wnt pathway. On the basis of these findings, the authors proposed that inversin serves as a molecular switch between Wnt signaling pathways and that fluid flow terminates canonical Wnt signaling. It remains to be determined whether this cell culture model system adequately represents the in vivo situation. The BATgal mouse (149), which allows visualization of β-catenin-dependent Wnt signaling, may provide more insights into the spatial and temporal activation of canonical Wnt signaling during renal development.

The oriented cell division in the kidney likely uses the same set or a subset of the PCP genes in other organs and species. Loss of Fat4 causes renal tubule dilation (150). Fat is a protocadherin that functions as a receptor for Dachsous in PCP signaling. Fat represses the transcription of Four-jointed, another transmembrane protein. Fat and Dachsous function upstream of Frizzled, although some studies suggest the former two proteins act in parallel with Frizzled PCP signaling. The Fat/Dachsous/Four-jointed group of proteins was originally identified in *Drosophila*. There are four Fat homologs, two Dachsous homologs, and one Four-jointed ortholog in vertebrates. Saburi et al. (150) observed a generalized loss of elongated tubules in Fat-deficient kidneys at E16.5. There were no significant differences in proliferation or apoptosis in *Fat4* mutants compared with wild type (150). However, that only 55% of mutant cells, in contrast to 92% of wild-type cells, have a mitotic angle smaller than $30°$ suggests that the tubular dilations in the *Fat4* mutants result from a loss of oriented cell division (150). *Fat4* likely acts in a partially redundant fashion with *Vang like-2* (*Vangl2*) because loss of one copy of *Vangl2* in *Fat4* mutants resulted in increased dilation of renal tubules compared with *Fat4* single homozygous mutants (see figure 5d in Reference 150). Interestingly, homozygous loss of both Four-jointed and *Fat4* enhanced the tubule dilation phenotype in the *Fat4* single mutants and resulted in hydronephrosis, double ureters, and the incidence of duplex kidneys, similar to that seen in *Foxc1* mutants (151) but never seen in *Fat4* single mutants, suggesting Four-jointed may affect *Fat4*-independent pathways in kidney development. The tubule dilation phenotype in *Fat4*-homozygous mutants is much milder than in the *Pkd1* mutants at the same developmental stage, suggesting that if PCP abnormality contributes, it may not be the sole

mechanism accounting for the cystic phenotype seen in the ADPKD animal models.

Polycystins and Planar Cell Polarity

To date the molecular events underlying oriented cell division and its control of tubular lumen size are unknown, as is the role of polycystins in PCP or the role of PCP in the pathogenesis of ADPKD. Interaction of the PC2 C-terminal C-tail with the N terminus of mDia1/Drf1 (mammalian Diaphanous or Diaphanous-related formin 1 protein) (152), however, suggests a potential role of PC2 in oriented cell division. mDia1 is a member of the RhoA GTPase–binding formin homology protein family, which participates in cytoskeletal organization, cytokinesis, and signal transduction. The mDia-PC2 interaction is more prevalent in dividing than in nondividing cells. Endogenous PC2 and mDia1 colocalize at the mitotic spindles of dividing cells (152). This finding is interesting in the context of abnormalities in PCP and oriented cell division in animal models of recessive PKD (146). PCP defects deserve to be carefully examined in orthologous models of human ADPKD genes.

SUMMARY AND FUTURE DIRECTIONS

In summary, polycystins are a family of membrane proteins divided into PC1 and PC2 topologies. PC1 subfamily members have 11 putative transmembrane-spanning segments and interact with G proteins and/or PC2 proteins. PC2 family members have six transmembrane-spanning domains and are Ca^{2+}-permeant ion channels of the TRP superfamily. Polycystins control tube lumen size through the regulation of a number of cell behaviors. PC1 and PC2 are key architects of tubule construction in development and are likely involved in tubular renovation in adult life. Only limited information is known about mammalian polycystin-like molecules at present. Future efforts are required to identify the downstream molecular events of polycystin-mediated Ca^{2+} and G protein signaling and how polycystin channels are regulated. The role of PCP in ADPKD remains to be determined. Two hits in *PKD1* or *PKD2* are sufficient to cause cysts in developing kidneys. A third genetic or nongenetic hit that promotes cell proliferation or activation of a renal developmental program is likely needed for rapid development of cysts in mature kidneys. Identification of the third hit(s) will improve our understanding of adult PKD pathogenesis and the development of therapeutic strategies.

DISCLOSURE STATEMENT

The author is not aware of any biases that might be perceived as affecting the objectivity of this review.

ACKNOWLEDGMENTS

The author wishes to thank the many colleagues whose work and discussions contributed to this review. In particular, the author thanks many of her prior trainees (particularly Weining Lu, Xingzhen Chen, Surya Nauli, and Robert Kolb) and current members of her laboratory (Ayumi Takakura and Xuefeng Su). The author apologizes in advance to the authors of many studies involving polycystins and polycystic kidney disease whose work was not directly discussed owing to space limitations. This work was supported by grants from the National Institutes of Health: DK40703, DK51050, DK074030, and DK53357.

LITERATURE CITED

1. Dalgaard OZ. 1957. Bilateral polycystic disease of the kidneys; a follow-up of two hundred and eighty-four patients and their families. *Acta Med. Scand. Suppl.* 328:1–255
2. Iglesias CG, Torres VE, Offord KP, Holley KE, Beard CM, Kurland LT. 1983. Epidemiology of adult polycystic kidney disease, Olmsted County, Minnesota: 1935–1980. *Am. J. Kidney Dis.* 2:630–39
3. Levy M, Feingold J. 2000. Estimating prevalence in single-gene kidney diseases progressing to renal failure. *Kidney Int.* 58:925–43
4. Igarashi P, Somlo S. 2002. Genetics and pathogenesis of polycystic kidney disease. *J. Am. Soc. Nephrol.* 13:2384–98
5. Ong AC, Harris PC. 2005. Molecular pathogenesis of ADPKD: The polycystin complex gets complex. *Kidney Int.* 67:1234–47
6. Reeders ST, Breuning MH, Davies KE, Nicholls RD, Jarman AP, et al. 1985. A highly polymorphic DNA marker linked to adult polycystic kidney disease on chromosome 16. *Nature* 317:542–44
7. Geng L, Segal Y, Peissel B, Deng N, Pei Y, et al. 1996. Identification and localization of polycystin, the PKD1 gene product. *J. Clin. Investig.* 98:2674–82
8. Geng L, Segal Y, Pavlova A, Barros EJ, Lohning C, et al. 1997. Distribution and developmentally regulated expression of murine polycystin. *Am. J. Physiol.* 272:F451–59
9. Ward CJ, Turley H, Ong AC, Comley M, Biddolph S, et al. 1996. Polycystin, the polycystic kidney disease 1 protein, is expressed by epithelial cells in fetal, adult, and polycystic kidney. *Proc. Natl. Acad. Sci. USA* 93:1524–28
10. Griffin MD, Torres VE, Grande JP, Kumar R. 1997. Vascular expression of polycystin. *J. Am. Soc. Nephrol.* 8:616–26
11. Ibraghimov-Beskrovnaya O, Dackowski WR, Foggensteiner L, Coleman N, Thiru S, et al. 1997. Polycystin: In vitro synthesis, in vivo tissue expression, and subcellular localization identifies a large membrane-associated protein. *Proc. Natl. Acad. Sci. USA* 94:6397–402
12. Foggensteiner L, Bevan AP, Thomas R, Coleman N, Boulter C, et al. 2000. Cellular and subcellular distribution of polycystin-2, the protein product of the *PKD2* gene. *J. Am. Soc. Nephrol.* 11:814–27

13. Torres VE, Cai Y, Chen X, Wu GQ, Geng L, et al. 2001. Vascular expression of polycystin-2. *J. Am. Soc. Nephrol.* 12:1–9

14. Ong AC, Harris PC, Biddolph S, Bowker C, Ward CJ. 1999. Characterisation and expression of the PKD-1 protein, polycystin, in renal and extrarenal tissues. *Kidney Int.* 55:2091–116

15. Markowitz GS, Cai Y, Li L, Wu G, Ward LC, et al. 1999. Polycystin-2 expression is developmentally regulated. *Am. J. Physiol.* 277:F17–25

16. Yoder BK, Hou X, Guay-Woodford LM. 2002. The polycystic kidney disease proteins, polycystin-1, polycystin-2, polaris, and cystin, are colocalized in renal cilia. *J. Am. Soc. Nephrol.* 13:2508–16

17. Luo Y, Vassilev PM, Li X, Kawanabe Y, Zhou J. 2003. Native polycystin 2 functions as a plasma membrane Ca^{2+}-permeable cation channel in renal epithelia. *Mol. Cell. Biol.* 23:2600–7

18. Nauli SM, Alenghat FJ, Luo Y, Williams E, Vassilev P, et al. 2003. Polycystins 1 and 2 mediate mechanosensation in the primary cilium of kidney cells. *Nat. Genet.* 33:129–37

19. Pazour GJ, San Agustin JT, Follit JA, Rosenbaum JL, Witman GB. 2002. Polycystin-2 localizes to kidney cilia and the ciliary level is elevated in orpk mice with polycystic kidney disease. *Curr. Biol.* 12:R378–80

20. McGrath J, Somlo S, Makova S, Tian X, Brueckner M. 2003. Two populations of node monocilia initiate left-right asymmetry in the mouse. *Cell* 114:61–73

21. Geng L, Okuhara D, Yu Z, Tian X, Cai Y, et al. 2006. Polycystin-2 traffics to cilia independently of polycystin-1 by using an N-terminal RVxP motif. *J. Cell Sci.* 119:1383–95

22. Bui-Xuan EF, Li Q, Chen XZ, Boucher CA, Sandford R, et al. 2006. More than colocalizing with polycystin-1, polycystin-L is in the centrosome. *Am. J. Physiol. Ren. Physiol.* 291:F395–406

23. Huan Y, van Adelsberg J. 1999. Polycystin-1, the PKD1 gene product, is in a complex containing E-cadherin and the catenins. *J. Clin. Investig.* 104:1459–68

24. Scheffers MS, Van Der Bent P, Prins F, Spruit L, Breuning MH, et al. 2000. Polycystin-1, the product of the polycystic kidney disease 1 gene, colocalizes with desmosomes in MDCK cells. *Hum. Mol. Genet.* 9:2743–50

25. Ong AC, Ward CJ, Butler RJ, Biddolph S, Bowker C, et al. 1999. Coordinate expression of the autosomal dominant polycystic kidney disease proteins, polycystin-2 and polycystin-1, in normal and cystic tissue. *Am. J. Pathol.* 154:1721–29

26. Cai Y, Maeda Y, Cedzich A, Torres VE, Wu G, et al. 1999. Identification and characterization of polycystin-2, the PKD2 gene product. *J. Biol. Chem.* 274:28557–65

27. Obermuller N, Gallagher AR, Cai Y, Gassler N, Gretz N, et al. 1999. The rat Pkd2 protein assumes distinct subcellular distributions in different organs. *Am. J. Physiol.* 277:F914–25

28. Pisitkun T, Shen RF, Knepper MA. 2004. Identification and proteomic profiling of exosomes in human urine. *Proc. Natl. Acad. Sci. USA* 101:13368–73

29. Nomura H, Turco AE, Pei Y, Kalaydjieva L, Schiavello T, et al. 1998. Identification of *PKDL*, a novel polycystic kidney disease 2-like gene whose murine homologue is deleted in mice with kidney and retinal defects. *J. Biol. Chem.* 273:25967–73

30. Chen XZ, Vassilev PM, Basora N, Peng JB, Nomura H, et al. 1999. Polycystin-L is a calcium-regulated cation channel permeable to calcium ions. *Nature* 401:383–86

31. Basora N, Nomura H, Berger UV, Stayner C, Guo L, et al. 2002. Tissue and cellular localization of a novel polycystic kidney disease-like gene product, polycystin-L. *J. Am. Soc. Nephrol.* 13:293–301

32. LopezJimenez ND, Cavenagh MM, Sainz E, Cruz-Ithier MA, Battey JF, Sullivan SL. 2006. Two members of the TRPP family of ion channels, *Pkd1l3* and *Pkd2l1*, are coexpressed in a subset of taste receptor cells. *J. Neurochem.* 98:68–77

33. Guo L, Schreiber TH, Weremowicz S, Morton CC, Lee C, Zhou J. 2000. Identification and characterization of a novel polycystin family member, polycystin-L2, in mouse and human: sequence, expression, alternative splicing, and chromosomal localization. *Genomics* 64:241–51

34. Chen Y, Zhang Z, Lv XY, Wang YD, Hu ZG, et al. 2008. Expression of *Pkd2l2* in testis is implicated in spermatogenesis. *Biol. Pharm. Bull.* 31:1496–500

35. Stayner C, Zhou J. 2001. Polycystin channels and kidney disease. *Trends Pharmacol. Sci.* 22:543–46

36. Koulen P, Cai Y, Geng L, Maeda Y, Nishimura S, et al. 2002. Polycystin-2 is an intracellular calcium release channel. *Nat. Cell Biol.* 4:191–97

37. Vassilev PM, Guo L, Chen XZ, Segal Y, Peng JB, et al. 2001. Polycystin-2 is a novel cation channel implicated in defective intracellular Ca^{2+} homeostasis in polycystic kidney disease. *Biochem. Biophys. Res. Commun.* 282:341–50

38. Streets AJ, Moon DJ, Kane ME, Obara T, Ong AC. 2006. Identification of an N-terminal glycogen synthase kinase 3 phosphorylation site which regulates the functional localization of polycystin-2 in vivo and in vitro. *Hum. Mol. Genet.* 15:1465–73

39. Kottgen M, Benzing T, Simmen T, Tauber R, Buchholz B, et al. 2005. Trafficking of TRPP2 by PACS proteins represents a novel mechanism of ion channel regulation. *EMBO J.* 24:705–16

40. Hidaka S, Konecke V, Osten L, Witzgall R. 2004. PIGEA-14, a novel coiled-coil protein affecting the intracellular distribution of polycystin-2. *J. Biol. Chem.* 279:35009–16

41. Sohara E, Luo Y, Zhang J, Manning DK, Beier DR, Zhou J. 2008. Nek8 regulates the expression and localization of polycystin-1 and polycystin-2. *J. Am. Soc. Nephrol.* 19:469–76

42. Hanaoka K, Qian F, Boletta A, Bhunia AK, Piontek K, et al. 2000. Co-assembly of polycystin-1 and -2 produces unique cation-permeable currents. *Nature* 408:990–94

43. Delmas P, Nomura H, Li X, Lakkis M, Luo Y, et al. 2002. Constitutive activation of G-proteins by polycystin-1 is antagonized by polycystin-2. *J. Biol. Chem.* 277:11276–83

44. Nauli SM, Rossetti S, Kolb RJ, Alenghat FJ, Consugar MB, et al. 2006. Loss of polycystin-1 in human cyst-lining epithelia leads to ciliary dysfunction. *J. Am. Soc. Nephrol.* 17:1015–25

45. Xu C, Rossetti S, Jiang L, Harris PC, Brown-Glaberman U, et al. 2007. Human ADPKD primary cyst epithelial cells with a novel, single codon deletion in the *PKD1* gene exhibit defective ciliary polycystin localization and loss of flow-induced Ca^{2+} signaling. *Am. J. Physiol. Ren. Physiol.* 292:F930–45

46. Grimm DH, Cai Y, Chauvet V, Rajendran V, Zeltner R, et al. 2003. Polycystin-1 distribution is modulated by polycystin-2 expression in mammalian cells. *J. Biol. Chem.* 278:36786–93

47. Pennekamp P, Karcher C, Fischer A, Schweickert A, Skryabin B, et al. 2002. The ion channel polycystin-2 is required for left-right axis determination in mice. *Curr. Biol.* 12:938–43

48. Parnell SC, Magenheimer BS, Maser RL, Rankin CA, Smine A, et al. 1998. The polycystic kidney disease-1 protein, polycystin-1, binds and activates heterotrimeric G-proteins in vitro. *Biochem. Biophys. Res. Commun.* 251:625–31

49. Yuasa T, Takakura A, Denker BM, Venugopal B, Zhou J. 2004. Polycystin-1L2 is a novel G-protein-binding protein. *Genomics* 84:126–38

50. Mengerink KJ, Moy GW, Vacquier VD. 2002. suREJ3, a polycystin-1 protein, is cleaved at the GPS domain and localizes to the acrosomal region of sea urchin sperm. *J. Biol. Chem.* 277:943–48

51. Qian F, Boletta A, Bhunia AK, Xu H, Liu L, et al. 2002. Cleavage of polycystin-1 requires the receptor for egg jelly domain and is disrupted by human autosomal-dominant polycystic kidney disease 1-associated mutations. *Proc. Natl. Acad. Sci. USA* 99:16981–86

52. Yu S, Hackmann K, Gao J, He X, Piontek K, et al. 2007. Essential role of cleavage of Polycystin-1 at G protein-coupled receptor proteolytic site for kidney tubular structure. *Proc. Natl. Acad. Sci. USA* 104:18688–93

53. Tsiokas L, Kim E, Arnould T, Sukhatme VP, Walz G. 1997. Homo- and heterodimeric interactions between the gene products of *PKD1* and *PKD2*. *Proc. Natl. Acad. Sci. USA* 94:6965–70

54. Qian F, Germino FJ, Cai Y, Zhang X, Somlo S, Germino GG. 1997. PKD1 interacts with PKD2 through a probable coiled-coil domain. *Nat. Genet.* 16:179–83

55. Delmas P, Nauli SM, Li X, Coste B, Osorio N, et al. 2004. Gating of the polycystin ion channel signaling complex in neurons and kidney cells. *FASEB J.* 18:740–42

56. Parnell SC, Magenheimer BS, Maser RL, Zien CA, Frischauf AM, Calvet JP. 2002. Polycystin-1 activation of c-Jun N-terminal kinase and AP-1 is mediated by heterotrimeric G proteins. *J. Biol. Chem.* 277:19566–72

57. Basavanna U, Weber KM, Hu Q, Ziegelstein RC, Germino GG, Sutters M. 2007. The isolated polycystin-1 COOH-terminal can activate or block polycystin-1 signaling. *Biochem. Biophys. Res. Commun.* 359:367–72

58. Kim E, Arnould T, Sellin L, Benzing T, Comella N, et al. 1999. Interaction between RGS7 and polycystin. *Proc. Natl. Acad. Sci. USA* 96:6371–76

59. Starremans PG, Li X, Finnerty PE, Guo L, Takakura A, et al. 2008. A mouse model for polycystic kidney disease through a somatic in-frame deletion in the 5′ end of *Pkd1*. *Kidney Int.* 73:1394–405

60. Gattone VH second, Wang X, Harris PC, Torres VE. 2003. Inhibition of renal cystic disease development and progression by a vasopressin V2 receptor antagonist. *Nat. Med.* 9:1323–26

61. Hanaoka K, Guggino WB. 2000. cAMP regulates cell proliferation and cyst formation in autosomal polycystic kidney disease cells. *J. Am. Soc. Nephrol.* 11:1179–87

62. Yamaguchi T, Pelling JC, Ramaswamy NT, Eppler JW, Wallace DP, et al. 2000. cAMP stimulates the in vitro proliferation of renal cyst epithelial cells by activating the extracellular signal-regulated kinase pathway. *Kidney Int.* 57:1460–71

63. Torres VE, Wang X, Qian Q, Somlo S, Harris PC, Gattone VH. 2004. Effective treatment of an orthologous model of autosomal dominant polycystic kidney disease. *Nat. Med.* 10:363–64

64. Ruggenenti P, Remuzzi A, Ondei P, Fasolini G, Antiga L, et al. 2005. Safety and efficacy of long-acting somatostatin treatment in autosomal-dominant polycystic kidney disease. *Kidney Int.* 68:206–16

65. Yang B, Sonawane ND, Zhao D, Somlo S, Verkman AS. 2008. Small-molecule CFTR inhibitors slow cyst growth in polycystic kidney disease. *J. Am. Soc. Nephrol.* 19:1300–10

66. Calvet JP. 2008. Strategies to inhibit cyst formation in ADPKD. *Clin. J. Am. Soc. Nephrol.* 3:1205–11

67. Gonzalez-Perret S, Kim K, Ibarra C, Damiano AE, Zotta E, et al. 2001. Polycystin-2, the protein mutated in autosomal dominant polycystic kidney disease (ADPKD), is a Ca^{2+}-permeable nonselective cation channel. *Proc. Natl. Acad. Sci. USA* 98:1182–87

68. Chen XZ, Segal Y, Basora N, Guo L, Peng JB, et al. 2001. Transport function of the naturally occurring pathogenic polycystin-2 mutant, R742X. *Biochem. Biophys. Res. Commun.* 282:1251–56

69. Cai Y, Anyatonwu G, Okuhara D, Lee KB, Yu Z, et al. 2004. Calcium dependence of polycystin-2 channel activity is modulated by phosphorylation at Ser^{812}. *J. Biol. Chem.* 279:19987–95

70. Feng S, Okenka GM, Bai CX, Streets AJ, Newby LJ, et al. 2008. Identification and functional characterisation of an N-terminal oligomerisation domain for polycystin-2. *J. Biol. Chem.* 283:28471–79

71. Bai CX, Giamarchi A, Rodat-Despoix L, Padilla F, Downs T, et al. 2008. Formation of a new receptor-operated channel by heteromeric assembly of TRPP2 and TRPC1 subunits. *EMBO Rep.* 9:472–79

72. Kottgen M, Buchholz B, Garcia-Gonzalez MA, Kotsis F, Fu X, et al. 2008. TRPP2 and TRPV4 form a polymodal sensory channel complex. *J. Cell Biol.* 182:437–47

73. Tsiokas L, Kim S, Ong EC. 2007. Cell biology of polycystin-2. *Cell. Signal.* 19:444–53

74. Ishimaru Y, Inada H, Kubota M, Zhuang H, Tominaga M, Matsunami H. 2006. Transient receptor potential family members PKD1L3 and PKD2L1 form a candidate sour taste receptor. *Proc. Natl. Acad. Sci. USA* 103:12569–74

75. Scholey JM. 2008. Intraflagellar transport motors in cilia: moving along the cell's antenna. *J. Cell Biol.* 180:23–29

76. Kozminski KG, Johnson KA, Forscher P, Rosenbaum JL. 1993. A motility in the eukaryotic flagellum unrelated to flagellar beating. *Proc. Natl. Acad. Sci. USA* 90:5519–23

77. Johnson KA, Rosenbaum JL. 1992. Polarity of flagellar assembly in *Chlamydomonas*. *J. Cell Biol.* 119:1605–11

78. Cole DG, Diener DR, Himelblau AL, Beech PL, Fuster JC, Rosenbaum JL. 1998. *Chlamydomonas* kinesin-II-dependent intraflagellar transport (IFT): IFT particles contain proteins required for ciliary assembly in *Caenorhabditis elegans* sensory neurons. *J. Cell Biol.* 141:993–1008

79. Piperno G, Siuda E, Henderson S, Segil M, Vaananen H, Sassaroli M. 1998. Distinct mutants of retrograde intraflagellar transport (IFT) share similar morphological and molecular defects. *J. Cell Biol.* 143:1591–601

80. Iomini C, Babaev-Khaimov V, Sassaroli M, Piperno G. 2001. Protein particles in *Chlamydomonas* flagella undergo a transport cycle consisting of four phases. *J. Cell Biol.* 153:13–24

81. Lucker BF, Behal RH, Qin H, Siron LC, Taggart WD, et al. 2005. Characterization of the intraflagellar transport complex B core: direct interaction of the IFT81 and IFT74/72 subunits. *J. Biol. Chem.* 280:27688–96

82. Jekely G, Arendt D. 2006. Evolution of intraflagellar transport from coated vesicles and autogenous origin of the eukaryotic cilium. *BioEssays* 28:191–98

83. Rosenbaum JL, Witman GB. 2002. Intraflagellar transport. *Nat. Rev. Mol. Cell Biol.* 3:813–25

84. Satir P, Christensen ST. 2007. Overview of structure and function of mammalian cilia. *Annu. Rev. Physiol.* 69:377–400

85. Pazour GJ, Dickert BL, Vucica Y, Seeley ES, Rosenbaum JL, et al. 2000. *Chlamydomonas IFT88* and its mouse homologue, polycystic kidney disease gene *Tg737*, are required for assembly of cilia and flagella. *J. Cell Biol.* 151:709–18

86. Lin F, Hiesberger T, Cordes K, Sinclair AM, Goldstein LS, et al. 2003. Kidney-specific inactivation of the KIF3A subunit of kinesin-II inhibits renal ciliogenesis and produces polycystic kidney disease. *Proc. Natl. Acad. Sci. USA* 100:5286–91

87. Praetorius HA, Spring KR. 2001. Bending the MDCK cell primary cilium increases intracellular calcium. *J. Membr. Biol.* 184:71–79

88. Forman JR, Qamar S, Paci E, Sandford RN, Clarke J. 2005. The remarkable mechanical strength of polycystin-1 supports a direct role in mechanotransduction. *J. Mol. Biol.* 349:861–71

89. Qian F, Wei W, Germino G, Oberhauser A. 2005. The nanomechanics of polycystin-1 extracellular region. *J. Biol. Chem.* 280:40723–30

90. Nauli SM, Kawanabe Y, Kaminski JJ, Pearce WJ, Ingber DE, Zhou J. 2008. Endothelial cilia are fluid shear sensors that regulate calcium signaling and nitric oxide production through polycystin-1. *Circulation* 117:1161–71

91. Tsiokas L, Arnould T, Zhu C, Kim E, Walz G, Sukhatme VP. 1999. Specific association of the gene product of PKD2 with the TRPC1 channel. *Proc. Natl. Acad. Sci. USA* 96:3934–39

92. Raychowdhury MK, McLaughlin M, Ramos AJ, Montalbetti N, Bouley R, et al. 2005. Characterization of single channel currents from primary cilia of renal epithelial cells. *J. Biol. Chem.* 280:34718–22

93. Liu W, Morimoto T, Woda C, Kleyman TR, Satlin LM. 2007. Ca^{2+} dependence of flow-stimulated K secretion in the mammalian cortical collecting duct. *Am. J. Physiol. Ren. Physiol.* 293:F227–35

94. Praetorius HA, Frokiaer J, Leipziger J. 2005. Transepithelial pressure pulses induce nucleotide release in polarized MDCK cells. *Am. J. Physiol. Ren. Physiol.* 288:F133–41

95. Jensen ME, Odgaard E, Christensen MH, Praetorius HA, Leipziger J. 2007. Flow-induced $[Ca^{2+}]_i$ increase depends on nucleotide release and subsequent purinergic signaling in the intact nephron. *J. Am. Soc. Nephrol.* 18:2062–70

96. Filipovic D, Sackin H. 1991. A calcium-permeable stretch-activated cation channel in renal proximal tubule. *Am. J. Physiol.* 260:F119–29

97. Hovater MB, Olteanu D, Hanson EL, Cheng NL, Siroky B, et al. 2008. Loss of apical monocilia on collecting duct principal cells impairs ATP secretion across the apical cell surface and ATP-dependent and flow-induced calcium signals. *Purinergic Signal.* 4:155–70

98. Cai Z, Xin J, Pollock DM, Pollock JS. 2000. Shear stress-mediated NO production in inner medullary collecting duct cells. *Am. J. Physiol. Ren. Physiol.* 279:F270–74

99. Liu W, Murcia NS, Duan Y, Weinbaum S, Yoder BK, et al. 2005. Mechanoregulation of intracellular Ca^{2+} concentration is attenuated in collecting duct of monocilium-impaired *orpk* mice. *Am. J. Physiol. Ren. Physiol.* 289:F978–88

100. Van Adelsberg J, Chamberlain S, D'Agati V. 1997. Polycystin expression is temporally and spatially regulated during renal development. *Am. J. Physiol.* 272:F602–9

101. Li X, Luo Y, Starremans PG, McNamara CA, Pei Y, Zhou J. 2005. Polycystin-1 and polycystin-2 regulate the cell cycle through the helix-loop-helix inhibitor Id2. *Nat. Cell Biol.* 7:1102–12

102. Woda CB, Miyawaki N, Ramalakshmi S, Ramkumar M, Rojas R, et al. 2003. Ontogeny of flow-stimulated potassium secretion in rabbit cortical collecting duct: functional and molecular aspects. *Am. J. Physiol. Ren. Physiol.* 285:F629–39

103. Satlin LM, Sheng S, Woda CB, Kleyman TR. 2001. Epithelial Na^+ channels are regulated by flow. *Am. J. Physiol. Ren. Physiol.* 280:F1010–18

104. Satlin LM, Carattino MD, Liu W, Kleyman TR. 2006. Regulation of cation transport in the distal nephron by mechanical forces. *Am. J. Physiol. Ren. Physiol.* 291:F923–31

105. Morimoto T, Liu W, Woda C, Carattino MD, Wei Y, et al. 2006. Mechanism underlying flow stimulation of sodium absorption in the mammalian collecting duct. *Am. J. Physiol. Ren. Physiol.* 291:F663–69

106. Garvin JL. 1990. Glucose absorption by isolated perfused rat proximal straight tubules. *Am. J. Physiol.* 259:F580–86

107. Dai LJ, Ritchie G, Kerstan D, Kang HS, Cole DE, Quamme GA. 2001. Magnesium transport in the renal distal convoluted tubule. *Physiol. Rev.* 81:51–84

108. Ma R, Li WP, Rundle D, Kong J, Akbarali HI, Tsiokas L. 2005. PKD2 functions as an epidermal growth factor-activated plasma membrane channel. *Mol. Cell. Biol.* 25:8285–98

109. Wang S, Zhang J, Nauli SM, Li X, Starremans PG, et al. 2007. Fibrocystin/polyductin, found in the same protein complex with polycystin-2, regulates calcium responses in kidney epithelia. *Mol. Cell. Biol.* 27:3241–52

110. Nauli SM, Zhou J. 2004. Polycystins and mechanosensation in renal and nodal cilia. *BioEssays* 26:844–56

111. Alenghat FJ, Nauli SM, Kolb R, Zhou J, Ingber DE. 2004. Global cytoskeletal control of mechanotransduction in kidney epithelial cells. *Exp. Cell Res.* 301:23–30

112. Chauvet V, Tian X, Husson H, Grimm DH, Wang T, et al. 2004. Mechanical stimuli induce cleavage and nuclear translocation of the polycystin-1 C terminus. *J. Clin. Investig.* 114:1433–43

113. Clapham DE, Runnels LW, Strubing C. 2001. The TRP ion channel family. *Nat. Rev. Neurosci.* 2:387–96

114. Threadgill DW, Dlugosz AA, Hansen LA, Tennenbaum T, Lichti U, et al. 1995. Targeted disruption of mouse EGF receptor: effect of genetic background on mutant phenotype. *Science* 269:230–34

115. Low SH, Vasanth S, Larson CH, Mukherjee S, Sharma N, et al. 2006. Polycystin-1, STAT6, and P100 function in a pathway that transduces ciliary mechanosensation and is activated in polycystic kidney disease. *Dev. Cell* 10:57–69

116. Grantham JJ. 1990. Polycystic kidney disease: neoplasia in disguise. *Am. J. Kidney Dis.* 15:110–16

117. Bhunia AK, Piontek K, Boletta A, Liu L, Qian F, et al. 2002. PKD1 induces p21^{waf1} and regulation of the cell cycle via direct activation of the JAK-STAT signaling pathway in a process requiring PKD2. *Cell* 109:157–68

118. Boletta A, Qian F, Onuchic LF, Bhunia AK, Phakdeekitcharoen B, et al. 2000. Polycystin-1, the gene product of *PKD1*, induces resistance to apoptosis and spontaneous tubulogenesis in MDCK cells. *Mol. Cell* 6:1267–73

119. Rockman SP, Currie SA, Ciavarella M, Vincan E, Dow C, et al. 2001. Id2 is a target of the β-catenin/T cell factor pathway in colon carcinoma. *J. Biol. Chem.* 276:45113–19

120. Kondo M, Cubillo E, Tobiume K, Shirakihara T, Fukuda N, et al. 2004. A role for Id in the regulation of TGF-β-induced epithelial-mesenchymal transdifferentiation. *Cell Death Differ.* 11:1092–101

121. Lasorella A, Noseda M, Beyna M, Yokota Y, Iavarone A. 2000. Id2 is a retinoblastoma protein target and mediates signalling by Myc oncoproteins. *Nature* 407:592–98

122. Kowanetz M, Valcourt U, Bergstrom R, Heldin CH, Moustakas A. 2004. Id2 and Id3 define the potency of cell proliferation and differentiation responses to transforming growth factor beta and bone morphogenetic protein. *Mol. Cell. Biol.* 24:4241–54

123. Bukanov NO, Smith LA, Klinger KW, Ledbetter SR, Ibraghimov-Beskrovnaya O. 2006. Long-lasting arrest of murine polycystic kidney disease with CDK inhibitor roscovitine. *Nature* 444:949–52

124. Wullschleger S, Loewith R, Hall MN. 2006. TOR signaling in growth and metabolism. *Cell* 124:471–84

125. Wahl PR, Serra AL, Le Hir M, Molle KD, Hall MN, Wuthrich RP. 2006. Inhibition of mTOR with sirolimus slows disease progression in Han:SPRD rats with autosomal dominant polycystic kidney disease (ADPKD). *Nephrol. Dial. Transplant.* 21:598–604

126. Hughes J, Ward CJ, Peral B, Aspinwall R, Clark K, et al. 1995. The polycystic kidney disease 1 (*PKD1*) gene encodes a novel protein with multiple cell recognition domains. *Nat. Genet.* 10:151–60

127. Kleymenova E, Ibraghimov-Beskrovnaya O, Kugoh H, Everitt J, Xu H, et al. 2001. Tuberin-dependent membrane localization of polycystin-1: a functional link between polycystic kidney disease and the TSC2 tumor suppressor gene. *Mol. Cell* 7:823–32

128. Shillingford JM, Murcia NS, Larson CH, Low SH, Hedgepeth R, et al. 2006. The mTOR pathway is regulated by polycystin-1, and its inhibition reverses renal cystogenesis in polycystic kidney disease. *Proc. Natl. Acad. Sci. USA* 103:5466–71

129. Qian Q, Du H, King BF, Kumar S, Dean PG, et al. 2008. Sirolimus reduces polycystic liver volume in ADPKD patients. *J. Am. Soc. Nephrol.* 19:631–38

130. Takakura A, Contrino L, Beck AW, Zhou J. 2008. Pkd1 inactivation induced in adulthood produces focal cystic disease. *J. Am. Soc. Nephrol.* 19:2351–63

131. Lantinga-van Leeuwen IS, Leonhard WN, Van Der Wal A, Breuning MH, de Heer E, Peters DJ. 2007. Kidney-specific inactivation of the *Pkd1* gene induces rapid cyst formation in developing kidneys and a slow onset of disease in adult mice. *Hum. Mol. Genet.* 16:3188–96

132. Piontek K, Menezes LF, Garcia-Gonzalez MA, Huso DL, Germino GG. 2007. A critical developmental switch defines the kinetics of kidney cyst formation after loss of Pkd1. *Nat. Med.* 13:1490–95

133. Lu W, Peissel B, Babakhanlou H, Pavlova A, Geng L, et al. 1997. Perinatal lethality with kidney and pancreas defects in mice with a targeted *Pkd1* mutation. *Nat. Genet.* 17:179–81

134. Lu W, Shen X, Pavlova A, Lakkis M, Babakhanlou H, et al. 2001. Comparison of *Pkd1*-targeted mutants reveals that loss of polycystin-1 causes cystogenesis and bone defects. *Hum. Mol. Genet.* 10:2385–96

135. Grantham JJ, Geiser JL, Evan AP. 1987. Cyst formation and growth in autosomal dominant polycystic kidney disease. *Kidney Int.* 31:1145–52

136. Nadasdy T, Laszik Z, Lajoie G, Blick KE, Wheeler DE, Silva FG. 1995. Proliferative activity of cyst epithelium in human renal cystic diseases. *J. Am. Soc. Nephrol.* 5:1462–68

137. Lanoix J, D'Agati V, Szabolcs M, Trudel M. 1996. Dysregulation of cellular proliferation and apoptosis mediates human autosomal dominant polycystic kidney disease (ADPKD). *Oncogene* 13:1153–60

138. Reeders ST. 1992. Multilocus polycystic disease. *Nat. Genet.* 1:235–37

139. Qian F, Watnick TJ, Onuchic LF, Germino GG. 1996. The molecular basis of focal cyst formation in human autosomal dominant polycystic kidney disease type I. *Cell* 87:979–87

140. Watnick TJ, Torres VE, Gandolph MA, Qian F, Onuchic LF, et al. 1998. Somatic mutation in individual liver cysts supports a two-hit model of cystogenesis in autosomal dominant polycystic kidney disease. *Mol. Cell* 2:247–51

141. Wu G, D'Agati V, Cai Y, Markowitz G, Park JH, et al. 1998. Somatic inactivation of Pkd2 results in polycystic kidney disease. *Cell* 93:177–88

142. Grantham JJ, Torres VE, Chapman AB, Guay-Woodford LM, Bae KT, et al. 2006. Volume progression in polycystic kidney disease. *N. Engl. J. Med.* 354:2122–30

143. Davenport JR, Watts AJ, Roper VC, Croyle MJ, van Groen T, et al. 2007. Disruption of intraflagellar transport in adult mice leads to obesity and slow-onset cystic kidney disease. *Curr. Biol.* 17:1586–94

144. Patel V, Li L, Cobo-Stark P, Shao X, Somlo S, et al. 2008. Acute kidney injury and aberrant planar cell polarity induce cyst formation in mice lacking renal cilia. *Hum. Mol. Genet.* 17:1578–90

145. Gong Y, Mo C, Fraser SE. 2004. Planar cell polarity signalling controls cell division orientation during zebrafish gastrulation. *Nature* 430:689–93

146. Fischer E, Legue E, Doyen A, Nato F, Nicolas JF, et al. 2006. Defective planar cell polarity in polycystic kidney disease. *Nat. Genet.* 38:21–23

147. Mitchell B, Jacobs R, Li J, Chien S, Kintner C. 2007. A positive feedback mechanism governs the polarity and motion of motile cilia. *Nature* 447:97–101

148. Simons M, Gloy J, Ganner A, Bullerkotte A, Bashkurov M, et al. 2005. Inversin, the gene product mutated in nephronophthisis type II, functions as a molecular switch between Wnt signaling pathways. *Nat. Genet.* 37:537–43

149. Maretto S, Cordenonsi M, Dupont S, Braghetta P, Broccoli V, et al. 2003. Mapping Wnt/β-catenin signaling during mouse development and in colorectal tumors. *Proc. Natl. Acad. Sci. USA* 100:3299–304

150. Saburi S, Hester I, Fischer E, Pontoglio M, Eremina V, et al. 2008. Loss of Fat4 disrupts PCP signaling and oriented cell division and leads to cystic kidney disease. *Nat. Genet.* 40:1010–15

151. Kume T, Deng K, Hogan BL. 2000. Murine forkhead/winged helix genes *Foxc1* (*Mf1*) and *Foxc2* (*Mfh1*) are required for the early organogenesis of the kidney and urinary tract. *Development* 127:1387–95

152. Rundle DR, Gorbsky G, Tsiokas L. 2004. PKD2 interacts and colocalizes with mDia1 to mitotic spindles of dividing cells: role of mDia1 in PKD2 localization to mitotic spindles. *J. Biol. Chem.* 279:29728–39

153. International Polycystic Kidney Disease Consortium. 1995. Polycystic kidney disease: the complete structure of the PKD1 gene and its protein. *Cell* 81:289–98

154. Hughes J, Ward CJ, Aspinwall R, Butler R, Harris PC. 1999. Identification of a human homologue of the sea urchin receptor for egg jelly: a polycystic kidney disease-like protein. *Hum. Mol. Genet.* 8:543–49

155. Yuasa T, Venugopal B, Weremowicz S, Morton CC, Guo L, Zhou J. 2002. The sequence, expression, and chromosomal localization of a novel polycystic kidney disease 1-like gene, *PKD1L1*, in human. *Genomics* 79:376–86

156. Li A, Tian X, Sung SW, Somlo S. 2003. Identification of two novel polycystic kidney disease-1-like genes in human and mouse genomes. *Genomics* 81:596–608

157. Mochizuki T, Wu G, Hayashi T, Xenophontos SL, Veldhuisen B, et al. 1996. *PKD2*, a gene for polycystic kidney disease that encodes an integral membrane protein. *Science* 272:1339–42

158. Schneider MC, Rodriguez AM, Nomura H, Zhou J, Morton CC, et al. 1996. A gene similar to PKD1 maps to chromosome 4q22: a candidate gene for PKD2. *Genomics* 38:1–4

159. Wu G, Hayashi T, Park JH, Dixit M, Reynolds DM, et al. 1998. Identification of *PKD2L*, a human PKD2-related gene: tissue-specific expression and mapping to chromosome 10q25. *Genomics* 54:564–68

160. Guo L, Chen M, Basora N, Zhou J. 2000. The human polycystic kidney disease 2-like (*PKDL*) gene: exon/intron structure and evidence for a novel splicing mechanism. *Mamm. Genome* 11:46–50

161. Veldhuisen B, Spruit L, Dauwerse HG, Breuning MH, Peters DJ. 1999. Genes homologous to the autosomal dominant polycystic kidney disease genes (*PKD1* and *PKD2*). *Eur. J. Hum. Genet.* 7:860–72

162. Lehtonen S, Ora A, Olkkonen VM, Geng L, Zerial M, et al. 2000. In vivo interaction of the adapter protein CD2-associated protein with the type 2 polycystic kidney disease protein, polycystin-2. *J. Biol. Chem.* 275:32888–93

163. Gallagher AR, Cedzich A, Gretz N, Somlo S, Witzgall R. 2000. The polycystic kidney disease protein PKD2 interacts with Hax-1, a protein associated with the actin cytoskeleton. *Proc. Natl. Acad. Sci. USA* 97:4017–22

164. Li Q, Montalbetti N, Shen PY, Dai XQ, Cheeseman CI, et al. 2005. α-Actinin associates with polycystin-2 and regulates its channel activity. *Hum. Mol. Genet.* 14:1587–603

165. Li Q, Dai Y, Guo L, Liu Y, Hao C, et al. 2003. Polycystin-2 associates with tropomyosin-1, an actin microfilament component. *J. Mol. Biol.* 325:949–62

166. Li Q, Shen PY, Wu G, Chen XZ. 2003. Polycystin-2 interacts with troponin I, an angiogenesis inhibitor. *Biochemistry* 42:450–57

167. Jurczyk A, Gromley A, Redick S, San Agustin J, Witman G, et al. 2004. Pericentrin forms a complex with intraflagellar transport proteins and polycystin-2 and is required for primary cilia assembly. *J. Cell Biol.* 166:637–43

168. Hu J, Bae YK, Knobel KM, Barr MM. 2006. Casein kinase II and calcineurin modulate TRPP function and ciliary localization. *Mol. Biol. Cell* 17:2200–11

169. Liang G, Li Q, Tang Y, Kokame K, Kikuchi T, et al. 2008. Polycystin-2 is regulated by endoplasmic reticulum-associated degradation. *Hum. Mol. Genet.* 17:1109–19

170. Bai CX, Kim S, Li WP, Streets AJ, Ong AC, Tsiokas L. 2008. Activation of TRPP2 through mDia1-dependent voltage gating. *EMBO J.* 27:1345–56

171. Liang G, Yang J, Wang Z, Li Q, Tang Y, Chen XZ. 2008. Polycystin-2 down-regulates cell proliferation via promoting PERK-dependent phosphorylation of eIF2α. *Hum. Mol. Genet.* 17:3254–62

172. Wu Y, Dai XQ, Li Q, Chen CX, Mai W, et al. 2006. Kinesin-2 mediates physical and functional interactions between polycystin-2 and fibrocystin. *Hum. Mol. Genet.* 15:3280–92

173. Kim I, Li C, Liang D, Chen XZ, Coffy RJ, et al. 2008. Polycystin-2 expression is regulated by a PC2-binding domain of intracellular portion of fibrocystin. *J. Biol. Chem.* 283:31559–66

174. Lu W, Fan X, Basora N, Babakhanlou H, Law T, et al. 1999. Late onset of renal and hepatic cysts in Pkd1-targeted heterozygotes. *Nat. Genet.* 21:160–61

175. Kim K, Drummond I, Ibraghimov-Beskrovnaya O, Klinger K, Arnaout MA. 2000. Polycystin 1 is required for the structural integrity of blood vessels. *Proc. Natl. Acad. Sci. USA* 97:1731–36

176. Boulter C, Mulroy S, Webb S, Fleming S, Brindle K, Sandford R. 2001. Cardiovascular, skeletal, and renal defects in mice with a targeted disruption of the *Pkd1* gene. *Proc. Natl. Acad. Sci. USA* 98:12174–79

177. Muto S, Aiba A, Saito Y, Nakao K, Nakamura K, et al. 2002. Pioglitazone improves the phenotype and molecular defects of a targeted *Pkd1* mutant. *Hum. Mol. Genet.* 11:1731–42

178. Herron BJ, Lu W, Rao C, Liu S, Peters H, et al. 2002. Efficient generation and mapping of recessive developmental mutations using ENU mutagenesis. *Nat. Genet.* 30:185–89

179. Lantinga-van Leeuwen IS, Dauwerse JG, Baelde HJ, Leonhard WN, van de Wal A, et al. 2004. Lowering of *Pkd1* expression is sufficient to cause polycystic kidney disease. *Hum. Mol. Genet.* 13:3069–77

180. Piontek KB, Huso DL, Grinberg A, Liu L, Bedja D, et al. 2004. A functional floxed allele of *Pkd1* that can be conditionally inactivated in vivo. *J. Am. Soc. Nephrol.* 15:3035–43

181. Wu G, Markowitz GS, Li L, D'Agati VD, Factor SM, et al. 2000. Cardiac defects and renal failure in mice with targeted mutations in *Pkd2*. *Nat. Genet.* 24:75–78

Subsystem Organization of the Mammalian Sense of Smell

Steven D. Munger,[1] Trese Leinders-Zufall,[2] and Frank Zufall[2]

[1]Department of Anatomy and Neurobiology, University of Maryland School of Medicine, Baltimore, Maryland 21201; email: smung001@umaryland.edu

[2]Department of Physiology, University of Saarland School of Medicine, 66421 Homburg, Germany; email: trese.leinders@uks.eu, frank.zufall@uks.eu

Annu. Rev. Physiol. 2009. 71:115–40

First published online as a Review in Advance on September 22, 2008

The *Annual Review of Physiology* is online at physiol.annualreviews.org

This article's doi: 10.1146/annurev.physiol.70.113006.100608

Copyright © 2009 by Annual Reviews. All rights reserved

0066-4278/09/0315-0115$20.00

Key Words

odor, pheromone, vomeronasal, G protein–coupled receptor

Abstract

The mammalian olfactory system senses an almost unlimited number of chemical stimuli and initiates a process of neural recognition that influences nearly every aspect of life. This review examines the organizational principles underlying the recognition of olfactory stimuli. The olfactory system is composed of a number of distinct subsystems that can be distinguished by the location of their sensory neurons in the nasal cavity, the receptors they use to detect chemosensory stimuli, the signaling mechanisms they employ to transduce those stimuli, and their axonal projections to specific regions of the olfactory forebrain. An integrative approach that includes gene targeting methods, optical and electrophysiological recording, and behavioral analysis has helped to elucidate the functional significance of this subsystem organization for the sense of smell.

INTRODUCTION

MOS: main olfactory system

MOE: main olfactory epithelium

OSN: olfactory sensory neuron

MOB: main olfactory bulb

Semiochemical: a chemosensory stimulus that communicates information between animals

To many, the nose appears to be a unitary organ with a single sensory role: to detect odors. However, the role of olfaction is broad in humans and other mammals. The sense of smell helps to identify food, to assay its quality, and to enhance its flavor. The activation of nasal chemosensory cells warns of potential toxins or pathogens. The olfactory system even detects information about reproductive status, gender, and genetic identity. In all these cases, the activation of chemosensory cells in the nasal cavity initiates a process of neural recognition that can influence behaviors, hormonal state, and mood.

How does the olfactory system accomplish so many diverse tasks? It has become increasingly clear that the concept of a single olfactory system is grossly oversimplified, even wrong. The olfactory system is actually composed of a number of subsystems, some well known and others only recently characterized (**Figure 1**) (1, 2). These subsystems may be anatomically segregated within the nasal cavity, and they each make distinct neural connections to regions of the olfactory forebrain. They are clearly distinguished by the receptors they express and the signaling mechanisms they employ to detect and transduce chemosensory stimuli. And they respond, sometimes quite specifically, to a plethora of diverse molecules that range from volatile odors to peptides and proteins. In this review, we discuss a number of olfactory subsystems in the mammalian nose. In particular, we emphasize exciting recent results that elucidate the chemosensory selectivity and transduction machinery of these subsystems. Those readers interested in a more comprehensive discussion of the main or accessory olfactory systems are encouraged to consult any of a number of reviews (e.g., References 3–12).

MAIN OLFACTORY SYSTEM

The main olfactory system (MOS) (**Figure 2**) consists of

1. the main olfactory epithelium (MOE), which contains ciliated olfactory sensory neurons (OSNs), microvillar cells of uncertain function, sustentacular cells that may serve a glia-like role, and populations of progenitor/stem cells that replenish this regenerating tissue;
2. the main olfactory bulb (MOB), a region of the forebrain innervated by axons of the OSNs and serving as the first processing center of olfactory information; and
3. higher olfactory centers that receive direct or indirect information from the MOB.

It is well established that the MOS is a broadly tuned odor sensor; it responds to thousands of volatile chemicals carrying information about the quality of food and the presence of pathogens, prey, predators, or potential mates (6). There is a growing recognition that the MOS is also responsive to semiochemicals that may elicit specific behaviors or hormonal responses (12). This functional heterogeneity suggests that the MOS contains several olfactory subsystems that serve specific chemosensory roles.

Indeed, distinct subpopulations of OSNs can be defined by their expression of specific chemosensory receptors, enzymes, and ion channels (7). Although the anatomical divisions of these OSN subpopulations are not clearly

---

Figure 1

Chemosensory subsystems in the mouse nose. Each is composed of heterogeneous cell populations. Abbreviations used: ACIII, adenylyl cyclase type III; AOB, accessory olfactory bulb; AOS, accessory olfactory system; CAII, carbonic anhydrase type II; CNG, cyclic nucleotide–gated channel; GG, Grueneberg ganglion; IP$_3$R3, inositol 1,4,5-trisphosphate receptor 3; MOB, main olfactory bulb; MOS, main olfactory system; N.D., not determined; OR, odor receptor; OSN, olfactory sensory neuron; PDE, phosphodiesterase; PLC, phospholipase C; TAAR, trace amine–associated receptor; V1R, type 1 vomeronasal receptor; V2R, type 2 vomeronasal receptor; TRPC2/6, transient receptor potential channel types C2 and C6; TRPM5, transient receptor potential channel type M5; VSN, vomeronasal sensory neuron.

System	Principal target	Signal transduction components	Stimuli
MOS			
Canonical OSNs	Glomeruli in general MOB	ORs, ACIII, $G\alpha_{olf}$, PDE1C2, PDE4A, CNGA2, CNGA4, CNGB1b	Volatile and nonvolatile (?) odor ligands
TAAR-expressing neurons	N.D.	TAARs, $G\alpha_{olf}$	Volatile amines
GC-D[+] neurons	Necklace glomeruli (dark circles, above)	GC-D, PDE2, CNGA3, CAII	Uroguanylin, guanylin, cues in urine, CO_2
TRP-expressing cells	TRPM5: glomeruli in medial, ventral, and lateral MOB	TRPM5, $G\gamma 13$, CNGA2, PLCβ2	TRPM5: 2,5-dimethylpyrazine, 2-heptanone
	TRPC2: N.D.	TRPC2	TRPC2: N.D.
	TRPC6: restricted to MOE	TRPC6, IP$_3$R3, PLCβ2	TRPC6: lilial, volatile odor ligands
V1R-expressing cells	N.D.	V1Rs	N.D.
AOS			
V1R-expressing VSNs	Rostral AOB	V1Rs, $G\alpha_{i2}$, TRPC2, PDE4A	Volatile phermones
V2R-expressing VSNs: H2-Mv[−]	Anterior part of caudal AOB	V2Rs, $G\alpha_o$, TRPC2	Genetically encoded ligands (peptides, proteins)
H2-Mv[+]	Posterior part of caudal AOB	V2Rs, $G\alpha_o$, TRPC2, H2-Mv	
OR-expressing VSNs	Rostral AOB	ORs, $G\alpha_{i2}$, TRPC2	General odor ligands (?)
SO			
Canonical OSNs	Posterior part of the ventromedial MOB	ORs, ACIII, $G\alpha_{olf}$, CNGA2	Volatile odors
GC-D[+] neurons	Necklace glomeruli (?)	GC-D, PDE2	N.D.
GG			
	Dorsocaudal MOB, near AOB and necklace glomeruli	TAARs V2R83, $G\alpha_o$, $G\alpha_{i2}$, ORs, $G\alpha_{olf}$	Alarm signal
Trigeminal system			
Solitary chemosensory cells α-gustducin[+]	Trigeminal nerve (ethmoid nerve)	T2Rs, α-gustducin, TRPM5, PLCβ2 or PLCγ13	General odor ligands (lilial, citral, geraniol), CO_2

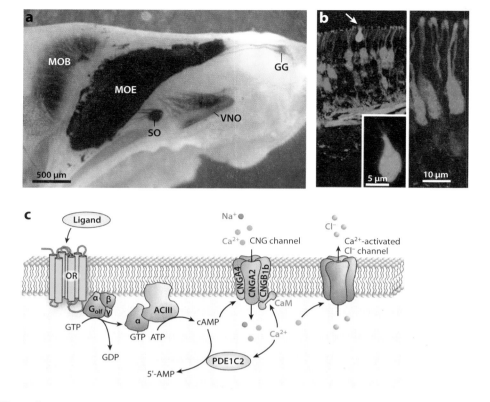

Figure 2

Organization of the sense of smell in mice. (*a*) Sagittal whole-mount view of the nasal cavity and forebrain of an *Omp-IRES-tau:LacZ* mouse (168) stained blue with X-Gal. Abbreviations used: MOB, main olfactory bulb; MOE, main olfactory epithelium; SO, septal organ of Masera; VNO, vomeronasal organ; GG, Grueneberg ganglion. Modified and reprinted from Reference 159 with permission from Wiley. (*b*) TRPM5 promoter–driven green fluorescent protein reveals microvillar (*arrowhead* and *inset*) and ciliated (*right panel*) TRPM5⁺ MOE cells. Modified and reprinted with permission from Reference 93. Copyright 2007, National Academy of Sciences. (*c*) Schematic representation of the cAMP second messenger cascade of canonical olfactory sensory neurons (OSNs). Abbreviations used: ACIII, adenylyl cyclase type III; CaM, calmodulin; CNGA2/A4/B1b, cyclic nucleotide–gated channel types A2, A4, and B1b; OR, odor receptor; PDE, phosphodiesterase. Odor activation of one of many hundred OR family members expressed in canonical OSNs initiates a cAMP-mediated signaling cascade that results in the depolarization of these cells.

delineated within the MOE, each subpopulation displays unique patterns of MOB innervation. For example, canonical OSNs express components of an adenosine 3′:5′-cyclic monophosphate (cAMP)-mediated signaling cascade to transduce odors (6), whereas MOE neurons expressing the guanylyl cyclase GC-D employ a guanosine 3′:5′-cyclic monophosphate (cGMP)-mediated sensory transduction mechanism (13–16). These two subpopulations of MOE sensory neurons make distinct con-

nections in the MOB (7), suggesting that these OSN subpopulations and their MOB targets are parts of specialized olfactory subsystems, each of which may process different subsets of chemosensory information.

Canonical Olfactory Sensory Neurons

A primary role of the MOS is to detect a wide array of odorants. Canonical OSNs are critical for this task. Each canonical OSN expresses one of hundreds of distinct OR-type odor receptors, utilizes a cAMP-mediated cascade to transduce odor stimuli, and sends odor information to the MOB. Thus, canonical OSNs serve as the general chemosensory cells of the MOS.

Cellular and molecular organization. Canonical OSNs are bipolar, ciliated neurons within the MOE. These cilia are the site of odor transduction and contain the receptors, enzymes, and ion channels required for this process. Each OSN sends a single, unbranched axon to the MOB, where the axon makes synaptic contact with interneurons and projection neurons within 1 of up to 2000 glomeruli (6).

The functional identity of these sensory cells is most clearly defined by their expression of OR-type odor receptors (17) (**Figure 2c**). ORs are members of the G protein–coupled receptor (GPCR) superfamily (8, 17). Although >1000 intact OR genes have been identified in mice (and ~350 in humans) (18–21), the vast majority have been defined as odor receptors on the basis of sequence homology alone. Thus, most ORs await functional annotation through deorphaning and localization to OSNs.

It appears that each canonical OSN expresses only a single OR gene (8). Those OSNs expressing a particular OR are restricted to only part of the MOE, although they are randomly distributed throughout that zone (8). Axons of OSNs expressing the same OR converge on a few glomeruli (usually two) within the MOB. This precise pattern of axonal targeting and convergence is influenced by many molecular factors, with the OR itself playing a critical, although as-yet-undefined, role in the process (8).

It was recognized more than two decades ago that most odors are transduced by a G protein–coupled, cAMP-mediated signaling mechanism (**Figure 2c**) that includes a stimulatory G protein, adenylyl cyclase, and a cAMP-gated ion channel (22–26). Three of the key molecules in the cascade were soon cloned: the G protein subunit $G\alpha_{olf}$ (27), a calcium-sensitive adenylyl cyclase (type III, or ACIII) (28), and the olfactory cyclic nucleotide–gated (CNG)

channel subunit CNGA2 (29). Two other subunits of the olfactory channel, CNGA4 (30, 31) and CNGB1b (32), were identified several years later. The protein(s) responsible for a Ca^{2+}-activated chloride conductance important for amplification of the odor response (33–35) has not been definitively identified, although intriguing candidates have been proposed (e.g., Reference 36). Proteins that may modulate the transduction cascade have also been identified: The guanine-exchange factor Ric-8B contributes to amplification of the odor signal through its actions on G proteins (37, 38), whereas two cyclic nucleotide phosphodiesterases (PDEs), PDE1C2 (39) and PDE4A (40), most likely participate in the adaptation or the termination of odor responses. Canonical OSNs also express olfactory marker protein (OMP) (41), which is critical for modulating the amplitude and kinetics of the odor response (42).

Chemoreceptive properties. Most OSNs recognize multiple odorants (6). ORs themselves are broadly tuned (8, 43, 44), and multiple ORs can respond to the same odorant, although usually with different efficacies (8). Thus, odors are encoded through use of a combinatorial strategy (5, 45). OSNs expressing the same OR converge on different sets of MOB glomeruli such that the activation of a group of glomeruli represents the presence of an individual odorant (46). The convergent innervation of individual glomeruli by OSNs expressing the same OR argues that the glomerulus, not the receptor, is the fundamental unit of odor coding.

All the ORs deorphaned to date respond to volatile odorants of diverse chemical classes (e.g., References 8, 43, 44, 47, and 48). However, not all odorants activate ORs; some odorants are antagonists (47, 49, 50). For example, undecanal inhibits the ability of bourgeonal to activate the human OR17-4, a receptor implicated in both olfactory function and sperm chemotaxis (49, 51), whereas an oxidatively dimerized isoeugenol derivative, but not isoeugenol itself, inhibits eugenol-dependent activation of mOR-EG (e.g., References 47 and

G protein–coupled receptors (GPCRs): characterized by seven transmembrane domains and the ability to couple ligand binding to G protein–mediated intracellular signaling. Most mammalian chemosensory receptors are GPCRs

Odorant: a compound that functions as an odor. Odors can be single odorants or mixtures of odorants

52). The potential for the presence of both agonist and antagonist odorants in complex odor mixtures provides an additional layer of complexity to the coding of odor stimuli.

It appears that interindividual differences in odor sensitivity and perception, including specific anosmias, are due largely to differences in the complement of OR genes. For example, human variation in sensitivity to, and perception of, the steroid odor androstenone (53) is associated with allelic variations of the OR7D4 receptor that affect receptor efficacy (54). The presence of large numbers of segregating pseudogenes in the olfactory genome (55) may also contribute to perceptual differences between individuals, such as differential sensitivity to isovaleric acid (56). The prevalence of interindividual differences in odor sensitivity and perception indicates that defining "normal" olfactory function may be problematic.

Signaling mechanisms. The molecular mechanisms of olfactory signaling are understood in some detail (6–8). The transduction process is initiated upon odor activation of one of many hundred OR family members. Members of this class of chemosensory receptor confer selective odor responsiveness on either OSNs (e.g., References 44 and 57) or heterologous cells (e.g., Reference 43). For example, OSNs expressing the mouse OR M71 respond to acetophenone and benzaldehyde. However, if the M71 receptor gene is replaced with a receptor gene encoding a well-characterized rat OR, I7, the odor selectivity of this same cell population is changed: These OSNs now respond to the I7 ligand octanal, but not to acetophenone or benzaldehyde (57). Thus, the OR dictates the stimulus tuning of the OSN.

Although ligands have been identified for several ORs, little is known about the key molecular determinants for ligand binding and selectivity, ligand-induced conformational changes, or G protein coupling (but see References 43, 44, 48, 50, 57, and 58). Thus, our understanding of the members of the largest GPCR family lags far behind that of many other receptors more amenable to structure-function analyses, such as rhodopsin and the β-adrenergic receptor (59). The biggest hurdle to more systematic structure-function studies is the ability to isolate large quantities of purified ORs. Unfortunately, the small number of OSNs expressing each OR proteins in the MOE makes biochemical purification of native ORs unrealistic. ORs express poorly in heterologous cells, even in the presence of chaperones or fusion tags geared to facilitate membrane targeting (43, 60), making in vitro strategies problematic as well. Overcoming the technical hurdles of OR expression and purification is critical if we are to understand how ORs recognize odors.

The preeminent role of a cAMP-mediated signaling cascade for the transduction of odors by canonical OSNs was confirmed by the characterization of gene-targeted mice in which each of the principal signaling molecules was deleted. $G\alpha_{olf}$ null mice exhibit a pronounced reduction in MOE responses, as assayed by electroolfactogram (EOG), to a variety of volatile odors (61). However, these responses are not completely abolished, possibly as a result of the expression of the partially redundant $G\alpha_s$ in the MOE (62). Similarly, the role of ACIII in odor transduction was confirmed by characterizing ACIII null mice (63): EOG responses to a number of odorants are completely absent in ACIII null mice, which also display deficits in olfaction-dependent learning.

The contribution of individual CNG channel subunits to the odor response is more complex. CNGA2 is necessary for the formation of a functional olfactory CNG channel in vitro, whereas the CNGA4 and CNGB1b subunits increase channel sensitivity to cyclic nucleotides (64). Indeed, deletion of *Cnga2* by gene targeting in mice (65) abolishes EOG responses to most odors tested, although responses to the volatile semiochemicals 2-heptanone and 2,5-dimethylpyrazine (66) and to the natriuretic peptides uroguanylin and guanylin (16) are maintained (see below). Both *Cnga4* and *Cngb1* null mice show reduced channel and OSN sensitivity to cyclic nucleotides (67, 68). *Cnga4* null mice also exhibit reduced behavioral sensitivity to odors (69) (behavioral tests of *Cngb1* null mice have not been reported).

The olfactory CNG channel is the principal site of Ca^{2+}/calmodulin (Ca^{2+}/CaM)-mediated odor adaptation in OSNs (64, 67, 70). Although CNGA2 homomeric channels are desensitized in vitro in the presence of Ca^{2+}/CaM, the specific contribution of the CNGA2 subunit to the mechanisms of odor adaptation is unresolved (64). *Cnga4* null mice show slower Ca^{2+}/CaM-mediated channel desensitization and defects in cellular odor adaptation to repeated or prolonged stimulation of OSNs (67). These mice are also profoundly impaired in their ability to discriminate olfactory stimuli in the presence of a background odor (69). Odor adaptation is also perturbed in the MOE of *Cngb1* null mice (68), although mutation of the Ca^{2+}/CaM binding site of this subunit affects adaptation to prolonged, but not repeated, odor stimulation (71). Together, these results indicate that all three subunits of the olfactory CNG channel play important, but distinct, roles in the process of odor adaptation.

Olfactory Sensory Neurons Expressing Trace Amine–Associated Receptors

Not all OSNs express OR-type odor receptors. Recently, a small family of trace amine–associated receptors (TAARs) has been identified in a subpopulation of OSNs that do not express ORs. These receptors, and the OSNs that express them, may be responsive to semiochemicals such as pheromones.

Cellular and molecular organization. Activation of the MOE by atypical olfactory stimuli such as volatile pheromones (66) and major histocompatibility complex (MHC)-related peptides (72) suggested that some OSNs might express members of other GPCR families. A systematic screening of OSN-enriched cDNA led to the amplification of several members of the TAAR family, a group of GPCRs more closely related to serotonin and dopamine receptors than to ORs (73, 74). Eight of nine mouse *Taar* genes are expressed in nonoverlapping subsets of OSNs in the MOE. TAAR-expressing OSNs do not appear to express ORs, although they do express $G\alpha_{olf}$ (74). Therefore, although TAAR-expressing OSNs may utilize a cAMP-mediated signaling cascade to transduce odors, they are distinct from the canonical OR-expressing OSNs.

Like canonical OSNs, $TAAR^+$ OSNs expressing the same receptor are zonally restricted within the MOE, although randomly distributed within those zones. Individual TAARs are expressed sparsely in the MOE ($\sim$1/1000 OSNs), again similar to the canonical OSNs (74). However, although TAAR-expressing OSNs share many similarities with canonical OSNs, it remains unknown if they observe the same properties of glomerular convergence, or if they preferentially target regions of the MOB. Therefore, it is unclear if there is combinatorial coding of TAAR ligands in the MOS.

Chemoreceptive properties. Although recordings from TAAR-expressing OSNs have not been reported, heterologous expression of olfactory TAARs has identified a number of ligands (74). As predicted (73), mouse TAARs (mTAARs) respond to biogenic amines, but not to related alcohols or amino acids (74). Of note are β-phenylethylamine (an mTAAR4 ligand), isoamylamine (an mTAAR3 ligand), and trimethylamine (an mTAAR5 ligand). Consistent with a role in detecting species-specific, urine-derived pheromones and other social cues, mTAAR5 responds to highly diluted (1/30,000) adult male mouse urine, but not to female urine, prepubescent male urine, or human male urine (74). Urine from either BALB/c or C57BL/6 mice, which vary in MHC haplotype, was equally effective, indicating that the mTAAR5 ligand is not related to MHC-linked genetic identity. Together, these results suggest that at least some TAARs may act as pheromone receptors. However, the conservation of TAARs in fish, chicken, mouse, and human (74–76) suggests that they serve a more general role as well.

Pheromones: a subclass of semiochemicals used for intraspecies communication and that elicit a behavioral or hormonal change in the recipient animal

SO: septal organ of
Masera

Signaling mechanisms. Little is known about signaling in TAAR-expressing OSNs. TAARs are coexpressed with $G\alpha_{olf}$, suggesting that they use the same transduction cascade as do canonical OSNs (74). Although TAARs can couple to cAMP signaling pathways in vitro (73, 74), this coupling has not been demonstrated in MOE neurons.

GC-D-Expressing Chemosensory Neurons and the Necklace Glomeruli

OSN subpopulations can also be differentiated on the basis of transduction mechanisms. $GC\text{-}D^+$ neurons, which express the receptor guanylyl cyclase GC-D, utilize a cGMP-mediated cascade to transduce chemosensory stimuli, including two natriuretic peptide hormones, uroguanylin and guanylin.

Cellular and molecular organization. The mammalian receptor guanylyl cyclases are a small family of peptide and orphan receptors (77). They contain three functional domains: an extracellular receptor domain, an intracellular regulatory domain, and an intracellular catalytic domain that generates the second messenger cGMP. One family member, GC-D, was identified by homology cloning from rat MOE (13). $GC\text{-}D^+$ neurons compose less than ~0.1% of MOE neurons, exhibit a typical OSN bipolar morphology, and can be found singly or in clusters within the MOE (13, 14, 78). The highest density of $GC\text{-}D^+$ neurons is in the dorsal recesses of the ectoturbinates, although they are also found on the endoturbinates and septum and in the septal organ of Masera (SO) (13, 78, 79).

$GC\text{-}D^+$ neurons do not express many of the transduction proteins found in canonical OSNs and implicated in cAMP-mediated olfactory transduction, including ACIII, $G\alpha_{olf}$, CNGA2, PDE1C2, and PDE4A (14, 15) (**Figure 3c**). However, these cells do express the cGMP-sensitive channel subunit CNGA3 and the cGMP-stimulated PDE2 (14, 15), suggesting that $GC\text{-}D^+$ neurons utilize a cGMP-mediated signaling cascade to transduce chemosensory

stimuli. PDE2 is localized throughout the $GC\text{-}D^+$ neurons, which permits the glomerular targets of $GC\text{-}D^+$ neurons to be easily identified (14). PDE2 immunolabeling of a small number of cholinesterase-positive glomeruli that ring the caudal MOB (14) indicated that $GC\text{-}D^+$ neurons specifically innervate a subset of atypical glomeruli known as the necklace glomeruli (80). Genetic labeling of GC-D neurons through gene targeting has supported this interpretation (16, 78, 81) (**Figure 3a**).

Chemoreceptive properties. The chemosensory stimuli to which $GC\text{-}D^+$ neurons respond were only recently determined. Two natriuretic peptides, uroguanylin and guanylin (82), elicited responses in the MOE of *Cnga2* null mice (16) (**Figure 3b**), indicating that the response to these stimuli was, as expected (14, 15, 83), independent of a canonical cAMP-mediated transduction cascade. These two peptides are highly effective stimuli: In EOG recordings from wild-type mice, $K_{1/2}$ values were as low as 66 pM. $GC\text{-}D^+$ neurons respond to stimulation with uroguanylin, guanylin, or dilute urine with an increase in action potential frequency and a rise in intracellular Ca^{2+} (16). However, two volatile semiochemicals found in urine that elicit c-fos activation in PDE2-positive glomeruli, 2-heptanone and 2,5-dimethylpyrazine (66), fail to activate $GC\text{-}D^+$ neurons themselves (16). $GC\text{-}D^+$ neurons are heterogeneous in their stimulus tuning: Although approximately one-half of $GC\text{-}D^+$ neurons respond to both peptides, the remainder respond to either uroguanylin or guanylin alone (16). Thus, $GC\text{-}D^+$ neurons function as receptors for uroguanylin, guanylin, and components of urine.

A contemporaneous study reached a very different conclusion: that GC-D neurons are sensitive CO_2 sensors (81). $GC\text{-}D^+$ neurons express the CO_2-catalyzing enzyme carbonic anhydrase type II (CAII) (81). Stimulation of labeled $GC\text{-}D^+$ neurons in a gene-targeted mouse expressing enhanced green fluorescent protein (EGFP) under the control of the *Gucy2d*

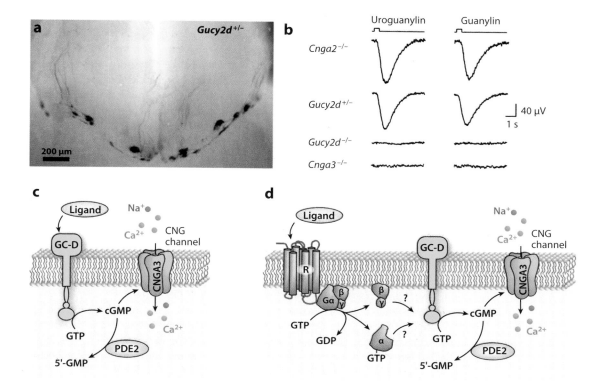

Figure 3

GC-D-expressing neurons and the necklace glomeruli. (*a*) Whole-mount X-Gal staining of paired olfactory bulbs from *Gucy2d-Mapt-lacZ* [+/−] mice showing stained necklace glomeruli (*ventral view*). (*b*) Stimulus-evoked field-potential responses to guanylin family peptides in the main olfactory epithelium (MOE) of *Cnga2*, *Gucy2d*, or *Cnga3* null mice. Panels *a* and *b* are reprinted with permission from Reference 16. Copyright 2007, National Academy of Sciences. (*c,d*) Possible signal transduction mechanisms in GC-D⁺ neurons. (*c*) GC-D itself may serve as the chemosensory receptor for these cells, or (*d*) GC-D activity may be modulated by an unknown GPCR (R). CNGA3 channel, cyclic nucleotide–gated channel type A3; Gα, -β, -γ, unknown G protein α, β, and γ subunits; PDE2, phosphodiesterase type 2.

promoter with CO_2 elicits a rise in intracellular Ca^{2+} that is blocked by a CA inhibitor. Olfactory bulb neurons associated with the necklace glomeruli are activated by CO_2 stimulation of the MOE (81). *Car2* null mice, which lack CAII, show reduced behavioral responses to CO_2 as compared with wild-type mice.

Further experiments are needed to resolve these two distinct functions. Some insights may be had from the observation that, although the *Gucy2d* ortholog is a pseudogene in most primates (including humans), it is intact in some prosimians and in dog, mouse, rat, and tree shrew (84). Thus, a common chemoreceptive role of GC-D⁺ neurons may be retained in these mammals.

Signaling mechanisms. GC-D⁺ neurons utilize a cGMP-mediated transduction cascade. They are unresponsive to the adenylyl cyclase activator forskolin but are stimulated by the membrane-permeant 8-bromo-cGMP or by PDE inhibitors (which elevate intracellular cyclic nucleotides) (16). Also, the CNG channel inhibitor l-*cis*-diltiazem attenuates uroguanylin and guanylin responses in the MOE (16). In contrast to the mammalian phototransduction mechanism, which responds to sensory stimulation with a decrease in intracellular cGMP and hyperpolarization of photoreceptor cells (85), GC-D$^+$ neurons respond to sensory stimulation with GC-D-dependent increases in intracellular cGMP and Ca^{2+}, membrane depolarization, and an increase in action potential firing (16). Thus, GC-D$^+$ neurons utilize an excitatory, cGMP-mediated signaling cascade to transduce peptide and urine stimuli. The mechanism by which CAII-dependent hydrolysis of CO$_2$ leads to action potential firing in GC-D$^+$ neurons remains unknown.

The nature of the peptide receptor in GC-D$^+$ neurons is also unclear. GC-D itself is an excellent candidate: Other members of the receptor guanylyl cyclase family respond to natriuretic peptides (82), and deletion of *Gucy2d* in mice abolishes responses to uroguanylin, guanylin, and urine (16). However, the three distinct tuning profiles of GC-D$^+$ neurons suggest that subpopulations of GC-D$^+$ neurons express distinct receptors, receptor complexes, or common receptors with distinct modifications or variants. Consistent with this view, rat GC-D responds to uroguanylin, but not to guanylin, in heterologous cells (86).

TRP Channel–Expressing Olfactory Sensory Neurons

The expression of several transient receptor potential (TRP) channel family members in subpopulations of cells in the MOE indicates further functional diversity in the MOS. However, it remains unclear whether olfactory TRP channels function as chemosensory receptors or transduction effectors, or whether they play some other role.

Cellular and molecular organization. Members of the TRP channel superfamily are structurally diverse and are found in numerous tissues (87). Three TRP channel isoforms, TRPC2, TRPC6, and TRPM5, are expressed in distinct subsets of MOE cells. TRPC2 is found in a small number of cells (<1%) in the adult and embryonic rat MOE; in adult MOE these cells appear restricted to the basal lay-

ers, suggesting that they are immature neurons (88). However, it is unclear if these cells innervate the central nervous system (CNS) or even if they are chemosensory neurons.

Expression of TRPC6 is restricted to a population of bipolar microvillar cells in the apical MOE that extend a process to, but not through, the basal lamina (89). The restriction of these cells to the MOE suggests that they may play a local role in chemosensory processing. Expression of neuropeptide Y in some TRPC6$^+$ cells suggests a role in development and/or regeneration (90, 91). TRPC6$^+$ cells express two components of a phosphoinositide (PI) signaling cascade, the inositol 1,4,5-trisphosphate receptor 3 (IP$_3$R3) and the effector enzyme phospholipase C β2 (PLCβ2), but do not express ACIII, CNGA2, or OMP, three key markers of canonical OSNs (89).

TRPM5 is expressed in two morphologically distinct cell types within the MOS: a population of solitary microvillar cells innervated by trigeminal nerve fibers (92; see below) and a large group of ciliated MOE neurons (93) (**Figure 2***b*). The ciliated TRPM5$^+$ OSNs are enriched in the ventrolateral MOE and project axons to the ventral, lateral, and medial MOB (93). Surprisingly, these neurons express components of both PI- and cAMP-mediated signaling cascades, including PLCβ2, the G protein subunit Gγ13, and the CNGA2 channel subunit (93). These neurons also express OMP (93).

Chemoreceptive properties. Some odorants, such as lilial and lyral, stimulate the production of IP$_3$ in MOE membrane preparations (e.g., Reference 94). However, the dependence of canonical OSN odor responses on an intact cAMP-mediated signaling cascade brings into question the relevance of IP$_3$-mediated odor transduction in the MOE (4, 61, 63, 65). TRPC6$^+$ cells respond to mixtures of IP$_3$-mediated odors and to the single odorant lilial (10 μM) with a rise in intracellular Ca^{2+} (89). If this Ca^{2+} increase is dependent on IP$_3$, such responses may at least partially account for the earlier biochemical results showing odor-dependent IP_3 increases in whole-MOE membrane preparations.

$TRPM5^+$ OSNs may respond to pheromones or other semiochemicals. For example, mouse urine, the putative pheromone 2,5-dimethylpyrazine, and the social cue (methylthio)methanethiol (MTMT) each elicit increased c-fos expression in periglomerular cells associated with MOB glomeruli receiving $TRPM5^+$ neuron innervation (93). The degree to which $TRPM5^+$ OSNs vary in their stimulus selectivity is unknown and awaits functional characterization of individual $TRPM5^+$ neurons.

Signaling mechanisms. The different subfamilies of TRP channels, including TRPC and TRPM isoforms, display distinct physiological properties and modes of activation (87). Both TRPC2 and TRPC6 are diacylglycerol-sensitive cation channels (87). Odor transduction in $TRPC6^+$ microvillar cells may be mediated by PI signaling; odor responses in these cells, which also express $PLC\beta2$ and IP_3R3, are at least partially independent of extracellular Ca^{2+} (89).

Odor transduction in $TRPM5^+$ OSNs may be more complex. They express components of both PI- and cAMP-mediated signaling cascades, but the role each cascade plays in cellular responses to odors is unclear. For example, EOG responses to the semiochemical 2,5-dimethylpyrazine, but not to the environmental odor lilial, are reduced in *Trpm5* null mice in the presence, but not in the absence, of an adenylyl cyclase inhibitor (93). Thus, both pathways may be required for the transduction of certain stimuli by $TRPM5^+$ OSNs.

ACCESSORY OLFACTORY SYSTEM

In addition to the MOS, many mammals possess an accessory olfactory system (AOS) (10, 95–98) (**Figure 4**). The AOS consists of

1. the vomeronasal organ (VNO), a chemoreceptive structure situated at the base of the nasal septum, which houses the microvillar vomeronasal sensory neurons (VSNs);
2. the accessory olfactory bulb (AOB), a region of the forebrain that receives synaptic input from the VNO and serves as the first processing center of vomeronasal information; and
3. higher olfactory centers that receive direct or indirect information from the AOB.

The AOS has attracted a great deal of attention over the past ten years because of a growing recognition of this system's essential role in chemical communication and the regulation of social behaviors (10, 95, 96, 99, 100). However, the traditional distinction that the MOS detects only volatile, environmental odorants and the AOS detects only nonvolatile pheromones is no longer valid. Both systems have considerable overlap in terms of the stimuli they detect and the effects that they mediate. Furthermore, both systems are capable of recognizing a wide variety of chemical signals and structures (10, 12).

CRANIAL NERVE ZERO: THE TERMINAL NERVE

The terminal nerve, or nervus terminalis, is present in all vertebrates whether or not they have a vomeronasal organ. Although the terminal nerve was first suggested to function as a chemosensory system in its own right, increasing evidence now points to a centrifugal, modulatory role (166). As the most anterior vertebrate cranial nerve, the terminal nerve extends between hypothalamic nuclei and the nasal cavity, reaching deep into the lamina propria surrounding the olfactory epithelium. The nerve consists of fibers that are highly heterogeneous in terms of neurochemically distinct phenotypes, including the expression of neuropeptide Y, gonadotropin-releasing hormone (GnRH), tyrosine hydroxylase, nitric oxide synthase, and several other molecules (167). With respect to neuropeptide Y, which plays an important role in the control of appetite and feeding and is released by stress, terminal nerve–derived peptide appears to modulate olfactory sensory neuron activity in a context-dependent manner, at least in lower vertebrates (166).

AOS: accessory olfactory system

VNO: vomeronasal organ

VSN: vomeronasal sensory neuron

AOB: accessory olfactory bulb

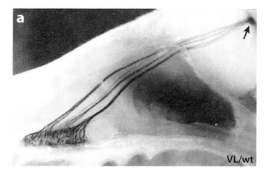

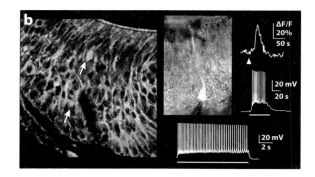

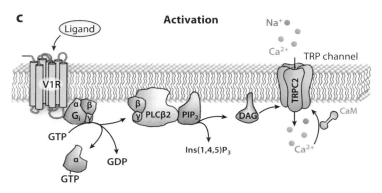

Figure 4

The accessory olfactory system. (*a*) Whole-mount X-Gal staining of a heterozygous *V1r2-IRES-tau-lacZ* (VL/wt) mouse reveals the distribution and axonal projection pattern of a subpopulation of vomeronasal sensory neurons (VSNs) expressing a single V1R. The AOB is indicated by the arrow. Reprinted from Reference 107, with permission from Elsevier. (*b*) Imaging and recording from identified VSNs. (*Left panel*) Confocal Ca^{2+} imaging in vomeronasal organ (VNO) slices showing VSNs responding to MHC class 1 peptides (*cyan, arrows*) are superimposed onto a protein expression map indicating the two epithelial layers (*red*, V1R$^+$ VSNs; *green*, V2R$^+$ VSNs). (*Right panel, counterclockwise from upper left*) Patch-clamp recording from a gene-targeted basal VSN expressing the V2R1b receptor (*white cell*). Current injection produces sustained action potential discharges that show spike-frequency adaptation (*bottom trace*). Ligand-induced responses in single VSNs, obtained by whole-cell current clamp recording (*middle trace*) or Ca^{2+} imaging (*upper trace*). Reprinted and modified from References 112, 118, and 120, with permission from AAAS, Elsevier, and the American Physiological Society, respectively. (*c*) Signal transduction mechanisms in V1R$^+$ VSNs. Activation of a V1r receptor initiates a G protein–coupled phospholipase C signaling cascade that results in Ca^{2+} entry and depolarization of the cells. Abbreviations used: CaM, calmodulin; DAG, diacylglycerol; Ins(1,4,5)P$_3$, inositol 1,4,5-trisphosphate; PIP$_2$, phosphatidylinositol-4,5-bisphosphate; PLCβ2, phospholipase C type β2; TRPC2, transient receptor potential channel canonical type 2.

The detection of molecular cues by the mouse VNO is mediated by independent subsystems that originate from VSN subpopulations residing in nonoverlapping apical and basal zones of the VNO neuroepithelium. VSNs of these two subdivisions are molecularly and functionally distinct. VSNs of the apical layer express members of the *V1r* family of vomeronasal receptor genes, whereas VSNs of the basal layer express members of the *V2r* family (8, 95). This spatial segregation correlates with the differential expression of two G protein subunits, Gα$_{i2}$ and Gα$_o$ (101, 102) (**Figure 4b**), and is maintained at the level of the AOB: VSN axons from the apical layer synapse in the anterior half of the AOB, whereas VSN axons from the basal layer synapse in the posterior half of the AOB. This segregation is at least partly maintained in higher levels of the CNS: Anterior and posterior divisions of the AOB each project to specific areas of the amygdala (103).

V1R-Expressing Vomeronasal Sensory Neurons

V1R$^+$ VSNs are narrowly tuned sensory neurons that detect a range of small natural ligands present in the urine of conspecifics. Each V1R$^+$ VSN expresses one V1R-type receptor, utilizes a phospholipase C (PLC)-mediated signaling cascade to transduce molecular cues, and sends information to the anterior aspect of the AOB. Thus, V1R$^+$ VSNs play a crucial role in chemical communication between members of the same species.

Cellular and molecular organization. The spatial segregation of the VNO into molecularly defined subsystems has important consequences for the sensing of structurally and functionally distinct sets of chemical stimuli and ultimately for the regulation of distinct sets of behavioral repertoires. V1R$^+$ VSNs are characterized by the expression of a given member of the V1R family, which consists of class A (rhodopsin-like) GPCRs (8, 95, 104). Database mining identified 308 *V1r* genes in the mouse genome, of which 191 appear to be intact (105). *V1r* genes are classified into 12 families, each containing between 1 and 30 members (105, 106). Transcription of *V1r* genes occurs in a monogenic and monoallelic manner (107). Humans have five intact *V1r* genes, at least some of which are expressed in the MOE (106). It is not yet clear whether V1R$^+$ MOE cells are displaced VSNs or whether they represent a unique cell type. Of all vertebrates surveyed thus far, the semiaquatic platypus has the largest *V1r* repertoire, with 270 intact genes and 579 pseudogenes (108). Several groups have used gene-targeted mouse lines, in which V1R$^+$ VSNs coexpress cellular markers, to determine the pattern of axonal projections to the AOB (107, 109). The results are complex and reveal a fundamentally different wiring logic as compared with canonical OSN-MOB projections. Individual mitral cells in the anterior AOB seem to receive information from multiple glomeruli associated with distinct, but possibly closely related, V1R$^+$ VSN populations (11, 110). This would indicate a considerable degree of integration of information at the level of the AOB.

Chemoreceptive properties. High-resolution fluorescence imaging techniques have been developed to investigate the activity of potential pheromone ligands in large neuronal populations of the VNO neuroepithelium (111). An acute coronal tissue slice preparation of the mouse VNO enables superimposition of neuronal activation maps onto protein or gene expression maps to identify the molecular identity of responsive VSNs (111, 112). Systematic analysis of the detection capabilities of individual VSNs established that V1R$^+$ VSNs function as highly sensitive pheromone detectors that recognize small organic pheromones present in the urine of conspecifics, such as the testosterone-dependent volatiles (*R,R*)-3,4-dehydro-*exo*-brevicomin (DHB) and (*S*)-2-*sec*-butyl-4,5-dihydrothiazole (SBT) (111). Neuronal responses showed highly selective tuning properties, and their specificity did not broaden as the stimulus concentration was increased (111). These functional properties predict that ablation of *V1r* genes would cause discrete deficits in the ability of the VNO to detect specific molecules. Indeed, this was observed following deletion of a cluster of 16 *V1r* genes (113). These *V1r*-deficient mice failed to show VSN responses to specific pheromonal cues, including 6-hydroxy-6-methyl-3-heptanone, *n*-pentylacetate, and isobutylamine (113). These mice also displayed alterations in social behaviors such as maternal aggression, thus establishing a role of V1Rs as pheromone receptors (113). In an alternative approach, green fluorescent protein–tagged VSNs that express the *V1rb2* gene responded to 2-heptanone, a response that was absent when the *V1rb2* gene was deleted (114). This indicates that V1Rb2 is a receptor for 2-heptanone, a urinary constituent that has a primer pheromonal effect extending the length of female estrous cycles (114).

Other imaging approaches used to analyze population responses in the VNO have produced contrasting findings with respect to the spatial distribution of responses to dilute urine in apical and basal VSNs (115, 116). One study concluded that urine responses are essentially confined to V1R$^+$ VSNs (116), whereas others showed evidence that subsets of V2R$^+$ VSNs respond to dilute urine or molecular cues present in urine (112, 115, 117). A complete understanding of chemoreception in V1R$^+$ VSNs will require the identification of ligand-receptor pairs for the entire V1R family.

Signaling mechanisms. Stimulation of V1R$^+$ VSNs elicits action potentials and elevates intracellular Ca^{2+} (111, 114). In patch-clamp recordings from identified V1Rb2$^+$ VSNs, depolarizing currents of only a few picoamperes are sufficient to produce repetitive firing (118). Low-threshold, regenerative Ca^{2+} spikes are responsible, in part, for driving action potential firing (118). There is good evidence that a PLC-mediated signaling cascade underlies primary signal transduction in these neurons (119) (**Figure 4c,d**), but genetic proof is required to firmly establish a critical role of Gα$_{i2}$ and PLC subtypes for the signal transduction mechanism. A key target of PLC activity is a 42-pS diacylglycerol-sensitive cation channel present in VSN dendrites (120). Activation of the diacylglycerol-sensitive channel is strongly impaired in mice exhibiting a targeted deletion of the transient receptor potential channel TRPC2 (120). *Trpc2*$^{-/-}$ mice also reveal a striking reduction in the electrical response to pheromonal ligands that activate V1R$^+$ cells (120, 121). Together, these results establish a direct link between PLC activity, gating of a TRPC2-dependent cation channel by diacylglycerol, and the sensory response in V1R$^+$ VSNs. Despite these advances, very little is known about the molecular architecture and subunit composition of the TRPC2 channel. Like the olfactory CNG channel, the TRPC2 channel is subject to

strong modulation by Ca^{2+}/CaM feedback, offering a powerful mechanism for pheromone adaptation in these cells (122). Spike-frequency adaptation of action potential bursts provides a second mechanism for regulating the temporal response properties of V1R$^+$ VSNs (118). Thus, multiple mechanisms exist in VSNs to mediate pheromone adaptation, in contrast to the previous belief that VSNs lack any form of sensory adaptation (123).

V2R-Expressing Vomeronasal Sensory Neurons

V2R$^+$ VSNs represent a second major class of sensory neurons in the VNO. Only very recently has it become possible to obtain functional information from these neurons. These investigations indicate that V2R$^+$ VSNs detect several families of peptide and protein pheromones that are critical for chemical communication and the regulation of social behaviors.

Cellular and molecular organization. *V2r* genes are class C GPCRs, characterized by a long extracellular N terminus (8, 3, 95). Nearly 300 *V2r* genes have been identified in the mouse genome, of which 61–120 are putatively functional (124, 125). Presently, all V2Rs are orphan receptors. They are grouped, according to sequence homology, into four families: A, B, C, and D (125, 126). V2R$^+$ VSNs show combinatorial coexpression of different V2Rs (126). Hence, these neurons seem to be an exception to the one neuron–one receptor rule for chemosensory cells.

V2R$^+$ VSNs express another multigene family, termed *H2-Mv*, containing nonclassical MHC class 1 genes (127, 128). Initially, it was thought that *H2-Mv* genes might function as subunits in a native receptor complex with V2Rs and might be required as escort molecules in the transport of V2Rs to the cell surface (128–130). However, a substantial fraction of mouse V2R$^+$ VSNs do not express any of the nine *H2-Mv* genes (131). These results reveal a novel compartmentalization of the basal layer of the

VNO neuroepithelium, with at least two distinct neuronal subpopulations: one expressing *H2-Mv* genes and the other not. V2R+/H2-Mv− cells are localized in the upper sublayer of the basal layer, i.e., the middle layer of the VNO neuroepithelium, a subdomain organization that is maintained at the level of the AOB (131). Whether this organizational feature underlies specific aspects of chemosensory processing is yet to be determined.

Chemoreceptive properties. V2R+ VSNs respond to several families of nonvolatile peptide and protein pheromones that require direct physical contact between the nose and the stimulus source for effective transmission (112, 117, 132). Fluorescence imaging of identified V2R+ VSNs in intact VNO tissue slices revealed a vast family of antigenic peptides—the MHC class 1 peptides—as sensory stimuli for these cells (112, 115) (**Figure 4*b***). It is not yet known whether VSN detection of MHC peptide ligands correlates with the expression of *H2-Mv* genes. Such MHC peptides, which are crucial in the context of immune surveillance, carry information about the genetic makeup of an individual (133). Hence, the sensing of MHC peptides can potentially serve as a self-referent genetic recognition mechanism whereby individuals compare their own MHC type with those of conspecifics (134). Indeed, behavioral studies in mice have shown that VSN detection of MHC peptides mediates the formation of a persistent memory that is required for mate recognition in the context of selective, odor-induced pregnancy termination (the Bruce effect) (112). Interestingly, MHC peptide ligands are also detected by cAMP-sensitive OSNs of the MOE (72). These results illustrate that the same molecular cues can be processed by distinct olfactory pathways (72) and likely with distinct functional consequences.

Additional stimuli for some V2R+ VSNs have been identified. A male-specific, 7-kDa peptide called ESP1, which is secreted from the extraorbital lacrimal gland, functions as a sensory cue for these neurons (132). ESP1 is encoded by a gene from a multigene family consisting of 38 members in mice (135). Field-potential recordings show that ESP1 elicits an electrical response in the VNO, whereas c-Fos activity measurements indicate that this response occurs in V2R+ VSNs (132, 135, 136). The exact role of ESP family peptides in mouse communication is still unclear, but the observation that ESP expression patterns differ between strains suggests that such expression patterns may transmit strain-specific information (135).

A third group of nonvolatile chemosensory stimuli detected by VSNs of the basal layer consists of the major urinary proteins (MUPs) (117). MUPs represent another polygenic and highly polymorphic set of proteins thought to be involved in multiple aspects of chemosensory communication, including identity recognition (137) and the induction of ovulation (138). New work (117) indicates that MUPs also act as male-male aggression pheromones that specifically stimulate $G\alpha_o^+$ VSNs.

Signaling mechanisms. Few data are available on the signaling properties of identified V2R+ VSNs, although it is clear that these VSNs respond to sensory stimulation with action potential generation (112, 135) and intracellular Ca^{2+} elevation (112, 115, 117, 132) (**Figure 4*b***). Patch-clamp analysis of gene-targeted VSNs expressing the V2R1b receptor, which are H2-Mv− (131), has investigated the mechanisms underlying action potential firing (118). These cells are capable of maintaining low-frequency persistent firing for tens to hundreds of seconds (**Figure 4*b***). This is interesting because long-term potentiation at the mitral-to-granule cell synapse in the AOB, which is thought to underlie pheromonal learning in the context of the Bruce effect, is effectively triggered by low-frequency, 10-Hz pulses applied for extended periods of time. Specific coupling of L-type voltage-gated Ca^{2+} channels and large-conductance Ca^{2+}-activated K^+ channels mediates persistent firing in V2R1b+ VSNs (118).

The TRPC2 channel is widely expressed in apical and basal layers of the VNO (88). Gene knockout studies firmly established this channel's critical role for VSN activation by urine and V1R$^+$ neuron–specific stimuli (121, 139), as well as for the regulation of a variety of social behaviors (121, 139, 140). Ca^{2+} responses to MUPs are reduced in basal VSNs of *Trpc2$^{-/-}$* mice (117). It came as a surprise, therefore, that deletion of *Trpc2* does not significantly influence the transduction of MHC peptides by V2R$^+$ VSNs (141). Likewise, memory formation in the context of the Bruce effect remains intact in *Trpc2$^{-/-}$* mice, despite a requirement for a fully functional VNO (141). Whether these findings reflect a TRPC2-independent transduction mechanism is not yet clear. Alternatively, Ca^{2+} flux through residual channels in *Trpc2$^{-/-}$* VSNs (120) may be sufficient to drive a secondary amplification mechanism and thus produce an excitatory response in these cells, not unlike the Ca^{2+}-activated Cl$^-$ conductance of canonical OSNs (see above). In any case, considerable differences in terms of transduction and signaling mechanisms appear to exist between distinct VSN populations, and systematic studies comparing the signaling properties of molecularly defined VSN subsets will be required to address these questions.

OR-Expressing Vomeronasal Sensory Neurons

An additional neuronal subpopulation in the VNO is defined by the expression of members of the OR gene family (142). RT-PCR analysis suggests that at least 44 different OR genes are expressed in the mouse VNO. OR$^+$ cells in the VNO also express TRPC2 and Gα_{i2}, are located in the apical layer, and project their axons to distinct glomeruli of the anterior subdomain of the AOB (142). On the basis of their dendritic morphology, i.e., the absence of cilia, OR$^+$ cells in the VNO resemble typical VSNs.

The biological role of these cells is presently unclear, although they may be responsible for the detection of environmental odorants by the VNO (142). It has been known since the 1970s that the VNO can detect a range of odorants that do not exhibit any known pheromonal functions (143–147). Like OSNs, odor-sensitive VSNs are activated by more than one odorant mixture (145), indicating that the breadth of tuning of these VSNs differs from that of narrowly tuned V1R$^+$ VSNs. However, it remains unknown whether odor-detecting VSNs are involved in the stimulation of innate behavioral responses.

SEPTAL ORGAN OF MASERA

Cellular and Molecular Organization

In addition to the MOE and VNO, mice and rats possess a small, isolated patch of sensory epithelium known as the septal organ of Masera (SO), which lies near the base of the nasal septum at the entrances to the nasopalatine ducts (**Figure 2a**). Despite its discovery decades ago, the functional role of the SO is still enigmatic. The SO sensory epithelium is composed of 1–3 layers of ciliated OSNs, compared with 6–8 layers in most regions of the MOE (1, 79, 148). The SO expresses 50–80 genes of the OR family, all of which are expressed in the MOE as well (149, 150). Greater than 90% of the SO cells express one of only nine ORs; there is no evidence that a single cell expresses more than one OR (149, 150). Like canonical OSNs, the vast majority of SO OSNs express OMP, ACIII, and Gα_{olf} (79). A small subset of SO OSNs express GC-D and PDE2 (78, 79). Whether additional OSN subpopulations found in the MOE also exist in the SO is not yet known.

OSNs in the SO project to a small subset of glomeruli in the MOB (1, 79, 151). These glomeruli are located in the posterior, ventromedial aspect of the bulb. Some glomeruli appear to be innervated exclusively by SO OSNs, whereas others seem to receive axonal input of OSNs from both the MOE and the SO (151).

Chemoreceptive Properties

Early field-potential recordings first demonstrated that SO OSNs respond at relatively low concentrations to several general odorants, including pentylacetate (152). More recently, patch-clamp recordings from individual knobs of SO OSNs have shown that these cells produce a sensory current in response to stimulation with known concentrations of diverse odorants, not unlike canonical OSNs (79, 153).

Signaling Mechanisms

Involvement of the cAMP second messenger system in SO odor transduction is supported by pharmacological evidence (79) and by genetic deletion of *Cnga2* (153). SO OSNs appear to serve dual functions as odor detectors and mechanical sensors because they respond to both chemical stimuli and mechanical stimuli (153). Remarkably, the mechanical responses appear to be mediated by the same cAMP-dependent pathway that is employed for signal transduction by chemical stimuli (153). The mechanical sensitivity of OSNs is also observed in the MOE and has been hypothesized to serve at least two functions: (*a*) to modulate sensory responses of OSNs with respect to airflow and (*b*) to synchronize rhythmic activity in the olfactory bulb with respiration (153).

GRUENEBERG GANGLION

Cellular and Molecular Organization

First described in 1973 (154), the Grueneberg ganglion (GG) was rediscovered just a few years ago (155–159). It consists of OMP^+ cells located at the dorsal tip of the nasal cavity, close to the opening of the naris. Its biological function is unknown. At a light-microscopic level, cells of the GG do not seem to possess prominent dendrites, cilia, or microvilli and lack direct access to the nasal lumen (1). Therefore, GG cells may detect gaseous or other highly membrane-permeant stimuli. Several chemosensory receptors found in the main and accessory olfactory systems, including the vomeronasal receptor V2R83 (160) and several TAARs (161), have been reported. However, it is unclear how sensory stimuli might access these receptors. GG cells project along the nasal septum and the medial surface of the MOB to reach the dorsal regions of the caudal MOB, near the AOB (155, 159). This region overlaps somewhat with that occupied by the necklace glomeruli, which are innervated by $GC-D^+$ neurons (see above). However, GG cells do not express GC-D (78), and the relationship between the caudal glomeruli innervated by GC-D neurons and the GG cells is unclear.

GG: Grueneberg ganglion

Chemoreceptive Properties

Physiological recordings from GG cells have not been reported. Consequently, no sensory stimuli detected by these cells are known, and the potential sensory function of the GG subsystem remains elusive (see note added in proof).

Signaling Mechanisms

On the basis of the expression of molecular markers, the GG appears to comprise cells of MOE-like and VNO-like molecular phenotypes (160). Signaling molecules found in canonical OSNs, such as ACIII and $G\alpha_{olf}$, are expressed by a few GG cells during the prenatal and perinatal stages (160). An antibody that recognizes members of the V2R2 family (126) labels a considerable number of cells in the GG, as do antibodies against $G\alpha_o$ and $G\alpha_i$ (160). By contrast, another study (158) concluded that the GG is unlikely to express ORs, V1Rs, V2Rs, or other typical elements of OSN and VSN signal transduction cascades.

TRIGEMINAL SYSTEM

Nasal chemosensation depends not only on the sense of smell but also on the activity of branches of the trigeminal nerve, which provide sensory innervation to the epithelia of the head (162). Inhalation of noxious or irritating chemical stimuli activates the trigeminal system and triggers protective reflexes such as apnea or sneezing. It was thought that receptors for trigeminal irritants are located exclusively on free nerve endings within the nasal epithelium (92, 93, 163, 164). However, new work has identified a large population of solitary chemosensory cells in mice and rats that reach the surface of the nasal epithelium, form synaptic contacts with trigeminal afferent nerve fibers, and respond to odorous irritants at high concentrations (92, 93, 163, 164). These cells were first characterized by the expression of T2R-type taste receptors, PLCβ2, and the G protein α-gustducin (163). More recently, these cells were also shown to contain TRPM5 (92, 93, 164) (**Figure 2b**). Ca^{2+} imaging of dissociated, genetically labeled $TRPM5^+$ solitary chemosensory cells demonstrated that they respond directly to a panel of volatile odorants at relatively high concentrations (92). It is not yet clear whether TRPM5 is directly involved in this response (92). Together, this work provides a new strategy for dissecting the mechanisms underlying the perception of irritating odors.

CONCLUSIONS

The cellular, molecular, and functional diversity of olfactory subsystems begs the question of why they are needed. The answer is most certainly multifaceted. Perhaps the most obvious reason for separate subsystems is that, because they express distinct families of chemosensory receptors, they expand the repertoire of chemicals that can be detected. The need for diverse transduction mechanisms is less clear but likely reflects both the coupling properties of the receptor itself and the kinetic and modulatory requirements associated with different classes of stimuli.

Some subsystems likely subserve species-specific roles. Although humans can respond quite well to diverse environmental odors and even to some semiochemicals, humans lack several olfactory subsystems present in lower mammals, including the entire AOS and GC-D^+ neurons of the MOS. These subsystems are particularly attuned to stimuli that are present in urine and glandular secretions and that may have been supplanted by many of the visual cues utilized for communication by higher primates.

Subsystems can also confer meaning to a stimulus through their connections to the CNS. The most dramatic example comes from mice in which olfactory signaling through the dorsal MOE and MOB has been disrupted (165). These mice no longer avoid odors associated with predators or spoiled food, even though they can still detect the stimuli, which stimulate OSNs in multiple regions of wild-type mouse MOE. Thus, the same odorant can evoke distinct perceptions, depending on the subset of OSNs it activates. The distinct CNS connections of the MOS and AOS further support a model in which olfactory subsystems help animals to extract information about the meaning of a stimulus, not just its identity.

SUMMARY POINTS

1. The mammalian olfactory system contains a diverse array of subsystems. These vary in the stimuli to which they respond, the cell types and molecules they employ to detect and transduce stimuli, and the connections they make with the CNS.

2. The use of gene targeting in mice has provided essential tools with which to identify, differentiate, and characterize olfactory subsystems.

3. The view that the main olfactory system is only a general odor sensor and the accessory olfactory system only a detector of semiochemicals such as pheromones is not valid.

4. Olfactory subsystems can allow for parallel processing of stimuli, thereby providing a means to extract different types of information from a single chemosensory cue.

FUTURE ISSUES

1. Although it is clear that olfactory subsystems utilize distinct molecular mechanisms to detect and transduce chemosensory stimuli, these mechanisms remain poorly understood. Identification of the key molecular players for each subsystem will provide invaluable tools for dissecting their biological function.

2. Elucidating the specific behavioral and physiological roles of each olfactory subsystem is required to understand how these subsystems work together to represent the sensory environment.

3. Little is known about how information from each olfactory subsystem is integrated within higher brain centers. Of particular interest is the question of how olfactory information is integrated with taste, somatosensory, and hormonal inputs, all of which can critically impact perception and motivation.

4. Although humans detect a complex olfactory world, they lack at least some subsystems present in rodents (e.g., the AOS). An understanding of the repertoire of human olfactory subsystems is needed if we are to understand fully the extent of the human chemosensory world.

DISCLOSURE STATEMENT

The authors are not aware of any biases that might be perceived as affecting the objectivity of this review.

ACKNOWLEDGMENTS

Work in the authors' laboratories is supported by the National Institute on Deafness and Other Communication Disorders and by the Deutsche Forschungsgemeinschaft. T.L.-Z. is a Lichtenberg Professor of the Volkswagen Foundation.

LITERATURE CITED

1. Breer H, Fleischer J, Strotmann J. 2006. The sense of smell: multiple olfactory subsystems. *Cell. Mol. Life Sci.* 63:1465–75
2. Ma M. 2007. Encoding olfactory signals via multiple chemosensory systems. *Crit. Rev. Biochem. Mol. Biol.* 42:463–80
3. Tirindelli R, Mucignat-Caretta C, Ryba NJ. 1998. Molecular aspects of pheromonal communication via the vomeronasal organ of mammals. *Trends Neurosci.* 21:482–86
4. Gold GH. 1999. Controversial issues in vertebrate olfactory transduction. *Annu. Rev. Physiol.* 61:857–71
5. Buck LB. 2000. The molecular architecture of odor and pheromone sensing in mammals. *Cell* 100:611–18
6. Firestein S. 2001. How the olfactory system makes sense of scents. *Nature* 413:211–18
7. Zufall F, Munger SD. 2001. From odor and pheromone transduction to the organization of the sense of smell. *Trends Neurosci.* 24:191–93
8. Mombaerts P. 2004. Genes and ligands for odorant, vomeronasal and taste receptors. *Nat. Rev. Neurosci.* 5:263–78
9. Ache BW, Young JM. 2005. Olfaction: diverse species, conserved principles. *Neuron* 48:417–30
10. Brennan PA, Zufall F. 2006. Pheromonal communication in vertebrates. *Nature* 444:308–15
11. Dulac C, Wagner S. 2006. Genetic analysis of brain circuits underlying pheromone signaling. *Annu. Rev. Genet.* 40:449–67

12. Zufall F, Leinders-Zufall T. 2007. Mammalian pheromone sensing. *Curr. Opin. Neurobiol.* 17:483–89

13. Fulle HJ, Vassar R, Foster DC, Yang RB, Axel R, Garbers DL. 1995. A receptor guanylyl cyclase expressed specifically in olfactory sensory neurons. *Proc. Natl. Acad. Sci. USA* 92:3571–75

14. Juilfs DM, Fulle HJ, Zhao AZ, Houslay MD, Garbers DL, Beavo JA. 1997. A subset of olfactory neurons that selectively express cGMP-stimulated phosphodiesterase (PDE2) and guanylyl cyclase-D define a unique olfactory signal transduction pathway. *Proc. Natl. Acad. Sci. USA* 94:3388–95

15. Meyer MR, Angele A, Kremmer E, Kaupp UB, Muller F. 2000. A cGMP-signaling pathway in a subset of olfactory sensory neurons. *Proc. Natl. Acad. Sci. USA* 97:10595–600

16. Leinders-Zufall T, Cockerham RE, Michalakis S, Biel M, Garbers DL, et al. 2007. Contribution of the receptor guanylyl cyclase GC-D to chemosensory function in the olfactory epithelium. *Proc. Natl. Acad. Sci. USA* 104:14507–12

17. Buck L, Axel R. 1991. A novel multigene family may encode odorant receptors: a molecular basis for odor recognition. *Cell* 65:175–87

18. Glusman G, Yanai I, Rubin I, Lancet D. 2001. The complete human olfactory subgenome. *Genome Res.* 11:685–702

19. Zozulya S, Echeverri F, Nguyen T. 2001. The human olfactory receptor repertoire. *Genome Biol.* 2:RESEARCH0018

20. Young JM, Friedman C, Williams EM, Ross JA, Tonnes-Priddy L, Trask BJ. 2002. Different evolutionary processes shaped the mouse and human olfactory receptor gene families. *Hum. Mol. Genet.* 11:535–46

21. Zhang X, Firestein S. 2002. The olfactory receptor gene superfamily of the mouse. *Nat. Neurosci.* 5:124–33

22. Pace U, Hanski E, Salomon Y, Lancet D. 1985. Odorant-sensitive adenylate cyclase may mediate olfactory reception. *Nature* 316:255–58

23. Pace U, Lancet D. 1986. Olfactory GTP-binding protein: signal-transducing polypeptide of vertebrate chemosensory neurons. *Proc. Natl. Acad. Sci. USA* 83:4947–51

24. Sklar PB, Anholt RR, Snyder SH. 1986. The odorant-sensitive adenylate cyclase of olfactory receptor cells. Differential stimulation by distinct classes of odorants. *J. Biol. Chem.* 261:15538–43

25. Anholt RR, Mumby SM, Stoffers DA, Girard PR, Kuo JF, Snyder SH. 1987. Transduction proteins of olfactory receptor cells: identification of guanine nucleotide binding proteins and protein kinase C. *Biochemistry* 26:788–95

26. Nakamura T, Gold GH. 1987. A cyclic nucleotide-gated conductance in olfactory receptor cilia. *Nature* 325:442–44

27. Jones DT, Reed RR. 1989. G_{olf}: an olfactory neuron specific-G protein involved in odorant signal transduction. *Science* 244:790–95

28. Bakalyar HA, Reed RR. 1990. Identification of a specialized adenylyl cyclase that may mediate odorant detection. *Science* 250:1403–6

29. Dhallan RS, Yau KW, Schrader KA, Reed RR. 1990. Primary structure and functional expression of a cyclic nucleotide-activated channel from olfactory neurons. *Nature* 347:184–87

30. Bradley J, Li J, Davidson N, Lester HA, Zinn K. 1994. Heteromeric olfactory cyclic nucleotide-gated channels: a subunit that confers increased sensitivity to cAMP. *Proc. Natl. Acad. Sci. USA* 91:8890–94

31. Liman ER, Buck LB. 1994. A second subunit of the olfactory cyclic nucleotide-gated channel confers high sensitivity to cAMP. *Neuron* 13:611–21

32. Bonigk W, Bradley J, Muller F, Sesti F, Boekhoff I, et al. 1999. The native rat olfactory cyclic nucleotide-gated channel is composed of three distinct subunits. *J. Neurosci.* 19:5332–47

33. Kleene SJ, Gesteland RC. 1991. Calcium-activated chloride conductance in frog olfactory cilia. *J. Neurosci.* 11:3624–29

34. Kurahashi T, Yau KW. 1993. Co-existence of cationic and chloride components in odorant-induced current of vertebrate olfactory receptor cells. *Nature* 363:71–74

35. Lowe G, Gold GH. 1993. Nonlinear amplification by calcium-dependent chloride channels in olfactory receptor cells. *Nature* 366:283–86

36. Pifferi S, Pascarella G, Boccaccio A, Mazzatenta A, Gustincich S, et al. 2006. Bestrophin-2 is a candidate calcium-activated chloride channel involved in olfactory transduction. *Proc. Natl. Acad. Sci. USA* 103:12929–34

37. Von Dannecker LE, Mercadante AF, Malnic B. 2005. Ric-8B, an olfactory putative GTP exchange factor, amplifies signal transduction through the olfactory-specific G-protein Gαolf. *J. Neurosci.* 25:3793–800

38. Kerr DS, Von Dannecker LE, Davalos M, Michaloski JS, Malnic B. 2008. Ric-8B interacts with Gα_{olf} and Gg$_{13}$ and colocalizes with Gα_{olf}, Gb$_1$ and G$_{13}$ in the cilia of olfactory sensory neurons. *Mol. Cell. Neurosci.* 38:341–48

39. Yan C, Zhao AZ, Bentley JK, Loughney K, Ferguson K, Beavo JA. 1995. Molecular cloning and characterization of a calmodulin-dependent phosphodiesterase enriched in olfactory sensory neurons. *Proc. Natl. Acad. Sci. USA* 92:9677–81

40. Cherry JA, Davis RL. 1995. A mouse homolog of *dunce*, a gene important for learning and memory in *Drosophila*, is preferentially expressed in olfactory receptor neurons. *J. Neurobiol.* 28:102–13

41. Keller A, Margolis FL. 1976. Isolation and characterization of rat olfactory marker protein. *J. Biol. Chem.* 251:6232–37

42. Buiakova OI, Baker H, Scott JW, Farbman A, Kream R, et al. 1996. Olfactory marker protein (OMP) gene deletion causes altered physiological activity of olfactory sensory neurons. *Proc. Natl. Acad. Sci. USA* 93:9858–63

43. Krautwurst D, Yau KW, Reed RR. 1998. Identification of ligands for olfactory receptors by functional expression of a receptor library. *Cell* 95:917–26

44. Zhao H, Ivic L, Otaki JM, Hashimoto M, Mikoshiba K, Firestein S. 1998. Functional expression of a mammalian odorant receptor. *Science* 279:237–42

45. Malnic B, Hirono J, Sato T, Buck LB. 1999. Combinatorial receptor codes for odors. *Cell* 96:713–23

46. Wachowiak M, Shipley MT. 2006. Coding and synaptic processing of sensory information in the glomerular layer of the olfactory bulb. *Semin. Cell Dev. Biol.* 17:411–23

47. Kajiya K, Inaki K, Tanaka M, Haga T, Kataoka H, Touhara K. 2001. Molecular bases of odor discrimination: reconstitution of olfactory receptors that recognize overlapping sets of odorants. *J. Neurosci.* 21:6018–25

48. Abaffy T, Matsunami H, Luetje CW. 2006. Functional analysis of a mammalian odorant receptor subfamily. *J. Neurochem.* 97:1506–18

49. Spehr M, Gisselmann G, Poplawski A, Riffell JA, Wetzel CH, et al. 2003. Identification of a testicular odorant receptor mediating human sperm chemotaxis. *Science* 299:2054–58

50. Shirokova E, Schmiedeberg K, Bedner P, Niessen H, Willecke K, et al. 2005. Identification of specific ligands for orphan olfactory receptors. G protein-dependent agonism and antagonism of odorants. *J. Biol. Chem.* 280:11807–15

51. Spehr M, Schwane K, Heilmann S, Gisselmann G, Hummel T, Hatt H. 2004. Dual capacity of a human olfactory receptor. *Curr. Biol.* 14: R832–83

52. Oka Y, Nakamura A, Watanabe H, Touhara K. 2004. An odorant derivative as an antagonist for an olfactory receptor. *Chem. Senses* 29:815–22

53. Wysocki CJ, Beauchamp GK. 1984. Ability to smell androstenone is genetically determined. *Proc. Natl. Acad. Sci. USA* 81:4899–902

54. Keller A, Zhuang H, Chi Q, Vosshall LB, Matsunami H. 2007. Genetic variation in a human odorant receptor alters odour perception. *Nature* 449:468–72

55. Menashe I, Man O, Lancet D, Gilad Y. 2003. Different noses for different people. *Nat. Genet.* 34:143–44

56. Menashe I, Abaffy T, Hasin Y, Goshen S, Yahalom V, et al. 2007. Genetic elucidation of human hyperosmia to isovaleric acid. *PLoS Biol.* 5:e284

57. Bozza T, Feinstein P, Zheng C, Mombaerts P. 2002. Odorant receptor expression defines functional units in the mouse olfactory system. *J. Neurosci.* 22:3033–43

58. Katada S, Hirokawa T, Oka Y, Suwa M, Touhara K. 2005. Structural basis for a broad but selective ligand spectrum of a mouse olfactory receptor: mapping the odorant-binding site. *J. Neurosci.* 25:1806–15

59. Lagerstrom MC, Schioth HB. 2008. Structural diversity of G protein-coupled receptors and significance for drug discovery. *Nat. Rev. Drug Discov.* 7:339–57

60. Saito H, Kubota M, Roberts RW, Chi Q, Matsunami H. 2004. RTP family members induce functional expression of mammalian odorant receptors. *Cell* 119:679–91

61. Belluscio L, Gold GH, Nemes A, Axel R. 1998. Mice deficient in G$_{olf}$ are anosmic. *Neuron* 20:69–81

62. Jones DT, Reed RR. 1987. Molecular cloning of five GTP-binding protein cDNA species from rat olfactory neuroepithelium. *J. Biol. Chem.* 262:14241–49

63. Wong ST, Trinh K, Hacker B, Chan GC, Lowe G, et al. 2000. Disruption of the type III adenylyl cyclase gene leads to peripheral and behavioral anosmia in transgenic mice. *Neuron* 27:487–97

64. Kaupp UB, Seifert R. 2002. Cyclic nucleotide-gated ion channels. *Physiol. Rev.* 82:769–824

65. Brunet LJ, Gold GH, Ngai J. 1996. General anosmia caused by a targeted disruption of the mouse olfactory cyclic nucleotide-gated cation channel. *Neuron* 17:681–93

66. Lin W, Arellano J, Slotnick B, Restrepo D. 2004. Odors detected by mice deficient in cyclic nucleotide-gated channel subunit A2 stimulate the main olfactory system. *J. Neurosci.* 24:3703–10

67. Munger SD, Lane AP, Zhong H, Leinders-Zufall T, Yau KW, et al. 2001. Central role of the CNGA4 channel subunit in Ca^{2+}-calmodulin-dependent odor adaptation. *Science* 294:2172–75

68. Michalakis S, Reisert J, Geiger H, Wetzel C, Zong X, et al. 2006. Loss of CNGB1 protein leads to olfactory dysfunction and subciliary cyclic nucleotide-gated channel trapping. *J. Biol. Chem.* 281:35156–66

69. Kelliher KR, Ziesmann J, Munger SD, Reed RR, Zufall F. 2003. Importance of the CNGA4 channel gene for odor discrimination and adaptation in behaving mice. *Proc. Natl. Acad. Sci. USA* 100:4299–304

70. Kurahashi T, Menini A. 1997. Mechanism of odorant adaptation in the olfactory receptor cell. *Nature* 385:725–29

71. Song Y, Cygnar KD, Sagdullaev B, Valley M, Hirsh S, et al. 2008. Olfactory CNG channel desensitization by Ca^{2+}/CaM via the B1b subunit affects response termination but not sensitivity to recurring stimulation. *Neuron* 58:374–86

72. Spehr M, Kelliher KR, Li XH, Boehm T, Leinders-Zufall T, Zufall F. 2006. Essential role of the main olfactory system in social recognition of major histocompatibility complex peptide ligands. *J. Neurosci.* 26:1961–70

73. Borowsky B, Adham N, Jones KA, Raddatz R, Artymyshyn R, et al. 2001. Trace amines: identification of a family of mammalian G protein-coupled receptors. *Proc. Natl. Acad. Sci. USA* 98:8966–71

74. Liberles SD, Buck LB. 2006. A second class of chemosensory receptors in the olfactory epithelium. *Nature* 442:645–50

75. Gloriam DE, Bjarnadottir TK, Yan YL, Postlethwait JH, Schioth HB, Fredriksson R. 2005. The repertoire of trace amine G-protein-coupled receptors: large expansion in zebrafish. *Mol. Phylogenet. Evol.* 35:470–82

76. Mueller JC, Steiger S, Fidler AE, Kempenaers B. 2008. Biogenic trace amine-associated receptors (TAARS) are encoded in avian genomes: evidence and possible implications. *J. Hered.* 99:174–76

77. Gibson AD, Garbers DL. 2000. Guanylyl cyclases as a family of putative odorant receptors. *Annu. Rev. Neurosci.* 23:417–39

78. Walz A, Feinstein P, Khan M, Mombaerts P. 2007. Axonal wiring of guanylate cyclase-D-expressing olfactory neurons is dependent on neuropilin 2 and semaphorin 3F. *Development* 134:4063–72

79. Ma M, Grosmaitre X, Iwema CL, Baker H, Greer CA, Shepherd GM. 2003. Olfactory signal transduction in the mouse septal organ. *J. Neurosci.* 23:317–24

80. Shinoda K, Shiotani Y, Osawa Y. 1989. "Necklace olfactory glomeruli" form unique components of the rat primary olfactory system. *J. Comp. Neurol.* 284:362–73

81. Hu J, Zhong C, Ding C, Chi Q, Walz A, et al. 2007. Detection of near-atmospheric concentrations of CO_2 by an olfactory subsystem in the mouse. *Science* 317:953–57

82. Forte LR Jr. 2004. Uroguanylin and guanylin peptides: pharmacology and experimental therapeutics. *Pharmacol. Ther.* 104:137–62

83. Baker H, Cummings DM, Munger SD, Margolis JW, Franzen L, et al. 1999. Targeted deletion of a cyclic nucleotide-gated channel subunit (OCNC1): biochemical and morphological consequences in adult mice. *J. Neurosci.* 19:9313–21

84. Young JM, Waters H, Dong C, Fulle HJ, Liman ER. 2007. Degeneration of the olfactory guanylyl cyclase D gene during primate evolution. *PLoS ONE* 2:e884

85. Fain GL. 2003. *Sensory Transduction*. Sunderland, MA: Sinauer Assoc.

86. Duda T, Sharma RK. 2008. ONE-GC membrane guanylate cyclase, a trimodal odorant signal transducer. *Biochem. Biophys. Res. Commun.* 367:440–45

87. Venkatachalam K, Montell C. 2007. TRP channels. *Annu. Rev. Biochem.* 76:387–417

88. Liman ER, Corey DP, Dulac C. 1999. TRP2: a candidate transduction channel for mammalian pheromone sensory signaling. *Proc. Natl. Acad. Sci. USA* 96:5791–96

89. Elsaesser R, Montani G, Tirindelli R, Paysan J. 2005. Phosphatidyl-inositide signaling proteins in a novel class of sensory cells in the mammalian olfactory epithelium. *Eur. J. Neurosci.* 21:2692–700

90. Hansel DE, Eipper BA, Ronnett GV. 2001. Neuropeptide Y functions as a neuroproliferative factor. *Nature* 410:940–44

91. Montani G, Tonelli S, Elsaesser R, Paysan J, Tirindelli R. 2006. Neuropeptide Y in the olfactory microvillar cells. *Eur. J. Neurosci.* 24:20–24

92. Lin W, Ogura T, Margolskee RF, Finger TE, Restrepo D. 2008. TRPM5-expressing solitary chemosensory cells respond to odorous irritants. *J. Neurophysiol.* 99:1451–60

93. Lin W, Margolskee R, Donnert G, Hell SW, Restrepo D. 2007. Olfactory neurons expressing transient receptor potential channel M5 (TRPM5) are involved in sensing semiochemicals. *Proc. Natl. Acad. Sci. USA* 104:2471–76

94. Boekhoff I, Tareilus E, Strotmann J, Breer H. 1990. Rapid activation of alternative second messenger pathways in olfactory cilia from rats by different odorants. *EMBO J.* 9:2453–58

95. Dulac C, Torello AT. 2003. Molecular detection of pheromone signals in mammals: from genes to behavior. *Nat. Rev. Neurosci.* 4:551–62

96. Halpern M, Martinez-Marcos A. 2003. Structure and function of the vomeronasal system: an update. *Prog. Neurobiol.* 70:245–318

97. Bigiani A, Mucignat-Caretta C, Montani G, Tirindelli R. 2005. Pheromone reception in mammals. *Rev. Physiol. Biochem. Pharmacol.* 154:1–35

98. Zufall F, Leinders-Zufall T, Puche A. 2008. Accessory olfactory system. In *The Senses: A Comprehensive Reference*, ed. AI Basbaum, A Kaneko, GM Shepherd, G Westheimer, pp. 783–814. San Diego: Academic

99. Luo M, Katz LC. 2004. Encoding pheromonal signals in the mammalian vomeronasal system. *Curr. Opin. Neurobiol.* 14:428–34

100. Broad KD, Keverne EB. 2008. More to pheromones than meets the nose. *Nat. Neurosci.* 11:128–29

101. Berghard A, Buck LB, Liman ER. 1996. Evidence for distinct signaling mechanisms in two mammalian olfactory sense organs. *Proc. Natl. Acad. Sci. USA* 93:2365–69

102. Jia C, Halpern M. 1996. Subclasses of vomeronasal receptor neurons: differential expression of G proteins ($G_{i\alpha 2}$ and $G_{o\alpha}$) and segregated projections to the accessory olfactory bulb. *Brain Res.* 719:117–28

103. Mohedano-Moriano A, Pro-Sistiaga P, Ubeda-Banon I, Crespo C, Insausti R, Martinez-Marcos A. 2007. Segregated pathways to the vomeronasal amygdala: differential projections from the anterior and posterior divisions of the accessory olfactory bulb. *Eur. J. Neurosci.* 25:2065–80

104. Dulac C, Axel R. 1995. A novel family of genes encoding putative pheromone receptors in mammals. *Cell* 83:195–206

105. Zhang X, Firestein S. 2007. Comparative genomics of odorant and pheromone receptor genes in rodents. *Genomics* 89:441–50

106. Rodriguez I, Mombaerts P. 2002. Novel human vomeronasal receptor-like genes reveal species-specific families. *Curr. Biol.* 12:R409–11

107. Rodriguez I, Feinstein P, Mombaerts P. 1999. Variable patterns of axonal projections of sensory neurons in the mouse vomeronasal system. *Cell* 97:199–208

108. Grus WE, Shi P, Zhang J. 2007. Largest vertebrate vomeronasal type 1 receptor gene repertoire in the semiaquatic platypus. *Mol. Biol. Evol.* 24:2153–57

109. Belluscio L, Koentges G, Axel R, Dulac C. 1999. A map of pheromone receptor activation in the mammalian brain. *Cell* 97:209–20

110. Wagner S, Gresser AL, Torello AT, Dulac C. 2006. A multireceptor genetic approach uncovers an ordered integration of VNO sensory inputs in the accessory olfactory bulb. *Neuron* 50:697–709

111. Leinders-Zufall T, Lane AP, Puche AC, Ma W, Novotny MV, et al. 2000. Ultrasensitive pheromone detection by mammalian vomeronasal neurons. *Nature* 405:792–96

112. Leinders-Zufall T, Brennan P, Widmayer P, Chandramani SP, Maul-Pavicic A, et al. 2004. MHC class I peptides as chemosensory signals in the vomeronasal organ. *Science* 306:1033–37

113. Del Punta K, Leinders-Zufall T, Rodriguez I, Jukam D, Wysocki CJ, et al. 2002. Deficient pheromone responses in mice lacking a cluster of vomeronasal receptor genes. *Nature* 419:70–74

114. Boschat C, Pelofi C, Randin O, Roppolo D, Luscher C, et al. 2002. Pheromone detection mediated by a V1r vomeronasal receptor. *Nat. Neurosci.* 5:1261–62

115. He J, Ma L, Kim S, Nakai J, Yu CR. 2008. Encoding gender and individual information in the mouse vomeronasal organ. *Science* 320:535–38

116. Holekamp TF, Turaga D, Holy TE. 2008. Fast three-dimensional fluorescence imaging of activity in neural populations by objective-coupled planar illumination microscopy. *Neuron* 57:661–72

117. Chamero P, Marton TF, Logan DW, Flanagan K, Cruz JR, et al. 2007. Identification of protein pheromones that promote aggressive behavior. *Nature* 450:899–902

118. Ukhanov K, Leinders-Zufall T, Zufall F. 2007. Patch-clamp analysis of gene-targeted vomeronasal neurons expressing a defined V1r or V2r receptor: ionic mechanisms underlying persistent firing. *J. Neurophysiol.* 98:2357–69

119. Zufall F, Ukhanov K, Lucas P, Liman ER, Leinders-Zufall T. 2005. Neurobiology of TRPC2: from gene to behavior. *Pflüg. Arch.* 451:61–71

120. Lucas P, Ukhanov K, Leinders-Zufall T, Zufall F. 2003. A diacylglycerol-gated cation channel in vomeronasal neuron dendrites is impaired in TRPC2 mutant mice: mechanism of pheromone transduction. *Neuron* 40:551–61

121. Leypold BG, Yu CR, Leinders-Zufall T, Kim MM, Zufall F, Axel R. 2002. Altered sexual and social behaviors in *trp2* mutant mice. *Proc. Natl. Acad. Sci. USA* 99:6376–81

122. Minke B. 2006. TRP channels and Ca^{2+} signaling. *Cell Calcium* 40:261–75

123. Holy TE, Dulac C, Meister M. 2000. Responses of vomeronasal neurons to natural stimuli. *Science* 289:1569–72

124. Yang H, Shi P, Zhang YP, Zhang J. 2005. Composition and evolution of the V2r vomeronasal receptor gene repertoire in mice and rats. *Genomics* 86:306–15

125. Young JM, Trask BJ. 2007. V2R gene families degenerated in primates, dog and cow, but expanded in opossum. *Trends Genet.* 23:212–15

126. Silvotti L, Moiani A, Gatti R, Tirindelli R. 2007. Combinatorial coexpression of pheromone receptors, V2Rs. *J. Neurochem.* 103:1753–63

127. Ishii T, Hirota J, Mombaerts P. 2003. Combinatorial coexpression of neural and immune multigene families in mouse vomeronasal sensory neurons. *Curr. Biol.* 13:394–400

128. Loconto J, Papes F, Chang E, Stowers L, Jones EP, et al. 2003. Functional expression of murine V2R pheromone receptors involves selective association with the M10 and M1 families of MHC class Ib molecules. *Cell* 112:607–18

129. Olson R, Huey-Tubman KE, Dulac C, Bjorkman PJ. 2005. Structure of a pheromone receptor-associated MHC molecule with an open and empty groove. *PLoS Biol.* 3:e257

130. Olson R, Dulac C, Bjorkman PJ. 2006. MHC homologs in the nervous system—They haven't lost their groove. *Curr. Opin. Neurobiol.* 16:351–57

131. Ishii T, Mombaerts P. 2008. Expression of nonclassical class I major histocompatibility genes defines a tripartite organization of the mouse vomeronasal system. *J. Neurosci.* 28:2332–41

132. Kimoto H, Haga S, Sato K, Touhara K. 2005. Sex-specific peptides from exocrine glands stimulate mouse vomeronasal sensory neurons. *Nature* 437:898–901

133. Boehm T, Zufall F. 2006. MHC peptides and the sensory evaluation of genotype. *Trends Neurosci.* 29:100–7

134. Villinger J, Waldman B. 2008. Self-referent MHC type matching in frog tadpoles. *Proc. Biol. Sci.* 275:1225–30

135. Kimoto H, Sato K, Nodari F, Haga S, Holy TE, Touhara K. 2007. Sex- and strain-specific expression and vomeronasal activity of mouse ESP family peptides. *Curr. Biol.* 17:1879–84

136. Touhara K. 2007. Molecular biology of peptide pheromone production and reception in mice. *Adv. Genet.* 59:147–71

137. Cheetham SA, Thom MD, Jury F, Ollier WE, Beynon RJ, Hurst JL. 2007. The genetic basis of individual-recognition signals in the mouse. *Curr. Biol.* 17:1771–77

138. More L. 2006. Mouse major urinary proteins trigger ovulation via the vomeronasal organ. *Chem. Senses* 31:393–401

139. Stowers L, Holy TE, Meister M, Dulac C, Koentges G. 2002. Loss of sex discrimination and male-male aggression in mice deficient for TRP2. *Science* 295:1493–500

140. Kimchi T, Xu J, Dulac C. 2007. A functional circuit underlying male sexual behavior in the female mouse brain. *Nature* 448:1009–14

141. Kelliher KR, Spehr M, Li XH, Zufall F, Leinders-Zufall T. 2006. Pheromonal recognition memory induced by TRPC2-independent vomeronasal sensing. *Eur. J. Neurosci.* 23:3385–90

142. Levai O, Feistel T, Breer H, Strotmann J. 2006. Cells in the vomeronasal organ express odorant receptors but project to the accessory olfactory bulb. *J. Comp. Neurol.* 498:476–90

143. Müller W. 1971. Vergleichende elektrophysiologische Untersuchungen an den Sinnesepithelien des Jacobsonschen Organs und der Nase von Amphibien (Rana), Reptilien (Lacerta) und Säugetieren (Mus). *Z. Vergl. Physiol.* 72:370–85

144. Tucker D. 1971. Nonolfactory responses from the nasal cavity: Jacobson's organ and the trigeminal system. In *Handbook of Sensory Physiology*, ed. H Autrum, R Jung, WR Loewenstein, DM MacKay, HL Teuber, pp. 152–81. Berlin/Heidelberg/New York: Springer-Verlag

145. Sam M, Vora S, Malnic B, Ma W, Novotny MV, Buck LB. 2001. Odorants may arouse instinctive behaviors. *Nature* 412:142

146. Trinh K, Storm DR. 2003. Vomeronasal organ detects odorants in absence of signaling through main olfactory epithelium. *Nat. Neurosci.* 6:519–25

147. Xu F, Schaefer M, Kida I, Schafer J, Liu N, et al. 2005. Simultaneous activation of mouse main and accessory olfactory bulbs by odors or pheromones. *J. Comp. Neurol.* 489:491–500

148. Weiler E, Farbman AI. 2003. The septal organ of the rat during postnatal development. *Chem. Senses* 28:581–93

149. Kaluza JF, Gussing F, Bohm S, Breer H, Strotmann J. 2004. Olfactory receptors in the mouse septal organ. *J. Neurosci. Res.* 76:442–52

150. Tian H, Ma M. 2004. Molecular organization of the olfactory septal organ. *J. Neurosci.* 24:8383–90

151. Levai O, Strotmann J. 2003. Projection pattern of nerve fibers from the septal organ: DiI-tracing studies with transgenic OMP mice. *Histochem. Cell Biol.* 120:483–92

152. Marshall DA, Maruniak JA. 1986. Masera's organ responds to odorants. *Brain Res.* 366:329–32

153. Grosmaitre X, Santarelli LC, Tan J, Luo M, Ma M. 2007. Dual functions of mammalian olfactory sensory neurons as odor detectors and mechanical sensors. *Nat. Neurosci.* 10:348–54

154. Grüneberg H. 1973. A ganglion probably belonging to the N. terminalis system in the nasal mucosa of the mouse. *Z Anat. Entwickl.* 140:39–52

155. Fuss SH, Omura M, Mombaerts P. 2005. The Grueneberg ganglion of the mouse projects axons to glomeruli in the olfactory bulb. *Eur. J. Neurosci.* 22:2649–54

156. Koos DS, Fraser SE. 2005. The Grueneberg ganglion projects to the olfactory bulb. *Neuroreport* 16:1929–32

157. Fleischer J, Hass N, Schwarzenbacher K, Besser S, Breer H. 2006. A novel population of neuronal cells expressing the olfactory marker protein (OMP) in the anterior/dorsal region of the nasal cavity. *Histochem. Cell Biol.* 125:337–49

158. Roppolo D, Ribaud V, Jungo VP, Luscher C, Rodriguez I. 2006. Projection of the Gruneberg ganglion to the mouse olfactory bulb. *Eur. J. Neurosci.* 23:2887–94

159. Storan MJ, Key B. 2006. Septal organ of Grüneberg is part of the olfactory system. *J. Comp. Neurol.* 494:834–44

160. Fleischer J, Schwarzenbacher K, Besser S, Hass N, Breer H. 2006. Olfactory receptors and signaling elements in the Grueneberg ganglion. *J. Neurochem.* 98:543–54

161. Fleischer J, Schwarzenbacher K, Breer H. 2007. Expression of trace amine-associated receptors in the Grueneberg ganglion. *Chem. Senses* 32:623–31

162. Bryant BP. 2000. Chemesthesis: the common chemical sense. In *The Neurobiology of Taste and Smell*, ed. TE Finger, WL Silver, D Restrepo, pp. 73–100. New York: Wiley

163. Finger TE, Bottger B, Hansen A, Anderson KT, Alimohammadi H, Silver WL. 2003. Solitary chemore-ceptor cells in the nasal cavity serve as sentinels of respiration. *Proc. Natl. Acad. Sci. USA* 100:8981–86

164. Kaske S, Krasteva G, Konig P, Kummer W, Hofmann T, et al. 2007. TRPM5, a taste-signaling transient receptor potential ion-channel, is a ubiquitous signaling component in chemosensory cells. *BMC Neurosci.* 8:49

165. Kobayakawa K, Kobayakawa R, Matsumoto H, Oka Y, Imai T, et al. 2007. Innate versus learned odour processing in the mouse olfactory bulb. *Nature* 450:503–8

166. Mousley A, Polese G, Marks NJ, Eisthen HL. 2006. Terminal nerve-derived neuropeptide Y modulates physiological responses in the olfactory epithelium of hungry axolotls (*Ambystoma mexicanum*). *J. Neurosci.* 26:7707–17

167. Von Bartheld CS. 2004. The terminal nerve and its relation with extrabulbar "olfactory" projections: lessons from lampreys and lungfishes. *Microsc. Res. Tech.* 65:13–24

168. Mombaerts P, Wang F, Dulac C, Chao SK, Nemes A, et al. 1996. Visualizing an olfactory sensory map. *Cell* 87:675–86

RELATED RESOURCES

1. Olfactory Receptor Database (ORDB) (**http://senselab.med.yale.edu/ordb/default.asp**)
2. The Human Olfactory Receptor Data Exploratorium (HORDE) (**http://bioportal.weizmann.ac.il/HORDE/**)
3. Touhara K, Vosshall LB. 2009. Sensing odorants and pheromones with chemosensory receptors. *Annu. Rev. Physiol.* 71:307–32
4. Baum MJ, Kelliher K. 2009. Complementary roles of main and accessory olfactory system processing of chemosignals in promoting mate recognition across mammalian phylogeny. *Annu. Rev. Physiol.* 71:141–60

NOTE ADDED IN PROOF

After acceptance of this manuscript, Brechbühl et al. (2008) reported that cells of the Grueneberg ganglion (GG) mediate alarm pheromone detection in mice. These chemicals, which remain unidentified in mammals, may signal stress, injury, or the presence of predators. This report also found that stimuli may gain access to GG neurons and their cilia though a keratinized epithelium that is permeable to hydrophilic substances.

Brechbühl J, Klaey M, Broillet M-C. 2008. Grueneberg ganglion cells mediate alarm pheromone detection in mice. *Science* 321:1092–95

Complementary Roles of the Main and Accessory Olfactory Systems in Mammalian Mate Recognition

Michael J. Baum[1] and Kevin R. Kelliher[2]

[1]Department of Biology, Boston University, Boston, Massachusetts 02215; email: baum@bu.edu

[2]Department of Biological Sciences, University of Idaho, Moscow, Idaho 83844; email: kelliher@uidaho.edu

Annu. Rev. Physiol. 2009. 71:141–60

First published online as a Review in Advance on September 25, 2008

The *Annual Review of Physiology* is online at physiol.annualreviews.org

This article's doi: 10.1146/annurev.physiol.010908.163137

Copyright © 2009 by Annual Reviews. All rights reserved

0066-4278/09/0315-0141$20.00

Key Words

pheromone, olfactory bulbs, medial amygdala, sex behavior, major histocompatibility complex

Abstract

We review studies conducted in mouse and ferret that have specified roles of both the main and the accessory olfactory nervous systems in the detection and processing of body odorants (e.g., urinary pheromones, extraorbital lacrimal gland secretions, major histocompatibility complex peptide ligands, and anal scent gland secretions) that play an essential role in sex discrimination and attraction between males and females leading to mate choice and successful reproduction. We also review literature that compares the forebrain processing of inputs from the two olfactory systems in the two sexes that underlies heterosexual partner preferences. Finally, we review experiments that raise the possibility that body odorants detected by the main olfactory system contribute to mate recognition in humans.

Main olfactory system: composed of the MOE that contains OSNs that project to the MOB. MOB mitral cells project to olfactory cortex and amygdala and include a recently identified projection to the medial (vomeronasal) amygdala

Accessory olfactory system: composed of the VNO neuroepithelium, which contains VSNs that project to the AOB. AOB mitral cells project to nuclei in the medial (vomeronasal) amygdala that in turn project to the BST, MPA, and VMH

Pheromone: a social chemosignal produced by one member of a species that regulates the endocrine function, behavior, or development of a conspecific (105). Signaler pheromones are chemosignals that specify reproductive status so as to contribute to mammalian mate recognition (106)

INTRODUCTION

The olfactory brain can be segregated into different classical and nonclassical components. Classical olfactory processing occurs through olfactory receptor neurons located in the main olfactory epithelium (MOE). In addition to the classical pathway, chemical cues that enter the nose can be detected by sensory neurons within the vomeronasal organ (VNO) epithelium, the septal organ, and the Grueneberg ganglion as well as by free nerve endings in the trigeminal system. Each of these larger systems comprises subsystems defined by differences in receptor mechanisms, signal transduction pathways, and physical location within the nasal cavity. To date, most research concerning mechanisms of mate recognition has focused on the detection of chemosensory cues by either the VNO neuroepithelium or, more recently, the MOE. Nothing is known about the role of the other above-mentioned olfactory systems in mate recognition. Therefore, this review concentrates on the roles of the main and the accessory olfactory systems in mammalian mate recognition. We focus on data for the mouse and ferret (species in which we have both conducted experiments) as well as for the human.

THE MAIN OLFACTORY SYSTEM

In all vertebrates the main olfactory system detects and processes the vast majority of chemical cues that enter the nasal cavity. This system is a chemical categorizer, first detecting a chemical and then cataloging it in the cortex for later recall. The role of the main olfactory system in the detection and processing of chemosensory cues that trigger innate, hard-wired physiological, or behavioral responses is still somewhat controversial but has recently received considerable attention (1–3).

Chemical cues are detected in the MOE after they bind to G protein–coupled receptors on the cilia of olfactory sensory neurons (OSNs). OSNs project their axons to the main olfactory bulb (MOB), where OSNs expressing a single type of olfactory receptor target one or two individual glomeruli (4, 5). Within the MOB, mitral cells that extend an apical dendrite to a single glomerulus in turn extend axons via the lateral olfactory tract to central brain regions including the anterior olfactory nucleus (AO), anterior cortical amygdala (ACo), olfactory turbercle (Tu), piriform cortex (Pir), and entorhinal cortex (Ent) (**Figure 1**). Recent results (details below) also show that MOB mitral cells project to the anterior medial amygdala (Me), where signals may be integrated with inputs coming from the VNO/accessory olfactory system. Inputs from the vomeronasal amygdala, which includes the Me and the posterodorsal Me (MePD), have potential access to limbic targets including the bed nucleus of the stria terminalis (BST), the medial preoptic area (MPA), and the ventromedial hypothalamus (VMH) (6, 7). Thus, odors detected by the MOE may activate neural circuits involved in cortical as well as limbic processing. Until recently, it was thought that the MOE detects only airborne volatile odors. In fact, murine OSNs can also respond to small nonvolatile major histocompatibility complex (MHC) peptide ligands that lead to changes in behavior (8) (more details below).

THE ACCESSORY OLFACTORY SYSTEM

Odorants emitted from other conspecifics (including classically defined pheromones) that are dissolved in nasal mucous gain access to the VNO via an active pumping mechanism (9). These nonvolatile odorants, in turn, bind G protein–coupled receptors located on vomeronasal sensory neurons (VSNs). VSNs extend their axons to the accessory olfactory bulb (AOB) (**Figure 1**). AOB mitral cells send their axons to the vomeronasal amygdala, which in turn, sends projections to several hypothalamic nuclei either directly or via the BST (7, 10). Choi et al. (11) have suggested that interactions between subnuclei of the Me and their projections to the VMH are important for the differential regulation of

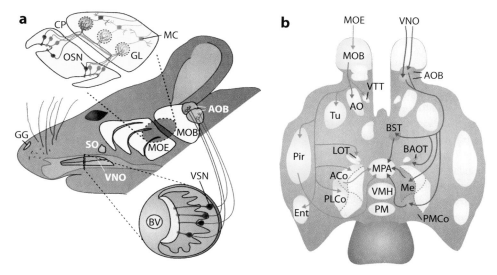

Figure 1

Diagrams showing the projection pathways of the main olfactory epithelium (MOE) and the vomeronasal organ (VNO). (*a*) Olfactory sensory neurons (OSN) in the MOE project through the cribiform plate (CP) to glomeruli (GL) in the main olfactory bulb (MOB). OSNs (*purple*, *blue*, and *orange*) expressing the same olfactory receptor genes converge on single GL in the MOB. MOB mitral cells (MC) extend an apical dendrite into abutting GL and extend axons to the forebrain. Vomeronasal sensory neurons (VSN) extend axons from the VNO to the accessory olfactory bulb (AOB). VSNs in the apical zone (*red*, *left*) of the VNO project to the rostral AOB; VSNs in the basal zone (*green*, *right*) of the VNO project to the caudal AOB. Additional abbreviations used: BV, blood vessel; GG, Grueneberg ganglion; SO, septal organ. (*b*) Forebrain projection pathways of the MOB (*orange*, *left*) and the AOB (*brown*, *right*). Abbreviations used: VTT, ventral tenia tecta; AO, anterior olfactory nucleus; Tu, olfactory turbercle; Pir, piriform cortex; Ent, entorhinal cortex; LOT, nucleus of the lateral olfactory tract; ACo, anterior cortical amygdala; PLCo, posterolateral cortical amygdala; BST, bed nucleus of the stria terminalis; BAOT, bed nucleus of the accessory olfactory tract; Me, medial amygdala; PMCo, posteromedial cortical amygdala; MPA, medial preoptic area; VMH, ventromedial hypothalamus; PM, premammillary nucleus. Adapted from Reference 104 with permission.

reproductive versus defensive behaviors. It has been presumed that the VNO preferentially responds to heavy-molecular-weight, nonvolatile molecules. However, VSNs respond in vitro to both volatile and nonvolatile chemosensory stimuli (12, 13) (more details are given, below).

THE MAIN OLFACTORY SYSTEM DETECTS VOLATILE ODORS IMPORTANT FOR MATE RECOGNITION IN MICE

Most common odorants activate a subset of MOE neurons that express one of up to ~1000 olfactory receptor genes (14). This same family of odorants receptors is also expressed in spermatozoa and may guide sperm to eggs that release chemical cues that bind to these receptors (15). The same family of olfactory receptors, expressed in the MOE, plays an important role in sex discrimination, sexual attraction, and mate recognition in female as well as in male mice.

An early study (16) showed that female mice were significantly more attracted to volatile odors emitted by urine of gonadally intact versus castrated males. This preference was displayed by females that had never had mating experience and was irrespective of stage of their estrous cycle. In a subsequent study (17), virgin female mice, regardless of stage of the estrous cycle, again preferred to investigate volatile odors emitted from gonadally intact versus castrated males. In addition, two volatile compounds isolated from gonad-intact male mouse urine, thiazoline and dehydrobrevicomin, when mixed with castrate male urine, made this urine as attractive to females as was gonad-intact male urine. Females were not attracted to these compounds unless they were dissolved in castrate male urine. Female mice preferred to approach urinary stimuli from normal males as opposed to males from which the preputial glands had been surgically removed (18). Two constituents of preputial gland secretions, α and β farnesene, when added to water, were preferentially approached by virgin female mice over water alone (19). Although volatile odors can gain access to the VNO, and these particular urinary compounds activate VSNs (13), there is no clear evidence that the accessory olfactory system alone can mediate mate recognition in response to these volatile odors.

One study reported that a highly volatile constituent of male urine, (methylthio)methanethiol (MTMT), enhanced the attractiveness of castrate male urine to sexually experienced female mice (20). In addition, females preferred to approach MTMT dissolved in water versus water alone. This study also identified a relatively small number of mitral cells in the ventral and lateral portions of the MOB of female mice that were electrically activated by urinary volatiles from both sexes. Approximately one-third of these urine-responsive neurons were also activated by MTMT. These experiments extend other studies (20–22) showing that volatile urinary odors from males with different MHC haplotypes stimulated c-fos expression in the periglomerular cells of different clusters of glomeruli located in the ventral MOB of female mice. Other investigators (23) also used functional magnetic resonance imaging (fMRI) to show that male urinary volatiles activated the ventral MOB of female subjects.

DAMAGE TO THE MAIN OLFACTORY EPITHELIUM DISRUPTS MATE RECOGNITION IN MICE

Several studies in which the function of the MOE was disrupted point to a critical role of the main olfactory system in heterosexual mate recognition in female mice. Thus, females given zinc sulfate lesions of the MOE lost their preference to investigate soiled bedding from a gonadally intact versus a castrated male (24). More recently (25), this result was extended to show that females with zinc sulfate lesions of the MOE lost both (a) their preference to approach volatile odors emitted from anesthetized, gonadally intact males versus castrated males and (b) their preference to approach urinary volatiles emitted from intact males. Likewise, selective destruction of the MOE in transgenic female mice, in which the expression of a bacterial nitroreductase gene was linked to the olfactory marker protein, eliminated subjects' ability to locate urinary odors from males (26). In another study (27) virgin female mice were exposed for seven days to volatile odorants emitted from the bedding of a dominant, as opposed to a subordinate, male. The females previously exposed to odors from a dominant male later preferred to approach volatile odors emitted from a dominant versus a subordinate male. In addition, there was significantly more incorporation of newly born neurons into the granule cell layer of the MOB in females previously exposed to dominant-male odors than in naive females. These results show that the main olfactory system plays an essential role in identifying volatile odors from potential male mating partners and in motivating females to seek out these conspecifics.

Sexually naive adult male mice preferred to approach urinary volatiles from females (estrous cycle stage not monitored) as opposed to water, whereas males with mating experience strongly preferred to approach urinary volatiles from females as opposed to either water or male urinary volatiles (28). Thus, in male mice, as in females, the ability to detect and prefer opposite-sex urinary volatiles appears to be hard wired, without any requirement for prior mating experience. Zinc sulfate destruction of MOE neurons caused sexually experienced male mice to stop displaying a preference to approach either volatile body odors or odors emitted from soiled bedding of an estrous versus an ovariectomized female (29). The VNO remained intact and able to convey input to the AOB in males with MOE lesions. Yet in the absence of a functional MOE, males showed no preference to seek out opposite-sex urinary volatiles. As stated above, volatile body odors can be detected by VNO sensory neurons (12, 13, 30); however, to date there is no clear evidence that odors detected in this manner motivate approach to opposite-sex conspecifics.

Volatile odorants that influence mate recognition in both sexes initially bind to olfactory receptor proteins expressed in MOE neurons that, in turn, activate a cAMP signaling pathway (31). In one study (32) male mice with a null mutation of the type 3 adenylyl cyclase (AC3−/−) showed neither behavioral (approach) nor MOE electrophysiological responses to urinary volatiles or to a putative mouse pheromone, 2-heptanone. Likewise, these AC3−/− males failed to display aggressive behavior toward a male intruder or sexual behavior toward a receptive female. A targeted null mutation of the cyclic nucleotide-gated channel subunit A2 (CNGA2) also eliminated the preference of sexually naive male mice to investigate female versus male urine (33). Like AC3−/− males, CNGA2 mutant males failed to display aggressive behavior toward a male intruder or mating behavior with an estrous female. Whereas CNGA2 null mutant males failed to prefer female over male urinary odors, mice with a null mutation of the CNGA2 retained MOE electrical responsiveness to several putative pheromones, including 2-heptanone (34). Also, in CNGA2 null mutant mice, pheromonal odors activated (stimulated c-fos expression in) a small set of MOB necklace glomeruli abutting the AOB. These necklace glomeruli are innervated by MOE neurons that express the cone subunit of the CNG channel (CNGA3) instead of CNGA2 and use a cGMP instead of a cAMP signaling system. In addition, another set of glomeruli located primarily in the ventral MOB remained responsive to putative pheromones in CNGA2 null mutant mice. Neurons innervating these glomeruli appear to use a phospholipase C (PLC) signaling pathway, which may involve the transient receptor potential channel M5 (TRPM5) (35). In male mice, as in females, exposure to urinary volatiles of male and female conspecifics activated (stimulated c-fos expression in periglomerular cells) overlapping but distinct clusters of glomeruli located in the ventral MOB (36). Finally, a new family of MOE odor receptor genes that encode trace amounts of volatile amines [trace amine–associated receptors (TARRs)] has been characterized in mice (37). HEK293 cells cotransfected with expression vectors encoding murine TAARs and a cAMP reporter gene (CRE-SEAP) were activated by several different volatile amines as well as by volatiles from adult male urine, but not by volatiles from castrate male or from female urine.

THE MAJOR HISTOCOMPATIBILITY COMPLEX INFLUENCES MURINE MATE CHOICE VIA THE MAIN OLFACTORY EPITHELIUM

Although sex discrimination and attraction to the opposite sex play critical roles in mate recognition, identification of individuals' immune status based on characteristics of body odorants also makes a contribution. Thus, as first shown by Yamazaki et al. (38), the adaptive immune system is directly involved in mate preference. These authors found that mice could discriminate between individuals with different MHC haplotypes and preferred to approach opposite-sex individuals with divergent MHC haplotypes. In the intervening 30 years, odor-based MHC mate preferences have been observed in many species, including humans (39, 40). The potential benefits of a mate preference based on MHC status include (*a*) inbreeding avoidance, leading to increased reproductive fitness; (*b*) increased MHC heterozygosity, resulting in offspring that are more resistant to multiple infections; and (*c*) provision of a moving target against pathogen escape, thereby decreasing the possibility that a pathogen may permanently avoid MHC immune recognition (41). If one or all of these hypotheses hold, then the recognition of individual MHC genotype likely plays an important role in mate preference. Studies in mice using a Y-maze or an olfactometer suggested that volatile urinary odors mediate discrimination of and preference for individuals with different MHC genotypes (38). Hypotheses about how the adaptive immune system produces a unique odor type have ranged from the breakdown products of the MHC class I molecules being excreted in urine to different patterns of body microflora producing unique patterns and concentrations of volatile body odors (42). Recently, Boehm & Zufall (39) proposed that the same peptide ligands that are used by the MHC class I molecules to recognize self versus nonself can also be detected by sensory neurons in the MOE and VNO. Although these peptide ligands are presumably nonvolatile molecules, they can activate OSNs in the MOE, leading male mice to prefer female urinary odors on the basis of the presence of peptides dissimilar to their own MHC genotype (8).

A COMPLEMENTARY ROLE FOR THE ACCESSORY OLFACTORY SYSTEM IN MURINE MATE RECOGNITION

Early anatomical data demonstrating a projection pathway from the VNO to the AOB (10, 43) and on to the Me and hypothalamus (see **Figure 1**) (7, 44) raised the possibility that this accessory olfactory system is the primary mediator of pheromonal actions on aspects of reproduction such as sex discrimination, attraction, and mate recognition (45). Early studies

(46, 47) showed that the ability of sexually naive male mice to display ultrasonic vocalizations after nasal contact with urine or soiled bedding from females was eliminated after VNO inputs to the AOB were destroyed. Males that were exposed to female odors on several occasions prior to VNO removal showed somewhat smaller deficits in ultrasonic vocalizations, implying that MOE inputs could partially compensate for the absence of VNO inputs (47). This view was further supported (48) by the observation that the ability of sexually experienced male mice to display ultrasonic vocalizations in response to female odors was partially attenuated by disruption of either main or accessory olfactory signaling but was totally eliminated by disruption of both olfactory systems. These results established the role of VNO signaling in the display of male-typical social behaviors toward same-sex conspecifics. However, none of this early work specifically identified a role of the VNO in males' ability to discriminate odor cues on the basis of conspecifics' sex or endocrine status.

DISRUPTION OF VOMERONASAL ORGAN SIGNALING IMPAIRS MOTIVATIONAL ASPECTS OF MATE RECOGNITION BUT NOT SEX DISCRIMINATION

Behavioral experiments exploring accessory olfactory system functions languished during the 1990s. However, the discovery (49) that the type 2 transient receptor potential channel (TRP2C) plays an essential role in the generation of action potentials in VNO sensory neurons following the detection of pheromonal stimuli via VR1 and/or VR2 receptors motivated the production of mice with a null mutation of this cation channel. VNO neurons of TRP2C mutant mice showed greatly reduced electrical responses to urine (50). Moreover, TRP2C mutant males displayed no aggressive behavior toward male intruders. These results supported earlier findings that VNO removal eliminated aggressive behavior otherwise displayed by VNO-intact males toward intruder males (51, 52). Instead of aggressive behavior, TRP2C males showed significantly more mounting behavior than did wild-type or heterozygous TRP2C mutant control males toward a male intruder. Finally, TRP2C mutant males indiscriminately displayed ultrasonic vocalizations toward female and male conspecifics, whereas control males vocalized only in the presence of a female. An independent study (53) obtained a similar set of results, including the observation that TRP2C male mice displayed normal mating behavior with an estrous female. However, these mutant males also attempted to mount male intruders instead of attacking them. More recently, it was reported (54) that female TRP2C mutant mice indiscriminately mounted and produced ultrasonic vocalizations toward either a male or a female intruder (presented on separate occasions). By contrast, wild-type and heterozygous TRP2C control females displayed low levels of these two types of behavior toward an intruder female. Wild-type females from which the VNO was surgically removed also showed mounting and ultrasonic vocalization toward a female intruder. All three studies (50, 53, 54) concluded that VNO signaling plays a critical role in sex discrimination and, by inference, in mate recognition.

Surprisingly, none of the three studies that characterized the social behaviors of TRP2C mutant male and female mice assessed subjects' ability to discriminate body odorants from males versus females or to show a preference to investigate urinary odors from animals of the two sexes in different endocrine conditions. In a related study (55), surgical removal of the VNO from sexually naive mice failed to disrupt their ability to discriminate volatile odors from males versus estrous females in habituation/dishabituation tests given in the home cage. VNO removal also failed to disrupt males' sexual behavior in tests with an estrous female. Furthermore, when presented simultaneously with an estrous female and a castrated

male (with urine from a gonadally intact male swabbed on the castrated male's back), VNO-lesioned males and VNO-intact males showed an equivalent preference to mount the estrous female, although both groups directed a minority (∼35%) of their mounts toward the castrated male. The only significant behavioral effect observed was that VNO removal eliminated the preference of both sexually naive mice and sexually experienced male mice to make nasal contact with urine from an estrous female versus a gonadally intact male. In another study (56) sexually naive male mice with intact versus lesioned VNOs showed an equivalent preference to make nasal contact with estrous urine versus water in the two goal boxes of a Y-maze. However, when the choice was between investigating male versus estrous female urine, VNO-intact controls spent significantly more time investigating female urine, whereas VNO-lesioned males spent an equivalent time investigating these two stimuli. Likewise, sexually naive VNO-intact females preferred to approach and nasally contact soiled bedding or urine from a male versus an estrous female in Y-maze tests, whereas VNO-lesioned females did not show such a preference (57). Both VNO-intact females and VNO-lesioned females preferred to approach volatile body odors (anesthetized stimulus animals in a goal box) or volatile urinary odors from a male as opposed to an estrous female when direct nasal access to the stimuli was prevented in Y-maze tests. Presumably, VNO signals generated when males or females come into direct nasal contact with opposite-sex urinary odorants motivate mice to remain in close proximity to these reproductively relevant stimuli. Contrary to the claims of papers describing the phenotype of TRP2C null mutant mice (50, 53, 54), there is no evidence that the accessory olfactory system plays an obligatory role in animals' ability to discriminate the sex of conspecifics on the basis of their pheromone profile. Instead, evidence already reviewed strongly implicates the main olfactory system in this critical first step in mate recognition.

DETECTION AND PROCESSING OF PHEROMONES BY THE ACCESSORY OLFACTORY SYSTEM IS SEXUALLY DIFFERENTIATED IN MICE

Several studies that have directly monitored the responses of murine VNO neurons to pheromones suggest that subjects' sex and endocrine status modulates VNO responsiveness. In one study (12) action potentials of VNO neurons were monitored in vitro with a flat array of 61 extracellular electrodes. Many VNO neurons from each sex responded selectively to different dilutions of male versus female urine, regardless of the individual identity of donors of each sex. In two other studies (58, 59) the expression of immediate early genes (IEGs) c-fos or EGR-1 were used to monitor the activation of murine VNO neurons in response to pheromones emitted from soiled bedding. Both studies revealed that male odorants stimulated significantly higher IEG expression in the female than in the male VNO, and showed that this effect was upregulated by administering estradiol to gonadectomized male and female subjects. In addition, castrated male subjects showed maximal IEG induction to female bedding odorants, and this effect was further enhanced by testosterone. In another study (60) a male mouse–specific peptide (exocrine gland–secreting peptide 1) isolated from the extraorbital lacrimal gland stimulated c-fos expression in sensory neurons located in the basal zone of the female's VNO. Pheromonal cues, detected by the murine VNO, activate a projection circuit that includes the AOB, the Me, and other targets in the hypothalamus. Exposure of male mice to soiled bedding from either male or female conspecifics stimulated IEG expression in the AOB (58, 59, 61). Likewise, nasal application of whole male urine or the low-molecular-weight (but not high-molecular-weight) component of male urine to female mice stimulated c-fos expression in the AOB (57). VNO removal attenuates AOB c-fos responses in both sexes (55, 57).

Male urine stimulates c-fos expression in AOB mitral cells that project to the Me, as shown by retrograde labeling after cholera toxin B (CTB) injections into the Me (62). Interestingly, CTB-Fos-labeled AOB mitral cells were significantly greater in gonadectomized, estradiol-primed females than in males. In another study (63) electrophysiological recordings were made from putative AOB mitral cells in freely behaving mice. These cells were only activated when subjects made direct physical contact with the bodies of stimulus mice—especially with the ano-genital or head/face regions, which are sources of urinary pheromones and exocrine gland–secreting peptide 1. Individual AOB mitral cells were preferentially activated by odors from one sex or mouse strain, raising the possibility that these cells respond to the selective activation of a single type of VNO receptor.

The ability of male body odors to stimulate c-fos expression in the Me, MePD, BST, and MPA is sexually differentiated; only females are responsive (59). This sexually differentiated response appears to be organized by perinatal androgen exposure in the male. Tfm (androgen receptor mutant) male mice showed a female-typical profile of Fos expression in these sites (64). By contrast, neonatal treatment of female mice with the androgen dihydrotestosterone defeminized (eliminated) the later ability of male bedding odors to induce c-fos responses in the BST and MPA (65).

INTEGRATED FUNCTIONING OF THE MAIN AND ACCESSORY OLFACTORY SYSTEMS IN MICE

Most investigators who study murine pheromone signaling accept the dogma that AOB mitral cells project to the vomeronasal medial amygdalar nuclei (Me and MePD), whereas a subset of MOB mitral cells projects to the olfactory amygdala [including the ACo and posterolateral cortical amygdala (PLCo)] (10, 43). Volatile odorants that have the potential to function as pheromones (e.g., MTMT and TAAR ligands) are preferentially detected by the MOE sensory neurons and are processed via the MOB. In contrast, nonvolatile pheromones [e.g., major urinary proteins (MUPs)] are dissolved in nasal mucous and gain access to the VNO lumen, where they activate VNO sensory neurons leading to information processing in the AOB. Several recent studies suggest that volatile pheromones derived from opposite-sex conspecifics selectively activate the AOB following their detection by the MOE and processing in the MOB. An early indication that such signaling may occur came from the observation (23) that in female mice a blood-oxygen-level-dependent (BOLD) fMRI signal was induced in the MOB immediately after exposure to male urinary volatiles whereupon a peak increase in the BOLD signal occurred in the females' AOB several minutes later. This result raises the possibility that volatile urinary odors, initially detected by the MOE, lead to an activation of the AOB. A study (36) confirmed this hypothesis, showing that ovariectomized female mice (given no replacement hormone) showed an increase in AOB mitral and granule layer c-fos expression in response to urinary volatiles from male, but not female, conspecifics. Interestingly, this response was sexually dimorphic in that castrated male mice (again, given no hormones) showed an increase in AOB c-fos expression in response to urinary volatiles from an estrous female, but not a male conspecific. In both sexes these respective odor-induced AOB c-fos responses were blocked by prior zinc sulfate lesioning of the MOE. Additional experiments showed that VNO-AOB signaling remained viable in females previously given intranasal zinc sulfate (nasal application of male urine stimulated significant AOB c-fos responses in these lesioned females). Thus, the absence of an AOB response to urinary volatiles in zinc sulfate–lesioned mice was likely not an artifact of inadvertent damage to the VNO.

There is evidence (66) that centrifugal inputs to the murine AOB originate in the Me and in the bed nucleus of the accessory olfactory tract (BAOT). A recent study (K.M. Martel & M.J. Baum, unpublished results) replicated this anatomical result after injection of the retrograde tracer CTB into the AOB.

In female mice that received AOB injections of CTB, the percentage of retrogradely labeled Me neurons that coexpressed c-fos was significantly increased in animals exposed to urinary volatiles from a male mouse as opposed to a female mouse, a cat, or food odors. This outcome is illuminated by other recent findings (N. Kang, M.J. Baum & J.A. Cherry, unpublished results) showing that a subset of mitral cells, located in the medial and ventral-central MOB, projects axons directly to the Me and MePD, as opposed to the more conventional MOB mitral projection to the ACo and PLCo. Again, after CTB injections into the Me of estrous female mice, exposure to urinary volatiles from male, but not female, conspecifics stimulated CTB/c-fos double labeling of MOB mitral cells that project to the Me. This input is perhaps responsible for the activation of the AOB seen in response to opposite-sex urinary volatiles. The functional significance of AOB activation that originates with MOE signaling remains unclear. Whether MOB-mediated activation of the AOB is a retrograde signaling mechanism or a means whereby the MOB gates the responsiveness of the AOB to opposite-sex odors remains to be determined. However, a recent study (67) showed that sexually naive male mice investigated volatiles emitted from an estrous female as opposed to those from a male when the stimuli were presented in sequence (regardless of order of presentation). A striking reduction in investigation of estrous female, but not of male, urinary volatiles was seen in male mice given either complete or incomplete bilateral lesions of the AOB. As in the case of VNO lesions (55), AOB lesions failed to disrupt males' mating behavior in tests with an estrous female.

These results suggest that the AOB/accessory olfactory system enhances the salience of the volatile opposite-sex pheromones in sexually naive animals, thereby increasing their motivation to investigate these cues. Some additional support for this view is provided by the observation (36) that female subjects showed enhanced c-fos responses in the MPA and nucleus accumbens shell in response to urinary volatiles from male, but not female, conspecifics. This latter region is a projection target of dopamine neurons in the ventral tegmental area that have been implicated in learning and reward functions.

SEXUALLY DIMORPHIC ODOR CUES CONTRIBUTE TO MATE RECOGNITION IN THE FERRET

An early study (68) of feral ferrets showed that males and females establish separate territories in which contact between adults of the two sexes occurs only during the breeding season. The absence of direct social interaction between male and female ferrets raises the possibility that olfactory cues play an important role in ferrets' ability to find a mate. In this respect we would expect to find that scent marks left behind by both male and female ferrets communicate information about conspecifics' sex and reproductive fitness. Clapperton and coworkers (69) used gas chromatography to demonstrate that the composition of anal scent gland secretions is sexually dimorphic in ferrets. Zhang et al. (70) later confirmed and extended these findings with the observation that the composition of volatile compounds contained in urine from male versus female ferrets also differed significantly. Another classic study (71) showed that ferrets display a repertoire of scent-marking behaviors, including urogenital wiping and flank rubbing. Observations of gonadally intact ferrets showed that during the spring breeding season males displayed significantly higher levels of both types of scent-marking behavior than did females. More recently Chang et al. (72) found that after adult gonadectomy and in the absence of any hormone replacement, male ferrets displayed significantly more urogenital wipes than did female ferrets. Furthermore, administration of either estradiol or testosterone augmented the display of this wiping behavior, again with males still displaying significantly higher levels of the behavior than did females. Taken together, these data suggest that some combination of anal scent gland and/or urinary odorants is available to communicate gender as well as breeding (endocrine) status of ferrets.

Clapperton and coworkers (69) provided some of the earliest evidence that ferrets' anal scent gland odorants attract the opposite sex. Thus, in Y-maze tests gonadally intact male and female ferrets were more likely initially to approach anal scent gland odorants from the opposite sex. More recently, Kelliher & Baum (73) monitored the amount of time that subjects spent in proximity to volatile odors emitted from anesthetized estrous female versus breeding male ferrets that were placed in the two arms of an air-tight Y-maze. First, in the absence of previous mating experience, gonadectomized male and female ferrets given either estradiol benzoate (EB) or testosterone propionate (TP) treatments showed no preference for opposite-versus same-sex odors. However, when subjects were retested after receiving mating experience, ovariectomized females treated with either EB or TP spent significantly more time in the vicinity of the Y-maze goal box containing an anesthetized male versus the female stimulus, whereas castrated males given EB or TP preferred the goal box containing an anesthetized female versus the male stimulus. The outcome of this study was corroborated in another study (72) that assessed the time that sexually experienced ferrets spent investigating blocks of wood that had previously been soiled (scent marked) by an estrous female versus a breeding male ferret. Gonadectomized male and female ferrets given no sex hormones showed equal preference for briefly investigating both wood blocks soiled by a male and those soiled by a female. However, following EB or TP treatment, ovariectomized females preferred to investigate blocks soiled with male odors. By contrast, castrated males given TP preferred to investigate blocks soiled with female odors. At first blush, these results imply that ferrets prefer to approach body odors from opposite-sex conspecifics only after they have received mating experience. However, in the study by Kelliher & Baum (73), in significantly more Y-maze trials sexually naive subjects (ovariectomized females treated with EB; castrated males treated with

TP) preferred to approach volatile odors emitted by an anesthetized opposite-sex ferret as opposed to volatile odors emitted by same-sex animals. (Subjects were placed in a start box prior to each trial in the maze.) This same method was used with sexually experienced male and female ferrets to assess the contribution of anal scent gland odorants to subjects' preference to approach opposite-sex body odorants (74). Gonadectomized male (TP-treated) and female (EB-treated) subjects preferred to approach the Y-maze goal box that emitted volatile odorants from opposite-sex anal scent glands; however, these same subjects persisted in their preference to approach a goal box containing an opposite-sex conspecific even if that ferret had been surgically descented. Thus, there appears to be a hard-wired preference among ferrets to approach opposite-sex volatile body odors that likely include a sexually dimorphic suite of compounds emitted from both urine and anal scent glands and/or additional putative odor sources such as lacrimal or skin glands. This preference is only revealed when adult ferrets have sex hormones circulating, and the preference is enhanced in both sexes after they receive mating experience.

THE MAIN OLFACTORY SYSTEM CONTROLS MATE RECOGNITION IN THE FERRET

A series of studies asked whether the main or the accessory olfactory systems play a preferential role in any contribution that pheromones from conspecifics make to heterosexual mate recognition. To the extent that ferrets, like other mammalian species, prefer to approach purely volatile odorants from opposite-sex conspecifics, it appears that any olfactory contribution to mate recognition may depend initially on odor detection and processing via the main olfactory system. Indeed, nasal contact with soiled bedding from opposite-sex conspecifics caused a significant stimulation of c-fos expression in MOB granule cells but not in the cell layer of the AOB (75). In a subsequent study (76) exposure to soiled bedding

from estrous female ferrets again stimulated c-fos expression on the MOB granule cell layer of gonadectomized male and female ferrets, and the magnitude of this response significantly increased after TP administration. Again, there was no odor-induced stimulation of AOB Fos responses in either male or female subjects. When gonadectomized, TP-treated male and female ferrets were exposed to soiled male bedding, and there was, again, a significant induction of Fos expression in MOB granule cells, but not in the AOB. In a parallel set of results (77), male and female ferrets taken at postnatal day 15 were placed in physical (nasal) contact with their anesthetized mother following a 4-h period of separation. Again, Fos responses were reliably seen in the MOB granule cell layer, but not in the AOB. In a final study (78) breeding ferrets of both sexes were exposed to volatile anal scent gland odorants from male versus female conspecifics in breeding condition. The presence of Fos-IR in the periglomerular cells of MOB glomeruli was taken as an index of glomerular activation (22). Groups of estrous female ferrets showed overlapping, although statistically distinct, profiles of glomerular activation in the ventral MOB 90 min after the onset of exposure to male versus female anal scents. A very similar profile of MOB glomerular activation was seen in breeding male subjects following exposure to these same odorants. There was no evidence in either sex that anal scents induced Fos in any cells of the AOB. These data suggest that pheromonal cues activate the main olfactory system in ferrets; however, the absence of an odor-induced Fos response in the AOB cannot be taken as definitive evidence that the ferret's VNO–accessory olfactory system is not activated by pheromones.

Experiments were undertaken selectively to disable either main or accessory olfactory function in ferrets to determine whether either procedure would disrupt heterosexual partner preference or mating behavior itself. An initial study (79) took advantage of the fact that the ferret (a carnivore), unlike rodent species, is able to breathe via the mouth after total occlusion of the nasal sinuses. Infusing dental impression material into the nares of gonadectomized (TP-treated) males and (EB-treated) females eliminated subjects' ability to use a peppermint odor to locate food in Y-maze tests and eliminated all MOB c-fos expression. Even prior to receiving mating experience, sham-occluded females preferred to approach volatile male body odors from an anesthetized male in the Y-maze goal box. Not surprisingly, nares-occluded (anosmic) females approached the volatile odors from male and female stimuli equally. Male ferrets are 30% larger than females. Yet when awake stimulus animals were placed behind a transparent barrier (with holes to allow the passage of odors and sounds) in the goal boxes, nares-occluded females still showed no preference to approach the male stimulus (in contrast to sham-occluded controls). When placed in a small compartment with a stud male ferret, the nares-occluded females, like the control females, readily displayed all aspects of sexually receptive sexual behavior. Later, when again allowed to choose in Y-maze tests between male and female odors alone, or between the suite of odor, visual, and auditory cues emitted from male versus female stimulus ferrets, nares-occluded females (in contrast to control females) displayed no preference for one stimulus over the other. This was even the case in a final series of Y-maze tests in which subjects were allowed to have a brief physical interaction with tethered male versus female stimulus animals on each trial. An identical profile of Y-maze results was obtained in male subjects following bilateral nares occlusion, except that control males in that experiment showed a significant preference to approach the odor stimuli emitted from the estrous female stimulus animals. Although these results alone did not definitively rule out a possible role for the accessory olfactory system in heterosexual mate recognition, they did establish in both sexes the essential role of the main olfactory system in this process.

More definitive evidence that the accessory olfactory system plays no role in the ability of female ferrets to identify and seek out volatile body odors from male conspecifics came from a study (80) in which surgical removal of the VNO from ovariectomized, EB-treated female ferrets failed to disrupt such a preference in Y-maze testing. By contrast, females from which the VNO had been removed spent significantly less time than VNO-intact control females investigating spots of male urine when they were first presented in a sequence of odor stimuli on a piece of filter paper adhered to the front of the subject's home cage. These same two groups of females investigated (made nasal contact with) same-sex urine stimuli for equal times. Both groups investigated either male or female anal scent gland odorants for equal times when they were first presented. It thus appears that VNO inputs generated by odorants emitted from male (opposite-sex) urine, but not anal scents, prolonged subjects' motivation to remain in close proximity to this stimulus. Therefore, whereas VNO-AOB inputs are not required for sex discrimination and heterosexual mate recognition, they may facilitate the motivation to remain in nasal contact with opposite-sex urinary stimuli. As in the case of disrupting main olfactory processing of odors, VNO destruction had no significant effect on the display of receptive sexual behaviors by female ferrets.

FOREBRAIN PROCESSING OF PHEROMONES IS SEXUALLY DIFFERENTIATED IN THE FERRET

Several studies suggest that body odors, detected and initially processed via the main olfactory system, are processed differently in male and female ferrets by projection pathways to the amygdala and on to the hypothalamus. Thus, exposure to soiled bedding from opposite-sex stimulus ferrets stimulated c-fos expression in the MOB granule cell layer (but not in the AOB) and in the medial amygdalar nucleus of both male and female subjects; however, only female subjects showed increased c-fos expression in the MPA and VMH (75). In a subsequent study (76) gonadectomized male and female ferrets (regardless of whether they received TP or oil vehicle) showed significant increases in c-fos expression in the MOB granule cell layer (but not in the AOB), Me, and BST after exposure to soiled female bedding. However, only female subjects showed a significant c-fos response in the MPA. In addition, exposure of gonadectomized, TP-treated male and female subjects to soiled male bedding again activated Fos in the MOB granule cell layer, Me, and BST of both sexes. In addition, however, odorants in soiled male bedding stimulated c-fos expression selectively in the MPA and VMH of female, but not male, subjects. Taken together, these results suggest that body odorants that are initially detected by the MOE and processed via the MOB, Me, and BST in both sexes are differentially processed in the hypothalamus as a function of subjects' sex and odor source (male versus female).

Additional studies suggest that the differential processing of body odorants in different subregions of the hypothalamus may determine the different preferences of male and female ferrets to approach volatile body odors emitted from potential mating partners (male versus female). Early studies (81–83) suggested that the ability of adult male ferrets to display male-typical neck grip, mount, and pelvic thrusting behaviors (that lead to penile intromission) depends on the perinatal actions of testosterone and/or estradiol in the brain. As explained above, gonadectomized, adult ferrets given sex hormones show a sex difference in their preference to approach male versus female stimulus animals (males prefer to approach females; females prefer to approach males). Perinatal treatment of female ferrets with testosterone masculinized their later partner preference profile just as it augmented females' capacity to show male-typical mating responses later in life (84). It is noteworthy that exposing female ferrets to soiled male bedding stimulated c-fos expression in a forebrain circuit leading from the MOB to the Me and terminating in the VMH (75, 76). The functional implication of this observation was revealed by a study (85) in which bilateral electrolytic lesions of the female ferret's VMH eliminated the females' motivation to approach volatile odors emitted from

an anesthetized male ferret housed behind an opaque barrier in the goal box of a Y-maze. Females with VMH lesions also showed a reduction in their motivation to approach and interact briefly with a tethered male in Y-maze tests. By contrast, other females with bilateral lesions of the MPA/AH (anterior hypothalamus) resembled sham-operated control females in displaying a strong preference to approach odors emitted from either an anesthetized male or a tethered male in Y-maze tests. These results suggest that odor inputs processed by the neural circuit from the Me to the VMH encode maleness in female ferrets.

The male ferret, like the male rat (86), possesses a sexually dimorphic structure in the medial MPA/AH. Ferrets have a cluster of large neurons that differentiates solely in males during the last quarter of the 41-day gestation in response to the action of estradiol formed via neural aromatization of testosterone (87, 88). Localized electrolytic lesions of the adult male ferret's sexually dimorphic nucleus failed to disrupt the display of male-typical mating behavior. By contrast, however, destruction of this nucleus in males caused them to display a female-typical profile of estradiol-induced approach responses to a tethered male kept in the goal box of an L-shaped maze (89). Likewise, placement of either excitotoxic (90) or electrolytic (91) lesions in the male ferrets' sexually dimorphic MPA/AH caused them to prefer to approach and interact sexually with a tethered same-sex (male) versus a female conspecific in T-maze tests. Finally, electrolytic lesions of the sexually dimorphic MPA/AH caused male ferrets to switch their preference from female to male body odors in Y-maze tests (92). Male ferrets with bilateral lesion damage to the male nucleus of the MPA/AH showed a female-typical Fos response in the MPA (a site rostral to the lesion) after exposure for 90 min to soiled male bedding.

Taken together, these results suggest that the female-typical default situation is for male odors to activate neurons in the VMH. These odor inputs originate primarily from the main olfactory system, although accessory olfactory inputs may also contribute to the females' motivation to remain in proximity to urinary odor deposits from a male. As a result of hypothalamic sexual differentiation, the male ferret develops a circuit that tonically blocks inputs of social odorant cues to this brain region. For reasons that are not understood, the male thus develops a preference to approach female-derived body odors. After destruction of the male's sexually dimorphic nucleus of the MPA/AH, he partially reverts to the female-typical profile of Fos responsiveness to male odors while showing a female-typical preference to seek out male, as opposed to female, body odors.

MATE RECOGNITION IN HUMANS: EVIDENCE FOR A POSSIBLE ROLE OF THE MAIN OLFACTORY SYSTEM

Several studies have addressed the actions of the volatile odorous steroid 4,16-androstadien-3-one (androstadienone), which is excreted in sweat from the axillary region of men and, to lesser extent, women (93). The MOE odorant receptor OR7D4 mediates responses to androstadienone, and single-nucleotide polymorphisms in the OR7D4 gene reduce humans' ability to detect this odorant (94), resulting in considerable variability in its detection threshold (95). Application of androstadienone to the upper lip exerted a subtle positive effect on mood in female subjects (96), enhanced the masculinity rating of men's faces in male subjects (97), and enhanced cortisol secretion in women (98). However, to date there is no report that androstadienone has any effect on at-titudes of women toward or preference for particular men or on genital arousal in women. The current consensus (99, 100) is that the VNO is not functional in adult humans, perhaps because there is no AOB to which VSNs can project. Thus, any pheromone signaling that occurs in humans presumably depends on pheromone detection by receptor neurons in the MOE and further processing via the MOB, the Me, and the hypothalamus. Savic and colleagues (101) used positron emission tomography (PET) scans to assess the distribution of forebrain sites activated by androstadienone as a function of genetic sex. Androstadienone augmented activity in the hypothalamus of self-identified heterosexual women, but not in the hypothalamus of heterosexual men. The PET method used in this study was not sufficiently sensitive to provide unambiguous resolution of the subnuclei of the hypothalamus that were activated in women by androstadienone. Even so, there is an apparent correspondence between this outcome in humans and previously described results from ferret (76) and mouse (59) in which putative male pheromones more strongly stimulated Fos expression in hypothalamic nuclei of females than in those of males. It has yet to be determined whether the male mouse gender signal, MTMT, is able to duplicate the effects of male mouse urine on the female-specific activation of forebrain c-fos expression. However, there is an obvious analogy between the behavioral (attractant) actions of MTMT in mice and a potential attractant role of androstadienone in humans. Finally, there are reports (102, 103) that women more positively rated body odors from men with dissimilar MHC type.

SUMMARY POINTS

1. The main olfactory system detects volatile pheromones important for sex discrimination and mate recognition in mice.

2. Major histocompatibility complex peptide ligands influence murine mate recognition after their detection by the main olfactory system.

3. The accessory olfactory system contributes to motivational aspects of mate recognition, including attraction to opposite-sex pheromones.

4. Detection and processing of pheromones by the accessory olfactory system are sexually differentiated in mice.

5. Sexually dimorphic processing of pheromones in the hypothalamus controls heterosexual partner preference in ferrets.

6. More research is needed to determine whether either the odorous androgen androstadienone or MHC peptide ligands contribute to mate recognition in humans after their detection and processing via the main olfactory system.

FUTURE ISSUES

1. Do OSNs that express murine TAAR receptors also express TRPM5 cation channels?

2. Do OSNs dedicated to detecting pheromones project to MOB glomeruli that provide direct access via abutting mitral cells to the medial (vomeronasal) amygdala?

3. Is the salience of opposite-sex pheromones that are initially detected by the MOE enhanced by further processing via the AOB prior to these inputs being passed back to the forebrain?

4. Which forebrain circuits bias males to seek out female pheromones and vice versa?

5. Do MHC peptides exist in humans, and are such peptides detected by the OSNs?

6. Do MHC molecules or androstadienone function as human pheromones that contribute to sexual attraction or mate recognition?

DISCLOSURE STATEMENT

The authors are not aware of any bias that might be perceived as affecting the objectivity of this review.

ACKNOWLEDGMENT

We acknowledge support from NIH grants HD044897 and HD21094 (M.J.B.) and CD006603 (K.R.K.).

LITERATURE CITED

1. Kobayakawa K, Kobayakawa R, Matsumoto H, Oka Y, Imai T, et al. 2007. Innate versus learned odour processing in the mouse olfactory bulb. *Nature* 450:503–8
2. Restrepo D, Arellano J, Oliva AM, Schaefer ML, Lin W. 2004. Emerging views on the distinct but related roles of the main and accessory olfactory systems in responsiveness to chemosensory signals in mice. *Horm. Behav.* 46:247–56
3. Shepherd GM. 2006. Behaviour: smells, brains and hormones. *Nature* 439:149–51
4. Ressler KJ, Sullivan SL, Buck LB. 1994. Information coding in the olfactory system: evidence for a stereotyped and highly organized epitope map in the olfactory bulb. *Cell* 79:1245–55

5. Vassar R, Chao SK, Sitcheran R, Nunez JM, Vosshall LB, Axel R. 1994. Topographic organization of sensory projections to the olfactory bulb. *Cell* 79:981–91

6. Dong HW, Petrovich GD, Swanson LW. 2001. Topography of projections from amygdala to bed nuclei of the stria terminalis. *Brain Res. Brain Res. Rev.* 38:192–246

7. Kevetter GA, Winans SS. 1981. Connections of the corticomedial amygdala in the golden hamster. I. Efferents of the "vomeronasal amygdala." *J. Comp. Neurol.* 197:81–98

8. Spehr M, Kelliher KR, Li XH, Boehm T, Leinders-Zufall T, Zufall F. 2006. Essential role of the main olfactory system in social recognition of major histocompatibility complex peptide ligands. *J. Neurosci.* 26:1961–70

9. Meredith M, Marques DM, O'Connell RO, Stern FL. 1980. Vomeronasal pump: significance for male hamster sexual behavior. *Science* 207:1224–26

10. Davis BJ, Macrides F, Youngs WM, Schneider SP, Rosene DL. 1978. Efferents and centrifugal afferents of the main and accessory olfactory bulbs in the hamster. *Brain Res. Bull.* 3:59–72

11. Choi GB, Dong HW, Murphy AJ, Valenzuela DM, Yancopoulos GD, et al. 2005. Lhx6 delineates a pathway mediating innate reproductive behaviors from the amygdala to the hypothalamus. *Neuron* 46:647–60

12. Holy TE, Dulac C, Meister M. 2000. Responses of vomeronasal neurons to natural stimuli. *Science* 289:1569–72

13. Leinders-Zufall T, Lane AP, Puche AC, Ma W, Novotny MV, et al. 2000. Ultrasensitive pheromone detection by mammalian vomeronasal neurons. *Nature* 405:792–96

14. Buck L, Axel R. 1991. A novel multigene family may encode odorant receptors: a molecular basis for odor recognition. *Cell* 65:175–87

15. Spehr M, Gisselmann G, Poplawski A, Riffell JA, Wetzel CH, et al. 2003. Identification of a testicular odorant receptor mediating human sperm chemotaxis. *Science* 299:2054–58

16. Scott JW, Pfaff DW. 1970. Behavioral and electrophysiological responses of female mice to male urine odors. *Physiol. Behav.* 5:407–11

17. Jemiolo B, Alberts J, Sochinski-Wiggins S, Harvey S, Novotny M. 1985. Behavioural and endocrine responses of female mice to synthetic analogues of volatile compounds in male urine. *Anim. Behav.* 33:1114–18

18. Bronson FH, Caroom D. 1971. Preputial gland of the male mouse; attractant function. *J. Reprod. Fertil.* 25:279–82

19. Jemiolo B, Xie TM, Novotny M. 1991. Socio-sexual olfactory preference in female mice: attractiveness of synthetic chemosignals. *Physiol. Behav.* 50:1119–22

20. Lin DY, Zhang SZ, Block E, Katz LC. 2005. Encoding social signals in the mouse main olfactory bulb. *Nature* 434:470–77

21. Schaefer ML, Yamazaki K, Osada K, Restrepo D, Beauchamp GK. 2002. Olfactory fingerprints for major histocompatibility complex-determined body odors II: relationship among odor maps, genetics, odor composition, and behavior. *J. Neurosci.* 22:9513–21

22. Schaefer ML, Young DA, Restrepo D. 2001. Olfactory fingerprints for major histocompatibility complex-determined body odors. *J. Neurosci.* 21:2481–87

23. Xu F, Schaefer M, Kida I, Schafer J, Liu N, et al. 2005. Simultaneous activation of mouse main and accessory olfactory bulbs by odors or pheromones. *J. Comp. Neurol.* 489:491–500

24. Lloyd-Thomas A, Keverne EB. 1982. Role of the brain and accessory olfactory system in the block to pregnancy in mice. *Neuroscience* 7:907–13

25. Keller M, Douhard Q, Baum MJ, Bakker J. 2006. Destruction of the main olfactory epithelium reduces female sexual behavior and olfactory investigation in female mice. *Chem. Senses* 31:315–23

26. Ma D, Allen ND, Van Bergen YC, Jones CM, Baum MJ, et al. 2002. Selective ablation of olfactory receptor neurons without functional impairment of vomeronasal receptor neurons in OMP-ntr transgenic mice. *Eur. J. Neurosci.* 16:2317–23

27. Mak GK, Enwere EK, Gregg C, Pakarainen T, Poutanen M, et al. 2007. Male pheromone-stimulated neurogenesis in the adult female brain: possible role in mating behavior. *Nat. Neurosci.* 10:1003–11

28. Nyby J, Day E, Bean N, Dahinden Z, Kerchner M. 1985. Male mouse attraction to airborne urinary odors of conspecifics and to food odors: effects of food deprivation. *J. Comp. Psychol.* 99:479–90

29. Keller M, Douhard Q, Baum MJ, Bakker J. 2006. Sexual experience does not compensate for the disruptive effects of zinc sulfate–lesioning of the main olfactory epithelium on sexual behavior in male mice. *Chem. Senses* 31:753–62

30. Sam M, Vora S, Malnic B, Ma W, Novotny MV, Buck LB. 2001. Neuropharmacology. Odorants may arouse instinctive behaviours. *Nature* 412:142

31. Firestein S. 2001. How the olfactory system makes sense of scents. *Nature* 413:211–18

32. Wang Z, Balet Sindreu C, Li V, Nudelman A, Chan GC, Storm DR. 2006. Pheromone detection in male mice depends on signaling through the type 3 adenylyl cyclase in the main olfactory epithelium. *J. Neurosci.* 26:7375–79

33. Mandiyan VS, Coats JK, Shah NM. 2005. Deficits in sexual and aggressive behaviors in *Cnga2* mutant mice. *Nat. Neurosci.* 8:1660–62

34. Lin W, Arellano J, Slotnick B, Restrepo D. 2004. Odors detected by mice deficient in cyclic nucleotide-gated channel subunit A2 stimulate the main olfactory system. *J. Neurosci.* 24:3703–10

35. Lin W, Margolskee R, Donnert G, Hell SW, Restrepo D. 2007. Olfactory neurons expressing transient receptor potential channel M5 (TRPM5) are involved in sensing semiochemicals. *Proc. Natl. Acad. Sci. USA* 104:2471–76

36. Martel KL, Baum MJ. 2007. Sexually dimorphic activation of the accessory, but not the main, olfactory bulb in mice by urinary volatiles. *Eur. J. Neurosci.* 26:463–75

37. Liberles SD, Buck LB. 2006. A second class of chemosensory receptors in the olfactory epithelium. *Nature* 442:645–50

38. Yamazaki K, Boyse EA, Mike V, Thaler HT, Mathieson BJ, et al. 1976. Control of mating preferences in mice by genes in the major histocompatibility complex. *J. Exp. Med.* 144:1324–35

39. Boehm T, Zufall F. 2006. MHC peptides and the sensory evaluation of genotype. *Trends Neurosci.* 29:100–7

40. Eggert F, Muller-Ruchholtz W, Ferstl R. 1998. Olfactory cues associated with the major histocompatibility complex. *Genetica* 104:191–97

41. Slev PR, Nelson AC, Potts WK. 2006. Sensory neurons with MHC-like peptide binding properties: disease consequences. *Curr. Opin. Immunol.* 18:608–16

42. Penn D, Potts W. 1998. MHC-disassortative mating preferences reversed by cross-fostering. *Proc. Biol. Sci.* 265:1299–306

43. Scalia F, Winans SS. 1975. The differential projections of the olfactory bulb and accessory olfactory bulb in mammals. *J. Comp. Neurol.* 161:31–55

44. Kevetter GA, Winans SS. 1981. Connections of the corticomedial amygdala in the golden hamster. II. Efferents of the "olfactory amygdala." *J. Comp. Neurol.* 197:99–111

45. Wysocki CJ. 1979. Neurobehavioral evidence for the involvement of the vomeronasal system in mammalian reproduction. *Neurosci. Biobehav. Rev.* 3:301–41

46. Bean NJ. 1982. Olfactory and vomeronasal mediation of ultrasonic vocalizations in male mice. *Physiol. Behav.* 28:31–37

47. Wysocki CJ, Nyby J, Whitney G, Beauchamp GK, Katz Y. 1982. The vomeronasal organ: primary role in mouse chemosensory gender recognition. *Physiol. Behav.* 29:315–27

48. Sipos ML, Wysocki CJ, Nyby JG, Wysocki L, Nemura TA. 1995. An ephemeral pheromone of female house mice: perception via the main and accessory olfactory systems. *Physiol. Behav.* 58:529–34

49. Liman ER, Corey DP, Dulac C. 1999. TRP2: a candidate transduction channel for mammalian pheromone sensory signaling. *Proc. Natl. Acad. Sci. USA* 96:5791–96

50. Stowers L, Holy TE, Meister M, Dulac C, Koentges G. 2002. Loss of sex discrimination and male-male aggression in mice deficient for TRP2. *Science* 295:1493–500

51. Bean NJ. 1982. Modulation of agonistic behavior by the dual olfactory system in male mice. *Physiol. Behav.* 29:433–37

52. Clancy AN, Coquelin A, Macrides F, Gorski RA, Noble EP. 1984. Sexual behavior and aggression in male mice: involvement of the vomeronasal system. *J. Neurosci.* 4:2222–29

53. Leypold BG, Yu CR, Leinders-Zufall T, Kim MM, Zufall F, Axel R. 2002. Altered sexual and social behaviors in *trp2* mutant mice. *Proc. Natl. Acad. Sci. USA* 99:6376–81

54. Kimchi T, Xu J, Dulac C. 2007. A functional circuit underlying male sexual behaviour in the female mouse brain. *Nature* 448:1009–14

55. Pankevich DE, Baum MJ, Cherry JA. 2004. Olfactory sex discrimination persists, whereas the preference for urinary odorants from estrous females disappears in male mice after vomeronasal organ removal. *J. Neurosci.* 24:9451–57

56. Pankevich DE, Cherry JA, Baum MJ. 2006. Effect of vomeronasal organ removal from male mice on their preference for and neural Fos responses to female urinary odors. *Behav. Neurosci.* 120:925–36

57. Keller M, Pierman S, Douhard Q, Baum MJ, Bakker J. 2006. The vomeronasal organ is required for the expression of lordosis behaviour, but not sex discrimination in female mice. *Eur. J. Neurosci.* 23:521–30

58. Halem HA, Baum MJ, Cherry JA. 2001. Sex difference and steroid modulation of pheromone-induced immediate early genes in the two zones of the mouse accessory olfactory system. *J. Neurosci.* 21:2474–80

59. Halem HA, Cherry JA, Baum MJ. 1999. Vomeronasal neuroepithelium and forebrain Fos responses to male pheromones in male and female mice. *J. Neurobiol.* 39:249–63

60. Kimoto H, Haga S, Sato K, Touhara K. 2005. Sex-specific peptides from exocrine glands stimulate mouse vomeronasal sensory neurons. *Nature* 437:898–901

61. Dudley CA, Moss RL. 1999. Activation of an anatomically distinct subpopulation of accessory olfactory bulb neurons by chemosensory stimulation. *Neuroscience* 91:1549–56

62. Kang N, Janes A, Baum MJ, Cherry JA. 2006. Sex difference in Fos induced by male urine in medial amygdala-projecting accessory olfactory bulb mitral cells of mice. *Neurosci. Lett.* 398:59–62

63. Luo M, Fee MS, Katz LC. 2003. Encoding pheromonal signals in the accessory olfactory bulb of behaving mice. *Science* 299:1196–201

64. Bodo C, Rissman EF. 2007. Androgen receptor is essential for sexual differentiation of responses to olfactory cues in mice. *Eur. J. Neurosci.* 25:2182–90

65. Bodo C, Rissman E. 2008. The androgen receptor is selectively involved in organization of sexually dimorphic social behaviors in mice. *Endocrinology* 149:4142–50

66. Barber PC. 1982. Adjacent laminar terminations of two centrifugal afferent pathways to the accessory olfactory bulb in the mouse. *Brain Res.* 245:215–21

67. Jakupovic J, Kang N, Baum MJ. 2008. Effect of bilateral accessory olfactory bulb lesions on volatile urinary odor discrimination and investigation as well as mating behavior in male mice. *Physiol. Behav.* 93:467–73

68. Moors S, Lavers R. 1981. Movements and home range of ferrets at Pukepuke Lagoon, New Zealand. *N. Z. J. Zool.* 8:413–23

69. Clapperton B, Minot E, Crump D. 1988. An olfactory recognition system in the ferret. *Anim. Behav.* 36:541–53

70. Zhang JX, Soini HA, Bruce KE, Wiesler D, Woodley SK, et al. 2005. Putative chemosignals of the ferret (*Mustela furo*) associated with individual and gender recognition. *Chem. Senses* 30:727–37

71. Clapperton B. 1989. Scent-marking behaviour of the ferret. *Anim. Behav.* 38:436–46

72. Chang YM, Kelliher KR, Baum MJ. 2000. Steroidal modulation of scent investigation and marking behaviors in male and female ferrets. *J. Comp. Psychol.* 114:401–7

73. Kelliher K, Baum M. 2002. Effect of sex steroids and coital experience on ferrets' preference for the smell, sight and sound of conspecifics. *Physiol. Behav.* 76:1–7

74. Cloe AL, Woodley SK, Waters P, Zhou H, Baum MJ. 2004. Contribution of anal scent gland and urinary odorants to mate recognition in the ferret. *Physiol. Behav.* 82:871–75

75. Wersinger SR, Baum MJ. 1997. Sexually dimorphic processing of somatosensory and chemosensory inputs to forebrain luteinizing hormone-releasing hormone neurons in mated ferrets. *Endocrinology* 138:1121–29

76. Kelliher KR, Chang YM, Wersinger SR, Baum MJ. 1998. Sex difference and testosterone modulation of pheromone-induced NeuronalFos in the ferret's main olfactory bulb and hypothalamus. *Biol. Reprod.* 59:1454–63

77. Chang YM, Kelliher KR, Baum MJ. 2001. Maternal odours induce Fos in the main but not the accessory olfactory bulbs of neonatal male and female ferrets. *J. Neuroendocrinol.* 13:551–60

78. Woodley SK, Baum MJ. 2004. Differential activation of glomeruli in the ferret's main olfactory bulb by anal scent gland odours from males and females: an early step in mate identification. *Eur. J. Neurosci.* 20:1025–32

79. Kelliher KR, Baum MJ. 2001. Nares occlusion eliminates heterosexual partner selection without disrupting coitus in ferrets of both sexes. *J. Neurosci.* 21:5832–40

80. Woodley SK, Cloe AL, Waters P, Baum MJ. 2004. Effects of vomeronasal organ removal on olfactory sex discrimination and odor preferences of female ferrets. *Chem. Senses* 29:659–69

81. Baum MJ. 1976. Effects of testosterone propionate administered perinatally on sexual behavior of female ferrets. *J. Comp. Physiol. Psychol.* 90:399–410

82. Baum MJ, Gallagher CA, Martin JT, Damassa DA. 1982. Effects of testosterone, dihydrotestosterone, or estradiol administered neonatally on sexual behavior of female ferrets. *Endocrinology* 111:773–80

83. Tobet SA, Baum MJ. 1987. Role for prenatal estrogen in the development of masculine sexual behavior in the male ferret. *Horm. Behav.* 21:419–29

84. Baum MJ, Erskine MS, Kornberg E, Weaver CE. 1990. Prenatal and neonatal testosterone exposure interact to affect differentiation of sexual behavior and partner preference in female ferrets. *Behav. Neurosci.* 104:183–98

85. Robarts DW, Baum MJ. 2007. Ventromedial hypothalamic nucleus lesions disrupt olfactory mate recognition and receptivity in female ferrets. *Horm. Behav.* 51:104–13

86. Gorski RA, Gordon JH, Shryne JE, Southam AM. 1978. Evidence for a morphological sex difference within the medial preoptic area of the rat brain. *Brain Res.* 148:333–46

87. Cherry JA, Basham ME, Weaver CE, Krohmer RW, Baum MJ. 1990. Ontogeny of the sexually dimorphic male nucleus in the preoptic/anterior hypothalamus of ferrets and its manipulation by gonadal steroids. *J. Neurobiol.* 21:844–57

88. Tobet SA, Zahniser DJ, Baum MJ. 1986. Differentiation in male ferrets of a sexually dimorphic nucleus of the preoptic/anterior hypothalamic area requires prenatal estrogen. *Neuroendocrinology* 44:299–308

89. Cherry JA, Baum MJ. 1990. Effects of lesions of a sexually dimorphic nucleus in the preoptic/anterior hypothalamic area on the expression of androgen- and estrogen-dependent sexual behaviors in male ferrets. *Brain Res.* 522:191–203

90. Paredes RG, Baum MJ. 1995. Altered sexual partner preference in male ferrets given excitotoxic lesions of the preoptic area/anterior hypothalamus. *J. Neurosci.* 15:6619–30

91. Kindon HA, Baum MJ, Paredes RJ. 1996. Medial preoptic/anterior hypothalamic lesions induce a female-typical profile of sexual partner preference in male ferrets. *Horm. Behav.* 30:514–27

92. Alekseyenko OV, Waters P, Zhou H, Baum MJ. 2007. Bilateral damage to the sexually dimorphic medial preoptic area/anterior hypothalamus of male ferrets causes a female-typical preference for and a hypothalamic Fos response to male body odors. *Physiol. Behav.* 90:438–49

93. Nixon A, Mallet AI, Gower DB. 1988. Simultaneous quantification of five odorous steroids (16-androstenes) in the axillary hair of men. *J. Steroid Biochem.* 29:505–10

94. Keller A, Zhuang H, Chi Q, Vosshall LB, Matsunami H. 2007. Genetic variation in a human odorant receptor alters odour perception. *Nature* 449:468–72

95. Lundstrom JN, Hummel T, Olsson MJ. 2003. Individual differences in sensitivity to the odor of 4,16-androstadien-3-one. *Chem. Senses* 28:643–50

96. Jacob S, Garcia S, Hayreh D, McClintock MK. 2002. Psychological effects of musky compounds: comparison of androstadienone with androstenol and muscone. *Horm. Behav.* 42:274–83

97. Kovacs G, Gulyas B, Savic I, Perrett DI, Cornwell RE, et al. 2004. Smelling human sex hormone-like compounds affects face gender judgment of men. *Neuroreport* 15:1275–77

98. Wyart C, Webster WW, Chen JH, Wilson SR, McClary A, et al. 2007. Smelling a single component of male sweat alters levels of cortisol in women. *J. Neurosci.* 27:1261–65

99. Brennan PA, Zufall F. 2006. Pheromonal communication in vertebrates. *Nature* 444:308–15

100. Meredith M. 2001. Human vomeronasal organ function: a critical review of best and worst cases. *Chem. Senses* 26:433–45

101. Savic I, Berglund H, Gulyas B, Roland P. 2001. Smelling of odorous sex hormone-like compounds causes sex-differentiated hypothalamic activations in humans. *Neuron* 31:661–68

102. Jacob S, McClintock MK, Zelano B, Ober C. 2002. Paternally inherited HLA alleles are associated with women's choice of male odor. *Nat. Genet.* 30:175–79

103. Wedekind C, Furi S. 1997. Body odour preferences in men and women: Do they aim for specific MHC combinations or simply heterozygosity? *Proc. Biol. Sci.* 264:1471–79

104. Spehr M, Spehr J, Ukhanov K, Kelliher KR, Leinders-Zufall T, Zufall F. 2006. Parallel processing of social signals by the mammalian main and accessory olfactory systems. *Cell Mol. Life Sci.* 63:1476–84

105. Karlson P, Luscher M. 1959. 'Pheromones': a new term for a class of biologically active substances. *Nature* 183:55–56

106. McClintock MK. 2002. Pheromones, odors, and vasanas: The neuroendocrinology of social chemosignals in humans and animals. In *Hormones, Brain and Behavior*, ed. DW Pfaff, AP Arnold, AM Etgen, SE Fahrbach, RT Rubin, pp. 797–870. San Diego: Elsevier

Pheromone Communication in Amphibians and Reptiles

Lynne D. Houck

Department of Zoology, Oregon State University, Corvallis, Oregon 97331;
email: houckl@science.oregonstate.edu

Annu. Rev. Physiol. 2009. 71:161–76

First published online as a Review in Advance on
October 10, 2008

The *Annual Review of Physiology* is online at
physiol.annualreviews.org

This article's doi:
10.1146/annurev.physiol.010908.163134

Copyright © 2009 by Annual Reviews.
All rights reserved

0066-4278/09/0315-0161$20.00

Key Words

chemical signals, vertebrates

Abstract

This selective review considers herpetological papers that feature the
use of chemical cues, particularly pheromones involved in reproductive interactions between potential mates. Primary examples include
garter snake females that attract males, lacertid lizards and the effects
of their femoral gland secretions, aquatic male newts that chemically
attract females, and terrestrial salamander males that chemically per-
suade a female to mate. Each case study spans a number of research
approaches (molecular, biochemical, behavioral) and is related to sensory processing and the physiological effects of pheromone delivery.
These and related studies show that natural pheromones can be identified, validated with behavioral tests, and incorporated in research on
vomeronasal functional response.

INTRODUCTION

Chemical communication is a flourishing area of physiological investigations. At a recent symposium on chemical senses, researchers addressed the genetic and physiological bases of signals that affected olfactory, central, and endocrine responses (1). Reports at this symposium revealed enormous gains in understanding the pathways of olfactory perception and the genetic basis of signal production in these systems. These reports, however, focused almost exclusively on model study systems, including *Drosophila*, *Caenorhabditis elegans*, house mice, and rats. In a separate event, an excellent recent review titled "Pheromonal Communication in Vertebrates" (2) quickly narrowed its focus to mammals.

The focus on mammals and model study systems is understandable. The momentum of recent breakthroughs has created an exciting atmosphere of research discoveries. At the same time, studies of amphibians and reptiles, particularly those documenting the nature and function of protein pheromone signals, have also progressed. These studies address modes of pheromone reception and the evolution of chemical signals across phylogenies of related species. Studies of salamanders, for example, have revealed extremely rapid molecular evolution of pheromone proteins, a situation not yet documented for mammals.

This brief review of pheromone communication in amphibians and reptiles centers on chemical signaling among conspecifics (members of the same species), with a particular emphasis on reproductive pheromones. This focus extends to signal biochemistry as well as the evolution and coevolution of signals and responses. With recent studies of protein pheromones revealing stunning signal variation (e.g., 3), it is time to consider nonmodel studies that can provide perspectives on such extraordinary change. Along these lines, three goals guide this review: (*a*) to consider recent studies of pheromone communication in reptiles and amphibians, (*b*) to provide a convenient and clear summary of the natural pheromones that

are (or easily can be made) available for sensory research, and (*c*) to encourage cross-discipline connections that will facilitate the exploration of similarities and differences in pheromone communication in all vertebrates.

This review is highly selective, not a compendium of herpetological communication via chemical cues. Examples include only chemical signaling among conspecifics; predator-prey interactions are excluded. A particular focus is on reproductive interactions, but topics range from the biochemical characterization of protein-based signals to the neural pathways of receiver response and to the nature of variation in pheromone communication in multiple species within a single genus.

Reptiles and amphibians do not begin to approach mammals in terms of sociality. Nonetheless, snakes, lizards, frogs, and salamanders still need to interact with conspecifics, defend a territory, and determine the sex and reproductive condition of potential mates. Each study of pheromone communication considered below provides information that includes either a new analysis, a framework for comparative differences across related species, or a novel pheromone.

REPTILES

In snakes and lizards, chemical communication with conspecifics is based primarily on protein-lipid signals received when the tongue makes contact with a substrate cue or a conspecific. The tongue acquires molecules that are transferred to the vomeronasal organ (VNO), where pheromone ligands bind with neural receptors (4–11). Lizard species with the capacity for chemical communication typically have femoral glands (on the ventral thigh) that produce lipid-protein secretions. These secretions are placed on a substrate and may reveal individual identity, reproductive condition, and other information.

Below, I consider three main examples of studies that have focused on multiple aspects related to reproductive pheromone interactions. First, however, a few shorter studies are briefly mentioned. These short studies note particular aspects of pheromone communication that are related to reproductive interactions.

Analysis of femoral gland secretions in desert iguanas (*Dipsosaurus dorsalis*) revealed individual-specific patterns of protein banding (12). Desert iguanas communicate by depositing femoral gland secretions on hard surfaces such as rocks. These secretions are composed of 80% protein and 20% lipid material. Individuals differed in gel banding patterns of femoral gland proteins. This paper was one of the first to examine lizard gland secretions for evidence of individual identity.

Supplementation of testosterone (T) increased the rate of femoral gland secretion in *Amphibolurus ornatus* (an agamid lizard from Australia) (13). In addition, castration eliminated femoral gland secretion. Other studies also concur that androgen is necessary for the development of lizard femoral glands as secondary sex characteristics (see citations in Reference 13).

Female wall lizards (*Podarcis hispanica*) could detect and discriminate between male femoral secretions of T-supplemented males versus controls (14). Analysis of secretions that were most attractive to females revealed particular proportions of certain compounds. Reports of this type of female preference are not common because such studies require individual biochemical analyses of secretions from multiple males.

The vomeronasal receptor cells of a turtle (*Pseudemys scripta*) contained high levels of galactosamine sugar residues that are not expressed by receptor cells of the olfactory mucosa (15). These specific sugar residues presumably function in the chemoreception and transduction of pheromone signals.

The short studies summarized above show that lizard pheromones can be used for identification of conspecific individuals and for confirmation of breeding levels of androgen in potential male mates. In addition, the variation in VNO sugar residues in a turtle supports the general separation of main and accessory olfactory receptors common in other vertebrates.

For the three studies of snakes and lizards considered below, the environmental and behavioral background of each study species is summarized. This background information reveals limitations and opportunities that provide a context in which to evaluate physiological and evolutionary circumstances that could affect pheromone signaling.

Snakes

Colubrid snakes are common and are tractable subjects for testing pheromone response. Almost all colubrid snakes are nonvenomous, and snakes kept under laboratory conditions are likely to respond naturally to odorants of significance. In chemosensory experiments, the rate of tongue flicking typically is taken as a measure of an animal's response to particular odorant cues.

Red-sided garter snakes (*Thamnophis sirtalis*). These garter snakes have a widespread distribution across the United States and thus inhabit a range of environments that vary greatly in the duration and timing of mating seasons. The studies discussed below focused on *T. sirtalis parietalis* snakes, which are at the northernmost limits of the *T. sirtalis* distribution.

Context. Red-sided garter snakes in Canada spend approximately eight months of the year hibernating in underground dens. The garter snakes emerge in May, and mating primarily occurs in the vicinity of the den. Males emerge first, and thousands may amass near the entrance of a particularly large den. The skin of a reproductive female is the source of a male-attracting pheromone. An emerging female contends with hundreds of males, each attracted and ready to mate with her. One male is successful at inserting a hemipene into the female's cloaca and transferring spermatozoa. Mating results in the male leaving a copulatory plug in the female's cloaca. This secretory plug apparently dissuades other males from pursuing the female, and she leaves the den area to forage for food. After most females disperse, the males also disperse to forage. In June, an adult male experiences a rise in testosterone levels that co-occurs with spermatogenesis. Mature sperm are stored until the beginning of the next mating season (the following year). This dissociation of spermatogenesis and mating permits a male to transfer sperm to a female immediately upon emergence from the den, rather than delaying mating for the 6–8 weeks (or more) required for spermatogenesis to occur.

Research. Mason and colleagues have conducted behavioral (16), hormonal (17, 18), and biochemical studies of garter snakes living in Canada at the northern edge of their distribution. Reproductive females emerging from a hibernaculum produce a skin compound that happens to act as a male-attracting pheromone. A male can identify a reproductive female simply by contacting her (or her odor trail) by tongue-flicking to make contact with the odor. This behavioral method of assaying the male response to the female was used to verify a synthesized version of the female's male-attracting pheromone (18, 19). To date, the nonvolatile organic pheromone produced by the female is one of only a few dozen vertebrate pheromones that have been chemically identified, synthesized, and shown to have full behavioral effects (20). Whittier & Tokarz (21) provided a detailed description of female mating tendency and the postcopulatory decline in female attractiveness.

Comments. Male-male competition for mates is extreme, and selection presumably favored males that identified a gravid female by responding to her skin odor. In this sense, the female pheromone may simply be a consequence of female reproductive condition; thus, this mate attractant is beneficial to the male but not necessarily advantageous for the female that is inundated by scores of potential mates. The extreme circumstances of the Canadian populations of garter snakes (the massive emergence of snakes at den sites as soon as weather first permits) do not typify the behavior of snakes living in warmer climates. In these warmer areas, in-dividuals may be more widely dispersed, and the mate-attracting pheromone in these populations may actually be beneficial to the female in obtaining a mate.

Lizards

Lacertid lizards are a speciose group (more than 30 genera) that is common in Europe, Asia, and Africa. Often known as wall lizards or rock lizards, lacertids typically have a small body size (up to 9 cm body length). Habitats range from forest and scrub woodlands to grasslands to rocky, arid areas. The relatively dry habitats promote the use of chemical cues that are applied to the substrate.

Iberian rock lizards (*Lacerta monticola* and *Lacerta vivipara*)

Context. The Iberian rock lizard occupies rocky habitats in mountains within the Iberian Peninsula. Males are promiscuous and do not contribute to parental care. Chemical cues released from the femoral pores are a major mode of communication in these lizards, but visual cues (e.g., body coloration and body size) are also important. Substrates that are scent-marked with secretions can function as territorial markers for individual males.

Research. Males can detect and discriminate between self-produced femoral pore secretions and the secretions of conspecific males; a male can also differentiate chemical odors of familiar and unfamiliar males (22). Femoral pore secretions are likely to affect female mate choice (23). Male head size and body size affect the outcome of male-male agonistic interactions, with larger males being dominant. Chemical cues from male femoral secretions reveal information about a male's identity and dominance status (24). The production of secretions is directly related to levels of circulating androgens.

The relative proportions of particular lipophilic chemical compounds found in male secretions are variable; thus, conspecifics may obtain individualized information from the producer of a scent mark (25). In fact, the chemical nature of a male's scent marks alone may be the basis of female mate choice (25): Males with a greater T cell immune response have higher proportions of two steroids (ergosterol and dehydrocholesterol) in their femoral secretions. A link between T cell immunity and a male pheromone would mean that a female attracted to such a male could benefit from this sire (26).

Comments. An extensive series of behavioral and chemical studies in these rock lizards ultimately focused on the hypothesis that a male's mating success might be linked to his production of a chemical signal that was costly to synthesize. The concept of pheromones as indicators of male quality was recently reviewed (27): These pheromones not only must be honest (i.e., accurately revealing a male's status), but also must be costly (i.e., energetically expensive). The multiple functions of the male rock lizard's chemical signal (territorial marker, female mate attractant, etc.), along with, e.g., variables of a male's size, territory location, and accessibility, add some doubt that femoral secretions can convey all this information. Additional experimental evidence (e.g., observing mating success after experimentally giving a weaker male a costly secretion) is necessary to prove that a signal is costly.

Swedish sand lizards (*Lacerta agilis*)

Context. As with the rock lizards, male sand lizards are not social, provide only sperm to the female, and do not provide parental care for offspring. A male will mate repeatedly and with more than one female.

Research. A male often guards a female after copulation (28). Male reproductive success is based on male-male aggression associated with guarding a territory and guarding females; a male can follow the scent trail of a female (29). Males have home ranges and prefer to mate with larger females, which tend to have larger clutches (30). The home range of a female is much smaller than that of the male (30).

Experiments tested whether a female would associate with a male that had a similar major histocompatibility complex (MHC) profile. In fact, females preferred to associate with males that were less similar in MHC loci (31).

Comments. Mate guarding may prevent remating during the time it takes for physiological changes (e.g., higher prostaglandin levels) to dampen the female's receptivity (cf. 21). Research has established that some mammals can avoid individuals with similar MHC odors (e.g., 32–34). Similarly, these female sand lizards also avoid males having a comparable MHC genotype. Additional MHC studies on other reptiles are scarce, as is knowledge of VNO processing of the MHC chemical signal.

Overview

Given the information above, many aspects of the physiological and behavioral components of pheromone response seem highly conserved between reptiles and mammals. Pheromone reception via the VNO is a common feature. Lizards and snakes would also be good candidates for tracing connections between natural pheromone stimulation of the VNO to identify particular receptor subsets. In addition, natural pheromone stimulation may help solve some of the mysteries of functional connectivity between the VNO and the hypothalamus. This basic level of similarity across vertebrate groups is also expected for amphibians, as discussed below.

AMPHIBIANS

Studies of chemical communication in amphibians are not common. Even fewer studies feature integrative research that spans molecular, behavioral, physiological, and comparative levels of inquiry, particularly research with a focus on reproductive pheromones. However, two salamander study systems are highly integrative, and these systems are described below. Before I turn to these detailed studies, I illustrate the scope of pheromone action currently documented for amphibian species.

A new male sex pheromone, splendipherin, was identified as a skin peptide in the Australian tree frog *Litoria splendida* (35). A 25-amino-acid peptide was isolated, chemically characterized, and tested to show that this peptide attracted females in water. A synthetic version of splendipherin also elicited female response. This peptide was the first anuran sex pheromone to be discovered and one of relatively few vertebrate pheromones to be fully biochemically and behaviorally tested (20).

Terrestrial toads (*Pseudophryne bibronii*) could discriminate between male and female conspecific substrate cues (36). Three different experiments supported cue discrimination. (*a*) In behavioral choice trials, female *P. bibronii* preferred substrate marked by either conspecific males or females, as opposed to a control (unmarked) substrate. In contrast, males preferred substrate marked by females and avoided substrate marked by other males (36). (*b*) In a second experiment, females preferentially followed a path having a substrate marked with male chemosignals (versus a path with no odor) to reach a calling male. Female response was enhanced with the combination of acoustic and chemical signals (36). (*c*) In a third experiment, a female odor stimulated an increase in male advertisement calls, but a male odor stimulated territorial calling (36). Studies of male and female odors that elicit very different male responses may provide a robust documentation of the varying neural consequences of pheromone stimulation.

A chemosignal in the dwarf African clawed frogs (*Hymenochirus* sp.) attracted mates (37). Females were attracted to chemical cues in water that had housed (*a*) adult males or (*b*) homogenized male breeding glands. Females did not respond to water-borne cues from other females or to water from males that had their breeding glands removed. In similar trials, males were not attracted to water that housed either females or other males. These behavioral trials are the first to demonstrate a mate-

attractant function for anuran breeding glands. Given the ease with which anurans can be used in laboratory studies, this chemosignal represents an example of a natural odorant waiting to be characterized and used to elucidate the molecular basis of VNO processing.

The architecture of the developing olfactory bulb in the anuran *Xenopus laevis* is highly similar to that of mammals (38). The origin, chemoarchitecture, and central connections of olfactory bulbs are highly conserved among mammals. This new information provides further evidence that tests of amphibian olfactory function may directly relate to mammalian studies.

Vomeronasal receptor genes were expressed in *X. laevis* (39). Genes for the *Xenopus* V2R receptor, along with genes for the G protein G_o, were expressed throughout the sensory epithelium of the VNO.

Axolotls (*Ambystoma mexicanum*) used chemical cues to assess conspecific sex and reproductive condition (40). In this fully aquatic salamander, whole-body odorants were used in behavioral tests. Males were active when exposed to female odorants, but females were not. Electro-olfactogram (EOG) recordings from the olfactory and vomeronasal epithelia produced stronger signals in axolotls responding to other-sex odorants. In males, EOG recordings from particular locations in the nasal epithelia differed in response to stimuli from gravid or from spawned females. The authors concluded that both the vomeronasal and the olfactory systems probably contribute to discrimination of conspecific sex and reproductive condition.

The brief summaries above reveal diverse approaches to the discernment of chemical cues in amphibian species. The following section reveals multilevel approaches for studying the effects of chemical cues in aquatic newts and terrestrial lungless salamanders. Both terrestrial and aquatic salamanders possess a vomeronasal system similar in physiology and morphology to those of other vertebrates (41). The VNO is not physically separated from the main olfactory epithelium, but VNO function and the presence of V2 receptors have been clearly demonstrated (e.g., 42, 43).

Aquatic Newts

Newts are a subgroup of salamanders within the Salamandridae. Newts have an aquatic courtship during which a male deposits a spermatophore in front of the female. The spermatophore is composed of a gelatinous base that supports an apical sperm cap. Sperm transfer occurs when the female moves over the spermatophore, lowers her cloaca, and lifts off after the sperm mass is lodged in her cloaca. Prior to sperm transfer, a male may increase the chances that a female will be inseminated by releasing courtship pheromones into the water.

Japanese newts (*Cynops* spp.): female-attracting pheromone

Context. Japanese newts in the genus *Cynops* have a biphasic life cycle, with adults returning to water to mate and oviposit. The mating season is relatively short, and females typically mate and oviposit soon after entering the breeding pond. During courtship interactions, the male will release a pheromone near the female. This pheromone is secreted by the abdominal gland, released through the cloaca, and transferred to the female via diffusion into the water. This chemical signal most likely functions primarily as a courtship pheromone (as defined in Reference 44) in that the male delivers the pheromone to a female that he is courting. However, the male pheromone can also act as a sex attractant because nearby females that are ready to mate are attracted and will swim toward the scent of the male pheromone.

Research. Studies of the *Cynops* pheromone address female behavioral response, hormonal effects on pheromone synthesis, pheromone variation across and within species, and the genetic basis for pheromone production.

1. The *Cynops* sex attractant, a peptide, is termed sodefrin (in *Cynops pyrrhogaster*) or silefrin (in *Cynops ensicauda*); each of these two peptides comprises 10 amino acid residues (45).

2. Molecular analysis of cDNA encoding these peptides showed that both are produced from precursor proteins. Multiple genes encode sodefrin and its variants (46).

3. Prolactin and androgen regulate pheromone synthesis (47).

4. Prolactin and estrogen enhance female responsiveness; arginine vasotocin (AVT) elicits the release of sodefrin into the water (48, 49).

5. Three types of AVT receptors were cloned in *C. pyrrhogaster*, and each was linked by sequence similarity to a corresponding mammalian AVT receptor. AVT receptor mRNA was expressed in various tissues, including male and female reproductive organs, the brain, and the pituitary (50).

6. An EOG recording from the vomeronasal area showed that females responded to sodefrin solutions in a dose-dependent manner (49).

7. From male *C. pyrrhogaster* in the Nara area of Japan, a sodefrin variant (aonirin) was identified. This variant attracted Nara females but was much less (or not) effective for female *C. pyrrhogaster* from other areas of Japan (51).

Comments. Earlier behavioral studies by Japanese colleagues revealed behavioral (sexual) isolation between species of *Cynops*. Some sexual isolation was also documented between different populations of one species, *C. pyrrhogaster* (52). The work cited above (45–51) confirms that pheromone specificity is critical for female response. Regional variation in female response to the male pheromone has apparently resulted in multiple amino acid changes in a protein pheromone that is only a decapeptide. The currently observed variation in female response may lead to increased reproductive isolation in future generations (52). If so, this isolation would parallel results of the speciation event that created separate pheromones for *C. pyrrhogaster* (sodefrin) and *C. ensicauda* (silifrin). Comparative studies of variation in measures of receptor response (e.g., binding affinities, signaling pathways) would presumably reveal differences in VNO reception as a basis for the species-specific female response.

Terrestrial Salamanders

Plethodontid salamanders are a group of nearly 400 species that are characterized by their lack of lungs. Most species are completely terrestrial: Development occurs within the egg such that the young emerges in the form of a miniature adult. Plethodontids also share the characteristic of having naso-labial grooves. When a salamander's upper lip contacts a moist substrate, these paired, narrow grooves provide transport of water-borne odorants (via capillary action) from the substrate to the nares. This action provides immediate access to odorant cues.

Red-legged salamanders (*Plethodon shermani*): persuasion by courtship pheromones

Context. In contrast to the aquatic newts that deliver pheromone via water currents, *Plethodon shermani* have a completely terrestrial life cycle: Mating and oviposition occur on land. Male courtship pheromones are produced from a submandibular (mental) gland (**Figure 1**). The male delivers gland secretions only to a female and only during courtship (**Supplemental Video 1**; follow the **Supplemental Material link** from the Annual Reviews home page at **http://www.annualreviews.org**). In fact, courtship pheromones have a persuasive function because these pheromones are delivered only if the female is not sufficiently receptive. In contrast to the relatively rapid mating described above for garter snakes in Canada, female *P. shermani* typically have low levels of receptivity because (*a*) the mating season lasts for at least 2–3 months; (*b*) a female typically

Figure 1

Pheromone delivery behavior in the plethodontid salamander *Plethodon shermani*. The male turns back to "slap" his submandibular gland on the female's snout. Pheromones delivered to the female enter the nasal cavities and are shunted laterally to the vomeronasal organ. Typically, a male delivers pheromone multiple times during the course of a courtship. **Supplemental Video 1** shows pheromone delivery when the male's gland contacts the female's nares. Pheromone delivery typically occurs when the male and female are engaged in a tail-straddling walk, as observed in the video. Although this tail-straddling walk indicates some courtship interest on the part of the female, she may leave the male at any time.

is inseminated only 2–3 times during the entire season (cf. 53); (*c*) a female can store sperm (in her spermatheca) for several months; (*d*) oviposition occurs several weeks or even months after the female's last insemination (54); and (*e*) the male does not clasp or restrain the female during courtship (55), so the female can (and does) leave a courting male at any point during their interactions (L.D. Houck, personal observations). In this context, a male typically assesses female behavioral response and then delivers courtship pheromones to a female that does not indicate high levels of receptivity (56, 57). Pheromone delivery occurs when the male "slaps" his mental (submandibular) gland on the female's snout (**Figure 1**; **Supplemental Video 1**). The female cooperates by raising her head so that her nares will be contacted by the male's mental gland. Slapping behavior delivers courtship pheromones directly to the female's nares and hence to her VNO (58). Delivery apparently increases the female's focus on the male and, indirectly, her level of receptivity. Thus, pheromone delivery is a form of chemical persuasion that is promoted by sexual selection (44).

Research. The analysis of this salamander pheromone signaling system has been pursued at multiple levels, from the molecular and biochemical analysis of pheromone components to behavioral tests of female response to pheromone isoforms. Salient results are summarized here.

1. A blend of proteinaceous pheromones is produced in an exocrine gland (the mental gland) on the male's chin. Secretions from the mental gland are applied when the male's chin contacts the female's nares during courtship (55; see **Figure 1**).

2. These pheromones activate sensory neurons of the female's VNO (42, 43; see **Figure 2**).

3. Male pheromones increase female sexual receptivity (tendency to mate) and thus differ from other vertebrate sex pheromones that function to attract

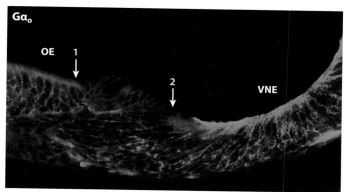

Figure 2

Immunocytochemical labeling of the chemosensory epithelium in *Plethodon shermani*. Vomeronasal epithelium (VNE) demonstrated intense labeling of the dendritic knob and microvillar surface following the application of antisera generated against the alpha subunit of the G_o protein ($G\alpha_o$). Labeling was also apparent in the membrane of the cell bodies. Olfactory epithelium (OE) did not show labeling of the dendritic knob region or cilia. The olfactory and vomeronasal epithelia are separated by a region of vascular connective tissue (the area between *arrows 1* and *2*). These data demonstrate that the transduction mechanism for chemosensory stimuli at the surface of the vomeronasal epithelium includes the use of the $G\alpha_o$ subunit.

mates or to coordinate mating partners (59–62).

4. Administration of pheromone extract to a female's nares enhances early gene expression (c-Fos) in areas of the brain (e.g., amygdala and preoptic areas, ventral hypothalamus, torus/dorsal tegmentum, and raphe median). Thus, pheromone stimulation of the VNO results in higher-level processing that likely leads to further physiological or behavioral effects (63).

5. The connectivity and cytoarchitecture of the ventral telencephalon were studied by labeling cell groups. The rostral part of the urodele amygdala corresponds to the central and basolateral amygdala, and the caudal part of the urodele amygdala corresponds to the cortical/medial amygdala of mammals (64).

6. The neuroanatomical distribution of AVT is widespread in the *P. shermani* brain, including sites in the amygdala, preoptic area, and ventral hypothalamus (65).

Table 1 Individual male profiles for isoforms (A, B, C) of the pheromone plethodontid receptivity factor (PRF) are presented for 22 *Plethodon shermani* salamanders from a single population[a]

Male	A	B	C	Male	A	B	C
1	0	20	80	12	52	18	30
2	9	50	41	13	52	11	37
3	12	66	22	14	53	37	10
4	24	65	11	15	53	33	14
5	26	34	40	16	56	7	37
6	31	0	69	17	56	0	44
7	32	33	35	18	58	4	38
8	32	19	49	19	59	0	41
9	34	22	44	20	64	16	20
10	36	41	23	21	65	9	26
11	42	38	20	22	71	5	24

[a]Percentages of these three isoforms were determined by HPLC analysis of mental gland extracts (cf. 66). Each of the 22 males had a unique percentage of the three isoforms. One male did not express isoform A, and three males did not express isoform B. In an unpublished preliminary study, individual males expressed the same isoform percentage across a multiweek period (R.C. Feldhoff, data not shown). Individual male values are listed in order of the amount of isoform A. Percentages were rounded to whole numbers.

7. The extract from a male's mental gland consists of multiple different proteins, only one of which has been demonstrated to increase female receptivity. This protein pheromone, plethodontid receptivity factor (PRF), is massively expressed in the male mental gland (66). PRF is a 22-kDa protein of 192 amino acids (59, 67).

8. PRF has three variants (isoforms) that together account for ~98% of all PRF proteins in the male's mental gland (see **Table 1**). One of these PRFs (isoform A) was synthesized and was as effective as the native pheromone in eliciting the female behavioral response (61). The substantial expression of isoforms B and C (see **Table 1**) indicates that these PRF proteins also affect female behavior.

9. In a broader investigation of variation in PRF, isoform sequences were obtained for additional species of *Plethodon* salamanders. Comparison of these nucleotide sequences revealed that PRF proteins have undergone rapid, selection-driven evolution over the past 27 million years (67, 68). This continual, rapid evolution is characterized by convergent evolutionary events, gene duplication, the production of pseudogenes, and the consistent maintenance of polymorphisms (67–69). The sexual selection process producing this evolutionary pattern in male courtship pheromones may be driven by coevolution between PRF isoforms and the reception of these isoforms in the female VNO.

Comments. Sexual selection has produced significant changes in the PRF protein pheromone, particularly between different lineages of terrestrial salamanders. This isoform variability is presumably paralleled by similar changes in neural responsiveness in the female VNO (cf. 70). The main question is how changes in the VNO are related to species differences in response to the PRF pheromone. At the population level, a female encounters male pheromones that are mixtures of three different PRF isoforms. Are there measurable differences among females in their VNO response to individual or multi-isoform pheromones?

Overview

Among amphibians, many species have reproductive pheromones that can be or have been chemically characterized. These natural pheromones are ideal for use in functional studies of VNO reception as well as in studies that pursue connections from the amygdala to the hypothalamus.

FUTURE DIRECTIONS

Many Natural Pheromones Are Available for Sensory Studies

Recent research has already documented chemical signals that have undergone measurable evolutionary change, including MHC, mouse urinary proteins (MUPs), and exocrine gland–secreting peptides (ESPs) (e.g., 33, 70, 71). These signals are ideal candidates for studies of sensory processing. Many additional candidates have already been identified in the non-mammalian literature (e.g., 31), and even more natural pheromones are likely to emerge. The *Cynops* newt research alone clearly shows how recent changes in a decapeptide are still altering reproductive interactions between populations of a newt species.

Interactive Effects of Multiple Signals

In signals other than pheromones, interactive effects between signals and signal components are well established. Sensory processing in anuran species provides many examples in which signals interact to affect receiver response (72). In gray tree frogs, for example, females prefer male calls with two spectral peaks over calls with a single high-frequency peak (73, 74). Behavioral response to multiple signals shows similar effects in reptiles. Female sagebrush lizards exposed to displays of both male aggression and courtship altered their physiologies and behaviors in response to combinations of these signals (75). Whether such interactions are mediated at the level of signal transduction or during later sensory processing, signal combinations clearly interact to affect receiver behavior.

Interactive effects may also be very common in the processing of pheromone signals. In arthropods, precise combinations and amounts of particular pheromone components are typically required for a response. Only this required combination will produce a response: Individual components, if presented alone, are ineffective (76). Indeed, most insect pheromones are multicomponent, with synergistic interaction among those components (77). In most vertebrates, physiological and behavioral responses to natural (but complex) chemical signals (e.g., rodent bedding, vaginal secretions) are usually studied in lieu of responses to individual pheromone components, presented singly or in combination. Consequently, interaction between individual pheromone components is a little-studied topic. Synergistic interaction is known, however, in the pheromone communication of mice. Two components isolated from the urine of male mice interact synergistically to mediate aggressive responses (78). At a broader level, many studies have established that pheromones interact with other sensory modalities to produce a particular behavioral response (e.g., 79). In hamsters, for example, deciphering information from a conspecific pheromone depends on the interaction between the conspecific odor and the neural "memory" of the spatial location of the pheromone (80).

The evolutionary dynamics observed for the PRF pheromone across multiple *Plethodon* species are characteristic of proteins that are embedded in sexual communication systems (67, 69). The PRF picture of rapid evolutionary change in pheromone proteins (driven by selection) is an unusual pattern found only in a small minority of signal proteins. Although rare, these rapidly evolving proteins are usually found in sexual communication systems: Proteins that mediate interactions between the sexes may evolve rapidly. Examples include gamete recognition proteins in abalone, sea urchins, and mice (81–84) and several pheromone-associated proteins in rodents (85). Thus, this type of coevolutionary process has often been invoked as an explanation for unusual protein evolution in sexual communication systems (81–88). The Holy Grail in this field is to determine whether receptors rapidly coevolve with their ligands. So far, such direct evidence has been produced only for the abalone lysin-receptor case (86–88). Current work with the *Plethodon* study system is focused on the molecular analysis of V2R receptors that respond to the PRF pheromone.

Receptors in the Vomeronasal Organ

The focus on VNO receptors responding to characterized pheromones resonates with other current research. In a recent paper on proteins that facilitate aggressive behavior in mice, results strongly supported the hypothesis that MUP proteins functioned as pheromone ligands to mediate male-male aggression (89). Also, work by Touhara (3) and colleagues (71) recently revealed the family of ESPs, which apparently includes a repertoire of VNO-specific ligands. Other studies have revealed expression patterns of VNO receptors that suggest that pheromone coding may be achieved by activation of a small subset of odor-specific receptor neurons (70, 90–93). Studies of biochemically characterized reproductive pheromones will provide additional insight into sensory function and are also promising candidates to address functional connectivity in other central sites.

SUMMARY POINTS

1. Research in pheromone communication could benefit from more input from nontraditional study organisms.

2. Studies of nontraditional pheromones can provide new approaches to testing vomeronasal function. Naturally existing variants of terrestrial salamander pheromones (isoforms of PRF) can help reveal mechanisms by which females are preferentially stimulated by particular variants.

3. Protein pheromones can evolve rapidly, and these evolutionary differences—particularly those that occur over relatively short timescales—raise questions about the nature of sensory response to variation in reproductive pheromones.

4. The potential interaction effects of pheromones having multiple components should be tested with existing pheromone systems.

5. The continued advances in sensory physiology provided by work on model systems will make available useful tools for testing hypotheses in nontraditional studies.

DISCLOSURE STATEMENT

The author is not aware of any biases that might be perceived as affecting the objectivity of this review.

ACKNOWLEDGMENTS

Thanks to Karen Kiemnec for help in preparing the manuscript. I appreciate permission to use **Figure 1** (by Stevan J. Arnold) and **Figure 2** (by Celeste Wirsig-Wiechmann). Stevan J. Arnold graciously supplied the two video clips of salamander behavior. Work was supported in part by grants from the National Science Foundation: IOS-0416724 to L.D. Houck and IOS-0416834 to R.C. Feldhoff.

LITERATURE CITED

1. Keystone Symp. 2007. *Chemical Senses: From Genes to Perception (A7), Jan. 21–27, Snowbird, UT.* http://www.keystonesymposia.org/Meetings/ViewPastMeetings.cfm?MeetingID=843

2. Brennen PA, Zufall F. 2006. Pheromonal communication in vertebrates. *Nature* 16:308–15

3. Touhara K. 2007. Molecular biology of peptide pheromone production and reception in mice. In *Genetics of Sexual Differentiation and Sexually Dimorphic Behaviors.* Vol. 59: *Advances in Genetics*, ed. D Yamamoto, pp. 147–71. Cambridge, MA: Elsevier

4. Burghardt GM. 1970. Chemical perception in reptiles. In *Communication by Chemical Signals*, ed. JW Johnston Jr, DG Moulton, A Turk, pp. 241–308. New York: Appleton-Century Crofts

5. Cooper WE. 1994. Chemical discrimination by tongue-flicking in lizards: a review with hypotheses on its origin and its ecological and phylogenetic relationships. *J. Chem. Ecol.* 20:439–87

6. Gans C, Crews D, eds. *Biology of the Reptilia.* Vol. 18: *Physiology E, Hormones, Brain, and Behavior.* Chicago: Univ. Chicago Press

7. Halpern M. 1992. Nasal chemical senses in reptiles: structure and function. See Ref. 6, pp. 423–523

8. Halpern M, Martínez-Marcos A. 2003. Structure and function of the vomeronasal system: an update. *Prog. Neurobiol.* 70:245–318

9. Martínez-Marcos A, Lanuza E, Halpern M. 2002. Neural substrates for processing chemosensory information in snakes. *Brain Res. Bull.* 57:543–46

10. Alberts AC, Phillips JA, Werner DI. 1993. Sources of intraspecific variability in the protein composition of lizard femoral gland secretions. *Copeia* 1993:775–81

11. Simon C. 1983. A review of lizard chemoreception. In *Lizard Ecology, Studies of a Model Organism*, ed. RB Huey, ER Pianka, TW Schoener, pp. 119–33. Cambridge, MA: Harvard Univ. Press

12. Alberts AC. 1990. Chemical properties of femoral gland secretions in the desert iguana *Dipsosaurus dorsalis.* *J. Chem. Ecol.* 16:13–25

13. Ferguson B, Bradshaw SD, Cannon JR. 1985. Hormonal control of femoral gland secretion in the lizard, *Amphibolurus ornatus. Gen. Comp. Endocrinol.* 57:371–76

14. Martín J, López P, Gabirot M, Pilz KM. 2007. Effects of testosterone supplementation on chemical signals of male Iberian wall lizards: consequences for female mate choice. *Behav. Ecol. Sociobiol.* 61:1275–82

15. Franceschini V, Lazzari M, Ciana F. Identification of surface glycoconjugates in the olfactory system of turtle. *Brain Res.* 725:81–87

16. Shine R, Mason RT. 2001. Courting male garter snakes (*Thamnophis sirtalis parietalis*) use multiple cues to identify potential mates. *Behav. Ecol. Sociobiol.* 49:465–73

17. Smith MT, Mason RT. 1997. Gonadotropin antagonist modulates courtship behavior in male red-sided garter snakes, *Thamnophis sirtalis parietalis. Physiol. Behav.* 61:137–43

18. Mason RT. 1992. Reptilian pheromones. See Ref. 6, pp. 114–228

19. Mason RT. 1993. Chemical ecology of the red-sided garter snake, *Thamnophis sirtalis parietalis. Brain Behav. Evol.* 41:261–68

20. Houck LD, Watts RA, Arnold SA, Kiemnec KM, Bowman KE, et al. 2008. A recombinant courtship pheromone affects sexual receptivity in a plethodontid salamander. *Chemical Senses* 33:623–31

21. Whittier JM, Tokarz RR. 1992. Physiological regulation of sexual behavior in female reptiles. See Ref. 6, pp. 24–69

22. Aragón P, López P, Martín J. 2001. Discrimination of femoral gland secretions from familiar and unfamiliar conspecifics by male Iberian rock-lizards, *Lacerta monticola. J. Herpetol.* 35:346–50

23. Aragón P, López P, Martín J. 2001. Chemosensory discrimination of familiar and unfamiliar conspecifics by lizards: implication of field spatial relationships between males. *Behav. Ecol. Sociobiol.* 50:128–33

24. Moreira PL, López P, Martín J. 2006. Femoral secretions and copulatory plugs convey chemical information about male identity and dominance status in Iberian rock lizards (*Lacerta monticola*). *Behav. Ecol. Sociobiol.* 60:166–74

25. López P, Amo L, Martín J. 2006. Reliable signaling by chemical cues of male traits and health state in male lizards, *Lacerta monticola. J. Chem. Ecol.* 32:439–87

26. Aragón P, López P, Martín J. 2006. Roles of male residence and relative size in the social behavior of Iberian rock lizards, *Lacerta monticola. Behav. Ecol. Sociobiol.* 59:762–69

27. Johansson BG, Jones TM. 2007. The role of chemical communication in mate choice. *Biol. Rev.* 82:265–89
28. Olsson M. 1994. Nuptial coloration in the sand lizard *Lacerta agilis*: an intrasexual selected cue to fighting ability. *Anim. Behav.* 48:607–13
29. Olsson M, Madsen T. 2001. Promiscuity in sand lizards (*Lacerta agilis*) and adder snakes (*Vipera berus*). *J. Hered.* 92:190–97
30. Olsson M. 1993. Male preference for large females and assortative mating for body size in the sand lizard. *Behav. Ecol. Sociobiol.* 32:37–41
31. Olsson M, Madsen T, Nordby J, Wapstra E, Ujvari B, Wittsell H. 2003. Major histocompatibility complex and mate choice in sand lizards. *Proc. Biol. Sci.* 270(Suppl. 2):S254–56
32. Penn DJ, Potts WK. 1999. The evolution of mating preferences and major histocompatibility complex genes. *Am. Nat.* 153:145–64
33. Cheetham SA, Thom MD, Jury F, Ollier WER, Beynon RJ, Hurst JL. 2007. The genetic basis of individual-recognition signals in the mouse. *Curr. Biol.* 17:1771–77
34. Kelliher KR. 2007. The combined role of the main olfactory and vomeronasal systems in social communication in mammals. *Horm. Behav.* 52:561–70
35. Wabnitz PA, Bowie JH, Tyler MJ, Wallace JC, Smith BP. 2000. Differences in the skin peptides of the male and female Australian tree frog *Litoria splendida*. *Eur. J. Biochem.* 267:269–75
36. Byrne PG, Keogh JS. 2007. Terrestrial toadlets use chemosignals to recognize conspecifics, locate mates and strategically adjust calling behaviour. *Anim. Behav.* 74:1155–62
37. Pearl CA, Cervantes M, Chan M, Ho U, Shoji R, Thomas EO. 2000. Evidence for a mate-attracting chemosignal in the dwarf African clawed frog, *Hymenochirus*. *Horm. Behav.* 38:67–74
38. Moreno N, Morona R, López JM, Dominguez L, Muñoz M, González A. 2008. Anuran olfactory bulb organization: embryology, neurochemistry and hodology. *Brain Res. Bull.* 75:241–45
39. Hagino-Yamagishi K, Moriya K, Kubo H, Wakabayashi Y, Isobe N, et al. 2004. Expression of vomeronasal receptor genes in *Xenopus laevis*. *J. Comp. Neurol.* 472:246–56
40. Park D, McGuire JM, Majchrzak AL, Ziobro JM, Eisthen HL. 2004. Discrimination of conspecific sex and reproductive condition using chemical cues in axolotls (*Ambystoma mexicanum*). *J. Comp. Physiol. A* 190:415–27
41. Eisthen H, Polese G. 2006. Evolution of vertebrate olfactory subsystems. In *Evolution of Nervous Systems*. Vol. 2: *Non-Mammalian Vertebrates*, ed. JH Kaas, pp. 355–406. Oxford, UK: Academic
42. Wirsig-Wiechmann CR, Houck LD, Feldhoff PW, Feldhoff RC. 2002. Pheromonal activation of vomeronasal neurons in plethodontid salamanders. *Brain Res.* 952:335–44
43. Wirsig-Wiechmann CR, Houck LD, Wood JM, Feldhoff PW, Feldhoff RC. 2006. Male pheromone protein components activate female vomeronasal neurons in the salamander *Plethodon shermani*. *BMC Neurosci.* 7:26
44. Arnold SJ, Houck LD. 1982. Courtship pheromones: evolution by natural and sexual selection. In *Biochemical Aspects of Evolutionary Biology*, ed. M Nitecki, pp. 173–211. Chicago: Univ. Chicago Press
45. Kikuyama S, Yamamoto K, Iwata T, Toyoda F. 2002. Peptide and protein pheromones and amphibians. *Comp. Biochem. Physiol. B Biochem. Mol. Biol.* 132:69–74
46. Iwata T, Conlon JM, Nakada T, Toyoda F, Yamamoto K, Kikuyama S. 2004. Processing of multiple forms of preprosodefrin in the abdominal gland of the red-bellied newt *Cynops pyrrhogaster*: regional and individual differences in preprosodefrin gene expression. *Peptides* 25:1537–43
47. Kikuyama S, Nakada T, Toyoda F, Iwata T, Yamamoto K, Conlon JM. 2005. Amphibian pheromones and endocrine control of their secretion. *Trends Comp. Endocrinol. Neurobiol.* 1040:123–30
48. Toyoda F, Tanaka S, Matsuda K, Kikuyama S. 1994. Hormonal control of response to and secretion of sex attractants in Japanese newts. *Physiol. Behav.* 55:569–76
49. Toyoda F, Kikuyama S. 2000. Hormonal influence on the olfactory response to a female-attracting pheromone, sodefrin, in the newt, *Cynops pyrrhogaster*. *Comp. Biochem. Physiol. B* 126:239–45
50. Hasunuma I, Sakai T, Nakada T, Toyoda F, Namiki H, Kikuyama S. 2007. Molecular cloning of three types of arginine vasotocin receptor in the newt, *Cynops pyrrhogaster*. *Gen. Comp. Endocrinol.* 151:252–58
51. Nakada T, Toyoda F, Iwata T, Yamamoto K, Conlon JM, et al. 2007. Isolation, characterization and bioactivity of a region-specific pheromone, [Val^8]sodefrin from the newt *Cynops pyrrhogaster*. *Peptides* 28:774–80

52. Kawamura T, Sawada S. 1959. On the sexual isolation among different species and local races of Japanese newts. *J. Sci. Hiroshima Univ. Ser. B Div. 1* 18:17–31

53. Adams EM. 2004. Reproductive strategies of the Ocoee Salamander, *Desmognathus ocoee*. PhD thesis. Oregon State Univ. 113 pp.

54. Houck LD, Sever DM. 1994. Role of the skin in reproduction and behaviour. In *Amphibian Biology*. Vol. 1: *The Integument*, ed. H Heatwole, GT Barthalmus, pp. 351–81. Chipping Norton, Aust.: Surrey Beatty & Sons

55. Houck LD, Arnold SJ. 2003. Courtship and mating. In *Phylogeny and Reproductive Biology of Urodela (Amphibia)*, ed. DM Sever, pp. 383–424. Enfield, NH: Sci. Publ.

56. Arnold SJ. 1976. Sexual behavior, sexual interference and sexual defense in the salamanders *Ambystoma maculatum*, *Ambystoma tigrinum*, and *Plethodon jordani*. *Zeit. Tierpsychol.* 42:247–300

57. Houck LD. 1986. The evolution of salamander courtship pheromones. In *Chemical Signals in Vertebrates*, Vol. 4, ed. Duvall D, Muller-Schwarze D, Silverstein RM. New York: Plenum Press. pp. 173–90

58. Dawley CM, Bass AH. 1989. Chemical access to the vomeronasal organs of a plethodontid salamander. *J. Morphol.* 200(2):163–74

59. Rollmann SM, Houck LH, Feldhoff RC. 1999. Proteinaceous pheromone affecting female receptivity in a terrestrial salamander. *Science* 285:1907–1909

60. Rollmann SM, Houck LH, Feldhoff RC. 2003. Conspecific and heterospecific pheromone effects on female receptivity. *Anim. Behav.* 66:857–61

61. Houck LD, Watts RA, Mead LS, Palmer CA, Arnold SA, et al. 2007. A new candidate pheromone, SPF, increases female receptivity in a salamander. In *Chemical Signals in Vertebrates*, Vol. 11, ed. J Hurst, R Beynon, SC Roberts, pp. 213–30. New York, New York: Springer

62. Houck LD. 1998. Integrative studies of amphibians: From molecules to mating. *Amer. Zool.* 38:108–17

63. Laberge F, Feldhoff RC, Feldhoff PW, Houck LD. 2008. Courtship pheromone-induced c-FOS-like immunolabeling in the female salamander brain. *Neuroscience* 151:329–39

64. Laberge F, Roth G. 2005. Connectivity and cytoarchitecture of the ventral telencephalon in the salamander *Plethodon shermani*. *J. Comp. Neurol.* 482:176–200

65. Hollis DM, Chu J, Walthers EA, Heppner BL, Searcy BT, Moore FL. 2005. Neuroanatomical distribution of vasotocin and mesotocin in two urodele amphibians (*Plethodon shermani* and *Taricha granulosa*) based on in situ hybridization histochemistry. *Brain Res.* 1035:1–12

66. Feldhoff RC, Rollmann SM, Houck LD. 1999. Chemical analyses of courtship pheromones in a plethodontid salamander. In *Advances in Chemical Signals in Vertebrates*, ed. RE Johnston, D Muller-Schwarze, PW Sorenson, pp. 117–25. New York: Kluwer Acad./Plenum

67. Watts RA, Palmer CA, Feldhoff RC, Feldhoff PW, Houck LD, et al. 2004. Stabilizing selection on behavior and morphology masks positive selection on the signal in a salamander pheromone signaling complex. *Mol. Biol. Evol.* 21:1032–41

68. Palmer CA, Houck LD. 2005. Responses to sex- and species-specific chemical signals in allopatric and sympatric salamander species. In *Chemical Signals in Vertebrates*, Vol. 10, ed. Mason RT, LeMaster MP, Müller-Schwarze D, pp. 32–41. New York: Kluwer Acad./Plenum

69. Palmer CA, Hollis DM, Watts RA, Houck LD, McCall MA, et al. 2007. Plethodontid modulating factor, a hypervariable salamander courtship pheromone in the three-finger protein superfamily. *FEBS J.* 274:2300–10

70. Dulac C, Torello AT. 2003. Molecular detection of pheromone signals in mammals: from genes to behaviour. *Nat. Rev. Neurosci.* 4:551–62

71. Kimoto H, Sato K, Nodari F, Haga S, Holy TE, Touhara K. 2007. Sex- and strain-specific expression and vomeronasal activity of mouse ESP family peptides. *Curr. Biol.* 17:1879–84

72. Ryan MJ, Rand AS. 2003. Sexual selection and female preference space: how female túngara frogs perceive and respond to complex population variation in acoustic mating signals. *Evolution* 57:2608–18

73. Gerhardt HC. 1992. Multiple messages in acoustic signals. *Sem. Neurosci.* 4:391–400

74. Gerhardt HC. 2005. Acoustic spectral preferences in two cryptic species of grey treefrogs: implications for mate choice and sensory mechanisms. *Anim. Behav.* 70:39–48

75. Kelso EC, Martins EP. 2008. Effects of two courtship display components on female reproductive behaviour and physiology in the sagebrush lizard. *Anim. Behav.* 18:75:639–46

76. Greenfield MD. 2002. *Signalers and Receivers: Mechanisms and Evolution of Arthropod Communication.* Oxford, UK: Oxford Univ. Press

77. Wyatt TD. 2003. *Pheromones and Animal Behaviour: Communication by Smell and Taste.* Cambridge Univ. Press

78. Novotny MV, Ma W, Zidek L, Daev E. 1999. Recent biochemical insights into puberty acceleration, estrus induction and puberty delay in the house mouse. In *Advances in Chemical Signals in Vertebrates,* ed. RE Johnston, D Muller-Schwarze, PW Sorenson, pp. 99–126. New York: Kluwer Acad./Plenum

79. Meredith M. 1998. Vomeronasal, olfactory, hormonal convergence in the brain: cooperation or coincidence? *Ann. N.Y. Acad. Sci.* 855:349–61

80. Mayeaux DJ, Johnston RE. 2004. Discrimination of social odors and their locations: role of lateral entorhinal area. *Physiol. Behav.* 82:653–62

81. Palumbi SR. 1999. All males are not created equal: Fertility differences depend on gamete recognition polymorphisms in sea urchins. *Proc. Natl. Acad. Sci. USA* 96:12632–37

82. Yang Z, Swanson WJ, Vacquier VD. 2000. Maximum-likelihood analysis of molecular adaptation in abalone sperm lysin reveals variable selective pressures among lineages and sites. *Mol. Biol. Evol.* 17:1446–55

83. McCartney MA, Lessios HA. 2004. Adaptive evolution of sperm bindin tracks egg incompatibility in neotropical sea urchins of the genus *Echinometra. Mol. Biol. Evol.* 21:732–45

84. Turner LM, Hoekstra HE. 2006. Adaptive evolution of fertilization proteins within a genus: variation in ZP2 and ZP3 in deer mice (*Peromyscus*). *Mol. Biol. Evol.* 23:1656–69

85. Emes RD, Beatson SA, Ponting CP, Goodstadt L. 2004. Evolution and comparative genomics of odorant- and pheromone-associated genes in rodents. *Genome Res.* 14:591–602

86. Galindo BE, Vacquier VD, Swanson WJ. 2003. Positive selection in egg receptor for abalone sperm lysin. *Proc. Natl. Acad. Sci. USA* 100:4639–43

87. Panhuis TM, Clark NL, Swanson WJ. 2006. Rapid evolution of reproductive proteins in abalone and *Drosophila. Philos. Trans. R. Soc. London Ser. B* 361:261–68

88. Levitan DR, Ferrell DL. 2007. Selection on gamete recognition proteins depends on sex, density, and genotype frequency. *Science* 312:267–69

89. Chamero P, Marton TF, Logan DW, Flanagan K, Cruz JR, et al. 2007. Identification of protein pheromones that promote aggressive behaviour. *Nature* 450:899–902

90. Galindo BE, Moy GW, Swanson WJ, Vacquier VD. 2003. Full-length sequence of VERL, the egg vitelline envelope receptor for abalone sperm lysin. *Gene* 288:111–17

91. Begun DJ, Lindfors HA. 2006. Rapid evolution of genomic *Acp* complement in the *melanogaster* subgroup of *Drosophila. Mol. Biol. Evol.* 22:2010–21

92. Dulac C. 2000. Sensory coding of pheromone signals in mammals. *Curr. Opin. Neurobiol.* 10:511–18

93. Luo M, Katz LC. 2004. Encoding pheromonal signals in the mammalian vomeronasal system. *Curr. Opin. Neurobiol.* 14:428–34

Transcriptional Control of Mitochondrial Biogenesis and Function

M. Benjamin Hock and Anastasia Kralli

Departments of Chemical Physiology and Cell Biology, The Scripps Research Institute, La Jolla, California 92037; email: kralli@scripps.edu

Annu. Rev. Physiol. 2009. 71:177–203

The *Annual Review of Physiology* is online at physiol.annualreviews.org

This article's doi: 10.1146/annurev.physiol.010908.163119

Copyright © 2009 by Annual Reviews. All rights reserved

0066-4278/09/0315-0177$20.00

Key Words

energy homeostasis, exercise training, nuclear receptor, PGC-1

Abstract

Mitochondria play central roles in energy homeostasis, metabolism, signaling, and apoptosis. Accordingly, the abundance, morphology, and functional properties of mitochondria are finely tuned to meet cell-specific energetic, metabolic, and signaling demands. This tuning is largely achieved at the level of transcriptional regulation. A highly interconnected network of transcription factors regulates a broad set of nuclear genes encoding mitochondrial proteins, including those that control replication and transcription of the mitochondrial genome. The same transcriptional network senses cues relaying cellular energy status, nutrient availability, and the physiological state of the organism and enables short- and long-term adaptive responses, resulting in adjustments to mitochondrial function and mitochondrial biogenesis. Mitochondrial dysfunction is associated with many human diseases. Characterization of the transcriptional mechanisms that regulate mitochondrial biogenesis and function can offer insights into possible therapeutic interventions aimed at modulating mitochondrial function.

Oxidative phosphorylation (OxPhos): the process of ATP generation by five inner mitochondrial membrane complexes that oxidize NADH and $FADH_2$—transferring electrons to molecular oxygen and pumping protons across the membrane—and use the ensuing electrochemical gradient to phosphorylate ADP

Mitochondrial transcription factor A (TFAM) and transcription factor B1/B2, mitochondrial (TFB1M/TFB2M): mitochondrial DNA–binding proteins that are encoded by the nuclear genome and that control transcription and replication of the mitochondrial genome

Nuclear respiratory factor-1 (NRF-1) and GA-binding protein (GABP or NRF-2): nuclear DNA–binding factors that regulate transcription of OxPhos and mitochondrial transcription/import genes

Peroxisome proliferator–activated receptors (PPARα, PPARδ, and PPARγ): nuclear receptors that regulate genes involved in lipid transport and metabolism in response to fatty acid–derived ligands

INTRODUCTION

Mitochondria are essential eukaryotic organelles that process glycolysis and lipolysis products to generate, via oxidative phosphorylation (OxPhos), the cellular energy carrier ATP. In addition, mitochondria contain enzymes critical for multiple biosynthetic processes—including lipid, cholesterol, nucleotide, heme, and steroid synthesis—and play important roles in amino acid metabolism and ion homeostasis. Mitochondria also signal, via reactive oxygen species (ROS) and Ca^{2+}, and are critical regulators of cell death pathways. Given their central bioenergetic, metabolic, and signaling roles, tight regulation of mitochondrial mass and mitochondrial function is vital. Notably, mitochondrial mass, function, and morphology differ significantly in different cell types and are dynamically regulated in response to a wide range of physiological cues (e.g., physical activity, nutrient availability, temperature, circadian cues, exposure to infectious agents).

Mitochondrial biogenesis is a complex process that requires the synthesis, import, and incorporation of proteins and lipids to the existing mitochondrial reticulum, as well as replication of the mitochondrial DNA (mtDNA). The mitochondrial proteome comprises ∼1100 to 1500 proteins (1). The vast majority of them are encoded by nuclear genes, and we refer to them in this review as simply mitochondrial genes. The mitochondrial genome encodes only 13 proteins, a small but essential group because all 13 are OxPhos components. Comparison of mitochondria across different tissues shows significant concordance between protein levels and mRNA levels (2), suggesting that mitochondrial mass in a cell is controlled largely, although not solely, at the level of transcription. Thus, mitochondrial biogenesis requires the coordinated transcription of the large number of mitochondrial genes in the nucleus, as well as of the fewer but essential genes in mitochondria. The coordination of the two genomes is achieved by nucleus-encoded mitochondrial proteins, such as TFAM, TFB1M, and TFB2M,

that control the transcription and replication of mtDNA and are induced in response to signals promoting mitochondrial biogenesis (3–5).

Mitochondrial biogenesis is a long-term adaptive response and is not always required to meet transiently increased energetic needs. Transient changes in energy demands can be met by increases in the expression of a subset of mitochondrial genes or of critical regulators and by the enhancement of mitochondrial function. Similarly, expression of subsets of mitochondrial genes is important for the specialized functions of mitochondria in different tissues (e.g., steroid synthesis in adrenal gland or cholesterol in liver) and different physiological states [e.g., expression of uncoupling proteins (UCPs) upon exposure to cold or after a meal, leading to increased thermogenesis]. Strikingly, ∼50% of the mitochondrial genes are expressed in a tissue-specific manner (2), suggesting that a large part of the mitochondrial proteome is dedicated to specialized functions.

The regulation of mitochondrial biogenesis and function presents a transcriptional challenge. Regulatory mechanisms must provide for the induction of the broad mitochondrial gene set and at the same time enable tissue- and signal-specific inductions of gene subsets. Pioneering studies by Scarpulla and colleagues started addressing this challenge by identifying transcription factors that recognize conserved motifs at the promoters of mitochondrial OxPhos genes, leading to the identification of nuclear respiratory factor (NRF)-1 and GA-binding protein (GABP) (also known as NRF-2) (5). In parallel, efforts in the nuclear receptor field to elucidate the function of orphan receptors led to the realization that the peroxisome proliferator–activated receptors (PPARs) control mitochondrial gene subsets with roles in fatty acid oxidation (FAO) and uncoupling (5, 6). A major breakthrough in our understanding of how the different gene subsets are coordinately regulated was the identification of PPARγ coactivator-1α (PGC-1α) as a transcriptional coactivator of NRF-1, GABP, and PPARs and the appreciation of the ability of PGC-1α to integrate physiological signals and to enhance mitochondrial biogenesis and oxidative function (7, 8). PGC-1α also led to the identification of related coactivators [PGC-1β and the PGC-1-related coactivator (PRC)] and other transcription factors [the nuclear receptors ERRs (estrogen-related receptors)] that function in the same or similar pathways (9, 10). This review discusses the transcriptional regulators that control mitochondrial biogenesis and function and the mechanisms by which these regulators sense energetic and metabolic demands associated with different physiological states.

TRANSCRIPTIONAL REGULATORS OF MITOCHONDRIAL BIOGENESIS AND FUNCTION

Overview

Expression of the large number of genes required for mitochondrial biogenesis and function is under the control of a network of nuclear DNA–binding transcription factors and coregulators (**Figure 1**). This network allows for broad and robust activation of the mitochondrial biogenesis program in response to varied physiological cues, as well as specialized tissue- or signal-specific modifications of mitochondrial gene expression and function. In this section, we first discuss the DNA-binding factors, which target overlapping but distinct sets of mitochondrial genes. Notably, each of these factors targets not only mitochondrial genes but also genes with nonmitochondrial functions, which poses the interesting question of how the factors sort out mitochondrial and non-mitochondrial roles. For each factor, we review mechanisms and signals that regulate their activity and expression so as to provide the context in which they contribute to mitochondrial gene expression. Next, we discuss the transcriptional coregulators that enhance or repress the activity of the DNA-binding factors. The ability of the coregulators to interact with multiple DNA-binding factors enables the integration of signals into the broad mitochondrial

gene expression program, as best illustrated for PGC-1α.

DNA-Binding Transcription Factors

Nuclear respiratory factor-1, regulator of OxPhos and mtDNA replication/transcription factors. NRF-1 was identified as a transcription factor binding to a conserved regulatory site of the *cytochrome c* promoter (11). NRF-1 binding sites are evolutionarily conserved in the proximal promoters of many mitochondrial genes (5). Accordingly, NRF-1 activates the expression of OxPhos components, mitochondrial transporters, and mitochondrial ribosomal proteins. In addition, NRF-1 regulates expression of *Tfam*, *Tfb1m*, and *Tfb2m* and thereby coordinates the increased expression of nuclear mitochondrial genes with increases in mtDNA replication and expression (5). NRF-1 may also affect expression of mitochondrial and metabolic genes via indirect mechanisms, e.g., by inducing expression of the transcription factor MEF2A, which activates *Cox* genes, *Glut4*, and *PGC-1α* (12).

Silencing of NRF-1 leads to a significant suppression of mitochondrial target genes, suggesting that endogenous NRF-1 is constitutively active and important for the basal expression of mitochondrial targets (13–15). Nevertheless, NRF-1 activity can also be regulated by phosphorylation and/or interactions with PGC-1α, PGC-1β, PRC, and cyclin D1. Phosphorylation of NRF-1 occurs upon exposure of quiescent fibroblasts to serum (which correlates with induction of *Cycs*) and exposure of hepatoma cells to oxidants (which leads to an NRF-1-dependent induction of *Tfam*). Depending on the context, phosphorylation affects NRF-1 translocation to the nucleus, DNA binding, and/or transcriptional activity (5). Physical interactions of the PGC-1 family members with NRF-1 enhance NRF-1-dependent gene expression (8, 16, 17). Finally, cyclin D1, which suppresses mitochondrial biogenesis, associates with NRF-1 and represses NRF-1 activity (14, 18).

Fatty acid oxidation (FAO) or β-oxidation: the mitochondrial degradation of fatty acids by a cycle of oxidation, hydration, oxidation, and thiolysis reactions, generating acetyl CoA and reducing equivalents (NADH and $FADH_2$)

PPARγ coactivators [PGC-1α, PGC-1β, and PGC-1-related coactivator (PRC)]: coactivators of ERRs, PPARs, NRF-1, GABP, and other transcription factors

Estrogen-related receptors (ERRα, ERRβ, and ERRγ): orphan nuclear receptors that regulate a broad set of mitochondrial genes

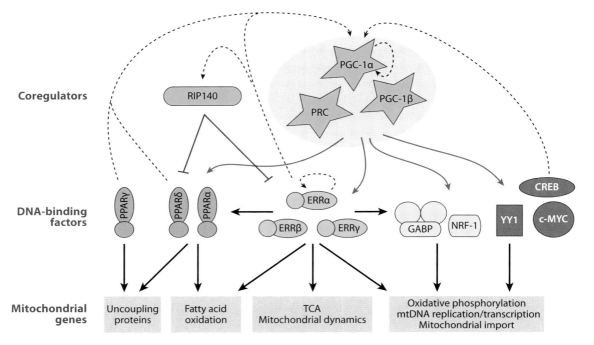

Figure 1

The transcriptional network that controls mitochondrial gene expression. DNA-binding factors regulate overlapping but distinct classes of mitochondrial genes (*solid black arrows*). The relative levels and activities of these factors in different tissues or in response to signals can thus endow mitochondria with specialized mitochondrial functions. Coregulators (PGC-1α, PGC-1β, PRC, and RIP140) interact with multiple DNA-binding factors to coordinate the regulation of multiple classes of mitochondrial genes (*blue arrows* indicate coactivation and *red bars* corepression). The relative levels and activities of the coregulators are major determinants of mitochondrial biogenesis. Several feed-forward and feedback loops (*dashed arrows*) control activity of the network. DNA-binding factors regulate expression of themselves [e.g., ERRα induces PPARα, NRF-1, GABPα, and its own promoter (30, 75)], as well as of coregulators (e.g., PPARγ, PPARδ, ERRγ, and CREB induce PGC-1α, whereas ERRα induces RIP140). These regulatory loops are likely tissue- and signal-specific [e.g., PPARα expression is reduced in BAT but not in WAT of ERRα null mice (10)]. Abbreviations used: BAT, brown adipose tissue; CREB, cAMP response element–binding protein; ERR, estrogen-related receptor; GABPα, GA-binding protein α; NRF-1, nuclear respiratory factor; PPAR, peroxisome proliferator–activated receptor; PGC-1, PPARγ coactivator-1; PRC, PGC-1-related coactivator; RIP140, receptor-interacting protein 140; WAT, white adipose tissue.

AMP-activated protein kinase (AMPK): a kinase that is activated in response to increases in cellular AMP:ATP ratio and that signals energetic stress by phosphorylating target proteins (e.g., acetyl-CoA carboxylase, PGC-1α)

Many signals known to induce mitochondrial biogenesis or respiratory function also induce NRF-1 expression, suggesting that NRF-1 is part of the energy-sensing pathway in mammalian cells. For example, NRF-1 expression is induced by electrical stimulation in cardiac myocytes, respiratory uncoupling in HeLa cells, PGC-1α overexpression in myotubes, serum activation in 3T3 fibroblasts, Ca^{2+} flux in skeletal myotubes, and etoposide stimulation of fibroblasts (5, 8, 19, 20). The induced NRF-1 levels are not just coincidental; dominant-negative NRF-1 inhibits PGC-1α-dependent mitochondrial biogenesis in my-

otubes (8). In vivo, NRF-1 expression in muscle is induced by exercise in rat and zebra fish (21–23). NRF-1 induction by exercise has not been seen in human studies, which may reflect species differences, the types of muscle tested, or the type and duration of physical activity (24, 25). Finally, NRF-1 expression was induced in the muscle of rats fed with a creatine analog that activates AMP-activated protein kinase (AMPK) and induces adaptations similar to those induced by exercise training (26). All together, two major signals have emerged as regulators of NRF-1 expression: increases in Ca^{2+} and activation of AMPK. Whether these signals regulate NRF-1 activity directly, and not simply NRF-1 expression, is not yet known.

In support of a role for NRF-1 as a critical transcription factor for expression of mitochondrial genes, NRF-1 null animals show early embryonic lethality, and NRF-1$^{-/-}$ blastocysts have reduced mtDNA content and mitochondrial membrane potential (27). Although NRF-1 seems necessary for mitochondrial biogenesis, its expression alone is not sufficient to drive this program. Transgenic overexpression of NRF-1 in muscle increases expression of select NRF-1 targets but does not enhance respiratory capacity, suggesting that activation of parallel transcription pathways must complement NRF-1 during exercise-induced muscle mitochondrial biogenesis (28).

Although identified as a transcriptional regulator of mitochondrial genes, NRF-1 also regulates many genes with nonmitochondrial functions and in particular genes with roles in cell-cycle control and proliferation (13). The breadth of NRF-1 function is the likely explanation for why NRF-1 null mice die at an earlier stage [embryonic day (E)3.5–6.5] than do mice lacking the mtDNA replication factor TFAM (E8.5–10.5). In addition, NRF-1 may have developmental functions, as shown for the *Drosophila* and zebra fish NRF-1 orthologs, which play roles in nervous system and muscle development (5).

GA-binding protein. Scarpulla and colleagues (29) identified GABP (also referred to as NRF-2) as an activator of the *CoxIV* promoter. It is a heterotetramer of two distinct and unrelated subunits: GABPα, which contains an ETS domain and serves as the DNA-binding subunit, and GABPβ (β1 or β2, encoded by two homologous genes), which contains a transcriptional activation domain. Functional GABP binding sites have been identified in the proximal promoters of many mitochondrial genes, including ones for OxPhos components, mitochondrial import, and *Tfam*, *Tfb1m*, and *Tfb2m* (which encode the mtDNA transcription factors) (5). Moreover, motifs with the consensus site for GABP, although

common to many promoters, are enriched in a set of coregulated OxPhos genes that show reduced expression in diabetes (30). Consistent with a role of GABP for OxPhos and other mitochondrial gene expression, knockdown of GABPα expression in cells leads to the reduced expression of all 10 nuclear-encoded *Cox* genes (as well as *Tfam*, *Tfb1m*, and the import machinery component *Tomm20*) and a 20% decrease in cellular COX activity (31).

The GABP protein integrates signaling information relevant to mitochondrial biogenesis and function. GABPα and GABPβ become phosphorylated in muscle cells treated with neuregulin, a factor that promotes expression of OxPhos genes (32, 33). In addition, phosphorylated GABP together with host-cell factor 1 (HCF1) recruit the transcriptional coactivators PGC-1α and PRC, which further enhance GABP-dependent transcription (33, 34). The third member of the PGC-1 family, PGC-1β, also interacts with HCF1 (17), suggesting that it may also coactivate GABP.

GABP expression is broad (35) and regulated by developmental and physiological signals that impact mitochondria. GABPα levels increase at times of mitochondrial biogenesis during brown adipose tissue (BAT) development in mice and during brown adipocyte differentiation in vitro (36). Likewise, GABPα is induced in myotubes by Ca^{2+} (19) and in skeletal muscle by exercise (22, 25). In liver, GABPα is induced by treatment with thyroid hormone, which enhances respiratory rate (37). Conversely, GABPα expression is reduced under pathological conditions in which mitochondrial gene expression is dysregulated, such as a rat model of congestive heart failure (38). The GABPα promoter contains regulatory binding sites for the nuclear receptor ERRα and for GABP itself, which enable transactivation of the promoter by PGC-1α (30). Thus, signals that activate PGC-1α, ERRα, or GABP are also likely to enhance GABPα expression.

Although often discussed as a transcription factor for mitochondrial genes, GABP has much wider functions, regulating cell-cycle, ribosomal, myeloid, and neuromuscular junction genes (39). Consistent with a wide range of targets, *Gabpa* disruption in mice results in preimplantation lethality (35), whereas immune-specific loss of GABPα reveals a critical role in B cell development and function (40). Mice lacking GABPα specifically in muscle show no overt muscle defects, with normal distribution and appearance of mitochondria, suggesting that GABP is not essential for muscle mitochondrial biogenesis (41, 42).

Peroxisome proliferator–activated receptors (PPARα, PPARγ, and PPARδ): regulators of lipid metabolism. PPARs are nuclear receptors that sense lipids and control lipid homeostasis. PPARα and PPARδ are primarily regulators of lipid oxidation, whereas PPARγ promotes lipid synthesis and storage. The three receptors have distinct tissue distributions and physiological functions. PPARα levels are highest in the liver, although also expressed strongly in the heart and BAT. PPARα promotes FAO and liver ketogenesis and is important for the response to fasting (6). PPARδ is expressed widely and is particularly abundant in skeletal muscle and heart. PPARδ, which has a broader function in oxidative metabolism than does PPARα, promotes glucose as well as lipid oxidation, enhances metabolic rate, and promotes the formation of oxidative fiber types in skeletal muscle (43, 44). PPARγ is most abundant in adipose tissue, where it promotes adipocyte differentiation and lipogenesis, and is present at lower levels in macrophages, muscle, and liver (6). In addition to endogenous ligands, PPARs are activated by synthetic ligands and drugs: PPARα ligands include the hypolipidemic fibrates, whereas PPARγ is the target of thiazolidinedione (TZD) class of insulin sensitizers (6).

PPARs act as heterodimers with retinoid X receptors (RXRs) to regulate a broad set of genes involved in lipid uptake, storage, and metabolism, including genes encoding mitochondrial FAO enzymes (6). Lipid uptake and metabolism provide substrates for mitochondrial oxidation and are thereby intimately related to mitochondrial function. PPARs also regulate the expression of genes encoding UCPs, i.e., transporters that reside in the inner mitochondrial membrane and play roles in thermogenesis, ROS production, and oxidative capacity. Thus, via their ability to regulate genes of lipid metabolism and mitochondrial UCPs, PPARs are poised to confer cell-type specialization to mitochondria and in particular to enable the use of lipids as high-energy sources for ATP production. Because PPAR ligands are endogenously produced (likely by lipolysis) at specific physiological states and in response to environmental signals (e.g., fasting, exposure to cold, exercise), PPARs also enable mitochondrial adaptation to changing energetic and metabolic needs. Importantly, these regulatory actions of PPARs are integrated with those of other regulators of mitochondrial biogenesis and function, such as NRF-1 and GABP, via PGC-1α and PGC-1β, which interact physically with PPARs and enhance their ability to induce target genes (7, 45, 46). Interestingly, the coactivation function of PGC-1α with PPARγ is gene-specific, suggesting that PGC-1α enhances a selective subset of PPAR targets and may thereby drive PPAR function to specific pathways (7).

In addition to their effects on lipid transport and metabolism, PPARγ and PPARδ promote mitochondrial biogenesis in a cell-type-specific manner. When one thinks of mitochondrial biogenesis, the focus is often on tissues rich in mitochondria, even though mitochondria are critical organelles for all cell types. White adipose tissue (WAT) is not particularly rich in mitochondria. Nevertheless, WAT mitochondrial biogenesis and activity respond dynamically to physiological signals. They become suppressed in animal models of diabetes and diet-induced obesity and are enhanced by treatment with the insulin-sensitizing PPARγ ligands (47–49). Treatment with the PPARγ ligand pioglitazone also increases mitochondrial biogenesis in subcutaneous adipose tissue in humans (50). The ability of PPARγ agonists to enhance mitochondrial biogenesis in adipocytes in vitro suggests that the effects are due to a cell-autonomous function of PPARγ in WAT and are not just a response to systemic changes brought about by PPARγ ligands in other tissues (49, 51). Notably, in all these studies PPARγ agonists induced the expression of endogenous PGC-1α, suggesting that PPARγ affects mitochondrial biogenesis indirectly by enhancing the transcription of PGC-1α. In support of this mechanism, Hondares et al. (52) have identified a functional PPAR response element (PPRE) in the PGC-1α promoter that determines PPARγ-dependent transcription in adipocytes (**Figure 2**). Because PGC-1α coactivates PPARγ, this element also enables PGC-1α to enhance its own expression in an autoregulatory fashion. Notably, these studies suggest that changes in mitochondrial biogenesis in WAT may underlie both diabetes-related pathogenesis and TZD-induced improvements in glycemic control. The PPARγ/PGC-1α-stimulated WAT mitochondrial activity may ameliorate symptoms of metabolic disease by increasing energy expenditure, mitochondrial capacity for lipogenesis, and/or the synthesis and secretion of adipokines. Finally, TZDs and PPARγ may increase not only PGC-1α but also PGC-1β expression (49).

Activation of PPARδ affects mitochondrial biogenesis and function in skeletal muscle. Treatment of mice with an agonist PPARδ ligand enhances muscle lipid uptake, FAO, and mitochondrial biogenesis; it also increases expression of UCPs, GLUT4, and PGC-1α (53). Similarly, transgenic mice expressing PPARδ or a constitutively active PPARδ-VP16 chimera specifically in muscle show enhanced expression of oxidative metabolism and uncoupling genes and a shift to more oxidative muscle fiber types (46, 54). Conversely, mice lacking PPARδ specifically in muscle show a decrease in mitochondrial gene expression and in oxidative capacity (55). Studies with PPARδ ligands in vitro indicate that these effects are largely due to PPARδ-regulated gene expression in muscle (56, 57). Although PPARδ acts directly on genes of lipid metabolism and UCPs, via characterized PPREs, there is no evidence so far for a direct PPARδ effect on mitochondrial biogenesis and OxPhos genes. Thus, it is possible that,

analogous to PPARγ in adipocytes, PPARδ impacts mitochondrial biogenesis via its ability to induce PGC-1α expression in muscle. In support of this, PPARδ ligands induce PGC-1α in muscle in vitro and in vivo, and mice lacking PPARδ show reduced levels of muscle PGC-1α expression (52, 53, 58). The same PPRE that mediates PPARγ-dependent induction of PGC-1α in adipocytes mediates PPARδ-dependent induction in myotubes (52, 55, 59).

Similar pathways of PPAR-dependent enhancement of PGC-1α and/or PGC-1β may take place in other tissues. PPARγ and PPARδ promote oxidative metabolism and mitochondrial biogenesis during alternative activation in macrophages (60, 61). PGC-1β, which is important for macrophage alternative activation (62), is expressed at reduced levels in PPARδ null macrophages (63). Finally, although no effect of PPARα or PPARα ligands on PGC-1 expression has yet been reported, given the central and common pathways of regulation and function of PPARα and PGC-1α in liver, we speculate that PPARα contributes to the fasting-induced expression of PGC-1α in this tissue. Importantly, these observations suggest that PPARs, primarily appreciated for their effects on lipid metabolism, may have a wider impact on mitochondrial biogenesis and function by acting as transcriptional regulators of PGC-1 coactivators.

Estrogen-related receptors (ERRα, ERRβ, and ERRγ): regulators of a broad mitochondrial program. ERRα, ERRβ, and ERRγ are members of the nuclear receptor superfamily and the most recent discoveries in the mitochondrial gene expression regulatory network. As their name implies, ERRs show sequence similarities to the estrogen receptor, particularly in the DNA-binding and ligand-binding domains (64). Despite this similarity, ERRs are not activated by estrogens or estrogen-like molecules and can attain constitutively active ligand-binding domain conformations in the absence of a ligand (65, 66). The transcriptional activity of ERRs is instead regulated via

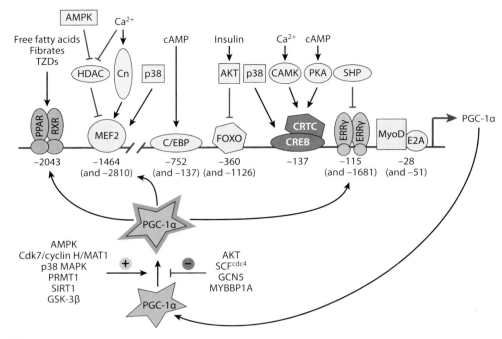

Figure 2

Positive and negative signals that regulate mitochondrial biogenesis and converge at the PGC-1α promoter and/or the PGC-1α protein. The upper part of the figure shows regulatory sites of the PGC-1α promoter, the corresponding DNA-binding transcription factors, and signals activating (*black arrows*) or repressing (*red bars*) these factors. The promoter is not drawn to scale; where multiple *cis*-regulatory elements have been identified, only one is shown; and numbering is based on Reference 145. PGC-1α can coactivate several transcription factors that bind the PGC-1α promoter (PPARs, MEF2, and ERRγ, indicated by *arrows* below the promoter). As a result, signals that enhance or repress PGC-1α protein activity (shown at *bottom* and indicated by the *black arrow/green plus sign* and *red bar/red minus sign*, respectively) can similarly affect PGC-1α transcription. Many of the factors and signals shown here are expressed or active in a tissue-specific manner, so this diagram does not represent PGC-1α regulation in all tissues. An alternative promoter, upstream to the one studied so far and shown here, was described recently (178). The alternative promoter drives the expression of a cDNA with an alternatively spliced first exon and is reported to be the one induced by exercise and cold in skeletal muscle and BAT, respectively (178).

physical interaction with coregulators, such as coactivators of the SRC and PGC-1 family or corepressors like receptor-interacting protein 140 (RIP140) (10, 64). ERRα, in particular, enhances gene expression only when partnered with the PGC-1 coactivators, which convert ERRα from a latent to a potent transcriptional activator (67–69). Conversely, ERRs in complex with RIP140 are thought to repress gene expression (70). ERRs are also subject to regulation by phosphorylation and sumoylation, which affect DNA binding, dimerization, and coactivator interaction (64). Although the func-

tional consequences of these modifications for the regulation of mitochondrial gene expression are yet to be determined, one activator of ERRα is neuregulin, which also promotes mitochondrial function (32).

ERRs bind as monomers, homodimers, or heterodimers of different ERRs to ERR response elements (ERREs) (64). The identification of one of the first ERREs as a regulatory site in the promoter of the FAO gene *Acadm* suggested that ERRs could be regulators of FAO. Recent studies have indeed established a role for ERRα, the best characterized of the

Receptor-interacting protein 140 (RIP140 or NRIP1): a corepressor of many nuclear receptors three ERRs, in the regulation of lipid oxidation genes, as well as a wider set of mitochondrial genes, including components of OxPhos, tricarboxylic acid (TCA) cycle, mitochondrial import, mitochondrial dynamics, and oxidative stress defenses (10, 64). The actions of ERRα are mediated via ERREs present at the regulatory regions of many mitochondrial genes (71–74). ERREs are significantly enriched in OxPhos and PGC-1α-regulated gene sets and are often found in association with NRF-1 or GABP sites (30, 72). ERRα also acts on mitochondrial gene expression through indirect mechanisms by regulating the expression of GAPBα and PPARα (30, 75). siRNA and pharmacological approaches have shown that endogenous ERRα is required for the ability of exogenously expressed PGC-1α to induce mitochondrial biogenesis and respiration (30, 71). Conversely, overexpression of ERRα in cardiac myocytes acts similarly to PGC-1α, increasing lipid oxidation rates (75). Expression of an ERRα-VP16 chimera, which no longer requires PGC-1α or PGC-1β to be transcriptionally active, can by itself induce mitochondrial biogenesis, suggesting that activation of ERR target genes is central to the regulation of mitochondrial function (71).

Consistent with a role in mitochondrial biogenesis and function, ERRs are expressed at high levels in tissues with high energetic demands (64). Moreover, ERRα, which is induced by PGC-1α via a positive autoregulatory loop (30, 69, 76), responds to signals central to the regulation of mitochondrial biogenesis or function, such as upon exposure to cold (in BAT and muscle), fasting (in liver), and exercise (in skeletal muscle) (25, 69, 77). Studies of ERRα null mice have provided further support and clarification on the physiological role of ERRα for mitochondrial function. The BAT of ERRα null mice shows a 40% decrease in mitochondrial content and oxidative capacity. Although mild, the energetic deficiency renders these mice unable to defend their body temperature when challenged with even mildly cold temperatures (13°C), despite normal expression of the thermogenic *Ucp1* (74). An en-

ergetic deficiency, without alteration in mitochondrial content, is also seen in the heart. The defect manifests as signs of heart failure when animals are challenged with increased workload (78). Finally, ERRα is necessary for increased rates of respiration and ROS production in macrophages stimulated with interferon β (IFNβ). Consequently, ERRα null macrophages have a decreased ability to clear pathogens, and ERRα null mice have impaired survival rates when infected with *Listeria* (73).

Changes in mitochondrial gene expression or oxidative capacity are also seen in other tissues of ERRα null mice, such as WAT, intestine, and skeletal muscle (75, 79, 80), supporting a wide role of ERRα in mitochondrial function. Nevertheless, ERRα null mice are viable and fertile, suggesting that the mitochondrial defects are mild. The mild phenotype can be interpreted in at least two not mutually exclusive ways. First, ERRα may simply be part of the mechanism conferring adaptation to tissue- and physiology-specific cues and may not be important for basal levels of mitochondrial gene expression. This would be consistent with the defects in ERRα null mice being most apparent in tissues with highest mitochondrial content and in states of increased energy demand. Second, loss of ERRα function may be compensated by increased activity of other ERRs or other factors with similar roles. Consistent with this idea, ERRα null mice show increased expression of ERRγ and PGC-1α in heart; ERRγ in skeletal muscle; and ERRγ, PPARα, and PGC-1β in adipose tissue (75; J.A. Villena & A. Kralli, unpublished data). Furthermore, recent studies show that ERRα and ERRγ target highly overlapping sets of mitochondrial genes in heart (72) and that ERRγ null mice, which die perinatally, have signs of mitochondrial dysfunction and decreased oxidative capacity in the heart (81). The extent to which ERRβ may compensate for the absence of ERRα is not yet clear.

Given the evidence that ERRα activates the expression of many mitochondrial genes, a striking phenotype of ERRα null mice is decreased adiposity and a resistance to high-fat diet–induced obesity (79). The underlying

Tricarboxylic acid (TCA) cycle: the mitochondrial matrix pathway that oxidizes acetyl CoA (as well as other products of amino acid metabolism) to produce ATP and reducing equivalents (NADH and $FADH_2$) for OxPhos

mechanism(s) for this phenotype is not yet clear but may include a decrease in lipid absorption in the intestine (80); a developmental role for ERRα in adipocyte differentiation and function (82); defective lipogenesis, possibly due to an energetic and/or TCA cycle deficiency (79); and increased energy expenditure due to the increased expression of compensating regulators, such as PGC-1α, PGC-1β, and ERRγ.

As seen with other transcription factors that regulate mitochondrial gene expression, ERRs also have roles as regulators of nonmitochondrial programs. ERR-regulated targets include genes with roles in Ca^{2+} homeostasis, contractile function, glucose metabolism, angiogenesis, lung maturation, ion channel and transporter expression, and other cellular processes (64). ERRβ in particular plays important roles in extraembryonic cells during development (83), in embryonic stem cell biology (84), and in the development of the endolymph-producing cells of the inner ear (85).

Other nuclear transcription factors: CREB, c-Myc, and YY1. Additional transcriptional regulators with broad biological functions contribute to the control of mitochondrial gene expression. The cAMP response element–binding (CREB) protein regulates *Cycs* expression; binding sites for CREB protein are present not only at the *Cycs* but also at the *Cox5a*, *Cox8a*, *Idh3g*, *Nnt*, and *Ucp1* genes (5). Increased cAMP signaling is associated with states of changing energetic demands (e.g., adrenergic stimulation in BAT upon exposure to cold, and the fasting response in liver), and CREB contributes to mitochondrial function both directly, by acting at specific mitochondrial genes, as well as indirectly, by inducing PGC-1α expression (86, 87).

An increasing number of studies support the role of c-Myc as a regulator of mitochondrial genes (5). Recent genome-wide association studies show c-Myc binding to 107 mitochondrial genes, including the mitochondrial DNA polymerase γ (88). Moreover, c-Myc may affect mitochondrial biogenesis via its ability to activate the expression of PGC-1β (89). Con-

sistent with the binding and gene expression studies, c-Myc null cells have diminished mitochondrial mass (90). The context in which c-Myc regulates mitochondrial genes is not yet clear. An interesting possibility is that c-Myc links mitochondrial biogenesis to cell growth and proliferation.

Finally, the transcription factor YY1 has been implicated in both positive and negative regulation of COX genes (5). Recent studies show that YY1 in muscle is in a complex with PGC-1α, enhancing mitochondrial gene expression and cellular respiration (91). The interaction of YY1 with PGC-1α requires the activity of mammalian target of rapamycin (mTOR), suggesting that YY1 integrates information from two nutrient-sensing pathways: PGC-1α, which relays signals of low cellular energy state (92, 93), and mTOR, which promotes cell growth in the presence of nutrients. The physiological state during which this mechanism becomes important for mitochondrial biogenesis is not clear.

Transcriptional Coregulators

PPARγ coactivator-1 family (PGC-1α, PGC-1β, and PRC): promoters of mitochondrial biogenesis programs. PGC-1 coactivators play important roles in the control of mitochondrial biogenesis and function by integrating physiological signals and coordinately enhancing the function of diverse transcription factors acting at mitochondrial genes. PGC-1α was first identified by Spiegelman and colleagues (7) as a protein interacting with PPARγ, selectively expressed in BAT, and induced by exposure to cold. PGC-1β and PRC were identified on the basis of their similarity to PGC-1α (16, 17, 94). The three coactivators regulate expression of a broad mitochondrial gene set and promote mitochondrial biogenesis (9). They also carry important functions outside the mitochondrial gene expression program (9), which are not discussed here.

The PGC-1 proteins share three molecular features that are important for the regulation of mitochondrial genes. First, they contain protein surfaces that enable interactions with NRF-1, GABP, PPARs, ERRs, and YY1 and are thereby recruited to target regulatory sites. These protein surfaces include leucine-rich motifs that mediate interactions with nuclear receptors (16, 45, 94, 95); a conserved DHDY motif that binds HCF and presumably enables interactions with GABP (17, 33, 34); and less well-characterized interfaces for NRF-1, YY1, and other transcription factors (8, 91, 96). Second, the three PGC-1 proteins share similar transcriptional activation domains that enable the enhancement of gene expression (16, 45, 94, 97, 98). The molecular mechanisms are best characterized for PGC-1α and include the ability to recruit the histone acetyltransferases CBP/p300 and Mediator (9). Finally, PGC-1 proteins contain sites of posttranslational modifications or interaction with regulatory proteins. Some of these sites are conserved in the three members, suggesting common mechanisms of regulation (99). In summary, the mode of PGC-1 action at mitochondrial genes seems deceptively simple: PGC-1 docks on transcription factors bound at their respective response elements and enables the recruitment of histone acetyltransferases and the Mediator complex, thereby enhancing transcription initiation and/or elongation. PGC-1 proteins also have domains proposed to regulate posttranscriptional steps, such as RNA splicing (100, 101); the importance of this mechanism for mitochondrial function has not been addressed.

Overexpression of PGC-1α and PGC-1β in many cell types induces mitochondrial biogenesis and enhances respiration, suggesting that the two coactivators are limiting for the mitochondrial gene expression program (9). The functional properties of mitochondria in PGC-1α- and PGC-1β-expressing cells differ in terms of coupling and oxidative stress defenses, suggesting that PGC-1α and PGC-1β induce similar but not identical programs (102). Differential effects of PGC-1α and PGC-1β may be due to selective preferences in associations with DNA-binding transcription factors and/or differences in the communication with the gen-

eral transcription machinery. Overexpression of PRC induces OxPhos genes, and knockdown of PRC decreases cytochrome oxidase activity; however, it is not clear if increased PRC expression is sufficient to induce mitochondrial biogenesis (34).

The central roles of PGC-1α and PGC-1β in mitochondrial gene expression and biogenesis have been demonstrated in mice with gain- and loss-of-function studies. Transgenic expression of PGC-1α or PGC-1β in skeletal muscle leads to an increase in mitochondrial content, increased expression of mitochondrial genes, and enhanced exercise performance (103–105). PGC-1α and PGC-1β activate expression of distinct muscle contractile proteins (PGC-1α promotes type IIa and type I, and PGC-1β promotes type IIx fibers), consistent with the notion that the two proteins carry some distinct physiological roles (103, 104). Conversely, loss of PGC-1α by genetic inactivation results in viable mice with modest but significant decreases in expression of mitochondrial genes; decreased mitochondrial enzymatic activities; and phenotypes of mild to moderate mitochondrial dysfunction, such as a failure to defend body temperature when exposed to cold, reduced capacity to sustain running, and energetic impairments in heart in response to β-adrenergic stimulation or cardiac pressure overload (106–109). Similarly, PGC-1β null or hypomorph mice show decreased mitochondrial gene expression and defects in thermogenesis and cardiac performance (73, 110, 111). Overall, mitochondrial biogenesis defects are subtle, with decreases in mitochondrial volume seen only in some tissues [e.g., skeletal muscle of PGC-1α (108) and PGC-1β hypomorph (111) mice].

The mild mitochondrial biogenesis defects seen in mice lacking just PGC-1α or PGC-1β suggest that PGC-1α and PGC-1β compensate for each other's loss in vivo. In support of this notion, the induction of mitochondrial biogenesis during in vitro brown adipocyte differentiation is not affected by the lack of a single PGC-1 but is abolished when both PGC-1α and PGC-1β are knocked down (112). Further

Sirtuin 1 (SIRT1): NAD$^+$-dependent protein deacetylase homologous to the yeast silent information regulator (Sir2) and implicated in caloric restriction and aging pathways

Histone deacetylase (HDAC): a family of enzymes that deacetylate histones and other proteins

confirmation is provided by the recent generation of double PGC-1α/PGC-1β null mice, which show severe reductions in BAT mitochondrial density, late fetal arrest in cardiac mitochondrial biogenesis, small hearts, reduced cardiac output and other signs of heart defects and die shortly after birth (113). These findings suggest that PGC-1α and PGC-1β are essential for the developmental program that drives high levels of mitochondrial biogenesis in tissues with high energy demands, such as BAT and heart, but not for basal levels of mitochondrial biogenesis (by comparison, *Tfam* null mice die at E10.5).

One of the most interesting aspects of PGC-1 biology is the potential of these coregulators to sense signals of energetic or metabolic needs and to relay such signals to changes in gene expression. PGC-1α has served as the prototype PGC-1 family member in understanding this role (9). Signaling information is to a large extent integrated at two levels: transcriptional regulation and posttranslational regulation (**Figure 2**). At the transcriptional level, both PGC-1α and PGC-1β are expressed in a tissue-selective manner, with high levels in tissues with high energy demands, suggesting that their transcription depends on tissue-specific developmental cues (9). Moreover, PGC-1α, but not PGC-1β, is highly inducible in response to signals of increased energy needs (e.g., in BAT and muscle upon exposure to cold, in liver upon fasting, in skeletal muscle in response to exercise), suggesting that PGC-1α plays a role in long-term adaptation to such needs (7–9). At the posttranslational level, PGC-1α activity is regulated via phosphorylation by mitogen-activated protein kinase (MAPK) p38, AKT, AMPK, and glycogen synthase kinase-3 (GSK-3) (93, 95, 99, 114, 115), (de)acetylation by GCN5 and Sirtuin 1 (SIRT1) (92, 116), arginine methylation by PRMT1 (117), ubiquitination by SCF^{cdc4} (99), and interaction with the repressor MYBBP1A (118). Several of these modifications are likely to affect PGC-1β activity as well because the PGC-1α target modification sites are conserved. The signaling pathways that regulate PGC-1α expression and

activity, and the physiological context in which these pathways act, are discussed in the next section.

Receptor-interacting protein 140: a brake on mitochondrial biogenesis. The nuclear receptor corepressor RIP140 functions as the antithesis of PGC-1 coactivators and acts as a transcriptional brake on mitochondrial biogenesis. Like PGC-1α/β, RIP140 interacts with a broad set of nuclear receptors (including ERRs and PPARs) via a series of LXXLL motifs (70, 119). However, RIP140 docking to nuclear receptors recruits additional corepressors, such as CtBP and histone deacetylases (HDACs), and leads to suppression of gene transcription. In vitro and in vivo studies support the role of RIP140 in mitochondrial function. Silencing of RIP140 in 3T3L1 cells leads to increased expression of many mitochondrial genes, including ones with roles in the TCA cycle, OxPhos, FAO, and organelle biogenesis (120). The ability of RIP140 to repress at least some of these genes depends on endogenous ERRα (120), indicating that the same transcription factor (i.e., ERRα) can mediate positive and negative effects on mitochondrial gene expression, depending on the cellular context and type of coregulator present. RIP140 null animals have increased oxygen consumption and expression of mitochondrial genes (121). Muscle-specific deletion of RIP140 results in increased mitochondrial volume and number of oxidative fibers, i.e., effects similar to ones seen in mice overexpressing PGC-1α (103, 122).

Even though RIP140 is expressed widely, there is some correlation of high RIP140 levels and low mitochondrial content. For example, RIP140 levels are higher in WAT than in BAT and higher in glycolytic than in oxidative muscle fibers (70). RIP140 expression is also induced by many nuclear receptors, including ERRα (123). The ERRα-mediated induction of RIP140 may serve as a mechanism that limits mitochondrial gene induction in amplitude and/or temporally. Like PGC-1α, RIP140 is regulated by protein modifications. Sumoylation (124) and acetylation (125) have both been proposed to enhance the repressive ability of RIP140. Interestingly, the same arginine methyltransferase, PRMT1, modifies RIP140 and PGC-1α (117, 126). PRMT1-mediated methylation enhances the activity of PGC-1α and suppresses that of RIP140, suggesting that PRMT1 can act as a switch in the cellular balance of PGC-1α versus RIP140. It will be interesting to define physiological signals that regulate PRMT1.

PHYSIOLOGICAL STATES THAT PROMOTE MITOCHONDRIAL BIOGENESIS

Overview

Energetic demands vary not only among cell types but also in different physiological states. Thus, gene expression programs of mitochondrial biogenesis and function are regulated in response to physiological signals that accompany increased demands for energy or energetic efficiency. One of the best-understood paradigms is endurance exercise training—in which increased mitochondrial biogenesis contributes to muscle performance (3, 127). Similarly, long-term cold exposure induces mitochondrial biogenesis in BAT of small animals, which enables a higher capacity for adaptive thermogenesis (128). Finally, and of particular interest to today's calorie-ridden society, caloric restriction induces mitochondrial biogenesis (4, 129). In this section, we first present an overview of the signals associated with exercise, exposure to cold, and caloric restriction and then review how each of these signals communicates to the transcription factors that control mitochondrial gene expression.

Endurance exercise activates signals that are associated with physical activity, such as muscle contraction–induced increases in cytoplasmic Ca^{2+}, as well as signals that relay energy deficits, such as AMPK (127) (**Figure 3**). Exercise also activates the sympathetic nervous system, leading to adrenergic stimulation and increased cAMP signaling. Ca^{2+}-derived signals, AMPK, cAMP, and other signals (e.g., nitric oxide) then activate the transcription factors controlling mitochondrial gene expression (5, 9, 130). Single bouts of exercise lead to transient increases in the levels of PGC-1α, PRC, NRF-1, GABP, ERRα, PPARδ, and downstream mitochondrial targets (22, 24, 25, 127, 131). The increase in expression of these transcription factors is likely mediated by post-translational modifications that enhance their activity, as proposed for PGC-1α (132, 133). Repeated bouts of exercise are necessary for long-term adaptive responses, including a stable increase in PGC-1α protein and increased mitochondrial biogenesis (3, 127).

Exposure to cold leads to the activation of the sympathetic nervous system, adrenergic stimulation of BAT, and increased cAMP and cAMP-dependent pathways [i.e., activation of protein kinase A (PKA), p38 MAPK, and transcription factors CREB/ATF2], which induce PGC-1α (87) (**Figure 3**). PGC-1α and CREB/ATF2 induce the thermogenic *Ucp1*, which uses the mitochondrial proton gradient to generate heat (7, 87). PGC-1α also induces *deiodinase 2*, which generates local thyroid hormone and further enhances mitochondrial function (112). As with exercise, acute short-term exposure to cold induces a small set of mitochondrial genes, whereas long-term exposure results in mitochondrial biogenesis.

Caloric restriction without malnutrition enhances mitochondrial biogenesis in rodents and humans (4, 134–136). The molecular signals implicated in caloric restriction, which likely imposes a demand for increased energetic efficiency, include increased SIRT1 activity, increased AMPK activity (possibly due to an increase in circulating adiponectin levels), and the induction of endothelial nitric oxide synthase (eNOS) (**Figure 3**). All three signals converge on PGC-1α by regulating PGC-1α activity and expression levels (92, 93, 137).

Induction of mitochondrial biogenesis is also seen in "stressed" cells, in which stress may be due to (*a*) an energetic deficiency, as seen in treatment of cells with uncoupling agents, (*b*) DNA damage, as seen in cells exposed to ionizing radiation or the drug etoposide, or (*c*) microtubule disruption (20, 133, 138–140).

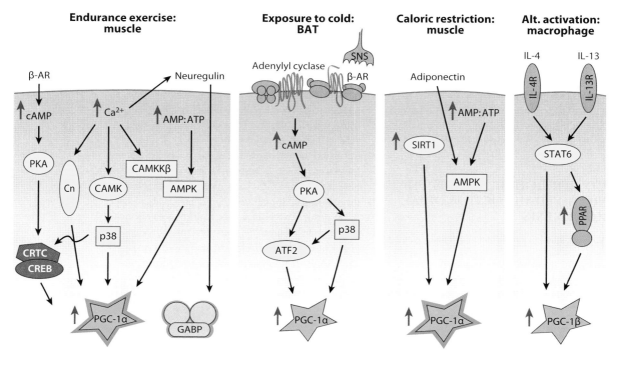

Figure 3

Diverse physiological signals regulate mitochondrial biogenesis in a tissue-specific manner. Shown are signaling pathways that induce mitochondrial biogenesis in skeletal muscle in response to endurance exercise or caloric restriction, in BAT in response to cold exposure, and in macrophages in response to signals promoting alternative activation. The signals enhance activity (*orange outlines*) and/or expression (*upward blue vertical arrows*) of transcriptional regulators PGC-1α, GABP, or PGC-1β.

An emerging theme from these studies is the repeated use of similar signals as in physiological states of mitochondrial biogenesis: AMPK, activated by uncoupling agents and the DNA damage–responsive kinase ATM (as in exercise and caloric restriction) (20, 139), and bursts of Ca²⁺, which mediate the uncoupling-induced response and retrograde signaling from mitochondria to nucleus (as in exercise) (139, 141).

Ca²⁺-Induced Pathways

Muscle contraction leads to Ca²⁺ bursts that signal to mitochondrial gene expression programs via the Ca²⁺-dependent phosphatase calcineurin, Ca²⁺/calmodulin-dependent kinases (including CAMKII, CAMKIV, and CAMKKβ), and p38 MAPK (127) (**Figure 3**).

Several lines of evidence support the roles of these molecules in mitochondrial biogenesis and function. First, caffeine treatment of cultured myotubes, which leads to Ca²⁺ release, activates p38 in a CAMK-dependent manner and mimics the effects of exercise, inducing PGC-1α, NRF-1, GABP, and TFAM (19). Inhibitors of CAMK or p38 block these effects (19, 142). Second, transgenic mice expressing constitutively active calcineurin or CAMKIV in muscle show increased mtDNA content, mitochondrial volume, and PGC-1α levels (143, 144). Third, the calcineurin, CAMK, and p38 signals converge at regulatory sites of the PGC-1α promoter, where they act on transcription factors of the MEF2 family (activated via calcineurin and p38) and the CREB/ATF2 family (activated by CAMKIV and p38) to induce PGC-1α expression (**Figure 2**) (145–148). A CAMK-dependent phosphorylation of HDAC5 may also export HDAC5 from the nucleus and relieve the inhibitory effect of HDAC5 on MEF2, thereby further enhancing PGC-1α expression (149).

In addition to inducing PGC-1α expression via the activation of MEF2 and CREB/ATF2, the MAPK p38 phosphorylates and enhances the activity of PGC-1α, thereby promoting PGC-1α to coactivate MEF2 and induce its own expression (**Figure 2**) (95, 114). Thus, in vivo, activation of the PGC-1α protein may precede the induced expression of PGC-1α (132). p38 also regulates stability of the PGC-1α protein. In cells with low SCF^cdc4 E3 ubiquitin ligase activity (e.g., 293 cells), phosphorylation of PGC-1α by p38 leads to increased protein stability (114). Conversely, in cells that have SCF^cdc4 and active GSK-3β, p38- and GSK-3β-mediated phosphorylation of PGC-1α promotes ubiquitination and ubiquitin-mediated degradation, thereby decreasing protein stability (99). The effects of GSK-3β and SCF^cdc4 on PGC-1α in muscle have not been examined. However, activation of p38 in myotubes exposed to palmitate leads to decreased PGC-1α expression, suggesting that p38 can act both positively and negatively, depending on cellular context (150).

AMPK: Cellular Energy Status Sensor

AMPK senses cellular energetic deficiencies as an increase in the AMP:ATP ratio and becomes activated in endurance exercise, caloric restriction, and other stressor-induced states. Several studies have established a role for AMPK in mitochondrial biogenesis and oxidative metabolism (127). Briefly, rodents treated with chemical activators of AMPK [e.g., 5′-aminoimidazole-4-carboxamide-1-β-ribofuranoside (AICAR)] or expressing constitutively active AMPK have enhanced muscle mitochondrial biogenesis, FAO, and expression of PGC-1α, NRF-1, and PPARα (26, 127, 151, 152). The ability of AMPK to induce PGC-1α in muscle is lost in PGC-1α null

mice, consistent with AMPK enhancing the activity of PGC-1α protein, which then promotes PGC-1α transcription (**Figure 2**). Indeed, AMPK phosphorylates PGC-1α and enhances PGC-1α activity (93). AMPK also phosphorylates and inactivates the MEF2-associated repressor HDAC5, thereby further enhancing PGC-1α transcription (**Figure 2**) (148, 153). PGC-1α is important for the AMPK-dependent induction of some mitochondrial genes, like *Cycs*, but not others, like *Ucp3* and *Pdk4*, suggesting that AMPK activates other transcription factors besides PGC-1α (93). Interestingly, AICAR-mediated activation of AMPK in mice enhances not just muscle mitochondrial biogenesis but also the capacity for exercise, suggesting a central and wide role of AMPK in the program induced by endurance exercise (154).

Finally, AMPK is activated by other hormones and signals that enhance mitochondrial biogenesis and function, such as adiponectin, leptin, thyroid hormone, and the DNA double-strand break–sensing kinase ATM, suggesting that AMPK plays a central role in multiple pathways that enhance mitochondrial biogenesis (20, 129, 155–157).

SIRT1: A Nutrient Deprivation Sensor

SIRT1 is a NAD$^+$-dependent deacetylase and the mammalian homolog of the yeast Sir2, which mediates effects of caloric restriction on yeast life span. Similar to the yeast Sir2, the mammalian SIRT1 is activated in states of nutrient deprivation, such as fasting and caloric restriction (158). SIRT1 deacetylates and activates PGC-1α, thereby reversing the effects of the acetyltransferase GCN5, which acetylates and represses PGC-1α (158). In muscle, SIRT1 promotes the deacetylation of PGC-1α and the increased expression of PGC-1α, ERRα and many mitochondrial genes, including *Tfam*, TCA cycle, OxPhos, and FAO genes (92). Increasing SIRT1 activity, by feeding mice the SIRT1 activator resveratrol, induces muscle mitochondrial biogenesis and enhances exercise performance, suggesting an important role of SIRT1 for mitochondrial function in vivo (159). Besides SIRT1, resveratrol may also activate AMPK (160). Thus, studies with specific SIRT1 activators will be necessary to define the in vivo role of SIRT1 (161). Finally, the effects of SIRT1 on PGC-1α may depend on cellular context because SIRT1 represses PGC-1α activity in PC12 cells (162). There are 13 acetylated lysines in PGC-1α (158); some of them may enhance, whereas others may repress, PGC-1α activity.

Other Signals: cAMP, Neuregulins, and Nitric Oxide

Several other signals have been implicated in the regulation of mitochondrial biogenesis. Induction of PGC-1α in muscle by exercise is blunted in mice lacking β-adrenergic receptors, indicating that adrenergic stimulation and cAMP signaling contribute to the adaptive responses (130) (**Figure 3**). Moreover, exercise and Ca^{2+} induce the expression of neuregulins, which activate ErbB tyrosine kinases (163). Neuregulins enhance expression of PPARδ, PGC-1α, and mitochondrial genes and increase oxidative capacity in muscle cells (164). Interestingly, neuregulins may act at multiple steps because the ErbB pathway also activates GABP and ERRα (33, 165). Another signal that impacts mitochondrial biogenesis is the gas nitric oxide (NO), which induces PGC-1α expression via a cGMP-dependent mechanism (166). Mice lacking eNOS, an NO-producing enzyme that is induced by caloric restriction, show mitochondrial defects in multiple tissues (135).

Molecular studies of PGC-1α have also identified regulators whose role in the physiological pathways of mitochondrial biogenesis is not yet clear. Some of these regulators affect mitochondrial function in vitro and in vivo. For example, the kinase cyclinH/Cdk7/MAT1 regulates the expression of PGC-1α and OxPhos genes in HIB1B cells, and mice lacking MAT1 in heart show decreased PGC-1α and PGC-1β activity and mitochondrial and energetic deficiencies (167, 168). The underlying mechanism

for this regulation is not known. Similarly, future studies need to elucidate the physiological contributions of other regulators, such as SCF^{cdc4} (which is regulated in neurons by oxidative stress), MYBBP1A, and PRMT1 (99, 117, 118).

(Patho)physiological States Associated with Decreased Mitochondrial Biogenesis

Decreased mitochondrial biogenesis and function are seen in aging, physical inactivity, obesity, and insulin resistance. They often parallel decreases in the expression of transcriptional regulators, like PGC-1α, PGC-1β, and NRF-1 (9, 169). One likely cause for the decreased mitochondrial biogenesis is simply reduced input in the positive signals discussed above. For example, aging is associated with a blunted AMPK stimulation by exercise, insulin resistance with decreased plasma adiponectin levels, obesity with increased cytokines that suppress eNOS expression, and so on (129, 169, 170). Signals that specifically repress the transcriptional program of mitochondrial biogenesis may include lipids. Lipid infusion or a high-fat diet in humans leads to decreased expression of PGC-1α, PGC-1β, and their mitochondrial targets (171, 172). Similarly, exposure to palmitate downregulates PGC-1α and mitochondrial gene expression, as well as oxygen consumption in C2C12 myotubes (150). However, increasing plasma free fatty acids and high-fat diets in rats induce PGC-1α and PPARβ/δ expression and mitochondrial content (173, 174). The conflicting results may be due to differences in fatty acid and diet composition, suggesting that it is the types rather than simply the levels of lipids that signal to mitochondrial gene expression programs. Finally, insulin can negatively regulate PGC-1α by two distinct mechanisms: first, by repressing FOXO-dependent PGC-1α expression (175, 176) and second, via AKT-dependent phosphorylation and repression of PGC-1α activity (115) (**Figure 2**). The significance of the AKT effect for mitochondrial biogenesis in muscle is not clear.

FEATURES OF THE TRANSCRIPTIONAL NETWORK AND THE REGULATING SIGNALS

One of the remarkable features of the network of transcription factors regulating mitochondrial biogenesis and function is the extensive use of feed-forward and feedback loops (**Figures 1** and **2**). Increases in one network component often enhance the expression of others. For example, PGC-1α not only coactivates ERRα, GABP, and NRF-1 but also increases their expression levels. ERRα enhances the expression of PPARα, GABP, and the negative regulator RIP140. PPARγ, PPARδ, ERRγ, and CREB induce the expression of PGC-1α. ERRα and PGC-1α induce their own expression. These regulatory loops are likely to be important in enhancing the amplitude of the mitochondrial gene response, as well as in limiting it temporally. Moreover, modeling of biological feedback loops suggests that such coupled positive and negative feedback systems are better able to respond faithfully to signals in the presence of noise (177).

A second interesting feature is the parallel activation of multiple signaling pathways in physiological states that enhance mitochondrial biogenesis and function (e.g., Ca^{2+} and AMPK in exercise, AMPK and SIRT1 in caloric restriction) (**Figure 3**). The importance of parallel pathways is nicely demonstrated in a recent study showing that, although pharmacological activation of PPARδ by itself does not enhance exercise performance, the combination of the PPARδ agonist with exercise training results in the synergistic enhancement of exercise performance (154). It seems reasonable to assume that a robust and specific induction of an expensive process, such as mitochondrial biogenesis, requires concurrent activation of more than one pathway, as well as multiple transcription factors.

SUMMARY POINTS

1. Mitochondrial biogenesis and function are dynamically regulated in a tissue- and signal-specific manner to enable cellular adaptation to energetic and metabolic demands.

2. Mitochondrial biogenesis in response to physiological signals is a long-term adaptive response and requires changes in the transcription of nuclear genome–encoded mitochondrial genes.

3. Short-term adaptation can be achieved by the increased expression of a subset of mitochondrial genes that enhance mitochondrial function without inducing mtDNA replication and organelle expansion.

4. Mitochondrial biogenesis and function are regulated by a transcriptional network comprising (*a*) DNA-binding factors that target distinct but overlapping sets of mitochondrial genes and (*b*) coregulators that integrate signals and coordinate the action of multiple DNA-binding factors. The network allows the concerted regulation of a broad mitochondrial gene set while also permitting tissue- and signal-specific expression patterns of subsets of mitochondrial genes.

5. A series of regulatory feed-forward and feedback loops among the transcription factors of the mitochondrial gene expression network allows robust and specific transcriptional responses to physiological signals.

6. Distinct physiological stimuli that induce mitochondrial biogenesis use common signals, including activity-dependent bursts of Ca^{2+} and the cellular energy status sensor AMPK. Cellular stress can induce the same signals.

7. The coactivator PGC-1α plays a central role in adaptive responses by integrating diverse signals that impact mitochondrial biogenesis and coordinating multiple DNA-binding factors to induce broad sets of mitochondrial genes.

FUTURE ISSUES

1. The central role of PGC-1α in integrating most if not all signals known to regulate mitochondrial biogenesis may largely reflect the attention allotted to this coactivator. Future studies on the mechanisms that regulate activity of PGC-1β, PRC, RIP140, and the downstream DNA-binding factors will likely elucidate contexts in which these other regulators integrate signals, and thus may decipher possible "codes" that determine activity of the transcriptional network.

2. The field has made major advances in characterizing the signals and mechanisms that induce mitochondrial biogenesis. Signals or states resulting in mitochondrial dysfunction are less well understood. Studies on the molecular mechanisms that may underlie decreased mitochondrial biogenesis and function in (patho)physiological states will provide insights into basic science questions regarding such decreases (e.g., do they have causal roles, or are they secondary to disease?), as well as suggest novel modes of intervention.

3. There are a large number of transcriptional regulators of mitochondrial biogenesis and function. Some of them seem, at first glance, to have similar roles. It will be important to understand the specific roles and contributions of individual regulators in different tissues and at different physiological states. This knowledge will help identify targets for intervention in cases of cell-type-specific mitochondrial dysfunction. It may also help in the design of drugs that spare tissues where increased mitochondrial function may not be safe.

DISCLOSURE STATEMENT

The authors are not aware of any biases that might be perceived as affecting the objectivity of this review.

ACKNOWLEDGMENTS

We apologize to the many investigators whose work could not be cited owing to space limitations. We thank Mari Gantner, Barbara Sullivan, and Enrique Saez for discussions and comments and the NIH for support (NIH grant DK64951 to A.K.).

LITERATURE CITED

1. Pagliarini DJ, Calvo SE, Chang B, Sheth SA, Vafai SB, et al. 2008. A mitochondrial protein compendium elucidates complex I disease biology. *Cell* 134:112–23
2. Mootha VK, Bunkenborg J, Olsen JV, Hjerrild M, Wisniewski JR, et al. 2003. Integrated analysis of protein composition, tissue diversity, and gene regulation in mouse mitochondria. *Cell* 115:629–40

3. Chow LS, Greenlund LJ, Asmann YW, Short KR, McCrady SK, et al. 2007. Impact of endurance training on murine spontaneous activity, muscle mitochondrial DNA abundance, gene transcripts, and function. *J. Appl. Physiol.* 102:1078–89

4. Civitarese AE, Carling S, Heilbronn LK, Hulver MH, Ukropcova B, et al. 2007. Calorie restriction increases muscle mitochondrial biogenesis in healthy humans. *PLoS Med.* 4:e76

5. Scarpulla RC. 2008. Transcriptional paradigms in mammalian mitochondrial biogenesis and function. *Physiol. Rev.* 88:611–38

6. Evans RM, Barish GD, Wang YX. 2004. PPARs and the complex journey to obesity. *Nat. Med.* 10:355–61

7. Puigserver P, Wu Z, Park CW, Graves R, Wright M, Spiegelman BM. 1998. A cold-inducible coactivator of nuclear receptors linked to adaptive thermogenesis. *Cell* 92:829–39

8. Wu Z, Puigserver P, Andersson U, Zhang C, Adelmant G, et al. 1999. Mechanisms controlling mitochondrial biogenesis and respiration through the thermogenic coactivator PGC-1. *Cell* 98:115–24

9. Handschin C, Spiegelman BM. 2006. Peroxisome proliferator-activated receptor γ coactivator 1 coactivators, energy homeostasis, and metabolism. *Endocr. Rev.* 27:728–35

10. Villena JA, Kralli A. 2008. ERRα: a metabolic function for the oldest orphan. *Trends Endocrinol. Metab.* 8:269–76

11. Virbasius CA, Virbasius JV, Scarpulla RC. 1993. NRF-1, an activator involved in nuclear-mitochondrial interactions, utilizes a new DNA-binding domain conserved in a family of developmental regulators. *Genes Dev.* 7:2431–45

12. Ramachandran B, Yu G, Gulick T. 2008. Nuclear respiratory factor 1 controls myocyte enhancer factor 2A transcription to provide a mechanism for coordinate expression of respiratory chain subunits. *J. Biol. Chem.* 283:11935–46

13. Cam H, Balciunaite E, Blais A, Spektor A, Scarpulla RC, et al. 2004. A common set of gene regulatory networks links metabolism and growth inhibition. *Mol. Cell* 16:399–411

14. Wang C, Li Z, Lu Y, Du R, Katiyar S, et al. 2006. Cyclin D1 repression of nuclear respiratory factor 1 integrates nuclear DNA synthesis and mitochondrial function. *Proc. Natl. Acad. Sci. USA* 103:11567–72

15. Dhar SS, Ongwijitwat S, Wong-Riley MT. 2008. Nuclear respiratory factor 1 regulates all ten nuclear-encoded subunits of cytochrome *c* oxidase in neurons. *J. Biol. Chem.* 283:3120–29

16. Andersson U, Scarpulla RC. 2001. Pgc-1-related coactivator, a novel, serum-inducible coactivator of nuclear respiratory factor 1-dependent transcription in mammalian cells. *Mol. Cell. Biol.* 21:3738–49

17. Lin J, Puigserver P, Donovan J, Tarr P, Spiegelman BM. 2002. Peroxisome proliferator-activated receptor γ coactivator 1β (PGC-1β), a novel PGC-1-related transcription coactivator associated with host cell factor. *J. Biol. Chem.* 277:1645–48

18. Sakamaki T, Casimiro MC, Ju X, Quong AA, Katiyar S, et al. 2006. Cyclin D1 determines mitochondrial function in vivo. *Mol. Cell. Biol.* 26:5449–69

19. Ojuka EO, Jones TE, Han DH, Chen M, Holloszy JO. 2003. Raising Ca^{2+} in L6 myotubes mimics effects of exercise on mitochondrial biogenesis in muscle. *FASEB J.* 17:675–81

20. Fu X, Wan S, Lyu YL, Liu LF, Qi H. 2008. Etoposide induces ATM-dependent mitochondrial biogenesis through AMPK activation. *PLoS ONE* 3:e2009

21. Murakami T, Shimomura Y, Yoshimura A, Sokabe M, Fujitsuka N. 1998. Induction of nuclear respiratory factor-1 expression by an acute bout of exercise in rat muscle. *Biochim. Biophys. Acta* 1381:113–22

22. Baar K, Wende AR, Jones TE, Marison M, Nolte LA, et al. 2002. Adaptations of skeletal muscle to exercise: rapid increase in the transcriptional coactivator PGC-1. *FASEB J.* 16:1879–86

23. McClelland GB, Craig PM, Dhekney K, Dipardo S. 2006. Temperature- and exercise-induced gene expression and metabolic enzyme changes in skeletal muscle of adult zebrafish (*Danio rerio*). *J. Physiol.* 577:739–51

24. Pilegaard H, Saltin B, Neufer PD. 2003. Exercise induces transient transcriptional activation of the PGC-1α gene in human skeletal muscle. *J. Physiol.* 546:851–58

25. Cartoni R, Leger B, Hock MB, Praz M, Crettenand A, et al. 2005. Mitofusins 1/2 and ERRα expression are increased in human skeletal muscle after physical exercise. *J. Physiol.* 567:349–58

26. Bergeron R, Ren JM, Cadman KS, Moore IK, Perret P, et al. 2001. Chronic activation of AMP kinase results in NRF-1 activation and mitochondrial biogenesis. *Am. J. Physiol. Endocrinol. Metab.* 281:E1340–46

27. Huo L, Scarpulla RC. 2001. Mitochondrial DNA instability and peri-implantation lethality associated with targeted disruption of nuclear respiratory factor 1 in mice. *Mol. Cell. Biol.* 21:644–54

28. Baar K, Song Z, Semenkovich CF, Jones TE, Han DH, et al. 2003. Skeletal muscle overexpression of nuclear respiratory factor 1 increases glucose transport capacity. *FASEB J.* 17:1666–73

29. Virbasius JV, Virbasius CA, Scarpulla RC. 1993. Identity of GABP with NRF-2, a multisubunit activator of cytochrome oxidase expression, reveals a cellular role for an ETS domain activator of viral promoters. *Genes Dev.* 7:380–92

30. Mootha VK, Handschin C, Arlow D, Xie X, St. Pierre J, et al. 2004. Errα and Gabpa/b specify PGC-1α-dependent oxidative phosphorylation gene expression that is altered in diabetic muscle. *Proc. Natl. Acad. Sci.* 101:6570–75

31. Ongwijitwat S, Liang HL, Graboyes EM, Wong-Riley MT. 2006. Nuclear respiratory factor 2 senses changing cellular energy demands and its silencing down-regulates cytochrome oxidase and other target gene mRNAs. *Gene* 374:39–49

32. Canto C, Pich S, Paz JC, Sanches R, Martinez V, et al. 2007. Neuregulins increase mitochondrial oxidative capacity and insulin sensitivity in skeletal muscle cells. *Diabetes* 56:2185–93

33. Handschin C, Kobayashi YM, Chin S, Seale P, Campbell KP, Spiegelman BM. 2007. PGC-1α regulates the neuromuscular junction program and ameliorates Duchenne muscular dystrophy. *Genes Dev.* 21:770–83

34. Vercauteren K, Gleyzer N, Scarpulla RC. 2008. PGC-1-related coactivator (PRC) complexes with HCF-1 and NRF-2β in mediating NRF-2(GABP)-dependent respiratory gene expression. *J. Biol. Chem.* 283:12102–11

35. Ristevski S, O'Leary DA, Thornell AP, Owen MJ, Kola I, Hertzog PJ. 2004. The ETS transcription factor GABPα is essential for early embryogenesis. *Mol. Cell. Biol.* 24:5844–49

36. Villena JA, Carmona MC, Rodríguez de la Concepción M, Rossmeisl M, Viñas O, et al. 2002. Mitochondrial biogenesis in brown adipose tissue is associated with differential expression of transcription regulatory factors. *Cell. Mol. Life Sci.* 59:1934–44

37. Rodríguez-Peña A, Escrivá H, Handler AC, Vallejo CG. 2002. Thyroid hormone increases transcription of GA-binding protein/nuclear respiratory factor-2 α-subunit in rat liver. *FEBS Lett.* 514:309–14

38. Garnier A, Fortin D, Delomenie C, Momken I, Veksler V, Ventura-Clapier R. 2003. Depressed mitochondrial transcription factors and oxidative capacity in rat failing cardiac and skeletal muscles. *J. Physiol.* 551:491–501

39. Yang ZF, Mott S, Rosmarin AG. 2007. The Ets transcription factor GABP is required for cell-cycle progression. *Nat. Cell Biol.* 9:339–46

40. Xue HH, Bollenbacher-Reilley J, Wu Z, Spolski R, Jing X, et al. 2007. The transcription factor GABP is a critical regulator of B lymphocyte development. *Immunity* 26:421–31

41. Jaworski A, Smith CL, Burden SJ. 2007. GA-binding protein is dispensable for neuromuscular synapse formation and synapse-specific gene expression. *Mol. Cell. Biol.* 27:5040–46

42. O'Leary DA, Noakes PG, Lavidis NA, Kola I, Hertzog PJ, Ristevski S. 2007. Targeting of the ETS factor Gabpα disrupts neuromuscular junction synaptic function. *Mol. Cell. Biol.* 27:3470–80

43. Burkart EM, Sambandam N, Han X, Gross RW, Courtois M, et al. 2007. Nuclear receptors PPARβ/δ and PPARα direct distinct metabolic regulatory programs in the mouse heart. *J. Clin. Investig.* 117:3930–39

44. Reilly SM, Lee CH. 2008. PPARδ as a therapeutic target in metabolic disease. *FEBS Lett.* 582:26–31

45. Vega RB, Huss JM, Kelly DP. 2000. The coactivator PGC-1 cooperates with peroxisome proliferator-activated receptor α in transcriptional control of nuclear genes encoding mitochondrial fatty acid oxidation enzymes. *Mol. Cell. Biol.* 20:1868–76

46. Wang YX, Zhang CL, Yu RT, Cho HK, Nelson MC, et al. 2004. Regulation of muscle fiber type and running endurance by PPARδ. *PLoS Biol.* 2:e294

47. Wilson-Fritch L, Nicoloro S, Chouinard M, Lazar MA, Chui PC, et al. 2004. Mitochondrial remodeling in adipose tissue associated with obesity and treatment with rosiglitazone. *J. Clin. Investig.* 114:1281–89

48. Choo HJ, Kim JH, Kwon OB, Lee CS, Mun JY, et al. 2006. Mitochondria are impaired in the adipocytes of type 2 diabetic mice. *Diabetologia* 49:784–91
49. Rong JX, Qiu Y, Hansen MK, Zhu L, Zhang V, et al. 2007. Adipose mitochondrial biogenesis is suppressed in db/db and high-fat diet-fed mice and improved by rosiglitazone. *Diabetes* 56:1751–60
50. Bogacka I, Xie H, Bray GA, Smith SR. 2005. Pioglitazone induces mitochondrial biogenesis in human subcutaneous adipose tissue in vivo. *Diabetes* 54:1392–99
51. Wilson-Fritch L, Burkart A, Bell G, Mendelson K, Leszyk J, et al. 2003. Mitochondrial biogenesis and remodeling during adipogenesis and in response to the insulin sensitizer rosiglitazone. *Mol. Cell. Biol.* 23:1085–94
52. Hondares E, Mora O, Yubero P, Rodriguez de la Concepción M, Iglesias R, et al. 2006. Thiazolidinediones and rexinoids induce peroxisome proliferator-activated receptor-coactivator (PGC)-1α gene transcription: An autoregulatory loop controls PGC-1α expression in adipocytes via peroxisome proliferator-activated receptor-γ coactivation. *Endocrinology* 147:2829–38
53. Tanaka T, Yamamoto J, Iwasaki S, Asaba H, Hamura H, et al. 2003. Activation of peroxisome proliferator-activated receptor δ induces fatty acid β-oxidation in skeletal muscle and attenuates metabolic syndrome. *Proc. Natl. Acad. Sci. USA* 100:15924–29
54. Luquet S, Lopez-Soriano J, Holst D, Fredenrich A, Melki J, et al. 2003. Peroxisome proliferator-activated receptor δ controls muscle development and oxidative capability. *FASEB J.* 17:2299–301
55. Schuler M, Ali F, Chambon C, Duteil D, Bornert JM, et al. 2006. PGC1α expression is controlled in skeletal muscles by PPARβ, whose ablation results in fiber-type switching, obesity, and type 2 diabetes. *Cell Metab.* 4:407–14
56. Dressel U, Allen TL, Pippal JB, Rohde PR, Lau P, Muscat GE. 2003. The peroxisome proliferator-activated receptor β/δ agonist, GW501516, regulates the expression of genes involved in lipid catabolism and energy uncoupling in skeletal muscle cells. *Mol. Endocrinol.* 17:2477–93
57. Holst D, Luquet S, Nogueira V, Kristiansen K, Leverve X, Grimaldi PA. 2003. Nutritional regulation and role of peroxisome proliferator-activated receptor δ in fatty acid catabolism in skeletal muscle. *Biochim. Biophys. Acta* 1633:43–50
58. Bastin J, Aubey F, Rotig A, Munnich A, Djouadi F. 2008. Activation of peroxisome proliferator-activated receptor pathway stimulates the mitochondrial respiratory chain and can correct deficiencies in patients' cells lacking its components. *J. Clin. Endocrinol. Metab.* 93:1433–41
59. Hondares E, Pineda-Torra I, Iglesias R, Staels B, Villarroya F, Giralt M. 2007. PPARδ, but not PPARα, activates PGC-1α gene transcription in muscle. *Biochem. Biophys. Res. Commun.* 354:1021–27
60. Lee CH, Kang K, Mehl IR, Nofsinger R, Alaynick WA, et al. 2006. Peroxisome proliferator-activated receptor δ promotes very low-density lipoprotein-derived fatty acid catabolism in the macrophage. *Proc. Natl. Acad. Sci. USA* 103:2434–39
61. Odegaard JI, Ricardo-Gonzalez RR, Goforth MH, Morel CR, Subramanian V, et al. 2007. Macrophage-specific PPARγ controls alternative activation and improves insulin resistance. *Nature* 447:1116–20
62. Vats D, Mukundan L, Odegaard JI, Zhang L, Smith KL, et al. 2006. Oxidative metabolism and PGC-1β attenuate macrophage-mediated inflammation. *Cell Metab.* 4:13–24
63. Kang K, Reilly SM, Karabacak V, Gangl MR, Fitzgerald K, et al. 2008. Adipocyte-derived Th2 cytokines and myeloid PPARδ regulate macrophage polarization and insulin sensitivity. *Cell Metab.* 7:485–95
64. Giguere V. 2008. Transcriptional control of energy homeostasis by the estrogen-related receptors (ERRs). *Endocr. Rev.* 29:677–96
65. Greschik H, Wurtz JM, Sanglier S, Bourguet W, van Dorsselaer A, et al. 2002. Structural and functional evidence for ligand-independent transcriptional activation by the estrogen-related receptor 3. *Mol. Cell* 9:303–13
66. Kallen J, Schlaeppi JM, Bitsch F, Filipuzzi I, Schilb A, et al. 2004. Evidence for ligand-independent transcriptional activation of the human estrogen-related receptor α (ERRβ): Crystal structure of ERRα ligand binding domain in complex with peroxisome proliferator-activated receptor coactivator-1α. *J. Biol. Chem.* 279:49330–37
67. Huss JM, Kopp RP, Kelly DP. 2002. Peroxisome proliferator-activated receptor coactivator-1α (PGC-1α) coactivates the cardiac-enriched nuclear receptors estrogen-related receptor-α and -γ. Identification of novel leucine-rich interaction motif within PGC-1α. *J. Biol. Chem.* 277:40265–74

68. Kamei Y, Ohizumi H, Fujitani Y, Nemoto T, Tanaka T, et al. 2003. PPARγ coactivator 1β/ERR ligand 1 is an ERR protein ligand, whose expression induces a high-energy expenditure and antagonizes obesity. *Proc. Natl. Acad. Sci. USA* 100:12378–83

69. Schreiber SN, Knutti D, Brogli K, Uhlmann T, Kralli A. 2003. The transcriptional coactivator PGC-1 regulates the expression and activity of the orphan nuclear receptor estrogen-related receptor α (ERRα). *J. Biol. Chem.* 278:9013–18

70. White R, Morganstein D, Christian M, Seth A, Herzog B, Parker MG. 2008. Role of RIP140 in metabolic tissues: connections to disease. *FEBS Lett.* 582:39–45

71. Schreiber SN, Emter R, Hock MB, Knutti D, Cardenas J, et al. 2004. The estrogen-related receptor α (ERRα) functions in PPARγ coactivator 1α (PGC-1α)-induced mitochondrial biogenesis. *Proc. Natl. Acad. Sci. USA* 101:6472–77

72. Dufour CR, Wilson BJ, Huss JM, Kelly DP, Alaynick WA, et al. 2007. Genome-wide orchestration of cardiac functions by the orphan nuclear receptors ERRα and γ. *Cell Metab.* 5:345–56

73. Sonoda J, Laganiere J, Mehl IR, Barish GD, Chong LW, et al. 2007. Nuclear receptor ERRα and coactivator PGC-1β are effectors of IFN-γ-induced host defense. *Genes Dev.* 21:1909–20

74. Villena JA, Hock MB, Chang WY, Barcas JE, Giguere V, Kralli A. 2007. Orphan nuclear receptor estrogen-related receptor α is essential for adaptive thermogenesis. *Proc. Natl. Acad. Sci. USA* 104:1418–23

75. Huss JM, Torra IP, Staels B, Giguere V, Kelly DP. 2004. Estrogen-related receptor α directs peroxisome proliferator-activated receptor α signaling in the transcriptional control of energy metabolism in cardiac and skeletal muscle. *Mol. Cell. Biol.* 24:9079–91

76. Laganiere J, Tremblay GB, Dufour CR, Giroux S, Rousseau F, Giguere V. 2004. A polymorphic autoregulatory hormone response element in the human estrogen-related receptor α (ERRα) promoter dictates peroxisome proliferator-activated receptor γ coactivator-1α control of ERRα expression. *J. Biol. Chem.* 279:18504–10

77. Ichida M, Nemoto S, Finkel T. 2002. Identification of a specific molecular repressor of the peroxisome proliferator-activated receptor γ coactivator-1α (PGC-1α). *J. Biol. Chem.* 277:50991–95

78. Huss JM, Imahashi K-I, Dufour CR, Weinheimer CJ, Courtois M, et al. 2007. The nuclear receptor ERRα is required for the bioenergetic and functional adaptation to cardiac pressure overload. *Cell Metab.* 6:25–37

79. Luo J, Sladek R, Carrier J, Bader JA, Richard D, Giguere V. 2003. Reduced fat mass in mice lacking orphan nuclear receptor estrogen-related receptor α. *Mol. Cell. Biol.* 23:7947–56

80. Carrier JC, Deblois G, Champigny C, Levy E, Giguere V. 2004. Estrogen-related receptor α (ERRα) is a transcriptional regulator of apolipoprotein A-IV and controls lipid handling in the intestine. *J. Biol. Chem.* 279:52052–58

81. Alaynick WA, Kondo RP, Xie W, He W, Dufour CR, et al. 2007. ERRγ directs and maintains the transition to oxidative metabolism in the postnatal heart. *Cell Metab.* 6:13–24

82. Ijichi N, Ikeda K, Horie-Inoue K, Yagi K, Okazaki Y, Inoue S. 2007. Estrogen-related receptor α modulates the expression of adipogenesis-related genes during adipocyte differentiation. *Biochem. Biophys. Res. Commun.* 358:813–18

83. Luo J, Sladek R, Bader JA, Matthyssen A, Rossant J, Giguere V. 1997. Placental abnormalities in mouse embryos lacking the orphan nuclear receptor ERR-β. *Nature* 388:778–82

84. Ivanova N, Dobrin R, Lu R, Kotenko I, Levorse J, et al. 2006. Dissecting self-renewal in stem cells with RNA interference. *Nature* 442:533–38

85. Chen J, Nathans J. 2007. Estrogen-related receptor β/NR3B2 controls epithelial cell fate and endolymph production by the stria vascularis. *Dev. Cell* 13:325–37

86. Herzig S, Long F, Jhala US, Hedrick S, Quinn R, et al. 2001. CREB regulates hepatic gluconeogenesis through the coactivator PGC-1. *Nature* 413:179–83

87. Cao W, Daniel KW, Robidoux J, Puigserver P, Medvedev AV, et al. 2004. p38 mitogen-activated protein kinase is the central regulator of cyclic AMP-dependent transcription of the brown fat uncoupling protein 1 gene. *Mol. Cell. Biol.* 24:3057–67

88. Kim J, Lee JH, Iyer VR. 2008. Global identification of Myc target genes reveals its direct role in mitochondrial biogenesis and its E-box usage in vivo. *PLoS ONE* 3:e1798

89. Zhang H, Gao P, Fukuda R, Kumar G, Krishnamachary B, et al. 2007. HIF-1 inhibits mitochondrial biogenesis and cellular respiration in VHL-deficient renal cell carcinoma by repression of C-MYC activity. *Cancer Cell* 11:407–20

90. Li F, Wang Y, Zeller KI, Potter JJ, Wonsey DR, et al. 2005. Myc stimulates nuclearly encoded mitochondrial genes and mitochondrial biogenesis. *Mol. Cell. Biol.* 25:6225–34

91. Cunningham JT, Rodgers JT, Arlow DH, Vazquez F, Mootha VK, Puigserver P. 2007. mTOR controls mitochondrial oxidative function through a YY1-PGC-1α transcriptional complex. *Nature* 450:736–40

92. Gerhart-Hines Z, Rodgers JT, Bare O, Lerin C, Kim SH, et al. 2007. Metabolic control of muscle mitochondrial function and fatty acid oxidation through SIRT1/PGC-1α. *EMBO J.* 26:1913–23

93. Jager S, Handschin C, St-Pierre J, Spiegelman BM. 2007. AMP-activated protein kinase (AMPK) action in skeletal muscle via direct phosphorylation of PGC-1α. *Proc. Natl. Acad. Sci. USA* 104:12017–22

94. Kressler D, Schreiber SN, Knutti D, Kralli A. 2002. The PGC-1-related protein PERC is a selective coactivator of estrogen receptor α. *J. Biol. Chem.* 277:13918–25

95. Knutti D, Kressler D, Kralli A. 2001. Regulation of the transcriptional coactivator PGC-1 via MAPK-sensitive interaction with a repressor. *Proc. Natl. Acad. Sci. USA* 98:9713–18

96. Vercauteren K, Pasko RA, Gleyzer N, Marino VM, Scarpulla RC. 2006. PGC-1-related coactivator: immediate early expression and characterization of a CREB/NRF-1 binding domain associated with cytochrome *c* promoter occupancy and respiratory growth. *Mol. Cell. Biol.* 26:7409–19

97. Puigserver P, Adelmant G, Wu Z, Fan M, Xu J, et al. 1999. Activation of PPARγ coactivator-1 through transcription factor docking. *Science* 286:1368–71

98. Knutti D, Kaul A, Kralli A. 2000. A tissue-specific coactivator of steroid receptors, identified in a functional genetic screen. *Mol. Cell. Biol.* 20:2411–22

99. Olson BL, Hock MB, Ekholm-Reed S, Wohlschlegel JA, Dev KK, et al. 2008. SCF^{Cdc4} acts antagonistically to the PGC-1α transcriptional coactivator by targeting it for ubiquitin-mediated proteolysis. *Genes Dev.* 22:252–64

100. Monsalve M, Wu Z, Adelmant G, Puigserver P, Fan M, Spiegelman BM. 2000. Direct coupling of transcription and mRNA processing through the thermogenic coactivator PGC-1. *Mol. Cell* 6:307–16

101. Thijssen-Timmer DC, Schiphorst MP, Kwakkel J, Emter R, Kralli A, et al. 2006. PGC-1α regulates the isoform mRNA ratio of the alternatively spliced thyroid hormone receptor α transcript. *J. Mol. Endocrinol.* 37:251–57

102. St-Pierre J, Lin J, Krauss S, Tarr PT, Yang R, et al. 2003. Bioenergetic analysis of peroxisome proliferator-activated receptor γ coactivators 1α and 1β (PGC-1α and PGC-1β) in muscle cells. *J. Biol. Chem.* 278:26597–603

103. Lin J, Wu H, Tarr PT, Zhang CY, Wu Z, et al. 2002. Transcriptional coactivator PGC-1α drives the formation of slow-twitch muscle fibres. *Nature* 418:797–801

104. Arany Z, Lebrasseur N, Morris C, Smith E, Yang W, et al. 2007. The transcriptional coactivator PGC-1β drives the formation of oxidative type IIX fibers in skeletal muscle. *Cell Metab.* 5:35–46

105. Calvo JA, Daniels TG, Wang X, Paul A, Lin J, et al. 2008. Muscle-specific expression of PPARγ coactivator-1α improves exercise performance and increases peak oxygen uptake. *J. Appl. Physiol.* 104:1304–12

106. Lin J, Wu P-H, Tarr PT, Lindenberg KS, St-Pierre J, et al. 2004. Defects in adaptive energy metabolism with CNS-linked hyperactivity in PGC-1α null mice. *Cell* 119:121–35

107. Arany Z, He H, Lin J, Hoyer K, Handschin C, et al. 2005. Transcriptional coactivator PGC-1α controls the energy state and contractile function of cardiac muscle. *Cell Metab.* 1:259–71

108. Leone TC, Lehman JJ, Finck BN, Schaeffer PJ, Wende AR, et al. 2005. PGC-1α deficiency causes multi-system energy metabolic derangements: muscle dysfunction, abnormal weight control and hepatic steatosis. *PLoS Biol.* 3:e101

109. Arany Z, Novikov M, Chin S, Ma Y, Rosenzweig A, Spiegelman BM. 2006. Transverse aortic constriction leads to accelerated heart failure in mice lacking PPAR-γ coactivator 1α. *Proc. Natl. Acad. Sci. USA* 103:10086–91

110. Lelliott CJ, Medina-Gomez G, Petrovic N, Kis A, Feldmann HM, et al. 2006. Ablation of PGC-1β results in defective mitochondrial activity, thermogenesis, hepatic function, and cardiac performance. *PLoS Biol.* 4:e369

111. Vianna CR, Huntgeburth M, Coppari R, Choi CS, Lin J, et al. 2006. Hypomorphic mutation of PGC-1β causes mitochondrial dysfunction and liver insulin resistance. *Cell Metab.* 4:453–64

112. Uldry M, Yang W, St-Pierre J, Lin J, Seale P, Spiegelman BM. 2006. Complementary action of the PGC-1 coactivators in mitochondrial biogenesis and brown fat differentiation. *Cell Metab.* 3:333–41

113. Lai L, Leone TC, Zechner C, Schaeffer PJ, Kelly SM, et al. 2008. Transcriptional coactivators PGC-1α and PGC-lβ control overlapping programs required for perinatal maturation of the heart. *Genes Dev.* 22:1948–61

114. Puigserver P, Rhee J, Lin J, Wu Z, Yoon JC, et al. 2001. Cytokine stimulation of energy expenditure through p38 MAP kinase activation of PPARγ coactivator-1. *Mol. Cell* 8:971–82

115. Li X, Monks B, Ge Q, Birnbaum MJ. 2007. Akt/PKB regulates hepatic metabolism by directly inhibiting PGC-1α transcription coactivator. *Nature* 447:1012–16

116. Lerin C, Rodgers JT, Kalume DE, Kim S-H, Pandey A, Puigserver P. 2006. GCN5 acetyltransferase complex controls glucose metabolism through transcriptional repression of PGC-1α. *Cell Metab.* 3:429–38

117. Teyssier C, Ma H, Emter R, Kralli A, Stallcup MR. 2005. Activation of nuclear receptor coactivator PGC-1α by arginine methylation. *Genes Dev.* 19:1466–73

118. Fan M, Rhee J, St-Pierre J, Handschin C, Puigserver P, et al. 2004. Suppression of mitochondrial respiration through recruitment of p160 myb binding protein to PGC-1α: modulation by p38 MAPK. *Genes Dev.* 18:278–89

119. Christian M, White R, Parker MG. 2006. Metabolic regulation by the nuclear receptor corepressor RIP140. *Trends Endocrinol. Metab.* 17:243–50

120. Powelka AM, Seth A, Virbasius JV, Kiskinis E, Nicoloro SM, et al. 2006. Suppression of oxidative metabolism and mitochondrial biogenesis by the transcriptional corepressor RIP140 in mouse adipocytes. *J. Clin. Investig.* 116:125–36

121. Leonardsson G, Steel JH, Christian M, Pocock V, Milligan S, et al. 2004. Nuclear receptor corepressor RIP140 regulates fat accumulation. *Proc. Natl. Acad. Sci. USA* 101:8437–42

122. Seth A, Steel JH, Nichol D, Pocock V, Kumaran MK, et al. 2007. The transcriptional corepressor RIP140 regulates oxidative metabolism in skeletal muscle. *Cell Metab.* 6:236–45

123. Nichol D, Christian M, Steel JH, White R, Parker MG. 2006. RIP140 expression is stimulated by estrogen-related receptor α during adipogenesis. *J. Biol. Chem.* 281:32140–47

124. Rytinki MM, Palvimo JJ. 2008. SUMOylation modulates the transcription repressor function of RIP140. *J. Biol. Chem.* 283:11586–95

125. Ho PC, Gupta P, Tsui YC, Ha SG, Huq M, Wei LN. 2008. Modulation of lysine acetylation-stimulated repressive activity by Erk2-mediated phosphorylation of RIP140 in adipocyte differentiation. *Cell Signal.* 20:1911–19

126. Mostaqul Huq MD, Gupta P, Tsai NP, White R, Parker MG, Wei LN. 2006. Suppression of receptor interacting protein 140 repressive activity by protein arginine methylation. *EMBO J.* 25:5094–104

127. Rockl KS, Witczak CA, Goodyear LJ. 2008. Signaling mechanisms in skeletal muscle: acute responses and chronic adaptations to exercise. *IUBMB Life* 60:145–53

128. Klingenspor M. 2003. Cold-induced recruitment of brown adipose tissue thermogenesis. *Exp. Physiol.* 88:141–48

129. Civitarese AE, Smith SR, Ravussin E. 2007. Diet, energy metabolism and mitochondrial biogenesis. *Curr. Opin. Clin. Nutr. Metab. Care* 10:679–87

130. Miura S, Kawanaka K, Kai Y, Tamura M, Goto M, et al. 2007. An increase in murine skeletal muscle peroxisome proliferator-activated receptor-γ coactivator-1α (PGC-1α) mRNA in response to exercise is mediated by β-adrenergic receptor activation. *Endocrinology* 148:3441–48

131. Russell AP, Hesselink MK, Lo SK, Schrauwen P. 2005. Regulation of metabolic transcriptional coactivators and transcription factors with acute exercise. *FASEB J.* 19:986–88

132. Wright DC, Han DH, Garcia-Roves PM, Geiger PC, Jones TE, Holloszy JO. 2007. Exercise-induced mitochondrial biogenesis begins before the increase in muscle PGC-1α expression. *J. Biol. Chem.* 282:194–99

133. Arany Z, Wagner BK, Ma Y, Chinsomboon J, Laznik D, Spiegelman BM. 2008. Gene expression-based screening identifies microtubule inhibitors as inducers of PGC-1α and oxidative phosphorylation. *Proc. Natl. Acad. Sci. USA* 105:4721–26

134. Lambert AJ, Wang B, Yardley J, Edwards J, Merry BJ. 2004. The effect of aging and caloric restriction on mitochondrial protein density and oxygen consumption. *Exp. Gerontol.* 39:289–95

135. Nisoli E, Tonello C, Cardile A, Cozzi V, Bracale R, et al. 2005. Calorie restriction promotes mitochondrial biogenesis by inducing the expression of eNOS. *Science* 310:314–17

136. Lopez-Lluch G, Hunt N, Jones B, Zhu M, Jamieson H, et al. 2006. Calorie restriction induces mitochondrial biogenesis and bioenergetic efficiency. *Proc. Natl. Acad. Sci. USA* 103:1768–73

137. Borniquel S, Valle I, Cadenas S, Lamas S, Monsalve M. 2006. Nitric oxide regulates mitochondrial oxidative stress protection via the transcriptional coactivator PGC-1α. *FASEB J.* 20:1889–91

138. Eaton JS, Lin ZP, Sartorelli AC, Bonawitz ND, Shadel GS. 2007. Ataxia-telangiectasia mutated kinase regulates ribonucleotide reductase and mitochondrial homeostasis. *J. Clin. Investig.* 117:2723–34

139. Rohas LM, St-Pierre J, Uldry M, Jager S, Handschin C, Spiegelman BM. 2007. A fundamental system of cellular energy homeostasis regulated by PGC-1α. *Proc. Natl. Acad. Sci. USA* 104:7933–38

140. Wagner BK, Kitami T, Gilbert TJ, Peck D, Ramanathan A, et al. 2008. Large-scale chemical dissection of mitochondrial function. *Nat. Biotechnol.* 26:343–51

141. Biswas G, Adebanjo OA, Freedman BD, Anandatheerthavarada HK, Vijayasarathy C, et al. 1999. Retrograde Ca²⁺ signaling in C2C12 skeletal myocytes in response to mitochondrial genetic and metabolic stress: a novel mode of inter-organelle crosstalk. *EMBO J.* 18:522–33

142. Wright DC, Geiger PC, Han DH, Jones TE, Holloszy JO. 2007. Calcium induces increases in peroxisome proliferator-activated receptor γ coactivator-1α and mitochondrial biogenesis by a pathway leading to p38 mitogen-activated protein kinase activation. *J. Biol. Chem.* 282:18793–99

143. Naya FJ, Mercer B, Shelton J, Richardson JA, Williams RS, Olson EN. 2000. Stimulation of slow skeletal muscle fiber gene expression by calcineurin in vivo. *J. Biol. Chem.* 275:4545–48

144. Wu H, Kanatous SB, Thurmond FA, Gallardo T, Isotani E, et al. 2002. Regulation of mitochondrial biogenesis in skeletal muscle by CaMK. *Science* 296:349–52

145. Handschin C, Rhee J, Lin J, Tarr PT, Spiegelman BM. 2003. An autoregulatory loop controls peroxisome proliferator-activated receptor γ coactivator 1α expression in muscle. *Proc. Natl. Acad. Sci. USA* 100:7111–16

146. McGee SL, Hargreaves M. 2004. Exercise and myocyte enhancer factor 2 regulation in human skeletal muscle. *Diabetes* 53:1208–14

147. Akimoto T, Pohnert SC, Li P, Zhang M, Gumbs C, et al. 2005. Exercise stimulates Pgc-1α transcription in skeletal muscle through activation of the p38 MAPK pathway. *J. Biol. Chem.* 280:19587–93

148. Akimoto T, Li P, Yan Z. 2008. Functional interaction of regulatory factors with the Pgc-1α promoter in response to exercise by in vivo imaging. *Am. J. Physiol. Cell Physiol.* 295:C288–92

149. Czubryt MP, McAnally J, Fishman GI, Olson EN. 2003. Regulation of peroxisome proliferator-activated receptor γ coactivator 1α (PGC-1α) and mitochondrial function by MEF2 and HDAC5. *Proc. Natl. Acad. Sci. USA* 100:1711–16

150. Crunkhorn S, Dearie F, Mantzoros C, Gami H, da Silva WS, et al. 2007. Peroxisome proliferator activator receptor γ coactivator-1 expression is reduced in obesity: potential pathogenic role of saturated fatty acids and p38 mitogen-activated protein kinase activation. *J. Biol. Chem.* 282:15439–50

151. Winder WW, Holmes BF, Rubink DS, Jensen EB, Chen M, Holloszy JO. 2000. Activation of AMP-activated protein kinase increases mitochondrial enzymes in skeletal muscle. *J. Appl. Physiol.* 88:2219–26

152. Lee WJ, Kim M, Park H-S, Kim HS, Jeon MJ, et al. 2006. AMPK activation increases fatty acid oxidation in skeletal muscle by activating PPARα and PGC-1. *Biochem. Biophys. Res. Commun.* 340:291–95

153. McGee SL, van Denderen BJ, Howlett KF, Mollica J, Schertzer JD, et al. 2008. AMP-activated protein kinase regulates GLUT4 transcription by phosphorylating histone deacetylase 5. *Diabetes* 57:860–67

154. Narkar VA, Downes M, Yu RT, Embler E, Wang YX, et al. 2008. AMPK and PPARδ agonists are exercise mimetics. *Cell* 134:405–15

155. Minokoshi Y, Kim YB, Peroni OD, Fryer LG, Muller C, et al. 2002. Leptin stimulates fatty-acid oxidation by activating AMP-activated protein kinase. *Nature* 415:339–43

156. Yoon MJ, Lee GY, Chung JJ, Ahn YH, Hong SH, Kim JB. 2006. Adiponectin increases fatty acid oxidation in skeletal muscle cells by sequential activation of AMP-activated protein kinase, p38 mitogen-activated protein kinase, and peroxisome proliferator-activated receptor α. *Diabetes* 55:2562–70

157. Irrcher I, Walkinshaw DR, Sheehan TE, Hood DA. 2008. Thyroid hormone (T3) rapidly activates p38 and AMPK in skeletal muscle in vivo. *J. Appl. Physiol.* 104:178–85

158. Rodgers JT, Lerin C, Gerhart-Hines Z, Puigserver P. 2008. Metabolic adaptations through the PGC-1α and SIRT1 pathways. *FEBS Lett.* 582:46–53

159. Lagouge M, Argmann C, Gerhart-Hines Z, Meziane H, Lerin C, et al. 2006. Resveratrol improves mitochondrial function and protects against metabolic disease by activating SIRT1 and PGC-1α. *Cell* 127:1109–22

160. Dasgupta B, Milbrandt J. 2007. Resveratrol stimulates AMP kinase activity in neurons. *Proc. Natl. Acad. Sci. USA* 104:7217–22

161. Milne JC, Lambert PD, Schenk S, Carney DP, Smith JJ, et al. 2007. Small molecule activators of SIRT1 as therapeutics for the treatment of type 2 diabetes. *Nature* 450:712–16

162. Nemoto S, Fergusson MM, Finkel T. 2005. SIRT1 functionally interacts with the metabolic regulator and transcriptional coactivator PGC-1α. *J. Biol. Chem.* 280:16456–60

163. Lebrasseur NK, Cote GM, Miller TA, Fielding RA, Sawyer DB. 2003. Regulation of neuregulin/ErbB signaling by contractile activity in skeletal muscle. *Am. J. Physiol. Cell Physiol.* 284:C1149–55

164. Canto C, Chibalin AV, Barnes BR, Glund S, Suarez E, et al. 2006. Neuregulins mediate calcium-induced glucose transport during muscle contraction. *J. Biol. Chem.* 281:21690–97

165. Ariazi EA, Kraus RJ, Farrell ML, Jordan VC, Mertz JE. 2007. Estrogen-related receptor α1 transcriptional activities are regulated in part via the ErbB2/HER2 signaling pathway. *Mol. Cancer Res.* 5:71–85

166. Nisoli E, Clementi E, Paolucci C, Cozzi V, Tonello C, et al. 2003. Mitochondrial biogenesis in mammals: the role of endogenous nitric oxide. *Science* 299:896–99

167. Sano M, Izumi Y, Helenius K, Asakura M, Rossi DJ, et al. 2007. Ménage-à-trois 1 is critical for the transcriptional function of PPARγ coactivator 1. *Cell Metab.* 5:129–42

168. Wu C, Delano DL, Mitro N, Su SV, Janes J, et al. 2008. Gene set enrichment in eQTL data identifies novel annotations and pathway regulators. *PLoS Genet.* 4:e1000070

169. Reznick RM, Zong H, Li J, Morino K, Moore IK, et al. 2007. Aging-associated reductions in AMP-activated protein kinase activity and mitochondrial biogenesis. *Cell Metab.* 5:151–56

170. Valerio A, Cardile A, Cozzi V, Bracale R, Tedesco L, et al. 2006. TNF-α downregulates eNOS expression and mitochondrial biogenesis in fat and muscle of obese rodents. *J. Clin. Investig.* 116:2791–98

171. Richardson DK, Kashyap S, Bajaj M, Cusi K, Mandarino SJ, et al. 2005. Lipid infusion decreases the expression of nuclear encoded mitochondrial genes and increases the expression of extracellular matrix genes in human skeletal muscle. *J. Biol. Chem.* 280:10290–97

172. Sparks LM, Xie H, Koza RA, Mynatt R, Hulver MW, et al. 2005. A high-fat diet coordinately downregulates genes required for mitochondrial oxidative phosphorylation in skeletal muscle. *Diabetes* 54:1926–33

173. Garcia-Roves P, Huss JM, Han DH, Hancock CR, Iglesias-Gutierrez E, et al. 2007. Raising plasma fatty acid concentration induces increased biogenesis of mitochondria in skeletal muscle. *Proc. Natl. Acad. Sci. USA* 104:10709–13

174. Hancock CR, Han D-H, Chen M, Terada S, Yasuda T, et al. 2008. High-fat diets cause insulin resistance despite an increase in muscle mitochondria. *Proc. Natl. Acad. Sci. USA* 105:7815–20

175. Daitoku H, Yamagata K, Matsuzaki H, Hatta M, Fukamizu A. 2003. Regulation of PGC-1 promoter activity by protein kinase B and the forkhead transcription factor FKHR. *Diabetes* 52:642–49

176. Southgate RJ, Bruce CR, Carey AL, Steinberg GR, Walder K, et al. 2005. PGC-1α gene expression is down-regulated by Akt-mediated phosphorylation and nuclear exclusion of FoxO1 in insulin-stimulated skeletal muscle. *FASEB J.* doi:10.1096/fj.05-3993fje

177. Kim D, Kwon YK, Cho KH. 2007. Coupled positive and negative feedback circuits form an essential building block of cellular signaling pathways. *Bioessays* 29:85–90

178. Miura S, Kai Y, Kamei Y, Ezaki O. 2008. Isoform-specific increases in murine skeletal muscle peroxisome proliferator-activated receptor-γ coactivator-1α (PGC-1α) mRNA in response to β2-adrenergic receptor activation and exercise. *Endocrinology* 149:4527–33

The Functions and Roles of the Extracellular Ca^{2+}–Sensing Receptor along the Gastrointestinal Tract

John P. Geibel[1] and Steven C. Hebert[2]

[1]Department of Surgery and [2]Department of Cellular and Molecular Physiology, Yale University School of Medicine, New Haven, Connecticut 06520; email: john.geibel@yale.edu

Annu. Rev. Physiol. 2009. 71:205–17

First published online as a Review in Advance on November 13, 2008

The *Annual Review of Physiology* is online at physiol.annualreviews.org

This article's doi: 10.1146/annurev.physiol.010908.163128

Copyright © 2009 by Annual Reviews. All rights reserved

0066-4278/09/0315-0205$20.00

Key Words

stomach, intestine, colon, electrolytes, secretion

Abstract

Digestion of food and normal salt and water homeostasis in the body require a functional digestive tract. Recently an increasing number of studies have demonstrated a role for the calcium-sensing receptor along the entire gastrointestinal tract and its role in normal gut physiology. Detailed studies have been performed on colonic fluid transport and gastric acid secretion. We have now demonstrated that the receptor can modulate fluid secretion and absorption along the intestine and can thereby be a potent target to prevent secretory diarrhea. Recent studies have demonstrated that organic nutrients such as polyamines and L-amino acids can act as agonists by allosterically modifying the receptor. Thus, the receptor may detect nutrient availability to epithelial cells along the gastrointestinal tract and may be involved in the coordinated rapid turnover of the intestinal epithelium. Furthermore, the receptor has been suggested as a link for the mechanisms leading to calcium uptake by the colon and may thus reduce the risk for colon cancer.

LOCALIZATION OF THE Ca²⁺-SENSING RECEPTOR ALONG THE GASTROINTESTINAL TRACT

The gastrointestinal tracts of amphibia (1, 2), birds (3), fish (4), and mammals (5–10), including humans (5, 11, 12), have all shown Ca^{2+}-sensing receptor (CaSR) transcripts and/or protein. To date, the earliest evolutionary expression of the receptor in the gastrointestinal tract goes back to cartilaginous fish (elasmobranchs), e.g., the dogfish shark (4). In cartilaginous and bony fish, the receptor is present on the apical surfaces of both the stomach and the intestine (4). In the marine environment the CaSR is thought to have evolved to support osmo-adaptation. This notion is supported by the general expression of CaSR in many other tissues outside the gastrointestinal tract that are involved in mono- and divalent ion transport in fish that live in a seawater environment rich in divalent minerals and sodium chloride (4, 13). This involvement of the CaSR in divalent and monovalent metabolism is repeated in mammals (e.g., the role of the receptor in fluid and electrolyte transport in the colon; see discussion below).

In the amphibian *Necturus maculosus*, CaSR expression was detected on the basal surface of gastric epithelial cells (1). In contrast, in the frog stomach CaSR expression was detected on apical membranes of the acid-secreting oxyntic cells (2). In chicken, *Gallus domesticus*, CaSR was identified in the duodenum, but to date the remainder of the gastrointestinal tissues has not been assayed (3). One of the most complete explorations for CaSR expression has taken place in mammals, along the majority of the gastrointestinal tract (5–10, 12). To date receptor transcripts and/or protein have been detected in stomach, small intestinal, and colonic mucosal epithelia, along with the underlying neural plexuses of Meissner and Auerbach. In addition the CaSR has also been detected in several human intestinal cell lines (T84, HT-29, Caco-2, FET, SW480, MOSER, and CBS) (8, 14–16)

and in primary cultures of human gastric mucosa (11, 17, 18).

In mammalian stomach, CaSR has been identified on both the apical and the basolateral membranes of human gastrin-secreting cells (G cells) (17, 18) and mucous-secreting cells (11), whereas in parietal cells the receptor has been detected only at the basolateral membrane (2, 6, 19). In contrast, in small intestine there is expression on both the apical and the basolateral membranes of villus cells (7). CaSR has also been detected on both the apical and the basolateral membranes of both surface and crypt epithelial cells in rat and human colons (5, 7). This expression pattern of CaSR immunostaining was also observed in rat proximal and distal colon (5). In human large intestine, CaSR was identified at the apical and the basolateral membranes of crypts and in certain enteroendocrine cells at the base of crypts (5, 12).

Ca²⁺-SENSING RECEPTOR IN THE STOMACH

Overview

For digestion to take place the stomach must generate large quantities of 0.16 N hydrochloric acid. To signal the stomach to secrete acid, a complex series of hormonal, neuronal, and/or paracrine/autocrine feedback regulatory mechanisms (20–23) that allow for the continued production of acid is activated. **Figure 1** shows a model of acid secretion of a parietal cell, the cell responsible for the generation of the concentrated acid. This model, a composite that summarizes the data from many laboratories (21, 22, 24–26), shows the flow of protons from the apical pole of the cell, where they are combined with secreted Cl$^-$ ions to produce the concentrated 0.1 N acid. In the resting cell the ion pumps remain off the surface of the apical pole and are found along the surface of the tubular vesicular elements. Following stimulation, the H$^+$,K$^+$-ATPase (proton pump) is trafficked to the apical surface, where it provides the substrate for the vectorial transport of protons (20, 22, 23, 27–29). In synchrony with the release of

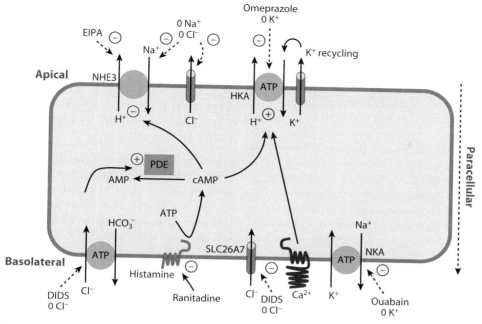

Figure 1

Schematic illustration of the parietal cell showing the known mechanisms of gastric acid secretion. Both the Ca^{2+}-sensing receptor (CaSR) and the histamine receptor are located on the basolateral membrane. HCl secretion results from H^+ extrusion via the H^+,K^+-ATPase (HKA) coupled with Cl^- secretion via an apical channel. Activation of the CaSR can stimulate the HKA in resting cells. In addition, CaSR stimulation of gastrin release from gastrin-secreting cells (G cells) leads to increases in histamine release and adds to HCl secretion. Abbreviations used: DIDS, 4,4′-di-isothiocyanato-2,2′-disulfostilbene; EIPA, 5-(N-ethyl-N-isopropyl)-amiloride; NHE3, N^+/H^+ exchanger 3; NKA, N^+,K^+-ATPase; PDE, phosphodiesterase.

acid, the chief cells at the base of the gland secrete pepsinogen, which the concentrated acid converts to pepsin. The combined enzymatic acid solution is propelled to the surface of the gland and into the interior of the stomach. This solution can then aid in the digestion and breakdown of proteins and prepares them for amino acid or peptide reabsorption in the intestine.

Classic models of acid secretion (**Figure 1**) involve activation of the gastric glands via either a neuronal route or a hormonal pathway, which leads to the release of gastrin from the G cells. Gastrin directly acts on the endocrine (ECL) cells of the gastric gland to release histamine. Histamine binds to receptors on the basolateral surface of the parietal cell, causing insertion of the proton pump (H^+,K^+-ATPase) into the apical membrane of the gland (30). As alluded to above, this process requires an active tubular-vesicular insertion mechanism to transport the H^+,K^+-ATPase into the apical region of the parietal cell (28). **Figure 1** illustrates that in conjunction with proton pump insertion and activation there is a simultaneous activation of an apical K^+ channel(s) that provides for the recycling of K^+. Such recycling of K^+ is necessary for its continued exchange with H^+ on the pump (20, 22, 24, 31). To produce HCl, a secondary action of histamine release is the activation and/or insertion of Cl^- channels into the apical membranes of parietal cells, which mediates synchronized Cl^- secretion to accompany H^+ secretion (20, 24, 25; see **Figure 1**). In summary, the acid secretory process requires (*a*) the insertion of proton pumps into the apical surface for H^+ secretion; (*b*) concurrent activation of apical Cl^- channels to mediate Cl⁻ secretion, which is essential for HCl formation; and (*c*) activation of apical K⁺ channels that are necessary for the recycling of K⁺ into the lumen of the gland, allowing for the maintenance of H⁺,K⁺-ATPase activity.

When there is unregulated HCl secretion due to unregulated cellular feedback or via increased output of acid at rest, many symptoms manifest themselves clinically. These range from mild heartburn to the potential for lesions and ulcerations of the gastric mucosa or the esophagus (21, 26). If gastric ulcers remain untreated, the outcome can be abdominal bleeding or cell hyperplasia, which, if recurrent, potentially results in tumor formation. Compounding the difficulty of characterizing this tissue, a complete analysis of cell function and feedback loops has been precluded owing to the inherent complexity of the tissue with its mixed collection of cell types (at least seven different types

of cells identified in the glands to date) as well as the complex physiology (different populations of glands with specific functions, along with a layer of surface cells that secrete a protective mucous gel layer, and bicarbonate).

THE ROLE OF Ca²⁺ AND THE Ca²⁺-SENSING RECEPTOR IN ACID SECRETION

Following hormonal or neuronal stimulation of acid secretion in the gastric gland, increased levels of cell Ca²⁺ have been detected (20, 21, 23, 36) (see **Figure 2**) and have been associated with the activation of the trafficking process of pumps to the membrane and with acid secretion (37). After this initial secretagogue-induced stimulation, intracellular Ca²⁺ levels fall, and acid secretion diminishes to basal levels.

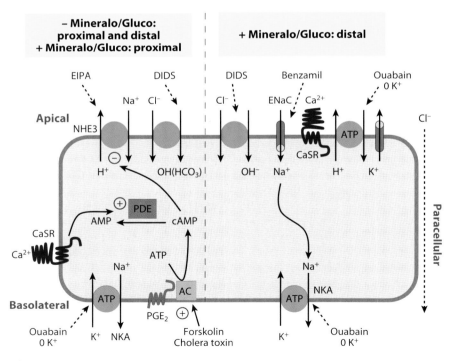

Figure 2

Schematic representation of the colonic crypt showing both resting and mineralocorticoid-induced fluid and electrolyte transport in both the proximal and distal colon. Mineral/Gluco denotes the localization of transport proteins in resting conditions and also following mineralocorticoid and/or glucocorticoid exposure.

The first report that identified functionally active CaSR in stomach focused on the amphibian *N. maculosus* (3). In this study activation of the basolateral receptor in *Necturus* gastric antrum by increasing extracellular Ca^{2+} concentration (or other receptor agonists like NPS-467 or neomycin) induced a rapid hyperpolarization and a decrease in resistance of the basolateral membrane. Results from circuit analysis suggested that the detected electrophysiological effects were due to activation of basolateral K^+ channels (3).

In rat stomach, when extracellular Ca^{2+} was increased in the absence of conventional extracellular secretagogues (e.g., histamine) (6), intracellular Ca^{2+} levels of parietal cells rapidly doubled. Furthermore, activation of the CaSR in the intact rat gastric gland by Ca^{2+} or Mg^{2+} or by the potent CaSR agonist Gd^{3+} elevated the rate of acid secretion via the apical H^+,K^+-ATPase (19). Inactivation of the CaSR by reducing the concentration of extracellular divalent minerals resulted in a downregulation in acid secretion even in the continued presence of potent secretagogues like histamine (19). Therefore, the CaSR plays an important modulatory role in regulating acid secretion in the mammalian gastric gland.

The frog (*Rana esculenta*) stomach shows immunostaining for the CaSR with an intracellular distribution in the acid-secreting oxyntic cells that colocalized with the plasma membrane Ca^{2+}-ATPase and the H^+,K^+-ATPase (2). Nearing et al. (4) suggested that the receptor may reside on the same tubular-vesicular network so that it could traffic to the apical membrane along with the H^+,K^+-ATPase. A physiological role for such an apical receptor has yet to be established.

CaSR has also been identified in human gastric tissues (11, 17, 18, 27, 28). When mucous epithelial cells were placed in culture, receptor activation resulted in a rapid elevation in intracellular Ca^{2+} and a proliferative response (11). In G cells, activation of the receptor leads to gastrin release, which is accompanied by activation of phospholipase C (17) and a concurrent rise in intracellular Ca^{2+} (17, 18). Histamine release caused by the CaSR-mediated secretion of gastrin by the G cells may account for the rebound acid secretion that occurs with the ingestion of Ca^{2+}-containing antacids. These data support the hypothesis that the CaSR in the stomach plays an important role in both acid secretion and mucosal repair. One possibility is that the receptor acts to modulate the rate of acid secretion in response to the body's requirement for increased Ca^{2+} absorption. As the demand for Ca^{2+} increases, more acid is produced, which may ionize more ingested Ca^{2+}, thereby allowing more ionized Ca^{2+} to be available for intestinal absorption. With a continued elevation in Ca^{2+}, gastric CaSR would become internalized, resulting in a downregulation in acid secretion. Secondary to this downregulation, the elevated serum Ca^{2+} may aid in repairing the mucosal layer prior to the next cycle of acid secretion.

Patients with Zollinger-Ellison syndrome (ZES) are characterized by ulcer disease of the upper gastrointestinal tract, increased gastrin secretion, and non-β-cell tumors of the pancreas (i.e., gastrinomas). Levels of gastrin secretion tend to correlate with the activity of frequently associated hyperparathyroidism (29). This correlation is consistent with gastrin secretion paralleling high parathyroid hormone (PTH)-driven elevations in plasma Ca^{2+}. Significant, although variable, CaSR expression has been detected in human gastrinomas (27, 28), suggesting a possible link with the receptor mediating the effect of extracellular Ca^{2+} on gastrin secretion. Consistent with this possibility are recent studies in which activation of the CaSR occurred by raising extracellular Ca^{2+}, resulting in a rapid increase in intracellular Ca^{2+} that was not altered by the Ca^{2+} channel blocker nifedipine (28).

Ca^{2+}-SENSING RECEPTOR ALONG THE INTESTINE

Overview

Although the CaSR is expressed in epithelial cells along the entire small and large intestines, detailed investigations of the receptor have been conducted only in the colon. For this reason discussion of the receptor is focused on the potential roles of the receptor in normal intestinal function, in diarrheal states, and in the effect of oral Ca^{2+} intake on the risk of colon cancer. Expression of the CaSR in nerve plexuses involved in smooth muscle function and coordination suggests a potential role of the CaSR in modulating intestinal motility. This latter role for the receptor may be important in coordinating the delivery of food (Ca^{2+}, amino acids, polyamines) and intestinal motility, thus maximizing nutrient absorption. CaSR activation of intestinal motility may also contribute to constipation associated with hypercalcemic states.

Fluid Transport in the Colon

The colon's major function is both to absorb and to secrete fluid, which contributes to the maintenance of normal salt and water homeostasis. The colon has extensive invaginations of the surface into crypts (**Figure 3**), with approximately 10% of the epithelial mass on the colonic surface and the remaining 90% in crypts. Although early studies had suggested that surface cells absorb and crypt cells secrete fluid in the colon, recent evidence has established that surface and crypt cells can both absorb and secrete fluid [see review for details (41)]. The direction of net fluid transport is determined by the relative magnitudes of the absorptive and secretory fluxes. Because more than 90% of the colonic epithelial surface area forms invaginations or crypts, these structures constitute the major functional unit of the colonic epithelium (42, 43).

In the absence of hormones, drugs, or other factors (i.e., under basal conditions), net fluid transport by colonic crypts is absorptive (44). However, upon exposure to cell-permeable cyclic AMP (cAMP) analogs, forskolin or other agents that activate adenylate cyclase, or modulators of cAMP metabolism such as phosphodiesterase (PDE) inhibitors, crypts shift the direction of net fluid transport to secretion (44). The addition of cAMP-generating hormones/factors such as 5-hydroxytryptophan or prostaglandin E_2 to the blood/interstitial surface of the crypt epithelium also increases fluid secretion by colonic crypts (42, 43). When cyclic nucleotides (cAMP or cGMP) modulate these fluid transport processes in the colon, the results—profound fluid and electrolyte losses with secretory diarrheas such as cholera (43)—can be devastating.

The Ca^{2+}-Sensing Receptor Can Modulate Colonic Fluid Transport

Numerous early physiological studies measuring Ca^{2+} fluxes in rat isolated colonic mucosa presented evidence that the colon has the capacity to respond to changes in extracellular Ca^{2+}. For example, the colon, similar to the small intestine, can absorb and/or secrete Ca^{2+} in response to changes in extracellular Ca^{2+} as well as modulations in 1,25-dihydroxy vitamin D_3 levels (30, 31). The latter observations indicate that the colonic mucosal epithelium

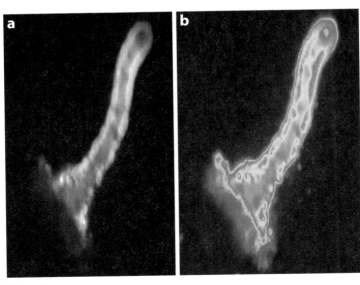

Figure 3

Rise in cell Ca^{2+} elicited by activation of the Ca^{2+}-sensing receptor in colonic crypts. Individual rat colonic crypts perfused in vitro and loaded with the Ca^{2+}-sensitive dye Fluo-4. (*a*) A phase contrast image of the gland at rest in 0.5 mM extracellular Ca^{2+}. (*b*) The crypt following exposure to 5 mM extracellular Ca^{2+} shows a uniform increase in cell Ca^{2+}.

is equipped with a Ca^{2+}-sensing mechanism. More recent studies have indicated that this divalent mineral–sensing mechanism in colonic epithelia is the CaSR. This conclusion is based on both immunolocalization of the receptor in apical and basolateral membranes and receptor function studies (6, 8, 9, 12).

Activation of the colonic CaSR by the divalent ion Ca^{2+}, the trivalent ion Gd^{3+}, or the antibiotic neomycin results in a rapid rise in intracellular Ca^{2+} in both surface and crypt cells (6). This rise in intracellular Ca^{2+} takes place within a few seconds, which is consistent with activation of the phosphatidylinositol–phospholipase C–inositol 1,4,5-trisphosphate (PI-PLC-IP_3) pathway by G protein–coupled cell membrane receptors (see **Figure 4**). Pretreatment of crypt cells with U-73122, a specific inhibitor of PI-PLC (see **Figure 4**), prevents the CaSR-mediated increases in intracellular Ca^{2+}. This drug induced PLC inhibition and confirmed that intracellular Ca^{2+} transients induced by CaSR agonists were due not to altered entry of extracellular Ca^{2+} into colonic epithelial cells but to receptor-mediated activation of the PI-PLC pathway. The release of Ca^{2+} from thapsigargin-sensitive cell stores causes this receptor-mediated increase in intracellular Ca^{2+} in colon (6).

To assess the role of the CaSR as a modulator of colonic fluid movement, perfused colonic crypts have been isolated by use of the in vitro microperfusion technique (6, 32). When crypts are maintained under basal conditions (i.e., in the absence of forskolin), crypts exhibit net fluid absorption (33).

If crypts are exposed to forskolin, there is a change from net fluid absorption to net fluid secretion (33). Activation of either the luminal or the basolateral CaSR by Ca^{2+} and/or a polyamine (spermine) results in a reversal of forskolin-stimulated fluid secretion (5, 32). To completely understand the mechanism of CaSR effects on cAMP-mediated fluid secretion will require further studies. On the basis of the information obtained following receptor activation on vasopressin-stimulated increases in cAMP in the kidney thick ascending limb of

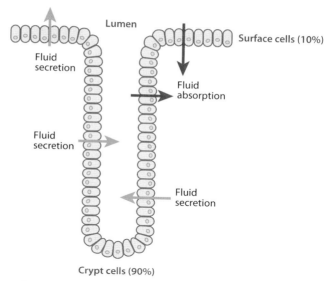

Figure 4

Diagrammatic representation of the mammalian crypt showing that both surface and crypt cells mediate fluid absorption and secretion. This model shows that both types of cells (surface and crypt) can either secrete fluid or absorb fluid. Both types of cells are sensitive to Ca^{2+}-sensing receptor (CaSR) activation.

Henle (34), we postulate a model of receptor action as shown in **Figure 5**. The pathway shown in **Figure 5** for modulating intestinal fluid secretion through the colonic CaSR could provide a novel method for the prevention or treatment of certain diarrheal diseases (i.e., cholera and other cAMP-associated diarrheal diseases).

POSSIBLE ROLES OF THE Ca^{2+}-SENSING RECEPTOR IN INTESTINAL EPITHELIAL CELL GROWTH, DIFFERENTIATION, AND NUTRIENT SENSING

Overview

The epithelial barrier of the colon and the small intestine is in a state of constant renewal. In the colon, cells proliferate and differentiate as they migrate from the base of the crypt (progenitor zone) to the surface. For this reason it has been thought that the cells at the base of the crypt are highly proliferative but are less differentiated, whereas cells at the surface of the colon

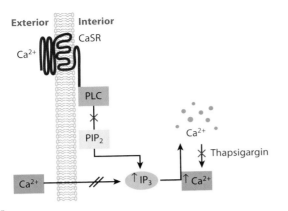

Figure 5

A model of Ca^{2+}-sensing receptor (CaSR) action. Stimulation of the CaSR on either the apical or the basolateral membranes leads to a rise in intracellular Ca^{2+} mediated by generation of inositol 1,4,5-trisphosphate (IP_3) via phospholipase C (PLC) and then the release of Ca^{2+} from a thapsigargin-sensitive pool. The CaSR-mediated rise in intracellular Ca^{2+} activates the Ca^{2+}/calmodulin-sensitive phosphodiesterases, which metabolize intracellular cAMP and thereby abrogate fluid secretion. PIP_2 denotes phosphatidylinositol bisphosphate.

in providing feedback information on nutrient delivery to intestinal cells are discussed below.

Role of the Ca^{2+}-Sensing Receptor in Intestinal Cell Growth and Differentiation

Dietary polyamine intake is essential for normal gastrointestinal tract cell growth and development in rats and mice (45–49). Hypothesized mechanisms for the prodifferentiation and anticancer effects of dietary Ca^{2+}/polyamines include (*a*) the formation of insoluble salts of Ca^{2+} with otherwise known tumorigenic fatty acids and bile salts and (*b*) the modification of the rates and/or fates of biologically active molecules such as nucleic acids, proteins, and phospholipids (50–52). Identification of CaSRs on the plasma membranes of both surface and crypt epithelial cells suggests that CaSR may mediate some of the effects of dietary Ca^{2+}, polyamines, and other nutrients on tissue modeling of the intestinal epithelia.

Enhanced levels of polyamines, in particular spermine, result in the generation of IP_3, which in turn raises intracellular Ca^{2+} and modulates forskolin-stimulated fluid secretion, consistent with activation of the colonic epithelial CaSR (32). Extracellular Ca^{2+} potentiates polyamine (spermine > spermidine > putrescine)-mediated enhancement of intracellular IP_3 and Ca^{2+} accumulation. Polyamines can also lower the EC_{50} for Ca^{2+}_o-mediated activation of the CaSR (32). These recent findings indicate that the colonic epithelial CaSR responds to polyamines and may use these compounds as a natural allosteric modulating substrate for itself.

The receptor also increases E-cadherin production and reduces β-catenin production; E-cadherin and β-cadherin are markers for intestinal differentiation in cultured intestinal cell lines (14, 53, 54). Activation of the CaSR in Caco-2 cells by raising extracellular Ca^{2+} increases thymidine incorporation into DNA, which is a measure of cell proliferation (16). Conversely, when Caco-2 cells are exposed to low extracellular Ca^{2+} concentrations, there is

are highly differentiated but are in a nonproliferative state. Should alteration of this highly regulated process occur, gross morphometric changes would result, which in turn could lead to the development of tumors. One potential role for the CaSR in colon is in the regulation of epithelial cell proliferation, differentiation, and development. This is suggested by recent observations that receptor activation reduces proliferation and induces differentiation of a number of different cell types. For example, recent data show that activation of the CaSR enhances cell differentiation in both mouse (35, 36) and human (37, 38) keratinocytes (39). Furthermore, activation of the CaSR in other cells modulates proliferation and inhibits apoptosis (40–42).

The ability of CaSR to modulate such a wide variety of effects in different cells may relate to the fact that the receptor not only responds to modulations in divalent minerals but also can detect changes in organic nutrients such as polyamines (43) and amino acids (44). Organic nutrients function primarily by altering the EC_{50} of the CaSR for Ca^{2+}, although weak direct agonist effects have been observed. Potential roles for these nutrients in coordinating protein and divalent mineral metabolism and

a protein kinase C (PKC)-dependent increase in c-*myc* proto-oncogene expression; activation of the CaSR by raising extracellular Ca^{2+} can stop this proproliferative effect (16). In keratinocytes and certain other cells, the CaSR also modulates the proliferation/differentiation response and alters the activities of mitogen-activated protein (MAP) kinases and tyrosine kinases associated with cell proliferation (16, 35–37, 42, 55–59). Taken together, these data support a potential role for the CaSR as a modulator of the cell proliferation and differentiation response in intestinal epithelial cells.

Role of Ca^{2+} and the Ca^{2+}-Sensing Receptor in Colon Cancer

Recent studies have shown that high dietary Ca^{2+} intake promotes colonic mucosal epithelial cell differentiation, decreases cell growth, and reduces risk for the development of colorectal cancer [see recent summaries (60, 61) and References 14, 15, 59, 62, and 63]. Cancer of the colon and rectum remains the second most frequently diagnosed malignancy in the United States in addition to the second most common cause of cancer-related death (>56,000 American deaths in 2007). Recent observations demonstrate that increasing dietary calcium markedly reduces the risk of developing colon adenomas. In particular, increases in dietary calcium intake (*a*) decrease the risk for colorectal cancer by threefold in men consuming 1400–1500 mg calcium per day [results of a 19-year prospective study of men working at the Western Electric Co. in Chicago (64)]; (*b*) significantly reduce colonic crypt cell proliferation and enhance markers of cell differentiation in human subjects at increased risk for colon cancer (63, 65, 66); and (*c*) reduce the risk of colorectal adenomas in humans [as concluded from the significant reduction in the recurrence of colorectal adenomas in a randomized, double-blind trial of 930 subjects (67)]. Moreover, long-term calcium supplementation significantly suppresses colonic cell proliferation in adenoma patients (62). In addition, virtually all rat studies have shown a marked reduction

in the incidence and number of carcinogen-induced colonic tumors following exposure to Ca^{2+} (see Reference 68 for a review). Activation of the CaSR in human carcinoma cell lines by elevating levels of extracellular Ca^{2+} promotes E-cadherin expression and suppresses β-catenin activation (14) and expression of other markers of cell differentiation (53, 54). Finally, there is a correlation between CaSR expression and the state of differentiation in human colon tumors (14). These observations provide a significant body of evidence that dietary Ca^{2+} reduces the risk of colon cancer and that the activation of CaSR mediates this effect.

Theory: Role of the Ca^{2+}-Sensing Receptor in Nutrient Sensing

The capability of the CaSR to respond to L-amino acids has been suggested as a possible link between protein ingestion and calcium metabolism (44). Evidence for this theory came

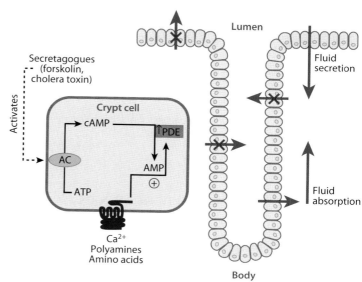

Figure 6

Mechanism for Ca^{2+}-sensing receptor (CaSR)-induced modulation of fluid secretion by the mammalian colonic crypt. Activation of either apical or basolateral receptors reduces cyclic AMP or cyclic GMP generated by modulators of adenylate cyclase (e.g., forskolin or cholera toxin, STa) by activating Ca^{2+}-dependent phosphodiesterases (PDE). AC denotes adenylate cyclase.

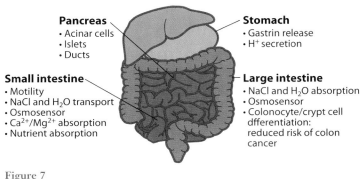

Pancreas
- Acinar cells
- Islets
- Ducts

Stomach
- Gastrin release
- H$^+$ secretion

Small intestine
- Motility
- NaCl and H$_2$O transport
- Osmosensor
- Ca^{2+}/Mg^{2+} absorption
- Nutrient absorption

Large intestine
- NaCl and H$_2$O absorption
- Osmosensor
- Colonocyte/crypt cell dfferentiation: reduced risk of colon cancer

Figure 7

Summary of the known physiological roles for the Ca^{2+}-sensing receptor in gastrointestinal biology.

from the direct relationship between dietary protein intake and renal Ca^{2+} excretion (69, 70). Acute ingestion of a high-protein diet increases urinary Ca^{2+} excretion (69, 70), whereas a low protein intake causes elevated PTH levels (71). The resulting increased urinary Ca^{2+} excretion associated with high protein intake appears related to elevations in intestinal Ca^{2+} absorption (69, 70), although this finding remains somewhat controversial (72). As discussed above, activation of the CaSR by Ca^{2+} stimulates gastric acid secretion, which in turn promotes acid digestion of proteins (together with peptidases). The resultant release of L-amino acids then promotes the absorption of Ca^{2+} in both the small

and the large intestines because of their synergistic activation of the CaSR.

SUMMARY

Figure 6 summarizes the presently known potential roles of the CaSR in gastrointestinal biology. On the basis of the unique properties of the CaSR in recognizing and responding to modulations in extracellular Ca^{2+} and nutrients, this receptor provides a logical potential mechanism to link dietary metabolism (i.e., the digestion of food and the associated nutrient absorption) with (*a*) nutrient availability with epithelial growth and differentiation; (*b*) divalent mineral and protein metabolism; (*c*) a reduction in risk of colon cancer with an increase in dietary Ca^{2+} uptake; and (*d*) salt, fluid, and nutrient homeostasis. In addition, the potential effects of CaSR activation by nutrients and intestinal motility, coupled to the demonstrated reduction in fluid secretion, may lead to an increase in nutrient-epithelial contact time for absorption. Finally, the potent ability of CaSR agonists to reduce cAMP-mediated fluid secretion by the colon has important implications for development of a novel oral therapy for cyclic nucleotide–dependent diarrheas such as cholera (**Figure 7**).

DISCLOSURE STATEMENT

The authors are not aware of any biases that might be perceived as affecting the objectivity of this review.

ACKNOWLEDGMENTS

This article is dedicated to my dear friend and colleague Steven C. Hebert, who passed away suddenly before the completion of this work. S.C.H. was supported by grants from the National Institutes of Health (DK 54999) and Amgen. J.P.G. is supported by grants from the National Institutes of Health (DK 17433), Amgen, and Astra Zeneca.

LITERATURE CITED

1. Cima RR, Cheng I, Klingensmith ME, Chattopadhyay N, Kifor O, et al. 1997. Identification and functional assay of an extracellular calcium-sensing receptor in *Necturus* gastric mucosa. *Am. J. Physiol.* 273:G1051–60
2. Caroppo R, Gerbino A, Debellis L, Kifor O, Soybel DI, et al. 2001. Asymmetrical, agonist-induced fluctuations in local extracellular [Ca^{2+}] in intact polarized epithelia. *EMBO J.* 20:6316–26

3. Diaz R, Hurwitz S, Chattopadhyay N, Pines M, Yang Y, et al. 1997. Cloning, expression, and tissue localization of the calcium-sensing receptor in chicken (*Gallus domesticus*). *Am. J. Physiol.* 273:R1008–16

4. Nearing J, Betka M, Quinn S, Hentschel H, Elger M, et al. 2002. Polyvalent cation receptor proteins (CaRs) are salinity sensors in fish. *Proc. Natl. Acad. Sci. USA* 99:9231–36

5. Cheng SX, Okuda M, Hall AE, Geibel JP, Hebert SC. 2002. Expression of calcium-sensing receptor in rat colonic epithelium: evidence for modulation of fluid secretion. *Am. J. Physiol. Gastrointest. Liver Physiol.* 283:G240–50

6. Cheng I, Qureshi I, Chattopadhyay N, Qureshi A, Butters RR, et al. 1999. Expression of an extracellular calcium-sensing receptor in rat stomach. *Gastroenterology* 116:118–26

7. Chattopadhyay N, Cheng I, Rogers K, Riccardi D, Hall A, et al. 1998. Identification and localization of extracellular Ca^{2+}-sensing receptor in rat intestine. *Am. J. Physiol.* 274:G122–30

8. Gama L, Baxendale-Cox LM, Breitwieser GE. 1997. Ca^{2+}-sensing receptors in intestinal epithelium. *Am. J. Physiol.* 273:C1168–75

9. Mitsuma T, Rhue N, Kayama M, Mori Y, Adachi K, et al. 1999. Distribution of calcium sensing receptor in rats: an immunohistochemical study. *Endocr. Regul.* 33:55–59

10. Butters RR Jr, Chattopadhyay N, Nielsen P, Smith CP, Mithal A, et al. 1997. Cloning and characterization of a calcium-sensing receptor from the hypercalcemic New Zealand white rabbit reveals unaltered responsiveness to extracellular calcium. *J. Bone Miner. Res.* 12:568–79

11. Rutten MJ, Bacon KD, Marlink KL, Stoney M, Meichsner CL, et al. 1999. Identification of a functional Ca^{2+}-sensing receptor in normal human gastric mucous epithelial cells. *Am. J. Physiol.* 277:G662–70

12. Sheinin Y, Kallay E, Wrba F, Kriwanek S, Peterlik M, Cross HS. 2000. Immunocytochemical localization of the extracellular calcium-sensing receptor in normal and malignant human large intestinal mucosa. *J. Histochem. Cytochem.* 48:595–602

13. Hentschel H, Nearing J, Harris HW, Betka M, Baum M, et al. 2003. Localization of Mg^{2+}-sensing shark kidney calcium receptor SKCaR in kidney of spiny dogfish, *Squalus acanthias*. *Am. J. Physiol. Ren. Physiol.* 285:F430–39

14. Chakrabarty S, Radjendirane V, Appelman H, Varani J. 2003. Extracellular calcium and calcium sensing receptor function in human colon carcinomas: promotion of E-cadherin expression and suppression of β-catenin/TCF activation. *Cancer Res.* 63:67–71

15. Kallay E, Bajna E, Wrba F, Kriwanek S, Peterlik M, Cross HS. 2000. Dietary calcium and growth modulation of human colon cancer cells: role of the extracellular calcium-sensing receptor. *Cancer Detect. Prev.* 24:127–36

16. Kallay E, Kifor O, Chattopadhyay N, Brown EM, Bischof MG, et al. 1997. Calcium-dependent c-*myc* proto-oncogene expression and proliferation of Caco-2 cells: a role for a luminal extracellular calcium-sensing receptor. *Biochem. Biophys. Res. Commun.* 232:80–83

17. Buchan AM, Squires PE, Ring M, Meloche RM. 2001. Mechanism of action of the calcium-sensing receptor in human antral gastrin cells. *Gastroenterology* 120:1128–39

18. Ray JM, Squires PE, Curtis SB, Meloche MR, Buchan AM. 1997. Expression of the calcium-sensing receptor on human antral gastrin cells in culture. *J. Clin. Investig.* 99:2328–33

19. Geibel JP, Wagner CA, Caroppo R, Qureshi I, Gloeckner J, et al. 2001. The stomach divalent ion-sensing receptor scar is a modulator of gastric acid secretion. *J. Biol. Chem.* 276:39549–52

20. Prinz C, Kajimura M, Scott D, Helander H, Shin J, et al. 1992. Acid secretion and the H,K ATPase of stomach. *Yale J. Biol. Med.* 65:577–96

21. Hirschowitz BI, Keeling D, Lewin M, Okabe S, Parsons M, et al. 1995. Pharmacological aspects of acid secretion. *Dig. Dis. Sci.* 40:S3–23

22. Hersey SJ, Sachs G. 1995. Gastric acid secretion. *Physiol. Rev.* 75:155–89

23. Sachs G, Shin JM, Briving C, Wallmark B, Hersey S. 1995. The pharmacology of the gastric acid pump: the H^+,K^+ ATPase. *Annu. Rev. Pharmacol. Toxicol.* 35:277–305

24. Cuppoletti J, Sachs G. 1984. Regulation of gastric acid secretion via modulation of a chloride conductance. *J. Biol. Chem.* 259:14952–9

25. Malinowska DH, Cuppoletti J, Sachs G. 1983. Cl^- requirement of acid secretion in isolated gastric glands. *Am. J. Physiol.* 245:G573–81

26. Banerjee S, El-Omar E, Mowat A, Ardill JE, Park RH, et al. 1996. Sucralfate suppresses *Helicobacter pylori* infection and reduces gastric acid secretion by 50% in patients with duodenal ulcer. *Gastroenterology* 110:717–24

27. Goebel SU, Peghini PL, Goldsmith PK, Spiegel AM, Gibril F, et al. 2000. Expression of the calcium-sensing receptor in gastrinomas. *J. Clin. Endocrinol. Metab.* 85:4131–37

28. Itami A, Kato M, Komoto I, Doi R, Hosotani R, et al. 2001. Human gastrinoma cells express calcium-sensing receptor. *Life Sci.* 70:119–29

29. Jansen JB, Lamers CB. 1982. Effect of changes in serum calcium on secretin-stimulated serum gastrin in patients with Zollinger-Ellison syndrome. *Gastroenterology* 83:173–78

30. Favus MJ, Kathpalia SC, Coe FL. 1981. Kinetic characteristics of calcium absorption and secretion by rat colon. *Am. J. Physiol.* 240:G350–54

31. Favus MJ, Kathpalia SC, Coe FL, Mond AE. 1980. Effects of diet calcium and 1,25-dihydroxyvitamin D3 on colon calcium active transport. *Am. J. Physiol.* 238:G75–78

32. Cheng SX, Geibel JP, Hebert SC. 2004. Extracellular polyamines regulate fluid secretion in rat colonic crypts via the extracellular calcium-sensing receptor. *Gastroenterology* 126:148–58

33. Singh SK, Binder HJ, Boron WF, Geibel JP. 1995. Fluid absorption in isolated perfused colonic crypts. *J. Clin. Investig.* 96:2373–79

34. De Jesus Ferreira MC, Bailly C. 1998. Extracellular Ca^{2+} decreases chloride reabsorption in rat CTAL by inhibiting cAMP pathway. *Am. J. Physiol.* 275:F198–203

35. Oda Y, Tu CL, Chang W, Crumrine D, Komuves L, et al. 2000. The calcium sensing receptor and its alternatively spliced form in murine epidermal differentiation. *J. Biol. Chem.* 275:1183–90

36. Oda Y, Tu CL, Pillai S, Bikle DD. 1998. The calcium sensing receptor and its alternatively spliced form in keratinocyte differentiation. *J. Biol. Chem.* 273:23344–52

37. Tu CL, Oda Y, Bikle DD. 1999. Effects of a calcium receptor activator on the cellular response to calcium in human keratinocytes. *J. Investig. Dermatol.* 113:340–45

38. Bikle DD. 2004. Vitamin D regulated keratinocyte differentiation. *J. Cell Biochem.* 92:436–44

39. Bikle DD, Ng D, Tu CL, Oda Y, Xie Z. 2001. Calcium- and vitamin D-regulated keratinocyte differentiation. *Mol. Cell. Endocrinol.* 177:161–71

40. Mailland M, Waelchli R, Ruat M, Boddeke HG, Seuwen K. 1997. Stimulation of cell proliferation by calcium and a calcimimetic compound. *Endocrinology* 138:3601–5

41. Lin KI, Chattopadhyay N, Bai M, Alvarez R, Dang CV, et al. 1998. Elevated extracellular calcium can prevent apoptosis via the calcium-sensing receptor. *Biochem. Biophys. Res. Commun.* 249:325–31

42. McNeil SE, Hobson SA, Nipper V, Rodland KD. 1998. Functional calcium-sensing receptors in rat fibroblasts are required for activation of SRC kinase and mitogen-activated protein kinase in response to extracellular calcium. *J. Biol. Chem.* 273:1114–20

43. Quinn SJ, Ye CP, Diaz R, Kifor O, Bai M, et al. 1997. The Ca^{2+}-sensing receptor: a target for polyamines. *Am. J. Physiol.* 273:C1315–23

44. Conigrave AD, Franks AH, Brown EM, Quinn SJ. 2002. L-Amino acid sensing by the calcium-sensing receptor: a general mechanism for coupling protein and calcium metabolism? *Eur. J. Clin. Nutr.* 56:1072–80

45. Buts JP, De Keyser N, Kolanowski J, Sokal E, Van Hoof F. 1993. Maturation of villus and crypt cell functions in rat small intestine. Role of dietary polyamines. *Dig. Dis. Sci.* 38:1091–98

46. Loser C, Eisel A, Harms D, Folsch UR. 1999. Dietary polyamines are essential luminal growth factors for small intestinal and colonic mucosal growth and development. *Gut* 44:12–16

47. Dufour C, Dandrifosse G, Forget P, Vermesse F, Romain N, Lepoint P. 1988. Spermine and spermidine induce intestinal maturation in the rat. *Gastroenterology* 95:112–16

48. Bardocz S, Duguid TJ, Brown DS, Grant G, Pusztai A, et al. 1995. The importance of dietary polyamines in cell regeneration and growth. *Br. J. Nutr.* 73:819–28

49. Osborne DL, Seidel ER. 1989. Microflora-derived polyamines modulate obstruction-induced colonic mucosal hypertrophy. *Am. J. Physiol.* 256:G1049–57

50. Seiler N, Delcros JG, Moulinoux JP. 1996. Polyamine transport in mammalian cells. An update. *Int. J. Biochem. Cell Biol.* 28:843–61

51. Pegg AE, McCann PP. 1982. Polyamine metabolism and function. *Am. J. Physiol.* 243:C212–21
52. Marton LJ, Pegg AE. 1995. Polyamines as targets for therapeutic intervention. *Annu. Rev. Pharmacol. Toxicol.* 35:55–91
53. Van Aken E, De Wever O, Correia da Rocha AS, Mareel M. 2001. Defective E-cadherin/catenin complexes in human cancer. *Virchows Arch.* 439:725–51
54. Wong NA, Pignatelli M. 2002. β-Catenin—a linchpin in colorectal carcinogenesis? *Am. J. Pathol.* 160:389–401
55. Yamaguchi T, Chattopadhyay N, Kifor O, Sanders JL, Brown EM. 2000. Activation of p42/44 and p38 mitogen-activated protein kinases by extracellular calcium-sensing receptor agonists induces mitogenic responses in the mouse osteoblastic MC3T3-E1 cell line. *Biochem. Biophys. Res. Commun.* 279:363–68
56. Yamaguchi T, Kifor O, Chattopadhyay N, Bai M, Brown EM. 1998. Extracellular calcium (Ca^{2+}_o)-sensing receptor in a mouse monocyte-macrophage cell line (J774): potential mediator of the actions of Ca^{2+}_o on the function of J774 cells. *J. Bone Miner. Res.* 13:1390–97
57. Yamaguchi T, Yamauchi M, Sugimoto T, Chauhan D, Anderson KC, et al. 2002. The extracellular calcium Ca^{2+}_o-sensing receptor is expressed in myeloma cells and modulates cell proliferation. *Biochem. Biophys. Res. Commun.* 299:532–38
58. Hobson SA, McNeil SE, Lee F, Rodland KD. 2000. Signal transduction mechanisms linking increased extracellular calcium to proliferation in ovarian surface epithelial cells. *Exp. Cell Res.* 258:1–11
59. Kifor O, MacLeod RJ, Diaz R, Bai M, Yamaguchi T, et al. 2001. Regulation of MAP kinase by calcium-sensing receptor in bovine parathyroid and CaR-transfected HEK293 cells. *Am. J. Physiol. Renal Physiol.* 280:F291–302
60. Bresalier RS. 1999. Calcium, chemoprevention, and cancer: a small step forward (a long way to go). *Gastroenterology* 116:1261–63
61. Lamprecht SA, Lipkin M. 2001. Cellular mechanisms of calcium and vitamin D in the inhibition of colorectal carcinogenesis. *Ann. N. Y. Acad. Sci.* 952:73–87
62. Rozen P, Lubin F, Papo N, Knaani J, Farbstein H, et al. 2001. Calcium supplements interact significantly with long-term diet while suppressing rectal epithelial proliferation of adenoma patients. *Cancer* 91:833–40
63. Holt PR, Atillasoy EO, Gilman J, Guss J, Moss SF, et al. 1998. Modulation of abnormal colonic epithelial cell proliferation and differentiation by low-fat dairy foods: a randomized controlled trial. *JAMA* 280:1074–79
64. Garland C, Shekelle RB, Barrett-Connor E, Criqui MH, Rossof AH, Paul O. 1985. Dietary vitamin D and calcium and risk of colorectal cancer: a 19-year prospective study in men. *Lancet* 1:307–9
65. Holt PR, Lipkin M, Newmark H. 1999. Calcium intake and colon cancer biomarkers. *JAMA* 281:1172–73
66. Ahnen DJ, Byers T. 1998. Proliferation happens. *JAMA* 280:1095–96
67. Baron JA, Beach M, Mandel JS, van Stolk RU, Haile RW, et al. 1999. Calcium supplements and colorectal adenomas. Polyp Prevention Study Group. *Ann. N. Y. Acad. Sci.* 889:138–45
68. Lipkin M, Newmark H. 1995. Development of clinical chemoprevention trials. *J. Natl. Cancer Inst.* 87:1275–77
69. Kerstetter JE, O'Brien KO, Insogna KL. 2003. Dietary protein, calcium metabolism, and skeletal homeostasis revisited. *Am. J. Clin. Nutr.* 78:S584–92
70. Kerstetter JE, O'Brien KO, Insogna KL. 1998. Dietary protein affects intestinal calcium absorption. *Am. J. Clin. Nutr.* 68:859–65
71. Kerstetter JE, Svastisalee CM, Caseria DM, Mitnick ME, Insogna KL. 2000. A threshold for low-protein-diet-induced elevations in parathyroid hormone. *Am. J. Clin. Nutr.* 72:168–73
72. Heaney RP. 2000. Dietary protein and phosphorus do not affect calcium absorption. *Am. J. Clin. Nutr.* 72:758–61

Neuroendocrine Control of the Gut During Stress: Corticotropin-Releasing Factor Signaling Pathways in the Spotlight

Andreas Stengel and Yvette Taché

Department of Medicine and CURE Digestive Diseases Research Center, Center for Neurobiology of Stress, University of California at Los Angeles, and VA Greater Los Angeles Healthcare System, Los Angeles, California 90073; email: ytache@mednet.ucla.edu

Annu. Rev. Physiol. 2009. 71:219–39

First published online as a Review in Advance on October 17, 2008

The *Annual Review of Physiology* is online at physiol.annualreviews.org

This article's doi: 10.1146/annurev.physiol.010908.163221

Copyright © 2009 by Annual Reviews. All rights reserved

0066-4278/09/0315-0219$20.00

Key Words

motility, CRF antagonists, colon, stomach, urocortin

Abstract

Stress affects the gastrointestinal tract as part of the visceral response. Various stressors induce similar profiles of gut motor function alterations, including inhibition of gastric emptying, stimulation of colonic propulsive motility, and hypersensitivity to colorectal distension. In recent years, substantial progress has been made in our understanding of the underlying mechanisms of stress's impact on gut function. Activation of corticotropin-releasing factor (CRF) signaling pathways mediates both the inhibition of upper gastrointestinal (GI) and the stimulation of lower GI motor function through interaction with different CRF receptor subtypes. Here, we review how various stressors affect the gut, with special emphasis on the central and peripheral CRF signaling systems.

INTRODUCTION

More than 70 years ago, Hans Selye (1) identified the gut, along with the endocrine and immune systems, as the primary target altered by a variety of chemical and physical challenges and pioneered the concept of stress as the "stereotyped biological response to any demand." Later, the term allostasis was introduced by Sterling & Eyer (2) to refer to the "maintenance of stability through change." Subsequently, McEwens (3) applied this concept to define stress as the physiological adaptation processes that maintain stability in times of internal or external challenges. Excessive stress can result in cumulative biological changes (known as allostatic load) and can alter adaptive mechanisms, resulting in an inefficient allostatic response and a permanent change in the basal levels of stress mediators (4). A perpetual imbalance between adaptation capacity and stressors can result in allostatic overload, leading to a state of illness (4) that may affect different body systems and induce development of functional bowel diseases (5). It is now appreciated that signaling pathways involving corticotropin-releasing factor (CRF) are altered by stress and contribute to functional bowel diseases (5).

STRESS AND THE GUT

In recent years, our understanding of the circuitries and biochemical coding involved in the stress response has increased tremendously (6). Many studies have used the immediate early gene c-*fos* protein (Fos) immunohistochemistry as a marker of neuronal activity, thereby identifying brain nuclei that respond to acute or chronic stress (7) and their relation to the autonomic regulation of gut function (8–11). CRF is the primary neurohormone involved in the hallmark response to stress: the activation of the hypothalamic-pituitary-adrenal (HPA) axis. CRF also acts as a neurotransmitter/neuromodulator to coordinate the behavioral, autonomic, and visceral efferent limbs of the stress response (12–14). Convergent findings support the involvement of CRF receptors in the brain and the gut as important mediators of acute or chronic stress–related alterations of gut function (14–16). Furthermore, environmental stressors seem to play a role in the development and/or exacerbation of functional bowel diseases, such as irritable bowel syndrome (IBS), which is characterized by altered bowel habits and visceral hypersensitivity (17–20). Growing preclinical and clinical reports indicate that increased central and peripheral CRF signaling may contribute to the development and maintenance of functional bowel disorders through the alteration of autonomic, enteric nervous, and immune system activity (5, 16).

THE CORTICOTROPIN-RELEASING FACTOR FAMILY AND ITS RECEPTORS

Mammalian Corticotropin-Releasing Factor and Urocortins

CRF, originally isolated by Vale and colleagues in 1981 (21), is a 41-amino-acid (aa) hypothalamic releasing peptide that stimulates the synthesis and release of adrenocorticotropic hormone and β-endorphin from the anterior pituitary. More recently, three other mammalian CRF-related peptides have been characterized: urocortin 1 (Ucn 1), a 40-aa peptide with 45% sequence identity with rat/human (r/h) CRF; urocortin 2 (Ucn 2); and urocortin 3 (Ucn 3) (22–25). Mouse Ucn 2 (mUcn 2) is a 38-aa peptide sharing 34% homology with r/h CRF and 42% with r/h Ucn 1 (24). However, the 38-aa peptide mUcn 3 shares only 26% and 21% homology to r/h CRF and r/h Ucn 1, respectively (25). Phylogenetic profiling of the CRF peptide family indicates that these four distinct genes—those encoding CRF, Ucn 1, Ucn 2, and Ucn 3—are highly conserved through evolution and can be traced back to invertebrates, indicating their important roles in survival and adaptation (26, 27).

Irritable bowel syndrome (IBS): a functional disease characterized by altered bowel habits and visceral pain

Urocortins (Ucns): mammalian CRF-related peptides

CRF$_1$ and CRF$_2$ Receptors

CRF and urocortins interact with two receptors, CRF$_1$ and CRF$_2$, which are encoded by two distinct genes exhibiting 70% sequence homology (28). The human and rat genomic structures of the CRF$_1$ receptor contain 14 and 13 exons, respectively. In most mammals, the active CRF$_1$ receptor protein results from transcription of all exons. In contrast, translation of all 14 exons in humans results in a human-specific, 444-aa protein named CRF$_{1b}$ that contains an extended first intracellular loop and exhibits impaired agonist binding and signaling properties (29, 30). The 415-aa protein CRF$_{1a}$ is the main functional CRF$_1$ variant resulting from the excision of exon 6. In addition to CRF$_{1a}$ and CRF$_{1b}$, the CRF$_1$ gene gives rise to multiple additional splice variants (1c, 1d, 1e, 1f, 1g, 1j, 1k, 1m, and 1n) that have neither a ligand binding site nor a signaling domain, although some variants modulate CRF and Ucn 1 actions (29). The expression of these CRF$_1$ isoforms is tissue specific and can vary with the tissues' functional activity as well as with environmental factors (29). For instance, the onset of labor is associated with an increased transcription of myometrial CRF$_{1a}$ gene and with differential up- and downregulation of other specific variants that may be implicated in the passage of quiescent to procontractile activity of the myometrium during labor (31). The expression and regulation of CRF$_1$ receptor variants under conditions of acute or chronic stress in the brain and the gut are still largely unexplored.

In humans, there are three functional splicing variants of the CRF$_2$ receptor (namely 2a, 2b, and 2c), whereas in other mammals only 2a and 2b are expressed (29, 32). The CRF$_2$ variants display a distinct expression profile in mammals (29, 32). The CRF$_2$ isoforms result from alternative splicing of exon 1 to exon 3. This splicing leads to structurally distinct N-terminal extracellular domains (34 aa for CRF$_{2a}$, 61 aa for CRF$_{2b}$, and 20 aa for CRF$_{2c}$), which are involved in ligand-receptor interaction (32, 33). In addition to the wild-type CRF$_{2a-1}$, five additional splice variants, CRF$_{2a-2}$ to CRF$_{2a-6}$, have been identified in the rat upper gut (34). Interestingly, the mouse CRF$_{2a}$ splice variant, originally identified in the brain as a soluble binding protein (sCRF$_{2a}$) for CRF and Ucn 1 (35), is also expressed in the rat upper gut (CRF$_{2a-6}$) (34).

CRF$_1$ and CRF$_2$ receptors show distinct affinities to CRF and related peptides (29, 32). Although CRF has a 10- to 40-fold higher affinity for the CRF$_1$ receptor than for the CRF$_2$ receptor, urocortins preferably signal through CRF$_2$ receptors. Ucn 1 binds with equal affinity to both CRF receptors and has a 100-fold higher affinity than CRF for the CRF$_2$ receptor (29, 36). In contrast, Ucn 2 and Ucn 3 show high selectivity for the CRF$_2$ receptor (29, 36) (**Figure 1**). Because none of the endogenous CRF ligands characterized thus far exclusively activate CRF$_1$ receptors, selective peptide CRF$_1$ agonists (namely cortagine and stressin$_1$-A) have recently been developed (36–38) (**Figure 1**). However, the molecular determinants that govern the binding of CRF-family peptides to their cognate receptors have been extensively characterized (29). The long N-terminal extracellular domain of CRF receptors primarily interacts with the C-terminal residues of CRF, whereas the N-terminal residues of CRF interact with the transmembrane region of the receptor, resulting in conformational changes that enable G protein activation (29).

In most tissues, signal transduction of CRF$_1$ and CRF$_2$ primarily involves coupling to the G$_s$–adenyl cyclase system, with subsequent cAMP generation and protein kinase A activation. In addition, CRF receptors, like most heptahelical G protein–coupled receptors, can interact with multiple G protein systems including G$_q$, G$_i$, G$_o$, G$_{il/2}$, and G$_z$ to relay signals to diverse intracellular effectors in an agonist- and tissue-specific manner (29). Thus, CRF receptors (*a*) may modulate various kinases, including phosphokinases A, B, and C, (*b*) can phosphorylate and activate mitogen-activated protein kinase (MAPK), in particular the ERK1/2 and p38/MAPK pathways, and

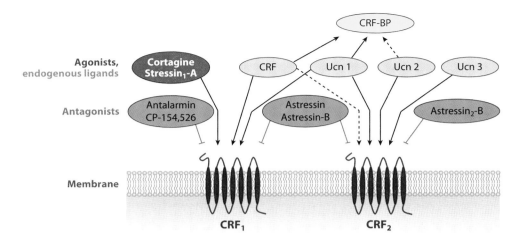

Figure 1

Schematic overview of preferential binding affinities and specificities of endogenous or synthetic corticotropin-releasing factor (CRF) receptor ligands. The CRF_1 and CRF_2 receptors share 70% homology and belong to the family of seven-transmembrane G protein–coupled receptors. BP, binding protein.

(*c*) can alter intracellular Ca^{2+} concentrations (29, 34).

Brain Distribution of Corticotropin-Releasing Factor Ligands and Receptors

The distribution of CRF-immunoreactive (ir) cells and fibers in the rat brain has been extensively described (39–41). The major brain areas expressing CRF messenger RNA (mRNA) and CRF-ir cells include the paraventricular nucleus (PVN) of the hypothalamus, the cerebral cortex, the amygdalar-hippocampal complex, and Barrington's nucleus in the dorsolateral pons. The PVN is the major site of CRF-containing cell bodies projecting to the median eminence. CRF neurons in the central amygdala project to the PVN, the locus coeruleus (LC), and the parabrachial nucleus, and neurons in the bed nucleus of stria terminalis project to the dorsal vagal complex (DVC). CRF-containing neurons in Barrington's nucleus contribute to the CRF innervation of the LC and the sacral spinal cord (40, 42).

CRF, Ucn 1, Ucn 2, and Ucn 3 distribution in the rat brain shows limited neuroanatomical overlap (43). In contrast to Ucn 1's widespread peripheral distribution, peptide expression in the brain is limited (22, 44, 45). The most prominent brain site of Ucn 1 expression and immunoreactivity is the Edinger-Westphal nucleus (44, 46). Moreover, Ucn 1–ir can be detected in the lateral superior olive; the olfactory bulb; the supraoptic nucleus (SON); the ventromedial hypothalamus (VMH); and magnocellular parts of the PVN, the lateral hypothalamic area, and the ambiguous nucleus; as well as the cranial nerve motor nuclei (facial and hypoglossal) (46, 47). The majority of Ucn 1–ir projections provide descending input to the brainstem, whereas ascending projections are restricted. In particular, Ucn 1–ir fibers project to CRF_2 receptor–containing nuclei including the lateral septum, the dorsal raphé, the interpeduncular nucleus, the nucleus of the solitary tract, and the area postrema (46). Ucn 2 gene expression is localized in the parvo- and magnocellular PVN, the SON, the arcuate nucleus of the hypothalamus, and the LC, as well as several cranial nerve motor nuclei (trigeminal, facial, and hypoglossal nuclei) and the ventral horn of the spinal cord (24, 48). Ucn 3 mRNA has been detected in the PVN, the amygdala (basomedial

Paraventricular nucleus (PVN): hypothalamic brain nucleus implicated in autonomic regulation of gastrointestinal functions

Locus coeruleus (LC): pontine catecholaminergic nucleus involved in physiological responses to stress

nucleus), and the basomedial nucleus of the stria terminalis (43, 48). Assessment of brain distribution of Ucn 2–ir fibers has been hampered by the lack of specific Ucn 2 antibodies. Ucn 3–ir fibers innervate the lateral septum, the amygdala (except for the central nucleus), the dorsal aspect of the VMH, and the dorsal raphé, as well as the area postrema (25, 48, 49).

In rat brain, the CRF_1 receptor is densely expressed in the forebrain and subcortical limbic structures in the septal region and amygdala. In the hypothalamus, expression is low under basal conditions but can be significantly upregulated during stress or following CRF application (50–52). Dense CRF_1 receptor representation also appears in the anterior and intermediate lobes of the pituitary, which supports its role in the activation of the HPA axis by CRF (53). In contrast, CRF_2 receptor expression in rodents is confined to subfornical structures, with high expression in the lateral septum, amygdala (with the exception of central nuclei), and hypothalamus (including high levels in the VMH and SON) (54). In the hindbrain, the dorsal raphé, area postrema, nucleus of the solitary tract, and chorionic plexus express the CRF_2 receptor (54). A close association has been found between Ucn 3–ir–terminal fields and expression of CRF_2 receptors in specific hypothalamic nuclei (49).

Corticotropin-Releasing Factor Receptor Antagonists and Binding Protein

Key to the understanding of the physiological role of the CRF signaling pathways is the early development of CRF receptor antagonists by Rivier et al. (55). The first of these antagonists to be developed are the nonselective CRF_1/CRF_2 receptor antagonists α-helical CRF_{9-41}, D-Phe^{12}CRF$_{12-41}$, astressin, and the long-acting astressin-B. Recently, two groups developed the peptide CRF_2 receptor antagonists antisauvagine-30, K41498, [D-Phe^{11}, His^{12}, Nle^{17}] sauvagine$_{11-40}$, and the more potent and long-acting analog astressin$_2$-B, which are competitive antagonists that bind

equally to the a, b, and c variants of CRF_2 receptor (35, 56, 57) (**Figure 1**). A common feature of these peptide antagonists is that they display poor penetrance into the brain when administered peripherally. In an early study, our group showed that an intravenous injection of astressin did not influence the inhibition of gastric transit in response to CRF injected into the cisterna magna, although the same dose blocked peripheral CRF–induced delay of gastric emptying (58). With regard to selective CRF_1 antagonists, pharmaceutical firms have developed a large number of high-affinity small hydrophobic molecules that can cross the blood-brain barrier. The impetus to create these molecules arose from their potential therapeutic application to curtail dysregulation of CRF-signaling pathways that may be relevant to the pathogenesis of human illnesses such as anxiety and depression, eating disorders, inflammatory diseases, substance abuse, preterm parturition, and functional bowel disorders (59–62). Among the CRF_1 antagonists most commonly used in experimental studies are CP-154,526, antalarmin, NBI-34041, NBI-30545, and NBI-35965 (60, 63).

In addition to the synthetic antagonists that can block CRF receptors, a 332-aa endogenous CRF binding protein (CRF-BP) has been isolated across different species (64, 65). The CRF-BP functions as an endogenous antagonist by sequestering CRF ligands and therefore modulating the access of CRF and related peptides to CRF receptors (64). Rat and human CRF and Ucn 1 display high (picomolar-range) affinity for the CRF-BP, whereas Ucn 2 shows a moderate (nanomolar-range) affinity and Ucn 3 displays no affinity (65) (**Figure 1**). The decrease in food intake and body weight, the anxiogenic-like behavior occurring in CRF-BP-deficient mice, and, conversely, the body weight gain in CRF-BP-overexpressing mice support the contention that CRF-BP can sequester CRF/Ucn 1 and modulate endogenous CRF/Ucn 1 biological actions (66, 67). Recent studies have identified distinct regions and residues of the CRF-BP

Astressin-B: long-acting CRF_1/CRF_2 receptor antagonist

Astressin$_2$-B: long-acting selective peptide CRF_2 receptor antagonist

that are responsible for r/h CRF and r/h Ucn 1 binding to CRF-BP. In particular, a single alanine mutation (R56A) in CRF-BP was effective in creating an Ucn 1–specific antagonist, thereby opening new venues for the design of selective CRF versus Ucn 1 antagonists to dissect their respective role in the stress response (68). In rat brain, CRF-BP immunoreactivity is prominently expressed in hypothalamic regions involved in the neuroendocrine and autonomic responses to stress, including the subdivision of the dorsal cap of the PVN projecting to the spinal cord (64, 69).

The isolation of CRF and endogenous selective CRF_2 receptor agonists, along with the development of selective CRF_1 agonists and antagonists and CRF_2 antagonists, provided essential tools to establish the pleiotropic actions of CRF and urocortins in the brain by acting at one or both CRF receptors. In particular, this pharmacological approach has allowed investigators to dissect the primary involvement of CRF_1 signaling pathways in the stress-related stimulation of the HPA axis, anxiogenic behavior, alterations in the autonomic nervous system activity, and visceral responses (5, 13, 70).

STRESS-RELATED ALTERATIONS OF GASTROINTESTINAL MOTILITY MEDIATED BY BRAIN CORTICOTROPIN-RELEASING FACTOR RECEPTORS

Various acute stressors most commonly delay gastric emptying in experimental animals as well as in healthy humans (12). In contrast, a range of stressors (e.g., anxiety, dichotomous listening, fear, intermittent hand immersion in cold water, and stressful interviews) increase colonic motility in healthy volunteers (12). Activation of propulsive colonic motor function has also been shown in rodents following exposure to diverse stressors such as open field tests, conditioned fear, loud sounds, restraint, cold exposure, water avoidance, inescapable foot or tail shocks, and central injection of interleukin-1 (12, 71–73). Consistent experimental evidence highlights the role of central CRF receptors in stress-related inhibition of gastric motor function and in stimulation of colonic propulsive motor activity, as shown by central injection of CRF and urocortins in nonstressed animals as well as by the injection of CRF antagonists under stress conditions.

Brain Corticotropin-Releasing Factor Receptors Mediate Stress-Related Inhibition of Gastric Motor Function

A number of studies have established that central injection [intracerebroventricularly (icv), intracisternally (ic), or into the fourth ventricle] of CRF, Ucn 1, Ucn 2, or nonmammalian CRF-related peptides such as sauvagine and urotensin I inhibits gastric emptying of acaloric liquid, caloric liquid, or solid meal and alters gastric motility. These changes include the suppression of propagative contractions, cyclic activity front, high-amplitude contractions, and the disruption of fasted pattern in species including rats, mice, and dogs (74). Central injection of α-helical CRF_{9-41}, D-$Phe^{12}CRF_{12-41}$, astressin, astressin-B, and $astressin_2$-B blocks the icv or ic CRF-, Ucn 1–, and Ucn 2–induced delay of gastric emptying and inhibition of motility in rats, mice, and dogs. However, selective CRF_1 antagonists have no effect, indicating that CRF's and urocortins' actions are primarily mediated by interaction with CRF_2 receptors (11, 73, 75–88). The PVN and DVC, which influence autonomic outflow to the stomach, have been identified as responsive brain nuclei for CRF-induced inhibition of gastric emptying and motility, whereas the LC seems not to be involved (89–93). The expression of CRF_{2a} receptor mRNA has been found in hypothalamic and brainstem nuclei, such as the PVN and DVC, as well as in limbic structures; this is consistent with the locations of the responsive sites (54).

The central action of CRF—inhibition of gastric transit—is independent of the activation of the HPA axis and is mediated by the autonomic nervous system, as shown by the persistence of the gastric response in hypophysectomized or adrenalectomized rats (75, 94). A number of reports have shown that (*a*) delay of gastric transit induced by icv or ic injection of CRF and Ucn 1 and (*b*) alteration of motility require the integrity of the vagus nerve in rats and dogs (75, 82, 88, 90, 92, 95–97). Only two studies reported different results that showed a primary involvement of the sympathetic nervous system (86, 94). However, the delay of gastric emptying induced by the ic injection of Ucn 2 is not altered by vagotomy but is mediated by sympathetic pathways and peripheral α-adrenergic receptors; this indicates that CRF ligands act through both vagal and sympathetic pathways to influence gastric function (88).

Pretreatment with CRF receptor antagonists has provided pharmacological evidence that brain CRF receptors are involved in stress-induced inhibition of gastric motor function. The ic, icv, or PVN injection of α-helical CRF_{9-41}, D-$Phe^{12}CRF_{12-41}$, astressin, or astressin-B blocks acute stress–induced delayed gastric emptying (12, 74). Stressors used in these studies fall under one of the following categories: (*a*) psychological/physical (swim stress, restraint), (*b*) visceral (abdominal surgery, trepanation, peritoneal irritation with intraperitoneal 0.6% acetic acid), immunological (intravenous or central injection of interleukin-1β), and (*c*) chemical (ether) (12, 74). Of interest is the demonstration that electroacupuncture normalizes both restraint- and ic-CRF-induced delay of gastric emptying, suggesting that the beneficial action of electroacupuncture under conditions of stress may be related to interference with brain CRF pathways (98). Also supportive of a role of brain CRF pathways is the demonstration that a variety of physical, immune, and psychological stressors (e.g., abdominal surgery, immobilization, forced swimming, and interleukins) activate CRF neurons and lead to a rapid increase in CRF gene transcription and upregulation of CRF mRNA in the PVN, which is the primary site of CRF synthesis (11, 99–102). The parvocellular division of the PVN contains neurons in the dorsal and ventral parvocellular caps that regulate autonomic outflow to the viscera (103), which is consistent with the role of CRF

signaling in the PVN to influence gastric motor function. Ucn 1 and 2 are also expressed in the PVN and are upregulated by various stressors (104, 105). However, their involvement in stress-related inhibition of gastric motor function remains to be clarified. So far, investigations in Ucn 1–deficient mice suggest that Ucn 1 does not play a primary role in heart rate increase and sympathetic activation in response to acute restraint, as monitored by epinephrine and norepinephrine levels (106).

The CRF receptor subtype(s) involved in mediating stress-related inhibition of gastric motor function has yet to be fully characterized. Because CRF and urocortins inhibit gastric emptying via a CRF_2-mediated pathway (73, 85, 88), it was expected that CRF_2 receptors would be primarily involved in stress-induced delay of gastric emptying. This has been demonstrated in one study, where ic injection of $astressin_2$-B blocked the restraint stress–induced delay of gastric emptying in rats (86). However, it is surprising that under conditions of surgical stress (abdominal surgery and cecal palpation), central CRF_1 receptors play a predominant role: CRF_1-knockout mice and wild-type animals injected centrally with a CRF_1 antagonist no longer develop the inhibition of gastric emptying following abdominal surgery and cecal palpation (107). Experimental data in which the central injection of CRF_1/CRF_2 antagonists or selective CRF receptor subtype antagonists blocked stress-related alterations of gastric propulsive motor function provide new insights into the role of brain CRF signaling pathways as underlying mechanisms involved in both acute postoperative gastric ileus (108) and alterations of gastric digestive function during disease states associated with cytokine release (109, 110). Furthermore, pharmacological evidence indicates that several centrally administered brain-gut peptides ultimately converge on brain CRF signaling as the downstream effector to induce an autonomic-mediated suppression of gastric propulsive motor function. For instance, the delayed gastric emptying induced by icv, fourth ventricle, or ic injection of glucagon-like peptide–1 (GLP-1), cocaine and amphetamine-regulated transcript (CART) peptide, or des-acyl ghrelin is blocked by pretreatment with CRF antagonists injected via the same route (111–113). However, central application of CRF receptor antagonists does not alter basal gastric emptying of a liquid nonnutrient or solid nutrient meal in rats, mice, and dogs, indicating that central CRF pathways do not regulate fasted and postprandial gastric propulsive motor function under basal conditions, but that they do gain importance when recruited under stress conditions (12, 76, 82, 114).

Are Brain Corticotropin-Releasing Factor Receptors Involved in the Mediation of Stress-Related Alterations of Small Intestinal Motor Function?

Compared with our understanding of gastric motor function, much less is known about the effects of stress on small intestinal propulsive activity and the role of brain CRF receptors (74). As in the stomach, acute psychological stress, as well as central injection of CRF or Ucn 1, inhibits duodenal and small intestinal transit and motility (suppression of the occurrence of myoelectric migrating complex, disruption of the fasted pattern of duodenal motor activity to fed pattern) in rats and dogs (71, 95, 96, 115–118). Likewise, the central action of CRF is mediated via vagal pathways and is independent of the HPA axis activation (71, 94, 116, 119). However, the slowing of small intestinal transit induced by icv injection of CRF is not as prominent as it is in the stomach; this is probably due to the lesser vagal innervation of the small intestine compared with the stomach (120). The involvement of central CRF receptors in mediating stress-induced inhibition of small intestinal transit has been little studied, and results are conflicting. The icv injection of α-helical CRF_{9-41} blocks restraint stress–induced slowing of small intestinal transit in male rats, whereas there is no such effect in female rats at a dose effective to block restraint-induced stimulation of colonic transit (71, 116).

Whether these divergences are sex-related or specific differences in experimental conditions needs to be clarified. Furthermore, the central CRF receptor subtype involved in the mediation of the stress-induced inhibition of small intestinal motor function has not been investigated.

Brain CRF_1 Receptors Mediate Stress-Related Stimulation of Colonic Motor Function

In contrast to its inhibitory effect on gastric and small intestinal transit, centrally injected CRF or Ucn 1 stimulates colonic transit and defecation and induces a pattern of cecocolonic myoelectrical activity characterized by clustered spike-bursts of long duration in freely moving female and male rats, mice, and gerbils (9, 71–73, 81, 91, 94, 121–123) (**Figure 2**). Convergent evidence has established that the stimulation of colonic motor function in response to central CRF and Ucn 1, as well as various stressors, is primarily mediated via central CRF_1 receptors (5, 74) (**Figure 2**). First, colonic propulsive motor activity in response to stress is mimicked by centrally administered CRF_1 receptor–preferential agonists such as ovine CRF, r/h CRF (56), and Ucn 1, whereas Ucn 2 and Ucn 3 are inactive in mice when centrally injected at a dose similar to that of CRF (73). Second, the central injection of peptide CRF_1/CRF_2 receptor antagonists and selective CRF_1 antagonists blocks the colonic motor stimulation induced by central injection of CRF or Ucn 1 or by various stressors (5, 124) (**Figure 2**). For instance, central injection of astressin, α-helical CRF_{9-41}, and D-$Phe^{12}CRF_{12-41}$ blocks the effects of central injection of CRF, Ucn 1, or interleukin-1β; wrap or partial restraint; water avoidance; morphine withdrawal; and colorectal distention–induced stimulation of colonic transit and defecation; as well as the increased frequency of colonic spike-bursts induced by conditioned fear stress in rodents (9, 71–73, 81, 84, 116, 123, 125–128). Furthermore, central or peripheral application of the selective CRF_1 antagonists

CP-154,526, CRA 1000, NBI-27914, NBI-35965, JTC-017, and antalarmin reduces the acceleration of colonic transit time caused by restraint, fecal pellet output induced by water avoidance, social stress, painful stimuli, and diarrhea resulting from morphine withdrawal in rodents. CRF_1-knockout mice show significantly less defecation in an open field test than do their wild-type littermates (5, 12, 129) (**Table 1**). Lastly, central injection of the selective CRF_2 antagonist astressin$_2$-B at a dose effective to block CRF_2-mediated action on gastric emptying does not prevent the colonic response to central injection of CRF in rodents (12). As described above for gastric motor function, central CRF_1 receptors are not involved in the basal and postprandial regulation of colonic motor function under nonstress conditions (5, 12, 73). Of interest, however, is the growing pharmacological evidence that a number of brain peptides influencing food intake, such as neuropeptide Y, GLP-1, CART, and ghrelin, act in the brain to stimulate colonic motor function by recruiting brain CRF signaling pathways (130–133).

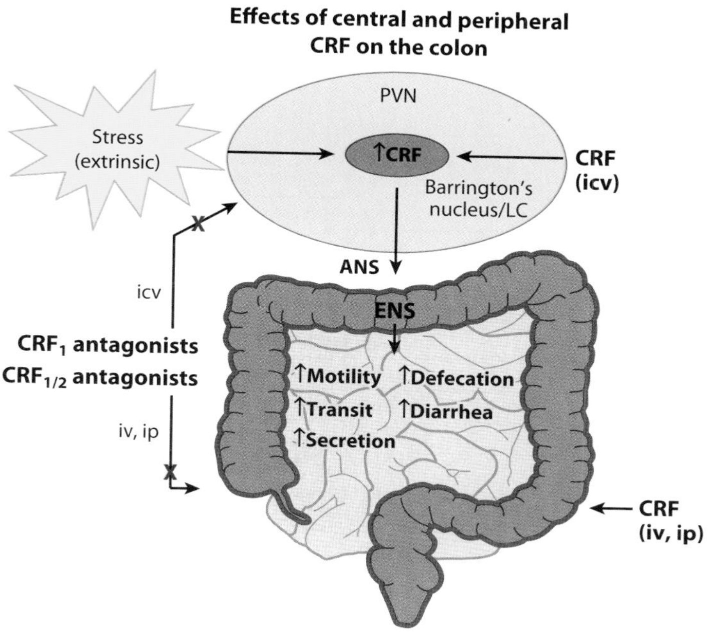

The central CRF- and stress-induced CRF_1-dependent stimulation of colonic motor function (transit and increased frequency of spike-burst activity) is not altered by blocking the activation of the HPA axis (72, 94). Rather, it is mediated by an increased parasympathetic outflow to the colon via vagal celiac branches innervating the proximal colon and via sacral parasympathetic fibers innervating the distal colon and rectum (72, 90, 94, 125, 126). Effector mechanisms within the colon involve parasympathetic-mediated activation of myenteric cholinergic and nitrergic neurons regulating the peristaltic reflex as well as serotonin (5-HT) acting on 5-HT$_3$ and 5-HT$_4$ receptors. This is supported by the demonstration that icv CRF-induced defecation is blocked by the peripheral administration of atropine and the 5-HT$_3$ antagonists ramosetron, ondansetron, azasetron, alosetron, and cilansetron, as well as the 5-HT$_4$ antagonist SB-204070, whereas the icv injection of these antagonists has no effects (123, 134–136). In addition, restraint stress or

Effects of central and peripheral CRF on the colon

Figure 2

Summary of corticotropin-releasing factor (CRF) actions on colonic function. Central and peripheral CRF stimulates various colonic functions, which recapitulate the effects of stress and are blocked by nonselective and CRF_1-selective CRF receptor antagonists. Abbreviations: ANS, autonomic nervous system; ENS, enteric nervous system; icv, intracerebroventricularly injected; ip, intraperitoneally injected; iv, intravenously injected; LC, locus coeruleus; PVN, paraventricular nucleus of the hypothalamus.

icv injection of CRF increases the 5-HT content in the feces of rat proximal colon (134). Taken together, these data suggest the cholinergic recruitment of peripheral serotonin pools from either enterochromaffin cells or enteric neurons by restraint stress and central CRF.

Consistent with the pharmacological evidence engaging the CRF_1 signaling pathways in the colonic response to stress, functional mapping and gene regulation studies support the PVN and LC/Barrington's complex in stress-related, CRF-mediated activation of colonic propulsive motor function (**Figure 2**). Tracing studies have shown that CRF neurons in the dorsal cap of the parvocellular part of the PVN have trans-synaptic connections to the colon (42). Also, CRF-synthesizing neurons in Barrington's nucleus project to the noradrenergic LC, as well as to the intermediolateral

Table 1 Blockade of acute stress–induced stimulation of colonic motor functions by selective CRF$_1$ antagonists[a]

Antagonist	Dose	Route	Species	Stress	Inhibition	Reference
Central administration						
NBI-27914	100 µg	Icv	Rats	Water avoidance	Defecation (67%)	84
NBI-35965	50 µg	Icv	Mice	Restraint	Defecation (100%)	73
Peripheral administration						
CP-154,526	20 mg kg^{-1}	Sc	Rats	Water avoidance	Defecation (55%)	153
CP-154,526	30 mg kg^{-1}	Sc	Rats	Morphine withdrawal	Diarrhea (50%)	127
CRA 1000	20 mg kg^{-1}	Sc	Mice	Morphine withdrawal	Diarrhea (50%)	171
CP-154,526	20 mg kg^{-1}	Sc	Rats	Partial restraint	Transit (55%)	157
NBI-35965	20 mg kg^{-1}	Sc	Rats	Water avoidance	Defecation (53%)	172
Antalarmin	20 mg kg^{-1}	Ip	Rats	Restraint	Defecation (49%)	173
JTC-017	10 mg kg^{-1}	Ip	Rats	Colon distention	Defecation (100%)	128
Antalarmin	20 mg kg^{-1}	Po	Monkeys	Social intruder	Defecation (40%)	174

[a]Abbreviations: CRF, corticotropin-releasing factor; Icv, intracerebroventricular; Ip, intraperitoneal; Po, per pos; Sc, subcutaneous.

column of the sacral spinal cord, which contains the sacral parasympathetic nucleus innervating the descending colon (42, 120). There are CRF-efferent fibers projecting directly from the PVN and Barrington's nucleus to the LC, as shown by anterograde tracing (137, 138). Neuroanatomical and functional studies have demonstrated that water avoidance stress activates the PVN and LC/Barrington's nuclei and CRF gene transcription in the PVN (9, 139), whereas icv injection of α-helical CRF$_{9-41}$ reduces Fos expression selectively in these hypothalamic and pontine nuclei in correlation with defecation score (9). Likewise, Lewis rats known to have a blunted hypothalamic CRF response to stress exposure (140) displayed a reduced activation of neurons in the PVN and sacral parasympathetic nucleus and showed a lower defecation response than did Fisher rats (126). Consistent with a role of CRF/CRF$_1$ signaling in the PVN, α-helical CRF$_{9-41}$ injected directly into the PVN blocks partial restraint– and water avoidance–induced stimulation of colonic transit and defecation, and various neurogenic and systemic stressors activate the transcription of the CRF$_1$ receptor gene in the PVN (50, 90, 125). CRF increases the firing rate of noradrenergic neurons in the LC and releases noradrenalin into the brain cortex, which results in arousal and anxiogenic behavior (137, 141). Therefore, CRF/CRF$_1$ signaling pathways in the PVN and LC may physiologically regulate the behavioral and autonomic responses to stress that influence colonic function as part of the brain-gut axis (91, 137). These pathways may play a role in diarrhea-predominant IBS patients with psychic comorbidities such as anxiety and depression (5).

STRESS-RELATED ALTERATIONS OF GASTROINTESTINAL MOTILITY MEDIATED BY PERIPHERAL CORTICOTROPIN-RELEASING FACTOR SIGNALING

Like many peptides once thought to be restricted to the brain and pituitary that were later detected in peripheral tissues, CRF ligands and receptors are widely expressed outside the brain in spinal cord and peripheral organs, including the gastrointestinal tract in animals and humans (12, 25, 142–146). This overlapping expression pattern of CRF ligands and receptors in the gastrointestinal tract gives rise to a local CRF signaling pathway that can act directly on the gut in either a paracrine or an autocrine manner (143, 144, 146–148). The potent actions of CRF ligands upon peripheral injection and CRF receptor blockade are consistent with the notion that peripheral activation of CRF_1 and CRF_2 signaling pathways may be part of the local effectors involved in gastric and colonic motor alterations induced by stress.

Peripheral Corticotropin-Releasing Factor Signaling Alters Gut Motor Function

When injected peripherally, CRF strongly alters gut motility and transit in several mammalian species including rodents, dogs, and humans (71, 75, 149–152). The iv or intraperitoneal (ip) injection of CRF or Ucn 1 in rats inhibits gastric emptying, delays small intestinal transit, and stimulates colonic transit and defecation with a potency similar to that of centrally injected (icv or ic) CRF (71, 152, 153). However, although the patterns of gut motor response are similar, distinct sites and mechanisms of action are involved in mediating the effects of centrally and peripherally injected CRF (58, 94, 96, 154). Ganglion blockade, which inhibits the icv CRF-induced delay of gastric emptying and acceleration of colonic transit, does not influence the ip CRF-induced alteration of gut transit (94). Likewise, sympathetic blockade, which prevents ic Ucn 2–induced delayed gastric emptying, does not influence iv Ucn 2–inhibitory action on gastric emptying in rats tested under otherwise-similar conditions (88). More importantly, peptide action can be reproduced in vitro in antral and colonic preparations. Studies performed on muscle strips of rat gastric antrum showed that perfusion of CRF, Ucn 1, or Ucn 2 decreases the amplitude of circular and longitudinal muscle contractions (148, 155). Moreover, in an isolated colonic rat preparation, CRF increased basal myoelectrical peristaltic activity (153, 156).

Convergent studies to characterize the CRF receptors involved in these processes have established that the delayed gastric emptying following peripheral injection of CRF, Ucn 1, Ucn 2, and Ucn 3 is mediated by CRF_2 receptors, whereas the stimulation of colonic motility after peripheral administration of CRF and Ucn 1 involves CRF_1 receptors located on the myen-

teric neurons of rats and mice (145). Ucn 2 (and, less potently, Ucn 3) injected peripherally delays gastric emptying of a solid or liquid meal without modifying distal colonic transit (152, 157). In contrast, ip injection of stressin$_1$-A induces defecation alone, without altering gastric emptying (38). In contrast, (a) ip or iv CRF and (b) Ucn 1 interacting with both CRF receptors inhibit gastric motor function while stimulating colonic propulsion and fecal pellet output in rats and mice (96, 152, 157). The use of selective CRF antagonists has shown that the peripheral injection of astressin$_2$-B and antisauvagine-30 blocks peripheral CRF- and Ucn 1–induced inhibition of gastric emptying without affecting the increase of distal colonic transit (152, 157). Conversely, the selective CRF_1 receptor antagonists CP-154,526 and NBI-27914 block peripheral CRF- or Ucn 1–induced stimulation of colonic motor function (clustered spike-burst activity, distal transit, defecation, and diarrhea) without influencing the delay of gastric emptying (152, 153, 157, 158). Consistent with a peripheral action within the gut, CRF receptors have been localized throughout the gastrointestinal tract in guinea pigs (147). In rat colon, CRF_1 receptor has been detected both at the gene level and by immunohistochemistry in goblet and stem cells of the crypts, as well as on surface epithelial cells, lamina propria, and, prominently, in the myenteric nervous plexus (144, 159). CRF_2 receptors are highly expressed at the gene and protein levels in the rat upper gut, including the esophagus and stomach (34, 148). Evidence for a pivotal role of the CRF_2 receptors, which are expressed on gastric myenteric neurons (148), comes from in vitro studies on gastric antral strips, in which CRF- and Ucn 2–induced reduction of spontaneous circular and longitudinal muscle contraction was tetrodotoxin dependent (148, 155).

Underlying alterations of motility linked with changes in gut transit have been characterized. In the upper gut, the iv injection of CRF decreases gastric intraluminal pressure in rats and inhibits motilin-induced jejunal migrating motor complex in dogs (155, 160). Likewise, in healthy humans, iv CRF reduces the basal fundic tone and stimulates the nonpropulsive postprandial duodenal motor activity as well as the pyloric and duodenal pressure wave (150, 161, 162). Ucn 1 injected intravenously disrupts the fasted motor pattern of gastroduodenal motility, which in conscious rats is replaced by the fed-like motor pattern (96). In contrast, when Ucn 1 is injected intravenously in ad libitum–fed animals, the fed motor pattern remains, but there is a decrease in antral and an increase in duodenal motor index (96). However, the increase in duodenal motility index is nonpropagative, as shown by the reduction of duodenal transit (96). In the colon, peripheral injection of CRF and Ucn 1 increases clustered spike-burst propagative activity in rats (153, 163). Clinical studies show that systemic application of CRF induces a colonic motility response that includes the occurrence of clustered contractions in the descending and sigmoid colon, which is more prominent in IBS patients than in healthy controls (151). The mediation of the CRF and Ucn 1 effect on colonic motor activity may involve a direct interaction with colonic myenteric neurons (159). Peripheral injection of CRF or of stressin$_1$-A induces robust Fos expression selectively in the myenteric ganglia of the colon; this expression can be blocked by peripheral application of astressin and CP-154,526 (159, 164). In addition, the activation takes place in cholinergic and nitrergic colonic myenteric neurons that are known to be involved in the peristaltic reflex and that bear CRF$_1$ receptors (144, 159).

Stress-Related Alteration of Gut Motor Function: Involvement of Peripheral Corticotropin-Releasing Factor Receptors

The functionality of the CRF signaling system in the gut during stress is supported by reports that peripherally injected peptide antagonists, namely α-helical CRF$_{9-41}$, D-Phe^{12}CRF$_{12-41}$, and astressin, block abdominal surgery–induced delay of gastric emptying (58, 165, 166). The inhibition of gastric emptying induced by acute wrap restraint stress is also blocked by peripheral application of a CRF$_2$ antagonist, whereas application of CP-154,526 has no effect (157). Likewise, the stimulation of distal colonic transit and fecal pellet output induced by acute wrap restraint or water avoidance stress is blocked or blunted by peripheral injection of α-helical CRF$_{9-41}$ or astressin (71, 123, 153, 167). However, peripheral CRF receptors seem not to be involved in the regulation of fasted and postprandial gut motor functions under basal conditions (71, 145, 153, 157). Researchers speculate that stress may recruit CRF ligands, which are expressed in the gut through autonomic alterations. For instance, CRF and Ucn 1 mRNA are detected in the submucosa and muscle layers, and immunoreactivity shows a cellular distribution in myenteric neurons, serotonin-containing enterochromaffin cells, and lamina propria cells of the mucosa in stomach and colon (148, 168, 169).

CONCLUSIONS

In summary, we have made major advances both in unraveling the components of CRF signaling pathways that encompass CRF, urocortins, CRF receptors, and CRF-BP and in mapping their expression in the brain and the gut. There are conclusive experimental data showing that activation of brain and colonic CRF$_1$ pathways by exogenous CRF or Ucn 1 or stress recapitulates cardinal features encountered during stress, including the stimulation of colonic motility, defecation/watery diarrhea, and visceral hypersensitivity (5). Selective CRF$_1$ antagonists abolish or reduce exogenous CRF- and stress-induced anxiogenic/depressive behavior, defective intestinal barrier, stimulation of colonic motility, myenteric neurons, mucus secretion, mast cell activation, defecation, diarrhea, and hyperalgesia (5, 170). Therefore, sustained activation of the CRF$_1$ system at central and/or peripheral sites may represent a key underlying mechanism whereby stress alters colonic function and can lead to stress-related functional bowel disorders such as IBS (15–20). In the upper gut, the brain and gastric CRF_2 signaling systems are more prominently involved in CRF ligands– and stress-related suppression of gastric motor function. Both central ligands (e.g., CRF and Ucn 1, Ucn 2, and Ucn 3) and stress inhibit propulsive gastric motor function through autonomic and enteric nervous system alterations.

Additional investigations on the stress-related regulation of CRF ligands and their receptors, including their variants in the gut and their mechanisms of action at the cellular level, may provide insight into new venues for effective therapies for patients suffering from stress-related functional bowel disorders.

SUMMARY POINTS

1. In recent years, our understanding of how stress alters the function of the gastrointestinal tract via the brain-gut axis has increased dramatically.

2. The discovery of the CRF peptide family; Ucn 1, Ucn 2, and Ucn 3; and the cloning of CRF receptor subtypes CRF_1, CRF_2, and CRF-BP; as well as the development of selective CRF_1 and CRF_2 antagonists, has provided relevant tools to characterize the primary implications of the brain CRF/CRF_1 signaling pathways in the endocrine, anxiogenic, autonomic, and visceral responses to stress.

3. Among the viscera, the gut functions are highly susceptible to the effects of stress, as has been shown by alterations of gut motility such as slowing of gastric transit and stimulation of colonic propulsive motor activity, along with altered intestinal barrier function.

4. Components of the CRF signaling system are expressed in brain nuclei influencing autonomic outflow to the viscera such as the PVN, the Barrington's nucleus/LC complex, and the DVC, along with myenteric, endocrine, and immune cells within the gut.

5. The stimulation of colonic motor function induced by various stressors is mediated by the brain (PVN, Barrington's nucleus/LC) and gut (enteric) CRF_1 signaling system, which contributes to the activation of the sacral parasympathetic and the colonic myenteric nervous systems, respectively. Such CRF_1 activation also contributes to the recruitment of colonic serotonin-containing enterochromaffin and mast cells.

6. The inhibition of gastric motor function by CRF ligands is mediated by activation of CRF_2 receptors both in the brain and in the stomach. CRF receptors are involved in the modulation of autonomic and gastric myenteric activity, which influences gastric function during stress.

7. Further studies on the regulation of the CRF signaling system in the gut in response to stress and specific mechanisms of action may provide the basis for effective therapeutic venues for stress-related functional bowel disorders.

DISCLOSURE STATEMENT

The authors are not aware of any biases that might be perceived as affecting the objectivity of this review.

ACKNOWLEDGMENTS

The authors were supported by National Institutes of Health grants R01 DK-33061 and DK 57238 (Y.T.), Center grant DK-41301 (Animal core, Y.T.), P50 AR 049550 (Y.T.), VA Career Scientist and Merit Award (Y.T.), and German Research Foundation grant STE 1765/1-1 (A.S.).

LITERATURE CITED

1. Selye H. 1998 [1936]. A syndrome produced by diverse nocuous agents. *J. Neuropsychiatr.* 10:230–31
2. Sterling P, Eyer J. 1981. Allostasis: a new paradigm to explain arousal pathology. In *Handbook of Life Stress, Cognition and Health*, ed. S Fisher, HS Reason. New York: Wiley
3. McEwen BS, Wingfield JC. 2003. The concept of allostasis in biology and biomedicine. *Horm. Behav.* 43:2–15
4. McEwen BS. 2000. Allostasis and allostatic load: implications for neuropsychopharmacology. *Neuropsychopharmacology* 22:108–24
5. Martinez V, Taché Y. 2006. CRF1 receptors as a therapeutic target for irritable bowel syndrome. *Curr. Pharm. Des.* 12:4071–88
6. Bale TL. 2005. Sensitivity to stress: dysregulation of CRF pathways and disease development. *Horm. Behav.* 48:1–10
7. Senba E, Ueyama T. 1997. Stress-induced expression of immediate early genes in the brain and peripheral organs of the rat. *Neurosci. Res.* 29:183–207
8. Taché Y, Garrick T, Raybould H. 1990. Central nervous system action of peptides to influence gastrointestinal motor function. *Gastroenterology* 98:517–28
9. Bonaz B, Taché Y. 1994. Water-avoidance stress-induced c-Fos expression in the rat brain and stimulation of fecal output: role of corticotropin-releasing factor. *Brain Res.* 641:21–28
10. Bonaz B, Taché Y. 1994. Induction of Fos immunoreactivity in the rat brain after cold-restraint-induced gastric lesions and fecal excretion. *Brain Res.* 652:56–64
11. Barquist E, Bonaz B, Martinez V, Rivier J, Zinner MJ, Taché Y. 1996. Neuronal pathways involved in abdominal surgery–induced gastric ileus in rats. *Am. J. Physiol.* 270:R888–94
12. Taché Y, Martinez V, Million M, Wang L. 2001. Stress and the gastrointestinal tract. III. Stress-related alterations of gut motor function: role of brain corticotropin-releasing factor receptors. *Am. J. Physiol. Gastrointest. Liver Physiol.* 280:G173–77
13. Bale TL, Vale WW. 2004. CRF and CRF receptors: role in stress responsivity and other behaviors. *Annu. Rev. Pharmacol. Toxicol.* 44:525–57
14. Caso JR, Leza JC, Menchen L. 2008. The effects of physical and psychological stress on the gastrointestinal tract: lessons from animal models. *Curr. Mol. Med.* 8:299–312
15. Taché Y, Bonaz B. 2007. Corticotropin-releasing factor receptors and stress-related alterations of gut motor function. *J. Clin. Invest.* 117:33–40
16. Fukudo S. 2007. Role of corticotropin-releasing hormone in irritable bowel syndrome and intestinal inflammation. *J. Gastroenterol.* 42(Suppl. 17):48–51
17. Mayer EA, Naliboff BD, Chang L, Coutinho SV. 2001. V. Stress and irritable bowel syndrome. *Am. J. Physiol. Gastrointest. Liver Physiol.* 280:G519–24
18. Mönnikes H, Tebbe JJ, Hildebrandt M, Arck P, Osmanoglou E, et al. 2001. Role of stress in functional gastrointestinal disorders. Evidence for stress-induced alterations in gastrointestinal motility and sensitivity. *Dig. Dis.* 19:201–11
19. Mulak A, Bonaz B. 2004. Irritable bowel syndrome: a model of the brain-gut interactions. *Med. Sci. Monit.* 10:RA55–62
20. Halpert A, Drossman D. 2005. Biopsychosocial issues in irritable bowel syndrome. *J. Clin. Gastroenterol.* 39:665–69
21. Vale W, Spiess J, Rivier C, Rivier J. 1981. Characterization of a 41-residue ovine hypothalamic peptide that stimulates secretion of corticotropin and β-endorphin. *Science* 213:1394–97
22. Vaughan J, Donaldson C, Bittencourt J, Perrin MH, Lewis K, et al. 1995. Urocortin, a mammalian neuropeptide related to fish urotensin I and to corticotropin-releasing factor. *Nature* 378:287–92
23. Hsu SY, Hsueh AJ. 2001. Human stresscopin and stresscopin-related peptide are selective ligands for the type 2 corticotropin-releasing hormone receptor. *Nat. Med.* 7:605–11
24. Reyes TM, Lewis K, Perrin MH, Kunitake KS, Vaughan J, et al. 2001. Urocortin II: a member of the corticotropin-releasing factor (CRF) neuropeptide family that is selectively bound by type 2 CRF receptors. *Proc. Natl. Acad. Sci. USA* 98:2843–48

25. Lewis K, Li C, Perrin MH, Blount A, Kunitake K, et al. 2001. Identification of urocortin III, an additional member of the corticotropin-releasing factor (CRF) family with high affinity for the CRF2 receptor. *Proc. Natl. Acad. Sci. USA* 98:7570–75

26. Lovejoy DA, Balment RJ. 1999. Evolution and physiology of the corticotropin-releasing factor (CRF) family of neuropeptides in vertebrates. *Gen. Comp. Endocrinol.* 115:1–22

27. Chang CL, Hsu SY. 2004. Ancient evolution of stress-regulating peptides in vertebrates. *Peptides* 25:1681–88

28. Perrin MH, Vale WW. 1999. Corticotropin releasing factor receptors and their ligand family. *Ann. N. Y. Acad. Sci.* 885:312–28

29. Hillhouse EW, Grammatopoulos DK. 2006. The molecular mechanisms underlying the regulation of the biological activity of corticotropin-releasing hormone receptors: implications for physiology and pathophysiology. *Endocr. Rev.* 27:260–86

30. Teli T, Markovic D, Hewitt ME, Levine MA, Hillhouse EW, Grammatopoulos DK. 2008. Structural domains determining signaling characteristics of the CRH-receptor type 1 variant R1β and response to PKC phosphorylation. *Cell. Signal.* 20:40–49

31. Markovic D, Vatish M, Gu M, Slater D, Newton R, et al. 2007. The onset of labor alters corticotropin-releasing hormone type 1 receptor variant expression in human myometrium: putative role of interleukin-1β. *Endocrinology* 148:3205–13

32. Hauger RL, Grigoriadis DE, Dallman MF, Plotsky PM, Vale WW, Dautzenberg FM. 2003. Int. Union Pharmacol. XXXVI. Current status of the nomenclature for receptors for corticotropin-releasing factor and their ligands. *Pharmacol. Rev.* 55:21–26

33. Catalano RD, Kyriakou T, Chen J, Easton A, Hillhouse EW. 2003. Regulation of corticotropin-releasing hormone type 2 receptors by multiple promoters and alternative splicing: identification of multiple splice variants. *Mol. Endocrinol.* 17:395–410

34. Wu SV, Yuan PQ, Wang L, Peng YL, Chen CY, Taché Y. 2007. Identification and characterization of multiple corticotropin-releasing factor type 2 receptor isoforms in the rat esophagus. *Endocrinology* 148:1675–87

35. Chen AM, Perrin MH, Digruccio MR, Vaughan JM, Brar BK, et al. 2005. A soluble mouse brain splice variant of type 2α corticotropin-releasing factor (CRF) receptor binds ligands and modulates their activity. *Proc. Natl. Acad. Sci. USA* 102:2620–25

36. Grace CR, Perrin MH, Cantle JP, Vale WW, Rivier JE, Riek R. 2007. Common and divergent structural features of a series of corticotropin releasing factor-related peptides. *J. Am. Chem. Soc.* 129:16102–14

37. Farrokhi CB, Tovote P, Blanchard RJ, Blanchard DC, Litvin Y, Spiess J. 2007. Cortagine: behavioral and autonomic function of the selective CRF receptor subtype 1 agonist. *CNS Drug Rev.* 13:423–43

38. Rivier J, Gulyas J, Kunitake K, DiGruccio M, Cantle JP, et al. 2007. Stressin₁-A, a potent corticotropin releasing factor receptor 1 (CRF1)-selective peptide agonist. *J. Med. Chem.* 50:1668–74

39. Swanson LW, Sawchenko PE, Lind RW. 1986. Regulation of multiple peptides in CRF parvocellular neurosecretory neurons: implications for the stress response. *Prog. Brain Res.* 68:169–90

40. Valentino RJ, Page ME, Luppi PH, Zhu Y, Van Bockstaele E, Aston-Jones G. 1994. Evidence for widespread afferents to Barrington's nucleus, a brainstem region rich in corticotropin-releasing hormone neurons. *Neuroscience* 62:125–43

41. De Souza EB. 1995. Corticotropin-releasing factor receptors: physiology, pharmacology, biochemistry and role in central nervous system and immune disorders. *Psychoneuroendocrinology* 20:789–819

42. Valentino RJ, Kosboth M, Colflesh M, Miselis RR. 2000. Transneuronal labeling from the rat distal colon: anatomic evidence for regulation of distal colon function by a pontine corticotropin-releasing factor system. *J. Comp. Neurol.* 417:399–414

43. Venihaki M, Sakihara S, Subramanian S, Dikkes P, Weninger SC, et al. 2004. Urocortin III, a brain neuropeptide of the corticotropin-releasing hormone family: modulation by stress and attenuation of some anxiety-like behaviours. *J. Neuroendocrinol.* 16:411–22

44. Morin SM, Ling N, Liu XJ, Kahl SD, Gehlert DR. 1999. Differential distribution of urocortin- and corticotropin-releasing factor–like immunoreactivities in the rat brain. *Neuroscience* 92:281–91

45. Martinez V, Wang L, Million M, Rivier J, Taché Y. 2004. Urocortins and the regulation of gastrointestinal motor function and visceral pain. *Peptides* 25:1733–44

46. Bittencourt JC, Vaughan J, Arias C, Rissman RA, Vale WW, Sawchenko PE. 1999. Urocortin expression in rat brain: evidence against a pervasive relationship of urocortin-containing projections with targets bearing type 2 CRF receptors. *J. Comp. Neurol.* 415:285–312

47. Kozicz T, Yanaihara H, Arimura A. 1998. Distribution of urocortin-like immunoreactivity in the central nervous system of the rat. *J. Comp. Neurol.* 391:1–10

48. Mano-Otagiri A, Shibasaki T. 2004. Distribution of urocortin 2 and urocortin 3 in rat brain. *J. Nippon Med. Sch.* 71:358–59

49. Li C, Vaughan J, Sawchenko PE, Vale WW. 2002. Urocortin III–immunoreactive projections in rat brain: partial overlap with sites of type 2 corticotrophin-releasing factor receptor expression. *J. Neurosci.* 22:991–1001

50. Bonaz B, Rivest S. 1998. Effect of a chronic stress on CRF neuronal activity and expression of its type 1 receptor in the rat brain. *Am. J. Physiol.* 275:R1438–49

51. Imaki T, Katsumata H, Miyata M, Naruse M, Imaki J, Minami S. 2001. Expression of corticotropin-releasing hormone type 1 receptor in paraventricular nucleus after acute stress. *Neuroendocrinology* 73:293–301

52. Konishi S, Kasagi Y, Katsumata H, Minami S, Imaki T. 2003. Regulation of corticotropin-releasing factor (CRF) type-1 receptor gene expression by CRF in the hypothalamus. *Endocr. J.* 50:21–36

53. Turnbull AV, Rivier C. 1997. Corticotropin-releasing factor (CRF) and endocrine responses to stress: CRF receptors, binding protein, and related peptides. *Proc. Soc. Exp. Biol. Med.* 215:1–10

54. Bittencourt JC, Sawchenko PE. 2000. Do centrally administered neuropeptides access cognate receptors? An analysis in the central corticotropin-releasing factor system. *J. Neurosci.* 20:1142–56

55. Rivier JE, Kirby DA, Lahrichi SL, Corrigan A, Vale WW, Rivier CL. 1999. Constrained corticotropin releasing factor antagonists (astressin analogues) with long duration of action in the rat. *J. Med. Chem.* 42:3175–82

56. Rühmann A, Bonk I, Lin CR, Rosenfeld MG, Spiess J. 1998. Structural requirements for peptidic antagonists of the corticotropin-releasing factor receptor (CRFR): development of CRFR2β-selective antisauvagine-30. *Proc. Natl. Acad. Sci. USA* 95:15264–69

57. Rivier J, Gulyas J, Kirby D, Low W, Perrin MH, et al. 2002. Potent and long-acting corticotropin releasing factor (CRF) receptor 2 selective peptide competitive antagonists. *J. Med. Chem.* 45:4737–47

58. Martinez V, Rivier J, Taché Y. 1999. Peripheral injection of a new corticotropin-releasing factor (CRF) antagonist, astressin, blocks peripheral CRF- and abdominal surgery–induced delayed gastric emptying in rats. *J. Pharmacol. Exp. Ther.* 290:629–34

59. Lanier M, Williams JP. 2002. Small molecule corticotropin-releasing factor antagonists. *Expert Opin. Ther. Patents* 12:1619–30

60. Zorrilla EP, Koob GF. 2004. The therapeutic potential of CRF1 antagonists for anxiety. *Expert Opin. Investig. Drugs* 13:799–828

61. Chen C. 2006. Recent advances in small molecule antagonists of the corticotropin-releasing factor type 1 receptor—focus on pharmacology and pharmacokinetics. *Curr. Med. Chem.* 13:1261–82

62. Ising M, Zimmermann US, Kunzel HE, Uhr M, Foster AC, et al. 2007. High-affinity CRF1 receptor antagonist NBI-34041: preclinical and clinical data suggest safety and efficacy in attenuating elevated stress response. *Neuropsychopharmacology* 32:1941–49

63. Seymour PA, Schmidt AW, Schulz DW. 2003. The pharmacology of CP-154526, a nonpeptide antagonist of the CRH1 receptor: a review. *CNS Drug Rev.* 9:57–96

64. Behan DP, De Souza EB, Lowry PJ, Potter E, Sawchenko P, Vale WW. 1995. Corticotropin releasing factor (CRF) binding protein: a novel regulator of CRF and related peptides. *Front. Neuroendocrinol.* 16:362–82

65. Westphal NJ, Seasholtz AF. 2006. CRH-BP: the regulation and function of a phylogenetically conserved binding protein. *Front. Biosci.* 11:1878–91

66. Lovejoy DA, Aubry JM, Turnbull A, Sutton S, Potter E, et al. 1998. Ectopic expression of the CRF-binding protein: minor impact on HPA axis regulation but induction of sexually dimorphic weight gain. *J. Neuroendocrinol.* 10:483–91

67. Karolyi IJ, Burrows HL, Ramesh TM, Nakajima M, Lesh JS, et al. 1999. Altered anxiety and weight gain in corticotropin-releasing hormone–binding protein-deficient mice. *Proc. Natl. Acad. Sci. USA* 96:11595–600

68. Huising MO, Vaughan JM, Shah SH, Grillot KL, Donaldson CJ, et al. 2008. Residues of corticotropin releasing factor–binding protein (CRF-BP) that selectively abrogate binding to CRF but not to urocortin 1. *J. Biol. Chem.* 283:8902–12

69. Henry BA, Lightman SL, Lowry CA. 2005. Distribution of corticotropin-releasing factor binding protein immunoreactivity in the rat hypothalamus: association with corticotropin-releasing factor-, urocortin 1- and vimentin-immunoreactive fibres. *J. Neuroendocrinol.* 17:135–44

70. Hauger RL, Risbrough V, Brauns O, Dautzenberg FM. 2006. Corticotropin releasing factor (CRF) receptor signaling in the central nervous system: new molecular targets. *CNS Neurol. Disord. Drug Targets* 5:453–79

71. Williams CL, Peterson JM, Villar RG, Burks TF. 1987. Corticotropin-releasing factor directly mediates colonic responses to stress. *Am. J. Physiol.* 253:G582–86

72. Gué M, Junien JL, Buéno L. 1991. Conditioned emotional response in rats enhances colonic motility through the central release of corticotropin-releasing factor. *Gastroenterology* 100:964–70

73. Martinez V, Wang L, Rivier J, Grigoriadis D, Taché Y. 2004. Central CRF, urocortins and stress increase colonic transit via CRF1 receptors while activation of CRF2 receptors delays gastric transit in mice. *J. Physiol.* 556:221–34

74. Taché Y, Million M. 2006. Central corticotropin-releasing factor and the hypothalamic–pituitary–adrenal axis in gastrointestinal physiology. In *Physiology of the Gastrointestinal Tract*, ed. LR Johnson, J Wood, pp. 791–816. Burlington, MA: Elsevier Academic

75. Taché Y, Maeda-Hagiwara M, Turkelson CM. 1987. Central nervous system action of corticotropin-releasing factor to inhibit gastric emptying in rats. *Am. J. Physiol.* 253:G241–45

76. Sheldon RJ, Qi JA, Porreca F, Fisher LA. 1990. Gastrointestinal motor effects of corticotropin-releasing factor in mice. *Regul. Pept.* 28:137–51

77. Taché Y, Barquist E, Stephens RL, Rivier J. 1991. Abdominal surgery– and trephination-induced delay in gastric emptying is prevented by intracisternal injection of CRF antagonist in the rat. *J. Gastrointest. Motil.* 3:19–25

78. Sütö G, Király A, Taché Y. 1994. Interleukin 1 β inhibits gastric emptying in rats: mediation through prostaglandin and corticotropin-releasing factor. *Gastroenterology* 106:1568–75

79. Smedh U, Uvnas-Moberg K, Grill HJ, Kaplan JM. 1995. Fourth ventricle injection of corticotropin-releasing factor and gastric emptying of glucose during gastric fill. *Am. J. Physiol.* 269:G1000–3

80. Coskun T, Bozkurt A, Alican I, Ozkutlu U, Kurtel H, Yegen BC. 1997. Pathways mediating CRF-induced inhibition of gastric emptying in rats. *Regul. Pept.* 69:113–20

81. Martinez V, Rivier J, Wang L, Taché Y. 1997. Central injection of a new corticotropin-releasing factor (CRF) antagonist, astressin, blocks CRF- and stress-related alterations of gastric and colonic motor function. *J. Pharmacol. Exp. Ther.* 280:754–60

82. Lee C, Sarna SK. 1997. Central regulation of gastric emptying of solid nutrient meals by corticotropin releasing factor. *Neurogastroenterol. Motil.* 9:221–29

83. Martinez V, Barquist E, Rivier J, Taché Y. 1998. Central CRF inhibits gastric emptying of a nutrient solid meal in rats: the role of CRF2 receptors. *Am. J. Physiol.* 274:G965–70

84. Martinez V, Taché Y. 2001. Role of CRF receptor 1 in central CRF-induced stimulation of colonic propulsion in rats. *Brain Res.* 893:29–35

85. Chen CY, Million M, Adelson DW, Martinez V, Rivier J, Taché Y. 2002. Intracisternal urocortin inhibits vagally stimulated gastric motility in rats: role of CRF2. *Br. J. Pharmacol.* 136:237–47

86. Nakade Y, Tsuchida D, Fukuda H, Iwa M, Pappas TN, Takahashi T. 2005. Restraint stress delays solid gastric emptying via a central CRF and peripheral sympathetic neuron in rats. *Am. J. Physiol. Regul. Integr. Comp. Physiol.* 288:R427–32

87. Nagata T, Uemoto M, Yuzuriha H, Asakawa A, Inui A, et al. 2005. Intracerebroventricularly administered urocortin inhibits gastric emptying in mice. *Int. J. Mol. Med.* 15:1041–43

88. Czimmer J, Million M, Taché Y. 2006. Urocortin 2 acts centrally to delay gastric emptying through sympathetic pathways while CRF and urocortin 1 inhibitory actions are vagal dependent in rats. *Am. J. Physiol. Gastrointest. Liver Physiol.* 290:G511–18

89. Heymann-Mönnikes I, Taché Y, Trauner M, Weiner H, Garrick T. 1991. CRF microinjected into the dorsal vagal complex inhibits TRH analog– and kainic acid–stimulated gastric contractility in rats. *Brain Res.* 554:139–44

90. Mönnikes H, Schmidt BG, Raybould HE, Taché Y. 1992. CRF in the paraventricular nucleus mediates gastric and colonic motor response to restraint stress. *Am. J. Physiol.* 262:G137–43

91. Mönnikes H, Schmidt BG, Tebbe J, Bauer C, Taché Y. 1994. Microinfusion of corticotropin releasing factor into the locus coeruleus/subcoeruleus nuclei stimulates colonic motor function in rats. *Brain Res.* 644:101–8

92. Lewis MW, Hermann GE, Rogers RC, Travagli RA. 2002. In vitro and in vivo analysis of the effects of corticotropin releasing factor on rat dorsal vagal complex. *J. Physiol.* 543:135–46

93. Smedh U, Moran TH. 2006. The dorsal vagal complex as a site for cocaine- and amphetamine-regulated transcript peptide to suppress gastric emptying. *Am. J. Physiol. Regul. Integr. Comp. Physiol.* 291:R124–30

94. Lenz HJ, Burlage M, Raedler A, Greten H. 1988. Central nervous system effects of corticotropin-releasing factor on gastrointestinal transit in the rat. *Gastroenterology* 94:598–602

95. Gué M, Fioramonti J, Frexinos J, Alvinerie M, Buéno L. 1987. Influence of acoustic stress by noise on gastrointestinal motility in dogs. *Dig. Dis. Sci.* 32:1411–17

96. Kihara N, Fujimura M, Yamamoto I, Itoh E, Inui A, Fujimiya M. 2001. Effects of central and peripheral urocortin on fed and fasted gastroduodenal motor activity in conscious rats. *Am. J. Physiol. Gastrointest. Liver Physiol.* 280:G406–19

97. Zhang H, Han T, Sun LN, Huang BK, Chen YF, et al. 2008. Regulative effects of essential oil from *Atractylodes lancea* on delayed gastric emptying in stress-induced rats. *Phytomedicine* 15:602–11

98. Iwa M, Nakade Y, Pappas TN, Takahashi T. 2006. Electroacupuncture elicits dual effects: Stimulation of delayed gastric emptying and inhibition of accelerated colonic transit induced by restraint stress in rats. *Dig. Dis. Sci.* 51:1493–500

99. Harbuz MS, Lightman SL. 1989. Responses of hypothalamic and pituitary mRNA to physical and psychological stress in the rat. *J. Endocrinol.* 122:705–11

100. Bonaz B, Plourde V, Taché Y. 1994. Abdominal surgery induces Fos immunoreactivity in the rat brain. *J. Comp. Neurol.* 349:212–22

101. Rivest S, Rivier C. 1994. Stress and interleukin-1β-induced activation of c-Fos, NGFI-B and CRF gene expression in the hypothalamic PVN: comparison between Sprague-Dawley, Fisher-344 and Lewis rats. *J. Neuroendocrinol.* 6:101–17

102. Yao M, Denver RJ. 2007. Regulation of vertebrate corticotropin-releasing factor genes. *Gen. Comp. Endocrinol.* 153:200–16

103. Swanson LW, Sawchenko PE. 1980. Paraventricular nucleus: a site for the integration of neuroendocrine and autonomic mechanisms. *Neuroendocrinology* 31:410–17

104. Tanaka Y, Makino S, Noguchi T, Tamura K, Kaneda T, Hashimoto K. 2003. Effect of stress and adrenalectomy on urocortin II mRNA expression in the hypothalamic paraventricular nucleus of the rat. *Neuroendocrinology* 78:1–11

105. Korosi A, Schotanus S, Olivier B, Roubos EW, Kozicz T. 2005. Chronic ether stress-induced response of urocortin 1 neurons in the Edinger-Westphal nucleus in the mouse. *Brain Res.* 1046:172–79

106. Wang X, Su H, Copenhagen LD, Vaishnav S, Pieri F, et al. 2002. Urocortin-deficient mice display normal stress-induced anxiety behavior and autonomic control but an impaired acoustic startle response. *Mol. Cell Biol.* 22:6605–10

107. Luckey A, Wang L, Jamieson PM, Basa NR, Million M, et al. 2003. Corticotropin-releasing factor receptor 1–deficient mice do not develop postoperative gastric ileus. *Gastroenterology* 125:654–59

108. Luckey A, Livingston E, Taché Y. 2003. Mechanisms and treatment of postoperative ileus. *Arch. Surg.* 138:206–14

109. Taché Y, Saperas E. 1993. Central actions of interleukin 1 on gastrointestinal function. In *Neurobiology of Cytokines Part B*, ed. EB De Souza, pp. 169–83. San Diego: Academic

110. Rivest S. 1995. Molecular mechanisms and neural pathways mediating the influence of interleukin-1 on the activity of neuroendocrine CRF motoneurons in the rat. *Int. J. Dev. Neurosci.* 13:135–46

111. Smedh U, Moran TH. 2003. Peptides that regulate food intake: separable mechanisms for dorsal hindbrain CART peptide to inhibit gastric emptying and food intake. *Am. J. Physiol. Regul. Integr. Comp. Physiol.* 284:R1418–26

112. Chen CY, Inui A, Asakawa A, Fujino K, Kato I, et al. 2005. Des-acyl ghrelin acts by CRF type 2 receptors to disrupt fasted stomach motility in conscious rats. *Gastroenterology* 129:8–25

113. Nakade Y, Tsukamoto K, Pappas TN, Takahashi T. 2006. Central glucagon like peptide-1 delays solid gastric emptying via central CRF and peripheral sympathetic pathway in rats. *Brain Res.* 1111:117–21

114. Taché Y, Martinez V, Million M, Rivier J. 1999. Corticotropin-releasing factor and the brain-gut motor response to stress. *Can. J. Gastroenterol.* 13:18–25A

115. Buéno L, Fioramonti J. 1986. Effects of corticotropin-releasing factor, corticotropin and cortisol on gastrointestinal motility in dogs. *Peptides* 7:73–77

116. Lenz HJ, Raedler A, Greten H, Vale WW, Rivier JE. 1988. Stress-induced gastrointestinal secretory and motor responses in rats are mediated by endogenous corticotropin-releasing factor. *Gastroenterology* 95:1510–17

117. Wittmann T, Crenner F, Angel F, Hanusz L, Ringwald C, Grenier JF. 1990. Long-duration stress. Immediate and late effects on small and large bowel motility in rat. *Dig. Dis. Sci.* 35:495–500

118. Kellow JE, Langeluddecke PM, Eckersley GM, Jones MP, Tennant CC. 1992. Effects of acute psychologic stress on small-intestinal motility in health and the irritable bowel syndrome. *Scand. J. Gastroenterol.* 27:53–58

119. Williams CL, Villar RG, Peterson JM, Burks TF. 1988. Stress-induced changes in intestinal transit in the rat: a model for irritable bowel syndrome. *Gastroenterology* 94:611–21

120. Taché Y. 2002. The parasympathetic nervous system in the pathophysiology of the gastrointestinal tract. In *Handbook of Autonomic Nervous System in Health and Disease*, ed. CL Bolis, J Licinio, S Govoni, pp. 463–503. New York: Dekker

121. Jimenez M, Buéno L. 1990. Inhibitory effects of neuropeptide Y (NPY) on CRF and stress-induced cecal motor response in rats. *Life Sci.* 47:205–11

122. Gué M, Tekamp A, Tabis N, Junien JL, Buéno L. 1994. Cholecystokinin blockade of emotional stress– and CRF-induced colonic motor alterations in rats: role of the amygdala. *Brain Res.* 658:232–38

123. Miyata K, Ito H, Fukudo S. 1998. Involvement of the 5-HT$_3$ receptor in CRH-induced defecation in rats. *Am. J. Physiol.* 274:G827–31

124. Taché Y, Martinez V, Wang L, Million M. 2004. CRF1 receptor signaling pathways are involved in stress-related alterations of colonic function and viscerosensitivity: implications for irritable bowel syndrome. *Br. J. Pharmacol.* 141:1321–30

125. Mönnikes H, Schmidt BG, Taché Y. 1993. Psychological stress–induced accelerated colonic transit in rats involves hypothalamic corticotropin-releasing factor. *Gastroenterology* 104:716–23

126. Million M, Wang L, Martinez V, Taché Y. 2000. Differential Fos expression in the paraventricular nucleus of the hypothalamus, sacral parasympathetic nucleus and colonic motor response to water avoidance stress in Fischer and Lewis rats. *Brain Res.* 877:345–53

127. Lu L, Liu D, Ceng X, Ma L. 2000. Differential roles of corticotropin-releasing factor receptor subtypes 1 and 2 in opiate withdrawal and in relapse to opiate dependence. *Eur. J. Neurosci.* 12:4398–404

128. Saito K, Kasai T, Nagura Y, Ito H, Kanazawa M, Fukudo S. 2005. Corticotropin-releasing hormone receptor 1 antagonist blocks brain-gut activation induced by colonic distention in rats. *Gastroenterology* 129:1533–43

129. Bale TL, Picetti R, Contarino A, Koob GF, Vale WW, Lee KF. 2002. Mice deficient for both corticotropin-releasing factor receptor 1 (CRFR1) and CRFR2 have an impaired stress response and display sexually dichotomous anxiety-like behavior. *J. Neurosci.* 22:193–99

130. Gulpinar MA, Bozkurt A, Coskun T, Ulusoy NB, Yegen BC. 2000. Glucagon-like peptide (GLP-1) is involved in the central modulation of fecal output in rats. *Am. J. Physiol. Gastrointest. Liver Physiol.* 278:G924–29

131. Tebbe JJ, Ortmann E, Schumacher K, Mönnikes H, Kobelt P, et al. 2004. Cocaine- and amphetamine-regulated transcript stimulates colonic motility via central CRF receptor activation and peripheral cholinergic pathways in fed, conscious rats. *Neurogastroenterol. Motil.* 16:489–96

132. Tebbe JJ, Mronga S, Tebbe CG, Ortmann E, Arnold R, Schäfer MK. 2005. Ghrelin-induced stimulation of colonic propulsion is dependent on hypothalamic neuropeptide Y1– and corticotrophin-releasing factor 1 receptor activation. *J. Neuroendocrinol.* 17:570–76

133. Nakade Y, Tsukamoto K, Iwa M, Pappas TN, Takahashi T. 2007. Glucagon-like peptide-1 accelerates colonic transit via central CRF and peripheral vagal pathways in conscious rats. *Auton. Neurosci.* 131:50–56

134. Nakade Y, Fukuda H, Iwa M, Tsukamoto K, Yanagi H, et al. 2007. Restraint stress stimulates colonic motility via central corticotropin-releasing factor and peripheral 5-HT3 receptors in conscious rats. *Am. J. Physiol. Gastrointest. Liver Physiol.* 292:G1037–44

135. Ataka K, Kuge T, Fujino K, Takahashi T, Fujimiya M. 2007. Wood creosote prevents CRF-induced motility via 5-HT3 receptors in proximal and 5-HT4 receptors in distal colon in rats. *Auton. Neurosci.* 133:136–45

136. Hirata T, Keto Y, Nakata M, Takeuchi A, Funatsu T, et al. 2008. Effects of serotonin 5-HT$_3$ receptor antagonists on CRF-induced abnormal colonic water transport and defecation in rats. *Eur. J. Pharmacol.* 587:281–84

137. Valentino RJ, Miselis RR, Pavcovich LA. 1999. Pontine regulation of pelvic viscera: pharmacological target for pelvic visceral dysfunctions. *Trends Pharmacol. Sci.* 20:253–60

138. Reyes BA, Valentino RJ, Xu G, Van Bockstaele EJ. 2005. Hypothalamic projections to locus coeruleus neurons in rat brain. *Eur. J. Neurosci.* 22:93–106

139. Kresse AE, Million M, Saperas E, Taché Y. 2001. Colitis induces CRF expression in hypothalamic magnocellular neurons and blunts CRF gene response to stress in rats. *Am. J. Physiol. Gastrointest. Liver Physiol.* 281:G1203–13

140. Sternberg EM, Young WS 3rd, Bernardini R, Calogero AE, Chrousos GP, et al. 1989. A central nervous system defect in biosynthesis of corticotropin-releasing hormone is associated with susceptibility to streptococcal cell wall–induced arthritis in Lewis rats. *Proc. Natl. Acad. Sci. USA* 86:4771–75

141. Koob GF. 1999. Corticotropin-releasing factor, norepinephrine, and stress. *Biol. Psychiatry* 46:1167–80

142. Kawahito Y, Sano H, Mukai S, Asai K, Kimura S, et al. 1995. Corticotropin releasing hormone in colonic mucosa in patients with ulcerative colitis. *Gut* 37:544–51

143. Chatzaki E, Murphy BJ, Wang L, Million M, Ohning GV, et al. 2004. Differential profile of CRF receptor distribution in the rat stomach and duodenum assessed by newly developed CRF receptor antibodies. *J. Neurochem.* 88:1–11

144. Chatzaki E, Crowe PD, Wang L, Million M, Taché Y, Grigoriadis DE. 2004. CRF receptor type 1 and 2 expression and anatomical distribution in the rat colon. *J. Neurochem.* 90:309–16

145. Taché Y, Perdue MH. 2004. Role of peripheral CRF signaling pathways in stress-related alterations of gut motility and mucosal function. *Neurogastroenterol. Motil.* 16(Suppl. 1):137–42

146. Porcher C, Juhem A, Peinnequin A, Sinniger V, Bonaz B. 2005. Expression and effects of metabotropic CRF1 and CRF2 receptors in rat small intestine. *Am. J. Physiol. Gastrointest. Liver Physiol.* 288:G1091–103

147. Liu S, Gao X, Gao N, Wang X, Fang X, et al. 2005. Expression of type 1 corticotropin-releasing factor receptor in the guinea pig enteric nervous system. *J. Comp. Neurol.* 481:284–98

148. Porcher C, Peinnequin A, Pellissier S, Meregnani J, Sinniger V, et al. 2006. Endogenous expression and in vitro study of CRF-related peptides and CRF receptors in the rat gastric antrum. *Peptides* 27:1464–75

149. Pappas T, Debas H, Taché Y. 1985. Corticotropin-releasing factor inhibits gastric emptying in dogs. *Regul. Pept.* 11:193–99

150. Mayer EA, Sytnik B, Reddy NS, Van Deventer G, Taché Y. 1992. Corticotropin releasing factor (CRF) increases postprandial duodenal motor activity in humans. *J. Gastrointest. Motil.* 4:53–60

151. Fukudo S, Nomura T, Hongo M. 1998. Impact of corticotropin-releasing hormone on gastrointestinal motility and adrenocorticotropic hormone in normal controls and patients with irritable bowel syndrome. *Gut* 42:845–49

152. Martinez V, Wang L, Rivier JE, Vale W, Taché Y. 2002. Differential actions of peripheral corticotropin-releasing factor (CRF), urocortin II, and urocortin III on gastric emptying and colonic transit in mice: role of CRF receptor subtypes 1 and 2. *J. Pharmacol. Exp. Ther.* 301:611–17

153. Maillot C, Million M, Wei JY, Gauthier A, Taché Y. 2000. Peripheral corticotropin-releasing factor and stress-stimulated colonic motor activity involve type 1 receptor in rats. *Gastroenterology* 119:1569–79

154. Taché Y. 1999. Cyclic vomiting syndrome: the corticotropin-releasing-factor hypothesis. *Dig. Dis. Sci.* 44:79S–86S

155. Raybould H, Koelbel CB, Mayer EA, Taché Y. 1990. Inhibition of gastric motor function by circulating corticotropin-releasing factor in anesthetized rats. *J. Gastrointest. Motil.* 2:265–72

156. Mancinelli R, Azzena GB, Diana M, Forgione A, Fratta W. 1998. In vitro excitatory actions of corticotropin-releasing factor on rat colonic motility. *J. Auton. Pharmacol.* 18:319–24

157. Million M, Maillot C, Saunders P, Rivier J, Vale W, Taché Y. 2002. Human urocortin II, a new CRF-related peptide, displays selective CRF$_2$-mediated action on gastric transit in rats. *Am. J. Physiol. Gastrointest. Liver Physiol.* 282:G34–40

158. Saunders PR, Maillot C, Million M, Taché Y. 2002. Peripheral corticotropin-releasing factor induces diarrhea in rats: role of CRF1 receptor in fecal watery excretion. *Eur. J. Pharmacol.* 435:231–35

159. Yuan PQ, Million M, Wu SV, Rivier J, Taché Y. 2007. Peripheral corticotropin releasing factor (CRF) and a novel CRF1 receptor agonist, stressin$_1$-A, activate CRF1 receptor expressing cholinergic and nitrergic myenteric neurons selectively in the colon of conscious rats. *Neurogastroenterol. Motil.* 19:923–36

160. Buéno L, Fargeas MJ, Gué M, Peeters TL, Bormans V, Fioramonti J. 1986. Effects of corticotropin-releasing factor on plasma motilin and somatostatin levels and gastrointestinal motility in dogs. *Gastroenterology* 91:884–89

161. Su YC, Doran S, Wittert G, Chapman IM, Jones KL, et al. 2002. Effects of exogenous corticotropin-releasing factor on antropyloroduodenal motility and appetite in humans. *Am. J. Gastroenterol.* 97:49–57

162. Van Den Elzen BD, Van Den Wijngaard RM, Tytgat GN, Boeckxstaens GE. 2007. Influence of corticotropin-releasing hormone on gastric sensitivity and motor function in healthy volunteers. *Eur. J. Gastroenterol. Hepatol.* 19:401–7

163. Maillot C, Wang L, Million M, Taché Y. 2003. Intraperitoneal corticotropin-releasing factor and urocortin induce Fos expression in brain and spinal autonomic nuclei and long lasting stimulation of colonic motility in rats. *Brain Res.* 974:70–81

164. Miampamba M, Maillot C, Million M, Taché Y. 2002. Peripheral CRF activates myenteric neurons in the proximal colon through CRF$_1$ receptor in conscious rats. *Am. J. Physiol. Gastrointest. Liver Physiol.* 282:G857–65

165. Barquist E, Zinner M, Rivier J, Taché Y. 1992. Abdominal surgery–induced delayed gastric emptying in rats: role of CRF and sensory neurons. *Am. J. Physiol.* 262:G616–20

166. Nozu T, Martinez V, Rivier J, Taché Y. 1999. Peripheral urocortin delays gastric emptying: role of CRF receptor 2. *Am. J. Physiol.* 276:G867–74

167. Castagliuolo I, Lamont JT, Qiu B, Fleming SM, Bhaskar KR, et al. 1996. Acute stress causes mucin release from rat colon: role of corticotropin releasing factor and mast cells. *Am. J. Physiol.* 271:G884–92

168. Harada S, Imaki T, Naruse M, Chikada N, Nakajima K, Demura H. 1999. Urocortin mRNA is expressed in the enteric nervous system of the rat. *Neurosci. Lett.* 267:125–28

169. Yuan PQ, Wu SV, Million M, Larauche MH, Wang L, Taché Y. 2008. Corticotropin releasing factor (CRF) in rat colon: expression, localization and regulation by stress. *Gastroenterology* 134(Suppl. 1):A291

170. Gareau MG, Silva MA, Perdue MH. 2008. Pathophysiological mechanisms of stress-induced intestinal damage. *Curr. Mol. Med.* 8:274–81

171. Funada M, Hara C, Wada K. 2001. Involvement of corticotropin-releasing factor receptor subtype 1 in morphine withdrawal regulation of the brain noradrenergic system. *Eur. J. Pharmacol.* 430:277–81

172. Million M, Grigoriadis DE, Sullivan S, Crowe PD, McRoberts JA, et al. 2003. A novel water-soluble selective CRF1 receptor antagonist, NBI 35965, blunts stress-induced visceral hyperalgesia and colonic motor function in rats. *Brain Res.* 985:32–42

173. Gabry KE, Chrousos GP, Rice KC, Mostafa RM, Sternberg E, et al. 2002. Marked suppression of gastric ulcerogenesis and intestinal responses to stress by a novel class of drugs. *Mol. Psychiatry* 7:474–83

174. Habib KE, Weld KP, Rice KC, Pushkas J, Champoux M, et al. 2000. Oral administration of a corticotropin-releasing hormone receptor antagonist significantly attenuates behavioral, neuroendocrine, and autonomic responses to stress in primates. *Proc. Natl. Acad. Sci. USA* 97:6079–84

Stem Cells, Self-Renewal, and Differentiation in the Intestinal Epithelium

Laurens G. van der Flier and Hans Clevers

Hubrecht Institute, Royal Netherlands Academy of Arts and Sciences & University Medical Center Utrecht, 3584 CT, Utrecht, The Netherlands; email: l.vanderflier@niob.knaw.nl; clevers@niob.knaw.nl

Annu. Rev. Physiol. 2009. 71:241–60

First published online as a Review in Advance on September 22, 2008

The *Annual Review of Physiology* is online at physiol.annualreviews.org

This article's doi:
10.1146/annurev.physiol.010908.163145

Copyright © 2009 by Annual Reviews.
All rights reserved

0066-4278/09/0315-0241$20.00

Key Words

Wnt, Lgr5, Notch, colorectal cancer

Abstract

The mammalian intestine is covered by a single layer of epithelial cells that is renewed every 4–5 days. This high cell turnover makes it a very attractive and comprehensive adult organ system for the study of cell proliferation and differentiation. The intestine is composed of proliferative crypts, which contain intestinal stem cells, and villi, which contain differentiated specialized cell types. Through the recent identification of Lgr5, an intestinal stem cell marker, it is now possible to visualize stem cells and study their behavior and differentiation in a much broader context. In this review we describe the identification of intestinal stem cells. We also discuss genetic studies that have helped to elucidate those signals important for progenitor cells to differentiate into one of the specialized intestinal epithelial cell types. These studies describe a genetic hierarchy responsible for cell fate commitment in normal gut physiology. Where relevant we also mention aberrant deregulation of these molecular pathways that results in colon cancer.

INTRODUCTION

The primary function of the intestinal tract is the digestion and absorption of nutrients. The intestinal lumen is lined with a specialized simple epithelium, which performs the primary functions of digestion and water and nutrient absorption and forms a barrier against luminal pathogens. The gut is anatomically divided into the small intestine and the colon. The small intestine can be subdivided into the duodenum, the jejunum, and the ileum. The intestinal epithelium is the most vigorously self-renewing tissue of adult mammals (1). **Figure 1a** shows the organization of the intestinal epithelium into crypts and villi. The four differentiated cell types that reside within the epithelium—goblet cells, enteroendocrine cells, Paneth cells, and enterocytes—are visualized through staining with specific markers (**Figure 1b–e**).

Proliferative cells reside in the crypts of Lieberkühn, epithelial invasions into the underlying connective tissue. The crypts harbor stem cells and their progeny, transit-amplifying cells. Transit-amplifying cells spend approximately two days in the crypt, in which they divide 4–5 times before they terminally differentiate into the specialized intestinal epithelial cell types.

In the small intestine, the surface area is dramatically enlarged through epithelial protrusions called villi. Three types of differentiated epithelial cells cover these villi: the absorptive enterocytes, mucous-secreting goblet cells, and hormone-secreting enteroendocrine cells. Three days after their terminal differentiation, the cells reach the tip of the villus, undergo spontaneous apoptosis, and are shed into the gut lumen (2). Paneth cells are unusual in that

Epithelium: a continuous sheet of tightly linked cells that lines both the surfaces (e.g., skin) and the inside cavities (e.g., intestine) of the body

Stem cells: undifferentiated cells, residing in a specific location (a niche) within a tissue, that can produce one or more differentiated cell types

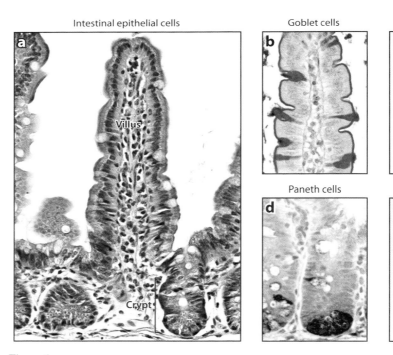

Figure 1

The intestinal epithelium. (*a*) H&E staining showing the morphology of the mouse intestine. The intestine is lined with a single layer of epithelial cells organized into invasions called the crypts of Lieberkühn and finger-like protrusions called villi. Immunohistochemical analysis for the main four differentiated cell types present in the intestinal epithelium: (*b*) periodic acid–Schiff (PAS) to stain goblet cells, (*c*) anti-synaptophysin to stain enteroendocrine cells, (*d*) lysozyme to stain Paneth cells, and (*e*) alkaline phosphatase to stain enterocytes.

they settle at the crypt bottoms and represent the only differentiated cells that escape the upward migration. Paneth cells have a function in innate immunity and antibacterial defense, to which ends they secrete bactericidal defensin peptides and lysozymes.

The modular organization of the epithelium of the small intestine and colon into crypts is globally comparable. Histologically, there are, however, two important differences between the two types of epithelia. The colon carries no villi but has a flat surface epithelium. Moreover, Paneth cells are absent in the colon.

This review focuses primarily on the physiological self-renewal of the mammalian adult small intestine. [For discussion of the development and patterning of the gut during embryogenesis, the reader is referred to excellent recent reviews (3–5).] In addition, we discuss the molecular pathways that play a role in the differentiation of stem cells into specialized epithelial cells along the crypt-villus axis in the adult intestine. Powerful genetic tools have been developed in recent years through the use of mouse transgenesis. As a consequence, the murine intestine has rapidly developed into the model of choice for the study of the intestinal epithelium. Most of the work described here is based on gain- or loss-of-function studies in genetically modified mice. The use of the Cre-LoxP system makes it possible to (in)activate genes in specific organs of interest (6). For conditional recombination in the intestinal epithelium, two Cre lines are commonly used. These are the lipophilic, xenobiotic, inducible *Cyp1a* promoter (P450) Cre line (7) and the tamoxifen-inducible villin-Cre-ER^{T2} mice (8). Upon induction, both these Cre lines are active in the intestinal epithelium, including in the intestinal stem cells.

PROLIFERATING INTESTINAL CELLS

The primary driving force behind the proliferation of epithelial cells in the intestinal crypts is the Wnt pathway. **Figure 2** shows a schematic representation of the Wnt signaling pathway.

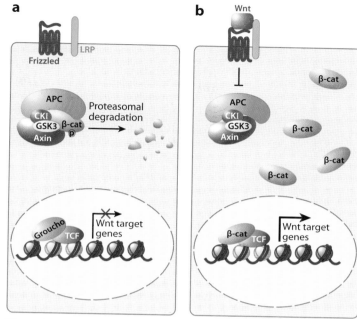

Figure 2

Schematic representation of the Wnt signaling pathway. (*a*) In the absence of Wnt stimulation, β-catenin (β-cat) levels in the cell are kept low through a dedicated destruction complex that consists of adenomatous polyposis coli (APC), casein kinase I (CKI), glycogen synthase kinase 3 (GSK3), and axin. This complex targets β-catenin for proteasomal degradation through phosphorylation. (*b*) An active Wnt signaling pathway. Wnt ligands bind to Frizzled receptors and low-density lipoprotein receptor–related protein (LRP) coreceptors. Consequently, the destruction complex no longer targets β-catenin for proteasomal degradation. The β-catenin level in the cell rises, resulting in the translocation of β-catenin into the nucleus. Here β-catenin replaces Groucho on T cell factor (TCF) transcription factors. β-Catenin/TCF form an active transcriptional complex, leading to the expression of Wnt target genes.

This pathway is highly conserved throughout the animal kingdom (9, 10). The central player in the canonical Wnt pathway is β-catenin. In the absence of a Wnt signal, β-catenin is targeted for proteasomal degradation through sequential phosphorylations occurring at its N terminus. A degradation complex, consisting of the tumor suppressors axin and adenomatous polyposis coli (APC) and the constitutively active kinases glycogen synthase kinase 3β and casein kinase I, regulates β-catenin phosphorylation status in a cell. When Wnt ligands signal through their Frizzled and low-density lipoprotein receptor–related protein (LRP) receptors, the destruction complex is inactivated

Cre-LoxP technology: a genetic method to control gene expression on the basis of short specific LoxP sequences that can site-specifically recombine through the Cre recombinase

Wnts: secreted (lipid modified) ligands for the Wnt signaling pathway; 19 different homologs exist in the mammalian genome

Canonical Wnt
pathway: regulates,
through a set of
evolutionarily highly
conserved proteins,
the interaction of
β-catenin with
TCF/LEF
transcription factors

APC: adenomatous
polyposis coli

TCF/LEF: T cell
factor/lymphocyte
enhancer factor

CRC: colorectal
cancer

in a not fully understood manner. As a result, β-catenin is no longer phosphorylated and accumulates in the cell. The coincident translocation of β-catenin into the nucleus results in the binding of β-catenin to transcription factors of the T cell factor/lymphocyte enhancer factor (TCF/LEF) family. TCF/LEF-β-catenin form an active transcriptional complex that activates target genes. In the absence of a Wnt signal, transcriptional repressors like Groucho bind TCF/LEF transcription factors (11, 12).

The Wnt pathway regulates its transcriptional target genes through TCF target sites located in promoters and/or enhancers. The optimal TCF binding site, AGATCAAAGG, is highly conserved between the four vertebrate TCF/LEF genes and *Drosophila* Tcf (13, 14). WNT/TCF reporter plasmids such as pTOPflash (15) are commonly used to measure Wnt pathway activation. These reporters consist of concatamers of the binding motif cloned upstream of a minimal promoter. A large variety of WNT/TCF target genes have been described since the discovery that this pathway represents the dominant force behind the proliferative activity of the healthy intestinal epithelium as well as behind colorectal cancer (CRC) (see below). A detailed list can be found at the Wnt home page hosted by R. Nusse (**http://www.stanford.edu/~rnusse/ pathways/targets.html**).

The advent of DNA array technology has made it possible to identify Wnt target genes on a genome-wide scale in situations in which the activity of the pathway can be manipulated (16–19). These studies show extensive overlap with microarray expression studies performed on crypt-derived RNA samples in mice (20) and humans (21). These studies have shown that intestine-specific Wnt target genes are expressed in proliferative crypt progenitors as well as in CRC cells.

Recent advances in chromatin analysis utilize a cross-linking technique to capture individual transcription factors bound on their cognate DNA motifs. Subsequent immunoprecipitation of the transcription factors and their associated chromatin reveals binding events

of these specific transcription factors on a genome-wide scale. The demonstration of the direct binding of TCF factors to regulatory elements in downstream genes makes it possible to distinguish whether a Wnt target is a direct or an indirect target and identifies the pertinent regulatory elements. Serial analysis of chromatin occupancy (SACO) identified 412 β-catenin-bound sites in HCT116 CRC cells (22). In another study, which used a genome-wide tiling array, analysis of TCF4-associated chromatin revealed almost 7000 TCF4-bound DNA elements in LS174T CRC cells. Testing of these TCF4-bound regions in luciferase-based reporter gene assays demonstrated that these genomic regions often behaved as Wnt-controlled enhancers or promoters (23). Strikingly, these two experimental studies compare well to each other but have hardly any overlap with DNA sites predicted through an algorithm-based Tcf4 binding site prediction, the enhancer element locator (EEL) bioinformatics tool (14).

Wnt target gene expression implies that the Wnt pathway is active in a gradient, with the highest activity at the crypt bottom. A recent study documents the expression pattern of all Wnts, Frizzleds, LRPs, Wnt antagonists, and TCFs in the intestinal epithelium, showing expression of multiple Wnts by the epithelial cells at the crypt bottom and implying active Wnt signaling (24).

Functional studies confirm that the Wnt pathway constitutes the master switch between proliferation and differentiation of the epithelial cells (17). Active Wnt signaling is essential for the maintenance of crypt progenitor compartments in the intestine. This is evidenced by mice lacking the Tcf4 transcription factor (25), by the conditional depletion of β-catenin from the intestinal epithelium (7, 26), and by transgenic inhibition of extracellular Wnt signaling through the secreted Dickkopf-1 Wnt inhibitor (27, 28). In all cases, a dramatic reduction of proliferative activity was observed. In the converse experiment, activating the Wnt pathway through transgenic expression of the Wnt agonist R-Spondin-1 resulted in a massive

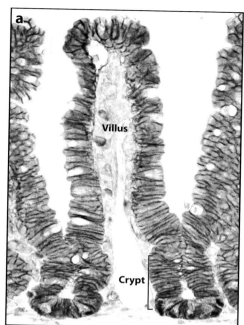 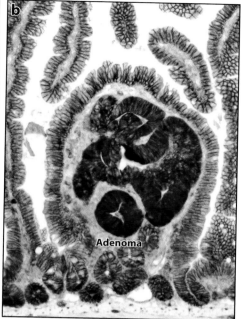

Figure 3

Active Wnt signaling in the intestine. Nuclear β-catenin is a hallmark for active Wnt signaling.
(*a*) β-Catenin staining in normal wild-type intestinal epithelium. On the villus only the membranes stain positive, and at the bottom of the crypts postmitotic Paneth cells also show nuclear β-catenin staining.
(*b*) Early adenoma tissue surrounded by wild-type crypt and villus epithelium in an Apc^{min} mouse shows elevated β-catenin levels in the adenoma cells, indicating hyperactive Wnt signaling.

hyperproliferation of intestinal crypts (29). Wnt signals in the crypt not only control proliferation of transit-amplifying progenitors but also are utilized by postmitotic Paneth cells for their terminal maturation (30). We discuss this topic in more detail below. **Figure 3** shows nuclear β-catenin staining, a hallmark for active Wnt signaling, in both the normal intestine (**Figure 3***a*) and an Apc mutant mouse adenoma (**Figure 3***b*).

Stem Cells

It has been known for decades that pluripotent stem cells fuel the proliferative activity of crypts (31, 32). Every crypt is commonly believed to contain approximately six independent stem cells; two schools of thought—the classic model and the stem cell zone model—define the exact identity of these stem cells. The small

intestinal crypt has been viewed as a tube of proliferating cells limited from below by Paneth cells. Since the late 1950s, the classic model has therefore proposed that the stem cells reside at position +4 relative to the crypt bottom. Terminally differentiated Paneth cells occupy the first three positions. Potten and colleagues have championed the +4 stem cell model; they have reported that DNA label–retaining cells reside specifically at this position (33). Additionally, these researchers have observed that the +4 cells are unusually radiation-sensitive, a property proposed to functionally protect the stem cell compartment from genetic damage (34). Damaged stem cells can in this model be replaced by the first 2–3 generations of transit-amplifying cells, which would fall back into the +4 position while regaining stem cell properties. The +4 cells are actively cycling. Label retention by the +4 cell would result from

Crypt base columnar
(CBC) cells: the
intestinal stem cells
found at the crypt
base; have Lgr5/Gpr49
expression

asymmetric segregation of old and new DNA strands (34, 35). Definitive proof of stemness requires that putative stem cells be experimentally linkable to their progeny. The current literature gives no insight into the identity of the cellular progeny of +4 cells. Therefore, the position of the +4 cells in the epithelial hierarchy is not clear. The second school of thought began with the identification more than 30 years ago of crypt base columnar (CBC) cells, which are small, undifferentiated, cycling cells squeezed in between the Paneth cells (20, 36–40). Originally on the basis of morphological considerations as well as, more recently, on clonal marking techniques, Leblond, Cheng, and Bjerknes have proposed that the CBC cells may represent the crypt stem cells.

One recent method to study clonal relations within crypts involves the use of inherited methylation patterns in single adult crypts. These studies have shown that methylation increases with aging, varies between crypts, and is mosaic within single crypts. Modeling of the data predicts multiple active stem cells in human adult colon crypts (41). These stem cells are proposed to give rise to new stem cells and transit-amplifying daughters in a stochastic manner. It is predicted that approximately 95% of the time a stem cell divides asymmetrically, resulting in one stem cell and one transit-amplifying daughter. In the remaining 5% of cases, a stem cell either becomes extinct (both daughter cells differentiate) or duplicates (both daughters remain as stem cells) (42). So, in each crypt stem cells can be lost and replaced randomly in time. Entire crypts can die out. Alternatively, the intestinal epithelium can repopulate itself through crypt fission (43).

Analyses of chimeric mice confirm the notion that crypts are polyclonal but over time become monoclonal (44). The villi receive epithelial cells from multiple crypts throughout life and therefore are by definition polyclonal. A number of human studies based on natural polymorphisms or somatic mutations have provided further proof that long-lived crypts are clonal (45–47). A study by Taylor and colleagues, in contrast, has shown that mutations

generating biochemical defects of cytochrome *c* oxidase create ribbons of mutant cells emanating from crypts, implying that several stem cells are simultaneously active (48). Such chimeric crypts may become monoclonal over time.

To definitively identify stem cells, one needs to be able to identify and/or mark such cells. Unfortunately, no specific marker has been available until very recently. Clonal marking of intestinal epithelial cells, through chemical mutagenesis studies, can be used to genetically mark intestinal epithelial cells by somatic mutations of the *Dlb-1* locus (49, 50). This technique was used to show that mutations in the intestinal epithelium occur in short-lived progenitors yielding one or two different cell types, as well as in long-lived cell progenitors capable of giving rise to all epithelial cells (38). Unfortunately, in this method it is not clear which cell sustains the first clonal mutation.

As described above, the Wnt pathway is the dominant force behind the proliferative activity of the intestinal epithelium both in its physiology and in CRC. For this reason, we began to study the target gene program activated by this pathway (17, 18). Through an in situ hybridization approach, we found that most of these target genes are expressed both in adenomas and in the proliferative wild-type crypts. However, a subset of these genes appeared to be restricted to a limited number of cells in the crypts (18).

One of these latter genes is *Lgr5/Gpr49*. Lgr5 is an orphan G protein–coupled receptor. Both in situ hybridization experiments as well as a LacZ knock-in allele reveal expression of Lgr5 to be restricted to the CBC cells, an observation confirmed by an enhanced green fluorescent protein (EGFP) knock-in allele (51). The EGFP in this knock-in allele is followed by CreERT2, a tamoxifen-inducible version of the Cre recombinase. A cross of this mouse strain with a strain in which the Rosa26R-LacZ reporter can be activated by Cre results in a genetic model in which tamoxifen induction genetically marks the CBC cells because the Cre induction irreversibly activates the genetic marker Rosa26R-LacZ by excision of a roadblock DNA sequence (52). **Figure 4a**

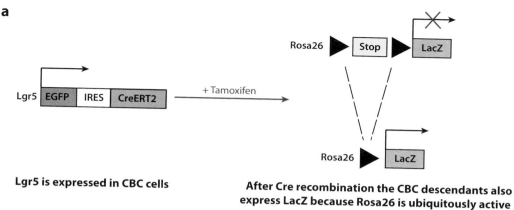

Lgr5 is expressed in CBC cells

After Cre recombination the CBC descendants also express LacZ because Rosa26 is ubiquitously active

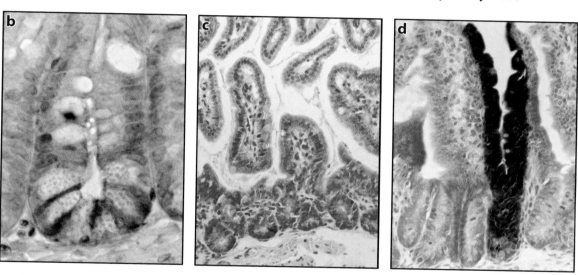

Figure 4

Lgr5 is an intestinal stem cell marker. (*a*) Genetic marking strategy to prove that Lgr5-expressing cells in the intestine are stem cells. Lgr5 knock-in mice, in which an EGFP-IRES-CreERT2 is cloned in frame with the ATG of the Lgr5 gene, are crossed with a strain in which the Rosa26R-LacZ reporter can be activated by Cre. The Rosa26R-LacZ strain by itself does not express LacZ because of a stop sequence in front of the LacZ gene. Tamoxifen-induced Cre activity in Lgr5-expressing cells will irreversibly activate LacZ by excision of the roadblock DNA sequence in the reporter strain. This genetic mark will be inherited by the daughter cells of Lgr5-positive cells because Rosa26 is ubiquitously expressed. (*b*) GFP staining of an Lgr5 knock-in allele showing that Lgr5 is expressed only in the crypt base columnar (CBC) cells that are located in between Paneth cells. (*c*) LacZ staining one day after tamoxifen induction in an Lgr5 EGFP-IRES-CreERT2 knock-in mouse crossed with a Rosa26R-LacZ reporter mouse. Single cells at the bottom of the crypts, which have not had time to divide, stain for LacZ. (*d*) LacZ staining six months after a single tamoxifen induction in an Lgr5 EGFP-IRES-CreERT2 knock-in mouse crossed with a Rosa26R-LacZ reporter mouse shows that crypts and adjacent villi are still lined with ribbons of LacZ-positive cells, proving that all these cells are descended from Lgr5-positive stem cells.

shows a schematic representation of this genetic lineage tracing experiment. The EGFP expression shows that Lgr5 is expressed only in CBC cells (**Figure 4***b*). Analysis at different time points after tamoxifen induction allows genetic tracing of the marked cells. Indeed, one day after Cre induction, only cells at the crypts base stain blue for LacZ (**Figure 4***c*). At later time points, blue ribbons of cells emanate from the crypts and run up the sides of the villi. These

Lineage tracing: the genetic marking of cells through the irreversibly excision of a roadblock DNA sequence through (inducible) Cre recombination, resulting in the expression of a marker protein (LacZ or a fluorescent protein)

Stem cell niche: the supporting microenvironment in which stem cells are found. In the intestine the niche is most likely formed by the subepithelial myofibroblasts

ribbons can still be observed six months after a single tamoxifen induction (**Figure 4d**). The clonal blue ribbons contain all four differentiated cell types of the intestine, proving that Lgr5 is a long-lived, pluripotent intestinal stem cell marker in both the small intestine and the colon (51).

Intestinal Stem Cell Niche

The epithelial homeostasis of the intestine is based on a delicate balance between self-renewal and differentiation, which must be maintained throughout life. The stemness of the CBC cells in the intestine is most likely not an intrinsic property but a nonautonomous feature and is best explained by a niche model (a niche is a limited microenvironment that supports stem cells) (53). The niche for the intestinal epithelium is probably formed by the tightly associated sheath of intestinal pericryptal fibroblasts, also called subepithelial myofibroblasts (54). These cells are believed to secrete various putative growth factors and cytokines that promote epithelial proliferation (55).

Bone morphogenetic protein (BMP) signals appear to be excluded from the crypt. BMP-2 and -4 ligands are expressed in the mesenchyme of villi (56, 57). BMPs, when bound to their receptors, transduce a signal from cytoplasm to nucleus through receptor-mediated phosphorylation of SMAD transcription factors (58). The BMP signaling pathway functions as a negative regulator of crypts. Conditional deletion of Bmp receptor 1A results in hyperproliferative crypts (59). Moreover, inhibition of BMP signaling on the villus through overexpression of the BMP inhibitor Noggin results in ectopic crypt formation (56). De novo crypt formation throughout the epithelium is also seen in juvenile polyposis patients, of which at least 50% carry germ-line mutations in one of the various components of the BMP signaling pathway (60–62).

Several mice that are mutant for genes expressed in the mesenchyme show an increase in proliferative cells in their intestinal epithelium, indicating that the proteins encoded by these genes work as negative regulators of crypt cell proliferation. Examples of these mesenchymal proteins are the forkhead homolog 6 (Fkh6) (63); the homeodomain transcription factor Nkx2-3 (64); and Epimorphin, a member of the syntaxin family of membrane-bound intracellular vesicle-docking proteins (65). These phenotypic effects are possibly mediated through the BMP signaling pathway because mRNA levels of both Bmp-2 and Bmp-4 are decreased in all three null mice.

Epithelium-to-mesenchyme interactions also take place in the opposite direction. Both the hedgehog pathway and the platelet-derived growth factor-A (Pdgf-A) and its receptor (Pdgfr-α) seem to play an important role hereby. Epithelial cells express the Sonic hedgehog (Shh) and Indian hedgehog (Ihh) ligands. These ligands signal to Patched 1 and Patched 2 receptors, which are expressed by the underlying mesenchyme (66). Inhibition of the hedgehog pathway through overexpression of a pan-hedgehog inhibitor results in the reduction of villi and a hyperproliferative epithelium (66). Shh and Ihh knockouts die shortly after birth, and for this reason only intestines from animals one day prior to birth have been analyzed. Both mutant mice show complex effects on embryonic gut development, with a reduced smooth muscle layer surrounding the gut and several other intestinal malformations (67). Pdgf-A, like the hedgehog ligands, is expressed by epithelial cells, although its receptor is expressed in the mesenchyme. Mice lacking Pdgf-A or Pdgfr-α develop fewer and misshapen villi (68).

CELL LINEAGE SPECIFICATION

The rare stem cells in the crypts give rise to a much larger pool of transit-amplifying cells. The transcription factor c-Myc is expressed by all cycling cells in the intestine and was the first identified target gene of the Wnt pathway in CRC cells (69). Conditional deletion of c-Myc in the intestinal epithelial results in a loss of c-Myc-deficient crypts (70). These crypts are rapidly replaced by c-Myc-proficient crypts through crypt fission of the latter within weeks. Myc$^{-/-}$ crypt cells remain in cycle, but they are smaller, cycle slower, and divide at a smaller size compared with wild-type crypts. Myc thus appears to be essential for the biosynthetic capacity of crypt progenitor cells to successfully progress through the cell cycle (70). Strikingly, although loss of Apc in the intestinal epithelium leads to the immediate transformation of the epithelium, this phenotype is fully rescued if c-Myc is simultaneously deleted. This implies that c-Myc plays a central role in the Wnt-driven target gene program in intestinal cancer (71).

Secretory Lineages versus Absorptive Lineage: The Notch Pathway

As discussed above, transit-amplifying cells terminally differentiate into one of the four principal epithelial cell lineages of the gastrointestinal tract. Three of these cell types—the goblet cells, the enteroendocrine cells, and the Paneth cells—belong to the secretory lineage. Absorptive enterocytes represent the fourth cell type. Besides these main four cell types, researchers have described some lesser-known cell types such as, e.g., M-cells and Brush cells (72, 73). In the following paragraphs, we focus on the molecular signals that regulate the fates of the four main cell types of the intestinal epithelium. Once determined, some lineages use reiteration of signaling pathways during terminal differentiation. Other lineages appear to utilize parallel signals that are important for full mature differentiation. **Figure 5** gives a schematic overview of the genetic hierarchy in cell lineage specification in the intestine.

The Notch pathway plays a central function in these intestinal cell fate decisions. Notch genes encode single transmembrane receptors that regulate a broad spectrum of cell fate decisions and differentiation processes during animal development (74). Interaction of one of the four Notch receptors with any one of five Notch ligands results in proteolytic cleavage of the receptor within the plane of the cell membrane. A key step in the cleavage process involves the activity of the gamma-secretase protease complex. The resulting free Notch intracellular domain (NICD) translocates into the nucleus. Here NICD binds to the transcription factor RBP-Jκ (CSL or CBF1) to activate target gene transcription (75).

Like Wnt signaling, the Notch pathway is essential to maintain the crypt compartment in its undifferentiated, proliferative state. Inhibition of the Notch pathway in the intestinal epithelium by conditional deletion of the CSL gene or through pharmacological gamma-secretase inhibitors results in the rapid and complete conversion of all epithelial cells into goblet cells (76, 77). The same effect is seen in intestinal tumors in Apc^{min} mice (78). Moreover, Cre-mediated lineage tracing of Notch1 activity in the intestine shows clonal ribbons of cells, indicating that Notch1 signaling is active in adult intestinal stem cells (79). Both the Notch1 and Notch2 receptors mediate Notch signals in the intestinal epithelium. These receptors work redundantly because only conditional inactivation of both these receptors in the gut results in the complete conversion of the proliferative crypt cells into postmitotic goblet cells (80). Gain of function through specific overexpression of a constitutively active Notch1 receptor in the intestinal epithelium results in the opposite effect, a depletion of goblet cells and a reduction in enteroendocrine and Paneth cell differentiation (81, 82). Thus, the Notch pathway controls absorptive versus secretory fate decisions in the intestinal epithelium. Of note, no mutational alterations in Notch signaling in intestinal tumorigenesis have been described.

A conserved feature of Notch signaling involves the regulation of the downstream effectors of Notch. Typically, active Notch signaling results in transcription of a first tier of genes of the Hairy/Enhancer of Split (Hes) class that encode trancriptional repressors. Hes repressors in turn repress transcription of a second tier of genes, typically basic helix-loop-helix (bHLH) transcription factors that, when derepressed, induce differentiation along specific lineages. In the intestine, the direct Notch target gene

Notch pathway: a highly conserved signaling pathway involved in cell-cell communication

bHLH: basic
helix-loop-helix

Hes1 represses transcription of the bHLH transcription factor Math1 (83). Hes1$^{-/-}$ animals are embryonic lethal, but intestines from these animals show an increase in Paneth, goblet, and enteroendocrine cells and a decrease in absorptive enterocytes (83, 84). Intestinal Math1 expression is required for commitment toward the secretory lineage because the epithelium of Math1 mutant mice is populated only by enterocytes (85). The zinc-finger transcriptional repressor Gfi1 is expressed in the secretory lineage of the intestine. Gfi1 is absent in

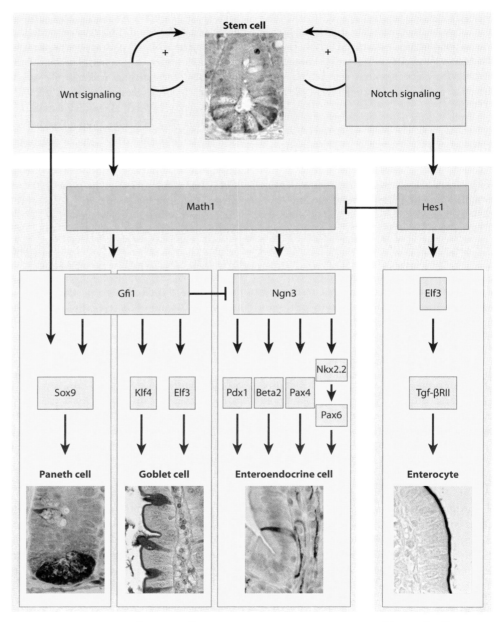

Math1$^{-/-}$ embryonic intestines, implying that it acts downstream of Math1. Gfi1$^{-/-}$ mice have no Paneth cells and display a clear reduction in the number of goblet cells. There is, however, an increase in the number of enteroendocrine cells (86). Mtgr1, a transcriptional corepressor, is required for secretory lineage maintenance. Mtgr1 null mice display a progressive depletion of the secretory lineage. Where Mtgr1 acts in the genetic hierarchy is not clear (for this reason Mtgr1 is not represented in **Figure 5**). The Mtgr1 null phenotype includes a loss of enteroendocrine cells, suggesting that Mtgr1 works upstream of Gfi1. However, Gfi1-positive progenitors are maintained in the absence of Mtgr1 (87).

The Wnt pathway also influences cell lineage specification of the intestine. Mouse models with impaired Wnt signaling show reductions of Math1-positive precursors in intestinal crypts. As a consequence, the secretory lineages are depleted, and the villi are lined mainly with enterocytes (7, 25, 28).

Goblet Cells

Goblet cells secrete protective mucins and trefoil proteins that are required for the movement and effective expulsion of gut contents, and provide protection against shear stress and chemical damage. The proportion of goblet cells among all epithelial cell types increases from the duodenum (~4%) to the descending colon (~16%) (88).

As discussed above, loss of function of Notch in the intestinal epithelium results in a massive conversion of epithelial cells into goblet cells (78, 89). Targeted deletion of kruppel-like factor 4 (Klf4), a zinc-finger transcription factor, results in the loss of goblet cells (90). Targeted inactivation of the most abundant secreted gastrointestinal mucin, Muc2, results in the absence of Alcian blue staining, a widely used goblet cell marker. Yet, the cells still express intestinal trefoil factor, suggesting that at least some aspects of the differentiation program of goblet cell lineage persist in Muc2$^{-/-}$ mice. Over time, Muc2 null animals frequently develop adenomas in the small intestine that progress to invasive adenocarcinomas (91).

Enteroendocrine Cells

Enteroendocrine, or neuroendocrine, cells coordinate gut functioning through specific peptide hormone secretion. There are up to 15 different subtypes of enteroendocrine cells defined by their morphology and expression of specific intestinal hormones or marker gene expression. Enteroendocrine cells are scattered as individual cells throughout the mucosa, representing approximately 1% of the cells lining the intestinal lumen (92).

As discussed above, the Notch pathway is important for the development of the secretory cell lineages of the intestine. Downstream of the Notch-Hes1-Math1 signaling cascade, Neurogenin3 (Ngn3) is required for endocrine cell fate specification. Mice homozygous for a null mutation in this bHLH transcription factor do not develop any intestinal endocrine cells (93). It is thought that, downstream of

Figure 5

Schematic overview of the genetic hierarchy of epithelial cell lineage commitment in the intestine. Intestinal stem cells proliferate under control of both the Wnt and the Notch pathways and can differentiate into all four differentiated cell types present in the intestinal epithelium. Math1 is required for the commitment to the secretory lineage. Gfi1 and Sox9 are responsible for differentiation into Paneth cells. Gfi1, kruppel-like factor 4 (Klf4), and E47-like factor 3 (Elf3) are necessary for goblet cell development. Neurogenin3 (Ngn3) is required for endocrine cell fate specification. Downstream of Ngn3, a set of transcription factors is responsible for the specification of the various enteroendocrine hormone–expressing cell types. Hairy/Enhancer of Split 1 (Hes1), through Elf3 and the transforming growth factor β type II receptor (Tgf-βRII), is responsible for differentiation into enterocytes of the absorptive lineage. The involvement of the various (transcription) factors is based on genetic experiments in mice and is discussed in more detail in the text.

Ngn3, a set of transcription factors is responsible for the specification of the various enteroendocrine hormone–expressing cell types. For secretin and cholecystokinin enteroendocrine cells, the bHLH transcription factor Beta2 (also called NeuroD) appears crucial; homozygous Beta2$^{-/-}$ mice do not have enteroendocrine cells expressing these particular hormones (94). Other genes that have been implicated in endocrine cell fate specification are the homeodomain transcription factors Pdx-1, Nkx2.2, Pax4, and Pax6 (95–98). Nkx2.2 appears to function upstream of Pax6 in enteroendocrine cell specification (95).

Paneth Cells

Paneth cells reside at the crypt base and have a function in innate immunity. They contain large apical secretory granules that contain specific proteins, including lysozymes, antimicrobials, and defensins, which likely relates to the abundance of the gastrointestinal flora. Paneth cells have a life expectancy of at least three weeks (36), much longer than that of their terminally differentiated villus counterparts. Unexpectedly, Paneth cells undergo active Wnt signaling, but they interpret these signals for their maturation and not for their proliferation (30). Labeling studies using ^{3}H-thymidine show that Paneth cells are formed around positions 5–7 of the crypt and subsequently migrate down to position 1. Thus, the Paneth cells at the bottom position have much bigger granules and are the cells showing the most degeneration (36). Paneth cells are the only differentiated intestinal epithelial cell type that migrates downward to the crypt bottom. This compartmentalization is achieved through the tyrosine kinase guidance receptors EphB2 and -3, both TCF targets. These tyrosine receptors generate repulsive forces when interacting with their ephrin-B ligands, which are expressed to high levels by villus cells. Paneth cells in both EphB3$^{-/-}$ or conditional ephrin-B1 knockout mice do not home to crypt bottoms but scatter along the crypt-villus axis (99, 100). The Frizzled-5 receptor is responsible for the Wnt activation in the Paneth cells because conditional deletion of this receptor results in EphB3-negative Paneth cells, scattered along the villus (30).

Sox9, a Wnt target (101), plays a role in the differentiation of Paneth cells in the intestinal epithelium. Conditional deletion of Sox9 in the intestinal epithelial cells results in a complete absence of Paneth cells. The Paneth cells are replaced by cycling, KI67-positive cells. Moreover, the crypts are wider in Sox9-deficient animals than in their wild-type counterparts (102, 103).

Absorptive Lineage: Enterocytes

Enterocytes, alternatively termed columnar cells, are highly polarized cells carrying an apical brush border that is responsible for absorbing and transporting nutrients across the epithelium. Enterocytes make up more than 80% of all intestinal epithelial cells.

As already mentioned above, Hes1$^{-/-}$ animals show a decrease in absorptive enterocytes (83). E47-like factor 3 (Elf3), a member of the Ets transcription family, appears to play a role in enterocyte and goblet cell differentiation. Mice homozygous for an Elf3 null mutation die shortly after birth and display poorly polarized enterocytes that have not reached maturity (104). The block in enterocyte differentiation in Elf3$^{-/-}$ mice can be genetically rescued through overexpression of transforming growth factor β type II receptor (Tgf-βRII) (105). Mice deficient for protein tyrosine kinase 6 (Ptk6) show a delay in expression of intestinal fatty acid binding protein, suggesting a role for Ptk6 in enterocyte differentiation (106).

COLORECTAL CANCER

Although Wnt signaling is very important in the normal physiology of the intestine, it was first characterized for its association with CRC, one of the most common cancers in industrialized countries. Activating mutations in the Wnt pathway initiate the overwhelming majority of CRC cases (107). These mutations either remove the tumor suppressors APC (108–110) or axin2 (111) or activate the proto-oncogene β-catenin (112, 113). As a common result, β-catenin accumulates in the nucleus and constitutively binds to the TCF/LEF transcription factors. Tcf4 is physiologically expressed in all epithelial cells of the intestine (114). The inappropriate formation of β-catenin/TCF4 complexes results in transcriptional activation of WNT/TCF4 target genes, initiating the transformation of intestinal epithelial cells (15, 112).

Fearon & Vogelstein (115) have proposed that CRC arises through an ordered sequence of mutations in what is called the adenoma-carcinoma sequence. Invariably, the initiating mutation occurs in the Wnt pathway, leading to the formation of benign yet long-lived adenomas. Subsequently, other mutations follow, e.g., in the Kras, Smad4, and p53 genes, ultimately resulting in metastasizing carcinomas (115).

Mouse Models

Activating Wnt pathway mutations result in adenoma formation in mice. A mutagenesis study discovered the first mouse strain with hereditary adenoma formation (116). This strain was called multiple intestinal neoplasia (Min) mice and turned out to harbor a heterozygous point mutation in the *Apc* gene. It is thus the murine counterpart of hereditary familial adenomatous polyposis in humans. Because of spontaneous mutation of the remaining wild-type *Apc* allele in intestinal epithelial cells, these Min mice develop multiple adenomas in the intestine within months (117). Several Apc knockout mouse models have been generated since, with different patterns of (extra)intestinal tumor formation (118). Conditional deletion of Apc in murine intestines leads to the rapid accumulation of nuclear β-catenin and the consequent transformation of the entire intestinal epithelium (16, 119, 120).

Another way of mutationally activating the Wnt pathway involves the stabilization of β-catenin. Conditional deletion of the third exon of β-catenin, which encodes the phosphorylated residues that control its stability, also re-sults in polyp formation in mouse intestine (121). At the molecular level, the various murine models are very good models for human adenoma formation, with one distinction: The adenomas in mice are preferably formed in the small intestine, whereas in humans they are almost exclusively found in the colon both in sporadic CRC as well as in hereditary familial adenomatous polyposis. The exact reason for this difference is unknown.

Cancer Stem Cells

CRC develops from a single epithelial cell, typically mutated in its APC gene, that gains the ability to proliferate aberrantly and eventually turn malignant. One central question in colon cancer biology still to be answered is the identity of the cells that sustain the initiating mutation. Can the initiating mutation occur in any cell, or is it mandatory that a stem cell be hit? And as soon as cells are transformed, is a stem/progenitor hierarchical organization established within the newly formed tumors? In a stochastic model, every tumor cell has equal tumor-initiating potential. However, according to the cancer stem cell theory, tumors are generated and maintained by a small defined subset of undifferentiated cells able to self-renew and differentiate into the bulk tumor population. These cells are termed tumor-initiating cells or cancer stem cells (122–124). Two recent publications favor the last model. Sorting human tumor cells on the basis of CD133 expression shows that the positive fraction has up to 200-fold more tumor-initiating potential in immunodeficient mice compared with CD133-negative fractions (125, 126). The interpretation of these data may be complicated by the fact that the assay involves xenotransplantation of human tumor cells into immunodeficient mice. A recent study in which lymphomas and leukemias of mouse origin were transplanted into histocompatible mice demonstrates that most tumor cells can seed tumor growth when not challenged by a species barrier. The low frequency of tumor-sustaining cells observed in xenotransplantation studies may reflect the limited ability of human tumor cells to adapt to growth in a foreign mouse milieu (127).

CONCLUDING REMARKS

In this review, we provide an overview and update of our current understanding of the biology of the intestinal epithelium. Recent genetic studies have created a wealth of new insights into the biology of the most rapidly self-renewing tissue of the adult mammalian body. Wnt and Notch, previously believed to uniquely control fate decisions during embryonic development, are the principal controllers of the proliferation/differentiation rheostat. Activating mutations in the Wnt cascade are the driving force behind colon cancer. A sophisticated genetic toolbox has been assembled to study in great detail the genes that control all aspects of the biology of the intestinal epithelium. These tools include a number of mouse strains that now allow the visualization, isolation, and genetic modification of the intestinal stem cells. Theories about the role of stem cells in cancer initiation and progression can now be tested.

SUMMARY POINTS

1. The intestine is lined with a rapidly self-renewing epithelium. It is organized into crypts (invaginations containing stem cells and transit-amplifying cells) and villi (protrusions that are covered with terminally differentiated cells).

2. The Wnt signaling pathway is the dominant force underlying the proliferative activity of the intestinal epithelium in both normal physiology and CRC development. The Notch pathway is needed to maintain proliferative progenitors and specific differentiated cell fates.

3. The proliferative crypt compartment of the intestine harbors CBC cells, pluripotent and long-lived stem cells that give rise to all differentiated cell types of the intestinal epithelium. The intestinal epithelium self-renews every 3–5 days.

4. Lgr5, an orphan G protein–coupled receptor, is the first definitive adult intestinal stem cell marker. Lineage tracing experiments using an inducible Cre knocked into the Lgr5 locus generate ribbons of epithelial cells of all types and genetically marked by LacZ expression for up to a year after a single Cre induction.

5. From stem cells in the crypts, a pool of transit-amplifying cells emanates. These cells will (after 2–3 days) terminally differentiate into one of the four principal epithelial cell lineages of the intestine.

6. The cancer stem cell concept proposes that tumors are generated and maintained by a small, defined subset of undifferentiated cells that are able to self-renew and differentiate into the bulk tumor population. These may be closely related to the CBC cells.

FUTURE ISSUES

1. With the identification of Lgr5 as the first definitive intestinal stem cell marker, it is now possible to visualize and isolate intestinal stem cells, study their behavior, and establish their stem cell–specific transcriptome. Novel stem cell markers should emerge from these studies.

2. The field direly lacks assays with which to study stem cells in culture or by transplantation. Such assays will be established in the near future.

3. Future efforts should be directed toward the definition of the crypt niche. It appears mandatory to identify unique marker genes for the niche cells to study their role in crypt biology and in cancer.

4. Through use of the inducible Cre recombinase integrated into the Lgr5 locus, it will be possible to genetically modify intestinal stem cells specifically for future studies. The same Cre knock-in can be used to mark Lgr5-positive cells within tumors and to study their potential cancer stem cell potential in situ.

5. An ever-increasing number of potential regulators of proliferation versus differentiation emerges from microarray studies in various models of intestinal biology. It is to be expected that—one by one—these regulators will be evaluated by what has become the standard approach, i.e., inducible deletion in the adult intestine.

DISCLOSURE STATEMENT

The authors are not aware of any biases that might be perceived as affecting the objectivity of this review.

ACKNOWLEDGMENTS

We thank Jeroen Kuipers, Nick Barker, and Hugo Snippert for their help with figure preparation.

LITERATURE CITED

1. Heath JP. 1996. Epithelial cell migration in the intestine. *Cell Biol. Int.* 20:139–46
2. Hall PA, Coates PJ, Ansari B, Hopwood D. 1994. Regulation of cell number in the mammalian gastrointestinal tract: the importance of apoptosis. *J. Cell Sci.* 107(Pt. 12):3569–77
3. Roberts DJ. 2000. Molecular mechanisms of development of the gastrointestinal tract. *Dev. Dyn.* 219:109–20
4. Beck F, Tata F, Chawengsaksophak K. 2000. Homeobox genes and gut development. *Bioessays* 22:431–41
5. Gregorieff A, Clevers H. 2005. Wnt signaling in the intestinal epithelium: from endoderm to cancer. *Genes Dev.* 19:877–90
6. Sauer B. 1998. Inducible gene targeting in mice using the Cre/lox system. *Methods* 14:381–92
7. Ireland H, Kemp R, Houghton C, Howard L, Clarke AR, et al. 2004. Inducible Cre-mediated control of gene expression in the murine gastrointestinal tract: effect of loss of β-catenin. *Gastroenterology* 126:1236–46
8. el Marjou F, Janssen KP, Chang BH, Li M, Hindie V, et al. 2004. Tissue-specific and inducible Cre-mediated recombination in the gut epithelium. *Genesis* 39:186–93
9. Clevers H. 2006. Wnt/β-catenin signaling in development and disease. *Cell* 127:469–80
10. Logan CY, Nusse R. 2004. The Wnt signaling pathway in development and disease. *Annu. Rev. Cell Dev. Biol.* 20:781–810
11. Cavallo RA, Cox RT, Moline MM, Roose J, Polevoy GA, et al. 1998. *Drosophila* Tcf and Groucho interact to repress Wingless signalling activity. *Nature* 395:604–8
12. Roose J, Molenaar M, Peterson J, Hurenkamp J, Brantjes H, et al. 1998. The *Xenopus* Wnt effector XTcf-3 interacts with Groucho-related transcriptional repressors. *Nature* 395:608–12

13. van de Wetering M, Cavallo R, Dooijes D, van Beest M, van Es J, et al. 1997. Armadillo coactivates transcription driven by the product of the *Drosophila* segment polarity gene *dTCF*. *Cell* 88:789–99

14. Hallikas O, Palin K, Sinjushina N, Rautiainen R, Partanen J, et al. 2006. Genome-wide prediction of mammalian enhancers based on analysis of transcription-factor binding affinity. *Cell* 124:47–59

15. Korinek V, Barker N, Morin PJ, van Wichen D, de Weger R, et al. 1997. Constitutive transcriptional activation by a β-catenin-Tcf complex in APC$^{-/-}$ colon carcinoma. *Science* 275:1784–87

16. Sansom OJ, Reed KR, Hayes AJ, Ireland H, Brinkmann H, et al. 2004. Loss of Apc in vivo immediately perturbs Wnt signaling, differentiation, and migration. *Genes Dev.* 18:1385–90

17. van de Wetering M, Sancho E, Verweij C, de Lau W, Oving I, et al. 2002. The β-catenin/TCF-4 complex imposes a crypt progenitor phenotype on colorectal cancer cells. *Cell* 111:241–50

18. van der Flier LG, Sabates-Bellver J, Oving I, Haegebarth A, De Palo M, et al. 2007. The intestinal Wnt/TCF signature. *Gastroenterology* 132:628–32

19. Gaspar C, Cardoso J, Franken P, Molenaar L, Morreau H, et al. 2008. Cross-species comparison of human and mouse intestinal polyps reveals conserved mechanisms in adenomatous polyposis coli (APC)-driven tumorigenesis. *Am. J. Pathol.* 172:1363–80

20. Stappenbeck TS, Mills JC, Gordon JI. 2003. Molecular features of adult mouse small intestinal epithelial progenitors. *Proc. Natl. Acad. Sci. USA* 100:1004–9

21. Kosinski C, Li VS, Chan AS, Zhang J, Ho C, et al. 2007. Gene expression patterns of human colon tops and basal crypts and BMP antagonists as intestinal stem cell niche factors. *Proc. Natl. Acad. Sci. USA* 104:15418–23

22. Yochum GS, McWeeney S, Rajaraman V, Cleland R, Peters S, Goodman RH. 2007. Serial analysis of chromatin occupancy identifies β-catenin target genes in colorectal carcinoma cells. *Proc. Natl. Acad. Sci. USA* 104:3324–29

23. Hatzis P, van der Flier LG, van Driel MA, Guryev V, Nielsen F, et al. 2008. Genome-wide pattern of TCF7L2/TCF4 chromatin occupancy in colorectal cancer cells. *Mol. Cell. Biol.* 28:2732–44

24. Gregorieff A, Pinto D, Begthel H, Destree O, Kielman M, Clevers H. 2005. Expression pattern of Wnt signaling components in the adult intestine. *Gastroenterology* 129:626–38

25. Korinek V, Barker N, Moerer P, van Donselaar E, Huls G, et al. 1998. Depletion of epithelial stem-cell compartments in the small intestine of mice lacking Tcf-4. *Nat. Genet.* 19:379–83

26. Fevr T, Robine S, Louvard D, Huelsken J. 2007. Wnt/β-catenin is essential for intestinal homeostasis and maintenance of intestinal stem cells. *Mol. Cell. Biol.* 27:7551–59

27. Kuhnert F, Davis CR, Wang HT, Chu P, Lee M, et al. 2004. Essential requirement for Wnt signaling in proliferation of adult small intestine and colon revealed by adenoviral expression of Dickkopf-1. *Proc. Natl. Acad. Sci. USA* 101:266–71

28. Pinto D, Gregorieff A, Begthel H, Clevers H. 2003. Canonical Wnt signals are essential for homeostasis of the intestinal epithelium. *Genes Dev.* 17:1709–13

29. Kim KA, Kakitani M, Zhao J, Oshima T, Tang T, et al. 2005. Mitogenic influence of human R-spondin1 on the intestinal epithelium. *Science* 309:1256–59

30. van Es JH, Jay P, Gregorieff A, van Gijn ME, Jonkheer S, et al. 2005. Wnt signalling induces maturation of Paneth cells in intestinal crypts. *Nat. Cell Biol.* 7:381–86

31. Bjerknes M, Cheng H. 2006. Intestinal epithelial stem cells and progenitors. *Methods Enzymol.* 419:337–83

32. Marshman E, Booth C, Potten CS. 2002. The intestinal epithelial stem cell. *Bioessays* 24:91–98

33. Potten CS, Kovacs L, Hamilton E. 1974. Continuous labelling studies on mouse skin and intestine. *Cell Tissue Kinet.* 7:271–83

34. Potten CS. 1977. Extreme sensitivity of some intestinal crypt cells to X and gamma irradiation. *Nature* 269:518–21

35. Potten CS, Owen G, Booth D. 2002. Intestinal stem cells protect their genome by selective segregation of template DNA strands. *J. Cell Sci.* 115:2381–88

36. Bjerknes M, Cheng H. 1981. The stem-cell zone of the small intestinal epithelium. I. Evidence from Paneth cells in the adult mouse. *Am. J. Anat.* 160:51–63

37. Bjerknes M, Cheng H. 1981. The stem-cell zone of the small intestinal epithelium. III. Evidence from columnar, enteroendocrine, and mucous cells in the adult mouse. *Am. J. Anat.* 160:77–91

38. Bjerknes M, Cheng H. 1999. Clonal analysis of mouse intestinal epithelial progenitors. *Gastroenterology* 116:7–14

39. Cheng H, Leblond CP. 1974. Origin, differentiation and renewal of the four main epithelial cell types in the mouse small intestine. V. Unitarian theory of the origin of the four epithelial cell types. *Am. J. Anat.* 141:537–61

40. Cheng H, Leblond CP. 1974. Origin, differentiation and renewal of the four main epithelial cell types in the mouse small intestine. I. Columnar cell. *Am. J. Anat.* 141:461–79

41. Nicolas P, Kim KM, Shibata D, Tavare S. 2007. The stem cell population of the human colon crypt: analysis via methylation patterns. *PLoS Comput. Biol.* 3:e28

42. Yatabe Y, Tavare S, Shibata D. 2001. Investigating stem cells in human colon by using methylation patterns. *Proc. Natl. Acad. Sci. USA* 98:10839–44

43. Cairnie AB, Millen BH. 1975. Fission of crypts in the small intestine of the irradiated mouse. *Cell Tissue Kinet.* 8:189–96

44. Schmidt GH, Winton DJ, Ponder BA. 1988. Development of the pattern of cell renewal in the crypt-villus unit of chimaeric mouse small intestine. *Development* 103:785–90

45. Fuller CE, Davies RP, Williams GT, Williams ED. 1990. Crypt restricted heterogeneity of goblet cell mucus glycoprotein in histologically normal human colonic mucosa: a potential marker of somatic mutation. *Br. J. Cancer* 61:382–84

46. Novelli M, Cossu A, Oukrif D, Quaglia A, Lakhani S, et al. 2003. X-inactivation patch size in human female tissue confounds the assessment of tumor clonality. *Proc. Natl. Acad. Sci. USA* 100:3311–14

47. Novelli MR, Williamson JA, Tomlinson IP, Elia G, Hodgson SV, et al. 1996. Polyclonal origin of colonic adenomas in an XO/XY patient with FAP. *Science* 272:1187–90

48. Taylor RW, Barron MJ, Borthwick GM, Gospel A, Chinnery PF, et al. 2003. Mitochondrial DNA mutations in human colonic crypt stem cells. *J. Clin. Investig.* 112:1351–60

49. Winton DJ, Blount MA, Ponder BA. 1988. A clonal marker induced by mutation in mouse intestinal epithelium. *Nature* 333:463–66

50. Winton DJ, Ponder BA. 1990. Stem-cell organization in mouse small intestine. *Proc. Biol. Sci.* 241:13–18

51. Barker N, van Es JH, Kuipers J, Kujala P, Van Den Born M, et al. 2007. Identification of stem cells in small intestine and colon by marker gene Lgr5. *Nature* 449:1003–7

52. Soriano P. 1999. Generalized *lacZ* expression with the ROSA26 Cre reporter strain. *Nat. Genet.* 21:70–71

53. Schofield R. 1978. The relationship between the spleen colony-forming cell and the haemopoietic stem cell. *Blood Cells* 4:7–25

54. Mills JC, Gordon JI. 2001. The intestinal stem cell niche: There grows the neighborhood. *Proc. Natl. Acad. Sci. USA* 98:12334–36

55. Powell DW, Mifflin RC, Valentich JD, Crowe SE, Saada JI, West AB. 1999. Myofibroblasts. II. Intestinal subepithelial myofibroblasts. *Am. J. Physiol.* 277:C183–201

56. Haramis AP, Begthel H, Van Den Born M, van Es J, Jonkheer S, et al. 2004. De novo crypt formation and juvenile polyposis on BMP inhibition in mouse intestine. *Science* 303:1684–86

57. Hardwick JC, Van Den Brink GR, Bleuming SA, Ballester I, Van Den Brande JM, et al. 2004. Bone morphogenetic protein 2 is expressed by, and acts upon, mature epithelial cells in the colon. *Gastroenterology* 126:111–21

58. Massague J, Seoane J, Wotton D. 2005. Smad transcription factors. *Genes Dev.* 19:2783–810

59. He XC, Zhang J, Tong WG, Tawfik O, Ross J, et al. 2004. BMP signaling inhibits intestinal stem cell self-renewal through suppression of Wnt-β-catenin signaling. *Nat. Genet.* 36:1117–21

60. Howe JR, Bair JL, Sayed MG, Anderson ME, Mitros FA, et al. 2001. Germline mutations of the gene encoding bone morphogenetic protein receptor 1A in juvenile polyposis. *Nat. Genet.* 28:184–87

61. Howe JR, Roth S, Ringold JC, Summers RW, Järvinen HJ, et al. 1998. Mutations in the *SMAD4/DPC4* gene in juvenile polyposis. *Science* 280:1086–88

62. Zhou XP, Woodford-Richens K, Lehtonen R, Kurose K, Aldred M, et al. 2001. Germline mutations in BMPR1A/ALK3 cause a subset of cases of juvenile polyposis syndrome and of Cowden and Bannayan-Riley-Ruvalcaba syndromes. *Am. J. Hum. Genet.* 69:704–11

63. Kaestner KH, Silberg DG, Traber PG, Schutz G. 1997. The mesenchymal winged helix transcription factor Fkh6 is required for the control of gastrointestinal proliferation and differentiation. *Genes Dev.* 11:1583–95

64. Pabst O, Zweigerdt R, Arnold HH. 1999. Targeted disruption of the homeobox transcription factor Nkx2-3 in mice results in postnatal lethality and abnormal development of small intestine and spleen. *Development* 126:2215–25

65. Wang Y, Wang L, Iordanov H, Swietlicki EA, Zheng Q, et al. 2006. Epimorphin$^{-/-}$ mice have increased intestinal growth, decreased susceptibility to dextran sodium sulfate colitis, and impaired spermatogenesis. *J. Clin. Investig.* 116:1535–46

66. Madison BB, Braunstein K, Kuizon E, Portman K, Qiao XT, Gumucio DL. 2005. Epithelial hedgehog signals pattern the intestinal crypt-villus axis. *Development* 132:279–89

67. Ramalho-Santos M, Melton DA, McMahon AP. 2000. Hedgehog signals regulate multiple aspects of gastrointestinal development. *Development* 127:2763–72

68. Karlsson L, Lindahl P, Heath JK, Betsholtz C. 2000. Abnormal gastrointestinal development in PDGF-A and PDGFR-α deficient mice implicates a novel mesenchymal structure with putative instructive properties in villus morphogenesis. *Development* 127:3457–66

69. He TC, Sparks AB, Rago C, Hermeking H, Zawel L, et al. 1998. Identification of c-MYC as a target of the APC pathway. *Science* 281:1509–12

70. Muncan V, Sansom OJ, Tertoolen L, Phesse TJ, Begthel H, et al. 2006. Rapid loss of intestinal crypts upon conditional deletion of the Wnt/Tcf-4 target gene c-*Myc*. *Mol. Cell Biol.* 26:8418–26

71. Sansom OJ, Meniel VS, Muncan V, Phesse TJ, Wilkins JA, et al. 2007. Myc deletion rescues Apc deficiency in the small intestine. *Nature* 446:676–79

72. Miller H, Zhang J, Kuolee R, Patel GB, Chen W. 2007. Intestinal M cells: the fallible sentinels? *World J. Gastroenterol.* 13:1477–86

73. Sbarbati A, Osculati F. 2005. A new fate for old cells: brush cells and related elements. *J. Anat.* 206:349–58

74. Artavanis-Tsakonas S, Rand MD, Lake RJ. 1999. Notch signaling: cell fate control and signal integration in development. *Science* 284:770–76

75. Tamura K, Taniguchi Y, Minoguchi S, Sakai T, Tun T, et al. 1995. Physical interaction between a novel domain of the receptor Notch and the transcription factor RBP-Jκ/Su(H). *Curr. Biol.* 5:1416–23

76. Milano J, McKay J, Dagenais C, Foster-Brown L, Pognan F, et al. 2004. Modulation of Notch processing by γ-secretase inhibitors causes intestinal goblet cell metaplasia and induction of genes known to specify gut secretory lineage differentiation. *Toxicol. Sci.* 82:341–58

77. Wong GT, Manfra D, Poulet FM, Zhang Q, Josien H, et al. 2004. Chronic treatment with the γ-secretase inhibitor LY-411,575 inhibits β-amyloid peptide production and alters lymphopoiesis and intestinal cell differentiation. *J. Biol. Chem.* 279:12876–82

78. van Es JH, van Gijn ME, Riccio O, Van Den Born M, Vooijs M, et al. 2005. Notch/γ-secretase inhibition turns proliferative cells in intestinal crypts and adenomas into goblet cells. *Nature* 435:959–63

79. Vooijs M, Ong CT, Hadland B, Huppert S, Liu Z, et al. 2007. Mapping the consequence of Notch1 proteolysis in vivo with NIP-CRE. *Development* 134:535–44

80. Riccio O, van Gijn ME, Bezdek AC, Pellegrinet L, van Es JH, et al. 2008. Loss of intestinal crypt progenitor cells owing to inactivation of both Notch1 and Notch2 is accompanied by derepression of CDK inhibitors p27^{Kip1} and p57^{Kip2}. *EMBO Rep.* 9:377–83

81. Fre S, Huyghe M, Mourikis P, Robine S, Louvard D, Artavanis-Tsakonas S. 2005. Notch signals control the fate of immature progenitor cells in the intestine. *Nature* 435:964–68

82. Stanger BZ, Datar R, Murtaugh LC, Melton DA. 2005. Direct regulation of intestinal fate by Notch. *Proc. Natl. Acad. Sci. USA* 102:12443–48

83. Jensen J, Pedersen EE, Galante P, Hald J, Heller RS, et al. 2000. Control of endodermal endocrine development by Hes-1. *Nat. Genet.* 24:36–44

84. Suzuki K, Fukui H, Kayahara T, Sawada M, Seno H, et al. 2005. Hes1-deficient mice show precocious differentiation of Paneth cells in the small intestine. *Biochem. Biophys. Res. Commun.* 328:348–52

85. Yang Q, Bermingham NA, Finegold MJ, Zoghbi HY. 2001. Requirement of Math1 for secretory cell lineage commitment in the mouse intestine. *Science* 294:2155–58

86. Shroyer NF, Wallis D, Venken KJ, Bellen HJ, Zoghbi HY. 2005. Gfi1 functions downstream of Math1 to control intestinal secretory cell subtype allocation and differentiation. *Genes Dev.* 19:2412–17

87. Amann JM, Chyla BJ, Ellis TC, Martinez A, Moore AC, et al. 2005. Mtgr1 is a transcriptional corepressor that is required for maintenance of the secretory cell lineage in the small intestine. *Mol. Cell. Biol.* 25:9576–85

88. Karam SM. 1999. Lineage commitment and maturation of epithelial cells in the gut. *Front. Biosci.* 4:D286–98

89. Crosnier C, Vargesson N, Gschmeissner S, Ariza-McNaughton L, Morrison A, Lewis J. 2005. Delta-Notch signalling controls commitment to a secretory fate in the zebrafish intestine. *Development* 132:1093–104

90. Katz JP, Perreault N, Goldstein BG, Lee CS, Labosky PA, et al. 2002. The zinc-finger transcription factor Klf4 is required for terminal differentiation of goblet cells in the colon. *Development* 129:2619–28

91. Velcich A, Yang W, Heyer J, Fragale A, Nicholas C, et al. 2002. Colorectal cancer in mice genetically deficient in the mucin Muc2. *Science* 295:1726–29

92. Schonhoff SE, Giel-Moloney M, Leiter AB. 2004. Minireview: development and differentiation of gut endocrine cells. *Endocrinology* 145:2639–44

93. Jenny M, Uhl C, Roche C, Duluc I, Guillermin V, et al. 2002. Neurogenin3 is differentially required for endocrine cell fate specification in the intestinal and gastric epithelium. *EMBO J.* 21:6338–47

94. Naya FJ, Huang HP, Qiu Y, Mutoh H, DeMayo FJ, et al. 1997. Diabetes, defective pancreatic morphogenesis, and abnormal enteroendocrine differentiation in BETA2/neuroD-deficient mice. *Genes Dev.* 11:2323–34

95. Desai S, Loomis Z, Pugh-Bernard A, Schrunk J, Doyle MJ, et al. 2008. Nkx2.2 regulates cell fate choice in the enteroendocrine cell lineages of the intestine. *Dev. Biol.* 313:58–66

96. Offield MF, Jetton TL, Labosky PA, Ray M, Stein RW, et al. 1996. PDX-1 is required for pancreatic outgrowth and differentiation of the rostral duodenum. *Development* 122:983–95

97. Hill ME, Asa SL, Drucker DJ. 1999. Essential requirement for Pax6 in control of enteroendocrine proglucagon gene transcription. *Mol. Endocrinol.* 13:1474–86

98. Larsson LI, St-Onge L, Hougaard DM, Sosa-Pineda B, Gruss P. 1998. Pax 4 and 6 regulate gastrointestinal endocrine cell development. *Mech. Dev.* 79:153–59

99. Batlle E, Henderson JT, Beghtel H, Van Den Born MM, Sancho E, et al. 2002. β-Catenin and TCF mediate cell positioning in the intestinal epithelium by controlling the expression of EphB/ephrinB. *Cell* 111:251–63

100. Cortina C, Palomo-Ponce S, Iglesias M, Fernandez-Masip JL, Vivancos A, et al. 2007. EphB-ephrin-B interactions suppress colorectal cancer progression by compartmentalizing tumor cells. *Nat. Genet.* 39:1376–83

101. Blache P, van de Wetering M, Duluc I, Domon C, Berta P, et al. 2004. SOX9 is an intestine crypt transcription factor, is regulated by the Wnt pathway, and represses the *CDX2* and *MUC2* genes. *J. Cell Biol.* 166:37–47

102. Bastide P, Darido C, Pannequin J, Kist R, Robine S, et al. 2007. Sox9 regulates cell proliferation and is required for Paneth cell differentiation in the intestinal epithelium. *J. Cell Biol.* 178:635–48

103. Mori-Akiyama Y, Van Den Born M, van Es JH, Hamilton SR, Adams HP, et al. 2007. SOX9 is required for the differentiation of Paneth cells in the intestinal epithelium. *Gastroenterology* 133:539–46

104. Ng AY, Waring P, Ristevski S, Wang C, Wilson T, et al. 2002. Inactivation of the transcription factor Elf3 in mice results in dysmorphogenesis and altered differentiation of intestinal epithelium. *Gastroenterology* 122:1455–66

105. Flentjar N, Chu PY, Ng AY, Johnstone CN, Heath JK, et al. 2007. TGF-βRII rescues development of small intestinal epithelial cells in Elf3-deficient mice. *Gastroenterology* 132:1410–19

106. Haegebarth A, Bie W, Yang R, Crawford SE, Vasioukhin V, et al. 2006. Protein tyrosine kinase 6 negatively regulates growth and promotes enterocyte differentiation in the small intestine. *Mol. Cell. Biol.* 26:4949–57

107. Fodde R, Brabletz T. 2007. Wnt/β-catenin signaling in cancer stemness and malignant behavior. *Curr. Opin. Cell Biol.* 19:150–58

108. Miyaki M, Konishi M, Kikuchi-Yanoshita R, Enomoto M, Igari T, et al. 1994. Characteristics of somatic mutation of the adenomatous polyposis coli gene in colorectal tumors. *Cancer Res.* 54:3011–20
109. Miyoshi Y, Ando H, Nagase H, Nishisho I, Horii A, et al. 1992. Germ-line mutations of the APC gene in 53 familial adenomatous polyposis patients. *Proc. Natl. Acad. Sci. USA* 89:4452–56
110. Powell SM, Zilz N, Beazer-Barclay Y, Bryan TM, Hamilton SR, et al. 1992. APC mutations occur early during colorectal tumorigenesis. *Nature* 359:235–37
111. Liu W, Dong X, Mai M, Seelan RS, Taniguchi K, et al. 2000. Mutations in AXIN2 cause colorectal cancer with defective mismatch repair by activating β-catenin/TCF signalling. *Nat. Genet.* 26:146–47
112. Morin PJ, Sparks AB, Korinek V, Barker N, Clevers H, et al. 1997. Activation of β-catenin-Tcf signaling in colon cancer by mutations in β-catenin or APC. *Science* 275:1787–90
113. Rubinfeld B, Robbins P, El-Gamil M, Albert I, Porfiri E, Polakis P. 1997. Stabilization of β-catenin by genetic defects in melanoma cell lines. *Science* 275:1790–92
114. Barker N, Huls G, Korinek V, Clevers H. 1999. Restricted high level expression of Tcf-4 protein in intestinal and mammary gland epithelium. *Am. J. Pathol.* 154:29–35
115. Fearon ER, Vogelstein B. 1990. A genetic model for colorectal tumorigenesis. *Cell* 61:759–67
116. Moser AR, Pitot HC, Dove WF. 1990. A dominant mutation that predisposes to multiple intestinal neoplasia in the mouse. *Science* 247:322–24
117. Su LK, Kinzler KW, Vogelstein B, Preisinger AC, Moser AR, et al. 1992. Multiple intestinal neoplasia caused by a mutation in the murine homolog of the APC gene. *Science* 256:668–70
118. Gaspar C, Fodde R. 2004. APC dosage effects in tumorigenesis and stem cell differentiation. *Int. J. Dev. Biol.* 48:377–86
119. Andreu P, Colnot S, Godard C, Gad S, Chafey P, et al. 2005. Crypt-restricted proliferation and commitment to the Paneth cell lineage following Apc loss in the mouse intestine. *Development* 132:1443–51
120. Shibata H, Toyama K, Shioya H, Ito M, Hirota M, et al. 1997. Rapid colorectal adenoma formation initiated by conditional targeting of the Apc gene. *Science* 278:120–23
121. Harada N, Tamai Y, Ishikawa T, Sauer B, Takaku K, et al. 1999. Intestinal polyposis in mice with a dominant stable mutation of the β-catenin gene. *EMBO J.* 18:5931–42
122. Dick JE. 2003. Breast cancer stem cells revealed. *Proc. Natl. Acad. Sci. USA* 100:3547–49
123. Reya T, Morrison SJ, Clarke MF, Weissman IL. 2001. Stem cells, cancer, and cancer stem cells. *Nature* 414:105–11
124. Wang JC, Dick JE. 2005. Cancer stem cells: lessons from leukemia. *Trends Cell Biol.* 15:494–501
125. O'Brien CA, Pollett A, Gallinger S, Dick JE. 2007. A human colon cancer cell capable of initiating tumour growth in immunodeficient mice. *Nature* 445:106–10
126. Ricci-Vitiani L, Lombardi DG, Pilozzi E, Biffoni M, Todaro M, et al. 2007. Identification and expansion of human colon-cancer-initiating cells. *Nature* 445:111–15
127. Kelly PN, Dakic A, Adams JM, Nutt SL, Strasser A. 2007. Tumor growth need not be driven by rare cancer stem cells. *Science* 317:337

Dendritic Spine Dynamics

D. Harshad Bhatt, Shengxiang Zhang,
and Wen-Biao Gan

Molecular Neurobiology Program, The Helen and Martin Kimmel Center for Biology and
Medicine at the Skirball Institute of Biomolecular Medicine, New York University School
of Medicine, New York, NY 10016; email: gan@saturn.med.nyu.edu

Annu. Rev. Physiol. 2009. 71:261–82

First published online as a Review in Advance on
November 13, 2008

The *Annual Review of Physiology* is online at
physiol.annualreviews.org

This article's doi:
10.1146/annurev.physiol.010908.163140

Copyright © 2009 by Annual Reviews.
All rights reserved

0066-4278/09/0315-0261$20.00

Key Words

dendritic filopodia, spine plasticity, spine stability, spine pathology, in
vivo imaging

Abstract

Dendritic spines are the postsynaptic components of most excitatory
synapses in the mammalian brain. Spines accumulate rapidly during
early postnatal development and undergo a substantial loss as animals
mature into adulthood. In past decades, studies have revealed that the
number and size of dendritic spines are regulated by a variety of gene
products and environmental factors, underscoring the dynamic nature
of spines and their importance to brain plasticity. Recently, in vivo time-
lapse imaging of dendritic spines in the cerebral cortex suggests that,
although spines are highly plastic during development, they are remarkably stable in adulthood, and most of them last throughout life.
Therefore, dendritic spines may provide a structural basis for lifelong
information storage, in addition to their well-established role in brain
plasticity. Because dendritic spines are the key elements for information
acquisition and retention, understanding how spines are formed and
maintained, particularly in the intact brain, will likely provide fundamental insights into how the brain possesses the extraordinary capacity
to learn and to remember.

INTRODUCTION

In 1888, Ramón y Cajal described a series of small protrusions extending from the dendrites of chicken Purkinje cells: "[T]he surface...appears bristling with points or short spines..." (1). Soon after Cajal's discovery of "espinas" (dendritic spines), ideas regarding dendritic spine dynamics and its contribution to the extent of neuronal structural immutability were proposed and debated. Cajal hypothesized that the function of the spine was to create more surface area on dendritic branches: "...[to] increase their receptive surface and establish more intimate contacts with axonal terminal arborizations" (2, 3). Although Cajal promulgated the idea that the connections between neurons would be heavily influenced both through hereditary means and on an experiential basis (3–5), his contemporaries, Demoor and Stefanowska, provided the first lines of experimental evidence to suggest that spines, or pyriform appendages, as Stefanowska referred to them, were able to change in size and shape during life and that such changes would impact neuronal connections to a variable degree (6–9).

With Sherrington's introduction of the concept of a synapse, the anatomical and physiological phenomena of neuronal connections were merged into a single term, providing more insight into the functional role of spines (10, 11). But Cajal, Sherrington, and the other investigators of their time were limited by the technical inability to verify their theories and definitively unite the anatomical dendritic spine with the physiological synapse, let alone their inability to explore spine dynamics and its functional implications. It was not until the late 1950s that electron microscopy (EM) identified the synapse and the dendritic spine as associated structures (12, 13). A couple of decades later, contractile actin was found to be ubiquitously present in dendritic spines, thus reinvigorating investigations into spine motility and its involvement in brain functions such as learning and memory, almost a century after spines were first discovered (14–17).

We now know that the vast majority of excitatory synapses are made on the heads of dendritic spines, which can be found on both excitatory and inhibitory neurons throughout the central nervous system and in an array of diverse species (18, 19). Dendritic spines are highly dynamic structures, particularly during postnatal development, when enormous numbers of synaptic connections are being rapidly made (20–29). As animals mature into adulthood, substantial changes in spine number and morphology may still occur during the learning process and under pathological conditions (30–41). Studies in the past several decades have shown that the structural and electrical properties of dendritic spines are critical for local signal integration and molecular compartmentalization (18, 42–46). Therefore, the dynamic properties of spines, including spine turnover and changes in spine shape and motility, are vital for the development and function of neural circuits. Consequently, abnormalities of spine structures and dynamics are intimately associated with disrupted synaptic, neuronal, and higher-order brain functions. Indeed, spine abnormalities have been found in many pathological conditions, including mental illnesses and age-related neurodegenerative diseases (31, 40).

Much of our understanding of dendritic spine dynamics, until very recently, has come predominantly from studies in fixed brain tissues, live neuronal cultures, and brain slices. Although studies of fixed or in vitro preparations have provided invaluable information on spine dynamics, it is necessary to examine changes of dendritic spines directly, in the living, intact brain, to fully appreciate the importance of spine dynamics to brain development and function. It has become possible only within the past few years to image individual dendritic spines over extended periods of time in the living mammalian brain. The advent of new technical advances, specifically, the use of two-photon microscopy (TPM), has allowed for the live imaging of fluorescently labeled synapses in the living cortex several hundred micrometers deep to the pial surface (47, 48). In addition, the expression of green fluorescent protein (GFP) and its spectral variants in specific cell types in the brain has permitted repeated imaging of individual synaptic structures in living animals (**Figure 1**) (23, 49–55). In recent years, investigators have applied TPM imaging in GFP/YFP (yellow fluorescent protein)-expressing transgenic animals to address some of the longstanding questions related to spine dynamics.

Despite our burgeoning ability to detect changes in spine shape and number, and the distance we have traveled since Cajal, key aspects of dendritic spine plasticity and function in living animals remain unclear. For example, how do spines form? How long does a spine last in an animal? How are the development and lifetime of spines regulated by the intrinsic genetic program and experience from the

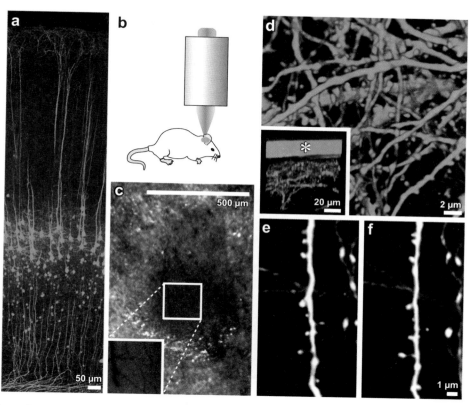

Figure 1

Long-term transcranial two-photon imaging of transgenic mice expressing yellow fluorescent protein (YFP). (*a*) Coronal sections of YFP H-line mice that show a subset of layer V pyramidal cells expressing YFP in the barrel cortex. The apical dendrites of layer V cells project to layer I and can be imaged with two-photon microscopy. (*b*) Diagram of transcranial two-photon imaging of an anesthetized mouse. (*c*) Dorsal view of a thinned-skull preparation. The skull above a cortical region to be imaged was thinned to ~20 μm in thickness in an area of ~200 μm in diameter. The inset shows the blood vasculature within the square, which can be used as a landmark for repeated imaging of the same neuronal structure. (*d*) Three-dimensional reconstruction of dendritic structures imaged through a thinned skull window. The inset shows a lateral view of a reconstructed imaging stack. The asterisk indicates the thinned skull, which is ~20 μm in thickness. (*e,f*) High-magnification images of dendritic and axonal structures taken three days apart, displaying that dendritic spines and axonal boutons can be identified over time in the same animal. Panel *a* is modified from Reference 54 with permission, panel *c* is adapted from Reference 92 with permission, and panels *e* and *f* are adapted from Reference 23 with permission.

external world? What is the role of abnormal spine plasticity in various brain diseases? With an increasing number of laboratories utilizing advanced imaging techniques and an expanse in transgenic and molecular technology, the answers to these questions are finally within reach. Owing to the flourish of studies of synapse and spine dynamics in the past decade, there are a good number of reviews on spine dynamics and function in the literature (18, 19, 40–45, 56–62). Here we thus focus the body of this discussion

on the most recent findings on spine plasticity in vivo, with an eye toward addressing some of the aforementioned queries and suggesting a platform for future directions.

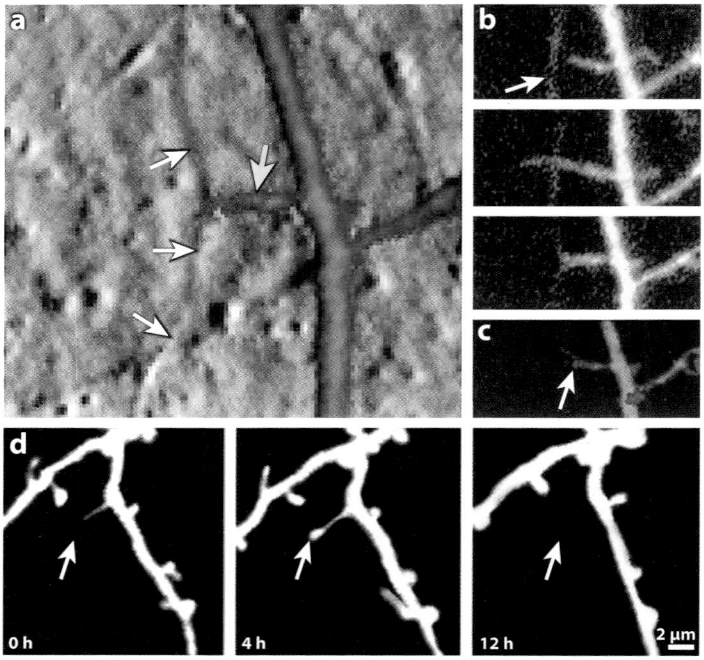

SPINOGENESIS

Dendritic spines appear early during development soon after dendritic processes are extended from neurons. From early stages of development, synaptic contacts already exist between spines and presynaptic axons, suggesting that the process of spine formation is intimately associated with the process of contact formation between neurons and the establishment of neural circuits. An important question pertaining to spine development is how spinogenesis and synaptogenesis are related to each other. The answer to this question would help forge our understanding of whether spine formation reflects some intrinsic properties of postsynaptic dendrites or if it is induced by extrinsic factors associated with presynaptic axonal terminals. Fundamentally, this kind of knowledge would shed light on the way in which we understand how synaptic connections are constructed in neural circuits.

Several lines of evidence suggest that dendritic filopodia, long and thin protrusions without bulbous heads, play a pivotal role in the initial stages of spinogenesis and synaptogenesis. First, during early development, when extensive spine formation occurs, filopodia are highly abundant and undergo rapid extension and retraction within minutes to hours (19–21, 29, 57, 63–65). Importantly, pioneering time-lapse observations from neuronal cultures and brain slices have revealed that these highly dynamic dendritic filopodia initiate contacts with presynaptic axons and are occasionally transformed into spines (**Figure 2**) (20, 29). Furthermore, EM studies indicate that some filopodia do not have any synaptic contacts with axons, whereas others make several synaptic contacts (66). In addition, filopodia and spines display various degrees of motility even while they are in contact with presynaptic axonal terminals (21, 67–69). Together, these findings

Figure 2

Dendritic filopodia in spinogenesis and synaptogenesis. (*a–c*) Dendritic filopodia initiate synaptic contacts with neighboring axons. (*a*) A lipophilic dye (FAST DiO)–labeled dendrite (*green*) of a cultured hippocampal neuron is overlaid onto a DIC image of the same field. White arrows indicate an axon near the dendrite. The yellow arrow indicates a filopodium extending from the dendritic shaft. The filopodium establishes a synaptic contact with the axon, as shown in panels *b* and *c*. (*b*) Time-lapse sequence of the filopodium shown in panel *a* at 0, 18, and 90 min. The white arrow indicates the axon shown in panel *a*. (*c*) Merged image showing FAST DiO–labeled dendrite (*green*) and FM 4–64–labeled presynaptic boutons (*red*). The white arrow indicates a bouton that formed at the filopodium contact site. (*d*) In vivo two-photon imaging shows that a filopodium (*white arrow*) first was converted to a spine-like protrusion within 4 h and disappeared by 12 h. Panels *a–c* are modified from Reference 29 with permission. Panel *d* is adapted from Reference 54 with permission.

suggest the following sequence of events during spine/synapse formation: Dendrites produce long and thin dendritic filopodia that exhibit dynamic growth, allowing them to sample some of the nearby axons. Choosing and capturing the appropriate presynaptic axon via activity-dependent or independent signaling would result in stabilization of the contact and maturation of the filopodia into dendritic spines. The absence of proper signals (or the presence of alternative ones) would result in the regression of the filopodia back into the dendritic shaft (19, 63, 65, 70–72).

The in vivo imaging of YFP-expressing mice showing that dendritic filopodia are indeed highly dynamic and can transform into spines provides more evidence consistent with the role of filopodia in spinogenesis (**Figure 2**) (23, 54). In young mice at one month of age, ~12% of dendritic protrusions in different cortical regions are filopodia, whereas the remaining are spines. Whereas most filopodia underwent rapid turnover within 4 h, ~15% of filopodia formed a bulbous head and persisted within the same observational period. Importantly, from the ~15% of these newly formed spine-like protrusions, ~40% persisted over 24 h, and ~20% lasted more than 48 h (54). These newly persistent protrusions are morphologically indistinguishable from preexisting spines. Thus, in vivo observations, in corroboration with findings from in vitro studies, indicate that a small percentage of filopodia are transformed into more stable thin or mushroom-like dendritic spines. These findings provide further evidence that filopodia are spine precursors acting as samplers of the local synaptic neighborhood. That most filopodia do not result in spines alludes to the possibility that forming and maintaining contacts between filopodia and their appropriate axonal partners are highly selective processes.

Although the aforementioned results give credence to the theory that dynamic filopodia find their presynaptic partners and mature into spines, other modes of spinogenesis may also occur and should be considered. EM studies have shown that the formation of shaft synapses precedes the formation of spine synapses during early developmental stages (19, 66). Furthermore, time-lapse imaging studies have found that spines can form directly from dendritic shafts rather than transitioning from filopodia (20). These observations raise the possibility that presynaptic axons recognize the shafts of dendrites and induce the postsynaptic cell to form a dendritic spine directly in apposition to the axonal terminal (19, 66). Such a mode of spine formation may occur particularly during early development, when highly plastic filopodia existing along axons may initiate contacts with postsynaptic dendrites (73–76). Moreover, spines may emerge independently of synaptic contact, acting as beacons for axonal terminals, which would then locate and recognize these preformed spines on dendrites and make targeted contacts. In agreement with such a scenario is the finding that in some parts of the brain, under certain conditions such as deafferentation, spinogenesis can occur in the absence of synaptogenesis (77, 78). Thus, although there is a strong correlation between the production of postsynaptic filopodia and dendritic spines and the establishment of connections with presynaptic axonal partners, the actual relationship between these two phenomena may vary with different developmental stages and under different experimental conditions.

Although the notion that dynamic filopodia find their partners and mature into spines remains a prominent idea and an important rule of spine development, it is still unclear whether filopodia are necessary precursors of all spines or if most spines are formed without filopodial intermediaries, directly from the dendritic shaft. If all spines were evolved from dynamic filopodia, it would suggest that synaptogenesis and spinogenesis are two inseparable processes, linked together teleologically. In this case, the determining factor for forming stable spines would be whether or not a given filopodium is able to make stable synaptic contacts with an axon. Alternatively, if spines were mainly formed directly from dendritic shaft synapses, synaptogenesis and spinogenesis would be two related but independent processes governed by distinctive mechanisms. The definitive answers to these questions require the use of combinatorial approaches for (a) simultaneously monitoring axons and dendrites at high temporal resolution as they are making contacts with each other and (b) subsequently determining whether such contacts are synaptic in nature either via highly specific synaptic markers or most incontrovertibly with EM identification. Utilizing such approaches, recent studies from developing hippocampal slices and mature cortex have shown that dendritic protrusions that have emerged from dendritic shafts form apparent synaptic contact with axonal terminals within a period of hours (79–81). These findings provide strong evidence in support of the view that dendritic filopodia play a major, if not exclusive, role in sampling the surrounding neuropil, initiating contacts with presynaptic terminals and forming dendritic spines. Because spine formation is associated with many rounds of filopodial extension and retraction and the newly formed spines are largely eliminated within days after they are formed (19, 20, 54, 57, 65), it appears that the formation of stable spines involves a protracted and highly selective process.

SPINE PRUNING

Shortly after Cajal initially described dendritic spines, he observed that the number of spines per unit length of a dendrite decreased toward the end of the postnatal development of the nervous system. He inferred this finding to be an indication that the connections between neurons were being altered through the removal of spines (3). This prediction, as with many of Cajal's predictions, was subsequently confirmed by more quantitative analyses of the number of synapses and spines in fixed brains from different species at varying developmental ages. Indeed, in the cerebral cortex of mammals, including that of humans, rapid synaptogenesis during early postnatal life is followed by a substantial (~50%) loss of synapses/spines that extends through adolescence (4, 25, 27, 82–85). In adulthood the number of spines remains relatively constant until aging-related loss of

synapses occurs (30, 35). These studies suggest that in late postnatal life, neuronal circuits undergo significant reorganization mediated in large part by synapse/spine elimination. However, these findings, taken predominantly from fixed postmortem tissues, were not able to determine the dynamic behavior of spines or thus the precise nature of cortical spine loss. That is, because spines are appearing and disappearing, the decrease in overall spine number may be due to an increase in the elimination of existing spines, the addition of fewer new spines, or some combination of both.

To address this issue, dendritic spines of layer V pyramidal cells were imaged in vivo with TPM techniques. Such studies found that in young adolescent mice (one month of age), 13–20% of spines were eliminated and 5–8% of spines were formed over a two-week interval in visual, barrel, primary motor, and frontal cortices, indicating a cortex-wide loss of spines during this developmental period (**Figure 3**) (23, 54). Consistent with this, examinations of spine dynamics in the visual and somatosensory cortices in ~1.5-month-old mice also show a net loss of spines over one or three weeks (86). Imaging dendrites from postnatal day 16 through day 26 in the barrel cortex showed that this net loss of spines could even have started prior to two weeks of age (87). Furthermore, from one to four months of age, a ~25% net loss of spines occurs as the result of a significantly higher rate of spine elimination compared with spine formation in both visual and barrel cortices (54). Together, these observations corroborate previous studies from fixed tissues showing that synaptic density in the mammalian cortex decreases substantially from infancy until puberty (4, 25, 27, 83). Importantly, because the degree of spine elimination is several-fold higher than that of spine formation during the period of net spine loss, these in vivo imaging studies indicate that the major reorganization of the cortex during late postnatal life involves the elimination of existing connections between neurons. The large loss of synapses occurs across different regions of the developing brain in humans and nonhuman primates, suggesting that pruning and sculpting early-established synaptic connections likely are processes fundamental to the maturation of the brain.

SPINE STABILITY

Although many of the initial investigations into spine plasticity have focused on development, a time when animals are undergoing remarkable changes in shape and behavior, little is understood about the degree of spine plasticity in adulthood. Dendritic spines possess an inherent plasticity, as is evidenced by significant and rapid changes in spine number in response to environmental challenges and under pathological conditions (30–32, 34–37, 40, 88). Nevertheless, the degree to which dendritic spines undergo remodeling in a normal and healthy adult brain remains a relative mystery. Uncovering the degree of spine mutability in size, shape, and number in the mature brain will provide important insights into how long-term information is stored (and lost) in neural circuits. For example, if most adult spines remained throughout life, memory and basic cortical functions could be stably maintained through synaptic connections that were established during development. In contrast, findings that adult spines were highly dynamic and showed a high degree of turnover over the lifetime of an animal would suggest that long-term information might instead be stored in a dynamic fashion in constantly and rapidly rewiring synaptic networks.

The first two studies using in vivo imaging of GFP/YFP-expressing dendritic spines generated fundamentally different views on the stability of spines in the adult mouse cortex (23, 53, 89, 90). One study showed that in the adult mouse primary visual cortex (>4 months of age), spines in apical dendrites from layer V pyramidal neurons are remarkably stable, with ~4% turnover per month (23). If all adult spines turned over at a similar rate and the average life span of a mouse were two to three years, a substantial proportion of adult spines would persist throughout a mouse's lifetime, thus providing a potential structural basis for long-lasting infor-

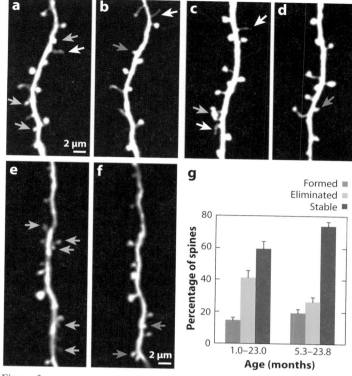

Figure 3

In vivo time-lapse imaging of dendritic spine dynamics and stability in the mouse cortex. (*a–d*) Repeated imaging of two dendritic branches from mice that are four to six weeks of age reveals spine elimination (*yellow arrows*) and formation (*red arrows*) as well as filopodium turnover (*white arrows*) in control (*a,b*) and sensory-deprived (*c,d*) barrel cortices. Long-term sensory deprivation via whisker trimming reduces spine elimination in young adolescent barrel cortex from four to six weeks of age. (*e,f*) Dendritic branches from a mouse imaged 18 months apart show that most spines remain at their same respective location. (*g*) Group data show the percentage of spines that are eliminated, formed, or stable over 18–22 months. One group of mice was imaged between 1 and 23.0 ± 2.6 months of age. Another group was imaged between 5.3 ± 1.0 and 23.8 ± 2.6 months of age. Panels *a–d* are modified from Reference 39 with permission, and panels *e–g* are from Reference 54 with permission.

mation storage. However, a different study in the mouse barrel cortex suggested that spines are highly plastic in adulthood, with ~20% of spine turnover within one day and ~40% spine turnover within eight days (53). Even those spines that persisted over eight days (classified as stable ones, ~60% of the total population) continued to be eliminated at a rate of ~16% over 22 days, which translates to a limited lifetime of this stable pool, ~120 days (53, 89).

Because spines are mostly short-lived and may turn over multiple times in the adult cortex, this study (53) implies that the rewiring of synaptic connections is a prominent form of neural plasticity not only during development but also in mature circuits. Such a scenario, although addressing the adaptability of the adult brain, does raise the question of how long-term memory is maintained in neural circuits that undergo such rapid and extensive rewiring.

Subsequent studies continue to show discrepancy over the degree of spine dynamism. Zuo et al. (54) showed that 3–5% of adult spines are formed and eliminated over two weeks in barrel, primary motor, and frontal cortices, a rate similar to that previously reported in the primary visual cortex. In addition, >70% of spines in adult barrel cortex were maintained over an 18-month interval, suggesting that the average lifetime of an individual adult spine may be ∼63 months, much longer than the entire life span of the animal (**Figure 3**). In another study, Holtmaat et al. (87) found that in six-month-old mouse barrel cortex, the percentage of spines that persisted for eight days or more was ∼73%, significantly larger than that in young adult mice (∼53% at 5–11 weeks of age). This suggests that the high spine turnover previously reported in barrel cortex (53) is partly due to the use of young adult mice (6–10 weeks of age) instead of mature ones. However, although this later study showed that most adult spines can be stable over weeks, more than 20% of total spines in six-month-old mice still turn over within four days, and ∼30% of spines turn over within one month in barrel and visual cortices (87). This work also showed that spines in barrel cortex had a significantly higher turnover rate than in visual cortex, a result different from that reported by Zuo et al. (54). In yet another study examining spine dynamics in the visual, somatosensory, and auditory cortices, Majewska et al. (86) found that ∼85% of spines are stable over three weeks in mice starting at postnatal day 40, which is in general agreement with the view that adult spines are largely stable.

Because the contrasting results lead to contradictory views on the structural plasticity of spines in the mature brain and have different implications for information storage and maintenance in neuronal circuits, it is important to identify which factor(s) is responsible for the disparity in previous reports. Although factors such as animal age and transgenic mouse line may contribute to dissimilar findings of spine dynamics, a recent study suggests that these differences may be due primarily to the use of an open-skull glass window versus a thinned-skull window for in vivo imaging. Using the same transgenic mouse line, Xu et al. examined the impact of cranial window type utilized during in vivo imaging on adult spine dynamics in barrel cortex (91, 92). These researchers found that spines are remarkably stable under thinned-skull windows (∼200 μm in diameter) but quite plastic under a large open-skull glass window (∼4–5 mm in diameter). In addition, the use of open-skull preparations leads to a substantial loss of spines within the first two weeks after surgery, followed by a high degree of spine turnover that lasts for at least an additional three to four weeks. Extensive glial activation was also found for at least one month after surgery under open-skull conditions, but not for thinned-skull preparations. These observations suggest that the discrepancy in adult spine dynamics is heavily, if not exclusively, influenced by the type of cranial window used for in vivo imaging (56, 91, 92). Nevertheless, open-skull preparations offer unique advantages for certain experiments (e.g., pharmacological perturbations, simultaneous in vivo imaging, and electrophysiological recordings) that cannot be easily performed under thinned-skull conditions, and thus the open-skull technique remains valuable. It is therefore important to develop methods that utilize open-skull windows while minimizing perturbations to the brain (86, 93).

Note that the above imaging studies were mainly from examinations of spines in apical dendrites of layer V pyramidal neurons in the cortex. It is difficult to say whether the observations of spine stability can be generalized to the entire dendritic tree of layer V pyramidal cells or to different neuronal types. Recent studies have shown only slightly different degrees of spine dynamics in different cell types (86, 87, 94). Experiments in other brain areas such as the hippocampus and the olfactory bulb have demonstrated high levels of stability of adult dendritic branches and spines over hours to days (95, 96). Imaging studies of inhibitory cortical neurons and adult-born neurons in the olfactory bulb have shown a small degree of changes in dendritic branches and spines in the mature brain (97, 98). Thus, in various cell types and cortical layers, a large percentage of dendritic spines likely persist for the duration of an animal's life.

Moreover, although the vast majority of spines may persist throughout adulthood, spine morphology undergoes change in the living cortex both during development and in adulthood (23, 54, 86, 87). Because spine size correlates with synaptic strength, changes in spine morphology indicate that synaptic strength can be modified without synapse turnover (24, 99). Such alterations may be important in the rapid plasticity seen in long-term potentiation and depression (42, 62, 99–106). Thus, although most adult spines remain stable in adulthood and provide a physical substrate for long-term information storage, a small degree of spine turnover and rapid changes in synaptic strength can occur throughout life and may underlie various forms of learning and plasticity in the mature brain.

EXPERIENCE-DEPENDENT SPINE FORMATION AND ELIMINATION

A fundamental feature of the nervous system is that experience profoundly affects patterns of neuronal connectivity and behavior throughout life. It is generally believed that a functional, mature nervous system is formed from an initial pool of imprecise synaptic connections by the selective establishment of some synapses and the elimination of others based on a combination of genetics and experience. A wide variety of experimental evidence has shown that experience/neuronal activity plays a critical role

in regulating synaptogenesis, particularly in developing neural circuits (22, 39, 42, 100, 107–118). For example, long-term sensory deprivation from birth continuing to young adulthood often reduces the number of synapses, whereas enriched environments during the same period increase dendritic branching and synapse number in various brain regions (32, 34, 101, 116, 117, 119–121). In cultured brain slices, induction of long-term potentiation or electric activity can lead to a rapid outgrowth of dendritic protrusions, an increase in spine motility, and synapse formation (21, 22, 28, 122). Conversely, induction of long-term depression and reduced neuronal activity cause shrinkage of existing dendritic spines, reduced spine motility, and a net loss in spine number (21, 102, 104, 123). In corroboration with these findings, in vivo imaging in young rat barrel cortex has shown that dendritic filopodia and spines are highly dynamic at approximately postnatal days 10–14 and that sensory deprivation via whisker trimming significantly reduces the motility of dendritic protrusions during a brief period (postnatal days 11–13) (50). Together, these studies underscore the pivotal importance of sensory experience and patterns of neuronal activity in the dynamics of dendritic spines and synaptic connections during early stages of development.

As evidence demonstrating the role of experience in synaptogenesis during early development accumulates, it remains unclear how experience is involved during the period of substantial loss of synaptic connections. Recently, time-lapse, two-photon imaging has been used to examine the effect of sensory experience on the rates of dendritic spine elimination and formation in mouse barrel cortex during late postnatal development. By trimming all the whiskers on one side of the mouse facial pad daily, Zuo et al. (39) showed that, in young adolescence, when extensive spine loss occurs, sensory deprivation over weeks preferentially reduces the rate of spine elimination rather than the rate of formation (**Figure 3**). Furthermore, as spines become increasingly stabilized in adulthood, sensory deprivation over a period of weeks has no significant effect on the rate of spine elimination or formation. Because sensory deprivation leads to a reduction in the net loss of spines during late postnatal development, experience-dependent sculpting of neuronal connectivity appears to be a major event in the reorganization of neural circuits during late postnatal development. These results resonate with studies of activity-dependent elimination of synaptic connections between spinal motor neurons and their muscle targets and between climbing fibers and Purkinje cells (113, 115, 124–126). In the peripheral nervous system, multiple motor neurons make synaptic contacts with individual muscle cells. This initial exuberance is then pruned by competition mediated by neuronal activity (111, 115, 124, 125, 127). Thus, experience- and activity-dependent sculpting of neuronal connectivity seems vital to the maturation of the mammalian nervous system.

All the aforementioned studies have revealed the fundamental role of experience in spine plasticity in the developing mammalian cortex. However, the degree to which experience modifies synaptic connectivity in the adult brain is still in question. Several recent studies have addressed this issue by in vivo two-photon imaging of dendritic spines after altering sensory inputs to the barrel cortex of GFP/YFP-expressing transgenic mice. Zuo et al. (39) showed that, although sensory deprivation by removing all the whiskers for two weeks had no significant effect on spine elimination and formation in adult barrel cortex (greater than four months), long-term deprivation for two months caused a small (~3%) but significant reduction in the rate of spine elimination in adult barrel cortex. Consistent with this result, the authors also showed that trimming every other whisker in a chessboard pattern on one side of the facial pad (chessboard deprivation) over two weeks preferentially reduced spine elimination without affecting the rate of spine formation in mice from four to six weeks of age, whereas similar manipulation had no significant impact on the rates of spine elimination and formation in adult mice (39). Thus, sensory experience

continues to sculpt synaptic connections in mature neural circuits, albeit to a lesser degree as compared with the effect of sensory experience on young adolescent barrel cortex. In another study, Trachtenberg et al. (53) found that in 5–10-week-old mice, chessboard trimming over a period of four days had no effect on overall spine density but caused a substantial increase in spine turnover. A more recent study showed that in mice between two to five months of age, chessboard trimming for up to 20 days continues to alter spine dynamics without impacting spine density in barrel cortex (94). Note again that different types of cranial windows were used for imaging experience-dependent spine dynamics and likely contributed to varying degrees of experience-dependent spine dynamics in the experiments described above. However, regardless of the differences in imaging techniques, the above studies demonstrate that under laboratory housing and sensory-deprived environments, spines are capable of forming and retracting, but only to a very limited degree, in adulthood. Such a small degree of experience-dependent spine remodeling in adult cortex may underlie the continuous adaptation of animals to their environment.

Many lines of evidence indicate that anatomical features such as dendrite- and axon-branching patterns in the developing mammals are highly sensitive to changes in the external environment and that this sensitivity decreases as animals mature into adulthood (21, 107, 108, 118, 128, 129). The studies described above suggest that experience-dependent spine plasticity also decreases as an animal's age increases. Specifically, there appears to be a temporal window after birth and before adulthood when spine plasticity and the malleability of the nervous system progressively diminish. Thus, one important extrapolation from studies on experience-dependent spine dynamics is that the period of heightened sensitivity to experience, termed the critical period, may be intimately related to the period of heightened spine plasticity. In addition, a small degree of spine turnover does occur and can be modified by experience in the mature brain. Therefore, the adult brain must retain some capacity, the extent and degree of which are yet to be determined, to form new synapses and rewire its circuitry outside of the critical period.

GENETIC CONTROL OF SPINE FORMATION AND ELIMINATION

One major focus of research in spine plasticity is to understand how and to what degree spine formation and elimination are regulated by intrinsic genetic programs. By removing or over-expressing individual genes, studies in the past two decades have made important progresses in identifying genes that are important for regulating the number and size of dendritic spines (40, 41, 58, 130). These genes encode a wide variety of proteins such as neurotransmitter receptors (131–136), adhesion molecules (130, 132, 137–141), postsynaptic density proteins (81, 142–150), protein kinases and phosphatases (151–153), and actin cytoskeleton and its regulatory elements (17, 43, 154–163). Because spine formation is intimately associated with synaptogenesis, genes participating in synapse formation generally have varying effects on the generation of dendritic filopodia and spines. The initial formation of spines can occur in the absence of neuronal activity. Even more dramatically, as seen in the cerebellum, spines of Purkinje neurons can form in the absence of axonal terminals (77, 78). Thus, it appears that some aspects of the ontogenesis of spines are genetically determined and do not rely on synaptic transmission.

One emerging scenario regarding the genetic control of spine plasticity is that various gene products converge to regulate actin polymerization and depolymerization in dendritic spines, leading to spine formation or elimination (40, 43, 58, 160, 162–165). Actin filaments form the main cytoskeleton of dendritic spines and underlie rapid spine motility. The Rho family of small GTPases, including RhoA, RhoB, and Rac, regulates the dynamics of the actin cytoskeleton and is an important contributor to spine formation and elimination. Over-

expression or suppression of these molecules as well as their interacting proteins results in changes in the density of dendritic filopodia and spines in vivo and in vitro (146, 166–169). In addition, neuronal activity can induce extensive remodeling of the actin cytoskeleton via calcium influx through glutamate receptors of the N-methyl-D-aspartate (NMDA) and the α-amino-3-hydroxy-5-methyl-4-isoxazole propionate (AMPA) subtypes, which are highly concentrated at the postsynaptic spines (21, 39, 158, 164, 170–172). Much evidence has shown that NMDA and AMPA receptor activation is essential for synapse formation and/or maintenance and that the calcium/calmodulin-dependent protein kinase II pathway at least partly mediates the effect of these receptors on actin dynamics (103, 173).

Although substantial progress has been made in the past decades on the gene regulation of spine development and plasticity, most of the studies so far have focused on the role of individual genes in regulating spine density from a single-time-point observation. Owing to the dynamic nature of spinogenesis, it would be important to combine molecular and time-lapse imaging approaches to determine the roles of different genes and signaling pathways in various phases of spine formation and maintenance. Furthermore, gene expression is dynamically regulated in vivo, and the lack of a specific gene's product is likely to alter not only the function of a particular gene but also patterns of neuronal activity. Thus, in living animals, better temporal and spatial control of gene expression in combination with improved in vivo imaging and electrophysiology are necessary to decipher the gene regulation of spine plasticity in the future. Such knowledge would be invaluable for understanding the pathogenic mechanisms underlying brain disorders, as discussed below.

DENDRITIC SPINE PATHOLOGY IN BRAIN DISORDERS

Because spine morphology and number are intimately linked to neuronal function, altered spine structures and derangements of the rules governing spine elimination and formation are likely to have diverse detrimental effects on neural circuits and to contribute to an array of cognitive impairments. Indeed, a number of psychiatric and neurological diseases are associated with substantial alterations in either spine morphology or density. In fragile X syndrome (FXS), for example, which is the most frequent form of inheritable mental retardation (174–176), spines are found in much higher density and display a more immature, long, and thin form. People with trisomy 21 (Down's syndrome) show decreased spine density in both the neocortex and hippocampus (177–180). Abnormal spine morphology and number also occur in other brain disorders, such as addiction, anxiety, and depression, that are often linked to environmental factors such as malnutrition, abnormal hormone levels, and chronic drug abuse (181–184). In animal models, repeated exposure to psychomotor stimulants such as amphetamine and cocaine causes increases in both the dendritic branching and spine density on apical dendrites of pyramidal cells in the prefrontal cortex as well as on medium spiny neurons of the nucleus accumbens (185, 186). The fact that substantial alterations in spine structure or density occur in various brain diseases strongly suggests that spine dynamics are well-balanced under normal circumstances and such a balance may be vulnerable to disruption through a variety of mechanisms under different pathological conditions.

Many psychiatric diseases may be viewed fundamentally as disorders of early development that either progress with age or do not manifest until early adulthood. Therefore, slight changes in spine plasticity may accumulate over time and lead to a substantial change in spine density, manifesting later in life or under specific environmental stressors. Consistent with such a view are postmortem studies on the brains of schizophrenic patients, which exhibit decreased spine density in neocortical pyramidal neurons, often in late adolescence (187–189). Important progress has been made in the past decades toward identifying

genes that may be responsible for psychiatric diseases and toward developing mouse models for mechanistic studies (190–194). For example, investigators have shown that FXS is caused by a mutation of the *fragile mental retardation 1* (*Fmr1*) gene located on the X chromosome (194–196). The *Fmr1* gene encodes a protein, the fragile X mental retardation protein (FMRP), which binds to many mRNA ligands and acts as a translational suppressor in different subcellular locations, including dendrites and dendritic spines (197). Similar alterations in dendritic spine morphology and density and overlapping behavioral phenotypes have been found in *Fmr1* knockout mice and human patients (193, 198). These findings suggest that such a mouse model is excellent for studying the mechanisms underlying the development and expression of FXS. The challenge in the future is to determine how the lack of certain gene products, such as Fmr1, leads to aberrant spine development and maintenance and how such structural abnormalities contribute to the functional deficits seen in disease states. The ability to image synaptic structure in living animals, with repeat sampling through development and aging, will be essential to monitoring changes in synaptic structure associated with disease symptoms.

Changes in dendritic spines have also been examined in other neurological disease models such as prion disease, seizures, and ischemia (95, 199–201). Recent in vivo imaging revealed a progressive loss of dendritic spines in a scrapie prion mouse model (199) and a rapid loss of spines in ischemia mouse models (37, 202–204). Although moderate reduction in blood flow is not associated with immediate damage to synaptic structure, a severe stroke can cause a rapid loss of spines and dendritic swelling (37, 203). Importantly, swelling dendrites and lost spines can recover after reperfusion of occluded vessels (37, 205). Recent in vivo imaging has also shown that dendritic arbors in the peri-infarct cortex undergo an extensive reorganization after stroke, including an increase in spine turnover and a recovery of spine density (206). Such enhanced synaptic and circuit remodeling

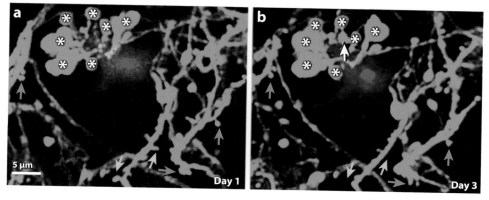

Figure 4

In vivo imaging of dendritic pathology in an Alzheimer's disease mouse model. The figure shows repeated imaging of cortical dendrites and axons (*green*) near an amyloid deposit (*red*) in cortical layer 1 of a mouse at six months of age. Panel *a* shows imaging at day one, and panel *b* shows imaging at day three. Although most spines (*blue arrows*) and varicosities (*asterisks*) were stable over two days, some structural changes [e.g., spine loss (*yellow arrows*) and varicosity formation (*white arrow*)] occurred near the deposits. Adapted from Reference 36 with permission.

may contribute to spontaneous recovery and behavioral improvement after ischemia.

Structural changes of dendritic spines also likely play an important role in the pathogenesis of many age-related neurodegenerative diseases. This is especially true in Alzheimer's disease (AD), in which the best correlate of cognitive dysfunction is thought to be the loss of synapses as seen by immunohistochemistry and EM in postmortem tissue (207–209). Amyloid peptide deposition, one of the pathological hallmarks of AD, is commonly associated with dystrophic neurites and synaptic loss (210, 211). By crossbreeding a transgenic mouse model of AD with mice expressing YFPs in subsets of neurons, researchers have studied the degree and time course of synaptic disruption associated with fibrillar amyloid deposits in vivo with TPM (**Figure 4**) (36). Time-lapse imaging over days to weeks revealed extensive formation and elimination of dendritic spines near the periphery of the amyloid plaques, suggesting that amyloid deposition triggers continuous remodeling of nearby neuronal structures. Furthermore, dendrites passing through or near fibrillar amyloid deposits all exhibit some degree of spine loss and reduction in shaft diam-

eter, whereas the axons in the vicinity of the deposits develop abnormally large varicosities. This finding implies that local dendritic and axonal abnormalities associated with amyloid deposits can eventually lead to synapse loss and the breakage of dendrites and axons (36, 212, 213). These studies suggest that the accumulation of fibrillar amyloid deposits leads to large-scale and permanent disruption of synaptic connections, contributing to the cognitive decline and memory loss seen in AD.

As outlined above, neurological disorders and neurodegenerative diseases are frequently associated with aberrant spine structure and plasticity in critical brain regions. Because dendritic spines are the major sites of neuronal connections in the brain, it is not surprising that changes in spine plasticity would have a significant impact on disease pathogenesis and progression. The important issue is to determine the mechanisms underlying abnormal spine dynamics and how alterations in spine plasticity contribute to the clinical symptoms and disease progression. The answers to such questions will undoubtedly provide important insight into the onset, progression, and treatment of psychiatric and neurological diseases.

DISCLOSURE STATEMENT

The authors are not aware of any biases that might be perceived as affecting the objectivity of this review.

ACKNOWLEDGMENTS

We thank Drs. Feng Pan and Guang Yang for many invaluable discussions. Support was provided by the National Institutes of Health to W.-B. Gan. D.H. Bhatt was supported by the residency program in the Department of Psychiatry, Weill Medical College of Cornell University. S. Zhang was supported by NYS Spinal Cord Injury Research Program Postdoctoral Fellowship.

LITERATURE CITED

1. Ramón y Cajal S. 1888. Estructura de los centros nerviosos de las aves. *Rev. Trim. Histol. Norm. Patol.* 1:1–10

2. Ramón y Cajal S. 1899. *Reglas y Consejos Sobre Investigación Biológica*. Madrid: Imprenta Fontanet

3. Ramón y Cajal S. 1899. *La Textura del Sistema Nervioso del Hombre y de los Vertebrados*. Madrid: Moya

4. De Felipe J, Marco P, Fairen A, Jones EG. 1997. Inhibitory synaptogenesis in mouse somatosensory cortex. *Cereb. Cortex* 7:619–34

5. Garcia-Lopez P, Garcia-Marin V, Freire M. 2007. The discovery of dendritic spines by Cajal in 1888 and its relevance in the present neuroscience. *Prog. Neurobiol.* 83:110–30

6. DeFelipe J. 2006. Brain plasticity and mental processes: Cajal again. *Nat. Rev. Neurosci.* 7:811–17

7. Demoor J. 1896. La plasticité morphologique des neurones cérébraux. *Arch. Biol. Brux.* 14:723–52

8. Stefanowska M. 1897. Les appendices terminaux des dendrites cérébraux et leur différents états physiologiques. *Ann. Soc. R. Sci. Med. Nat. Brux.* 6:351–407

9. Stefanowska M. 1897. Sur les appendices des dendrites. *Bull. Soc. R. Sci. Med. Nat. Brux.* 55:76–81

10. Foster M, Sherrington CS. 1897. *A Textbook of Physiology*. Part III. *The Central Nervous System*. London: MacMillan & Co.

11. Shepherd GM, Erulkar SD. 1997. Centenary of the synapse: from Sherrington to the molecular biology of the synapse and beyond. *Trends Neurosci.* 20:385–92

12. Gray EG. 1959. Electron microscopy of synaptic contacts on dendrite spines of the cerebral cortex. *Nature* 183:1592–93

13. Gray EG. 1959. Axo-somatic and axo-dendritic synapses of the cerebral cortex: an electron microscope study. *J. Anat.* 93:420–33

14. Blomberg F, Cohen RS, Siekevitz P. 1977. The structure of postsynaptic densities isolated from dog cerebral cortex. II. Characterization and arrangement of some of the major proteins within the structure. *J. Cell Biol.* 74:204–25

15. Crick F. 1982. Do dendritic spines twitch? *Trends Neurosci.* 5:44–46

16. Fifkova E, Delay RJ. 1982. Cytoplasmic actin in neuronal processes as a possible mediator of synaptic plasticity. *J. Cell Biol.* 95:345–50

17. Matus A, Ackermann M, Pehling G, Byers HR, Fujiwara K. 1982. High actin concentrations in brain dendritic spines and postsynaptic densities. *Proc. Natl. Acad. Sci. USA* 79:7590–94

18. Shepherd GM. 1996. The dendritic spine: a multifunctional integrative unit. *J. Neurophysiol.* 75:2197–210

19. Yuste R, Bonhoeffer T. 2004. Genesis of dendritic spines: insights from ultrastructural and imaging studies. *Nat. Rev. Neurosci.* 5:24–34

20. Dailey ME, Smith SJ. 1996. The dynamics of dendritic structure in developing hippocampal slices. *J. Neurosci.* 16:2983–94

21. Dunaevsky A, Tashiro A, Majewska A, Mason C, Yuste R. 1999. Developmental regulation of spine motility in the mammalian central nervous system. *Proc. Natl. Acad. Sci. USA* 96:13438–43

22. Engert F, Bonhoeffer T. 1999. Dendritic spine changes associated with hippocampal long-term synaptic plasticity. *Nature* 399:66–70

23. Grutzendler J, Kasthuri N, Gan WB. 2002. Long-term dendritic spine stability in the adult cortex. *Nature* 420:812–16

24. Harris KM, Stevens JK. 1989. Dendritic spines of CA 1 pyramidal cells in the rat hippocampus: serial electron microscopy with reference to their biophysical characteristics. *J. Neurosci.* 9:2982–97

25. Huttenlocher PR. 1990. Morphometric study of human cerebral cortex development. *Neuropsychologia* 28:517–27

26. Matsuzaki M, Honkura N, Ellis-Davies GC, Kasai HM. 2004. Structural basis of long-term potentiation in single dendritic spines. *Nature* 429:761–66

27. Rakic P, Bourgeois JP, Eckenhoff MF, Zecevic N, Goldman-Rakic PS. 1986. Concurrent overproduction of synapses in diverse regions of the primate cerebral cortex. *Science* 232:232–35

28. Toni N, Buchs PA, Nikonenko I, Bron CR, Muller D. 1999. LTP promotes formation of multiple spine synapses between a single axon terminal and a dendrite. *Nature* 402:421–25

29. Ziv NE, Smith SJ. 1996. Evidence for a role of dendritic filopodia in synaptogenesis and spine formation. *Neuron* 17:91–102

30. Duan H, Wearne SL, Rocher AB, Macedo A, Morrison JH, Hof PR. 2003. Age-related dendritic and spine changes in corticocortically projecting neurons in macaque monkeys. *Cereb. Cortex* 13:950–61

31. Fiala JC, Spacek J, Harris KM. 2002. Dendritic spine pathology: cause or consequence of neurological disorders? *Brain Res. Brain Res. Rev.* 39:29–54

32. Greenough WT, Volkmar FR, Juraska JM. 1973. Effects of rearing complexity on dendritic branching in frontolateral and temporal cortex of the rat. *Exp. Neurol.* 41:371–78

33. Kleim JA, Vij K, Ballard DH, Greenough WT. 1997. Learning-dependent synaptic modifications in the cerebellar cortex of the adult rat persist for at least four weeks. *J. Neurosci.* 17:717–21

34. Knott GW, Quairiaux C, Genoud C, Welker E. 2002. Formation of dendritic spines with GABAergic synapses induced by whisker stimulation in adult mice. *Neuron* 34:265–73

35. Terry RD, Masliah E, Salmon DP, Butters N, DeTeresa R, et al. 1991. Physical basis of cognitive alterations in Alzheimer's disease: Synapse loss is the major correlate of cognitive impairment. *Ann. Neurol.* 30:572–80

36. Tsai J, Grutzendler J, Duff K, Gan WB. 2004. Fibrillar amyloid deposition leads to local synaptic abnormalities and breakage of neuronal branches. *Nat. Neurosci.* 7:1181–83

37. Zhang S, Boyd J, Delaney K, Murphy TH. 2005. Rapid reversible changes in dendritic spine structure in vivo gated by the degree of ischemia. *J. Neurosci.* 25:5333–38

38. Zuo Y, Lin A, Chang P, Gan WB. 2005. Development of long-term dendritic spine stability in diverse regions of cerebral cortex. *Neuron* 46:181–89

39. Zuo Y, Yang G, Kwon E, Gan WB. 2005. Long-term sensory deprivation prevents dendritic spine loss in primary somatosensory cortex. *Nature* 436:261–65

40. Calabrese B, Wilson MS, Halpain S. 2006. Development and regulation of dendritic spine synapses. *Physiology (Bethesda)* 21:38–47

41. Carlisle HJ, Kennedy MB. 2005. Spine architecture and synaptic plasticity. *Trends Neurosci.* 28:182–87

42. Yuste R, Bonhoeffer T. 2001. Morphological changes in dendritic spines associated with long-term synaptic plasticity. *Annu. Rev. Neurosci.* 24:1071–89

43. Matus A. 2000. Actin-based plasticity in dendritic spines. *Science* 290:754–58

44. Segal M. 2002. Changing views of Cajal's neuron: the case of the dendritic spine. *Prog. Brain Res.* 136:101–7

45. Alvarez VA, Sabatini BL. 2007. Anatomical and physiological plasticity of dendritic spines. *Annu. Rev. Neurosci.* 30:79–97

46. Yuste R, Majewska A, Holthoff K. 2000. From form to function: calcium compartmentalization in dendritic spines. *Nat. Neurosci.* 3:653–59

47. Denk W, Strickler JH, Webb WW. 1990. Two-photon laser scanning fluorescence microscopy. *Science* 248:73–76

48. Theer P, Hasan MT, Denk W. 2003. Two-photon imaging to a depth of 1000 μm in living brains by use of a Ti:Al$_2$O$_3$ regenerative amplifier. *Opt. Lett.* 28:1022–24

49. Feng G, Mellor RH, Bernstein M, Keller-Peck C, Nguyen QT, et al. 2000. Imaging neuronal subsets in transgenic mice expressing multiple spectral variants of GFP. *Neuron* 28:41–51

50. Lendvai B, Stern EA, Chen B, Svoboda K. 2000. Experience-dependent plasticity of dendritic spines in the developing rat barrel cortex in vivo. *Nature* 404:876–81

51. Sin WC, Haas K, Ruthazer ES, Cline HT. 2002. Dendrite growth increased by visual activity requires NMDA receptor and Rho GTPases. *Nature* 419:475–80

52. Stettler DD, Yamahachi H, Li W, Denk W, Gilbert CD. 2006. Axons and synaptic boutons are highly dynamic in adult visual cortex. *Neuron* 49:877–87

53. Trachtenberg JT, Chen BE, Knott GW, Feng G, Sanes JR, et al. 2002. Long-term in vivo imaging of experience-dependent synaptic plasticity in adult cortex. *Nature* 420:788–94

54. Zuo Y, Lin A, Chang P, Gan WB. 2005. Development of long-term dendritic spine stability in diverse regions of cerebral cortex. *Neuron* 46:181–89

55. Zuo Y, Yang G, Kwon E, Gan WB. 2005. Long-term sensory deprivation prevents dendritic spine loss in primary somatosensory cortex. *Nature* 436:261–65

56. Grutzendler J, Gan WB. 2006. Two-photon imaging of synaptic plasticity and pathology in the living mouse brain. *NeuroRx* 3:489–96

57. Portera Cailliau C, Yuste R. 2001. On the function of dendritic filopodia. *Rev. Neurol.* 33:1158–66

58. Tada T, Sheng M. 2006. Molecular mechanisms of dendritic spine morphogenesis. *Curr. Opin. Neurobiol.* 16:95–101

59. Nimchinsky EA, Sabatini BL, Svoboda K. 2002. Structure and function of dendritic spines. *Annu. Rev. Physiol.* 64:313–53

60. Hayashi Y, Majewska AK. 2005. Dendritic spine geometry: functional implication and regulation. *Neuron* 46:529–32

61. McKinney RA. 2005. Physiological roles of spine motility: development, plasticity and disorders. *Biochem. Soc. Trans.* 33:1299–302

62. Bonhoeffer T, Yuste R. 2002. Spine motility. Phenomenology, mechanisms, and function. *Neuron* 35:1019–27

63. Portera-Cailliau C, Pan DT, Yuste R. 2003. Activity-regulated dynamic behavior of early dendritic protrusions: evidence for different types of dendritic filopodia. *J. Neurosci.* 23:7129–42

64. Tashiro A, Dunaevsky A, Blazeski R, Mason CA, Yuste R. 2003. Bidirectional regulation of hippocampal mossy fiber filopodial motility by kainate receptors: a two-step model of synaptogenesis. *Neuron* 38:773–84

65. Jontes JD, Smith SJ. 2000. Filopodia, spines, and the generation of synaptic diversity. *Neuron* 27:11–14

66. Fiala JC, Feinberg M, Popov V, Harris KM. 1998. Synaptogenesis via dendritic filopodia in developing hippocampal area CA1. *J. Neurosci.* 18:8900–11

67. Deng J, Dunaevsky A. 2005. Dynamics of dendritic spines and their afferent terminals: Spines are more motile than presynaptic boutons. *Dev. Biol.* 277:366–77

68. Konur S, Yuste R. 2004. Imaging the motility of dendritic protrusions and axon terminals: roles in axon sampling and synaptic competition. *Mol. Cell. Neurosci.* 27:427–40

69. Dunaevsky A, Blazeski R, Yuste R, Mason C. 2001. Spine motility with synaptic contact. *Nat. Neurosci.* 4:685–86

70. Korkotian E, Segal M. 2001. Regulation of dendritic spine motility in cultured hippocampal neurons. *J. Neurosci.* 21:6115–24

71. Lohmann C, Bonhoeffer T. 2008. A role for local calcium signaling in rapid synaptic partner selection by dendritic filopodia. *Neuron* 59:253–60

72. Richards DA, Mateos JM, Hugel S, de Paola V, Caroni P, et al. 2005. Glutamate induces the rapid formation of spine head protrusions in hippocampal slice cultures. *Proc. Natl. Acad. Sci. USA* 102:6166–71

73. Hua JY, Smear MC, Baier H, Smith SJ. 2005. Regulation of axon growth in vivo by activity-based competition. *Nature* 434:1022–26

74. Meyer MP, Smith SJ. 2006. Evidence from in vivo imaging that synaptogenesis guides the growth and branching of axonal arbors by two distinct mechanisms. *J. Neurosci.* 26:3604–14

75. Cline H, Haas K. 2008. The regulation of dendritic arbor development and plasticity by glutamatergic synaptic input: a review of the synaptotrophic hypothesis. *J. Physiol.* 586:1509–17

76. Wu GY, Zou DJ, Rajan I, Cline H. 1999. Dendritic dynamics in vivo change during neuronal maturation. *J. Neurosci.* 19:4472–83

77. Cesa R, Morando L, Strata P. 2005. Purkinje cell spinogenesis during architectural rewiring in the mature cerebellum. *Eur. J. Neurosci.* 22:579–86

78. Bravin M, Morando L, Vercelli A, Rossi F, Strata P. 1999. Control of spine formation by electrical activity in the adult rat cerebellum. *Proc. Natl. Acad. Sci. USA* 96:1704–9

79. Knott GW, Holtmaat A, Wilbrecht L, Welker E, Svoboda K. 2006. Spine growth precedes synapse formation in the adult neocortex in vivo. *Nat. Neurosci.* 9:1117–24

80. Nagerl UV, Kostinger G, Anderson JC, Martin KA, Bonhoeffer T. 2007. Protracted synaptogenesis after activity-dependent spinogenesis in hippocampal neurons. *J. Neurosci.* 27:8149–56

81. Marrs GS, Green SH, Dailey ME. 2001. Rapid formation and remodeling of postsynaptic densities in developing dendrites. *Nat. Neurosci.* 4:1006–13

82. Huttenlocher PR. 1979. Synaptic density in human frontal cortex—developmental changes and effects of aging. *Brain Res.* 163:195–205

83. Markus EJ, Petit TL. 1987. Neocortical synaptogenesis, aging, and behavior: lifespan development in the motor-sensory system of the rat. *Exp. Neurol.* 96:262–78

84. Rakic P, Bourgeois JP, Goldman-Rakic PS. 1994. Synaptic development of the cerebral cortex: implications for learning, memory, and mental illness. *Prog. Brain Res.* 102:227–43

85. Lubke J, Albus K. 1989. The postnatal development of layer VI pyramidal neurons in the cat's striate cortex, as visualized by intracellular Lucifer yellow injections in aldehyde-fixed tissue. *Brain Res. Dev. Brain Res.* 45:29–38

86. Majewska AK, Newton JR, Sur M. 2006. Remodeling of synaptic structure in sensory cortical areas in vivo. *J. Neurosci.* 26:3021–29

87. Holtmaat AJ, Trachtenberg JT, Wilbrecht L, Shepherd GM, Zhang X, et al. 2005. Transient and persistent dendritic spines in the neocortex in vivo. *Neuron* 45:279–91

88. Kirov SA, Petrak LJ, Fiala JC, Harris KM. 2004. Dendritic spines disappear with chilling but proliferate excessively upon rewarming of mature hippocampus. *Neuroscience* 127:69–80

89. Ottersen OP, Helm PJ. 2002. How hardwired is the brain? *Nature* 420:751–52

90. Meyer MP, Niell CM, Smith SJ. 2003. Brain imaging: How stable are synaptic connections? *Curr. Biol.* 13:R180–82

91. Xu HT, Pan F, Yang G, Gan WB. 2007. Choice of cranial window type for in vivo imaging affects dendritic spine turnover in the cortex. *Nat. Neurosci.* 10:549–51

92. Pan F, Gan WB. 2008. Two-photon imaging of dendritic spine development in the mouse cortex. *Dev. Neurobiol.* 68:771–78

93. Nishiyama H, Fukaya M, Watanabe M, Linden DJ. 2007. Axonal motility and its modulation by activity are branch-type specific in the intact adult cerebellum. *Neuron* 56:472–87

94. Holtmaat A, Wilbrecht L, Knott GW, Welker E, Svoboda K. 2006. Experience-dependent and cell-type-specific spine growth in the neocortex. *Nature* 441:979–83

95. Mizrahi A, Crowley JC, Shtoyerman E, Katz LC. 2004. High-resolution in vivo imaging of hippocampal dendrites and spines. *J. Neurosci.* 24:3147–51

96. Mizrahi A, Katz LC. 2003. Dendritic stability in the adult olfactory bulb. *Nat. Neurosci.* 6:1201–7

97. Lee WC, Huang H, Feng G, Sanes JR, Brown EN, et al. 2006. Dynamic remodeling of dendritic arbors in GABAergic interneurons of adult visual cortex. *PLoS Biol.* 4:e29

98. Mizrahi A. 2007. Dendritic development and plasticity of adult-born neurons in the mouse olfactory bulb. *Nat. Neurosci.* 10:444–52

99. Matsuzaki M, Honkura N, Ellis-Davies GC, Kasai H. 2004. Structural basis of long-term potentiation in single dendritic spines. *Nature* 429:761–66

100. Buonomano DV, Merzenich MM. 1998. Cortical plasticity: from synapses to maps. *Annu. Rev. Neurosci.* 21:149–86

101. Grossman AW, Churchill JD, Bates KE, Kleim JA, Greenough WT. 2002. A brain adaptation view of plasticity: Is synaptic plasticity an overly limited concept? *Prog. Brain Res.* 138:91–108

102. Nagerl UV, Eberhorn N, Cambridge SB, Bonhoeffer T. 2004. Bidirectional activity-dependent morphological plasticity in hippocampal neurons. *Neuron* 44:759–67

103. Segal M. 2001. Rapid plasticity of dendritic spine: hints to possible functions? *Prog. Neurobiol.* 63:61–70

104. Zhou Q, Homma KJ, Poo MM. 2004. Shrinkage of dendritic spines associated with long-term depression of hippocampal synapses. *Neuron* 44:749–57

105. Kauer JA, Malenka RC, Nicoll RA. 1988. A persistent postsynaptic modification mediates long-term potentiation in the hippocampus. *Neuron* 1:911–17

106. Oliet SH, Malenka RC, Nicoll RA. 1997. Two distinct forms of long-term depression coexist in CA1 hippocampal pyramidal cells. *Neuron* 18:969–82

107. Wiesel TN. 1982. The postnatal development of the visual cortex and the influence of environment. *Biosci. Rep.* 2:351–77

108. Hubel DH, Wiesel TN, LeVay S. 1977. Plasticity of ocular dominance columns in monkey striate cortex. *Philos. Trans. R. Soc. London Ser. B* 278:377–409

109. Bailey CH, Kandel ER. 1993. Structural changes accompanying memory storage. *Annu. Rev. Physiol.* 55:397–426

110. Fox K. 2002. Anatomical pathways and molecular mechanisms for plasticity in the barrel cortex. *Neuroscience* 111:799–814

111. Gan WB, Lichtman JW. 1998. Synaptic segregation at the developing neuromuscular junction. *Science* 282:1508–11

112. Gordon JA, Stryker MP. 1996. Experience-dependent plasticity of binocular responses in the primary visual cortex of the mouse. *J. Neurosci.* 16:3274–86

113. Kakizawa S, Yamasaki M, Watanabe M, Kano M. 2000. Critical period for activity-dependent synapse elimination in developing cerebellum. *J. Neurosci.* 20:4954–61

114. Katz LC, Shatz CJ. 1996. Synaptic activity and the construction of cortical circuits. *Science* 274:1133–38

115. Walsh MK, Lichtman JW. 2003. In vivo time-lapse imaging of synaptic takeover associated with naturally occurring synapse elimination. *Neuron* 37:67–73

116. Winkelmann E, Brauer K, Klutz K. 1977. [Spine density of lamina V pyramidal cells in the visual cortex of laboratory rats after lengthy dark exposure]. *J. Hirnforsch.* 18:21–28 (In German)

117. Winkelmann E, Brauer K, Werner L. 1976. [Studies on the spine density in lamina V pyramidal cells of the visual cortex in young and subadult rats after dark-rearing and destruction of the dorsal nucleus in the lateral geniculate body]. *J. Hirnforsch.* 17:489–500 (In German)

118. Zhang LI, Tao HW, Holt CE, Harris WA, Poo M. 1998. A critical window for cooperation and competition among developing retinotectal synapses. *Nature* 395:37–44

119. Fiala BA, Joyce JN, Greenough WT. 1978. Environmental complexity modulates growth of granule cell dendrites in developing but not adult hippocampus of rats. *Exp. Neurol.* 59:372–83

120. Valverde F. 1967. Apical dendritic spines of the visual cortex and light deprivation in the mouse. *Exp. Brain Res.* 3:337–52

121. Valverde F. 1971. Rate and extent of recovery from dark rearing in the visual cortex of the mouse. *Brain Res.* 33:1–11

122. Maletic-Savatic M, Malinow R, Svoboda K. 1999. Rapid dendritic morphogenesis in CA1 hippocampal dendrites induced by synaptic activity. *Science* 283:1923–27

123. Wallace W, Bear MF. 2004. A morphological correlate of synaptic scaling in visual cortex. *J. Neurosci.* 24:6928–38

124. Lichtman JW, Colman H. 2000. Synapse elimination and indelible memory. *Neuron* 25:269–78

125. Lichtman JW, Magrassi L, Purves D. 1987. Visualization of neuromuscular junctions over periods of several months in living mice. *J. Neurosci.* 7:1215–22

126. Hashimoto K, Kano M. 2005. Postnatal development and synapse elimination of climbing fiber to Purkinje cell projection in the cerebellum. *Neurosci. Res.* 53:221–28

127. Kasthuri N, Lichtman JW. 2004. Structural dynamics of synapses in living animals. *Curr. Opin. Neurobiol.* 14:105–11

128. Chapman B, Jacobson MD, Reiter HO, Stryker MP. 1986. Ocular dominance shift in kitten visual cortex caused by imbalance in retinal electrical activity. *Nature* 324:154–56

129. Crair MC, Gillespie DC, Stryker MP. 1998. The role of visual experience in the development of columns in cat visual cortex. *Science* 279:566–70

130. Arikkath J, Reichardt LF. 2008. Cadherins and catenins at synapses: roles in synaptogenesis and synaptic plasticity. *Trends Neurosci.* 31:487–94

131. Passafaro M, Nakagawa T, Sala C, Sheng M. 2003. Induction of dendritic spines by an extracellular domain of AMPA receptor subunit GluR2. *Nature* 424:677–81

132. Saglietti L, Dequidt C, Kamieniarz K, Rousset MC, Valnegri P, et al. 2007. Extracellular interactions between GluR2 and N-cadherin in spine regulation. *Neuron* 54:461–77

133. Alvarez VA, Ridenour DA, Sabatini BL. 2007. Distinct structural and ionotropic roles of NMDA receptors in controlling spine and synapse stability. *J. Neurosci.* 27:7365–76

134. Ultanir SK, Kim JE, Hall BJ, Deerinck T, Ellisman M, Ghosh A. 2007. Regulation of spine morphology and spine density by NMDA receptor signaling in vivo. *Proc. Natl. Acad. Sci. USA* 104:19553–58

135. Kauer JA, Malenka RC, Nicoll RA. 1988. NMDA application potentiates synaptic transmission in the hippocampus. *Nature* 334:250–52

136. Luscher C, Xia H, Beattie EC, Carroll RC, von Zastrow M, et al. 1999. Role of AMPA receptor cycling in synaptic transmission and plasticity. *Neuron* 24:649–58

137. Bozdagi O, Shan W, Tanaka H, Benson DL, Huntley GW. 2000. Increasing numbers of synaptic puncta during late-phase LTP: N-cadherin is synthesized, recruited to synaptic sites, and required for potentiation. *Neuron* 28:245–59

138. Togashi H, Abe K, Mizoguchi A, Takaoka K, Chisaka O, Takeichi M. 2002. Cadherin regulates dendritic spine morphogenesis. *Neuron* 35:77–89

139. Abe K, Chisaka O, Van Roy F, Takeichi M. 2004. Stability of dendritic spines and synaptic contacts is controlled by αN-catenin. *Nat. Neurosci.* 7:357–63

140. Furutani Y, Matsuno H, Kawasaki M, Sasaki T, Mori K, Yoshihara Y. 2007. Interaction between telencephalin and ERM family proteins mediates dendritic filopodia formation. *J. Neurosci.* 27:8866–76

141. Matsuno H, Okabe S, Mishina M, Yanagida T, Mori K, Yoshihara Y. 2006. Telencephalin slows spine maturation. *J. Neurosci.* 26:1776–86

142. Ehrlich I, Klein M, Rumpel S, Malinow R. 2007. PSD-95 is required for activity-driven synapse stabilization. *Proc. Natl. Acad. Sci. USA* 104:4176–81

143. Xu W, Schluter OM, Steiner P, Czervionke BL, Sabatini B, Malenka RC. 2008. Molecular dissociation of the role of PSD-95 in regulating synaptic strength and LTD. *Neuron* 57:248–62

144. Okabe S, Miwa A, Okado H. 2001. Spine formation and correlated assembly of presynaptic and postsynaptic molecules. *J. Neurosci.* 21:6105–14

145. Dalva MB, Takasu MA, Lin MZ, Shamah SM, Hu L, et al. 2000. EphB receptors interact with NMDA receptors and regulate excitatory synapse formation. *Cell* 103:945–56

146. Penzes P, Beeser A, Chernoff J, Schiller MR, Eipper BA, et al. 2003. Rapid induction of dendritic spine morphogenesis by trans-synaptic ephrinB-EphB receptor activation of the Rho-GEF kalirin. *Neuron* 37:263–74

147. Irie F, Yamaguchi Y. 2004. EPHB receptor signaling in dendritic spine development. *Front. Biosci.* 9:1365–73

148. Murai KK, Nguyen LN, Irie F, Yamaguchi Y, Pasquale EB. 2003. Control of hippocampal dendritic spine morphology through ephrin-A3/EphA4 signaling. *Nat. Neurosci.* 6:153–60

149. Chen L, Chetkovich DM, Petralia RS, Sweeney NT, Kawasaki Y, et al. 2000. Stargazin regulates synaptic targeting of AMPA receptors by two distinct mechanisms. *Nature* 408:936–43

150. Schnell E, Sizemore M, Karimzadegan S, Chen L, Bredt DS, Nicoll RA. 2002. Direct interactions between PSD-95 and stargazin control synaptic AMPA receptor number. *Proc. Natl. Acad. Sci. USA* 99:13902–7

151. Pak DT, Sheng M. 2003. Targeted protein degradation and synapse remodeling by an inducible protein kinase. *Science* 302:1368–73

152. Calabrese B, Halpain S. 2005. Essential role for the PKC target MARCKS in maintaining dendritic spine morphology. *Neuron* 48:77–90

153. Dunah AW, Hueske E, Wyszynski M, Hoogenraad CC, Jaworski J, et al. 2005. LAR receptor protein tyrosine phosphatases in the development and maintenance of excitatory synapses. *Nat. Neurosci.* 8:458–67

154. Hering H, Sheng M. 2003. Activity-dependent redistribution and essential role of cortactin in dendritic spine morphogenesis. *J. Neurosci.* 23:11759–69

155. Nakagawa T, Engler JA, Sheng M. 2004. The dynamic turnover and functional roles of α-actinin in dendritic spines. *Neuropharmacology* 47:734–45

156. Ryu J, Liu L, Wong TP, Wu DC, Burette A, et al. 2006. A critical role for myosin IIb in dendritic spine morphology and synaptic function. *Neuron* 49:175–82

157. Tada T, Simonetta A, Batterton M, Kinoshita M, Edbauer D, Sheng M. 2007. Role of Septin cytoskeleton in spine morphogenesis and dendrite development in neurons. *Curr. Biol.* 17:1752–58

158. Ackermann M, Matus A. 2003. Activity-induced targeting of profilin and stabilization of dendritic spine morphology. *Nat. Neurosci.* 6:1194–200

159. Iki J, Inoue A, Bito H, Okabe S. 2005. Bi-directional regulation of postsynaptic cortactin distribution by BDNF and NMDA receptor activity. *Eur. J. Neurosci.* 22:2985–94

160. Sekino Y, Kojima N, Shirao T. 2007. Role of actin cytoskeleton in dendritic spine morphogenesis. *Neurochem. Int.* 51:92–104

161. Takahashi H, Sekino Y, Tanaka S, Mizui T, Kishi S, Shirao T. 2003. Drebrin-dependent actin clustering in dendritic filopodia governs synaptic targeting of postsynaptic density-95 and dendritic spine morphogenesis. *J. Neurosci.* 23:6586–95

162. Fischer M, Kaech S, Knutti D, Matus A. 1998. Rapid actin-based plasticity in dendritic spines. *Neuron* 20:847–54

163. Oertner TG, Matus A. 2005. Calcium regulation of actin dynamics in dendritic spines. *Cell Calcium* 37:477–82

164. Fischer M, Kaech S, Wagner U, Brinkhaus H, Matus A. 2000. Glutamate receptors regulate actin-based plasticity in dendritic spines. *Nat. Neurosci.* 3:887–94

165. Feng J, Yan Z, Ferreira A, Tomizawa K, Liauw JA, et al. 2000. Spinophilin regulates the formation and function of dendritic spines. *Proc. Natl. Acad. Sci. USA* 97:9287–92

166. Tolias KF, Bikoff JB, Kane CG, Tolias CS, Hu L, Greenberg ME. 2007. The Rac1 guanine nucleotide exchange factor Tiam1 mediates EphB receptor-dependent dendritic spine development. *Proc. Natl. Acad. Sci. USA* 104:7265–70

167. Newey SE, Velamoor V, Govek EE, Van Aelst L. 2005. Rho GTPases, dendritic structure, and mental retardation. *J. Neurobiol.* 64:58–74

168. Ma XM, Wang Y, Ferraro F, Mains RE, Eipper BA. 2008. Kalirin-7 is an essential component of both shaft and spine excitatory synapses in hippocampal interneurons. *J. Neurosci.* 28:711–24

169. Rabiner CA, Mains RE, Eipper BA. 2005. Kalirin: a dual Rho guanine nucleotide exchange factor that is so much more than the sum of its many parts. *Neuroscientist* 11:148–60

170. Tian L, Stefanidakis M, Ning L, Van Lint P, Nyman-Huttunen H, et al. 2007. Activation of NMDA receptors promotes dendritic spine development through MMP-mediated ICAM-5 cleavage. *J. Cell Biol.* 178:687–700

171. Blackstone C, Sheng M. 2002. Postsynaptic calcium signaling microdomains in neurons. *Front. Biosci.* 7:d872–85

172. De Roo M, Klauser P, Mendez P, Poglia L, Muller D. 2008. Activity-dependent PSD formation and stabilization of newly formed spines in hippocampal slice cultures. *Cereb. Cortex* 18:151–61

173. Wu GY, Cline HT. 1998. Stabilization of dendritic arbor structure in vivo by CaMKII. *Science* 279:222–26

174. Hinton VJ, Brown WT, Wisniewski K, Rudelli RD. 1991. Analysis of neocortex in three males with the fragile X syndrome. *Am. J. Med. Genet.* 41:289–94

175. Irwin SA, Patel B, Idupulapati M, Harris JB, Crisostomo RA, et al. 2001. Abnormal dendritic spine characteristics in the temporal and visual cortices of patients with fragile-X syndrome: a quantitative examination. *Am. J. Med. Genet.* 98:161–67

176. Wisniewski KE, Segan SM, Miezejeski CM, Sersen EA, Rudelli RD. 1991. The fra(X) syndrome: neurological, electrophysiological, and neuropathological abnormalities. *Am. J. Med. Genet.* 38:476–80

177. Takashima S, Becker LE, Armstrong DL, Chan F. 1981. Abnormal neuronal development in the visual cortex of the human fetus and infant with Down's syndrome. A quantitative and qualitative Golgi study. *Brain Res.* 225:1–21

178. Marin-Padilla M. 1976. Pyramidal cell abnormalities in the motor cortex of a child with Down's syndrome. A Golgi study. *J. Comp. Neurol.* 167:63–81

179. Takashima S, Iida K, Mito T, Arima M. 1994. Dendritic and histochemical development and ageing in patients with Down's syndrome. *J. Intellect. Disabil. Res.* 38(Pt. 3):265–73

180. Ferrer I, Gullotta F. 1990. Down's syndrome and Alzheimer's disease: dendritic spine counts in the hippocampus. *Acta Neuropathol.* 79:680–85

181. Benitez-Bribiesca L, De la Rosa-Alvarez I, Mansilla-Olivares A. 1999. Dendritic spine pathology in infants with severe protein-calorie malnutrition. *Pediatrics* 104:e21

182. Chen Y, Dube CM, Rice CJ, Baram TZ. 2008. Rapid loss of dendritic spines after stress involves derangement of spine dynamics by corticotropin-releasing hormone. *J. Neurosci.* 28:2903–11

183. Robinson TE, Gorny G, Mitton E, Kolb B. 2001. Cocaine self-administration alters the morphology of dendrites and dendritic spines in the nucleus accumbens and neocortex. *Synapse* 39:257–66

184. Shors TJ, Chua C, Falduto J. 2001. Sex differences and opposite effects of stress on dendritic spine density in the male versus female hippocampus. *J. Neurosci.* 21:6292–97

185. Robinson TE, Kolb B. 1999. Alterations in the morphology of dendrites and dendritic spines in the nucleus accumbens and prefrontal cortex following repeated treatment with amphetamine or cocaine. *Eur. J. Neurosci.* 11:1598–604

186. Norrholm SD, Bibb JA, Nestler EJ, Ouimet CC, Taylor JR, Greengard P. 2003. Cocaine-induced proliferation of dendritic spines in nucleus accumbens is dependent on the activity of cyclin-dependent kinase-5. *Neuroscience* 116:19–22

187. Glantz LA, Lewis DA. 2000. Decreased dendritic spine density on prefrontal cortical pyramidal neurons in schizophrenia. *Arch. Gen. Psychiatry* 57:65–73

188. Sweet RA, Henteleff RA, Zhang W, Sampson AR, Lewis DA. 2008. Reduced dendritic spine density in auditory cortex of subjects with schizophrenia. *Neuropsychopharmacology.* In press

189. Garey LJ, Ong WY, Patel TS, Kanani M, Davis A, et al. 1998. Reduced dendritic spine density on cerebral cortical pyramidal neurons in schizophrenia. *J. Neurol. Neurosurg. Psychiatry* 65:446–53

190. Amir RE, Van Den Veyver IB, Wan M, Tran CQ, Francke U, Zoghbi HY. 1999. Rett syndrome is caused by mutations in X-linked MECP2, encoding methyl-CpG-binding protein 2. *Nat. Genet.* 23:185–88

191. Chen RZ, Akbarian S, Tudor M, Jaenisch R. 2001. Deficiency of methyl-CpG binding protein-2 in CNS neurons results in a Rett-like phenotype in mice. *Nat. Genet.* 27:327–31

192. Guy J, Hendrich B, Holmes M, Martin JE, Bird A. 2001. A mouse *Mecp2*-null mutation causes neurological symptoms that mimic Rett syndrome. *Nat. Genet.* 27:322–26

193. The Dutch-Belgian Fragile X Consortium. 1994. *Fmr1* knockout mice: a model to study fragile X mental retardation. *Cell* 78:23–33

194. Verkerk AJ, Pieretti M, Sutcliffe JS, Fu YH, Kuhl DP, et al. 1991. Identification of a gene (*FMR-1*) containing a CGG repeat coincident with a breakpoint cluster region exhibiting length variation in fragile X syndrome. *Cell* 65:905–14

195. Fu YH, Kuhl DP, Pizzuti A, Pieretti M, Sutcliffe JS, et al. 1991. Variation of the CGG repeat at the fragile X site results in genetic instability: resolution of the Sherman paradox. *Cell* 67:1047–58

196. Oberle I, Rousseau F, Heitz D, Kretz C, Devys D, et al. 1991. Instability of a 550-base pair DNA segment and abnormal methylation in fragile X syndrome. *Science* 252:1097–102

197. Dolen G, Bear MF. 2008. Role for metabotropic glutamate receptor 5 (mGluR5) in the pathogenesis of fragile X syndrome. *J. Physiol.* 586:1503–8

198. Bakker CE, Oostra BA. 2003. Understanding fragile X syndrome: insights from animal models. *Cytogenet. Genome Res.* 100:111–23

199. Fuhrmann M, Mitteregger G, Kretzschmar H, Herms J. 2007. Dendritic pathology in prion disease starts at the synaptic spine. *J. Neurosci.* 27:6224–33

200. Rensing N, Ouyang Y, Yang XF, Yamada KA, Rothman SM, Wong M. 2005. In vivo imaging of dendritic spines during electrographic seizures. *Ann. Neurol.* 58:888–98

201. Zeng LH, Xu L, Rensing NR, Sinatra PM, Rothman SM, Wong M. 2007. Kainate seizures cause acute dendritic injury and actin depolymerization in vivo. *J. Neurosci.* 27:11604–13

202. Zhang ZG, Zhang L, Ding G, Jiang Q, Zhang RL, et al. 2005. A model of mini-embolic stroke offers measurements of the neurovascular unit response in the living mouse. *Stroke* 36:2701–4

203. Zhang S, Murphy TH. 2007. Imaging the impact of cortical microcirculation on synaptic structure and sensory-evoked hemodynamic responses in vivo. *PLoS Biol.* 5:e119

204. Enright LE, Zhang S, Murphy TH. 2007. Fine mapping of the spatial relationship between acute ischemia and dendritic structure indicates selective vulnerability of layer V neuron dendritic tufts within single neurons in vivo. *J. Cereb. Blood Flow Metab.* 27:1185–200

205. Murphy TH, Li P, Betts K, Liu R. 2008. Two-photon imaging of stroke onset in vivo reveals that NMDA-receptor independent ischemic depolarization is the major cause of rapid reversible damage to dendrites and spines. *J. Neurosci.* 28:1756–72

206. Brown CE, Li P, Boyd JD, Delaney KR, Murphy TH. 2007. Extensive turnover of dendritic spines and vascular remodeling in cortical tissues recovering from stroke. *J. Neurosci.* 27:4101–9

207. Baloyannis SJ, Manolidis SL, Manolidis LS. 1992. The acoustic cortex in Alzheimer's disease. *Acta Otolaryngol. Suppl.* 494:1–13

208. Davies CA, Mann DM, Sumpter PQ, Yates PO. 1987. A quantitative morphometric analysis of the neuronal and synaptic content of the frontal and temporal cortex in patients with Alzheimer's disease. *J. Neurol. Sci.* 78:151–64

209. Masliah E, Terry RD, DeTeresa RM, Hansen LA. 1989. Immunohistochemical quantification of the synapse-related protein synaptophysin in Alzheimer disease. *Neurosci. Lett.* 103:234–39

210. Knowles RB, Wyart C, Buldyrev SV, Cruz L, Urbanc B, et al. 1999. Plaque-induced neurite abnormalities: implications for disruption of neural networks in Alzheimer's disease. *Proc. Natl. Acad. Sci. USA* 96:5274–79

211. Phinney AL, Deller T, Stalder M, Calhoun ME, Frotscher M, et al. 1999. Cerebral amyloid induces aberrant axonal sprouting and ectopic terminal formation in amyloid precursor protein transgenic mice. *J. Neurosci.* 19:8552–59

212. Spires TL, Meyer-Luehmann M, Stern EA, McLean PJ, Skoch J, et al. 2005. Dendritic spine abnormalities in amyloid precursor protein transgenic mice demonstrated by gene transfer and intravital multiphoton microscopy. *J. Neurosci.* 25:7278–87

213. Spires-Jones TL, Meyer-Luehmann M, Osetek JD, Jones PB, Stern EA, et al. 2007. Impaired spine stability underlies plaque-related spine loss in an Alzheimer's disease mouse model. *Am. J. Pathol.* 171:1304–11

Endocannabinoid Signaling and Long-Term Synaptic Plasticity

Boris D. Heifets and Pablo E. Castillo

Dominick P. Purpura Department of Neuroscience, Albert Einstein College of Medicine, Bronx, New York 10461; email: pcastill@aecom.yu.edu

Annu. Rev. Physiol. 2009. 71:283–306

First published online as a Review in Advance on November 13, 2008

The *Annual Review of Physiology* is online at physiol.annualreviews.org

This article's doi:
10.1146/annurev.physiol.010908.163149

Copyright © 2009 by Annual Reviews.
All rights reserved

0066-4278/09/0315-0283$20.00

Key Words

cannabinoid, CB1, LTD, LTP, synaptic transmission, STDP, learning, drug addiction, Parkinson's disease

Abstract

Endocannabinoids (eCBs) are key activity-dependent signals regulating synaptic transmission throughout the central nervous system. Accordingly, eCBs are involved in neural functions ranging from feeding homeostasis to cognition. There is great interest in understanding how exogenous (e.g., cannabis) and endogenous cannabinoids affect behavior. Because behavioral adaptations are widely considered to rely on changes in synaptic strength, the prevalence of eCB-mediated long-term depression (eCB-LTD) at synapses throughout the brain merits close attention. The induction and expression of eCB-LTD, although remarkably similar at various synapses, are controlled by an array of regulatory influences that we are just beginning to uncover. This complexity endows eCB-LTD with important computational properties, such as coincidence detection and input specificity, critical for higher CNS functions like learning and memory. In this article, we review the major molecular and cellular mechanisms underlying eCB-LTD, as well as the potential physiological relevance of this widespread form of synaptic plasticity.

INTRODUCTION

eCB: endocannabinoid

LTD: long-term depression

AEA: anandamide

2-AG: 2-arachidonyl glycerol

Endocannabinoid system: encompasses eCBs, their synthetic and degradative enzymes, eCB transporters, and cannabinoid receptors

Depolarization-induced suppression of inhibition/ excitation (DSI/ DSE): transient (typically <1 min) suppression of inhibitory (DSI) or excitatory (DSE) synaptic transmission induced by brief (~seconds) postsynaptic depolarization

Retrograde signaling by endocannabinoids (eCBs) has emerged as a major theme in the study of synaptic plasticity. Neuronal activity releases these neuromodulators, which activate the presynaptic type 1 cannabinoid receptor [CB1R, a G protein–coupled receptor (GPCR)], suppressing neurotransmitter release at both excitatory and inhibitory synapses in a short- and a long-term manner (1–6). Examples of eCB-mediated long-term depression (eCB-LTD) have been reported throughout the brain. The discovery of eCB-LTD demonstrates that eCBs can have a long-term impact on neural function and has certainly expanded our view of the role of eCB signaling in the central nervous system (CNS). As a result of the past 10 years of research on eCB-signaling and synaptic plasticity, eCBs have become the most prominent example of retrograde signaling molecules in the CNS. Moreover, eCB-LTD is by now one of the best examples of presynaptic forms of long-term plasticity. Since our previous review of eCB-mediated synaptic plasticity (3), a number of exciting developments have prompted a fresh appraisal of this field. Thorough review articles on eCB signaling and synaptic transmission have recently been published (5, 6). As the focus of a growing number of studies, eCB-LTD now warrants its own dedicated review. In this article, we review the major developments regarding the induction, expression, computational properties, developmental regulation, and physiological relevance of eCB-LTD.

OVERVIEW OF ENDOCANNABINOID-MEDIATED LONG-TERM DEPRESSION

Endocannabinoids and the CB1 Receptor

Endogenous cannabinoids, or endocannabinoids (eCBs), are so called because of their close relationship to exogenous cannabinoids, the most famous being Δ^9-tetrahydrocannabinol (THC), a major psychoactive component of marijuana (7). One of the most surprising discoveries came from studies revealing a unique, previously unknown receptor for THC within the mammalian central nervous system (CNS). The CB1R was ultimately isolated, cloned, and characterized as a GPCR that signals via the $\alpha_{i/o}$ family of G proteins (8, 9). Soon after, investigators discovered that CB1Rs not only bind THC but also have an endogenous ligand; anandamide (AEA) was the first of several eCBs isolated, among which 2-arachidonyl glycerol (2-AG) is most prevalent in brain (10, 11). The physiological relevance of the eCB system quickly emerged: Neuronal activity was identified as a potent stimulus for eCB release, and eCBs were found to act as retrograde messengers, regulating synaptic transmission through presynaptic CB1Rs (1). Since 2001, eCBs have been identified as triggers for short- and long-term plasticity at synapses throughout the brain (3), in agreement with the ubiquitous expression of CB1Rs (12, 13).

Activation of the CB1R, and subsequent long-term reduction of transmitter release at the same synapse, defines eCB-LTD. CB1Rs also mediate short-term plasticity, exemplified by depolarization-induced suppression of inhibition or excitation (DSI or DSE, respectively). DSI/DSE are well characterized in the hippocampus and cerebellum and have been observed in other brain areas (2, 5). CB1Rs can engage a wide range of effector molecules, including (but not limited to) voltage-dependent Ca^{2+} channels (VDCCs), K^+ channels, protein kinase A (PKA), and mitogen-activated protein kinase (MAPK) (recently reviewed in Reference 14).

Endocannabinoid-Mediated Long-Term Depression: A Widespread Phenomenon in the Brain

The first evidence implicating eCB signaling in LTD emerged in 2002 at excitatory synapses in dorsal striatum (15). Since then, eCB-LTD has been reported in several other brain structures such as the nucleus accumbens (16), amygdala (17–19), hippocampus

Table 1 eCB-LTD is a widespread phenomenon in the brain[a]

Brain structure	Synapse	Induction protocol	Reference(s)
Neocortex			
Visual	Excitatory inputs, L5 pyramidal cell pairs	STDP (postsynaptic bursts)	26, 27
		STDP and LFS	28
	Excitatory inputs, L4 → L2/3 pyramidal neurons (immature visual cortex)	TBS	29
Somatosensory (barrel cortex)	Excitatory inputs to L2/3 pyramidal neurons	STDP (postsynaptic bursts)	30, 31
Prefrontal	L2/3 → L5/6	Moderate 10 Hz stimulation for 10 min	32
Hippocampus	Inhibitory inputs to CA1 pyramidal cells	HFS, TBS	20–22, 24, 25
	Excitatory inputs to CA1 pyramidal cells (immature hippocampus)	HFS	23
Amygdala	Inhibitory inputs to basolateral amygdala	LFS	17–19
Dorsal striatum	Excitatory inputs to medium spiny neurons	LFS, STDP	15, 39, 49, 50
Nucleus accumbens	Excitatory inputs to medium spiny neurons	Moderate 13 Hz stimulation for 10 min	16, 142
Cerebellum	Excitatory inputs to stellate interneurons	Four bouts of 25 stimuli at 30 Hz, delivered at 0.33 Hz	33
Ventral tegmental area (VTA)	Inhibitory inputs to dopamine neurons	Moderate 10 Hz stimulation for 5 min	34
Dorsal cochlear nucleus	Excitatory inputs to cartwheel cells	STDP	35
Superior colliculus	Inhibitory inputs to tectal neurons in vitro	HFS	36

[a]Examples of eCB-LTD were confirmed by use of CB1R antagonists, CB1R knockout mice, or both. Abbreviations used: HFS, high-frequency stimulation (100 Hz); LFS, low-frequency stimulation (1 Hz); STDP, spike timing–dependent plasticity; TBS, theta burst stimulation.

(20–25), visual cortex (26–29), somatosensory cortex (30, 31), prefrontal cortex (32), cerebellum (33), ventral tegmental area (VTA) (34), brain stem (35), and superior colliculus (36). As shown in **Table 1**, eCB-LTD is a widely expressed phenomenon in the brain that can be observed at both excitatory and inhibitory synapses. This prevalence strongly suggests that eCB-LTD may be a fundamental mechanism for making long-term changes to neural circuits and behavior.

INDUCTION OF ENDOCANNABINOID-MEDIATED LONG-TERM DEPRESSION

Strong similarities in the pattern of eCB-LTD induction and expression are evident at both excitatory and inhibitory synapses from brainstem to cortex (3). The main objectives of this section are to define (*a*) synaptic events that trigger eCB production/release, (*b*) how eCB production, release, and degradation may be regulated, and (*c*) which presynaptic events are required for successful induction of eCB-LTD.

Synaptic Events Triggering Endocannabinoid-Mediated Synaptic Plasticity

eCB-LTD induction typically begins with a transient increase in activity at glutamatergic afferents and a concomitant release of eCBs from a target (postsynaptic) neuron (**Figure 1**). eCBs then travel backward (retrogradely) across the synapse, activating CB1Rs on the presynaptic terminals of either the original afferent (homosynaptic eCB-LTD) or nearby afferents (heterosynaptic eCB-LTD) (3). Mounting evidence over the past few years indicates that eCB-LTD induction requires presynaptic

**Spike timing–
dependent plasticity
(STDP):** an
associative plasticity
paradigm in which the
precise temporal
relationship between
presynaptic and
postsynaptic activity
determines the
direction (LTP or
LTD) and the
magnitude of the
resulting change in
synaptic strength

t-LTP/t-LTD:
LTP/LTD induced by
spike timing protocols

LTP: long-term
potentiation

activity of the target afferent, independent of its role in triggering eCB release (see below).

Induction protocols differ widely across examples of eCB-LTD (**Table 1**). Some forms of eCB-LTD are induced by the tetanic stimulation of afferents, an approach used extensively in the study of synaptic plasticity. A number of induction protocols effectively produce eCB-LTD, from 100 pulses at 1 Hz to 100 Hz, or the more patterned theta burst stimulation (TBS). These afferent-only induction protocols for eCB-LTD have not been rigorously compared at most synapses, but at least for eCB-LTD at hippocampal inhibitory synapses, induction is effective over a broad range of frequencies (24, 37). eCB-LTD has also been found by repetitively firing presynaptic and postsynaptic neurons at fixed intervals with respect to each other. This induction protocol can yield spike timing–

dependent plasticity (STDP), in which the order and interval of the two spikes dictate the direction [i.e., t-LTD or t-LTP (long-term potentiation)] and magnitude of plasticity (for a recent review, see Reference 38). At this time, several instances of STDP feature a mechanistically distinct t-LTD and t-LTP, with only the former being CB1R dependent (eCB-t-LTD). eCB-t-LTD's presynaptic locus of expression

---

Figure 1

Schematic summary of the endocannabinoid-mediated long-term depression (eCB-LTD) induction mechanism. One of the most common initial steps of induction is the activation of postsynaptic group I metabotropic glutamate (Glu) receptors (mGluR-I), following repetitive activation of excitatory inputs. These receptors couple to phospholipase C (PLC) via $G\alpha_{q/11}$ subunits and promote diacylglycerol (DAG) formation [from phosphatidylinositol (PI)], which is converted into the eCB 2-arachidonyl glycerol (2-AG) by diacylglycerol lipase (DGL). 2-AG is then released from the postsynaptic neuron by a mechanism that presumably requires an eCB membrane transporter (EMT) and binds presynaptic type 1 cannabinoid receptors (CB1Rs). Postsynaptic Ca^{2+} can contribute to eCB mobilization either by stimulating PLC or in a PLC-independent, uncharacterized manner. This Ca^{2+} rise can be through voltage-dependent Ca^{2+} channels (VDCC) activated by action potentials (e.g., during spike timing–dependent protocols), through NMDARs, or released from the endoplasmic reticulum (ER), e.g., by the PLC product, inositol 1,4,5-trisphosphate (IP_3). In some synapses, induction of eCB-LTD by afferent-only stimulation protocols can occur independently of postsynaptic Ca^{2+}. At the presynaptic terminal, the CB1R inhibits adenylyl cyclase (AC) via $G\alpha_{i/o}$, reducing protein kinase A (PKA) activity. Induction of eCB-LTD may also require a presynaptic Ca^{2+} rise through presynaptic VDCCs or NMDARs (not shown) or release from Ca^{2+} internal stores. Activation of the Ca^{2+}-sensitive phosphatase calcineurin (CaN), in conjunction with the reduction of PKA activity, shifts the kinase/phosphatase activity balance, thereby promoting dephosphorylation of a presynaptic target (T) that mediates a long-lasting reduction of transmitter release. For clarity, eCB-LTD mediated by anandamide (AEA), and the contribution of other GPCRs in mobilizing eCBs, is not shown.

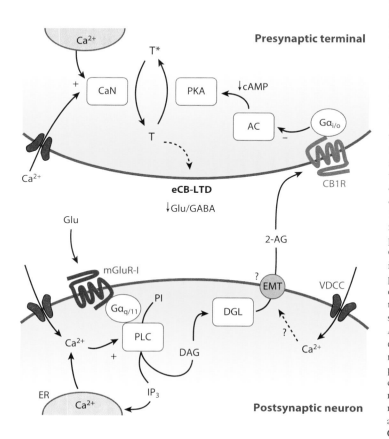

is indistinguishable from those forms induced with afferent-only stimulation protocols, described above. A given synapse may well be able to support eCB-LTD induced by both afferent-only and spike timing–dependent protocols (15, 39).

Stimuli for Endocannabinoid Production

Activity-dependent release of eCBs from the postsynaptic cell is the common trigger of all forms of eCB-LTD. Extensive studies have established that two separate processes, neurotransmitter release and neuronal depolarization, mediate eCB release (for comprehensive reviews on this topic, see References 5 and 40). For example, glutamate can stimulate eCB release through activating type I metabotropic glutamate receptors (mGluR-I), which can initiate eCB synthesis on demand through effectors of the $G\alpha_{q/11}$ subunit (41, 42). Neuronal depolarization is also a potent trigger of eCB release through a less understood Ca^{2+}-dependent process that does not depend on G proteins (1, 40).

Metabotropic receptor–dependent endocannabinoid release. Studies of neural tissue conducted in vivo or in acutely prepared brain slices indicate that eCB release need not be linked to special metabotropic receptors or a particular route of Ca^{2+} entry. In addition to mGluR-I, metabotropic dopamine (type 2; D_2), muscarinic acetylcholine (type 1/3; M_1/M_3), serotonin (type 2; $5\text{-}HT_2$), orexin, and cholecystokinin receptors are all effective stimuli for eCB release (43–48). Most of these GPCRs signal through $G\alpha_{q/11}$, engaging phospholipase C (PLC) and diacylglycerol lipase (DGL), which form the eCB 2-AG (**Figure 1**). However, alternative mechanisms coupling these receptors to eCB release have also been proposed. For example, in basolateral amygdala, mGluR1 appears to stimulate AEA production through a pathway involving adenylyl cyclase (AC) and PKA (18). Despite the marked differences in eCB synthetic route and even eCB identity, eCB-

LTD is expressed in a nearly identical fashion in amygdala and hippocampus (19), reinforcing the notion that eCB release by any means is a core component of eCB-LTD.

D_2 receptors (D_2Rs) differ from the other metabotropic receptors named above: They typically signal through the $G\alpha_{i/o}$ pathway and stimulate the formation of AEA, rather than 2-AG, although the precise synthetic pathway is not known (40, 43). D_2Rs are required for corticostriatal eCB-LTD (39, 49, 50), and several lines of evidence from slice electrophysiology suggest a role for AEA in this process (50, 51). In part because of the paucity of pharmacological tools for manipulating AEA biosynthesis, it has been difficult to determine the precise mechanism by which D_2Rs mediate eCB-LTD. D_2Rs may directly stimulate eCB release from medium spiny neurons (MSNs), or, as recently suggested, D_2Rs may exert an indirect effect on eCB release via cholinergic interneurons in dorsal striatum (52). However, a recent study has shown that the presence or absence of D_2Rs on MSNs determines whether eCB-LTD can be induced at an afferent glutamatergic synapse (50), owing to a D_2R-mediated facilitation of eCB release (49). Indeed, using spike timing–dependent induction protocols, researchers have recently reported that corticostriatal synapses onto D_1-expressing neurons (a population largely distinct from D_2R-expressing MSNs) can undergo eCB-LTD, provided induction occurs in the presence of a D_1R antagonist (39). Although these corticostriatal afferents are capable of expressing CB1R agonist–induced LTD (49, 53), specific postsynaptic factors may thus impose intrinsic limitations on eCB-LTD.

Calcium-dependent endocannabinoid release. Although not universally required, Ca^{2+}-dependent mechanisms play a pivotal role in many forms of eCB-LTD (15, 16, 26, 28, 30–33, 35, 54). Like metabotropic signaling pathways, which converge through multiple pathways onto eCB release to ultimately trigger eCB-LTD, Ca^{2+}-dependent eCB release also has a similarly broad set of initiating mechanisms. Ca^{2+} influx through NMDARs, L-type and T-type VDCCs, and Ca^{2+} release from intracellular stores have all been reported to drive eCB release in brain slices (30, 31, 54–57). The source of postsynaptic Ca^{2+} differs among individual examples of eCB-LTD. Although no specific mode of Ca^{2+} rise stands as a universal requirement, it is not clear to what extent different Ca^{2+} sources can compensate for each other.

Integrating Signals for Endocannabinoid Release

The metabotropic and Ca^{2+}-driven mechanisms can operate independently—mGluR-I can release substantial amounts of eCB even under strong Ca^{2+}-buffering conditions (18, 20, 42, 58)—but they can also act synergistically (25, 49, 59–61). Some of the enzymes mediating this cooperativity have been characterized. For example, the PLCβ isoform is critical for mGluR-I-triggered release of 2-AG (61). $G\alpha_{q/11}$, through PLCβ, provides the substrate for DGL, leading to 2-AG formation (62). In CA1 pyramidal cells of the hippocampus, activity of the isoform PLCβ1 is significantly enhanced by increased Ca^{2+} concentrations, which allows for a potential coincidence detection mechanism in the postsynaptic neuron, gating eCB release and possibly eCB-LTD (61). Another isoform of this enzyme, PLCβ4, mediates a similar process in the cerebellum at parallel fiber–to–Purkinje cell synapses (59, 63). Here, activation of mGluR-I renders eCB release more sensitive to intracellular Ca^{2+} levels, raised by either direct Purkinje cell depolarization or stimulation of powerful climbing-fiber inputs to the Purkinje cell. Although eCB-LTD has not been assessed in transgenic animals lacking the various PLCβ enzymes, pharmacological blockade of PLC effectively blocks eCB-LTD in hippocampus, VTA, and prefrontal cortex (20, 32, 34, 64). It remains to be seen whether similar synergy of eCB release between metabotropic and Ca^{2+}-driven signals occurs with enzymes in the AEA-synthetic pathway, as suggested for phospholipase D (40).

Spatial constraints on endocannabinoid production. Investigations into PLCβ have revealed a compelling picture of how the temporal pattern of synaptic input can amplify or constrain eCB production. In principle, this mechanism by PLCβ can also provide some spatial specificity to the eCB signal. Neuronal depolarization presumably triggers eCB synthesis/release wherever Ca^{2+} influx occurs; only active synaptic inputs engage the metabotropic facilitation of the eCB signal, whereas silent afferents do not. Even so, eCBs can trigger both homo- and heterosynaptic plasticity. How else, then, might a synapse limit the spatial impact of eCB release?

Another strategy neurons may employ to provide spatial specificity to the eCB signal is to confine the Ca^{2+} signal itself, thereby restricting the lateral extent of eCB release. This hypothesis has been explored in the context of eCB-LTD at cerebellar parallel fiber–to–stellate cell synapses (33). Stellate cells are especially notable for their absence of structural specializations, such as the spine head, which in other neurons effectively compartmentalizes second-messenger signals (65). Rather, the Ca^{2+} signal is biochemically "confined" through interactions with the Ca^{2+}-buffering protein parvalbumin as well as a generalized reduction of small-molecule mobility within the dendrite (33). These mechanisms can isolate the eCB release generated by a single glutamatergic synapse from its neighboring synapse, 15 μm away.

The spatial distribution of eCB-synthetic machinery may also impact synaptic transmission. As discussed above, DGL plays a key role in mGluR-dependent eCB production and several forms of eCB-LTD. Anatomical studies have revealed differential subcellular localization with respect to the synapse in various brain structures (66–69). The relevance of these findings to eCB-LTD induction remains unexplored.

Endocannabinoid degradation. Enzymes regulating the eCB lifetime in the synaptic cleft are also potentially significant targets for regulating eCB-LTD induction. The degradation of 2-AG and AEA largely depends on the enzymes monoacylglycerol lipase (MGL) and fatty acid amide hydrolase (FAAH), respectively (40). Other enzymes may also participate in eCB catabolism, although with less specificity (e.g., COX-2) (70, 71). In hippocampus, where 2-AG is very likely to be the dominant functional eCB, electron microscopic studies reveal a predominantly presynaptic pattern of MGL expression, supporting a role for MGL in regulating the effective concentration of eCBs near their site of action (72). Consistent with this idea, MGL inhibition can transform a subthreshold tetanus into an effective stimulus for eCB-LTD in the prefrontal cortex (32), mirroring similar results with FAAH inhibitors in eCB-LTD in neocortex, amygdala, and dorsal striatum (18, 26, 50). At present, it is not known if MGL or FAAH activity is dynamically regulated in neurons, although some data have also shown that elevated intracellular Ca^{2+} levels can inhibit MGL activity in microglia (73). Regulating MGL in this way offers a potential mechanism whereby presynaptic activity, and Ca^{2+} within the terminal, may hold sway over the duration of CB1R activation and possibly the induction of eCB-LTD.

A regulated endocannabinoid efflux step? Several aspects of eCB release are still poorly understood but may prove to be important points of control over eCB-LTD. A recent study has suggested that postsynaptic activity can regulate efflux of newly synthesized eCBs. Adermark & Lovinger (74) loaded postsynaptic striatal MSNs with AEA or 2-AG and found that pairs of test stimuli (0.1 Hz) could induce a CB1R-dependent suppression of excitatory transmission, independent of mGluR-I activation, postsynaptic Ca^{2+}, or postsynaptic membrane potential. Because afferent stimulation did not modulate direct presynaptic inhibition by a CB1R agonist, the authors hypothesized that afferent activity regulates a late step in eCB mobilization, prior to CB1R activation. Accordingly, postsynaptic loading with

blockers of the putative eCB membrane transporter (EMT) prevented the synaptic depression induced by combined postsynaptic AEA application and afferent activation (74). Indeed, Lovinger and coworkers have previously shown that postsynaptic loading of EMT blockers prevents eCB-LTD in dorsal striatum (75), an observation also confirmed by other researchers in somatosensory cortex (31). Whether postsynaptic activity governs other forms of eCB-LTD through the putative EMT is unknown; this type of regulation may be a synapse-specific phenomenon. In contrast to the situation at excitatory synapses, postsynaptically loaded eCBs suppress inhibitory synapses onto MSNs regardless of afferent activity (74).

Presynaptic Activity and Endocannabinoid-Mediated Long-Term Depression

The foregoing discussion primarily focuses on presynaptic (afferent) activity as a regulator of eCB release. However, several lines of evidence suggest that eCB-LTD requires presynaptic activity as a cofactor for induction, playing a role that is independent of eCB release and CB1R activation. The requirement for coincident activation of CB1Rs and presynaptic activity provides an additional mechanism to ensure synapse specificity such that only active fibers undergo eCB-LTD.

Spike timing–dependent endocannabinoid-mediated long-term depression and presynaptic NMDARs. In a series of elegant experiments using cell-pair recordings in L5 of visual cortex, Sjöström and coworkers (26) provided the first evidence that presynaptic activation of NMDARs is required for the induction of eCB-LTD, here induced by spike timing–dependent protocols (or eCB-t-LTD). They showed that exogenous activation of CB1Rs, a manipulation that short-cuts eCB release from the postsynaptic pyramidal neuron, induces LTD only if the presynaptic neuron is activated at a relatively high frequency, and importantly, they showed that this LTD is blocked by

Endocannabinoid mobilization: an inclusive term that encompasses both endocannabinoid production and release

the NMDAR antagonist D-APV (26). Sjöstrom et al. (26) postulated that coincident activation of presynaptic NMDA and CB1 receptors is required for the induction of eCB-t-LTD. Other groups studying somatosensory cortex (30, 31) and visual cortex (28, 76) have independently provided further support for the presynaptic NMDAR requirement, showing that bath application of NMDAR antagonists blocks eCB-LTD, but disrupting postsynaptic NMDAR function with either hyperpolarization or the inclusion of MK-801 in the postsynaptic recording pipette does not affect the induction of eCB-LTD. A recent study in somatosensory cortex has provided direct evidence that presynaptic NMDARs are required for t-LTD induction. Indeed, loading presynaptic neurons with MK-801 in synaptically coupled cell pairs selectively abolished t-LTD without affecting t-LTP (77). An open question is how presynaptic NMDARs and CB1Rs interact to induce a long-lasting reduction in transmitter release. It is tempting to postulate that Ca^{2+} influx through presynaptic NMDARs may be required to activate some metabolic process at the presynaptic terminal. Alternatively, presynaptic NMDARs may signal in a Ca^{2+}-independent manner. These possibilities remain to be tested.

CB1Rs are insufficient for endocannabinoid-mediated long-term depression, and presynaptic activity is required. The involvement of presynaptic NMDARs may represent a unique aspect linking the various forms of eCB-t-LTD (26, 28, 31). However, the fact that several examples of eCB-LTD are NMDAR-independent immediately raises the question of whether presynaptic Ca^{2+} entry, or even presynaptic activity, is a general requirement of eCB-LTD induction. Experiments performed with eCB-t-LTD showed that CB1R agonists alone do not trigger long-term plasticity without coincident presynaptic activity (26, 31). CB1R activation alone is similarly unable to trigger eCB-LTD in hippocampus, dorsal striatum, and VTA, where afferent-only induction protocols are typically used (22, 34, 53, 64, 75, 78–80).

Data from hippocampus and striatum show that eCB-LTD requires minutes of CB1R activation after a brief induction stimulus (20, 75), suggesting that another induction signal, such as presynaptic activity, may be integrated during that period to induce eCB-LTD. This issue has been investigated in some detail for eCB-LTD in dorsal striatum and at CA1 inhibitory synapses.

Recent evidence shows that eCB-LTD in the dorsal striatum requires low-frequency presynaptic activity during the period of CB1R activation (53). Singla et al. (53) propose that this dual requirement confers synapse specificity on eCB-LTD under conditions of, e.g., eCB spillover. This finding is consistent with a previous report showing that CB1R activation alone cannot induce striatal eCB-LTD (75). The requirement for presynaptic activity fits well with the fact that CB1R activation for more than 5 min after the induction tetanus is also necessary for eCB-LTD in the hippocampus (20) and striatum (75). In these studies, the role of afferent stimulation may have been either (*a*) to supply activity at the test synapse or (*b*) to release a cofactor crucial for induction. This latter concern is not trivial, given that striatal eCB-LTD may involve a complex interplay between dopamine receptors and cholinergic interneurons, in addition to events at the glutamatergic corticostriatal synapse (52, 81). Our own experiments in the hippocampus using interneuron–pyramidal cell pairs provide direct experimental evidence that eCBs and presynaptic activity at the test synapse interact to induce eCB-LTD. Under conditions of maximal eCB release (i.e., minutes of mGluR-I activation), we found that firing these interneurons produces eCB-LTD, whereas holding neurons silent during eCB release effectively prevents induction of eCB-LTD (78).

Timing of afferent activity: milliseconds versus minutes. Most evidence indicates that eCB-LTD induction requires presynaptic activity, in addition to CB1R activation. The timing requirements for presynaptic activity, however, seem to differ significantly among synapses. There is some evidence to suggest that these timing requirements reflect signal integration at the presynaptic terminal, rather than special conditions needed for eCB release. In L5 pyramidal cells of visual cortex, even manipulations that extended the lifetime of eCBs in the synaptic cleft did not remove spike timing dependence; rather, they only broadened the time window during which post-before-prefiring could trigger eCB-LTD (26). Similarly, in the dorsal cochlear nucleus, blocking the t-LTP component broadened the effective time window for eCB-LTD induction without changing the required order of spikes (35). It is tempting to speculate that the requirement of a presynaptic NMDAR, which occurs exclusively in eCB-t-LTD, confers a timing property (82). eCB-t-LTD mediates coincidence detection on the order of milliseconds. For eCB-t-LTD in somatosensory cortex, this strict timing requirement may reflect its proposed physiological role in rodent whisker desensitization (83). In hippocampus and dorsal striatum, eCB-LTD seems to integrate aggregate events occurring over minutes to yield eCB-LTD. At present, the significance of these timing requirements is not known, but in principle they allow for an associativity between synaptic events on two very different timescales.

Presynaptic calcium and endocannabinoid-mediated long-term depression. As suggested previously, the known involvement of presynaptic NMDARs in some forms of eCB-t-LTD may indicate a requirement for presynaptic Ca^{2+} in the induction of eCB-LTD. Our data (78) and those of Singla et al. (53), although indirect, support such a role for presynaptic Ca^{2+}. In experiments both in dorsal striatum (53) and in hippocampal inhibitory synapses, several methods of disrupting presynaptic Ca^{2+} dynamics all blocked eCB-LTD, strongly suggesting that presynaptic spikes (whether from interneurons or glutamatergic cortical afferents) regulate this form of plasticity by raising Ca^{2+} at the nerve terminal. In addition, induc-

tion of eCB-LTD at hippocampal synapses also requires activation of the Ca^{2+}-sensitive phosphatase calcineurin (78). Together, these observations lead us to propose a mechanism in which firing, by raising Ca^{2+} levels at the presynaptic terminal, activates calcineurin, thereby shifting the balance of kinase and phosphatase activity in the presynaptic terminal, finally inducing eCB-LTD (see below). At the moment, it is unclear why induction of eCB-LTD requires activation of presynaptic NMDARs at some synapses whereas presynaptic Ca^{2+} influx via VDCCs is sufficient for induction at other synapses.

In conclusion, most evidence indicates that both pre- and postsynaptic activity can control eCB-LTD induction. These regulatory mechanisms endow eCB-LTD with at least two forms of associativity, as summarized in **Figure 2**. In the postsynaptic compartment, glutamate release and activity-dependent Ca^{2+} rise facilitate eCB mobilization, a process mediated by PLC. Presynaptic associativity involves both CB1R activation by eCBs and presynaptic firing, which increases Ca^{2+} concentration, thus engaging long-term suppression of transmitter release.

EXPRESSION MECHANISMS

The molecular mechanisms of eCB-LTD's expression and maintenance are just beginning to emerge. It was recognized early on that CB1R activation is necessary for induction but not for expression of eCB-LTD. Accordingly, washing in a CB1R antagonist after eCB-LTD is established fails to reverse plasticity in all synapses in which this manipulation has been tested (20, 26, 75). Here, we discuss what processes downstream of CB1R activation can account for the long-term reduction of transmitter release reported during eCB-LTD. Changes in the amount of transmitter release can result from modifications of the presynaptic action potential–induced Ca^{2+} influx and/or the downstream release machinery, as well as changes in excitability of the afferent axon.

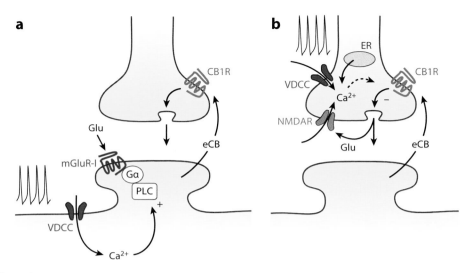

Figure 2

Potential mechanisms of associativity involved in endocannabinoid-mediated long-term depression (eCB-LTD) induction. (*a*) The postsynaptic compartment. The postsynaptic neuron can integrate action potential firing [which promotes Ca^{2+} rise via voltage-dependent Ca^{2+} channels (VDCCs)] and synaptic release of glutamate (Glu) (which activates mGluR-I to facilitate eCB mobilization and eCB-LTD induction). Other Ca^{2+} sources (e.g., NMDARs and Ca^{2+} internal stores) may also contribute. In this model, phospholipase C (PLC) operates as a coincidence detector (61). (*b*) The presynaptic compartment. The presynaptic terminal can also integrate two signals, eCBs [which activate presynaptic type 1 cannabinoid receptor (CB1Rs)] and presynaptic firing (which promotes a Ca^{2+} rise via VDCCs and NMDARs). Ca^{2+} stores [endoplasmic reticulum (ER)] may also contribute to this process. The presynaptic NMDAR may operate as a coincidence detector during a brief time window (26). In addition, the activity-dependent Ca^{2+} rise that occurs during minutes of CB1R activation likely promotes dephosphorylation of a presynaptic target downstream of the CB1R (78).

Role of cAMP/PKA Cascade in Endocannabinoid-Mediated Long-Term Depression

One of the most striking requirements distinguishing eCB-LTD induction from eCB-dependent short-term depression such as DSI/DSE is the extended duration of CB1R activity (several minutes) (3). This relatively long induction period suggests that CB1Rs may engage a different signaling process than the one mediating transient suppression of transmitter release. We have recently investigated this possibility in the hippocampus, where inhibitory synapses can undergo both short- and long-term eCB-mediated plasticity (DSI and I-LTD, respectively) (20). Here, one can directly compare the expression mechanisms un-

I-LTD: long-term depression at inhibitory synapses

derlying these two forms of depression at synapses impinging on the same target cell (19).

Compelling evidence indicates that the transient suppression of transmitter release occurring during DSI/DSE is mainly due to CB1R-dependent inhibition of presynaptic VDCCs (84–87), a process likely mediated by the $G_{\beta\gamma}$ subunits (84). The relatively fast decay of DSI (<1 min) may reflect the transient profile of eCB production and CB1R engagement. We hypothesized that the increased time requirement for eCB-dependent I-LTD induction could reflect the recruitment of the $\alpha_{i/o}$ effector limb of the CB1R G protein signaling cascade, which inhibits the cAMP/PKA pathway (88, 89). Indeed, we found that inhibiting PKA activity with selective blockers, or raising cAMP levels by continuous activation of AC, prevents I-LTD without affecting DSI (19). More recently, we reported that presynaptic phosphatase activity may complement the CB1R-dependent inhibition of PKA during eCB-LTD induction. Specifically, inhibiting the Ca^{2+}-activated phosphatase calcineurin fully blocked I-LTD, once again leaving DSI intact (78). In addition, postsynaptic blockade of PKA and calcineurin had no effect on either I-LTD or DSI, supporting the idea that these kinase and phosphatase activities occur presynaptically. Altogether these findings indicate that (a) the signaling pathways downstream CB1Rs that mediate transient versus long-lasting depression of transmitter release differ, and (b) a CB1R-mediated reduction of cAMP/PKA activity underlies I-LTD.

The involvement of the cAMP/PKA pathway in eCB-LTD also occurs in other brain structures. As in the hippocampus, inhibiting PKA or raising cAMP levels interferes with eCB-LTD at glutamatergic synapses in the nucleus accumbens (90). In addition, protein kinase inhibitors reportedly prevent striatal eCB-LTD (91), although in this case PKA may also be required postsynaptically for AEA production, as suggested for eCB-LTD in the amygdala (18). The cAMP/PKA cascade may also mediate the eCB-LTD recently reported at excitatory hippocampal synapses in neonatal rats (23). Thus, most evidence suggests that CB1R-dependent inhibition of the cAMP/PKA signaling pathway is a critical step in eCB-LTD at both excitatory and inhibitory synapses. Whether this pathway also mediates eCB-LTD in other brain areas remains to be tested; other signaling pathways downstream of CB1Rs may also participate.

Endocannabinoid-Mediated Long-Term Depression Expressed by Changes in the Release Machinery

Modulation of the cAMP/PKA pathway has been previously implicated in presynaptic forms of LTP and LTD (for a review, see Reference 92). A common theme in these forms of plasticity is that their expression mechanism relies on changes in the transmitter release machinery (93–95). In particular, the active-zone protein RIM1α, which is phosphorylated by PKA in vitro (94), is a demonstrated requirement for PKA-dependent LTP at several excitatory synapses, including synapses from mossy fibers to CA3 pyramidal cells (93), from parallel fibers to Purkinje cells (93, 94), and from the Schaffer collateral axons to CA1 pyramidal neurons (96). Using RIM1α knockout mice, we have recently found that I-LTD in the hippocampus and amygdala also requires RIM1α (19). In addition, the reduction of mIPSC (miniature inhibitory postsynaptic current) frequency (but not amplitude) mediated by a CB1R agonist was markedly reduced both in these mice and in the presence of a PKA blocker, altogether suggesting that CB1Rs and PKA can target the release machinery in a RIM1α-dependent manner. Thus, changes in the release machinery via cAMP/PKA signaling and RIM1α may represent a general mechanism underlying presynaptic forms of long-term plasticity (LTP and LTD) at both excitatory and inhibitory synapses.

According to the PKA/RIM1α model of presynaptic plasticity, a reduction of PKA activity (e.g., via CB1R activation) enables RIM1α dephosphorylation, which results in a depression of transmitter release. Conversely, increases in PKA activity and RIM1α phosphorylation lead to LTP (19, 93, 94, 97). A previous report suggested that phosphorylation of RIM1α at serine 413 is necessary for presynaptic LTP in parallel fiber synapses formed in vitro by cultured cerebellar neurons (94). To test whether RIM1α phosphorylation/dephosphorylation are indeed required for PKA-dependent forms of plasticity reported in acute brain slices, Südhof and coworkers (98) have recently generated knock-in mice in which serine 413 is mutated to alanine. Surprisingly, mossy fiber LTP, cerebellar LTP, and hippocampal I-LTD are all normal in these transgenic mice (98). These findings strongly suggest that, although the RIM1α complex is essential for presynaptic long-term plasticity,

IPSC: inhibitory postsynaptic current

Miniature inhibitory postsynaptic current (mIPSC): a spontaneous, action potential–independent, inhibitory synaptic current that reflects quantal release from a presynaptic terminal

EPSC: excitatory postsynaptic current

Miniature excitatory postsynaptic current (mEPSC): a spontaneous, action potential–independent, excitatory synaptic current that reflects quantal release from a presynaptic terminal

Fiber volley: extracellular potential representing the action potential firing of presynaptic fibers

PKA regulates this type of plasticity by a mechanism distinct from RIM1α phosphorylation at serine 413. Indeed, PKA-dependent modulation of transmitter release can occur as a result of the phosphorylation/dephosphorylation of several other presynaptic proteins involved in exocytosis (for a review, see Reference 99). Future studies to identify the nature of such a presynaptic protein(s) in PKA/RIM1α-dependent forms of plasticity, including eCB-LTD, are warranted.

Endocannabinoid-Mediated Long-Term Depression Expressed by Changes in Voltage-Dependent Calcium Channels

Manzoni and coworkers (90) have recently reported that blockade of P/Q-type, but not of L- or N-type, VDCCs occludes eCB-LTD at excitatory synapses in the nucleus accumbens. Previous work by this group showed that these synapses can undergo an mGluR2/3-dependent form of LTD that also depends on the cAMP/PKA pathway and whose expression mechanism likely involves a selective reduction in the contribution of P/Q-type VDCCs to glutamate release (100). Importantly, mGluR2/3-dependent LTD and eCB-LTD mutually occlude, supporting the idea that both forms of plasticity may share a common mechanism (101). Together, these observations suggest that a PKA-dependent reduction of presynaptic Ca^{2+} influx via P/Q-type VDCCs may underlie both forms of LTD. The involvement of P/Q-type VDCCs in the expression of a presynaptic (although eCB-independent) form of LTD has been previously demonstrated at mossy fiber synapses onto stratum lucidum interneurons in the hippocampus (102). Indeed, using two-photon Ca^{2+} imaging from anatomically defined mossy fiber terminals, Pelkey and coworkers (102) have directly shown that expression of mGluR7-dependent LTD is due to a long-lasting reduction of presynaptic Ca^{2+} influx via P/Q-type VDCCs. Whether eCB-LTD in nucleus accumbens or other brain structures is also associated with a long-term reduction of presy-

naptic Ca^{2+} signals has not been examined. Selective depression of P/Q-type VDCC function has also been reported in amphetamine-induced, CB1R-dependent LTD of excitatory synaptic transmission in the amygdala (103). Interestingly, this form of plasticity is accompanied by a change in the frequency, but not in the amplitude, of miniature excitatory postsynaptic currents (mEPSCs), suggesting a second (parallel) mechanism of expression downstream Ca^{2+} influx. Whether eCB-LTD at a given presynaptic terminal can express both RIM1α-dependent and P/Q-type VDCC-dependent mechanisms remains unknown.

Endocannabinoid-Mediated Long-Term Depression Expressed by Changes in Presynaptic Excitability

A third expression mechanism has been suggested by a recent study describing a heterosynaptic eCB-LTD of glutamatergic synaptic transmission in the developing CA1 area of the hippocampus (23). This form of eCB-LTD is associated with a decrease of fiber volley amplitude, suggesting a reduction of presynaptic excitability. Importantly, LTD of both synaptic transmission and fiber volley amplitude is mimicked by a CB1R agonist and blocked by a CB1R antagonist. Moreover, K$^+$ channel blockers impair depression of fiber volleys (23). These observations strongly suggest that heterosynaptic eCB-LTD at excitatory synapses in the immature hippocampus is likely due to a CB1R-dependent reduction of presynaptic excitability through activation of K$^+$ channels at presynaptic fibers. It will be interesting to see whether a similar expression mechanism also underlies the heterosynaptic eCB-LTD recently reported at glutamatergic synapses in the immature visual cortex (29).

DEVELOPMENTAL REGULATION OF ENDOCANNABINOID-MEDIATED LONG-TERM DEPRESSION

The refinement of neural circuits during postnatal development is believed to involve activity-dependent changes in synaptic strength. Presumably paralleling the ongoing changes in pre- and postsynaptic properties, induction mechanisms of synaptic plasticity adjust during development. Consistent with this notion, several recent studies have shown that eCB-LTD can be developmentally regulated. From a mechanistic perspective, regulation of eCB-LTD likely reflects developmental changes in eCB signaling (e.g., eCB production/degradation, CB1R number/function). For example, the probability that high-frequency stimulation (HFS) can induce striatal LTD increases in brain slices from rats during the third and fourth postnatal weeks (104), a change that has been associated with a developmental increase in AEA levels in striatal tissue (51). In the hippocampus, the magnitude of eCB-mediated I-LTD is reportedly lower in adult rats [postnatal day (P)75–110] versus adolescents (P28–35), in tandem with the reduced ability of a CB1R agonist to suppress inhibitory neurotransmission (22). It is unclear whether this developmental regulation reflects differences in CB1R number and/or function at GABAergic terminals, although binding studies have reported maximal CB1R expression levels at P30–40 (105).

As described above, glutamatergic synapses in the immature hippocampus (P2–10) can express a form of heterosynaptic eCB-LTD, reportedly associated with a reduction of presynaptic excitability (23). This form of plasticity becomes less prominent during development and cannot be induced in the mature hippocampus. At present, the mechanism of this developmental change is unknown. A similar heterosynaptic eCB-LTD has recently been shown in the immature visual cortex (29). HFS of excitatory inputs to L2/3 neurons induces heterosynaptic eCB-LTD in most mice at P7–20, but not at P35–41. Interestingly, brain-derived neurotrophic factor (BDNF), which may be released from strongly activated presynaptic sites, appears to prevent the induction of homosynaptic eCB-LTD. Presumably, an interaction between the CB1R and TrkB (the high-affinity receptor for BDNF) at active presynaptic terminals leads to the blockade of CB1R signaling. A developmental upregulation of BDNF expression, diffusion of BDNF in cortical tissues, and a concomitant reduction in CB1R expression have been proposed as potential mechanisms underlying the developmental regulation of heterosynaptic eCB-LTD in visual cortex (29).

Developmental changes in any of the major pre- and postsynaptic determinants of eCB-LTD may regulate induction and expression. In visual cortex, developmental loss of presynaptic NMDARs occurs abruptly at the onset of the critical period for receptive field plasticity (>P23 in rats) (76). During the peak of the critical period for ocular dominance (P23–30), t-LTD in L2/3 of visual cortex requires activation of postsynaptic NMDARs (76). Given that most studies on eCB-dependent t-LTD in neocortex have been performed in young rodents and show a requirement for presynaptic NMDARs, future studies will be necessary to determine whether older animals retain this form of plasticity.

METAPLASTICITY OF ENDOCANNABINOID-MEDIATED LONG-TERM DEPRESSION

Metaplasticity, the plasticity of synaptic plasticity, refers to an enduring change in the ability of neurons or synapses to generate synaptic plasticity (for a recent review, see Reference 106). eCBs can mediate metaplasticity by triggering long-term suppression of inhibitory transmission (i.e., I-LTD), thereby facilitating subsequent induction of LTP at excitatory inputs (18, 37). Recent work suggests that eCB signaling itself can be subject to plasticity. Indeed, repetitive stimulation in CA1 produces a long-term upregulation of eCB signaling, as indicated by an enduring enhancement of DSI magnitude (24, 107). In one study, DSI potentiation was triggered by low-frequency stimulation (LFS) (24), whereas in another study induction required strong HFS, as well as the coactivation of mGluRs, AMPA/kainate receptors, and

HFS: high-frequency stimulation (~100 Hz)

LFS: low-frequency stimulation (~1 Hz)

CB1Rs (107). Although the mechanism underlying DSI potentiation induced by LFS has not been explored, an upregulation of CB1R number likely mediates eCB plasticity induced by strong HFS (107). This mechanism has previously been implicated in the DSI potentiation that occurs following febrile seizures in rats (108). A subsequent study by Alger and coworkers (25) has shown that the long-lasting enhancement of DSI can be triggered by brief activation of mGluR-I. This study showed that the transient postsynaptic Ca^{2+} rise that occurs during a single episode of DSI can facilitate subsequent mGluR-I-dependent mobilization of eCBs and the induction of I-LTD (25). The Ca^{2+}-dependent process that causes this remarkable form of metaplasticity is unknown. In summary, these examples underscore that eCB signaling, and as a result eCB-mediated plasticity, can be dynamically regulated. Because some induction protocols can trigger simultaneous LTP and LTD, regulation of eCB signaling can be important in modulating the balance between the two.

FUNCTIONAL RELEVANCE OF ENDOCANNABINOID-MEDIATED LONG-TERM DEPRESSION

Reflecting the widespread distribution of the CB1R in the CNS (12, 13), exogenous ligands for CB1R impact CNS functions ranging broadly from homeostatic control of feeding to associative memory formation; some of these CB1R ligands have also been associated with neuropsychiatric disorders. However, ascribing specific roles for eCB-LTD in these behaviors has been challenging. Here, we discuss the most promising avenues of investigation linking eCBs and eCB-LTD to normal and pathological physiology.

Sensory Deprivation

LTD is a leading model to explain the synaptic weakening that occurs at cortical synapses after a period of sensory deprivation (109–111; but see 112, 113). Studies in both somatosensory and visual cortices have revealed that this LTD-like phenomenon mechanistically resembles eCB-LTD reported in vitro. Feldman and coworkers have recently shown that whisker removal weakens L4-L2/3 synapses in rat barrel cortex by decreasing presynaptic function (114), mirroring the expression mechanism of eCB-t-LTD described at these synapses (31) (see above). Similarly, monocular deprivation (MD) depresses visually evoked responses within multiple layers of visual cortex (115, 116). Bear and coworkers found that a previously described LTD at L4-L2/3 synapses in vitro (117), thought to contribute to this MD-induced synaptic depression, is CB1R dependent (28). Most notably, in brain slices prepared after whisker deprivation or MD, eCB-t-LTD is abolished, suggesting that sensory deprivation may occlude eCB-t-LTD and therefore may operate through this synaptic mechanism (28, 114).

Data from genetic and pharmacological disruption of CB1Rs support a role for eCB-LTD in sensory-driven cortical plasticity. Anatomical analysis of the barrel cortex in CB1–/– mice revealed abnormal barrel maps (118). During a brief period of MD, systemic application of the CB1R antagonist AM251 fully prevented the depression of inputs to L2/3 of visual cortex in vivo, the same area that expresses eCB-LTD (28, 119). Whether eCB-t-LTD drives the loss of cortical responses to deprived sensory inputs remains to be determined. In addition, short-term plasticity mediated by eCBs, such as DSI/DSE (120, 121), may also contribute to these forms of cortical plasticity.

Associative Learning

Long-term changes in synaptic strength are also believed to underlie associative memory formation in hippocampus and amygdala (122). The first studies to show a role for the eCB/CB1R system in hippocampus- and amygdala-specific learning tasks (the Morris water maze and cued fear conditioning, respectively) used CB1–/– mice or systemically applied CB1R antagonists

(17, 123). These and subsequent studies found that eCBs and CB1Rs were specifically required for extinguishing an established fear or spatial memory but were not needed to acquire these types of memories (17, 123–126). In addition, FAAH inhibitors, which enhance eCB signaling, facilitated extinction of cue-shock associations as well as memories of a hidden platform's former position in the Morris water maze (126, 127). Although these findings may point to an underlying involvement of eCB-LTD, a number of eCB-mediated effects on synaptic transmission have been noted in these two main structures. eCB-LTD has been described at inhibitory synapses in CA1 and basolateral amygdala (17, 20) and, in developing hippocampus, at excitatory synapses (23). Hippocampal synapses also express DSI and DSE (1, 3). In addition, in both hippocampus and amygdala, eCB-LTD modulates excitability and the induction of LTP at excitatory synapses, a process widely thought to underlie memory acquisition (18, 20, 37). Marsicano and colleagues (128) have developed a promising strategy to address the contribution of CB1Rs at excitatory and inhibitory terminals in the context of epilepsy, generating a line of transgenic mice with selective deletions of the CB1R in either forebrain GABAergic interneurons or principal neurons. Identifying unique molecular determinants of eCB-LTD versus DSI/DSE (such as RIM1α) will further aid in resolving these issues.

Parkinson's Disease

Parkinson's disease (PD) is a neurodegenerative disorder characterized by hallmark motor symptoms such as rigidity, bradykinesia/akinesia, and resting tremor, as well as the progressive loss of the dopaminergic neurons in the substantia nigra pars compacta. In addition to dopamine depletion, PD is also thought to involve a reactive increase in corticostriatal excitatory drive in the basal ganglia via the indirect (striatopallidal) pathway, a circuit identified by MSNs exclusively expressing the D_2R rather than the D_1R (129). Accordingly, strategies to treat these symptoms are aimed not only at replacing the missing dopamine but also at reducing transmission through the indirect pathway. As a D_2R-dependent mechanism that suppresses corticostriatal transmission (4, 39, 50, 130), striatal eCB-LTD is a potential therapeutic target in PD. In fact, in vivo administration of the FAAH inhibitor URB597 reduces parkinsonian motor deficits in dopamine-depleted animal models of PD, but only when URB597 is administered together with a D_2R agonist, which presumably triggers eCB release (43, 50). A direct demonstration that eCBs mediate this effect awaits confirmation. Interestingly, striatal eCB-LTD was also abolished in these animal models but rescued in the presence of URB597 or the D_2R agonist quinpirole (50). MSNs express high levels of mGluR5 (but not mGluR1) (68), whose activation can mobilize eCB release (42) and trigger chemical eCB-LTD at striatopallidal MSNs (50; see also 131). It is therefore conceivable that reduced eCB production, particularly in the indirect pathway, may be an important mechanism contributing to the pathophysiology of PD, presumably by eliminating eCB-LTD. Therefore, manipulations of this pathway by modulating eCB production may provide an alternative approach for the treatment of striatal-based brain disorders, such as PD.

Drug Addiction

Like PD, drug addiction is a robust behavioral phenomenon particularly amenable to analysis in animals because its major pathological features are reproducible across species. The rewarding properties of exogenous cannabinoids are well known (132, 133). However, the process of addiction not only involves the acute, rewarding effects of a drug but also may involve long-term neural adaptations induced by the drug of abuse (134, 135). These adaptations are thought to include changes of synaptic strength, especially in the VTA, a midbrain area densely populated by dopaminergic neurons, and the nucleus accumbens, which receives prominent projections from the VTA (136–138). At least one such behavioral adaptation, sensitization to the rewarding properties of cocaine, requires eCBs and CB1Rs, on the basis of experiments with CB1R antagonists and CB1R–/– mice (recently reviewed in Reference 139). In addition, in vivo exposure to drugs of abuse, including cocaine, can modulate eCB-LTD in both VTA and nucleus accumbens (34, 140–142). Repetitive cocaine exposure in vivo fully occludes eCB-LTD at GABAergic synapses in VTA (34). This reduced GABAergic inhibition has been linked to a facilitation of LTP at nearby excitatory synapses in VTA (143), believed to be critical for inducing behavioral sensitization (136). A similar process has been described in nucleus accumbens, where a single dose of cocaine abolishes eCB-LTD at excitatory synapses (140). However, diminished mGluR5-driven eCB release, rather than true occlusion of eCB-LTD at the presynaptic terminal, appears to account for that effect. The role of the eCB/CB1R signaling system in addiction appears to vary depending on the specific drug of abuse (139), indicating a need to examine the interaction of other drugs of abuse with eCB-LTD in these critical brain areas.

FUTURE DIRECTIONS AND CONCLUDING REMARKS

eCBs are now well established as mediators of synaptic plasticity. Extensive work demonstrates that eCB-LTD is a widespread phenomenon in the brain and is expressed by both excitatory and inhibitory synapses. As the list of synapses expressing eCB-LTD grows, so does our understanding of the induction/expression mechanisms and computational power of this form of plasticity. eCBs also mediate short-term plasticity, and by regulating inhibitory transmission, they can modulate the induction of eCB-independent forms of plasticity such as LTP (3). All these actions combined clearly highlight the unique position of eCBs as regulators of synaptic function.

In addition to eCB-LTD, eCBs have been implicated in other forms of long-term plas-

ticity. For example, in the cerebellum, Safo & Regehr (144) found that a well-studied form of postsynaptically expressed LTD between parallel fibers and Purkinje cells depends on retrograde release of 2-AG and activation of presynaptic CB1Rs. The authors suggest that a second (anterograde) signal is interposed between terminals bearing the CB1R and the Purkinje cell, where LTD is ultimately expressed. Interestingly, a similar situation has been observed in the VIIIth nerve to Mauthner cell mixed (electrical and chemical) synapses in goldfish (145), in which presynaptic CB1Rs stimulate dopamine release from immunocytochemically identified dopaminergic terminals, triggering a postsynaptic LTP of both electrical and chemical neurotransmission. This CB1R-dependent regulation of a second neuromodulator may also account for an eCB-dependent LTP of excitatory transmission described in lamprey spinal cord (146). These cases possibly represent a novel class of eCB-mediated plasticity; however, they also join an existing literature indicating that eCB-dependent processes modulate an otherwise eCB-independent form of plasticity (3). Here, the familiar retrograde eCB-CB1R signaling system forms a component of a more complex induction process. The most striking departure from eCB-LTD is the novel functional role for CB1Rs, which, rather than suppressing neurotransmitter release, stimulate release of a key factor for inducing synaptic plasticity. Other novel roles for CB1R in long-term plasticity may come to light, given evidence of functional CB1Rs at postsynaptic sites in neocortex (147).

Although protein synthesis is a known requirement in some forms of LTD (148, 149), thus far only one study has directly examined the contribution of protein synthesis in eCB-LTD. Yin et al. (150) found that inhibition of protein translation, but not that of protein transcription, prevents striatal eCB-LTD, implicating new protein synthesis in this form of plasticity. The authors provide evidence that rapid protein translation, likely occurring at the presynaptic axon/terminal, is critical for the expression of striatal LTD. Future studies will have to determine whether presynaptic protein translation is a general requirement for the maintenance of eCB-LTD in other brain structures. It will also be important to identify the presynaptic signaling pathways (e.g., Ca^{2+}, PKA) that promote local protein synthesis, as well as the nature of the newly synthesized protein(s) required for the long-lasting reduction in transmitter release.

Recent progress in eCB signaling and eCB-mediated synaptic plasticity provides new targets for developing experimental tools and potential therapies for neuropsychiatric disorders. The existing literature on eCBs and synaptic plasticity raises several important questions as well. For example, it is unclear how long eCB-LTD can persist. No study has yet addressed whether structural alterations of the synapse and active zone accompany eCB-LTD in the maintenance phase. The reversal of eCB-LTD is another critically underexamined issue that has been extensively explored in other forms of plasticity. In the laboratory, induction paradigms are commonly selected on the basis of their effectiveness rather than their physiological significance. Further research will have to determine whether the two induction paradigms commonly used to trigger eCB-LTD can occur in vivo. Of equal importance will be elucidating the functional and computational roles of eCB-LTD. Given the widespread importance of the eCB-CB1R signaling pathway in both synaptic physiology and behavior, we predict tremendous interest and progress in this field in the coming years.

SUMMARY POINTS

1. Endogenous cannabinoids (eCBs) can have a long-term impact in neural function by mediating eCB-LTD. This form of plasticity has now been observed in several brain structures at both excitatory and inhibitory synapses.

2. Despite the variety of stimulation protocols and experimental conditions used for its induction, nearly every synapse studied still expresses eCB-LTD as a long-lasting reduction of neurotransmitter release. eCB-LTD is by now one of the best examples of presynaptic forms of long-term plasticity.

3. The emerging molecular details of induction and expression of eCB-LTD at various synapses reveal many points of regulation. Important determinants of eCB-LTD induction are the enzymes involved in eCB synthesis, degradation, and (perhaps) release, as well as the cAMP/PKA signaling pathway downstream of CB1Rs.

4. CB1R activation per se is not sufficient for eCB-LTD induction at a number of synapses. Rather, the presynaptic terminal seems to integrate multiple signals, such as presynaptic firing, to generate eCB-LTD. At several synapses, presynaptic regulatory mechanisms of eCB-LTD induction translate into key computational properties, such as synapse specificity and associativity.

DISCLOSURE STATEMENT

The authors are not aware of any biases that might be perceived as affecting the objectivity of this review.

ACKNOWLEDGMENTS

We wish to thank all members of the Castillo lab and all scientists whose data are reviewed in this article. We apologize to all the investigators whose work could not be cited owing to space limitations. The authors are supported by the National Institutes of Health/National Institute on Drug Abuse grants, a National Institutes of Health training grant (to B.D.H.), the Irma T. Hirschl Career Scientist Award (to P.E.C.), and the National Alliance for Research on Schizophrenia and Depression.

LITERATURE CITED

1. Freund TF, Katona I, Piomelli D. 2003. Role of endogenous cannabinoids in synaptic signaling. *Physiol. Rev.* 83:1017–66
2. Alger BE. 2002. Retrograde signaling in the regulation of synaptic transmission: focus on endocannabinoids. *Prog. Neurobiol.* 68:247–86
3. Chevaleyre V, Takahashi KA, Castillo PE. 2006. Endocannabinoid-mediated synaptic plasticity in the CNS. *Annu. Rev. Neurosci.* 29:37–76
4. Gerdeman GL, Lovinger DM. 2003. Emerging roles for endocannabinoids in long-term synaptic plasticity. *Br. J. Pharmacol.* 140:781–89
5. Kano M, Ohno-Shosaku T, Hashimotodani Y, Uchigashima M, Watanabe M. 2008. Endocannabinoid-mediated control of synaptic transmission. *Physiol. Rev.* In press
6. Lovinger DM. 2008. Presynaptic modulation by endocannabinoids. *Handb. Exp. Pharmacol.* 184:435–77
7. Gaoni Y, Mechoulam R. 1964. Isolation, structure and partial synthesis of an active constituent of hashish. *J. Am. Chem. Soc.* 86:1646–47
8. Matsuda LA, Lolait SJ, Brownstein MJ, Young AC, Bonner TI. 1990. Structure of a cannabinoid receptor and functional expression of the cloned cDNA. *Nature* 346:561–64
9. Devane WA, Dysarz FA 3rd, Johnson MR, Melvin LS, Howlett AC. 1988. Determination and characterization of a cannabinoid receptor in rat brain. *Mol. Pharmacol.* 34:605–13
10. Devane WA, Breuer A, Sheskin T, Jarbe TU, Eisen MS, Mechoulam R. 1992. A novel probe for the cannabinoid receptor. *J. Med. Chem.* 35:2065–69
11. Stella N, Schweitzer P, Piomelli D. 1997. A second endogenous cannabinoid that modulates long-term potentiation. *Nature* 388:773–78
12. Herkenham M, Lynn AB, Little MD, Johnson MR, Melvin LS, et al. 1990. Cannabinoid receptor localization in brain. *Proc. Natl. Acad. Sci. USA* 87:1932–36
13. Tsou K, Brown S, Sanudo-Pena MC, Mackie K, Walker JM. 1998. Immunohistochemical distribution of cannabinoid CB1 receptors in the rat central nervous system. *Neuroscience* 83:393–411
14. Howlett AC. 2005. Cannabinoid receptor signaling. *Handb. Exp. Pharmacol.* 168:53–79
15. Gerdeman GL, Ronesi J, Lovinger DM. 2002. Postsynaptic endocannabinoid release is critical to long-term depression in the striatum. *Nat. Neurosci.* 5:446–51
16. Robbe D, Kopf M, Remaury A, Bockaert J, Manzoni OJ. 2002. Endogenous cannabinoids mediate long-term synaptic depression in the nucleus accumbens. *Proc. Natl. Acad. Sci. USA* 99:8384–88
17. Marsicano G, Wotjak CT, Azad SC, Bisogno T, Rammes G, et al. 2002. The endogenous cannabinoid system controls extinction of aversive memories. *Nature* 418:530–34
18. Azad SC, Monory K, Marsicano G, Cravatt BF, Lutz B, et al. 2004. Circuitry for associative plasticity in the amygdala involves endocannabinoid signaling. *J. Neurosci.* 24:9953–61
19. Chevaleyre V, Heifets BD, Kaeser PS, Sudhof TC, Castillo PE. 2007. Endocannabinoid-mediated long-term plasticity requires cAMP/PKA signaling and RIM1α. *Neuron* 54:801–12
20. Chevaleyre V, Castillo PE. 2003. Heterosynaptic LTD of hippocampal GABAergic synapses: a novel role of endocannabinoids in regulating excitability. *Neuron* 38:461–72
21. Lafourcade CA, Alger BE. 2007. Distinctions among $GABA_A$ and $GABA_B$ responses revealed by calcium channel antagonists, cannabinoids, opioids, and synaptic plasticity in rat hippocampus. *Psychopharmacology* 198:539–49

22. Kang-Park MH, Wilson WA, Kuhn CM, Moore SD, Swartzwelder HS. 2007. Differential sensitivity of GABA$_A$ receptor-mediated IPSCs to cannabinoids in hippocampal slices from adolescent and adult rats. *J. Neurophysiol.* 98:1223–30

23. Yasuda H, Huang Y, Tsumoto T. 2008. Regulation of excitability and plasticity by endocannabinoids and PKA in developing hippocampus. *Proc. Natl. Acad. Sci. USA* 105:3106–11

24. Zhu PJ, Lovinger DM. 2007. Persistent synaptic activity produces long-lasting enhancement of endo-cannabinoid modulation and alters long-term synaptic plasticity. *J. Neurophysiol.* 97:4386–89

25. Edwards DA, Zhang L, Alger BE. 2008. Metaplastic control of the endocannabinoid system at inhibitory synapses in hippocampus. *Proc. Natl. Acad. Sci. USA* 105:8142–47

26. Sjöström PJ, Turrigiano GG, Nelson SB. 2003. Neocortical LTD via coincident activation of presynaptic NMDA and cannabinoid receptors. *Neuron* 39:641–54

27. Sjöström PJ, Turrigiano GG, Nelson SB. 2004. Endocannabinoid-dependent neocortical layer-5 LTD in the absence of postsynaptic spiking. *J. Neurophysiol.* 92:3338–43

28. Crozier RA, Wang Y, Liu CH, Bear MF. 2007. Deprivation-induced synaptic depression by distinct mechanisms in different layers of mouse visual cortex. *Proc. Natl. Acad. Sci. USA* 104:1383–88

29. Huang Y, Yasuda H, Sarihi A, Tsumoto T. 2008. Roles of endocannabinoids in heterosynaptic long-term depression of excitatory synaptic transmission in visual cortex of young mice. *J. Neurosci.* 28:7074–83

30. Nevian T, Sakmann B. 2006. Spine Ca^{2+} signaling in spike-timing-dependent plasticity. *J. Neurosci.* 26:11001–13

31. Bender VA, Bender KJ, Brasier DJ, Feldman DE. 2006. Two coincidence detectors for spike timing-dependent plasticity in somatosensory cortex. *J. Neurosci.* 26:4166–77

32. Lafourcade M, Elezgarai I, Mato S, Bakiri Y, Grandes P, Manzoni OJ. 2007. Molecular components and functions of the endocannabinoid system in mouse prefrontal cortex. *PLoS ONE* 2:e709

33. Soler-Llavina GJ, Sabatini BL. 2006. Synapse-specific plasticity and compartmentalized signaling in cerebellar stellate cells. *Nat. Neurosci.* 9:798–806

34. Pan B, Hillard CJ, Liu QS. 2008. Endocannabinoid signaling mediates cocaine-induced inhibitory synaptic plasticity in midbrain dopamine neurons. *J. Neurosci.* 28:1385–97

35. Tzounopoulos T, Rubio ME, Keen JE, Trussell LO. 2007. Coactivation of pre- and postsynaptic signaling mechanisms determines cell-specific spike-timing-dependent plasticity. *Neuron* 54:291–301

36. Henneberger C, Redman SJ, Grantyn R. 2007. Cortical efferent control of subcortical sensory neurons by synaptic disinhibition. *Cereb. Cortex* 17:2039–49

37. Chevaleyre V, Castillo PE. 2004. Endocannabinoid-mediated metaplasticity in the hippocampus. *Neuron* 43:871–81

38. Caporale N, Dan Y. 2008. Spike timing-dependent plasticity: a Hebbian learning rule. *Annu. Rev. Neurosci.* 31:25–46

39. Shen W, Flajolet M, Greengard P, Surmeier DJ. 2008. Dichotomous dopaminergic control of striatal synaptic plasticity. *Science* 321:848–51

40. Piomelli D. 2003. The molecular logic of endocannabinoid signalling. *Nat. Rev. Neurosci.* 4:873–84

41. Varma N, Carlson GC, Ledent C, Alger BE. 2001. Metabotropic glutamate receptors drive the endo-cannabinoid system in hippocampus. *J. Neurosci.* 21:RC188;1–5

42. Jung KM, Mangieri R, Stapleton C, Kim J, Fegley D, et al. 2005. Stimulation of endocannabinoid formation in brain slice cultures through activation of group I metabotropic glutamate receptors. *Mol. Pharmacol.* 68:1196–202

43. Giuffrida A, Parsons LH, Kerr TM, Rodriguez de Fonseca F, Navarro M, Piomelli D. 1999. Dopamine activation of endogenous cannabinoid signaling in dorsal striatum. *Nat. Neurosci.* 2:358–63

44. Best AR, Regehr WG. 2008. Serotonin evokes endocannabinoid release and retrogradely suppresses excitatory synapses. *J. Neurosci.* 28:6508–15

45. Kim J, Isokawa M, Ledent C, Alger BE. 2002. Activation of muscarinic acetylcholine receptors enhances the release of endogenous cannabinoids in the hippocampus. *J. Neurosci.* 22:10182–91

46. Haj-Dahmane S, Shen RY. 2005. The wake-promoting peptide orexin-B inhibits glutamatergic trans-mission to dorsal raphé nucleus serotonin neurons through retrograde endocannabinoid signaling. *J. Neurosci.* 25:896–905

47. Földy C, Lee SY, Szabadics J, Neu A, Soltesz I. 2007. Cell type-specific gating of perisomatic inhibition by cholecystokinin. *Nat. Neurosci.* 10:1128–30

48. Ohno-Shosaku T, Matsui M, Fukudome Y, Shosaku J, Tsubokawa H, et al. 2003. Postsynaptic M1 and M3 receptors are responsible for the muscarinic enhancement of retrograde endocannabinoid signalling in the hippocampus. *Eur. J. Neurosci.* 18:109–16

49. Kreitzer AC, Malenka RC. 2005. Dopamine modulation of state-dependent endocannabinoid release and long-term depression in the striatum. *J. Neurosci.* 25:10537–45

50. Kreitzer AC, Malenka RC. 2007. Endocannabinoid-mediated rescue of striatal LTD and motor deficits in Parkinson's disease models. *Nature* 445:643–47

51. Ade KK, Lovinger DM. 2007. Anandamide regulates postnatal development of long-term synaptic plasticity in the rat dorsolateral striatum. *J. Neurosci.* 27:2403–9

52. Wang Z, Kai L, Day M, Ronesi J, Yin HH, et al. 2006. Dopaminergic control of corticostriatal long-term synaptic depression in medium spiny neurons is mediated by cholinergic interneurons. *Neuron* 50:443–52

53. Singla S, Kreitzer AC, Malenka RC. 2007. Mechanisms for synapse specificity during striatal long-term depression. *J. Neurosci.* 27:5260–64

54. Adermark L, Lovinger DM. 2007. Combined activation of L-type Ca^{2+} channels and synaptic transmission is sufficient to induce striatal long-term depression. *J. Neurosci.* 27:6781–87

55. Isokawa M, Alger BE. 2006. Ryanodine receptor regulates endogenous cannabinoid mobilization in the hippocampus. *J. Neurophysiol.* 95:3001–11

56. Beierlein M, Regehr WG. 2006. Local interneurons regulate synaptic strength by retrograde release of endocannabinoids. *J. Neurosci.* 26:9935–43

57. Ohno-Shosaku T, Hashimotodani Y, Ano M, Takeda S, Tsubokawa H, Kano M. 2007. Endocannabinoid signalling triggered by NMDA receptor-mediated calcium entry into rat hippocampal neurons. *J. Physiol.* 584:407–18

58. Galante M, Diana MA. 2004. Group I metabotropic glutamate receptors inhibit GABA release at interneuron-Purkinje cell synapses through endocannabinoid production. *J. Neurosci.* 24:4865–74

59. Brenowitz SD, Regehr WG. 2005. Associative short-term synaptic plasticity mediated by endocannabinoids. *Neuron* 45:419–31

60. Ohno-Shosaku T, Shosaku J, Tsubokawa H, Kano M. 2002. Cooperative endocannabinoid production by neuronal depolarization and group I metabotropic glutamate receptor activation. *Eur. J. Neurosci.* 15:953–61

61. Hashimotodani Y, Ohno-Shosaku T, Tsubokawa H, Ogata H, Emoto K, et al. 2005. Phospholipase Cβ serves as a coincidence detector through its Ca^{2+} dependency for triggering retrograde endocannabinoid signal. *Neuron* 45:257–68

62. Jung KM, Astarita G, Zhu C, Wallace M, Mackie K, Piomelli D. 2007. A key role for diacylglycerol lipase-α in metabotropic glutamate receptor-dependent endocannabinoid mobilization. *Mol. Pharmacol.* 72:612–21

63. Maejima T, Oka S, Hashimotodani Y, Ohno-Shosaku T, Aiba A, et al. 2005. Synaptically driven endocannabinoid release requires Ca^{2+}-assisted metabotropic glutamate receptor subtype 1 to phospholipase Cβ4 signaling cascade in the cerebellum. *J. Neurosci.* 25:6826–35

64. Edwards DA, Kim J, Alger BE. 2006. Multiple mechanisms of endocannabinoid response initiation in hippocampus. *J. Neurophysiol.* 95:67–75

65. Goldberg JH, Tamas G, Aronov D, Yuste R. 2003. Calcium microdomains in aspiny dendrites. *Neuron* 40:807–21

66. Katona I, Urban GM, Wallace M, Ledent C, Jung KM, et al. 2006. Molecular composition of the endocannabinoid system at glutamatergic synapses. *J. Neurosci.* 26:5628–37

67. Yoshida T, Fukaya M, Uchigashima M, Miura E, Kamiya H, et al. 2006. Localization of diacylglycerol lipase-α around postsynaptic spine suggests close proximity between production site of an endocannabinoid, 2-arachidonoyl-glycerol, and presynaptic cannabinoid CB1 receptor. *J. Neurosci.* 26:4740–51

68. Uchigashima M, Narushima M, Fukaya M, Katona I, Kano M, Watanabe M. 2007. Subcellular arrangement of molecules for 2-arachidonoyl-glycerol-mediated retrograde signaling and its physiological contribution to synaptic modulation in the striatum. *J. Neurosci.* 27:3663–76

69. Matyas F, Urban GM, Watanabe M, Mackie K, Zimmer A, et al. 2008. Identification of the sites of 2-arachidonoylglycerol synthesis and action imply retrograde endocannabinoid signaling at both GABAergic and glutamatergic synapses in the ventral tegmental area. *Neuropharmacology* 54:95–107

70. Fowler CJ. 2007. The contribution of cyclooxygenase-2 to endocannabinoid metabolism and action. *Br. J. Pharmacol.* 152:594–601

71. Kim J, Alger BE. 2004. Inhibition of cyclooxygenase-2 potentiates retrograde endocannabinoid effects in hippocampus. *Nat. Neurosci.* 7:697–98

72. Gulyas AI, Cravatt BF, Bracey MH, Dinh TP, Piomelli D, et al. 2004. Segregation of two endocannabinoid-hydrolyzing enzymes into pre- and postsynaptic compartments in the rat hippocampus, cerebellum and amygdala. *Eur. J. Neurosci.* 20:441–58

73. Witting A, Walter L, Wacker J, Moller T, Stella N. 2004. P2×7 receptors control 2-arachidonoylglycerol production by microglial cells. *Proc. Natl. Acad. Sci. USA* 101:3214–19

74. Adermark L, Lovinger DM. 2007. Retrograde endocannabinoid signaling at striatal synapses requires a regulated postsynaptic release step. *Proc. Natl. Acad. Sci. USA* 104:20564–69

75. Ronesi J, Gerdeman GL, Lovinger DM. 2004. Disruption of endocannabinoid release and striatal long-term depression by postsynaptic blockade of endocannabinoid membrane transport. *J. Neurosci.* 24:1673–79

76. Corlew R, Wang Y, Ghermazien H, Erisir A, Philpot BD. 2007. Developmental switch in the contribution of presynaptic and postsynaptic NMDA receptors to long-term depression. *J. Neurosci.* 27:9835–45

77. Rodriguez-Moreno A, Paulsen O. 2008. Spike timing-dependent long-term depression requires presynaptic NMDA receptors. *Nat. Neurosci.* 11:744–45

78. Heifets BD, Chevaleyre V, Castillo PE. 2008. Interneuron activity controls endocannabinoid-mediated presynaptic plasticity through calcineurin. *Proc. Natl. Acad. Sci. USA* 105:10250–55

79. Yin HH, Lovinger DM. 2006. Frequency-specific and D2 receptor-mediated inhibition of glutamate release by retrograde endocannabinoid signaling. *Proc. Natl. Acad. Sci. USA* 103:8251–56

80. Huang CC, Lo SW, Hsu KS. 2001. Presynaptic mechanisms underlying cannabinoid inhibition of excitatory synaptic transmission in rat striatal neurons. *J. Physiol.* 532:731–48

81. Centonze D, Gubellini P, Pisani A, Bernardi G, Calabresi P. 2003. Dopamine, acetylcholine and nitric oxide systems interact to induce corticostriatal synaptic plasticity. *Rev. Neurosci.* 14:207–16

82. Duguid I, Sjöstrom PJ. 2006. Novel presynaptic mechanisms for coincidence detection in synaptic plasticity. *Curr. Opin. Neurobiol.* 16:312–22

83. Jacob V, Brasier DJ, Erchova I, Feldman D, Shulz DE. 2007. Spike timing-dependent synaptic depression in the in vivo barrel cortex of the rat. *J. Neurosci.* 27:1271–84

84. Wilson RI, Kunos G, Nicoll RA. 2001. Presynaptic specificity of endocannabinoid signaling in the hippocampus. *Neuron* 31:453–62

85. Varma N, Brager D, Morishita W, Lenz RA, London B, Alger B. 2002. Presynaptic factors in the regulation of DSI expression in hippocampus. *Neuropharmacology* 43:550–62

86. Kreitzer AC, Regehr WG. 2001. Retrograde inhibition of presynaptic calcium influx by endogenous cannabinoids at excitatory synapses onto Purkinje cells. *Neuron* 29:717–27

87. Lenz RA, Wagner JJ, Alger BE. 1998. N- and L-type calcium channel involvement in depolarization-induced suppression of inhibition in rat hippocampal CA1 cells. *J. Physiol.* 512(Pt. 1):61–73

88. Childers SR, Deadwyler SA. 1996. Role of cyclic AMP in the actions of cannabinoid receptors. *Biochem. Pharmacol.* 52:819–27

89. Howlett AC, Qualy JM, Khachatrian LL. 1986. Involvement of Gi in the inhibition of adenylate cyclase by cannabimimetic drugs. *Mol. Pharmacol.* 29:307–13

90. Mato S, Lafourcade M, Robbe D, Bakiri Y, Manzoni OJ. 2008. Role of the cyclic-AMP/PKA cascade and of P/Q-type Ca^{++} channels in endocannabinoid-mediated long-term depression in the nucleus accumbens. *Neuropharmacology* 54:87–94

91. Calabresi P, Pisani A, Mercuri NB, Bernardi G. 1994. Post-receptor mechanisms underlying striatal long-term depression. *J. Neurosci.* 14:4871–81

92. Huang CC, Chen YL, Liang YC, Hsu KS. 2002. Role for cAMP and protein phosphatase in the presynaptic expression of mouse hippocampal mossy fibre depotentiation. *J. Physiol.* 543:767–78

93. Castillo PE, Schoch S, Schmitz F, Sudhof TC, Malenka RC. 2002. RIM1α is required for presynaptic long-term potentiation. *Nature* 415:327–30

94. Lonart G, Schoch S, Kaeser PS, Larkin CJ, Sudhof TC, Linden DJ. 2003. Phosphorylation of RIM1α by PKA triggers presynaptic long-term potentiation at cerebellar parallel fiber synapses. *Cell* 115:49–60

95. Tzounopoulos T, Janz R, Sudhof TC, Nicoll RA, Malenka RC. 1998. A role for cAMP in long-term depression at hippocampal mossy fiber synapses. *Neuron* 21:837–45

96. Huang YY, Zakharenko SS, Schoch S, Kaeser PS, Janz R, et al. 2005. Genetic evidence for a protein-kinase-A-mediated presynaptic component in NMDA-receptor-dependent forms of long-term synaptic potentiation. *Proc. Natl. Acad. Sci. USA* 102:9365–70

97. Kaeser PS, Sudhof TC. 2005. RIM function in short- and long-term synaptic plasticity. *Biochem. Soc. Trans.* 33:1345–49

98. Kaeser PS, Kwon HB, Blundell J, Chevaleyre V, Morishita W, Malenka RC, et al. 2008. RIM1 phosphorylation at serine 413 by protein kinase A is not required for presynaptic long-term potentiation and learning. *Proc. Natl. Acad. Sci. USA* 105:14680–85

99. Seino S, Shibasaki T. 2005. PKA-dependent and PKA-independent pathways for cAMP-regulated exocytosis. *Physiol. Rev.* 85:1303–42

100. Robbe D, Alonso G, Chaumont S, Bockaert J, Manzoni OJ. 2002. Role of P/Q-Ca^{2+} channels in metabotropic glutamate receptor 2/3-dependent presynaptic long-term depression at nucleus accumbens synapses. *J. Neurosci.* 22:4346–56

101. Mato S, Robbe D, Puente N, Grandes P, Manzoni OJ. 2005. Presynaptic homeostatic plasticity rescues long-term depression after chronic Δ$_9$-tetrahydrocannabinol exposure. *J. Neurosci.* 25:11619–27

102. Pelkey KA, Topolnik L, Lacaille JC, McBain CJ. 2006. Compartmentalized Ca^{2+} channel regulation at divergent mossy-fiber release sites underlies target cell-dependent plasticity. *Neuron* 52:497–510

103. Huang YC, Wang SJ, Chiou LC, Gean PW. 2003. Mediation of amphetamine-induced long-term depression of synaptic transmission by CB1 cannabinoid receptors in the rat amygdala. *J. Neurosci.* 23:10311–20

104. Partridge JG, Tang KC, Lovinger DM. 2000. Regional and postnatal heterogeneity of activity-dependent long-term changes in synaptic efficacy in the dorsal striatum. *J. Neurophysiol.* 84:1422–29

105. Rodriguez de Fonseca F, Ramos JA, Bonnin A, Fernandez-Ruiz JJ. 1993. Presence of cannabinoid binding sites in the brain from early postnatal ages. *Neuroreport* 4:135–38

106. Abraham WC. 2008. Metaplasticity: tuning synapses and networks for plasticity. *Nat. Rev. Neurosci.* 9:387

107. Chen K, Neu A, Howard AL, Földy C, Echegoyen J, et al. 2007. Prevention of plasticity of endocannabinoid signaling inhibits persistent limbic hyperexcitability caused by developmental seizures. *J. Neurosci.* 27:46–58

108. Chen K, Ratzliff A, Hilgenberg L, Gulyas A, Freund TF, et al. 2003. Long-term plasticity of endocannabinoid signaling induced by developmental febrile seizures. *Neuron* 39:599–611

109. Glazewski S, Fox K. 1996. Time course of experience-dependent synaptic potentiation and depression in barrel cortex of adolescent rats. *J. Neurophysiol.* 75:1714–29

110. Rittenhouse CD, Shouval HZ, Paradiso MA, Bear MF. 1999. Monocular deprivation induces homosynaptic long-term depression in visual cortex. *Nature* 397:347–50

111. Allen CB, Celikel T, Feldman DE. 2003. Long-term depression induced by sensory deprivation during cortical map plasticity in vivo. *Nat. Neurosci.* 6:291–99

112. Hensch TK. 2005. Critical period plasticity in local cortical circuits. *Nat. Rev. Neurosci.* 6:877–88

113. Chklovskii DB, Mel BW, Svoboda K. 2004. Cortical rewiring and information storage. *Nature* 431:782–88

114. Bender KJ, Allen CB, Bender VA, Feldman DE. 2006. Synaptic basis for whisker deprivation-induced synaptic depression in rat somatosensory cortex. *J. Neurosci.* 26:4155–65

115. Trachtenberg JT, Trepel C, Stryker MP. 2000. Rapid extragranular plasticity in the absence of thalamocortical plasticity in the developing primary visual cortex. *Science* 287:2029–32

116. Daw NW, Fox K, Sato H, Czepita D. 1992. Critical period for monocular deprivation in the cat visual cortex. *J. Neurophysiol.* 67:197–202

117. Kirkwood A, Dudek SM, Gold JT, Aizenman CD, Bear MF. 1993. Common forms of synaptic plasticity in the hippocampus and neocortex in vitro. *Science* 260:1518–21

118. Deshmukh S, Onozuka K, Bender KJ, Bender VA, Lutz B, et al. 2007. Postnatal development of cannabinoid receptor type 1 expression in rodent somatosensory cortex. *Neuroscience* 145:279–87

119. Liu CH, Heynen AJ, Shuler MG, Bear MF. 2008. Cannabinoid receptor blockade reveals parallel plasticity mechanisms in different layers of mouse visual cortex. *Neuron* 58:340–45

120. Bodor AL, Katona I, Nyiri G, Mackie K, Ledent C, et al. 2005. Endocannabinoid signaling in rat somatosensory cortex: laminar differences and involvement of specific interneuron types. *J. Neurosci.* 25:6845–56

121. Fortin DA, Trettel J, Levine ES. 2004. Brief trains of action potentials enhance pyramidal neuron excitability via endocannabinoid-mediated suppression of inhibition. *J. Neurophysiol.* 92:2105–12

122. Martin SJ, Grimwood PD, Morris RG. 2000. Synaptic plasticity and memory: an evaluation of the hypothesis. *Annu. Rev. Neurosci.* 23:649–711

123. Varvel SA, Lichtman AH. 2002. Evaluation of CB1 receptor knockout mice in the Morris water maze. *J. Pharmacol. Exp. Ther.* 301:915–24

124. Varvel SA, Anum EA, Lichtman AH. 2005. Disruption of CB1 receptor signaling impairs extinction of spatial memory in mice. *Psychopharmacology* 179:863–72

125. Kamprath K, Marsicano G, Tang J, Monory K, Bisogno T, et al. 2006. Cannabinoid CB1 receptor mediates fear extinction via habituation-like processes. *J. Neurosci.* 26:6677–86

126. Chhatwal JP, Davis M, Maguschak KA, Ressler KJ. 2005. Enhancing cannabinoid neurotransmission augments the extinction of conditioned fear. *Neuropsychopharmacology* 30:516–24

127. Varvel SA, Wise LE, Niyuhire F, Cravatt BF, Lichtman AH. 2007. Inhibition of fatty-acid amide hydrolase accelerates acquisition and extinction rates in a spatial memory task. *Neuropsychopharmacology* 32:1032–41

128. Monory K, Massa F, Egertova M, Eder M, Blaudzun H, et al. 2006. The endocannabinoid system controls key epileptogenic circuits in the hippocampus. *Neuron* 51:455–66

129. Graybiel AM. 2000. The basal ganglia. *Curr. Biol.* 10:R509–11

130. Calabresi P, Saiardi A, Pisani A, Baik JH, Centonze D, et al. 1997. Abnormal synaptic plasticity in the striatum of mice lacking dopamine D2 receptors. *J. Neurosci.* 17:4536–44

131. Oueslati A, Breysse N, Amalric M, Kerkerian-Le Goff L, Salin P. 2005. Dysfunction of the cortico-basal ganglia-cortical loop in a rat model of early parkinsonism is reversed by metabotropic glutamate receptor 5 antagonism. *Eur. J. Neurosci.* 22:2765–74

132. Justinova Z, Goldberg SR, Heishman SJ, Tanda G. 2005. Self-administration of cannabinoids by experimental animals and human marijuana smokers. *Pharmacol. Biochem. Behav.* 81:285–99

133. Gardner EL. 2005. Endocannabinoid signaling system and brain reward: emphasis on dopamine. *Pharmacol. Biochem. Behav.* 81:263–84

134. Nestler EJ. 2004. Molecular mechanisms of drug addiction. *Neuropharmacology* 47(Suppl. 1):24–32

135. Koob GF, Le Moal M. 1997. Drug abuse: hedonic homeostatic dysregulation. *Science* 278:52–58

136. Kauer JA, Malenka RC. 2007. Synaptic plasticity and addiction. *Nat. Rev. Neurosci.* 8:844–58

137. Hyman SE, Malenka RC, Nestler EJ. 2006. Neural mechanisms of addiction: the role of reward-related learning and memory. *Annu. Rev. Neurosci.* 29:565–98

138. Kalivas PW, Nakamura M. 1999. Neural systems for behavioral activation and reward. *Curr. Opin. Neurobiol.* 9:223–27

139. Solinas M, Goldberg SR, Piomelli D. 2008. The endocannabinoid system in brain reward processes. *Br. J. Pharmacol.* 154:369–83

140. Fourgeaud L, Mato S, Bouchet D, Hemar A, Worley PF, Manzoni OJ. 2004. A single in vivo exposure to cocaine abolishes endocannabinoid-mediated long-term depression in the nucleus accumbens. *J. Neurosci.* 24:6939–45

141. Mato S, Chevaleyre V, Robbe D, Pazos A, Castillo PE, Manzoni OJ. 2004. A single in-vivo exposure to Δ_9THC blocks endocannabinoid-mediated synaptic plasticity. *Nat. Neurosci.* 7:585–86

142. Hoffman AF, Oz M, Caulder T, Lupica CR. 2003. Functional tolerance and blockade of long-term depression at synapses in the nucleus accumbens after chronic cannabinoid exposure. *J. Neurosci.* 23:4815–20

143. Liu QS, Pu L, Poo MM. 2005. Repeated cocaine exposure in vivo facilitates LTP induction in midbrain dopamine neurons. *Nature* 437:1027–31

144. Safo PK, Regehr WG. 2005. Endocannabinoids control the induction of cerebellar LTD. *Neuron* 48:647–59

145. Cachope R, Mackie K, Triller A, O'Brien J, Pereda AE. 2007. Potentiation of electrical and chemical synaptic transmission mediated by endocannabinoids. *Neuron* 56:1034–47

146. Kyriakatos A, El Manira A. 2007. Long-term plasticity of the spinal locomotor circuitry mediated by endocannabinoid and nitric oxide signaling. *J. Neurosci.* 27:12664–74

147. Bacci A, Huguenard JR, Prince DA. 2004. Long-lasting self-inhibition of neocortical interneurons mediated by endocannabinoids. *Nature* 431:312–16

148. Huang F, Chotiner JK, Steward O. 2005. The mRNA for elongation factor 1α is localized in dendrites and translated in response to treatments that induce long-term depression. *J. Neurosci.* 25:7199–209

149. Huber KM, Kayser MS, Bear MF. 2000. Role for rapid dendritic protein synthesis in hippocampal mGluR-dependent long-term depression. *Science* 288:1254–57

150. Yin HH, Davis MI, Ronesi JA, Lovinger DM. 2006. The role of protein synthesis in striatal long-term depression. *J. Neurosci.* 26:11811–20

Sensing Odorants and Pheromones with Chemosensory Receptors

Kazushige Touhara[1] and Leslie B. Vosshall[2]

[1]Department of Integrated Biosciences, The University of Tokyo, Chiba, 277-8562 Japan; email: touhara@k.u-tokyo.ac.jp

[2]Howard Hughes Medical Institute, Laboratory of Neurogenetics and Behavior, The Rockefeller University, New York, NY 10065; email: leslie@mail.rockefeller.edu

Annu. Rev. Physiol. 2009. 71:307–32

The *Annual Review of Physiology* is online at physiol.annualreviews.org

This article's doi: 10.1146/annurev.physiol.010908.163209

Copyright © 2009 by Annual Reviews. All rights reserved

0066-4278/09/0315-0307$20.00

Key Words

vomeronasal organ, olfaction, behavior, G protein–coupled receptors, olfactory bulb, olfactory sensory neuron

Abstract

Olfaction is a critical sensory modality that allows living things to acquire chemical information from the external world. The olfactory system processes two major classes of stimuli: (*a*) general odorants, small molecules derived from food or the environment that signal the presence of food, fire, or predators, and (*b*) pheromones, molecules released from individuals of the same species that convey social or sexual cues. Chemosensory receptors are broadly classified, by the ligands that activate them, into odorant or pheromone receptors. Peripheral sensory neurons expressing either odorant or pheromone receptors send signals to separate odor- and pheromone-processing centers in the brain to elicit distinct behavioral and neuroendocrinological outputs. General odorants activate receptors in a combinatorial fashion, whereas pheromones activate narrowly tuned receptors that activate sexually dimorphic neural circuits in the brain. We review recent progress on chemosensory receptor structure, function, and circuitry in vertebrates and invertebrates from the point of view of the molecular biology and physiology of these sensory systems.

INTRODUCTION

What Is an Odorant?
What Is a Pheromone?

Odorant: a volatile chemical compound, usually of molecular weight 300 or smaller, that activates olfactory neurons and induces the percept of an odor

Pheromone: a substance released by a member of the same species that elicits a stereotyped behavior and/or endocrinological response in another member of the same species. Pheromones can be proteins, small molecules, or a combination thereof

Conspecifics: members of the same animal species

An odorant is a volatile chemical compound with a molecular weight of lower than ~300 that humans or other animals perceive as odorous via the olfactory system. The number of possible odorant molecules that exist on earth is unknown, but essentially all living things such as plants, insects, animals, and microbes emit smells—purposefully or as by-products of metabolism or waste excretion. There are also inorganic sources of odorants: Sulfur dioxide and some metals have a distinct odor. All these odorants can potentially be exploited by animals to improve their chances for survival.

How many odorants are there, and how many can humans detect and distinguish? The scientific and popular literature is full of claims that the number of odorants that exist and that we can detect ranges from hundreds to thousands or even hundreds of thousands. Meanwhile, most laboratory studies in the field of olfaction use a few dozen odorants from a standard set of approximately 500 chemicals (1). The most oft-cited statistic is that humans can distinguish approximately 10,000 different odors. In his engaging book on smell (2), Avery Gilbert painstakingly traces the familiar 10,000 odors claim back through the twentieth-century scientific literature to a series of questionable assumptions in a theoretical model developed in 1927 by Ernest C. Crocker and Lloyd F. Henderson. Therefore, no one appears to have catalogued the exact number of known smells or how many such smells we can perceive. Regardless of exactly how many different odorants there may be, or how many a given animal can detect and discriminate, these small molecules are clearly important for animal survival, allowing animals to locate food and to avoid predators and environmental dangers such as fire.

In contrast to general odorants, a pheromone is defined as a specific substance that is secreted by an individual and received by a second individual of the same species, or conspecific, to induce a specific reaction such as a stereotyped behavior or endocrinological change (3) [see also the book by Wyatt (4) for an excellent review on the topic of pheromones]. The term was originally coined on the basis of the identification of a volatile sex attractant, bombykol, which is released by the female silk moth *Bombyx mori* and elicits the full sequence of sexual behaviors in male moths (5). The definition avoids any use of the terms odor or odorant because a pheromone does not have to be odorous or volatile as long as the signal is a chemical substance that is transferred between conspecifics. Pheromones can be nonvolatile substances with a molecular weight of larger than a few hundred, including relatively large organic compounds, peptides, and proteins.

In some cases, the distinctions between general odorants and pheromones are blurred. For instance, a pheromone can be an odorous compound, and an odorant can be a pheromone. As such, a pheromone released and utilized by one species can be produced as a general odorant for a second species but can be an informative odor that a third species uses to predate the second species. General odorants can also induce behavioral or physiological changes in a fashion similar to a pheromone. Insects such as moths and flies are attracted by the smell of flowers and plants, which induces stereotyped feeding behavior. We similarly experience physiological effects of general odorants elicited by trees, plants, and flowers, an effect on which the aromatherapy industry is based. These behavioral and physiological effects, however, are not categorized as pheromonal effects because the active chemical substances are not derived from conspecifics. There is an ongoing trend in the chemosensory field to extend the definition of pheromones beyond the strict definition of Karlson & Luscher (3) because there is increasing evidence that pheromones play much more diverse functional roles than are included in the above definition (4). Thus, a pheromone could be more broadly defined as a substance that is utilized for intraspecies communication even though it does not elicit obvious behavioral or endocrinological changes. We list below some possible criteria for a more inclusive definition of a pheromone:

1. Pheromones are released by one individual and received by conspecifics;
2. pheromones themselves can send information about sex, strain, and species to the receiver; and
3. pheromonal effects must be meaningful or informative for the species.

As molecular biologists, chemical ecologists, and neurobiologists continue to work together to understand the link between pheromones and behavior, the field is likely to arrive at an optimal definition of a pheromone in the future.

How Are Odorants and Pheromones Detected?

If one assumes that there are thousands of important odorants, how do animals recognize such a large diversity of chemical cues? Titus Lucretius Carus, a Roman poet and philosopher, proposed in 50 BCE that a large variety of odors exists because each odorant possesses a unique structure (6). In the mid-twentieth century, Amoore (7) formalized this concept as the stereospecific receptor theory, which attempted to explain the molecular mechanisms underlying the remarkable olfactory sensing system. The receptor theory postulates that there are many receptor sites for odorants and that odor perception occurs only when the structure of an odorant molecule and the structure of its binding site match. Buck & Axel (8) discovered in 1991 a large multigene family encoding receptor proteins for odorants in rat; this finding was later extended to all vertebrates studied. Thus, the large number and diversity of olfactory (or odorant) receptors (ORs) confirm the essential features of Amoore's receptor theory. Although we still cannot predict what odorants will bind to a particular OR, or how a particular odorant will smell, the stereochemical receptor theory remains the dominant theory in the field. Alternate theories proposed to explain odor perception, including vibrational, puncturing, radiational, and absorption theories, are hotly debated (9, 10) but remain unproven.

Odorants and pheromones are detected by olfactory sensory neurons in the olfactory system. Mammals usually detect general odorants by the nasal olfactory epithelium via the main olfactory system. Rodents and a number of other nonprimate species possess a secondary olfactory system called the vomeronasal pathway, which detects signals via the vomeronasal organ (VNO), located at the bottom of the nasal cavity. The appearance of the VNO coincided with the acquisition of the lung respiratory system during the Cambrian explosion (11), and the VNO became genetically vestigial during the evolution of the primate lineage (12). Similarly, most insects also have two olfactory organs, the antenna and the maxillary palp, although insects differ in the extent to which each organ is dedicated to general odorants, pheromones, or even nonolfactory cues such as taste and mechanical stimuli. Recent evidence suggests that the labial palps, typically thought to be exclusively taste organs, can also sense odors (13).

In this review, we summarize the current knowledge of chemosensory receptors for odorants and pheromones, point out what is common and what is different in chemosensing mechanisms between invertebrate and vertebrate animals, and discuss how sexually dimorphic responses to chemosignals are encoded in the brain.

CHEMOSENSORY RECEPTOR GENES

Vertebrate Olfactory Receptors

The vertebrate OR genes encode a large family of seven-transmembrane-domain G protein–coupled receptors (GPCRs) that play a role in recognizing odorant molecules in the olfactory epithelium (14). OR proteins were classified as members of the GPCR superfamily because of the presence of structural features common to GPCRs (8, 15) and because the ORs couple to and activate heterotrimeric G proteins (16, 17). The size of the OR family was estimated to be at least several hundred genes when the

OR: olfactory receptor or odorant receptor

VNO: vomeronasal organ

GPCR: G protein–coupled receptor

VR: vomeronasal receptor

family was originally identified in the rat (8). In the past several years, whole-genome sequencing projects have allowed for a comprehensive analysis of the OR gene family in diverse species and have made it possible to study the genomic structure and distribution of OR genes from various organisms (18). In mammals, the OR repertoire comprises ∼800–1500 members, whereas fish have a relatively small OR family of ∼100 genes (19). Thus, the OR gene expansion likely occurred when animals shifted from aqueous to terrestrial environments.

In every vertebrate in which the expression of OR genes has been examined, each olfactory sensory neuron expresses only a single member of the OR gene family. This is hypothesized to be important for olfactory coding, such that a given population of olfactory neurons responds to a restricted number of odorants and communicates ligand binding to central olfactory circuits. The mechanism by which an olfactory neuron selects a single OR, and represses the expression of all remaining ORs in the genome, remains a fascinating and controversial question (20–23).

A significant portion of the OR gene family has been pseudogenized in vertebrates, leading to the loss of a large fraction of the potentially functional ORs in a given species. Hominoids possess high pseudogene content in their ORs (∼50%), whereas ∼20% of mouse and dog OR genes and 25–35% of primate OR genes are pseudogenes (19, 24, 25). The OR family in the lineage of many species appears to be undergoing a rapid molecular evolution by tandem gene duplication and pseudogenization. The fraction of pseudogenes has increased during the evolution from rodents, monkeys, and humans, suggesting that the reduced function of the sense of smell correlates with the loss of functional OR genes. Indeed, in whale and dolphin, animals for which the auditory system is dominant over the olfactory system, 70–80% of OR genes appear to be pseudogenes (Y. Go, personal communication). Furthermore, OR genes are highly polymorphic, as suggested by human leukocyte antigen–linked OR genes (26, 27) and dog OR genes (28). Recent studies have

revealed that single-nucleotide polymorphisms in human OR genes account in part for individual differences in the ability to detect specific odors (29, 30).

Vertebrate Vomeronasal Receptors

The VNO expresses vomeronasal receptors (VRs), comprising two distinct subfamilies of GPCRs, the V1Rs and V2Rs (31–36). The vomeronasal sensory epithelium in the VNO can be divided into apical and basal layers, which express V1R- and V2R-type receptors, respectively. V1Rs and V2Rs begin to be expressed after birth (31), whereas ORs begin to be expressed during embryogenesis (37). The segregation of the vomeronasal epithelium into two layers occurs during the postnatal period (32), consistent with the development of the VNO as an organ that functions in adult communication.

The V1R family consists of a total of 308 sequences in mice, 187 of which appear to encode full-length open reading frames (38). Most of the gene family is located on chromosomes 6, 7, and 13, and like ORs, V1Rs are encoded by a single exon (39, 40). In humans, more than 90% of the V1R genes have been converted to pseudogenes, so only five V1R genes appear to be intact (41, 42). V1Rs do not show significant homology to the ORs, but they are weakly related to the T2R family of bitter receptors (31). The expression of V1Rs is restricted to the apical neuroepithelium in mice (43); a given vomeronasal neuron expresses just one member of the V1R family, as previously documented for the main olfactory system. In humans, who lack a functional VNO, V1Rs are expressed in the olfactory epithelium (41). The genomic structure and expression patterns of V1Rs appear to have undergone rapid changes during the process of the evolution (44).

In mice, the V2R family comprises 121 intact genes out of a total of 279 genes, including pseudogenes (38, 45). Several years were needed for the comprehensive genomic analysis of the V2R family because of the complex intron/exon structures of V2Rs (33–35). V2Rs possess a long extracellular N terminus that is common to calcium-sensing and metabotropic glutamate receptors and the T1R family of sweet and umami receptors (18). Intact V2R genes have not been found in the human genome. There appear to be only ∼20 V2R pseudogenes in humans, suggesting that V2R genes have changed more dramatically than V1R genes. The expression of V2Rs is restricted to the basal neuroepithelium in mice (33–35), and as in the case of ORs and V1Rs, the one cell–one receptor rule seems conserved (46). One notable exception to the one-receptor-per-cell rule is a small subfamily of V2R genes, the V2R2 genes, that appear to be broadly coexpressed with other V2Rs (47). The function of the V2R2s in the V2R2-expressing portion of the VNO remains a mystery.

Insect Olfactory Receptors and Gustatory Receptors

The initial description of vertebrate ORs by Buck & Axel (8) rapidly led to the identification of ORs in essentially every vertebrate species studied. However, all attempts to isolate insect homologs of vertebrate GPCR-type ORs selectively expressed in olfactory tissues failed. Instead, three groups used the unbiased approach of plus-minus screening and genome mining to identify candidate ORs in the fruit fly *Drosophila melanogaster* (48–50). The insect OR genes, comprising 62 members in the fruit fly, encode a novel family of seven-transmembrane-domain receptor proteins selectively expressed in subsets of olfactory neurons in the insect olfactory organs, the antennae and maxillary palps.

Subsequent to the discovery of *Drosophila* ORs, a family of divergent seven-trans-membrane-domain receptor genes, distantly related to the ORs, was isolated (51). This gene family was named the gustatory receptor (GR) family because many of the GR genes were expressed in taste organs such as the labial palps (51–53).

The number of chemosensory receptor genes is smaller in insects than in most mam-

mals and more closely approaches the number found in fish. Genomic analysis has identified 62 ORs and 68 GRs in *D. melanogaster* (54), 79 ORs and 72 GRs in the malaria vector mosquito *Anopheles gambiae* (55), 170 ORs and 13 GRs in the honeybee *Apis mellifera* (56), 131 ORs and 88 GRs in the yellow fever and dengue virus vector *Aedes aegyptii* (57), 341 ORs (58) and 62 GRs (59) in the beetle *Tribolium castaneum*, and, most recently, 66 ORs and 14 GRs in the silk moth *B. mori* (International Silkworm Genome Consortium, submitted).

The numbers of ORs and GRs and the relative ratios of these related chemosensory receptor gene families differ from insect to insect, suggesting that sociosexual behavior and lifestyle may have positively influenced these gene families during the insect evolution. For instance, the honeybee has a severely reduced repertoire of taste receptors, comprising only a handful of putative sugar receptors, and an expanded OR gene family. Robertson & Wanner (56) suggested that the absence of bitter taste receptors in the honeybee could be explained by the cooperative relationship these pollinators have with host plants, which would not produce the bitter alkaloids that would harm bees and thus would not need to be detected by bitter-sensing bee GRs. Fewer chemosensory receptor genes in insects than in vertebrates are pseudogenes, with the exception of the *Tribolium* beetle genome, in which many OR and GR pseudogenes exist. To examine the molecular evolution and dynamics of OR genes, ORs from more than 10 *Drosophila* species were identified, and phylogenetic analysis was performed (60). The insect OR genes appear to be evolutionarily more stable than the vertebrate OR genes, but there are some lineage-specific gene duplications and losses that are reflected by positive selection possibly because of environmental and behavioral differences between species.

Insect ORs and GRs do not show any significant sequence similarity to vertebrate ORs and also lack homology to chemosensory GPCRs in vertebrates such as the typical DRY and NPXXY amino acid motifs. Furthermore, both computer prediction and experimental analysis

GR: gustatory receptor

suggest that the insect ORs and GRs possess a transmembrane topology distinct from that of GPCRs, with the N terminus located intracellularly and the C terminus located extracellularly (61–63).

Another feature that distinguishes insect ORs from their molecularly distinct mammalian OR counterparts is that a highly conserved OR, called Or83b after its name in *D. melanogaster*, is coexpressed with other conventional ORs in single olfactory neurons (64–66). Or83b is an essential constant subunit of the heteromultimeric insect OR that forms a receptor complex with the variable ligand-binding ORs (62, 65, 67). We discuss the functional consequences of these differences in the structure and expression profile between vertebrate and invertebrate ORs (see Insect Olfactory Receptors subsection in Chemosensory Signal Transduction section, below). Later in this review, we continue to develop the concept that chemosensory receptors appear to have arisen independently in the evolution of vertebrate and invertebrate species.

CHEMOSENSORY RECEPTOR FUNCTION: LIGAND-RECEPTOR PAIRS

Mouse Olfactory Receptors

When originally cloned by Buck & Axel (8), the OR genes were hypothesized to encode the receptor proteins underlying the molecular basis of smell. Functional proof that ORs recognize odorants has been assessed by both homologous and heterologous expression systems that seek to link a given OR with its cognate odor ligand(s). Olfactory neurons themselves provide the homologous expression system: homologous because they possess the appropriate cellular machinery for OR expression and the transmission of odorant signals. Thus, olfactory neurons identified to express a certain OR of interest respond to the cognate ligands for the OR (68–71).

ORs have also been functionally expressed in heterologous expression systems such as mammalian cell lines and *Xenopus laevis* frog oocytes, making it possible to assess OR responsiveness to odorants (16, 17, 72–76). Thus, in HEK293 cells coexpressing tagged ORs and the promiscuous G protein $G_{\alpha15}$, Ca^{2+} responses were observed when the cells were stimulated with their ligands (16, 17, 73, 74). Without coexpression of $G_{\alpha15}$, ORs activate endogenous $G_{\alpha s}$ upon ligand stimulation, resulting in an increase in cAMP levels in various mammalian cell lines including HEK293, COS-7, and CHO-K1 cells (16, 72–74). A luciferase-reporter assay system using the *zif268* promoter allows luminescent detection of cAMP increases upon stimulation with an odorant (72–74, 77).

Although convenient for rapid and high-throughput expression outside of native olfactory neurons, the expression of ORs in heterologous cells is tricky because ORs appear to be inefficiently translocated to the plasma membrane. In some cases, adding the N-terminal leader sequence of rhodopsin, another GPCR, resulted in improved targeting of functional ORs to the plasma membrane and in successful odorant-response recording in a heterologous system such as HEK293 cells (16, 17, 74, 75).

To improve functional heterologous expression, several groups have searched for cofactors such as protein chaperones and other proteins to facilitate and stabilize cell-surface expression of ORs. The one-transmembrane-domain protein RTP1 expressed in olfactory neurons enhances cell-surface expression of ORs, and some ORs have been deorphanized in cell lines coexpressing RTP1 (77). In addition, Ric8B, a putative guanine nucleotide exchange factor for $G_{\alpha s}$ and $G_{\alpha olf}$, promotes efficient signal transduction of ORs (78). Moreover, coexpression of myristylated Ric8A, a Ric8B homolog that acts on $G_{\alpha q}$-type G proteins, greatly enhances $G_{\alpha15}$-mediated odorant signaling (79). These efforts to enhance cell-surface expression and signal transduction of ORs have greatly facilitated the process of deorphanizing ORs in the past few years (74).

What can we learn from the comparative analysis of OR structure and function and regions that are variable versus those that are conserved? ORs possess seven hydrophobic transmembrane domains, a disulfide bond between the conserved cysteines in the extracellular loops, a conserved glycosylation site in the N-terminal region (80), and several amino acid sequence patterns that are conserved in the OR family (81, 82). These conserved motifs likely contribute to proper folding and membrane trafficking of ORs to the plasma membrane so that ORs can function in binding odorants and coupling to the appropriate G proteins. In contrast, the transmembrane regions are relatively variable and may play a role in forming an odorant-binding pocket, such that variability in these domains allows ORs to cover a wide variety of odorant molecules in the ligand recognition spectra (83–88).

Touhara and coworkers (89) undertook a systematic experimental approach to decipher the OR odorant-binding site. They employed functional analysis of site-directed mutants and ligand docking simulation studies, revealing that most of the critical residues involved in odorant recognition are hydrophobic and located within the binding pocket formed by transmembrane (TM) domains TM3, TM5, and TM6 (89). Furthermore, the accuracy of the binding model was validated by the fact that single-amino-acid changes caused predictable changes in agonist and antagonist specificity (89). These results allowed us to conclude that vertebrate ORs recognize the size, shape, and functional group of an odorant, using both hydrogen bonds and hydrophobic interactions in the odor-binding pocket formed by transmembrane helices.

As more and more vertebrate ORs have been paired with ligands, the results have revealed a combinatorial coding strategy in which each odorant is recognized by a subset of ORs that is unique for the odorant (90). ORs that recognize many odorants with a wide range of structures are defined as broadly tuned or generalist receptors. This overlapping coding strategy may represent a molecular basis for the discriminative power of the olfactory system. In contrast, some ORs detect certain odorant molecules with high specificity and are referred to as nar-rowly tuned or specialist receptors. Such narrowly tuned receptors may mediate signals that activate specialized circuits in the brain, resulting in discrete behaviors or neuroendocrinological changes reminiscent of a pheromone effect. As discussed above, the differences between pheromones and odorants are in part semantic and relate to the source and effect of the stimulus. With recent advances in annotating OR genes in diverse species and in the efficient functional expression of ORs in heterologous cells, it seems feasible that the complete odorant-OR matrix will eventually be constructed for several species. Such information will reveal how broad the universe of odorant molecules recognized by each species is and which odorant-OR combinations elicit innate reactions that are essential for individuals to survive in various environments.

Mouse Vomeronasal Receptors

VRs are receptors that mark two neural compartments of the VNO, but how good is the evidence that either V1Rs or V2Rs function as pheromone receptors? In fact, recent evidence for V1Rs as pheromone receptors has emerged (18, 91). First, a mouse V1R expressed ectopically in vomeronasal sensory neurons responds to urine (92), a bodily substance shown to contain a number of molecules that elicit pheromonal effects in rodents. Second, mutant mice lacking a cluster of V1Rs fail to respond to some volatile pheromones and show decreased aggressiveness and sexual behavior (93). Third, V1Rb2 expressed in vomeronasal neurons responds to 2-heptanone, one of the components of urine that cause extension of the estrous cycle (94). Together with evidence from calcium imaging of intact VNO epithelial slices showing that vomeronasal neurons in the apical layer respond to various volatile pheromones (95), these findings suggest that V1Rs serve as sensors for volatile pheromones.

Vomeronasal neurons, however, can also detect nonpheromonal odorants (96), implying that V1R-expressing neurons may not be restricted to the detection of species-specific

TM: transmembrane

ESP: exocrine gland–secreting peptide

AOB: accessory olfactory bulb

volatile pheromones for mating but also may perceive non-species-related signals such as those from prey, predators, and the physical environment. Five V1Rs expressed in the human olfactory epithelium responded strongly to C9-C10 aliphatic alcohols or aldehydes in a combinatorial fashion (97). Whether these human receptors function as pheromone receptors and what ligands activate them remain to be elucidated. Although the available data are relatively limited, it seems that the strategy for V1Rs to recognize volatile pheromones or odorants is similar to that of ORs. Thus, how a combinatorial code of V1R activation is processed in the brain to elicit innate behavior or neuroendocrine changes remains to be studied.

V2Rs have a long N-terminal region related to the N-terminal region found in metabotropic glutamate receptors and calcium sensors, and therefore potential ligands for V2Rs have been thought to be nonvolatile pheromones such as peptides and proteins. Candidate ligands for V2Rs include three polypeptide ligands that have been implicated in VNO function: major urinary proteins (MUPs), major histocompatibility complex (MHC) peptides, and the exocrine gland–secreting peptide (ESP) family (98).

Biochemical studies in rats have demonstrated that some MUPs activate vomeronasal neurons in the basal layer, which express $G_{\alpha o}$ and V2Rs (99). In a recent paper, Stowers and coworkers (100) demonstrated that highly purified MUPs in the absence of volatile urine components activated VNO neurons derived from the basal zone. Interestingly, the synthetic MUPs alone sufficed to elicit aggressive behavior in male mice, confirming that MUPs fulfill the basic definition of a pheromone. However, there have been some contradictions in the various MUP studies. One study showed that MUPs activate isolated V2R-expessing basal neurons in calcium imaging (100), whereas other studies showed that urine, which contains MUPs, does not elicit electrical responses in the VNO or neural spikes in V2R-expressing neurons (101). The basis for these differences is not currently known.

For MHC peptides, calcium-imaging studies in mice indicated that vomeronasal sensory neurons responding to the peptides are located in the basal neuroepithelium, where V2Rs are expressed (102, 103). Moreover, these peptides sufficed to induce the physiological effect of pregnancy block, a pheromonal behavior in which exposure to a foreign male blocks implantation of embryos sired by a familiar male (102). These peptides, however, also activate the main olfactory epithelium, which does not express VRs (104).

ESP1 is a male-specific peptide that activates V2R neurons in four different assay systems. First, ESP1 induced c-Fos expression, a transcriptional readout of neuronal activation, both in V2R neurons (105) and in the accessory olfactory bulb (AOB) (S. Haga & K. Touhara, unpublished observations). Second, ESP1 activated V2R-expressing neurons as assayed by calcium imaging (K. Touhara & C.R. Yu, unpublished observation). Third, ESP1 evoked potentials in the vomeronasal epithelium as measured in electro-vomeronasogram (EVG) recordings (105). Fourth, ESP1 activated neural spikes as measured with electrical recordings of vomeronasal neurons (106). Although these studies do not provide direct evidence for pheromone-V2R interactions, the results support the idea that V2R-expressing neurons respond to nonvolatile pheromones.

Further studies were performed to identify which V2R(s) is expressed in vomeronasal neurons that recognize ESP1. Vomeronasal neurons were double-labeled with an anti-c-Fos antibody and an in situ RNA probe for V2Rs. Of 12 different RNA probes, one of them, V2Rp, which was designed to hybridize with five highly homologous V2Rp genes, clearly recognized c-Fos-positive neurons that responded to ESP1 (105). In contrast, none of the other probes overlapped with c-Fos. Further studies using specific probes to discriminate the five V2Rp genes revealed that a single V2R, named V2Rp5, was expressed in all the c-Fos neurons induced by exposure to ESP1 (107). V2Rp5-expressing neurons tagged with DsRed respond to ESP1, clearly demonstrating that V2Rp5 is a receptor for ESP1 (S. Haga, Y. Yoshihara & K. Touhara, unpublished observation). These results provide evidence that each V2R possesses a narrow ligand spectrum and thus is responsible for the detection of a specific peptide pheromone. Multielectrode recordings of the spike firing rate of vomeronasal neurons also demonstrated that sensory neurons, each of which expresses a single type of V2R, showed responses to the specific ESP ligand, and therefore the V2R neurons are narrowly tuned (106). This putative narrow tuning of the VNO contrasts with the combinatorial strategy observed in the main olfactory system.

Insect Olfactory Receptors

Initial attempts to characterize insect OR ligands were carried out by a strategy similar to that used for vertebrate ORs. *Drosophila* Or43a was expressed in a homologous system, in this case in the antenna (108), or in a heterologous system, *Xenopus* oocytes, which led to the identification of ligands for Or43a, including benzaldehyde and cyclohexanone (109). Later, researchers carried out a large-scale analysis of the *Drosophila* antennal ORs by genetically introducing individual ORs into an antennal neuron lacking the endogenous ligand-binding OR but retaining the Or83b coreceptor (110). Responses of 24 ORs (out of a total of 62 ORs) expressed in the adult fly olfactory system to 110 odorants revealed combinatorial receptor codes for odorants, similar to those in the vertebrate olfactory system (111, 112). Twenty-five ORs were detected in the larva, and 14 showed larva-specific expression, whereas 11 ORs were expressed in both larvae and adults (113, 114). Most larval ORs are also broadly tuned but can be divided into two general classes of sensitivity: aromatic or aliphatic compounds (114, 115). This suggests that the vast majority of insect ORs are generalist-type sensors that detect a variety of odorous molecules from food sources.

As we have seen, vertebrates use distinct receptor gene families to detect putative pheromones, but what detection strategy have insects evolved to respond to pheromones? Two independent studies were carried out to identify an insect pheromone receptor that specifically recognizes moth sex pheromones (116–119). Both studies were based on the assumption that a sex-pheromone receptor would be expressed sex-specifically in the male antenna. Indeed, two male-specific members of the insect OR gene family, named *BmOR1* and *BmOR3*, were discovered (116, 117). These were shown to function as sensors for bombykol and bombykal, two pheromone components in the silk moth *B. mori*, respectively (117). The function of these two proteins as pheromone sensors was confirmed by heterologous expression in *Xenopus* oocytes by coexpressing each receptor with BmOR2, the *B. mori* Or83b ortholog (117). The binding site for bombykol seems to reside in BmOR1, not the BmOR2 coreceptor, because Or83b orthologs from different species act as a functional partner to BmOR1 (117). BmOR1 ectopically expressed in *Drosophila* antennae also responded to bombykol (120), further confirming that BmOR1 is the pheromone-binding site. A second study showed that sex-pheromone receptors from a different moth species (*Heliothis virescens*), HR13/15/16, are also expressed male-specifically and recognize female-producing pheromone compounds (118).

After these initial reports of moth pheromone receptors, other studies linking identified insect pheromones to receptors were published. A male drone–biased OR, AmOr11, in honeybees was shown to be the receptor for the queen retinue pheromone 9-*oxo*-2-decenoic acid by coexpression of AmOr11 together with AmOr2, the honeybee Or83b ortholog, in *Xenopus* oocytes (121). The *Drosophila* aggregation and sex pheromone 11-*cis*-vaccenyl acetate (cVA) is also recognized by an OR named Or67d, which functions together with Or83b (122, 123). These results suggest that insects have selected pheromone receptors from a repertoire of ORs, rather than creating a new family of specific pheromone receptors. This is in contrast to vertebrate species in which structurally distinct chemosensory

cVA: 11-*cis*-vaccenyl acetate

receptor families, the V1Rs and V2Rs, play a role in pheromone detection.

Although insects use members of the OR family to detect pheromones, there is recent evidence that cVA-sensing neurons in *Drosophila* are specialized and that additional accessory factors are crucial to detecting pheromones. These include the CD36 homolog sensory neuron membrane protein (SNMP) and the pheromone-binding protein LUSH. SNMP is a two-transmembrane-domain protein that is required to tune the spontaneous activity of the Or67/Or83b complex (124, 125), and LUSH is increasingly viewed as a critical component of the ligand recognition of the cVA pheromone (126). It will be interesting to see if vertebrate pheromone receptors similarly use specific cofactors to recognize pheromone ligands.

CHEMOSENSORY SIGNAL TRANSDUCTION

Vertebrate Olfactory Receptors

The odorant signal received by ORs is converted to an electrical signal in olfactory sensory neurons (127). In vertebrates, the first step in olfactory signal transduction is the activation of the G protein $G_{\alpha olf}$ by odorant-bound activated ORs. Some amino acids in the C-terminal domain and third intracellular loop of ORs appear to be involved in coupling and activation of $G_{\alpha olf}$ (80, 128). Unlike the case of rhodopsin, signal amplification does not occur in the OR–G protein activation cycle (129). Instead, Ric8B, a guanine nucleotide exchange factor, enhances the accumulation of $G_{\alpha olf}$ at the cell membrane, thus improving the efficiency of OR coupling to $G_{\alpha olf}$ (78). The activated $G_{\alpha olf}$ in turn stimulates adenylyl cyclase III, resulting in a cAMP increase (130). The elevated levels of cAMP interact with and gate a cyclic nucleotide–gated (CNG) channel, leading to cation influx and to depolarization of the receptor neuron. Furthermore, calcium activates a chloride channel that leads to an efflux of Cl^-, contributing to the amplification of the sensory depolarization. Bestrophin-2 is a candidate for a molecular component of the olfactory calcium-activated chloride channel (131, 132).

Gene knockout mice lacking any of the three molecular components (i.e., $G_{\alpha olf}$, adenylyl cyclase III, or CNG channel) failed to respond to odorants, suggesting that the cAMP cascade is dominant in transmitting the odorant signal in olfactory neurons (133–135). However, investigators later demonstrated that CNGA2 knockout mice could detect a subset of odorants including 2-heptanone and 2,5-dimethylpyrazine, which are constituents of urine (136). Two additional olfactory pathways have been proposed: One involves membrane-bound guanylyl cyclase and phosphodiesterase type 2 (137–139), and another involves phospholipase C and TRPM5 channel (140). Multiple pathways in olfactory neurons may allow mice to respond readily to complex biological signals for social and sexual communication.

The activated olfactory neuron must return to the steady state in a process known as desensitization. As is the case for other GPCRs, G protein–coupled ORs may be phosphorylated upon odorant binding by protein kinases such as protein kinase A (PKA) and G protein–coupled receptor kinase (GRK), resulting in desensitization (141–143). Indeed, knockout mice lacking GRK showed slower recovery of cAMP increases, implicating the involvement of GRK in OR desensitization (144). In addition, PKA-mediated phosphorylation and subsequent internalization of ORs along with β-arrestin led to a reduction of the number of ORs in the membrane (145). Although direct evidence for OR phosphorylation has not been obtained, desensitization mechanisms likely exist at the level of ORs. In addition, elevated intracellular Ca^{2+} that enters the cell via the influx through CNG channels causes these channels to downregulate themselves and thus close (146–148). Ca^{2+} also negatively regulates adenylyl cyclase activity (149). These Ca^{2+}-mediated negative feedback mechanisms allow activated olfactory neurons to return to the steady state and to prepare for the next odor stimulus.

Vertebrate Vomeronasal Receptors

The mouse vomeronasal epithelium is divided into two functionally distinct layers: The neurons in the apical layer coexpress V1Rs and the G protein $G_{\alpha i2}$, and the neurons in the basal layer coexpress V2Rs and $G_{\alpha o}$ (32). V1Rs are likely to couple $G_{\alpha i2}$ and to stimulate $G_{\beta\gamma}$-mediated calcium signaling, whereas V2Rs are likely to couple to $G_{\alpha o}$ and also to stimulate a calcium signaling cascade. Indeed, mouse V1rb2, one member of the V1R subfamily, elicits an inhibitory signal for the cAMP cascade and $G_{\beta\gamma}$-mediated calcium increases upon stimulation with 2-heptanone via coupling with $G_{\alpha i2}$ in HeLa cells coexpressing CNG channels (97). $G_{\beta\gamma}$-mediated activation of phospholipase C resulted in the production of inositol-1,4,5-triphosphate (IP3), diacylglycerol (DAG), and polyunsaturated fatty acids, all of which have been implicated in vomeronasal sensory signaling (150–152). In contrast, human V1Rs couple to stimulatory G proteins such as $G_{\alpha s}$ and $G_{\alpha olf}$ and stimulate the cAMP pathway in HeLa/CNG cells (97). This observation is consistent with the fact that human V1Rs are coexpressed with $G_{\alpha olf}$ in the human olfactory epithelium. The VRs in various species appear to have evolved so that each VR can couple to the appropriate G proteins in the expressed neurons.

The primary transduction channel downstream of these products is thought to be the transient receptor potential channel TRPC2, which is strongly expressed in both apical and basal compartments of the VNO (153). Indeed, involvement of TRPC2 in VNO-mediated pheromone signaling was obtained by analysis of knockout mice whose sociosexual behavior was impaired (154–156). Caution is warranted in interpreting such results because expression of VRs in the VNO of TRPC2 knockout mice appears to be downregulated. Thus, a primary effect of TRPC2 function in transduction versus a secondary effect of TRPC function on the cellular viability of the VNO needs to be disentangled (155). The mechanisms underlying activation of TRPC2 via a VR/PLC pathway are currently unknown, and additional ion channels may be involved in producing the pheromone-evoked conductance of vomeronasal sensory neurons. Recently, we have shown that V2Rp5-mediated ESP1 signal is completely impaired in TRPC2 knockout mice, indicating that TRPC2 exists downstream of VRs (S. Haga & K. Touhara, manuscript in preparation).

Insect Olfactory Receptors

Vertebrate chemosensory signal transduction, as we discuss above, utilizes familiar signaling elements downstream of canonical GPCRs. To what extent do insects share canonical GPCR signal transduction mechanisms with vertebrates? Because insect OR genes encode proteins that span the membrane seven times, most scientists assume that insect ORs, like ORs in all other species, are seven-transmembrane-domain GPCRs. This assumption seems reasonable because the vast majority of proteins with seven transmembrane domains are in fact GPCRs [the light-activated ion channel channelrhodopsin is a notable exception (157)]. Accordingly, it is generally thought that the insect olfactory system utilizes a signal transduction mechanism similar to that in the vertebrate olfactory system: G protein activation and subsequent second-messenger production lead to an increase in dendritic membrane conductance and the generation of action potentials followed by membrane depolarization via ion channel opening. How strong is the evidence that G protein signaling is part of the primary signal transduction cascade in insect olfaction?

Here we review the existing evidence for and against metabotropic coupling of insect ORs to G protein signaling pathways. Electrophysiological experiments in insect antennae suggested that olfactory neurons have multiple types of ion channels sensitive to various second messengers such as IP3, DAG, cGMP, and calcium (158–160). Biochemical analysis also indicated that IP3 is produced upon odorant or pheromone simulation of insect moth antennae (161). However, none of these observational data really proved that such second messengers were necessary for insect olfaction. Genetic approaches using *Drosophila* made it possible to knock down, knock out, or overexpress various elements of canonical G protein signaling cascades. Overexpression of the *dunce* adenylyl cyclase produced dominant effects on olfactory behavior, leading Alcorta and colleagues (162) to conclude that cAMP signaling is crucial for fly smell. Carlson and colleagues (163, 164) used mutants defective in phospholipase signaling to reach the opposite conclusion: Insect ORs use phospholipase C signaling pathways. Genetic knockdown of G_q protein subunits resulted in reduced olfactory responses in electrophysiological recordings from antennae (165) and in odor-evoked behavior (166). None of the phenotypes obtained to date appear to be as strong as any of the mouse knockouts of primary olfactory signal transduction elements reviewed above (see Vertebrate Olfactory Receptors subsection in this section, above). Thus, the mechanisms mediating insect olfactory transduction are controversial, and the evidence for the involvement of G protein–mediated second-messenger molecules in insect olfaction remains equivocal and confusing.

Several groups have tried to clarify matters by expressing insect ORs in heterologous cells in a reductionist approach to understand what signaling elements are necessary and sufficient for insect ORs to respond to their cognate odor ligands. Initial reports using heterologous expression suggested that insect ORs functioned in heterologous cells without the coreceptor encoded by Or83b in *Drosophila* and its orthologs in other insects. Thus, the odor responsiveness of *Drosophila* Or43a was reconstituted in oocytes without Or83b (109). A *Bombyx* pheromone receptor, BmOR1, responded to bombykol in the presence of $BmG_{\alpha q}$ without BmOR2, the *Bombyx* Or83b ortholog (116). BmOR1 and HR13 were functionally expressed when $G_{\alpha 15}$ was cotransfected in HEK293 cells (167). These results suggested that the ligand-binding subunit of the insect OR could function as a GPCR in vitro, but these in vitro results were at odds with functional mechanisms in vivo in which the complex formation of conventional ORs and Or83b was required for insect OR function (65). Indeed, the response amplitude obtained by $BmOR1/BmG_{\alpha q}$ signaling in the absence of the BmOR2 coreceptor (116) was 100 times lower than that observed for BmOR1/BmOR2 (117). This led our groups and others to speculate that G protein–mediated pathways via the ligand-binding OR subunit contribute little to insect smell and that instead an unknown signaling cascade mediated by the OR/Or83b complex predominates in vivo.

The first evidence for atypical signal transduction characteristics of insect ORs came from studies of the silk moth bombykol receptor BmOR1. The BmOR1/BmOR2 complex expressed in *Xenopus* oocytes elicited a nonselective cation channel activity previously unobserved in oocyte membranes (117). This finding led to the intriguing hypothesis that the insect OR/Or83b complex forms ionotropic-type ligand-gated channels. Indeed, not only in oocytes, but also in mammalian cells such as HeLa and HEK293 cells, ligand-binding ORs from silk moth, fruit fly, and mosquito (i.e., BmOR1, Or47a, and GPROR2) combined with the orthologous Or83b coreceptor from these insect species exhibited similar ion channel properties in evoking nonselective ion conductance upon odorant or pheromone stimulation (168). Outside-out patch-clamp recordings of membranes expressing insect OR/Or83b complexes showed transient unitary currents, which were independent of intracellular factors such as cAMP, IP3, DAG, ATP, and GTP. Whole-cell currents and influx of extracellular calcium persisted in the presence of general inhibitors of G protein signaling (168), a result also confirmed by Newcomb and colleagues (169). The latency of current responses was much faster than that of G protein–coupled olfactory responses, and moreover, ion selectivity of the insect OR complex was dependent on subunit combination (168). Taken together, we concluded that insect ORs form heteromeric complexes and that the complex itself evokes currents directly upon ligand stimulation.

This is by no means the end of this intriguing story because Hansson's group (170) used a similar approach to examine signaling downstream of the fruit fly Or22a/Or83b complex in heterologous cells and reached somewhat different conclusions. Wicher et al. (170), like Sato et al. (168), noticed the same rapid, ionotropic current that was independent of G proteins, but characterized a later metabotropic current induced by cyclic nucleotide–dependent signaling via $G_{\alpha s}$. On the basis of their data, Wicher et al. (170) concluded that Or22a couples to G proteins and that the Or83b coreceptor functions as a CNG channel. Sato et al. (168) observed that some insect OR complexes exhibit a small ligand-independent sensitivity to cyclic nucleotides such as cAMP and cGMP; these investigators observed no clear increase in cAMP levels upon ligand stimulation (168). Thus, there is continued controversy as to whether cyclic nucleotides modulate activity of the OR/OR83b ion channel complex independently of ligand signal transduction or whether they are involved in primary olfactory signal transduction.

Despite the differences between the two recent papers discussed above (168, 170), both groups propose that the insect ORs have the capacity to act as ligand-gated ion channels. Like channelrhodopsin, these seven-transmembrane proteins thus represent an exception to the rule that seven-transmembrane-domain proteins are always GPCRs, and instead are proposed to be seven-transmembrane-domain proteins that form a ligand-gated ion channel complex. No known ion channel pore or selectivity filter motifs have been found in insect ORs, and therefore a molecular basis for the channel activity is unknown. Small deletions in the sixth transmembrane domain of Or83b that produced a GYG motif found in potassium channel selectivity filters altered the ion selectivity of the Or22a/Or83b complex (170), but the details of how these proteins form a cation-selective pore remain to be elucidated. Although an untested hypothesis, it seems reasonable to propose that insect GRs, proteins that are in the same receptor superfamily as

the ORs and that sense bitter and sweet tastants as well as carbon dioxide may also share an ionotropic coupling mechanism with the insect ORs. Support for this idea can be found in an earlier paper that documented ionotropic sugar-gated currents on the blowfly antenna (171). If future data confirm this proposal, the insect chemosensory gene family may represent the largest single family of ion channels in any genome.

How and why did insects acquire such an olfactory signaling mechanism completely different from that in the vertebrate olfactory system? The speed of G protein–mediated transduction utilized in the vertebrate olfactory system is relatively slow [50–100 ms for vertebrate ORs (172) versus 18–25 ms for insect ORs (168)]. The fast responses to the olfactory environment via odor-gated ion channels may be an advantage in the evolution of flying insects, such as flies and moths, that need to find mating partners and food sources while flying. Another possible advantage for using ionotropic receptors is to avoid energy consumption in second-messenger cascades using ATP and GTP in small cellular compartments like the olfactory dendrites. However, ionotropic coupling does come with a cost. Direct receptor activation without G protein amplification is not as efficient in generating depolarizing membrane potential changes unless the receptor molecules are highly concentrated in dendritic membranes. Indeed, immunohistochemical studies have shown that the insect ORs are highly enriched in membranes (65, 110).

The OR/OR83b complex is spontaneously active in outside-out patch membranes (168), which appears to account for previous observations that olfactory sensory neurons exhibit both ligand-activated and ligand-inhibited electrical activity in vivo (110, 111). Insect species have fewer ORs in comparison with vertebrate species. We speculate that these bipolar characteristics derived from the ionotropic properties of insect ORs may provide a greater variety of temporal response profiles and patterns, compensating for the small numbers of ORs. The distinct signal transduction

Glomerulus:
describes the
organization of the
first olfactory neuropil
in both vertebrates and
invertebrates. This
spherical structure
represents the
convergence of
incoming olfactory
sensory neurons as
well as the dendrites of
second-order
projection neurons
and processes of local
interneurons

OB: olfactory bulb

mechanisms between invertebrates and vertebrates may reflect different evolutionary and environmental pressures on the olfactory systems in these different animals.

INFORMATION TRANSMISSION TO THE BRAIN

Olfactory Receptor Circuits in Mammals and Insects

In mammalian species, the olfactory bulb (OB) is the first brain region that relays neural signals of olfactory sensory neurons to secondary neurons, called mitral/tufted cells, that in turn send their axons to the central olfactory system (173, 174). The OB corresponds to the antennal lobe in insects, and the two olfactory processing centers are organized in a similar manner (175). Individual odorants activate distinct subsets of ORs, resulting in the construction of a glomerular activation pattern that is unique for each odorant in a stereotyped region of the OB called an odor map (176–178). Different odorants elicit different glomerular activity patterns, but structurally related odorants activate similar sets of glomeruli because similar odorants are recognized by similar sets of ORs in the olfactory epithelium. Even for the same odorant, different concentrations result in different patterns in a way such that more glomeruli are recruited by higher concentrations of odorants (179). In mice, spatial and functional mapping of OR-defined glomeruli has revealed that the positional relationship of glomeruli varies considerably between individual mice and even between the left and right OBs in the same animal (179). Unlike the stereotyped odor map, precise bulbar OR maps appear to differ between individuals, and the odor map can be truly described only by repeated examinations using many animals. Odor maps that change according to odor concentration have also been documented in insect species (180, 181).

Mitral and tufted cells that synapse with olfactory sensory neurons in glomeruli project their axons to the olfactory cortex through the lateral olfactory tract (182). Classical neuroanatomical tracing studies have revealed that the regions in the primary olfactory cortex receiving input from the OB include the anterior olfactory nucleus (AON), taenia tecta (TT), olfactory tubercle (OT), piriform cortex (PYR), anterior cortical amygdaloid nucleus (ACN), posterolateral cortical amygdaloid nucleus (PLCN), and entorhinal cortex (EC) (183, 184). In insects, the secondary neurons are projection neurons that project the axons to higher brain regions of the mushroom body and the lateral horn of the protocerebrum. It is not yet entirely clear whether the orderly spatial representations created in olfactory bulb or antennal lobe glomeruli are preserved in higher brain regions or whether the representations are distributed. Cortical representations of odor information will be an important field for future study.

In mammals, each of the cortical subregions projects information to various areas in the brain (185). One such region is the orbitofrontal cortex, which is a prominent site of olfactory processing. Functional imaging studies using fMRI (functional magnetic resonance imaging) or PET (positron emission tomography) made it possible to identify specific areas in which olfactory processing takes place upon odorant stimulation (186, 187).

By targeting of transneuronal tracers to gonadotropin-releasing hormone (GnRH) neurons in transgenic mice, the medial preoptic area–anterior hypothalamus (MPOA-AH), in which many GnRH neurons are located, was shown to receive olfactory input from various cortical areas including the TT, the AON, and the PYR (188). In addition, a virus-mediated approach successfully labeled the olfactory-hypothalamus pathway (189). Trimethyl-thiazoline, an odorant that induces the freezing response in mice, activates the hypothalamic-pituitary-adrenal pathway via the bed nucleus of the stria terminalis (BST) (190). Neuroanatomical, genetic, and imaging approaches are beginning to suggest that hard-wired circuits for inducing innate behavior or reproductive changes exist in the main olfactory system. It will be of interest to study this circuitry at finer scales of resolution.

Vomeronasal Receptor Circuits

VNO-mediated chemosensory signals are sent to the AOB, which then projects to the basal forebrain, which contains regions important for reproductive and mating behavior (11, 43, 184, 191). To examine the projection pattern of vomeronasal sensory neurons to the AOB, researchers generated various transgenic and knock-in mice in which specific VR-expressing neurons could be visualized. V1R- and V2R-expressing neurons project to the rostral and the caudal regions of the AOB, respectively, and form several converged glomeruli wherein the sensory neuronal axon terminals synapse with the secondary neurons (46, 192). The secondary neurons then project to the BST, the BAOT, the nucleus of the accessory olfactory tract, and the medial and posteromedial cortical nuclei of the amygdala (MEA and PMCO, respectively), from which the information is relayed to the MPOA-AH; behavioral and endocrine responses ensue (184, 193, 194). Because the MPOA contains many GnRH or luteinizing hormone–releasing hormone (LHRH) neurons, the VNO inputs to the amygdala-hypothalamus pathway may regulate reproductive function. Recently, expression of a targeted transneuronal tracer in transgenic mice was used to visualize the VNO-AOB neuronal circuit, which connects to GnRH/LHRH neurons (188). Although one study suggested that this input is negligible (189), VNO signaling appears to be processed at the basal forebrain area, which may eventually lead to gender recognition and stimulation of mating and aggressive behaviors.

The emerging view of the higher-order neural circuitry in the vomeronasal system and main olfactory system in mice suggests that the underlying mechanisms for processing general odorant-induced behavioral or endocrinological effects are shared with those for pheromone-induced innate behavior or reproductive changes via the vomeronasal system. As discussed above (see subsection: What Is an Odorant? What Is a Pheromone?), whether the observed effect is defined as pheromonal or not is solely dependent on the information possessed by the chemical and whether it represents meaningful communication between conspecific individuals.

Sexual Dimorphism in Olfactory Circuits

Many pheromones elicit sexually dimorphic behaviors. A simple means to afford such sex-specific effects is to build sensory systems that are sexually dimorphic at various levels of the circuit from input to output: the pheromone, the receptor, and/or higher brain circuitry. In some insects, sexual dimorphism is extreme at all levels. Female moths produce female-specific pheromones that are detected by male-specific pheromone receptors, and this information is transmitted to a male-specific region of the antennal lobe, resulting in specific sex behavior. This mechanism is clearly documented in the *B. mori* silk moth: The male-specific OR BmOR1 recognizes bombykol, the sex pheromone that only female silk moths produce (116, 117), and male neurons expressing BmOR1 project to the male-specific macroglomerular complex in the antennal lobe (195). In this case, the sexual dimorphism is ensured at the three distinct levels.

In contrast, cVA is a male-specific pheromone that elicits aggregation behavior in both male and female *Drosophila* but induces different courtship behavior between male and female. cVA is detected by Or67d (122, 123), which is expressed in both male and female flies. Careful analysis of the higher-brain circuits upstream of Or67d input indicated that sexually dimorphic behavior elicited by cVA may be mediated by a sexually dimorphic neural circuit in the lateral horn (196).

Sexual dimorphism affording sex-specific pheromone perception is also inferred to exist in the mouse brain, but neuroanatomical demonstration of this idea is not yet available. For example, 2,5-dimethylpyrazine, a compound in female urine, delays the onset of puberty and induces longer estrous cycles in female mice (197, 198) but decreases testosterone levels in male mice (see discussion in Reference 199). 3,4-Dehydro-*exo*-brevicomin, 2-*sec*-butyl-4,5-dihydrothiazole, 6-hydroxy-6-methyl-3-heptanone, and α- and β-farnesenes from male urine or the preputial gland accelerate the onset of puberty and induce estrous in

HUMAN PHEROMONES

The importance of pheromones to animals is unquestionable, but what is the evidence for and against the existence of human pheromones? In 1971, Martha McClintock described the biological phenomenon of menstrual synchrony and suppression in human females (207). Axial secretions from one woman were observed to change menstrual cycle timing when they were painted on the upper lip of another woman. More than 35 years later, the compound(s) responsible for this effect remains unknown, and even the existence of the effect is controversial among psychologists (despite the fact that the concept of synchronized estrus is well-accepted in rodents).

Other studies have shown that odorous steroids such as androstenone and androstadienone, derived from male axial secretions, can selectively influence mood, physiological state, and neural activity in humans and that sexual orientation affects responses to these male-derived odors (205, 208). The ability to detect androstenone and androstadienone differs enormously between people and is in part genetically determined (209). Polymorphisms in a single human odorant receptor were recently linked to perceptual variation in androstenone and androstadienone perception (29).

The strict definition of a pheromone may have to be relaxed if this field is to progress (see discussion in subsection: What Is an Odorant? What Is a Pheromone?). Humans are complex creatures who use multiple sensory inputs in evaluating a potential mate. It is therefore unlikely that the kind of innate responses one sees in a male silk moth smelling bombykol can be expected in a human male. Nevertheless, the identification of a human pheromone that reliably modulates human sexual and social behavior would be of obvious clinical and commercial importance. Better chemistry to identify compounds produced by humans in different reproductive states, as well as more quantitative readouts of the effects of human-derived compounds on human physiology and behavior, are needed. If these two experimental improvements can be achieved, this area will certainly be one to watch in the years to come.

female mice (200–202). These male odors also induce aggressive behavior in male mice (203). Although the production of these pheromones is sex biased, there are no validated cases of any V1Rs or V2Rs that are sex-specifically expressed (but see Reference 33). Therefore, we can assume that both male and female mice have the same set of receptors for pheromones and that different behavioral output is established by sex-specific neural circuits in the brain. Support for this idea comes from an experiment in which researchers observed a sexually dimorphic distribution of c-Fos-labeled neurons in the brain of mice stimulated with soiled bedding or body odors of the opposite sex (204). This partial sexual dimorphism is also seen for the mouse ESP1 peptide pheromone. ESP1 is secreted in male mouse tears but is received by both male and female vomeronasal sensory neurons that express the ESP1 receptor, V2Rp5 (107). Expression of V2Rp5 is not sex specific, but ESP1 and the higher-brain neural circuit appear to be sex specific (S. Haga & K. Touhara, unpublished observation).

The existence of pheromonal communication in humans is hotly debated (see side bar: Human Pheromones). One clue indicating that humans are sensitive to social signals came from functional imaging studies that demonstrated that the male-derived odor androstadienone elicits a sex-specific pattern of hypothalamic activity that also depends on the sexual orientation of the subject (205, 206). Thus, sexually dimorphic mechanisms for the detection of physiologically important pheromones and odorant substances appear to be conserved between invertebrates and vertebrates, and this dimorphism can be established at different levels from chemical substances to receptors and finally to the brain.

CONCLUSIONS

In this review, we examine how odorant and pheromone signals are distinguished and discriminated, focusing our discussion on three molecular and cellular levels in chemosensory pathways: a chemosignaling molecule (odorant or pheromone), a receptor (OR, VR), and a neural circuit to the brain. The production of chemosensory signals and the expression of the receptor molecules are often carefully regulated to produce species- or sex-specific signals. Pheromonal effects are sometimes observed in both male and female individuals but in a sexually dimorphic fashion. In this case, sex specificity is likely encoded at the level of neural circuits in the brain. By comparing the olfactory systems of invertebrates such as flies and moths, and vertebrates such as mice and humans, we note both common strategies for integrating olfactory signals as well as completely different signal transduction mechanisms.

Although olfaction is a primitive sensory modality, the ligand-binding receptor molecules and associated signal transduction mechanisms are rapidly changing in each evolutionary lineage. We suggest that such rapid evolution originates in positive pressure from the unique environment and the lifestyle of a given species. Extreme differences in the underlying mechanisms of OR transduction between insect and mammal suggest that other, even more novel ORs and signal transduction mechanisms may remain to be discovered. Such diversity of chemosensory receptors and signaling systems may account for species-specific differences in olfaction. Deciphering the molecular and physiological mechanisms of olfactory systems in various animal species, including aqueous and terrestrial animals, would expand the view of various chemosensory systems. The eventual goal of the field would be to understand how animals developed such sophisticated and individually tailored chemical communication strategies that afford optimal survival of each organism.

SUMMARY POINTS

1. General odors and pheromones derive from different sources and elicit distinct behaviors and physiological effects when detected by an animal.

2. General odors and pheromones bind to structurally different receptors in different gene families in mammals and to different receptors in the same gene family in insects.

3. Vertebrate and insect olfactory receptors are structurally unrelated and signal through different pathways.

4. General odors in general are encoded in a combinatorial fashion, whereas pheromones in general appear to activate a labeled-line pathway in which pheromone input leads to a direct behavioral or physiological output.

FUTURE ISSUES

1. At the time of this writing, the details of insect chemosensory signal transduction are still highly controversial. Do G proteins figure in insect chemosensory signal transduction, or do the involved receptors signal independently of G proteins?

2. Are there pheromones for individual and species recognition in mammals? If so, what are they, and how are they processed by the brain?

3. In humans and nonhuman primates, which lack a functional VNO, does the main olfactory system perform the dual roles of general odor and pheromone detection?

4. Do humans use chemical signals to communicate? If so, what is the nature of these pheromones, and how are they received?

DISCLOSURE STATEMENT

The authors are not aware of any biases that might be perceived as affecting the objectivity of this review.

ACKNOWLEDGMENTS

Work in our laboratories is supported by grants from the Program for Promotion of Basic Research Activities for Innovative Biosciences, Japan (PROBRAIN) and a Grant-in-Aid for Scientific Research on Priority Areas from the Ministry of Education, Culture, Sports, Science, and Technology (MEXT) of Japan (K.T.); by the National Institutes of Health (NIH) and the Foundation for the National Institutes of Health through the Grand Challenges in Global Health Initiative (L.B.V.); and by joint grants to K.T. and L.B.V. from the NIH U.S.-Japan BRCP and the JSPS Japan-U.S. Cooperative Science Program.

LITERATURE CITED

1. Haddad R, Khan R, Takahashi YK, Mori K, Harel D, Sobel N. 2008. A metric for odorant comparison. *Nat. Methods* 5:425–29
2. Gilbert AN. 2008. *What the Nose Knows: The Science of Scent in Everyday Life*. New York: Crown
3. Karlson P, Luscher M. 1959. Pheromones: a new term for a class of biologically active substances. *Nature* 183:55–56
4. Wyatt TD. 2003. *Pheromones and Animal Behavior: Communication by Smell and Taste*. Oxford, UK: Oxford Univ. Press
5. Butenandt A, Beckmann R, Stamm D, Hecker E. 1959. Uber den Sexuallockstoff des Seidenspinners *Bombyx mori*, Reindarstellung und Konstitution. *Z. Naturforsch.* 14b:283–84
6. Lucretius. *On the Nature of Things: De Rerum Natura*. Transl. AM Esolen, 1995. Baltimore: Johns Hopkins Univ. Press. (From Latin)
7. Amoore JE. 1963. Stereochemical theory of olfaction. *Nature* 198:271–72
8. Buck L, Axel R. 1991. A novel multigene family may encode odorant receptors: a molecular basis for odor recognition. *Cell* 65:175–87
9. Keller A, Vosshall LB. 2004. A psychophysical test of the vibration theory of olfaction. *Nat. Neurosci.* 7:337–38
10. Burr C. 2002. *The Emperor of Scent*. New York: Random House
11. Halpern M, Martinez-Marcos A. 2003. Structure and function of the vomeronasal system: an update. *Prog. Neurobiol.* 70:245–318
12. Liman ER, Innan H. 2003. Relaxed selective pressure on an essential component of pheromone transduction in primate evolution. *Proc. Natl. Acad. Sci. USA* 100:3328–32
13. Kwon HW, Lu T, Rutzler M, Zwiebel LJ. 2006. Olfactory responses in a gustatory organ of the malaria vector mosquito *Anopheles gambiae*. *Proc. Natl. Acad. Sci. USA* 103:13526–31
14. Buck LB. 2004. The search for odorant receptors. *Cell* 116:S117–19
15. Strader CD, Fong TM, Tota MR, Underwood D. 1994. Structure and function of G protein-coupled receptors. *Annu. Rev. Biochem.* 63:101–32
16. Kajiya K, Inaki K, Tanaka M, Haga T, Kataoka H, Touhara K. 2001. Molecular bases of odor discrimination: reconstitution of olfactory receptors that recognize overlapping sets of odorants. *J. Neurosci.* 21:6018–25
17. Krautwurst D, Yau KW, Reed RR. 1998. Identification of ligands for olfactory receptors by functional expression of a receptor library. *Cell* 95:917–26
18. Mombaerts P. 2004. Genes and ligands for odorant, vomeronasal and taste receptors. *Nat. Rev. Neurosci.* 5:263–78
19. Niimura Y, Nei M. 2005. Evolutionary dynamics of olfactory receptor genes in fishes and tetrapods. *Proc. Natl. Acad. Sci. USA* 102:6039–44

20. Lomvardas S, Barnea G, Pisapia DJ, Mendelsohn M, Kirkland J, Axel R. 2006. Interchromosomal interactions and olfactory receptor choice. *Cell* 126:403–13

21. Serizawa S, Miyamichi K, Nakatani H, Suzuki M, Saito M, et al. 2003. Negative feedback regulation ensures the one receptor-one olfactory neuron rule in mouse. *Science* 302:2088–94

22. Nishizumi H, Kumasaka K, Inoue N, Nakashima A, Sakano H. 2007. Deletion of the core-H region in mice abolishes the expression of three proximal odorant receptor genes in *cis*. *Proc. Natl. Acad. Sci. USA* 104:20067–72

23. Fuss SH, Omura M, Mombaerts P. 2007. Local and *cis* effects of the H element on expression of odorant receptor genes in mouse. *Cell* 130:373–84

24. Gilad Y, Wiebe V, Przeworski M, Lancet D, Paabo S. 2004. Loss of olfactory receptor genes coincides with the acquisition of full trichromatic vision in primates. *PLoS Biol.* 2:E5

25. Quignon P, Giraud M, Rimbault M, Lavigne P, Tacher S, et al. 2005. The dog and rat olfactory receptor repertoires. *Genome Biol.* 6:R83

26. Ehlers A, Beck S, Forbes SA, Trowsdale J, Volz A, et al. 2000. MHC-linked olfactory receptor loci exhibit polymorphism and contribute to extended HLA/OR-haplotypes. *Genome Res.* 10:1968–78

27. Eklund AC, Belchak MM, Lapidos K, Raha-Chowdhury R, Ober C. 2000. Polymorphisms in the HLA-linked olfactory receptor genes in the Hutterites. *Hum. Immunol.* 61:711–17

28. Tacher S, Quignon P, Rimbault M, Dreano S, Andre C, Galibert F. 2005. Olfactory receptor sequence polymorphism within and between breeds of dogs. *J. Hered.* 96:812–16

29. Keller A, Zhuang H, Chi Q, Vosshall LB, Matsunami H. 2007. Genetic variation in a human odorant receptor alters odour perception. *Nature* 449:468–72

30. Menashe I, Abaffy T, Hasin Y, Goshen S, Yahalom V, et al. 2007. Genetic elucidation of human hyperosmia to isovaleric acid. *PLoS Biol.* 5:e284

31. Dulac C, Axel R. 1995. A novel family of genes encoding putative pheromone receptors in mammals. *Cell* 83:195–206

32. Berghard A, Buck LB. 1996. Sensory transduction in vomeronasal neurons: evidence for Gαo, Gαi2, and adenylyl cyclase II as major components of a pheromone signaling cascade. *J. Neurosci.* 16:909–18

33. Herrada G, Dulac C. 1997. A novel family of putative pheromone receptors in mammals with a topographically organized and sexually dimorphic distribution. *Cell* 90:763–73

34. Matsunami H, Buck LB. 1997. A multigene family encoding a diverse array of putative pheromone receptors in mammals. *Cell* 90:775–84

35. Ryba NJ, Tirindelli R. 1997. A new multigene family of putative pheromone receptors. *Neuron* 19:371–79

36. Tirindelli R, Mucignat-Caretta C, Ryba NJ. 1998. Molecular aspects of pheromonal communication via the vomeronasal organ of mammals. *Trends Neurosci.* 21:482–86

37. Sullivan SL, Bohm S, Ressler KJ, Horowitz LF, Buck LB. 1995. Target-independent pattern specification in the olfactory epithelium. *Neuron* 15:779–89

38. Shi P, Zhang J. 2007. Comparative genomic analysis identifies an evolutionary shift of vomeronasal receptor gene repertoires in the vertebrate transition from water to land. *Genome Res.* 17:166–74

39. Zhang X, Rodriguez I, Mombaerts P, Firestein S. 2004. Odorant and vomeronasal receptor genes in two mouse genome assemblies. *Genomics* 83:802–11

40. Shi P, Bielawski JP, Yang H, Zhang YP. 2005. Adaptive diversification of vomeronasal receptor 1 genes in rodents. *J. Mol. Evol.* 60:566–76

41. Rodriguez I, Greer CA, Mok MY, Mombaerts P. 2000. A putative pheromone receptor gene expressed in human olfactory mucosa. *Nat. Genet.* 26:18–19

42. Rodriguez I, Mombaerts P. 2002. Novel human vomeronasal receptor-like genes reveal species-specific families. *Curr. Biol.* 12:R409–11

43. Dulac C, Torello AT. 2003. Molecular detection of pheromone signals in mammals: from genes to behaviour. *Nat. Rev. Neurosci.* 4:551–62

44. Grus WE, Shi P, Zhang YP, Zhang J. 2005. Dramatic variation of the vomeronasal pheromone receptor gene repertoire among five orders of placental and marsupial mammals. *Proc. Natl. Acad. Sci. USA* 102:5767–72

45. Young JM, Trask BJ. 2007. V2R gene families degenerated in primates, dog and cow, but expanded in opossum. *Trends Genet.* 23:212–15

46. Del Punta K, Puche A, Adams NC, Rodriguez I, Mombaerts P. 2002. A divergent pattern of sensory axonal projections is rendered convergent by second-order neurons in the accessory olfactory bulb. *Neuron* 35:1057–66

47. Martini S, Silvotti L, Shirazi A, Ryba NJ, Tirindelli R. 2001. Co-expression of putative pheromone receptors in the sensory neurons of the vomeronasal organ. *J. Neurosci.* 21:843–48

48. Vosshall LB, Amrein H, Morozov PS, Rzhetsky A, Axel R. 1999. A spatial map of olfactory receptor expression in the *Drosophila* antenna. *Cell* 96:725–36

49. Clyne PJ, Warr CG, Freeman MR, Lessing D, Kim J, Carlson JR. 1999. A novel family of divergent seven-transmembrane proteins: candidate odorant receptors in *Drosophila*. *Neuron* 22:327–38

50. Gao Q, Chess A. 1999. Identification of candidate *Drosophila* olfactory receptors from genomic DNA sequence. *Genomics* 60:31–39

51. Clyne PJ, Warr CG, Carlson JR. 2000. Candidate taste receptors in *Drosophila*. *Science* 287:1830–34

52. Scott K, Brady R Jr, Cravchik A, Morozov P, Rzhetsky A, et al. 2001. A chemosensory gene family encoding candidate gustatory and olfactory receptors in *Drosophila*. *Cell* 104:661–73

53. Dunipace L, Meister S, McNealy C, Amrein H. 2001. Spatially restricted expression of candidate taste receptors in the *Drosophila* gustatory system. *Curr. Biol.* 11:822–35

54. Robertson HM, Warr CG, Carlson JR. 2003. Molecular evolution of the insect chemoreceptor gene superfamily in *Drosophila melanogaster*. *Proc. Natl. Acad. Sci. USA* 100(Suppl. 2):14537–42

55. Hill CA, Fox AN, Pitts RJ, Kent LB, Tan PL, et al. 2002. G protein-coupled receptors in *Anopheles gambiae*. *Science* 298:176–78

56. Robertson HM, Wanner KW. 2006. The chemoreceptor superfamily in the honey bee, *Apis mellifera*: expansion of the odorant, but not gustatory, receptor family. *Genome Res.* 16:1395–403

57. Kent LB, Walden KK, Robertson HM. 2008. The Gr family of candidate gustatory and olfactory receptors in the yellow-fever mosquito *Aedes aegypti*. *Chem. Senses* 33:79–93

58. Engsontia P, Sanderson AP, Cobb M, Walden KK, Robertson HM, Brown S. 2008. The red flour beetle's large nose: an expanded odorant receptor gene family in *Tribolium castaneum*. *Insect Biochem. Mol. Biol.* 38:387–97

59. Abdel-Latief M. 2007. A family of chemoreceptors in *Tribolium castaneum* (Tenebrionidae: Coleoptera). *PLoS ONE* 2:e1319

60. McBride CS. 2007. Rapid evolution of smell and taste receptor genes during host specialization in *Drosophila sechellia*. *Proc. Natl. Acad. Sci. USA* 104:4996–5001

61. Wistrand M, Kall L, Sonnhammer EL. 2006. A general model of G protein-coupled receptor sequences and its application to detect remote homologs. *Protein Sci.* 15:509–21

62. Benton R, Sachse S, Michnick SW, Vosshall LB. 2006. Atypical membrane topology and heteromeric function of *Drosophila* odorant receptors in vivo. *PLoS Biol.* 4:e20

63. Lundin C, Kall L, Kreher SA, Kapp K, Sonnhammer EL, et al. 2007. Membrane topology of the *Drosophila* OR83b odorant receptor. *FEBS Lett.* 581:5601–4

64. Krieger J, Klink O, Mohl C, Raming K, Breer H. 2003. A candidate olfactory receptor subtype highly conserved across different insect orders. *J. Comp. Physiol. A* 189:519–26

65. Larsson MC, Domingos AI, Jones WD, Chiappe ME, Amrein H, Vosshall LB. 2004. *Or83b* encodes a broadly expressed odorant receptor essential for *Drosophila* olfaction. *Neuron* 43:703–14

66. Jones WD, Nguyen TA, Kloss B, Lee KJ, Vosshall LB. 2005. Functional conservation of an insect odorant receptor gene across 250 million years of evolution. *Curr. Biol.* 15:R119–21

67. Neuhaus EM, Gisselmann G, Zhang W, Dooley R, Stortkuhl K, Hatt H. 2005. Odorant receptor heterodimerization in the olfactory system of *Drosophila melanogaster*. *Nat. Neurosci.* 8:15–17

68. Zhao H, Ivic L, Otaki JM, Hashimoto M, Mikoshiba K, Firestein S. 1998. Functional expression of a mammalian odorant receptor. *Science* 279:237–42

69. Touhara K, Sengoku S, Inaki K, Tsuboi A, Hirono J, et al. 1999. Functional identification and reconstitution of an odorant receptor in single olfactory neurons. *Proc. Natl. Acad. Sci. USA* 96:4040–45

70. Bozza T, Feinstein P, Zheng C, Mombaerts P. 2002. Odorant receptor expression defines functional units in the mouse olfactory system. *J. Neurosci.* 22:3033–43

71. Grosmaitre X, Vassalli A, Mombaerts P, Shepherd GM, Ma M. 2006. Odorant responses of olfactory sensory neurons expressing the odorant receptor MOR23: a patch clamp analysis in gene-targeted mice. *Proc. Natl. Acad. Sci. USA* 103:1970–75

72. Katada S, Nakagawa T, Kataoka H, Touhara K. 2003. Odorant response assays for a heterologously expressed olfactory receptor. *Biochem. Biophys. Res. Commun.* 305:964–69

73. Touhara K, Katada S, Nakagawa T, Oka Y. 2006. Ligand screening of olfactory receptors. In *G Protein-Coupled Receptors: Structure, Function, and Ligand Screening*, ed. T Haga, S Takeda, pp. 85–109. Boca Raton, FL: CRC Press

74. Touhara K. 2007. Deorphanizing vertebrate olfactory receptors: recent advances in odorant-response assays. *Neurochem. Int.* 51:132–39

75. Wetzel CH, Oles M, Wellerdieck C, Kuczkowiak M, Gisselmann G, Hatt H. 1999. Specificity and sensitivity of a human olfactory receptor functionally expressed in human embryonic kidney 293 cells and *Xenopus laevis* oocytes. *J. Neurosci.* 19:7426–33

76. Abaffy T, Matsunami H, Luetje CW. 2006. Functional analysis of a mammalian odorant receptor subfamily. *J. Neurochem.* 97:1506–18

77. Saito H, Kubota M, Roberts RW, Chi Q, Matsunami H. 2004. RTP family members induce functional expression of mammalian odorant receptors. *Cell* 119:679–91

78. Von Dannecker LE, Mercadante AF, Malnic B. 2006. Ric-8B promotes functional expression of odorant receptors. *Proc. Natl. Acad. Sci. USA* 103:9310–14

79. Yoshikawa K, Touhara K. 2008. Myr-Ric-8A enhances $G\alpha15$-mediated Ca^{2+} response of vertebrate olfactory receptors. *Chem. Senses.* In press

80. Katada S, Tanaka M, Touhara K. 2004. Structural determinants for membrane trafficking and G protein selectivity of a mouse olfactory receptor. *J. Neurochem.* 90:1453–63

81. Zhang X, Firestein S. 2002. The olfactory receptor gene superfamily of the mouse. *Nat. Neurosci.* 5:124–33

82. Zozulya S, Echeverri F, Nguyen T. 2001. The human olfactory receptor repertoire. *Genome Biol.* 2:RE-SEARCH0018

83. Man O, Gilad Y, Lancet D. 2004. Prediction of the odorant binding site of olfactory receptor proteins by human-mouse comparisons. *Protein Sci.* 13:240–54

84. Pilpel Y, Lancet D. 1999. The variable and conserved interfaces of modeled olfactory receptor proteins. *Protein Sci.* 8:969–77

85. Singer MS, Weisinger-Lewin Y, Lancet D, Shepherd GM. 1996. Positive selection moments identify potential functional residues in human olfactory receptors. *Receptors Channels* 4:141–47

86. Singer MS. 2000. Analysis of the molecular basis for octanal interactions in the expressed rat 17 olfactory receptor. *Chem. Senses* 25:155–65

87. Floriano WB, Vaidehi N, Goddard WA 3rd, Singer MS, Shepherd GM. 2000. Molecular mechanisms underlying differential odor responses of a mouse olfactory receptor. *Proc. Natl. Acad. Sci. USA* 97:10712–16

88. Floriano WB, Vaidehi N, Goddard WA 3rd. 2004. Making sense of olfaction through predictions of the 3-D structure and function of olfactory receptors. *Chem. Senses* 29:269–90

89. Katada S, Hirokawa T, Oka Y, Suwa M, Touhara K. 2005. Structural basis for a broad but selective ligand spectrum of a mouse olfactory receptor: mapping the odorant-binding site. *J. Neurosci.* 25:1806–15

90. Malnic B, Hirono J, Sato T, Buck LB. 1999. Combinatorial receptor codes for odors. *Cell* 96:713–23

91. Rodriguez I. 2004. Pheromone receptors in mammals. *Horm. Behav.* 46:219–30

92. Hagino-Yamagishi K, Matsuoka M, Ichikawa M, Wakabayashi Y, Mori Y, Yazaki K. 2001. The mouse putative pheromone receptor was specifically activated by stimulation with male mouse urine. *J. Biochem. (Tokyo)* 129:509–12

93. Del Punta K, Leinders-Zufall T, Rodriguez I, Jukam D, Wysocki CJ, et al. 2002. Deficient pheromone responses in mice lacking a cluster of vomeronasal receptor genes. *Nature* 419:70–74

94. Boschat C, Pelofi C, Randin O, Roppolo D, Luscher C, et al. 2002. Pheromone detection mediated by a V1r vomeronasal receptor. *Nat. Neurosci.* 5:1261–62

95. Leinders-Zufall T, Lane AP, Puche AC, Ma W, Novotny MV, et al. 2000. Ultrasensitive pheromone detection by mammalian vomeronasal neurons. *Nature* 405:792–96

96. Sam M, Vora S, Malnic B, Ma W, Novotny MV, Buck LB. 2001. Neuropharmacology. Odorants may arouse instinctive behaviours. *Nature* 412:142

97. Shirokova E, Raguse JD, Meyerhof W, Krautwurst D. 2008. The human vomeronasal type-1 receptor family—detection of volatiles and cAMP signaling in HeLa/Olf cells. *FASEB J.* 22:1416–25

98. Touhara K. 2007. Molecular biology of peptide pheromone production and reception in mice. *Adv. Genet.* 59:147–71

99. Krieger J, Schmitt A, Lobel D, Gudermann T, Schultz G, et al. 1999. Selective activation of G protein subtypes in the vomeronasal organ upon stimulation with urine-derived compounds. *J. Biol. Chem.* 274:4655–62

100. Chamero P, Marton TF, Logan DW, Flanagan K, Cruz JR, et al. 2007. Identification of protein pheromones that promote aggressive behaviour. *Nature* 450:899–902

101. Nodari F, Hsu FF, Fu X, Holekamp TF, Kao LF, et al. 2008. Sulfated steroids as natural ligands of mouse pheromone-sensing neurons. *J. Neurosci.* 28:6407–18

102. Leinders-Zufall T, Brennan P, Widmayer P, S PC, Maul-Pavicic A, et al. 2004. MHC class I peptides as chemosensory signals in the vomeronasal organ. *Science* 306:1033–37

103. He J, Ma L, Kim S, Nakai J, Yu CR. 2008. Encoding gender and individual information in the mouse vomeronasal organ. *Science* 320:535–38

104. Spehr M, Kelliher KR, Li XH, Boehm T, Leinders-Zufall T, Zufall F. 2006. Essential role of the main olfactory system in social recognition of major histocompatibility complex peptide ligands. *J. Neurosci.* 26:1961–70

105. Kimoto H, Haga S, Sato K, Touhara K. 2005. Sex-specific peptides from exocrine glands stimulate mouse vomeronasal sensory neurons. *Nature* 437:898–901

106. Kimoto H, Sato K, Nodari F, Haga S, Holy TE, Touhara K. 2007. Sex- and strain-specific expression and vomeronasal activity of mouse ESP family peptides. *Curr. Biol.* 17:1879–84

107. Haga S, Kimoto H, Touhara K. 2007. Molecular characterization of vomeronasal sensory neurons responding to the male-specific peptide: sex communication in mice. *Pure Appl. Chem.* 79:775–83

108. Stortkuhl KF, Kettler R. 2001. Functional analysis of an olfactory receptor in *Drosophila melanogaster*. *Proc. Natl. Acad. Sci. USA* 98:9381–85

109. Wetzel CH, Behrendt HJ, Gisselmann G, Stortkuhl KF, Hovemann B, Hatt H. 2001. Functional expression and characterization of a *Drosophila* odorant receptor in a heterologous cell system. *Proc. Natl. Acad. Sci. USA* 98:9377–80

110. Dobritsa AA, Van Der Goes van Naters W, Warr CG, Steinbrecht RA, Carlson JR. 2003. Integrating the molecular and cellular basis of odor coding in the *Drosophila* antenna. *Neuron* 37:827–41

111. Hallem EA, Ho MG, Carlson JR. 2004. The molecular basis of odor coding in the *Drosophila* antenna. *Cell* 117:965–79

112. Hallem EA, Carlson JR. 2006. Coding of odors by a receptor repertoire. *Cell* 125:143–60

113. Fishilevich E, Domingos AI, Asahina K, Naef F, Vosshall LB, Louis M. 2005. Chemotaxis behavior mediated by single larval olfactory neurons in *Drosophila*. *Curr. Biol.* 15:2086–96

114. Kreher SA, Kwon JY, Carlson JR. 2005. The molecular basis of odor coding in the *Drosophila* larva. *Neuron* 46:445–56

115. Kreher SA, Mathew D, Kim J, Carlson JR. 2008. Translation of sensory input into behavioral output via an olfactory system. *Neuron* 59:110–24

116. Sakurai T, Nakagawa T, Mitsuno H, Mori H, Endo Y, et al. 2004. Identification and functional characterization of a sex pheromone receptor in the silkmoth *Bombyx mori*. *Proc. Natl. Acad. Sci. USA* 101:16653–58

117. Nakagawa T, Sakurai T, Nishioka T, Touhara K. 2005. Insect sex-pheromone signals mediated by specific combinations of olfactory receptors. *Science* 307:1638–42

118. Krieger J, Grosse-Wilde E, Gohl T, Dewer YM, Raming K, Breer H. 2004. Genes encoding candidate pheromone receptors in a moth (*Heliothis virescens*). *Proc. Natl. Acad. Sci. USA* 101:11845–50

119. Krieger J, Grosse-Wilde E, Gohl T, Breer H. 2005. Candidate pheromone receptors of the silkmoth *Bombyx mori*. *Eur. J. Neurosci.* 21:2167–76

120. Syed Z, Ishida Y, Taylor K, Kimbrell DA, Leal WS. 2006. Pheromone reception in fruit flies expressing a moth's odorant receptor. *Proc. Natl. Acad. Sci. USA* 103:16538–43

121. Wanner KW, Nichols AS, Walden KK, Brockmann A, Luetje CW, Robertson HM. 2007. A honey bee odorant receptor for the queen substance 9-*oxo*-2-decenoic acid. *Proc. Natl. Acad. Sci. USA* 104:14383–88

122. Kurtovic A, Widmer A, Dickson BJ. 2007. A single class of olfactory neurons mediates behavioural responses to a *Drosophila* sex pheromone. *Nature* 446:542–46

123. Ha TS, Smith DP. 2006. A pheromone receptor mediates 11-*cis*-vaccenyl acetate-induced responses in *Drosophila*. *J. Neurosci.* 26:8727–33

124. Benton R, Vannice KS, Vosshall LB. 2007. An essential role for a CD36-related receptor in pheromone detection in *Drosophila*. *Nature* 450:289–93

125. Jin X, Ha TS, Smith DP. 2008. SNMP is a signaling component required for pheromone sensitivity in *Drosophila*. *Proc. Natl. Acad. Sci. USA* 105:10996–1001

126. Laughlin JD, Ha TS, Jones DN, Smith DP. 2008. Activation of pheromone-sensitive neurons is mediated by conformational activation of pheromone-binding protein. *Cell* 133:1255–65

127. Buck LB. 1996. Information coding in the vertebrate olfactory system. *Annu. Rev. Neurosci.* 19:517–44

128. Kato A, Katada S, Touhara K. 2008. Amino acids involved in conformational dynamics and G-protein coupling of an odorant receptor: targeting gain-of-function mutation. *J. Neurochem.* 107:1261–70

129. Bhandawat V, Reisert J, Yau KW. 2005. Elementary response of olfactory receptor neurons to odorants. *Science* 308:1931–34

130. Sklar PB, Anholt RR, Snyder SH. 1986. The odorant-sensitive adenylate cyclase of olfactory receptor cells. Differential stimulation by distinct classes of odorants. *J. Biol. Chem.* 261:15538–43

131. Pifferi S, Pascarella G, Boccaccio A, Mazzatenta A, Gustincich S, et al. 2006. Bestrophin-2 is a candidate calcium-activated chloride channel involved in olfactory transduction. *Proc. Natl. Acad. Sci. USA* 103:12929–34

132. Tsunenari T, Nathans J, Yau KW. 2006. Ca^{2+}-activated Cl^- current from human bestrophin-4 in excised membrane patches. *J. Gen. Physiol.* 127:749–54

133. Belluscio L, Gold GH, Nemes A, Axel R. 1998. Mice deficient in G_{olf} are anosmic. *Neuron* 20:69–81

134. Brunet LJ, Gold GH, Ngai J. 1996. General anosmia caused by a targeted disruption of the mouse olfactory cyclic nucleotide-gated cation channel. *Neuron* 17:681–93

135. Wong ST, Trinh K, Hacker B, Chan GC, Lowe G, et al. 2000. Disruption of the type III adenylyl cyclase gene leads to peripheral and behavioral anosmia in transgenic mice. *Neuron* 27:487–97

136. Lin W, Arellano J, Slotnick B, Restrepo D. 2004. Odors detected by mice deficient in cyclic nucleotide-gated channel subunit A2 stimulate the main olfactory system. *J. Neurosci.* 24:3703–10

137. Fulle HJ, Vassar R, Foster DC, Yang RB, Axel R, Garbers DL. 1995. A receptor guanylyl cyclase expressed specifically in olfactory sensory neurons. *Proc. Natl. Acad. Sci. USA* 92:3571–75

138. Juilfs DM, Fulle HJ, Zhao AZ, Houslay MD, Garbers DL, Beavo JA. 1997. A subset of olfactory neurons that selectively express cGMP-stimulated phosphodiesterase (PDE2) and guanylyl cyclase-D define a unique olfactory signal transduction pathway. *Proc. Natl. Acad. Sci. USA* 94:3388–95

139. Meyer MR, Angele A, Kremmer E, Kaupp UB, Muller F. 2000. A cGMP-signaling pathway in a subset of olfactory sensory neurons. *Proc. Natl. Acad. Sci. USA* 97:10595–600

140. Lin W, Margolskee R, Donnert G, Hell SW, Restrepo D. 2007. Olfactory neurons expressing transient receptor potential channel M5 (TRPM5) are involved in sensing semiochemicals. *Proc. Natl. Acad. Sci. USA* 104:2471–76

141. Lefkowitz RJ. 2004. Historical review: a brief history and personal retrospective of seven-transmembrane receptors. *Trends Pharmacol. Sci.* 25:413–22

142. Dawson TM, Arriza JL, Jaworsky DE, Borisy FF, Attramadal H, et al. 1993. β-Adrenergic receptor kinase-2 and β-arrestin-2 as mediators of odorant-induced desensitization. *Science* 259:825–29

143. Schleicher S, Boekhoff I, Arriza J, Lefkowitz RJ, Breer H. 1993. A β-adrenergic receptor kinase-like enzyme is involved in olfactory signal termination. *Proc. Natl. Acad. Sci. USA* 90:1420–24

144. Peppel K, Boekhoff I, McDonald P, Breer H, Caron MG, Lefkowitz RJ. 1997. G protein-coupled receptor kinase 3 (GRK3) gene disruption leads to loss of odorant receptor desensitization. *J. Biol. Chem.* 272:25425–28

145. Mashukova A, Spehr M, Hatt H, Neuhaus EM. 2006. β-Arrestin2-mediated internalization of mammalian odorant receptors. *J. Neurosci.* 26:9902–12

146. Kurahashi T, Shibuya T. 1990. Ca^{2+}-dependent adaptive properties in the solitary olfactory receptor cell of the newt. *Brain Res.* 515:261–68

147. Zufall F, Shepherd GM, Firestein S. 1991. Inhibition of the olfactory cyclic nucleotide gated ion channel by intracellular calcium. *Proc. Biol. Sci.* 246:225–30

148. Kurahashi T, Menini A. 1997. Mechanism of odorant adaptation in the olfactory receptor cell. *Nature* 385:725–29

149. Wei J, Zhao AZ, Chan GC, Baker LP, Impey S, et al. 1998. Phosphorylation and inhibition of olfactory adenylyl cyclase by CaM kinase II in neurons: a mechanism for attenuation of olfactory signals. *Neuron* 21:495–504

150. Lucas P, Ukhanov K, Leinders-Zufall T, Zufall F. 2003. A diacylglycerol-gated cation channel in vomeronasal neuron dendrites is impaired in TRPC2 mutant mice: mechanism of pheromone transduction. *Neuron* 40:551–61

151. Spehr M, Hatt H, Wetzel CH. 2002. Arachidonic acid plays a role in rat vomeronasal signal transduction. *J. Neurosci.* 22:8429–37

152. Zufall F, Ukhanov K, Lucas P, Leinders-Zufall T. 2005. Neurobiology of TRPC2: from gene to behavior. *Pflüg. Arch.* 451:61–71

153. Liman ER, Corey DP, Dulac C. 1999. TRP2: a candidate transduction channel for mammalian pheromone sensory signaling. *Proc. Natl. Acad. Sci. USA* 96:5791–96

154. Leypold BG, Yu CR, Leinders-Zufall T, Kim MM, Zufall F, Axel R. 2002. Altered sexual and social behaviors in *trp2* mutant mice. *Proc. Natl. Acad. Sci. USA* 99:6376–81

155. Stowers L, Holy TE, Meister M, Dulac C, Koentges G. 2002. Loss of sex discrimination and male-male aggression in mice deficient for TRP2. *Science* 295:1493–500

156. Kimchi T, Xu J, Dulac C. 2007. A functional circuit underlying male sexual behaviour in the female mouse brain. *Nature* 448:1009–14

157. Nagel G, Ollig D, Fuhrmann M, Kateriya S, Musti AM, et al. 2002. Channelrhodopsin-1: a light-gated proton channel in green algae. *Science* 296:2395–98

158. Zufall F, Hatt H. 1991. Dual activation of a sex pheromone-dependent ion channel from insect olfactory dendrites by protein kinase C activators and cyclic GMP. *Proc. Natl. Acad. Sci. USA* 88:8520–24

159. Zufall F, Hatt H, Keil TA. 1991. A calcium-activated nonspecific cation channel from olfactory receptor neurons of the silkmoth *Antheraea polyphemus*. *J. Exp. Biol.* 161:455–68

160. Kaissling KE. 1996. Peripheral mechanisms of pheromone reception in moths. *Chem. Senses* 21:257–68

161. Krieger J, Breer H. 2003. Transduction mechanisms of olfactory sensory neurons. In *Insect Pheromone Biochemistry and Molecular Biology*, ed. GJ Blomquist, RG Vogt, pp. 593–607. London: Elsevier Acad.

162. Gomez-Diaz C, Martin F, Alcorta E. 2004. The cAMP transduction cascade mediates olfactory reception in *Drosophila melanogaster*. *Behav. Genet.* 34:395–406

163. Riesgo-Escovar JR, Woodard C, Carlson JR. 1994. Olfactory physiology in the *Drosophila* maxillary palp requires the visual system gene *rdgB*. *J. Comp. Physiol. A* 175:687–93

164. Riesgo-Escovar J, Raha D, Carlson JR. 1995. Requirement for a phospholipase C in odor response: overlap between olfaction and vision in *Drosophila*. *Proc. Natl. Acad. Sci. USA* 92:2864–68

165. Kain P, Chakraborty TS, Sundaram S, Siddiqi O, Rodrigues V, Hasan G. 2008. Reduced odor responses from antennal neurons of $G_q\alpha$, phospholipase $C\beta$, and *rdgA* mutants in *Drosophila* support a role for a phospholipid intermediate in insect olfactory transduction. *J. Neurosci.* 28:4745–55

166. Kalidas S, Smith DP. 2002. Novel genomic cDNA hybrids produce effective RNA interference in adult *Drosophila*. *Neuron* 33:177–84

167. Grosse-Wilde E, Gohl T, Bouche E, Breer H, Krieger J. 2007. Candidate pheromone receptors provide the basis for the response of distinct antennal neurons to pheromonal compounds. *Eur. J. Neurosci.* 25:2364–73

168. Sato K, Pellegrino M, Nakagawa T, Nakagawa T, Vosshall LB, Touhara K. 2008. Insect olfactory receptors are heteromeric ligand-gated ion channels. *Nature* 452:1002–6

169. Smart R, Kiely A, Beale M, Vargas E, Carraher C, et al. 2008. *Drosophila* odorant receptors are novel seven transmembrane domain proteins that can signal independently of heterotrimeric G proteins. *Insect Biochem. Mol. Biol.* 38:770–80

170. Wicher D, Schafer R, Bauernfeind R, Stensmyr MC, Heller R, et al. 2008. *Drosophila* odorant receptors are both ligand-gated and cyclic-nucleotide-activated cation channels. *Nature* 452:1007–11

171. Murakami M, Kijima H. 2000. Transduction ion channels directly gated by sugars on the insect taste cell. *J. Gen. Physiol.* 115:455–66

172. Firestein S, Shepherd GM, Werblin FS. 1990. Time course of the membrane current underlying sensory transduction in salamander olfactory receptor neurones. *J. Physiol.* 430:135–58

173. Mori K, Nagao H, Yoshihara Y. 1999. The olfactory bulb: coding and processing of odor molecule information. *Science* 286:711–15

174. Shepherd GM, Chen WR, Greer CA. 2004. Olfactory bulb. In *The Synaptic Organization of the Brain*, ed. GM Shepherd, pp. 165–216. New York: Oxford Univ. Press

175. Hildebrand JG, Shepherd GM. 1997. Mechanisms of olfactory discrimination: converging evidence for common principles across phyla. *Annu. Rev. Neurosci.* 20:595–631

176. Xu F, Greer CA, Shepherd GM. 2000. Odor maps in the olfactory bulb. *J. Comp. Neurol.* 422:489–95

177. Leon M, Johnson BA. 2003. Olfactory coding in the mammalian olfactory bulb. *Brain Res. Brain Res. Rev.* 42:23–32

178. Mori K, Takahashi YK, Igarashi KM, Yamaguchi M. 2006. Maps of odorant molecular features in the mammalian olfactory bulb. *Physiol. Rev.* 86:409–33

179. Oka Y, Katada S, Omura M, Suwa M, Yoshihara Y, Touhara K. 2006. Odorant receptor map in the mouse olfactory bulb: in vivo sensitivity and specificity of receptor-defined glomeruli. *Neuron* 52:857–69

180. Joerge J, Kuttner A, Galizia G, Menzel R. 1997. Representations of odours and odour mixtures visualized in the honeybee brain. *Nature* 387:285–88

181. Wang JW, Wong AM, Flores J, Vosshall LB, Axel R. 2003. Two-photon calcium imaging reveals an odor-evoked map of activity in the fly brain. *Cell* 112:271–82

182. Shipley MT, Ennis M. 1996. Functional organization of olfactory system. *J. Neurobiol.* 30:123–76

183. Prices JL. 1990. Olfactory system. In *The Human Nervous System*, ed. G Paxinos, pp. 979–1001. San Diego: Academic

184. Meredith M. 1998. Vomeronasal, olfactory, hormonal convergence in the brain. Cooperation or coincidence? *Ann. N. Y. Acad. Sci.* 855:349–61

185. Zelano C, Sobel N. 2005. Humans as an animal model for systems-level organization of olfaction. *Neuron* 48:431–54

186. de Araujo IE, Rolls ET, Velazco MI, Margot C, Cayeux I. 2005. Cognitive modulation of olfactory processing. *Neuron* 46:671–79

187. Anderson AK, Christoff K, Stappen I, Panitz D, Ghahremani DG, et al. 2003. Dissociated neural representations of intensity and valence in human olfaction. *Nat. Neurosci.* 6:196–202

188. Boehm U, Zou Z, Buck LB. 2005. Feedback loops link odor and pheromone signaling with reproduction. *Cell* 123:683–95

189. Yoon H, Enquist LW, Dulac C. 2005. Olfactory inputs to hypothalamic neurons controlling reproduction and fertility. *Cell* 123:669–82

190. Kobayakawa K, Kobayakawa R, Matsumoto H, Oka Y, Imai T, et al. 2007. Innate versus learned odour processing in the mouse olfactory bulb. *Nature* 450:503–8

191. Keverne EB. 2004. Importance of olfactory and vomeronasal systems for male sexual function. *Physiol. Behav.* 83:177–87

192. Wagner S, Gresser AL, Torello AT, Dulac C. 2006. A multireceptor genetic approach uncovers an ordered integration of VNO sensory inputs in the accessory olfactory bulb. *Neuron* 50:697–709

193. von Campenhausen H, Mori K. 2000. Convergence of segregated pheromonal pathways from the accessory olfactory bulb to the cortex in the mouse. *Eur. J. Neurosci.* 12:33–46

194. Brennan PA, Kendrick KM. 2006. Mammalian social odours: attraction and individual recognition. *Philos. Trans. R. Soc. London Ser. B* 361:2061–78

195. Kanzaki R, Soo K, Seki Y, Wada S. 2003. Projections to higher olfactory centers from subdivisions of the antennal lobe macroglomerular complex of the male silkmoth. *Chem. Senses* 28:113–30

196. Datta SR, Vasconcelos ML, Ruta V, Luo S, Wong A, et al. 2008. The *Drosophila* pheromone cVA activates a sexually dimorphic neural circuit. *Nature* 452:473–77

197. Jemiolo B, Novotny M. 1993. Long-term effect of a urinary chemosignal on reproductive fitness in female mice. *Biol. Reprod.* 48:926–29

198. Ma W, Miao Z, Novotny MV. 1998. Role of the adrenal gland and adrenal-mediated chemosignals in suppression of estrus in the house mouse: the Lee-Boot effect revisited. *Biol. Reprod.* 59:1317–20

199. Koyama S. 2004. Primer effects by conspecific odors in house mice: a new perspective in the study of primer effects on reproductive activities. *Horm. Behav.* 46:303–10

200. Jemiolo B, Harvey S, Novotny M. 1986. Promotion of the Whitten effect in female mice by synthetic analogs of male urinary constituents. *Proc. Natl. Acad. Sci. USA* 83:4576–79

201. Ma W, Miao Z, Novotny MV. 1999. Induction of estrus in grouped female mice (*Mus domesticus*) by synthetic analogues of preputial gland constituents. *Chem. Senses* 24:289–93

202. Novotny MV, Jemiolo B, Wiesler D, Ma W, Harvey S, et al. 1999. A unique urinary constituent, 6-hydroxy-6-methyl-3-heptanone, is a pheromone that accelerates puberty in female mice. *Chem. Biol.* 6:377–83

203. Novotny M, Harvey S, Jemiolo B, Alberts J. 1985. Synthetic pheromones that promote intermale aggression in mice. *Proc. Natl. Acad. Sci. USA* 82:2059–61

204. Halem HA, Cherry JA, Baum MJ. 1999. Vomeronasal neuroepithelium and forebrain Fos responses to male pheromones in male and female mice. *J. Neurobiol.* 39:249–63

205. Savic I, Berglund H, Gulyas B, Roland P. 2001. Smelling of odorous sex hormone-like compounds causes sex-differentiated hypothalamic activations in humans. *Neuron* 31:661–68

206. Savic I, Berglund H, Lindstrom P. 2005. Brain response to putative pheromones in homosexual men. *Proc. Natl. Acad. Sci. USA* 102:7356–61

207. McClintock MK. 1971. Menstrual synchorony and suppression. *Nature* 229:244–45

208. Wyart C, Webster WW, Chen JH, Wilson SR, McClary A, et al. 2007. Smelling a single component of male sweat alters levels of cortisol in women. *J. Neurosci.* 27:1261–65

209. Wysocki CJ, Beauchamp GK. 1984. Ability to smell androstenone is genetically determined. *Proc. Natl. Acad. Sci. USA* 81:4899–902

RELATED RESOURCES

1. Two databases provide molecular and genomic information on chemosensory receptors: The Human Olfactory Receptor Data Explore (HORDE) (**http://bioportal.weizmann. ac.il/HORDE/**) and Olfactory Receptor DataBase (ORDB) (**http://senselab.med.yale. edu/ordb/default.asp**)

2. Bargmann CI. 2006. Comparative chemosensation from receptors to ecology. *Nature* 444:295–301

3. Brennan PA, Zufall F. 2006. Pheromonal communication in vertebrates. *Nature* 444:308–15

4. Vosshall LB, Stocker RF. 2007. Molecular architecture of smell and taste in *Drosophila*. *Annu. Rev. Neurosci.* 30:505–33

5. Wilson RI, Mainen ZF. 2006. Early events in olfactory processing. *Annu. Rev. Neurosci.* 29:163–201

Signaling at Purinergic P2X Receptors

Annmarie Surprenant and R. Alan North

Faculty of Life Sciences, University of Manchester, Manchester M13 9PT, United Kingdom;
email: a.surprenant@manchester.ac.uk, alan.north@manchester.ac.uk

Annu. Rev. Physiol. 2009. 71:333–59

First published online as a Review in Advance on
October 13, 2008

The *Annual Review of Physiology* is online at
physiol.annualreviews.org

This article's doi:
10.1146/annurev.physiol.70.113006.100630

Copyright © 2009 by Annual Reviews.
All rights reserved

0066-4278/09/0315-0333$20.00

Key Words

ATP, ligand-gated ion channels, transgenic mice, structure-function, mutagenesis, cell signaling

Abstract

P2X receptors are membrane cation channels gated by extracellular ATP. Seven P2X receptor subunits ($P2X_{1-7}$) are widely distributed in excitable and nonexcitable cells of vertebrates. They play key roles in inter alia afferent signaling (including pain), regulation of renal blood flow, vascular endothelium, and inflammatory responses. We summarize the evidence for these and other roles, emphasizing experimental work with selective receptor antagonists or with knockout mice. The receptors are trimeric membrane proteins: Studies of the biophysical properties of mutated subunits expressed in heterologous cells have indicated parts of the subunits involved in ATP binding, ion permeation (including calcium permeability), and membrane trafficking. We review our current understanding of the molecular properties of P2X receptors, including how this understanding is informed by the identification of distantly related P2X receptors in simple eukaryotes.

AMPA: α-amino-3-hydroxy-5-methylisoxazole-4-propionic acid

EPSP: excitatory postsynaptic potential

INTRODUCTION

P2X receptors came to our attention as the receptor involved in transmitting the actions of ATP released from sympathetic nerves onto the smooth muscles of the vas deferens and small intestinal arterioles (1). These receptors were distinguished from G protein–coupled metabotropic P2Y receptors (2). The historical development of thinking as to P2X receptors, and the initial associated skepticism, has been recently reviewed (3).

The cloning of P2X receptor subunit cDNAs in 1994 (4, 5) led directly to two major avenues of progress. The first is the demonstration that the proteins have a very widespread tissue distribution in vertebrates and indeed in eukaryotes. This finding in turn has led to the development of receptor antagonists and of mice in which receptor subunits have been genetically deleted; the use of such tools has indicated wide-ranging and hitherto unexpected functional roles for P2X receptors. Second, P2X receptors do not resemble other ion channels at the molecular level. This has spurred many studies to determine the molecular details of the operation of these receptors in the cell membrane. This review focuses on these two areas of research.

PHYSIOLOGICAL ROLES FOR P2X RECEPTORS

Central Nervous System

P2X receptors are widely distributed on central nervous system neurons and glia at the mRNA (6–8) and protein (8, 9) levels. At the protein level, the inadequacy of several of the existing antibodies has handicapped progress, particularly in the case of the $P2X_7$ receptor (10, 11). Two approaches dominate the literature: the use of either slices of brain tissue or cells dissociated and maintained in culture. The first includes studies of the effects of exogenous ATP and analogs to nerve cells and glia, measuring either the direct effects on membrane currents or intracellular calcium (postsynaptic effects) or the alteration of transmitter release (presyn-

aptic effects). The second complementary approach is to deduce roles for endogenous ATP in intercellular signaling by the use of pharmacological antagonists or knockout mice.

Postsynaptic effects. A careful immunohistochemical study using the postembedding technique for electron microscopy showed that $P2X_4$ and $P2X_6$ subunits are found in perisynaptic locations on the hippocampal CA1 pyramidal and cerebellar Purkinje cells (9). There are several reports of residual excitatory postsynaptic potentials (EPSPs) in the presence of high concentrations of blockers of AMPA (α-amino-3-hydroxy-5-methylisoxazole-4-propionic acid), kainate, and NMDA subtypes of glutamate receptor. These EPSPs are rather small in amplitude, and their attribution to ATP is based on blockade by antagonists of limited specificity [habenula (12), hippocampus (13, 14), somatosensory cortex (15–17)]. Pankratov et al. (17) concluded that in the mouse cortex, some glutamate-containing vesicles also contained ATP, whereas others did not. In the lateral hypothalamus of the mouse (and neonatal chick), GABA and ATP are co-released, and each evokes its own postsynaptic current that can be distinguished pharmacologically (18). There is a clear need for a systemic comparison of the properties of such EPSPs between wild-type mice and mice lacking P2X receptor subunits. In the case of the $P2X_4$ knockout mouse, long-term potentiation in the hippocampus is of reduced amplitude as compared with the wild-type animals, but the underlying cellular mechanism for this remains unclear (19).

Direct postsynaptic effects on neurons of exogenously applied ATP have been described in many parts of the central nervous system and are not reviewed here (see References 8, 20, and 21). More recently, there have been reports of ATP actions on glial cells that are attributable to P2X receptors. Fellin et al. (22) reported ATP-evoked currents in hippocampal astrocytes, although Jabs et al. (23) did not. Astrocytes in the mouse cortex respond to ATP with inward currents exhibiting many of the properties of the P2X$_{1/5}$ heteromer, and single-cell RT-PCR shows the predominant expression of the two subunits (24).

Presynaptic effects. Presynaptic effects of ATP have been reported at several sites in the central nervous system. The most prevalent effect is an increase in the spontaneous release of glutamate, as detected by increased frequency of spontaneously occurring excitatory postsynaptic currents (EPSCs) in the presence of tetrodotoxin (21). For example, Khakh & Henderson (25) found that ATP enhances glutamate release onto neurons in the motor nucleus of the fifth cranial nerve in the rat but that the receptor involved has properties different from those of the cell bodies of the sensory input neurons that give rise to these presynaptic terminals (located in the mesencephalic nucleus of the fifth nerve) (26). The presynaptic facilitatory effect is rather specific in the hippocampus: ATP facilitates glutamate release at synapses onto inhibitory interneurons, but not at synapses onto CA1 pyramidal cells. This effect was not seen in mice lacking the P2X$_2$ subunit (27).

In the nucleus tractus solitarius, calcium entry through presynaptic P2X receptors elicits glutamate release sufficient to drive postsynaptic cell firing (28). This effect does not involve voltage-gated calcium channels or tetrodotoxin-sensitive sodium channels. The spontaneous EPSCs elicited by ATP are larger in average amplitude than those occurring spontaneously, suggesting that P2X receptor–induced calcium entry evokes the release of a distinct subpopulation of glutamate-containing vesicles. αβmeATP (α,β-methylene adenosine 5′-triphosphate) (≥30 μM) mimics the effect of ATP, and TNP-ATP [2′,3′-O-(2,4,6-trinitrophenyl) adenosine 5′-triphosphate] (IC$_{50}$ ≈ 1 μM) blocks these effects. These results are consistent with the involvement of P2X$_1$- or P2X$_3$-containing receptors but do not completely exclude other subtypes. It will be important to determine the molecular basis through which P2X receptors appear to access a pool of releasable glutamate distinct from that

which occurs spontaneously. Release of GABA by presynaptic P2X receptors has also been described in the cerebellum (29), and spontaneous release of glycine release has been observed in the dorsal horn of the spinal cord (30).

ATP depolarizes pituitary gonadotrophs, and the properties of the corresponding inward current conform to those expected of P2X$_2$ and/or P2X$_4$ subtypes (31). This results in a release of luteinizing hormone. In the neurohypophysis, ATP elicits inward currents on vasopressin terminals, but not on oxytocin-releasing terminals (32, 33). The response has features most similar to those expected for P2X$_2$ receptors. In both the anterior pituitary and posterior pituitary, the overall physiological context of these ATP effects remains obscure.

Afferent Signaling

Action potentials in afferent nerve fibers are often initiated by the release of a specific chemical from specialized sensory cells that then acts on receptors near the peripheral terminals of the nerve. ATP has been implicated as one such chemical.

Taste. Recent work has indicated that homomeric and/or heteromeric P2X$_2$ and P2X$_3$ receptors are essential players in taste transduction. A single mammalian taste bud consists of approximately 100 cells of four distinct types. Type I supporting cells wrap around type II and type III cells to form the characteristic taste bud clusters. Type II cells are the taste receptors expressing distinct G protein–coupled receptors specific for the five differentiated tastes (sweet, sour, salty, bitter, and umami), but type II cells do not synapse with afferent nerves within the taste buds. Presynaptic type III cells synapse, via 5-HT$_3$ neurotransmission, onto afferent taste nerves. The physiology of type IV cells is not clear, although they are known to function in the regeneration of taste cells (34, 35).

Single-cell RT-PCR of isolated mouse taste bud cells found P2X$_2$, P2X$_4$, and P2X$_7$ subunits (36). Immunohistochemical studies using well-characterized antibodies localized P2X$_2$ to

EPSC: excitatory postsynaptic current

αβmeATP: α,β-methylene adenosine 5′-triphosphate

TNP-ATP: 2′,3′-O-(2,4,6-trinitrophenyl) adenosine 5′-triphosphate

SNARE: a multiprotein scaffold between exocytotic vesicle and plasma membrane that controls vesicle/plasma membrane fusion and transmitter release

afferent nerve fibers and type III presynaptic cells, localized P2X$_3$ to afferent fibers only, and showed no P2X$_4$ protein in taste bud cells or nerve fibers (36–38). ATP release in response to tastants has been measured by luciferase assays from isolated segments of lingual epithelia (39) and also from single isolated type II taste cells via a biosensor assay (40, 41). Release of ATP from these cells is not via classic SNARE-based exocytosis, and both Huang et al. (40) and Romanov et al. (41) have suggested that release is via a hemichannel-like pore made up of proteins of the pannexin or connexin family. Finger et al. (39) recorded afferent firing from both chorda tympani and glossopharyngeal gustatory nerve trunks in mice lacking both P2X$_2$ and P2X$_3$ receptors. These researchers found an almost complete absence of nerve response to tastants applied to the oral cavity; responses to mechanical and other chemical stimuli were not altered. Moreover, behavioral responses to sweeteners, glutamate, and bitter substances were diminished or absent in these double-knockout mice. Finger et al. (39) also found moderately diminished neural and behavioral responses to tastants in single-knockout (P2X$_3$ or P2X$_2$) mice. Taken together, these studies provide convincing evidence for the role of ATP as the primary neurotransmitter in taste perception, acting on homomeric P2X$_2$ receptors, homomeric P2X$_3$ receptors, and heteromeric P2X$_{2/3}$ receptors.

However, ATP neurotransmission in mouse taste buds (**Figure 1**) may be atypical. Rong et al. (42) used an isolated, intra-arterially perfused rat tongue with separated gustatory (chorda tympani) and general sensory (lingual) nerves. They observed P2X$_2$ receptor and P2X$_3$ receptor responses only from lingual nerves that did not respond to tastants applied to the tongue, whereas chorda tympani nerves responded robustly to salty or acidic solutions but not to ATP or ATP analogs. There may be technical difficulties in the use of an isolated tongue preparation that can account for the contrasting results, or this study may serve as a caution against the overwhelming domination of mouse tissue and mouse models in physiology research today.

Hayato et al. (36) have demonstrated the presence of functional P2X$_7$ receptors on subpopulations of taste bud cells, although the specific cell type has not been ascertained. Taste bud cells are among the shortest-lived of all mammalian cells, having turnover rates of only several days, whereby old taste cells are removed via apoptosis. Because prolonged P2X$_7$ receptor activation leads to apoptotic cell death, this demonstration of functional P2X$_7$ receptors in taste buds indicates a possible role for P2X$_7$ receptors in taste cell regeneration.

Hearing. Immunohistochemical and electrophysiological studies have been carried out in mouse, rat, and guinea pig cochlea (reviewed in References 43 and 44). These studies have provided evidence for ATP as a neurotransmitter

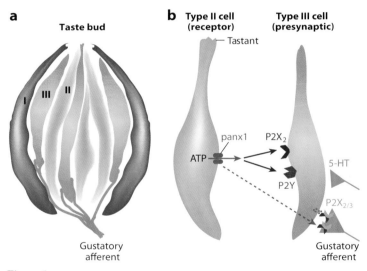

Figure 1

P2X$_2$ and P2X$_3$ receptors are involved in taste transduction. (*a*) Diagram of a single taste bud indicating three of the four types of taste cells: type I supporting cells, type II receptor cells, and type III presynaptic cells; type IV basal cells are not shown (35). (*b*) In response to tastants, specific type II cells release ATP via pannexin-1 (panx1) hemichannels (40, 41). The released ATP acts in a paracrine fashion to activate P2X$_2$ and/or G protein–coupled P2Y receptors in type III cells; this leads to increased intracellular calcium, which evokes classical synaptic release of ATP onto presynaptic P2X$_{2/3}$ receptors on afferent fibers (39). ATP released from type II cells may also act in a paracrine manner to activate presynaptic afferent P2X$_{2/3}$ receptors (*dashed arrow*). Depolarization of type III cells by ATP also leads to the release of 5-HT onto gustatory afferents (35).

or neuromodulator in afferent (but not efferent) auditory transmission. Patch-clamp recordings from inner and outer hair cells show ATP-gated currents that are likely mediated by P2X$_2$ receptors, whereas neuronal spiral ganglion cells show P2X$_3$ receptor–mediated inward currents (43–48). The electrophysiological data correspond to the immunolocalization of these receptor subunits. Housley and coworkers have amassed a large body of data that implicates P2X receptors, primarily P2X$_2$, in modulation of the endocochlear potential (43–45). However, there are currently no publications regarding the hearing phenotype of mice lacking P2X$_2$ and/or P2X$_3$ receptors or studies using selective P2X$_3$ antagonists to hearing behavior in animals or humans.

Chemoreceptors. The carotid body contains glomus cells that release a primary afferent transmitter onto primary afferent nerves in response to hypoxia. The identity of this transmitter (as also the primary hypoxia sensor) has been the subject of long-standing debate (49). Rong et al. (50) reported that ATP is released from type 1 glomus cells in response to hypoxia, that ATP activates primary afferent nerves (probably through a P2X$_{2/3}$ heteromeric receptor), and that the excitation of primary afferent nerves by hypoxia is much reduced in mice lacking the P2X$_2$ subunit (see Reference 51).

Bladder. Accumulating evidence indicates a role for P2X receptors in the initiation of primary afferent signaling in hollow viscera and particularly the urinary bladder. Distension of the bladder results in ATP release from urothelial cells (52). ATP and αβmeATP increase the excitability of bladder afferent nerves (53). Bladder afferent nerves express P2X$_3$ receptor subunits, and in the anesthetized P2X$_3$ knockout mouse the bladder must be distended to a greater volume than in wild-type mouse to initiate the voiding reflex (54, 55).

Pain. ATP acting at P2X receptors has been implicated in pain sensation at both periph-

eral and central sites within the nervous system. ATP applied within the skin evokes pain (56), and P2X$_3$ subunits are expressed on a subset of primary afferent neurons implicated in pain sensation (57, 58). Three lines of evidence implicate P2X$_3$ receptor subunits in some aspects of neuropathic and inflammatory pain perception: Some, although by no means all, pain-related behavior is blocked by antagonists selective for P2X$_3$ subunit–containing receptors (59), by reduction in P2X$_3$ subunit expression by RNA knockdown (60), and by genetic elimination of the P2X$_3$ subunits (reviewed in detail in Reference 61).

A role for P2X$_4$ receptors in neuropathic (but not inflammatory) pain has more recently been proposed. P2X$_4$ subunit expression is increased in microglia of the dorsal horn of the spinal cord following spinal nerve ligation, a commonly used model for neuropathic pain. The mechanical allodynia resulting from the ligation was reduced by intrathecal TNP-ATP and by intrathecal antisense oligonucleotides that somewhat reduced P2X$_4$ receptor expression (62). Conversely, intrathecal administration of cultured microglia that had been stimulated with ATP could induce allodynia in naive rats. It was proposed that spinal nerve ligation activates dorsal horn microglia by a mechanism involving P2X$_4$ receptors and that they release brain-derived nerve growth factor (BDNF). BDNF then acts on spinal cord interneurons to stimulate the expression of the potassium-chloride cotransporter KCC2, and the resultant reduction of intracellular chloride converts the usually inhibitory action of GABA into excitation (63). One difficulty with the attribution of these effects to the activation of P2X$_4$ receptors is that TNP-ATP has several nonspecific actions and is in any event a weak blocker of P2X$_4$ receptors (64). It will be important to repeat these studies in mice in which the P2X$_4$ and P2X$_7$ receptors have been knocked out, or with currently available P2X$_7$ receptor antagonists.

More recently, P2X$_7$ receptors have also been implicated in pain perception. Disruption of the P2X$_7$ receptor gene markedly reduces both neuropathic and chronic inflammatory

BDNF: brain-derived nerve growth factor

UVEC: umbilical
vein endothelial cell

IL-1β: interleukin-1
beta

IL-1Ra: a naturally
occurring protein that
binds to IL-1 receptor
without activating it
and prevents binding
of IL-1β

pain in mice (65), and antagonists with high selectivity for P2X$_7$ receptors strongly inhibit neuropathic pain in rats (66). The mechanism of these actions is as yet unclear but presumably involves the rearrangement of sensory processing pathways and decreased release of proinflammatory cytokines from microglia in the spinal cord.

Cardiovascular System

P2X receptors have widespread expression throughout tissues of the cardiovascular system (6). Our understanding of their function in two tissues—(a) the endothelium and its role in the control of blood pressure and (b) the platelet with respect to blood coagulation—has advanced rapidly in recent years.

Vascular endothelium. Vascular endothelium cells coexpress P2X$_4$ and P2X$_7$ receptors: mRNA and protein levels for the former are typically 5–20-fold higher than for the latter (67). The heterogeneity of vascular endothelia should be kept in mind when functional roles of any endothelial protein are discussed. One of the clearest examples of this heterogeneity is a microarray study performed by Chi et al. (68) on 53 different human endothelial cell lines obtained from various large and small arteries and veins; distinct patterns of gene expression were detected on the basis of not only tissue site, developmental stage, and artery versus vein type but also specific functions such as left/right asymmetry. Nevertheless, in endothelial cells increased intracellular calcium directly leads to nitric oxide (NO) production and release with consequent smooth muscle vasodilation; therefore, calcium assays (using Fura2 or Indo1) are confidently used in endothelial cells as indirect readouts of NO activity (69). Human and bovine umbilical vein endothelial cells (UVECs) have been the most commonly used preparations for in vitro studies of purinoceptor function in endothelia.

Ando and colleagues (70) have presented evidence for autocrine activation of P2X$_4$ receptors in human and bovine UVECs via shear stress–induced release of ATP from endothelial cells. The key evidence for P2X$_4$ receptor involvement is that antisense oligonucleotides that decreased protein levels abrogated shear stress–induced calcium transients and that heterologous expression of P2X$_4$ receptors in human embryonic kidney 293 cells resulted in a shear stress–induced calcium transient. Surprisingly, the ATP-induced responses blocked by P2X$_4$ receptor antisense oligonucleotide treatment were observed only with low ATP concentrations (<1 μM), concentrations that do not activate homomeric human or rodent P2X$_4$ receptors or heteromeric rat P2X$_{4/6}$ receptors (71).

In contrast, Wilson et al. (67) found no evidence for P2X$_4$ receptor–mediated membrane currents or calcium transients in human UVECs even under conditions (interferon-γ and TNFα stimulation) that upregulated P2X$_4$ receptor mRNA and protein by 10–75-fold; these investigators found that >95% of total P2X$_4$ protein is intracellular. The predominantly intracellular localization of the P2X$_4$ receptor has been documented in several other cell types and appears to be the default expression pattern of this receptor (72, 73). After activation by interferon-γ and TNFα, these human UVECs released small amounts of proinflammatory interleukin-1 beta (IL-1β) in response to P2X$_7$ receptor stimulation but simultaneously released anti-inflammatory IL-1Ra; the net effect was anti-inflammatory.

Although these in vitro studies of P2X receptor function in UVECs may seem contradictory, Ando and colleagues have provided conclusive evidence for P2X$_4$ receptor involvement in endothelial function in studies, using wild-type and P2X$_4$ receptor knockout mice (74). First, they found that flow-induced increases in calcium and NO production in cultured endothelia obtained from lung microvessels were virtually absent in preparations from P2X$_4$ receptor knockout mice and that these responses were regained with the reintroduction of P2X$_4$ receptor by adenoviral constructs. Second, these researchers found that in vivo vasodilation of preconstricted cremaster arterioles induced by intra-arterial ATP (but not acetylcholine), and the associated increased blood flow, was decreased in knockout mice; the decrease (30–100%) depended on ATP concentration and flow rate. A similar inhibition of ATP-induced vasodilation occurred in isolated mesenteric arteries from knockout mice. Third, P2X$_4$ receptor knockout mice had significantly higher blood pressure and lower basal plasma concentrations of nitrites and nitrates (an indicator of NO levels) than did wild-type controls. Fourth, Ando and coworkers found that the resting carotid artery diameter of P2X$_4$ receptor knockout mice was significantly smaller and that their smooth muscle layers were thicker compared with wild type. Moreover, these investigators used the carotid artery ligation to elicit neointimal depositions and long-term decreased vessel diameter, which is commonly employed as an animal model of atherosclerotic disease; P2X$_4$ receptor knockout mice showed no further decrease in vessel diameter. Chamberlain et al. (75) performed similar carotid artery ligation experiments on P2X$_7$ receptor knockout mice and found no differences in responses, thus ruling out P2X$_7$ receptor involvement in this animal model of vascular disease. Taken overall, the vascular phenotype of P2X$_4$ receptor knockout mice resembles that seen in endothelial nitric oxide synthase (eNOS) knockout mice (74), supporting the idea that the critical functional role of P2X$_4$ receptors in mouse vascular endothelia resides in the modulation of NO production and release. This finding has implications for the possible use of P2X$_4$ receptor antagonists in atherosclerosis.

Thrombosis. The study of ADP acting on P2Y receptors on platelets has a long history that has brought considerable clinical utility. The main players are the P2Y$_1$ and P2Y$_{12}$ receptors; the latter is blocked by the antithrombotic agents such as clopidogrel (76). Platelets also express P2X$_1$ receptors, and their activation by αβmeATP induces a rapid calcium influx associated with a transient shape change (77, 78). The bleeding time of P2X$_1$ receptor knockout mice is normal, suggesting little impairment of normal hemostasis; however, thrombosis associated with injury of the walls of small arterioles is reduced in the knockout mice (79). Thus, the P2X$_1$ receptor may play a role in thrombus formation under conditions of high shear stress, suggesting that a P2X$_1$-selective antagonist may be particularly effective in inhibiting thrombosis in conditions of stenosis, in which shear stress is high, while having little effect on normal hemostasis, in which shear stress is less (76).

Respiratory System

An important recent development has been the discovery that P2X receptors expressed by the ciliated epithelial cells of the bronchi can control ciliary beat frequency and thus the clearance of mucus from the airways. The subject has obvious importance for a range of lung diseases, including cystic fibrosis. A second area of substantive progress has been the central control of respiration. Sophisticated assays of ATP release combined with electrophysiological recordings have strongly supported a role for P2X receptors in the ventral surface of the medulla.

Ciliated epithelia. Ciliated epithelia play a critical role throughout the airways in clearing mucous secretions. Agents that increase ciliary beating, such as β$_2$-adrenergic and P2Y$_2$ receptor agonists, provide key therapies in several acute and chronic respiratory diseases (80). ATP increases ciliary beating in ciliated epithelia from nose, trachea, and other airway epithelia in several species via activation of both P2Y and P2X receptors (80, 81). Silberberg and coworkers have determined the properties of P2X responses in rabbit airway ciliary epithelia (82, 83). They have concluded that a novel P2X receptor, which they term P2X$_{cilia}$, underlies the ionotropic purinergic response. The functional responses reported for rabbit P2X$_{cilia}$ are the same as those reported for rodent P2X$_7$ receptors in most respects. These responses include agonist dose responses, divalent cation sensitivity, facilitation of response

N-methyl-D-glucamine (NMDG): the largest soluble cation (molecular weight 195.2) used in biophysical studies to estimate channel pore size

during prolonged agonist application, and inhibition by extracellular sodium ions. The key differences are that the rabbit cilia do not show uptake of dyes (such as propidium), do not show increases in *N*-methyl-D-glucamine (NMDG) permeability with prolonged agonist application, and have currents that are potentiated by zinc (1 μM) (83). Because dye-permeable pore formation and zinc inhibition are typical properties of $P2X_7$ receptors, these workers ruled out (homomeric) $P2X_7$ receptor involvement. However, the dye/NMDG-permeable properties of $P2X_7$ receptor function depend on the density of $P2X_7$ receptor protein expression (84) and/or the presence of other associated interacting proteins (85), and zinc modulation of many ligand-gated ion channels, including P2X receptors, commonly shows biphasic potentiation and inhibition of agonist-evoked responses depending on zinc concentration (21). The complete concentration dependence of the action of zinc in airway epithelial cells remains to be determined.

The Silberberg group (82, 83) also found that ivermectin potentiates the amplitude of the ATP-induced cationic current. Because ivermectin potentiation is a selective characteristic of $P2X_4$ receptors (21), this research group's general conclusion was that $P2X_{cilia}$ most likely resulted from a heteromeric $P2X_{4/7}$ receptor. However, ivermectin potentiated only the amplitude, and not the duration, of the ATP response in rabbit ciliary epithelia, whereas studies on ectopic or endogenous $P2X_4$ receptors have also shown a marked prolongation of the duration of $P2X_4$ receptor responses. We have recently observed that ivermectin significantly potentiates the amplitude of mouse (but not rat) $P2X_7$ receptor–induced current (J. Sim, unpublished observations) in a manner similar to that observed by Silberberg and colleagues in rabbit cilia.

Central control of respiration. Dale and colleagues (86) have cleverly developed an enzymatic/amperometric electrode to measure local ATP concentration. In one application of this device (87), they measured ATP concentrations in the ventral medulla during hypoxia. In vivo, they detected ATP release only in those discrete regions of the medullary surface known to correspond to classical CO_2 chemosensitivity. Slices of medulla maintained in vitro also released ATP in response to increased CO_2. Higher concentrations of CO_2 were required to stimulate respiratory activity in the presence of P2X receptor antagonists. Moreover, exogenous ATP itself mimicked the effect of increased CO_2 levels. These findings led the authors to suggest that ATP released in response to increased CO_2 acts on P2X receptors expressed on the dendrites of ventral respiratory column neurons and thus determines the adaptive responses in breathing (51, 87). This implies a key role for ATP released from peripheral chemosensors to initiate action potentials in primary afferent nerves (see above section, Afferent Signaling) and a very similar role for ATP in the ventral medulla to drive the central respiratory response. In both cases, an understanding of the mechanism of ATP release is urgently needed.

Genitourinary System

ATP acting at P2X receptors was first recognized as the efferent transmitter process from sympathetic nerves to the vas deferens (3), and more recently it has become clear that ATP released from urothelial cells can activate afferent nerves in a bladder distention reflex. More recent work implicates P2X receptors in the initial stage of urine formation, namely in the feedback mechanism for maintaining urinary filtration at the glomerulus in a wide range of physiological conditions.

Tuboglomerular feedback. Strong evidence now exists for a critical role of $P2X_1$ receptors in renal autoregulation. Autoregulation is the process whereby the kidney is able to maintain constant renal blood flow and glomerular filtration rate over changes in arterial pressure from approximately 80 to 180 mm Hg in the absence of changes in sympathetic input and/or vasoactive substances (88).

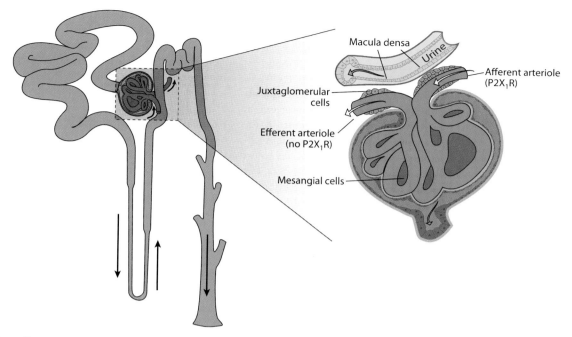

Figure 2

P2X$_1$ receptors are involved in renal autoregulation via tubuloglomerular feedback. To the left is a single nephron, with the glomerular apparatus enlarged on the right. Changes in tubular fluid flow rate occur in response to changes in glomerular filtration rate or the rate of reabsorption in the proximal tubule. Changes in glomerular filtration rate are sensed by macula densa cells that release ATP onto P2X$_1$ receptors on afferent arteriolar smooth muscle, triggering an increase in afferent arteriolar resistance. This reduces pressure in the glomerular capillaries and decreases glomerular filtration (i.e., tubuloglomerular feedback).

This autoregulation is due to both intrinsic myogenic response of smooth muscle and tubuloglomerular feedback, and P2X$_1$ receptors appear to be involved in the latter process. Tubuloglomerular feedback involves the juxtaglomerular apparatus, consisting of the macula densa, the glomerulus, and both efferent and afferent arterioles of the same nephron (**Figure 2**) (89). Apical macula densa cells are exquisitely sensitive to increases in NaCl concentration in the distal tubular fluid; as little as 5–10 mEq l^{-1} leads to afferent arteriolar vasoconstriction via tubuloglomerular feedback (90). How does this occur? Bell and colleagues (91, 92) used P2X$_2$ receptor–expressing PC12 cells as ATP biosensors. When they placed whole-cell patch-clamped (or Fura2-loaded) PC12 cells into contact with rabbit macula densa cells, they recorded (P2X$_2$-activated) currents (or calcium transients) when

luminal NaCl was increased by 5- to 10-fold; the responses corresponded to the release of ~10 μM ATP. Smaller changes in NaCl concentration were not examined in this assay. However, the Navar group made in vivo microdialysis measurements of canine renal interstitial ATP (and adenosine) in response to changes in renal arterial pressure within the autoregulatory range (90, 93). They found a tight correlation between interstitial concentrations of ATP (but not adenosine) and tubuloglomerular feedback in response to small stepwise changes in renal arterial pressure. Taken together, these results show conclusively that ATP is released from the macula densa in response to luminal changes in NaCl. Although the molecular mechanism of the ATP release remains unknown, it is currently hypothesized that this ATP release triggers tubuloglomerular feedback.

However, there is considerable debate as to how such tubuloglomerular feedback is achieved. On one side of the debate is the adenosine A_1 receptor hypothesis: Specific ectonucleotidases hydrolyze released ATP to adenosine with subsequent activation of vasoconstrictor A_1 receptors on afferent arterioles (94). On the other side of the debate is the P2X$_1$ receptor hypothesis: Released ATP directly activates vasoconstrictor P2X$_1$ receptors on afferent arteriolar smooth muscle (89). The main support for the former hypothesis consists of the following. (*a*) The specific ectonucleotidases required to break down ATP to adenosine are highly expressed in the juxtaglomerular apparatus, (*b*) specific A_1 receptor antagonists inhibit or abolish tubuloglomerular feedback, and (*c*) A_1 receptor knockout mice do not exhibit tubuloglomerular feedback (94–97). Equally strong support for the latter model comes from findings that (*a*) P2X$_1$ receptors are expressed only on afferent arterioles but not on efferent arterioles or other kidney cells, (*b*) a selective P2X$_1$ receptor antagonist (NF279) abolishes tubuloglomerular feedback, and (*c*) P2X$_1$ receptor knockout mice do not exhibit tubuloglomerular feedback (89, 98, 99). Interpretations of results from blockade of A_1 receptor with antagonists or knockout mice may be more complicated than for results from similar blockade of P2X$_1$ receptors because of the widespread expression of A_1 receptors in the kidney and renal vasculature and because of the concomitant vasodilatory actions of the similarly widespread expression of A_2 receptor throughout the renal vasculature (100). It is also likely that either or both ATP and adenosine underlie tubuloglomerular feedback, depending on subpopulations of nephrons and their individual metabolic state; there are more than 300,000 individual nephrons in a rodent kidney. Moreover, subtle differences in in vivo microdialysis and other in vitro techniques and protocols among laboratories may also contribute to experimental differences. This A_1 receptor/P2X$_1$ receptor puzzle may be resolved by exchange of the knockout mice among the relevant groups and parallel experiments in the same laboratory using both sets of mice, along with a more quantitative comparison of the kinetics and pharmacology of inhibition of tubuloglomerular feedback by A_1 and P2X$_1$ receptor antagonists.

Renal fibrosis. The abnormal accumulation of extracellular matrix (i.e., fibrosis) is associated with all chronic renal diseases. Renal mesangial cells are primarily responsible for extracellular matrix deposition via the secretion of transforming growth factor β (TGFβ), which leads to increased accumulation of fibronectin and other extracellular matrix proteins (101). Thus, mesangial cell dysfunction is a key contributor to interstitial fibrosis and, as such, has attracted recent interest in regard to the potential roles purinoceptors may play in normal and diseased mesangial cells.

Schulze-Lohoff et al. (102) first showed that P2X$_7$ receptor activation induces mesangial cell apoptosis. Harada et al. (103) showed that ATP exerts dual actions on isolated rat mesangial cells; P2Y$_2$ receptor and/or P2Y$_4$ receptor activation stimulated mesangial cell proliferation, and P2X$_7$ receptor stimulation with BzATP [2′,3′-*O*-(4-benzoyl-benzoyl) adenosine 5′-triphosphate] led to increased apoptosis. Soloni et al. (104) confirmed these findings and found that P2X$_7$ receptor stimulation increased secretion of TGFβ and subsequent fibronectin levels, effects that were significantly enhanced when cells were grown in a high-glucose medium (mimicking hyperglycemic conditions). Goncalves et al. (105) used an animal model of interstitial fibrosis, unilateral ureteral obstruction, in both wild-type and P2X$_7$ receptor knockout mice. This group found that the classical features of unilateral ureteral obstruction (high macrophage invasion into the interstitium, high TGFβ secretion, increased mesangial cell apoptosis, and interstitial fibrosis) were all significantly abrogated in the P2X$_7$ receptor knockout mice. Taken together, these results suggest a physiologically significant role for P2X$_7$ receptors in renal fibrosis and may point to the therapeutic potential of P2X$_7$ receptor antagonists in the treatment of chronic renal diseases.

However, Soloni et al. (106) used human mesangial cells from normal segments of kidneys obtained at nephrectomy and repeated experiments they had carried out on rat mesangial cells, with surprising results. They found that ATP clearly induced significant apoptosis in human mesangial cells, but could find no evidence for the presence or involvement of P2X$_7$ receptor; moreover, their pharmacological data were not readily compatible with the involvement of known human P2Y or P2X receptors. Only high concentrations of ATP (>1 mM), but not of BzATP, induced the apoptotic response, suggesting that ATP breakdown to adenosine may have been responsible; it has been shown that apoptosis of a hepatic epithelial cell line by ATP and ADP is due to the activation of adenosine A$_3$ receptors in response to hydrolysis of these nucleotides to adenosine (107). In any event, results from this study serve as caution against overreliance on, and overinterpretation of, results obtained from rodent tissues and cells.

Gastrointestinal System

The nonadrenergic, noncholinergic inhibitory transmission to the smooth muscle of the intestine was one of the very first demonstrations of a functional role for extracellular ATP (reviewed in References 1 and 3). Further, quite distinct roles in the gastrointestinal tract are now emerging. These include accessory structures such as salivary glands, liver, and exocrine pancreas. They also include signaling between nerve cells of the enteric nervous system within the gut wall.

Exocrine glands. Luminal ATP has long been known to activate purine receptors on several secretory epithelia with the consequent secretion of chloride, potassium, bicarbonate, or hormonal peptides (80, 81). The majority of studies on nucleotide-induced epithelial secretion show clearly that the P2Y family of purinocep-

tors, particularly the P2Y$_2$ receptor, is primarily responsible (80, 81). Additional involvement of P2X$_7$ receptors and/or P2X$_4$ receptors has also been suggested, but the interpretation is confounded by marked species differences in the physiological and pharmacological properties of these two receptors. Submandibular gland acinar and ductal cells from rat and mouse have been the most studied. Patch-clamp recordings show both P2X$_4$ receptor– and P2X$_7$ receptor–mediated currents, whereas Fura2 intracellular calcium measurements show predominantly P2X$_7$ receptor–mediated responses, with a very small contribution via P2X$_4$ receptors (<5% of total calcium response). Stimulation of P2X$_7$ receptors, but not of P2X$_4$ receptors, induces kallikrein secretion from ductal cells (108). Dehaye and colleagues (108–110) have built up a strong case for P2X$_7$ receptor signaling to salivary secretion through the modulation of phospholipid signaling processes, particularly phospholipase A2 and phospholipase D. The strongest evidence comes from a study comparing responses from P2X$_7$ receptor wild-type and knockout mice, in which (*a*) ATP no longer induced any phospholipid signaling, (*b*) the ATP-induced calcium signal was decreased by ~90–95%, and (*c*) the saliva from the knockout mice showed a significantly decreased potassium content (110). This study thus directly demonstrated the involvement of P2X$_7$ receptor in salivary secretions, but the evidence for phospholipid signaling as the underlying mechanism remains correlative.

The difficulty of matching P2X phenotype in epithelia with P2X protein expression is highlighted in a recent study on human gallbladder and rat liver bile duct cells, in which ATP and BzATP potently stimulated chloride secretion when applied to the apical surfaces of these epithelia (111). In these bile ducts, P2X$_4$ but not P2X$_7$ receptors were expressed by RT-PCR, immunoblot, and immunohistochemistry. However, the properties of the secretory response and whole-cell patch clamp recordings were more typical of activating P2X$_7$ than P2X$_4$ receptors. The recent availability of

BzATP: 2′,3′-O-(4-benzoyl-benzoyl) adenosine 5′-triphosphate highly selective and potent rodent and human $P2X_7$ receptor antagonists may resolve some of these discrepancies.

Intestinal motility. Afferents traveling by the vagus nerve or the dorsal root ganglia signal intestinal activity to the central nervous system. The role of P2X receptors on such nerves is described above (see Afferent Signaling section). In contrast, considerable evidence exists for a key role for P2X receptors in signaling within the enteric nervous system that underlies coordinated movements of the intestine. The principal intrinsic afferent neuron of the enteric nervous system is the AH cell (112). AH cells (which can be identified by immunoreactivity for calbindin D28K) express $P2X_2$ and $P2X_3$ subunits; there is evidence that the afferent terminals of these neurons, in the mucosal layer and in the muscle layers, are activated by ATP through P2X receptors, but the subtype has not been conclusively identified (113). Another main class of neuron, the descending interneurons on the pathway to the circular muscle motoneurons, also expresses $P2X_2$ subunits; in these cells ATP substantially contributes to the fast EPSP when presynaptic fibers are stimulated electrically, as shown by comparing wild-type and knockout mice (114). Peristalsis of the small intestine recorded in vitro is clearly impaired in the ileum taken from $P2X_3$ receptor knockout mice, suggesting most likely that the afferent neurons are less sensitive to the initiating distension (115).

Immune System

$P2X_1$, $P2X_4$, and $P2X_7$ receptor proteins coexist in most immune cells, including mast cells, B and T lymphocytes, monocytes, macrophages, microglia, and osteoclasts (8). Cation currents and calcium influx with the features expected for each of these P2X receptors have been demonstrated in all these immune cells. However, only $P2X_7$ receptors are thus far established as having physiological roles in the immune system. There is a long history of $P2X_7$ receptor involvement in inflammation via the unique ability of the $P2X_7$ receptor to induce the rapid activation of caspase-1 with subsequent release of the proinflammatory cytokine IL-1β from activated macrophage and microglia (116, 117). The underlying mechanism(s) by which $P2X_7$ receptor activation leads to caspase-1 activation and IL-1β release remains a matter of considerable debate and study, but abrogation of inflammatory responses by inhibition of $P2X_7$ receptors is directly linked to concomitant decreased levels of IL-1β in plasma or in the area of injury (116, 117).

Two lines of mice have been bred with disrupted $P2X_7$ receptors. In one [Glaxo mice (65)] 5′ exons were disrupted by a neomycin-resistant vector. In the other [Pfizer mice (118)] the gene was disrupted in the portion encoding the cytoplasmic tail, leaving open the possibility of expression of a truncated receptor. In both lines, ATP did not induce the release of IL-1β or other proinflammatory members of the IL-1 family (IL-18 and IL-1α) from endotoxin-activated macrophages. Moreover, mechanical hypersensitivity (allodynia) in response to adjuvant injection into the paw, as well as in response to partial nerve ligation, was absent, and levels of IL-1β from the area of paw inflammation systemically were reduced. Recently, similar results have been obtained from wild-type mice treated with the selective $P2X_7$ antagonist A740003 (119). Researchers have described other potent and selective $P2X_7$ receptor antagonists that show either human or rodent specificity (119, 120). One such molecule (AZD9056) produces clinically relevant improvements in patient assessments of symptoms, and reduced swollen and tender joints, in a Phase II clinical trial of 75 patients with active rheumatoid arthritis treated for one month (121). These early results hold promise that $P2X_7$ receptor antagonists in inflammation may be the first therapeutic benefits to emerge from P2X receptor research.

Killing intracellular mycobacteria. Stimulation of macrophage $P2X_7$ receptors can lead to the death of several types of intracellular bacteria, and the killing of mycobacterium tuberculosis is of much clinical relevance (122). The killing of intracellular bacilli directly parallels P2X$_7$ receptor–mediated macrophage apoptosis (122). Because IL-1β leads to apoptosis of surrounding cells, the direct cause of macrophage apoptosis and death of intracellular pathogens may be the P2X$_7$ receptor–induced release of IL-1β. Studies with RAW 264.7 murine macrophages have ruled out this possibility. These macrophages do not show caspase-1 activation/IL-1β release be-cause they lack a key adaptor protein (ASC) required to form the multiprotein complex (the inflammasome; see **Figure 3**) (123). Yet these macrophages express high levels of P2X$_7$ receptor, and prolonged activation leads to macrophage apoptosis and concomitant killing of intracellular pathogens (123, 124). It is not known whether macrophage apoptosis is necessary for killing of intracellular mycobacteria. However, the P2X$_7$ receptor–mediated apoptotic process is associated with the formation of

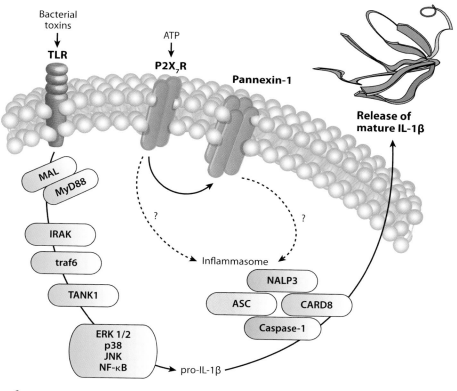

Figure 3

P2X$_7$ receptor (P2X$_7$R) and the NALP3 inflammasome. An initial inflammatory insult (e.g., bacterial endotoxins) activates Toll-like receptors (TLRs) on monocytes, macrophages, and microglia to initiate the classical nuclear factor-κB (NF-κB) cascade and the synthesis of pro-interleukin-1β (pro-IL-1β). However, in the absence of a secondary stimulus, little or no processing or release of IL-1β occur (116, 117). High concentrations of extracellular ATP, present at sites of inflammation, activate a P2X$_7$R/pannexin-1 protein complex (172). This leads to the activation of caspase-1, with the subsequent formation of a multiprotein intracellular complex, the NALP3-type inflammasome (173). Activated caspase-1 within the inflammasome scaffold now cleaves pro-IL-1β, and the release of mature, bioactive IL-1β follows. Blockade of either P2X$_7$R or pannexin-1 independent of P2X$_7$R prevents rapid IL-1β processing and release (117, 172). Mechanisms by which P2X$_7$R/pannexin-1 activation leads to caspase-1 activation and the release of mature IL-1β remain uncertain and are likely to be tissue/cell and stimulus specific.

large intracellular vacuoles and with the fusion of lysosomes with the bacilli-containing phagosomes, thus providing a means for killing the mycobacteria via lysosomal enzymes (122, 124, 125).

Osteoclasts. Several studies have implicated P2X$_7$ receptors in the formation of multinucleated giant cells in granulomatous diseases and in normal and pathological formation of multinucleated osteoclasts in bone (126). In one study, the Pfizer P2X$_7$ receptor knockout mouse showed abnormal bone formation and resorption in the form of a reduced periosteal circumference and decreased cortical bone content (127). In another study using Glaxo P2X$_7$R knockout mice, there was no difference between wild-type and knockout mice in the ability of mononuclear phagocytes to generate multinucleated osteoclasts (128). The P2X$_7$ receptor antagonist AZ11657312 significantly reduced bone resorption, synovial inflammation, and the number of osteoclastic giant cells (129). Thus, although it has not yet been determined whether P2X$_7$ receptor alters bone metabolism via alterations in the formation of multinucleated giant cells, recent studies provide strong evidence for its role in modulating the function of osteoclasts in bone.

P2X$_7$ polymorphisms and immune function. The human P2X$_7$ receptor is associated with more than 500 single-nucleotide polymorphisms, some of which render the P2X$_7$ receptor nonfunctional. Wiley and colleagues (124, 130) have examined five specific P2X$_7$ receptor polymorphisms known to result in nonfunctional P2X$_7$ receptors when examined in heterologous expression systems. In a cohort of more than 700 mainly Caucasian adults, these investigators found that 51% showed the wild-type gene at the five alleles they examined, 40% were heterozygous at one of these alleles, and 25% of the heterozygous adults showed reduced functional readouts compared with wild type. Approximately 3% of the population was homozygous for T357S/E496A without any functional P2X$_7$ receptor response

in their blood-derived macrophages (130). Macrophages obtained from individuals consistently show "high, inducible, or no" response to P2X$_7$ receptor stimulation (125, 130), and specific polymorphisms in the P2X$_7$ receptor gene may explain these observations. The molecular identity of individual P2X$_7$ receptors may of course alter their susceptibility to newly developed therapeutic antagonists.

MOLECULAR PROPERTIES OF P2X RECEPTORS

Direct structural approaches with single-particle electron microscopy and atomic force microscopy strongly support the view that the P2X receptor channel forms as a trimer (131, 132). This stoichiometry was originally based on polyacrylamide gel electrophoresis under nondenaturing conditions (133) and is supported by measurements of single-channel kinetics (134), or currents evoked by ATP through channels formed from concatenated subunits with reporter mutations (21, 135).

In the past few years, deduced P2X receptor sequences have become available from the genomes of several simple eukaryotes. Where these have been shown to function as ATP-gated ion channels, the limited sequence relatedness helps to identify those residues or parts of the molecule that are most critical for function. **Figure 4** illustrates this for five P2X receptors. These are green algae (*Ostreococcus tauri*), choanoflagellate (*Monosiga brevicollis*), slime mold (*Dictyostelium discoideum*), trematode worm (*Schistosoma mansoni*), and the vertebrate *Rattus norvegicus* (21, 136–138). The first two are single-celled organisms that are phylogenetically close to the origins of the divergence of plant and animal cells, the third has a life cycle with both single-celled and multicellular components, the fourth is a segmented worm, and the fifth is man.

Several points can be taken from this comparison of sequences of functional receptors. First, only relatively few residues, notably tyrosines, lysines, and glycines, are completely conserved. In almost all cases, these conserved

N terminus, TM1, post-TM1

```
rP2X2   ----MVRRLARGCWSAFWDYETPKVIVVRNRRLGFVHRMVQLLILLYFVWYVFIVQKSYQDSETGPESSIITKVKG 072
schis   ---------MVKGIAVLFEYETPKLVQISNIKIGVTQRLLQLVILIYVVCWVMIYEKGYQENDI-AKSAVTTKVKG 066
choan   MAASGFWGSIQQGIYSMLEYDTLKTVHIRSKKVGLIFRLLQITILAYVVGYGIIYQKGYQEVDS-AVSTVLTKVKG 075
algae   ---------------MGLSYTTQKSVVIRSWRLGALYYGLVAVVLAYVGFFLVYVERGYQRTSR-AVGNIGLKVKG 075
dicty   -------MGFSFDWDDIFQYSTVKIVRIRDRRLGILHLSFLVGIVAYIVVYSAIIKKGYLFTEV-PIGSVRTSLKG 078
```

Ectodomain

```
rP2X2   ---------------ITMSEDKVWDVEEYVKPPEGGSVVSIITRIEVTPSQTLG-----------TCPESMRVHSST 123
schis   -------VGFTNFSHIPGIGMRSWDVADYIVPPLGNNALFVITNLVKTERQSLS-----------KCQESSWVPEAA 133
choan   IAITCDNTDISSMNDCVPGDLRVWDTPDYIKPAQENDAFFVVSNSIQTSKQTQRAEGWDEDPAA--PVTGSASAFN 149
algae   ---------QATLRDATTGATLVYDANDLVMYEPSG---FFIATALATTLQARG-----------RCPGMDEDET 112
dicty   -------PNTFASNLTYCSNQQHNGSTYPFTPLECNY-WDEQLALFPVGQDSTF-------TCTTRVRLSKQEAN 128

rP2X2   CHSDD-DCIAGQLDMQGNGIRTGHCVPYYHG-----DSKTCEVSAWCPVEDG---TSDNH------FLGKMAPNFT 184
schis   CYKDS-DCKPYFISHLGNGAHTGKCIIKPGN-----DIGSCEIYSWCPLENDTLPLGRKS------FLFPMVYNYT 189
choan   CTSDA-DCPR--FATSRNGALTGECNT---------TTERCRIYGWGPVESKDEDDRATTDGLFYARHMPAVKNFT 213
algae   CTDAS-ACVVGTFSPS--GRMTGQCVATALKDEDGKVVKRCEVEGWCPGEPEKDEVTVLEN---------VGNFT 175
dicty   CNFTDPTCKFVDEPGSAKN------------------------------------------IYIADIESFT 132

rP2X2   ILIKNSIHYPKFKFSKGNIASQKSDY-LKHCTFDQDS--DPYCPIFRLGFIVEKAGEN-------------FTEL 243
schis   LLIKNDINFEKFGIHRRNIQNWASKKFLRTCLYNKTDPENRFCPIFQFGTIFEEANVD--------------QSIF 251
choan   VYIKNTVFFQRFGAKFG-STDESDKVDVYTCTWSPTG-LERHCPIFKIDTILNEAGIT-----------DFENQA 275
algae   VFTRISVEFPGIPDEDGEGNMLWTNLNGTKP-TLGWNLWTINDLLESGGMSVKEVARK--VTPYTLKEVARKGWDG 275
dicty   ILIDHTMYASSSGSQFNAVDLHGYILNQDGDEVQIDANGTSIGVSGKPDIMTIGQLLSFGGVSLDQASPV--LDQA 224

rP2X2   AHKGGVIGVIINWNCDLDLSESECNPKYSFRRLDPKYVPAS-SGYNFRFAKYYKINGTTTT--------------- 303
schis   IS-GGVIGIDIDWKCDLDWDVQYCNPTYSFRRLDDAHAKIA-SGFNFRYAHFYSENGTNY--------------- 309
choan   MRNGALITIQVNYDCNLDSSAHTCSPTYKFTRLD-TKSDLS-AGYNFRFANYQIDPPA---------------- 331
algae   RVDVEF-------DCNLDRGIDACAPKTPYTLKQVMHPNTLSEGFNIRWISGQNVGEPSAQAGVVYSNETANGPGKDV 307
dicty   SPVDSNVSIR--YDGVVLFVFITYSNTYTYSTSDFKYVYSVQQIANTIYDVPETIILESIHS-----------284
```

Pre-TM2, TM2, post-TM2/C terminus

```
rP2X2   RTLIKAYGIRIDVIVHGQAGKFSLIPTIINLATALTSIGVGSFLCDWILLTFMN-KNKLYSHKKFDKVRTPKHPS 377
schis   RDLIKAYGIRFVIHVSGEAGKFHLLPLTMNIGSGLALLGLAPTVCDIIALNLLR-SRDIYQRAKFETIAEEQAHL 383
choan   RDLYKVYGLRFVFVVSGTAGRFSMVPLLVALGSGLGLLGLATVIADLLVTKCIR-NANVYYGLKYQVVDEEDIDR 405
algae   RLLVKGYGPRIRFEVTGVGRKFDWLTLSTTVGAGVAFLGIASLVVNAVMMYCSGPKSKQYESWLFAEFHDTPYGS 382
dicty   RLLYKRHGIRVIFIQTGTIGSFHFQTLLLTLVSGLGLLAVATTVVDQLAIRLLP-QRKSYSSLKFQVTESMSNPM 358
```

Figure 4

Conservation of P2X receptors among disparate organisms. Sequence alignment of five P2X receptors that form functional ATP-gated channels when expressed in HEK cells. Overlines denote proposed transmembrane domains. Red denotes residues conserved among five sequences; blue denotes residues similar among five sequences (similar are IVLM, FYWH, KRH, EQDN). Sequences (with Genbank accessions) are rP2X2, *Rattus norvegicus* P2X$_2$ (P49653); schis, *Schistosoma mansoni* (CAH04147); choan, *Monosiga brevicollis* (EDQ92249); algae, *Ostreococcus tauri* (CAL54489); dicty, *Dictyostelium discoideum* (XP_645378). Further C termini (5–95 residues) are divergent and not shown.

residues have previously been shown to be essential for function in mammalian receptors (21, 135). Second, the highly conserved residues are in or close to the two transmembrane domains. This applies to the highly conserved intracellular Y-X-X-X-K motif that is found both ~10 residues before the start of transmembrane domain 1 (TM1) and ~10 residues after the end of transmembrane domain 2 (TM2). Such a positioning pattern also applies to the highly con-

served extracellular positively charged amino acids that are located ~15 residues after the end of TM1 (Lys[69] and Lys[71] in P2X$_2$) as well as ~20 residues before the start of TM2 (Arg[304] and Lys[308]). Third, the great bulk of the ectodomain is not conserved. This includes the ten cysteine residues. The lack of conservation extends also to several residues that, on the basis of mutagenesis in vertebrate receptors, had been considered to contribute to an ATP-binding site (see 135). The key residues that had been proposed to represent homology with the ATP-binding site of class II tRNA synthases (139) are also mostly missing in the simpler P2X receptors.

N Terminus

The most striking feature of the N terminus of the P2X receptor family is the Y-X-T-X[KR] motif that ends some 10 residues prior to the start of TM1. The T-X-K conforms to the protein kinase C consensus, and phosphorylation of Thr^{18} in the $P2X_2$ receptor has been demonstrated for $P2X_2$ (140) but not for $P2X_1$ receptors (141); mutations at this site alter peak currents and desensitization at the $P2X_1$ receptor (141). In the $P2X_2$ receptor, replacement of Tyr^{16} by cysteine prevents channel function, although such a substitution is tolerated at Thr^{18} or Lys^{20} (142).

Transmembrane Domain 1

Each residue of TM1 has been systematically substituted by cysteine [$P2X_2$ (142–144)], alanine [$P2X_2$ (145, 146)], or tryptophan [$P2X_4$ (147)]. The results indicate in general that (*a*) TM1 is very likely α-helical, (*b*) residues Gly^{30} and Tyr^{43} ($P2X_2$ numbering) are completely conserved and critical for function, (*c*) TM1 likely moves outward with respect to TM2 during channel gating (148), and (*d*) residues in TM1 seem not to contribute directly to the permeation pathway (146).

Post–Transmembrane Domain 1

The 20 amino acids immediately following TM1 do not show strong conservation (**Figure 4**) except for Tyr^{55} and Gln^{56}, which are essential for receptor function (142, 148), and the KΦKG motif Lys^{69} and Lys^{71} in the $P2X_2$ receptor, where Φ is any hydrophobic amino acid (142). Abundant evidence now supports the view that one or both of these Lys residues play a critical role in ATP binding. The three main lines of evidence are as follows.

(*a*) Changing the charge on the nearby Ile^{67} can affect ATP potency in a manner that is consistent with a direct electrostatic effect on the binding of a negatively charged agonist (142), (*b*) comparing the effects of ATP with those of the partial agonist BzATP suggest a direct binding rather than a gating role (149), and (*c*) Lys^{69} is also required for channel activation by TNP-ATP at the constitutively active $P2X_2$ receptor carrying a T339S mutation (150). Experiments with coexpressed $P2X_2$ subunits, in which one contains Lys^{69} and the other does not (i.e., $P2X_2$[K69A]), strongly suggest that two rather than three such lysines are required for agonist activation of the channel and, moreover, that the ATP-binding site likely forms between lysines provided by two different subunits (151).

Ectodomain

Figure 4 illustrates that most parts of the ectodomain that are conserved among vertebrates are not found in the *Dictyostelium* sequence; this includes the ten cysteine residues. This indicates that much of the ectodomain is not critical for function as an ATP-gated channel and presumably subserves other roles such as interactions with proteins provided by the matrix or by neighboring cells. The pattern of disulfide bond formation has been deduced. However, although the cysteines play a role in transport to the cell membrane, none of the cysteines are critical for channel function (152, 153). Zinc potentiates currents evoked by ATP at $P2X_2$ receptors. By using concatenated cDNAs to express subunits joined in tandem, Hume and colleagues (154) showed that His^{120} and His^{213} form the zinc-binding site and that different subunits contribute to the compositions of these two histidines. When these histidines are replaced by cysteine, an intersubunit disulfide bond can form. Similar intersubunit ectopic disulfide bonds between pairs of cysteines introduced into the $P2X_1$ receptor (Lys^{68} and Phe^{291}) have been demonstrated (155), as was first shown for cysteines at the outer ends of the transmembrane domains of the $P2X_2$ receptor [Val^{48} to Ile^{328} (148)]. One conserved motif is present in all known sequences, including the *Dictyostelium* sequence: [EQP]-[NDS]-F-T-Φ-Φ-Φ-[KD]-NH]-[STN] (where Φ = any hydrophobic residue). The FT (Phe[183]-Thr[184] in P2X$_2$) residues may play a role in stacking against the adenine moiety of ATP (149, 156). A similar role has also been ascribed to the NFR motif (Asn[288]-Phe[289]-Arg[290] in P2X$_2$) (156), but this seems less likely given that these residues are missing from the *Dictyostelium* sequence (137).

Pre–Transmembrane Domain 2

This region extends from the highly conserved Arg[304] (P2X$_2$) to the beginning of TM2 (around Ile[328]). The first half of this region shows particularly high sequence conservation (**Figure 4**). The lysine residue at position 308 (in P2X$_2$; position 309 in P2X$_1$ and position 313 in P2X$_4$) is required for channel function and has been implicated in ATP binding (21, 135, 142, 149, 157). For the P2X$_{2/3}$ heteromer, Wilkinson et al. (158) suggested that the ATP-binding site may have contributions from Lys[69] of one subunit and Lys[308] of another. Other work suggests a more fundamental role for Lys[308] in channel gating. Yan et al. (157) favor a role for this residue (Lys[313] in P2X$_4$ receptor) in contributing to intersubunit interactions that accompany gating, rather than contributing directly to binding. Cao et al. (150) have reported that this residue is critical for channel activation in a slightly modified channel (P2X$_2$[T339S]) that opens spontaneously in the complete absence of any agonist. P2X$_2$[T339S] is also activated by suramin, but P2X$_2$[T339S,K308A] is not; this finding is also best interpreted to indicate a key role for Lys[308] in channel gating. With the exception of Arg[304] and Lys[308], substitution of any individual residue in pre-TM2 by cysteine (in P2X$_1$) does not markedly affect the ability of ATP to evoke inward currents (159). In most cases, these cysteines were accessible to labeling with MTSEA-biotin, suggesting their exposure on the protein surface. Yan et al. (157) studied alanine substitutions (in P2X$_4$) and have speculated that the pre-

TM2 segment functions as a "signal transduction module" from ATP-binding site to channel gate. Taken together, it appears that this highly conserved region may subserve roles other than ATP binding and channel gating.

Transmembrane Domain 2

The consensus view from cysteine scanning (21, 135), alanine scanning (145), and tryptophan scanning (147) is that TM2 does not cross the membrane as an uninterrupted α-helix. Cysteine is tolerated at almost all positions through TM2 when substituted individually (160, 161). Tryptophan is tolerated in the outer half of TM2, and the positions causing the greatest perturbation of function (actually an increase in sensitivity to ATP) align along one surface of a helix (P2X$_4$) (147). Near the center of TM2, tryptophan is not tolerated at Ala[349], Leu[352], Cys[353], Asp[354], Val[357], and Leu[358] (P2X$_4$; positions correspond to P2X$_2$ Gly[344], Leu[347], Cys[348], Asp[349], Leu[352], and Leu[353], respectively), suggesting to these authors (147) "a tightly packed and relatively delicate structure that poorly tolerates substitution" at this position in the protein.

Early studies with cysteine scanning and the effects of methanethiosulfonates had indicated that the gate might form in the middle of TM2 [between Thr[339] and Asp[349] on the P2X$_2$ receptor (160)], and studies on the permeability of the P2X receptor now support this interpretation. Migita et al. (162) identified Thr[339] and Ser[340] of the P2X$_2$ receptor as having the greatest effect on cation permeability. More recently, the fraction of the inward current carried by calcium ions (*Pf%*) has been determined by combination of measurement of charge with measurement of change in calcium concentration (163). Measurements of *Pf%* also implicate residues Thr[339] and Ser[340] as the main deep determinants of calcium flux. Calcium *Pf%* varies among homomeric P2X receptors; P2X$_1$ and P2X$_4$ receptors show the highest values [11–15%, similar to calcium *Pf%* for NMDA receptors; see Egan & Khakh (163)]. The difference between these two receptors and others in the family may result in part from the calcium-concentrating effect of fixed negative charges provided by glutamate residues near the extracellular ends of both TM1 and TM2 (164).

The Asp residue at position 349 ($P2X_2$) is completely conserved among P2X receptors other than in the algae *O. tauri* (138). Replacement of Asn by Asp in the algae receptor causes a 50% increase in relative calcium permeability. However, the reverse mutation in $P2X_2$ receptors does not change relative calcium permeability (138, 162). Thus, this aspartate residue does not appear to play a major role in the determination of calcium permeability. In contrast, it does have a key role in the intramembrane interhelix interactions required for channel assembly ($P2X_5$) (165).

C Terminus

The cytoplasmic tails of six of the P2X receptors retain considerable sequence relatedness for the first 25 amino acids (**Figure 4**), but relatedness thereafter disappears completely. A YXXXK motif, beginning ~10 residues from the cytoplasmic end on TM2 (or 28 in $P2X_7$; see below), is very highly conserved among all P2X receptors, including those from single-celled organisms that have poorly conserved ectodomains. This lends a symmetry to P2X receptor sequences, the conserved region of which both begins and ends with YXXXK motifs in the cytoplasmic regions (**Figure 4**). Rassendren and colleagues (166) have shown that the Tyr and the Lys of this motif are required to stabilize the P2X receptor subunit in the plasma membrane; in heteromeric $P2X_{2/3}$ receptors the motif needs to be present on only one of the partners. The $P2X_7$ receptor differs from the other six in this region by the presence of an 18-amino-acid cysteine-rich insertion. Removal of this insertion prevents the $P2X_7$ receptor from undergoing the increase in NMDG permeability that normally occurs when the ATP application is continued for several seconds. Deletion of this cysteine-rich segment has no effect on the properties of the channel in normal extra-cellular sodium concentrations, and it does not affect the ability of the $P2X_7$ receptor to drive uptake of fluorescent dyes (167).

To the C-terminal side of the YXXXK motif, the sequences of P2X receptors are subunit specific. For the $P2X_2$ subunit, the C terminus binds to Fe65 (168), a multidomain protein first identified by its ability to bind to β-amyloid precursor protein. Both proteins colocalize to the postsynaptic membranes of excitatory synapses in the hippocampus. Coexpression in oocytes showed that the presence of Fe65 slows the rate of pore dilation (increase in NMDG permeability) that occurs when the ATP application is applied for several minutes (see Reference 21). $P2X_4$ subunits have a motif (YXXGΦ) that directly binds the μ2 subunit of the AP2 adaptor protein (73). This interaction is responsible for $P2X_4$ receptor endocytosis, despite the presence of a nearby YXXΦ motif that is commonly involved in the binding of AP2 to other proteins.

$P2X_7$ receptors interact with several partners, including cytoskeletal proteins (85). One of the interactors is receptor protein tyrosine phosphatase β, which has a clear effect on the currents through the $P2X_7$ receptor when it is activated repeatedly (85). Other partners include epithelial membrane proteins (169). Motifs have been identified in conserved regions of the $P2X_7$ receptor C terminus that resemble sequences in the SH3 domain–binding proteins *Mycoplasma genitalium* cytadherence accessory protein HMW3 and *Caenorhabditis elegans* protein C18H2.1 (170). However, the most intriguing binding partner may be the bacterial endotoxin lipopolysaccharide (LPS) (170). A motif very close to the tail of the $P2X_7$ sequence (residues 573 to 590) is well conserved among all known $P2X_7$ subunits and closely resembles a sequence in soluble LPS-binding protein. Bertics and his colleagues showed that a peptide corresponding to this motif (rat $P2X_7$ 573 to 590: CRWRIRKEF-PKTQGQYSG) directly binds FITC-labeled LPS, as does the LPS-binding protein itself. The same two basic amino acids in the $P2X_7$

peptide and in LPS-binding protein are essential for LPS binding (170), and mutation of these two residues impairs both receptor trafficking and channel function in HEK293 cells (171). It will be important to test the possibility that this region of the $P2X_7$ receptor is directly responsible for those actions of bacterial LPS that are independent of Toll-like receptor 4 (TLR4).

CONCLUDING THOUGHTS AND FUTURE ISSUES

The field continues to be limited by the availability of agonists, antagonists, and modulators with high selectivity for one or another P2X receptor, but there have been notable advances. Particularly in the cases of $P2X_1$, $P2X_3$, and $P2X_7$ receptors, there are now antagonists available with nanomolar affinity. There remain two important caveats to their application. The first is the issue of species specificity, which seems particularly important to take into account for $P2X_7$ antagonists. The second is the problem associated with actions at receptors other than P2X receptors. For example, suramin analogs and ivermectin have several such effects. Physiologists studying the functional roles of P2X receptors would welcome

further progress in the development of such compounds.

The availability of mice lacking $P2X_1$, $P2X_2$, $P2X_3$, $P2X_4$, and $P2X_7$ receptors has impacted the discovery of new functional roles. The difficulties here are those familiar to all such approaches—the species specificity and the possibilities of compensatory changes. Tissue-specific and inducible knock-in and knockout constructs would be most valuable.

At the molecular level, progress continues to be made with respect to our understanding of the permeation pathway and the ATP-binding site. The former will benefit from single-channel recordings combined with mutagenesis, and the latter from the sequencing of receptor fragments cross-linked to appropriate agonists or antagonists. Similar mass spectroscopic approaches need to be further extended to identify the range of proteins that interact directly with P2X receptors. These will lead us to an improved understanding of the subcellular trafficking and localization of P2X receptors, in which context it will be important to seek further evidence for intracellular functional roles such as those reported for *Dictyostelium*. And finally, high-resolution structure of a crystallized receptor would of course provide the field with its greatest impetus.

LPS: bacterial lipopolysaccharide

SUMMARY POINTS

1. $P2X_1$ receptors are neurotransmitter receptors at sympathetically innervated smooth muscle and play functional roles on platelets and in the juxtaglomerular apparatus of the kidney.

2. $P2X_2$ and $P2X_3$ subunits (and $P2X_{2/3}$ heteromeric receptors) are necessary for the initiation of sensory signaling in pathways subserving taste, chemoreception, visceral distension, and neuropathic pain.

3. $P2X_4$ receptors are involved in the function of vascular endothelium. Mice lacking the $P2X_4$ receptor gene are hypertensive and have smaller-diameter arteries. Microglial $P2X_4$ receptors in the spinal cord have also been implicated in neuropathic pain.

4. Activation of $P2X_7$ receptors drives the release of proinflammatory cytokines from macrophages primed with lipopolysaccharides. $P2X_7$ antagonists appear beneficial in the inflammation of rheumatoid arthritis.

5. In the central nervous system there are widespread presynaptic and postsynaptic actions of ATP, but the importance of P2X signaling remains unclear at the cellular level.

6. P2X receptors in a simple eukaryote are intracellular and involved in osmoregulation. Thus, functions on intracellular membranes in mammalian cells may be worth seeking. The distant sequence relatedness of these primitive P2X receptors also indicates that substantial parts of the receptor ectodomain are not needed for receptor function.

7. Single-channel recordings and mutagenesis are identifying the regions of the P2X receptor involved in ion permeation, ATP binding, and trafficking to the cell surface, but higher-resolution structural information would greatly accelerate this work.

DISCLOSURE STATEMENT

The authors are not aware of any biases that might be perceived as affecting the objectivity of this review.

ACKNOWLEDGMENTS

This work was supported by The Wellcome Trust and the BBSRC.

LITERATURE CITED

1. Burnstock G. 2007. Physiology and pathophysiology of purinergic neurotransmission. *Physiol. Rev.* 87:659–797
2. Burnstock G, Kennedy C. 1985. Is there a basis for distinguishing two types of P2-purinoceptor? *Gen. Pharmacol.* 16:433–40
3. **Burnstock G. 2006. Historical review: ATP as a neurotransmitter. *Trends Pharmacol. Sci.* 27:166–76.**
4. Valera S, Hussy N, Evans RJ, North RA, Surprenant A, et al. 1994. A new class of ligand-gated ion channel defined by P2X receptor for extracellular ATP. *Nature* 371:516–19
5. Brake AJ, Wagenbach MJ, Julius D. 1994. New structural motif for ligand-gated ion channels defined by an ionotropic ATP receptor. *Nature* 371:519–23
6. Collo G, North RA, Kawashima E, Merlo-Pich E, Neidhart S, et al. 1996. Cloning of P2X$_5$ and P2X$_6$ receptors and the distribution and properties of an extended family of ATP-gated ion channels. *J. Neurosci.* 16:2495–507
7. Collo G, Neidhart S, Kawashima E, Kosco-Vilbois M, North RA, Buell G. 1997. Tissue distribution of the P2X$_7$ receptor. *Neuropharmacology* 36:1277–83
8. Burnstock G, Knight GE. 2004. Cellular distribution and functions of P2 receptor subtypes in different systems. *Int. Rev. Cytol.* 240:31–304
9. Rubio ME, Soto F. 2001. Distinct localization of P2X receptors at excitatory postsynaptic specializations. *J. Neurosci.* 21:641–53
10. Sim JA, Young MT, Sung HY, North RA, Surprenant A. 2004. Reanalysis of P2X$_7$ receptor expression in rodent brain. *J. Neurosci.* 24:6307–14
11. Anderson CM, Nedergaard M. 2006. Emerging challenges of assigning P2X$_7$ receptor function and immunoreactivity in neurons. *Trends Neurosci.* 29:257–62
12. Edwards FA, Gibb AJ, Colquhoun D. 1992. ATP receptor-mediated synaptic currents in the central nervous system. *Nature* 359:144–46
13. Pankratov Y, Castro E, Miras-Portugal MT, Krishtal O. 1998. A purinergic component of the excitatory postsynaptic current mediated by P2X receptors in the CA1 neurons of the rat hippocampus. *Eur. J. Neurosci.* 10:3898–902
14. Mori M, Heuss C, Gähwiler BH, Gerber U. 2001. Fast synaptic transmission mediated by P2X receptors in CA3 pyramidal cells of rat hippocampal slice cultures. *J. Physiol.* 535:115–23

3. Engaging and authoritative personal perspective on the initiation and progress of the field of purine receptor biology.

15. Pankratov Y, Lalo U, Krishtal O, Verkhratsky A. 2002. Ionotropic P2X purinoreceptors mediate synaptic transmission in rat pyramidal neurones of layer II/III of somato-sensory cortex. *J. Physiol.* 542:529–36

16. Pankratov Y, Lalo U, Krishtal O, Verkhratsky A. 2003. P2X receptor-mediated excitatory synaptic currents in somatosensory cortex. *Mol. Cell. Neurosci.* 24:842–49

17. Pankratov Y, Lalo U, Verkhratsky A, North RA. 2007. Quantal release of ATP in mouse cortex. *J. Gen. Physiol.* 129:257–65

18. Jo YH, Role LW. 2002. Coordinate release of ATP and GABA at in vitro synapses of lateral hypothalamic neurons. *J. Neurosci.* 22:4794–804

19. Sim JA, Chaumont S, Jo J, Ulmann L, Young MT, et al. 2006. Altered hippocampal synaptic potentiation in P2X$_4$ knock-out mice. *J. Neurosci.* 26:9006–9

20. Sperlágh B, Vizi ES, Wirkner K, Illes P. 2006. P2X receptors in the nervous system. *Prog. Neurobiol.* 78:327–46

21. North RA. 2002. Molecular physiology of P2X receptors. *Physiol. Rev.* 82:1013–67

22. Fellin T, Pozzan T, Carmignoto G. 2006. Purinergic receptors mediate two distinct glutamate release pathways in hippocampal astrocytes. *J. Biol. Chem.* 281:4274–84

23. Jabs R, Matthias K, Grote A, Grauer M, Seifert G, Steinhäuser C. 2007. Lack of P2X receptor mediated currents in astrocytes and GluR type glial cells of the hippocampal CA1 region. *Glia* 55:1648–55

24. Lalo U, Pankratov Y, Wichert SP, Rossner MJ, North RA, et al. 2008. P2X$_1$ and P2X$_5$ subunits form the functional P2X receptor in mouse cortical astrocytes. *J. Neurosci.* 28:5473–80

25. Khakh BS, Henderson G. 1998. ATP receptor-mediated enhancement of fast excitatory neurotransmitter release in the brain. *Mol. Pharmacol.* 54:372–78

26. Patel MK, Khakh BS, Henderson G. 2001. Properties of native P2X receptors in rat trigeminal mesencephalic nucleus neurones: lack of correlation with known, heterologously expressed P2X receptors. *Neuropharmacology* 40:96–105

27. Khakh BS, Gittermann D, Cockayne DA, Jones A. 2003. ATP modulation of excitatory synapses onto interneurons. *J. Neurosci.* 23:7426–37

28. Shigetomi E, Kato F. 2004. Action potential-independent release of glutamate by Ca^{2+} entry through presynaptic P2X receptors elicits postsynaptic firing in the brainstem autonomic network. *J. Neurosci.* 24:3125–35

29. Donato R, Rodrigues RJ, Takahashi M, Tsai MC, Soto D, et al. 2008. GABA release by basket cells onto Purkinje cells, in rat cerebellar slices, is directly controlled by presynaptic purinergic receptors, modulating Ca^{2+} influx. *Cell Calcium* 44:521–32

30. Rhee JS, Wang ZM, Nabekura J, Inoue K, Akaike N. 2000. ATP facilitates spontaneous glycinergic IPSC frequency at dissociated rat dorsal horn interneuron synapses. *J. Physiol.* 524:471–83

31. Zemkova H, Yan Z, Liang Z, Jelinkova I, Tomic M, Stojilkovic SS. 2007. Role of aromatic and charged ectodomain residues in the P2X$_4$ receptor functions. *J. Neurochem.* 102:1139–50

32. Troadec JD, Thirion S, Nicaise G, Lemos JR, Dayanithi G. 1998. ATP-evoked increases in [Ca^{2+}]$_i$ and peptide release from rat isolated neurohypophysial terminals via a P2X$_2$ purinoceptor. *J. Physiol.* 511:89–103

33. Knott TK, Velázquez-Marrero C, Lemos JR. 2005. ATP elicits inward currents in isolated vasopressinergic neurohypophysial terminals via P2X$_2$ and P2X$_3$ receptors. *Pflüg. Arch.* 450:381–89

34. Yoshida R, Yasumatsu K, Shigemura N, Ninomiya Y. 2006. Coding channels for taste perception: information transmission from taste cells to gustatory nerve fibers. *Arch. Histol. Cytol.* 69:233–42

35. Roper SD. 2007. Signal transduction and information processing in mammalian taste buds. *Pflüg. Arch. Eur. J. Physiol.* 454:759–76

36. Hayato R, Ohtubo Y, Yoshii K. 2007. Functional expression of inotropic purinergic receptors on mouse taste bud cells. *J. Physiol.* 584:473–88

37. Bo X, Alavi A, Xiang Z, Oglesby I, Ford A, Burnstock G. 1999. Localization of ATP-gated P2X$_2$ and P2X$_3$ receptor immunoreactive nerves in rat taste buds. *Neuroreport* 10:1107–11

38. Kataoka S, Toyono T, Seta Y, Toyoshima K. 2006. Expression of ATP-gated P2X$_3$ receptors in rat gustatory papillae and taste buds. *Arch. Histol. Cytol.* 69:281–88

39. **Finger TE, Danilova V, Barrows J, Bartel DL, Vigers AJ, et al. 2005. ATP signaling is crucial for communication from taste buds to gustatory nerves.** *Science* 310:1495–99

40. Huang YJ, Maruyama Y, Dvoryanchikov G, Pereira E, Chaudhari N, Roper SD. 2007. The role of pannexin 1 hemichannels in ATP release and cell-cell communication in mouse taste buds. *Proc. Natl. Acad. Sci. USA* 104:6436–41

41. Romanov RA, Rogachevskaja OA, Bystrova MF, Jiang P, Margolskee RF, Kolesnikov SS. 2007. Afferent neurotransmission mediated by hemichannels in mammalian taste cells. *EMBO J.* 26:657–67

42. Rong W, Burnstock G, Spyer KM. 2000. P2X purinoceptor-mediated excitation of trigeminal lingual nerve terminals in an in vitro intra-arterially perfused rat tongue preparation. *J. Physiol.* 524:891–902

43. Housley GD, Jagger DJ, Greenwood D, Raybould NP, Salih SG, et al. 2002. Purinergic regulation of sound transduction and auditory neurotransmission. *Audiol. Neurootol.* 7:55–61

44. Housley GD, Marcotti W, Navaratnam D, Yamoah EN. 2006. Hair cells—beyond the transducer. *J. Membr. Biol.* 209:89–118

45. Munoz DJB, Thorne PR, Housley GD. 1999. P2X receptor-mediated changes in cochlear potentials arising from exogenous adenosine 5′-triphosphate in endolymph. *Hear. Res.* 138:56–64

46. Salih SG, Jagger DJ, Housley GD. 2002. ATP-gated currents in rat primary auditory neurones in situ arise from a heteromultimeric P2X receptor subunit assembly. *Neuropharmacology* 42:386–95

47. Lee JH, Chiba T, Marcus DC. 2001. P2X$_2$ receptor mediates stimulation of parasensory cation absorption by cochlear outer sulcus ells and vestibular transitional cells. *J. Neurosci.* 21:9168–74

48. Dulon D, Jagger DJ, Lin X, Davis RL. 2006. Neuromodulation in the spiral ganglion: shaping signals from the organ of Corti to the CNS. *J. Memb. Biol.* 209:167–75

49. Prabhakar NR. 2000. Oxygen sensing by the carotid body chemoreceptors. *J. Appl. Physiol.* 88:2287–95

50. Rong W, Gourine AV, Cockayne DA, Xiang Z, Ford AP, et al. 2003. Pivotal role of nucleotide P2X$_2$ receptor subunit of the ATP-gated ion channel mediating ventilatory responses to hypoxia. *J. Neurosci.* 23:11315–21

51. Spyer KM, Dale N, Gourine AV. 2004. ATP is a key mediator of central and peripheral chemosensory transduction. *Exp. Physiol.* 89:53–59

52. Ferguson DR, Kennedy I, Burton TJ. 1997. ATP is released from rabbit urinary bladder epithelial cells by hydrostatic pressure changes—a possible sensory mechanism? *J. Physiol.* 505:503–11

53. Yu Y, de Groat WC. 2008. Sensitization of pelvic afferent nerves in the in vitro rat urinary bladder-pelvic nerve preparation by purinergic agonists and cyclophosphamide pretreatment. *Am. J. Physiol. Ren. Physiol.* 294:F1146–56

54. **Cockayne DA, Hamilton SG, Zhu QM, Dunn PM, Zhong Y, et al. 2000. Urinary bladder hyporeflexia and reduced pain-related behaviour in P2X$_3$-deficient mice.** *Nature* 407:1011–15

55. Vlaskovska M, Kasakov L, Rong W, Bodin P, Bardini M, et al. 2001. P2X$_3$ knock-out mice reveal a major sensory role for urothelially released ATP. *J. Neurosci.* 21:567–77

56. Hamilton SG, Warburton J, Bhattacharjee A, Ward J, McMahon SB. 2000. ATP in human skin elicits a dose-related pain response which is potentiated under conditions of hyperalgesia. *Brain* 123:1238–46

57. Lewis C, Neidhart S, Holy C, North RA, Buell G, Surprenant A. 1995. Coexpression of P2X$_2$ and P2X$_3$ receptor subunits can account for ATP-gated currents in sensory neurons. *Nature* 377:432–35

58. Chen CC, Akopian AN, Sivilotti L, Colquhoun D, Burnstock G, Wood JN. 1995. A P2X purinoceptor expressed by a subset of sensory neurons. *Nature* 377:428–31

59. Jarvis MF, Burgard EC, McGaraughty S, Honore P, Lynch K, et al. 2002. A-317491, a novel potent and selective non-nucleotide antagonist of P2X$_3$ and P2X$_{2/3}$ receptors, reduces chronic inflammatory and neuropathic pain in the rat. *Proc. Natl. Acad. Sci. USA* 99:17179–84

60. Honore P, Kage K, Mikusa J, Watt AT, Johnston JF, et al. 2002. Analgesic profile of intrathecal P2X$_3$ antisense oligonucleotide treatment in chronic inflammatory and neuropathic pain states in rats. *Pain* 99:11–19

61. Wirkner K, Sperlagh B, Illes P. 2007. P2X$_3$ receptor involvement in pain states. *Mol. Neurobiol.* 36:165–83

62. Tsuda M, Shigemoto-Mogami Y, Koizumi S, Mizokoshi A, Kohsaka S, et al. 2003. P2X$_4$ receptors induced in spinal microglia gate tactile allodynia after nerve injury. *Nature* 424:778–83

63. Trang T, Beggs S, Salter MW. 2006. Purinoceptors in microglia and neuropathic pain. *Pflüg. Arch.* 452:645–52

64. Virginio C, Robertson G, Surprenant A, North RA. 1998. Trinitrophenyl-substituted nucleotides are potent antagonists selective for P2X$_1$, P2X$_3$, and heteromeric P2X$_{2/3}$ receptors. *Mol. Pharmacol.* 53:969–73

65. Chessell IP, Hatcher JP, Bountra C, Michel AD, Hughes JP, et al. 2005. Disruption of the P2X$_7$ purinoceptor gene abolishes chronic inflammatory and neuropathic pain. *Pain* 114:386–96

66. Honore P, Donnelly-Roberts D, Namovic MT, Hsieh G, Zhu CZ, et al. 2006. A-740003 [N-(1-{[(cyanoimino)(5-quinolinylamino) methyl]amino}-2,2-dimethylpropyl)-2-(3,4-dimethoxyphenyl)acetamide], a novel and selective P2X$_7$ receptor antagonist, dose-dependently reduces neuropathic pain in the rat. *J. Pharmacol. Exp. Ther.* 319:1376–85

67. Wilson HL, Varcoe RW, Stokes L, Holland KL, Francis SE, et al. 2007. P2X receptor characterization and IL-1/IL-rA release from human endothelial cells. *Br. J. Pharmacol.* 151:96–108

68. Chi JT, Chang HY, Haraldsen G, Jahnsen FL, Troyanskaya OG, et al. 2003. Endothelial cell diversity revealed by global expression profiling. *Proc. Natl. Acad. Sci. USA* 100:10623–28

69. Tran QK, Watanabe H. 2006. Calcium signaling in the endothelium. *Handb. Exp. Pharmacol.* 176:145–87

70. Yamamoto K, Korenaga R, Kamiya A, Qi Z, Sokabe M, Ando J. 2000. P2X$_4$ receptors mediate ATP-induced calcium influx in human vascular endothelial cells. *Am. J. Physiol.* 27:H285–92

71. Garcia-Guzman M, Soto F, Gomez-Hernandez JM, Lund P, Stuhmer W. 1997. Characterization of recombinant human P2X$_4$ receptor reveals pharmacological differences to the rat homologue. *Mol. Pharmacol.* 51:109–18

72. Bobanovic LK, Royle SJ, Murrell-Lagnado RD. 2002. P2X receptor trafficking in neurons is subunit specific. *J. Neurosci.* 22:4814–24

73. Royle SJ, Qureshi OS, Bobanović LK, Evans PR, Owen DJ, Murrell-Lagnado RD. 2005. Non-canonical YXXGΦ endocytic motifs: recognition by AP2 and preferential utilization in P2X$_4$ receptors. *J. Cell Sci.* 118:3073–80

74. Yamamoto K, Sokabe T, Matsumoto T, Yoshimura K, Shibata M, et al. 2006. Impaired flow-dependent control of vascular tone and remodeling in P2X$_4$-deficient mice. *Nat. Med.* 12:133–37

75. Chamberlain J, Evans D, King A, Dewberry R, Dower S, et al. 2006. Interleukin-1β and signaling of interleukin-1 in vascular wall and circulating cells modulates the extent of neointimal formation in mice. *Am. J. Pathol.* 168:1396–403

76. Gachet C. 2008. P2 receptors, platelet function and pharmacological implications. *Thromb. Haemost.* 99:466–72

77. Rolf MG, Brearley CA, Mahaut-Smith MP. 2001. Platelet shape change evoked by selective activation of P2X$_1$ purinoceptors with α,β-methylene ATP. *Thromb. Haemost.* 85:303–8

78. Rolf MG, Mahaut-Smith MP. 2002. Effects of enhanced P2X$_1$ receptor Ca^{2+} influx on functional responses in human platelets. *Thromb. Haemost.* 88:495–502

79. Hechler B, Lenain N, Marchese P, Vial C, Heim V, et al. 2003. A role of the fast ATP-gated P2X$_1$ cation channel in thrombosis of small arteries in vivo. *J. Exp. Med.* 198:661–67

80. Leipziger J. 2003. Control of epithelial transport via luminal P2 receptors. *Am. J. Physiol.* 284:F419–32

81. Schwiebert EM, Zsembery A. 2003. Extracellular ATP as a signalling molecule for epithelial cells. *Biochim. Biophys. Acta* 1615:7–32

82. Ma W, Korngreen A, Uzlaner N, Priel Z, Silberberg SD. 1999. Extracellular sodium regulates airway ciliary motility by inhibiting a P2X receptor. *Nature* 400:894–97

83. Ma W, Korngreen A, Weil S, Cohen EB, Priel A, et al. 2006. Pore properties and pharmacological features of the P2X receptor channel in airway ciliated cells. *J. Physiol.* 571:503–17

84. Fujiwara Y, Kubo Y. 2004. Density-dependent changes of the pore properties of the P2X$_2$ receptor channel. *J. Physiol.* 558:31–43

85. Kim M, Jiang LH, Wilson HL, North RA, Surprenant A. 2001. Proteomic and functional evidence for a P2X$_7$ receptor signaling complex. *EMBO J.* 20:6347–58

86. Llaudet E, Hatz S, Droniou M, Dale N. 2005. Microelectrode biosensor for real-time measurement of ATP in biological tissue. *Anal. Chem.* 77:3267–73

87. Gourine AV, Llaudet E, Dale N, Spyer KM. 2005. Release of ATP in the ventral medulla during hypoxia in rats: role in hypoxic ventilatory response. *J. Neurosci.* 25:1211–18

74. P2X$_4$R knockout mice provide striking evidence for the role of P2X$_4$R in endothelial cell function.

88. Schnermann J, Levine DZ. 2003. Paracrine factors in tubuloglomerular feedback: adenosine, ATP and nitric oxide. *Annu. Rev. Physiol.* 65:501–29

89. Guan Z, Osmond DA, Inscho EW. 2007. P2X receptors as regulators of the renal microvasculature. *Trends Pharmacol. Sci.* 28:646–52

90. Nishiyama A, Majid DSA, Taher KA, Miyatake A, Navar LG. 2000. Relation between renal interstitial ATP concentrations and autoregulation-mediated changes in renal vascular resistance. *Circ. Res.* 86:656–62

91. Komlosi P, Fintha A, Bell PD. 2005. Renal cell-to-cell communication via extracellular ATP. *Physiology* 20:86–90

92. Bell PD, Lapointe J, Sabirov R, Hayashi S, Peti-Peterdi J, et al. 2003. Macula densa cell signaling involves ATP release through a maxi anion channel. *Proc. Natl. Acad. Sci. USA* 100:4322–27

93. Nishiyama A, Majid DSA, Walker M, Miyatake A, Navar LG. 2001. Renal interstitial ATP responses to changes in arterial pressure during alterations in tubuloglomerular feedback activity. *Hypertension* 37(Pt. 2):753–59

94. Castrop H. 2007. Mediators of tubuloglomerular feedback regulation of glomerular filtration: ATP and adenosine. *Acta Physiol.* 189:3–14

95. Sun D, Samuelson LC, Yang T, Huang Y, Paliege A, et al. 2001. Mediation of tubuloglomerular feedback by adenosine: evidence from mice lacking adenosine 1 receptors. *Proc. Natl. Acad. Sci. USA* 98:9983–88

96. Brown R, Ollerstam A, Johansson B, Skott O, Gebre-Mehin S, et al. 2001. Abolished tubuloglomerular feedback and increased plasma rennin in adenosine A1 receptor-deficient mice. *Am. J. Physiol. Regul. Integr. Comp. Physiol.* 281:1362–67

97. Ren Y, Garvin JL, Liu R, Carretero OA. 2004. Role of macula densa adenosine triphosphate (ATP) in tubuloglomerular feedback. *Kidney Int.* 66:1479–85

98. Zhao X, Cook AK, Field M, Edwards B, Zhang S, et al. 2005. Impaired Ca^{2+} signaling attenuates P2X receptor-mediated vasoconstriction of afferent arterioles in angiotensin II hypertension. *Hypertension* 46:562–68

99. Inscho EW, Cook AK, Imig JD, Vial C, Evans RJ. 2003. Physiological role for P2X$_1$ receptors in renal microvascular autoregulatory behavior. *J. Clin. Investig.* 112:895–905

100. Feng M, Navar LG. 2007. Adenosine A2 receptor activation attenuates afferent arteriolar autoregulation during adenosine receptor saturation in rats. *Hypertension* 50:744–49

101. Hillman KA, Burnstock G, Unwin RJ. 2005. The P2X$_7$ ATP receptor in the kidney: a matter of life or death? *Nephron Exp. Nephrol.* 101:24–30

102. Schulze-Lohoff E, Hugo C, Rost S, Arnold S, Gruber A, et al. 1998. Extracellular ATP causes apoptosis and necrosis of cultured mesangial cells via P2Z/P2X$_7$ receptors. *Am. J. Physiol.* 275:F962–71

103. Harada H, Chan CM, Loesch A, Unwin R, Burnstock G. 2000. Induction of proliferation and apoptotic cell death via P2Y and P2X receptors, respectively, in rat glomerular mesangial cells. *Kidney Int.* 57:949–58

104. Solini A, Iacobini C, Ricci C, Chiozzi P, Amadio L, et al. 2005. Purinergic modulation of mesangial extracellular matrix production: role in diabetic and other glomerular diseases. *Kidney Int.* 67:875–85

105. Goncalves RG, Gabrich L, Rosario A, Takiya CM, Ferreira ML, et al. 2006. The role of purinergic P2X$_7$ receptors in the inflammation and fibrosis of unilateral ureteral obstruction in mice. *Kidney Int.* 70:1599–606

106. Solini A, Santini E, Chimenti D, Chiozzi P, Pratesi F, et al. 2007. Multiple P2X receptors are involved in the modulation of apoptosis in human mesangial cells: evidence for a role of P2X$_4$. *Am. J. Physiol.* 292:F1537–47

107. Wen LT, Knowles AF. 2003. Extracellular ATP and adenosine induce cell apoptosis of human hepatoma Li-7A cells via the A3 adenosine receptor. *Br. J. Pharmacol.* 140:1009–18

108. Alola E, Perez-Etxebarria A, Kabre E, Fogarty DJ, Metioui M, et al. 1998. Activation by P2X$_7$ agonists of two phospholipases A$_2$ (PLA$_2$) in ductal cells of rat submandibular gland. *J. Biol. Chem.* 273:30208–17

109. Garcia-Marcos M, Pochet S, Marino A, Dehaye JP. 2006. P2X$_7$ and phospholipid signaling: the search of the "missing link" in epithelial cells. *Cell. Signal.* 18:2098–104

110. Pochet S, Garcia-Marcos M, Seil M, Otto A, Marino A, Dehaye JP. 2007. Contribution of two ionotropic purinergic receptors to ATP responses in submandibular gland ductal cells. *Cell. Signal.* 19:2155–64

99. P2X$_1$R knockout mice provide convincing evidence for their essential role in renal autoregulation via tubuloglomerular feedback.

105. Using P2X$_7$R knockout mice, this study provides evidence for the role of P2X$_7$R in collagen deposition and renal fibrosis in response to ureteral obstruction.

111. Doctor RB, Matzakos T, McWilliams R, Johnson S, Feranchak AP, et al. 2005. Purinergic regulation of cholangiocyte secretion: identification of a novel role for P2X receptors. *Am. J. Physiol.* 288:G779–86

112. Furness JB, Jones C, Nurgali K, Clerc N. 2004. Intrinsic primary afferent neurons and nerve circuits within the intestine. *Prog. Neurobiol.* 72:143–64

113. Galligan JJ. 2004. Enteric P2X receptors as potential targets for drug treatment of irritable bowel syndrome. *Br. J. Pharmacol.* 141:1294–302

114. Ren J, Bian X, DeVries M, Schnegelsberg B, Cockayne DA, et al. 2003. $P2X_2$ subunits contribute to fast synaptic excitation in myenteric neurons of the mouse small intestine. *J. Physiol.* 552:809–21

115. Bian X, Ren J, DeVries M, Schnegelsberg B, Cockayne DA, et al. 2003. Peristalsis is impaired in the small intestine of mice lacking the $P2X_3$ subunit. *J. Physiol.* 551:309–22

116. Ferrari D, Pizzirani C, Adinolfi E, Lemoli RM, Curti A, et al. 2006. The $P2X_7$ receptor: a key player in IL-1 processing and release. *J. Immunol.* 176:3877–83

117. Di Virgilio F. 2007. Liaisons dangereuses: $P2X_7$ and the inflammasome. *Trends Pharmacol. Sci.* 28:465–72

118. Solle M, Labasi J, Perregaux DG, Stam E, Petrushova N, et al. 2001. Altered cytokine production in mice lacking $P2X_7$ receptors. *J. Biol. Chem.* 276;125–32

119. King BF. 2007. Novel $P2X_7$ receptor antagonists ease the pain. *Br. J. Pharmacol.* 151:565–67

120. Donnelly-Roberts DL, Jarvis MF. 2007. Discovery of $P2X_7$ receptor-selective antagonists offers new insights into $P2X_7$ receptor function and indicates a role in chronic pain states. *Br. J. Pharmacol.* 151:571–79

121. McInnes IB, Snell NJ, Perrett JH, Parmar H, Wang MM, et al. 2007. Results of a Phase II clinical trial of a novel $P2X_7$ receptor antagonist, AZD9056, in patients with active rheumatoid arthritis (CREATE study). *Am. College Rheumatol.* Abstr. 2085

122. Placido R, Auricchio G, Falzoni S, Battistini L, Colizzi V, et al. 2006. $P2X_7$ purinergic receptors and extracellular ATP mediate apoptosis of human monocytes/macrophages infected with *Mycobacterium tuberculosis* reducing the intracellular bacterial viability. *Cell. Immunol.* 244:10–18

123. Pelegrin P, Barroso-Gutierrez C, Surprenant A. 2008. $P2X_7$ receptor differentially couples to distinct release pathways for IL-1β in mouse macrophage. *J. Immunol.* 180:7147–57

124. Saunders BM, Fernando L, Sluyter R, Britton WJ, Wiley JS. 2003. A loss-of-function polymorphism in the human $P2X_7$ receptor abolishes ATP-mediated killing of mycobacteria. *J. Immunol.* 171:5442–46

125. Fairbairn IP, Stober CB, Kumararatne DS, Lammas DA. 2001. ATP-mediated killing of intracellular mycobacteria by macrophages is a $P2X_7$-dependent process inducing bacterial death by phagosome-lysosome fusion. *J. Immunol.* 167:3300–7

126. Steinberg TH, Hiken JF. 2007. P2 receptors in macrophage fusion and osteoclast formation. *Purinergic Signal.* 3:53–57

127. Ke HZ, Qi H, Weidema AF, Zhang Q, Panupinthu N, et al. 2003. Deletion of the $P2X_7$ nucleotide receptor reveals its regulatory roles in bone formation and resorption. *Mol. Endocrinol.* 17:1356–67

128. Gartland A, Buckley KA, Hipskind RA, Perry MJ, Tobias JH, et al. 2003. Multinucleated osteoclast formation in vivo and in vitro by $P2X_7$ receptor-deficient mice. *Crit. Rev. Eukaryot. Gene Expr.* 13:243–53

129. Cruwys S, Midha A, Rendall E, McCormick M, Nicol A, et al. 2007. Antagonism of $P2X_7$ receptor attenuates joint destruction in a model of arthritis. *Am. Coll. Rheumatol.* Abstr. 1772

130. Sherman AN, Sluyter R, Fernando SL, Clarke A, Dao-Ung LP, et al. 2006. A Thr^{357} to Ser polymorphism in homozygous and compound heterozygous subjects causes absent or reduced $P2X_7$ function and impairs ATP-induced mycobacterial killing by macrophages. *J. Biol. Chem.* 281:2079–86

131. Mio K, Kubo Y, Ogura T, Yamamoto T, Sato C. 2005. Visualization of the trimeric $P2X_2$ receptor with a crown-capped extracellular domain. *Biochem. Biophys. Res. Commun.* 337:998–1005

132. Barrera NP, Ormond SJ, Henderson RM, Murrell-Lagnado RD, Edwardson JM. 2005. Atomic force microscopy imaging demonstrates that $P2X_2$ receptors are trimers but that $P2X_6$ receptor subunits do not oligomerize. *J. Biol. Chem.* 280:10759–65

133. Nicke A, Bäumert HG, Rettinger J, Eichele A, Lambrecht G, et al. 1998. $P2X_1$ and $P2X_3$ receptors form stable trimers: a novel structural motif of ligand-gated ion channels. *EMBO J.* 17:3016–28

134. Ding S, Sachs F. 1999. Single channel properties of $P2X_2$ purinoceptors. *J. Gen. Physiol.* 113:695–720

121. The first drug targeting a P2XR to make it to Phase II clinical trials, with initial positive results for inflammatory disease.

135. Roberts JA, Vial C, Digby HR, Agboh KC, Wen H, et al. 2006. Molecular properties of P2X receptors. *Pflüg. Arch.* 452:486–500

136. Agboh KC, Webb TE, Evans RJ, Ennion SJ. 2004. Functional characterization of a P2X receptor from *Schistosoma mansoni*. *J. Biol. Chem.* 279:41650–57

137. Fountain SJ, Parkinson K, Young MT, Cao L, Thompson CR, North RA. 2007. An intracellular P2X receptor required for osmoregulation in *Dictyostelium discoideum*. *Nature* 448:200–3

138. Fountain SJ, Cao L, Young MT, North RA. 2008. Permeation properties of a P2X receptor in the green algae *Ostreococcus tauri*. *J. Biol. Chem.* 283:15122–26

139. Freist W, Verhey JF, Stuhmer W, Gauss DH. 1998. ATP binding site of P2X channel proteins: structural similarities with class II aminoacyl tRNA-synthetases. *FEBS Lett.* 434:61–65

140. Boué-Grabot E, Archambault V, Séguéla P. 2000. A protein kinase C site highly conserved in P2X subunits controls the desensitization kinetics of P2X$_2$ ATP-gated channels. *J. Biol. Chem.* 275:10190–95

141. Ennion SJ, Evans RJ. 2002. P2X$_1$ receptor subunit contribution to gating revealed by a dominant negative PKC mutant. *Biochem. Biophys. Res. Commun.* 291:611–16

142. Jiang LH, Rassendren F, Surprenant A, North RA. 2000. Identification of amino acid residues contributing to the ATP-binding site of a purinergic P2X receptor. *J. Biol. Chem.* 275:34190–96

143. Haines WR, Migita K, Cox JA, Egan TM, Voigt MM. 2001. The first transmembrane domain of the P2X receptor subunit participates in the agonist-induced gating of the channel. *J. Biol. Chem.* 276:32793–98

144. Haines WR, Voigt MM, Migita K, Torres GE, Egan TM. 2001. On the contribution of the first transmembrane domain to whole-cell current through an ATP-gated ionotropic P2X receptor. *J. Neurosci.* 21:5885–92

145. Li Z, Migita K, Samways DS, Voigt MM, Egan TM. 2004. Gain and loss of channel function by alanine substitutions in the transmembrane segments of the rat ATP-gated P2X$_2$ receptor. *J. Neurosci.* 24:7378–86

146. Samways DS, Migita K, Li Z, Egan TM. 2008. On the role of the first transmembrane domain in cation permeability and flux of the ATP-gated P2X$_2$ receptor. *J. Biol. Chem.* 283:5110–17

147. Silberberg SD, Chang TH, Swartz KJ. 2005. Secondary structure and gating rearrangements of transmembrane segments in rat P2X$_4$ receptor channels. *J. Gen. Physiol.* 125:347–59

148. Jiang LH, Rassendren F, Spelta V, Surprenant A, North RA. 2001. Amino acid residues involved in gating identified in the first membrane-spanning domain of the rat P2X$_2$ receptor. *J. Biol. Chem.* 276:14902–8

149. Roberts JA, Evans RJ. 2004. ATP binding at human P2X$_1$ receptors. Contribution of aromatic and basic amino acids revealed using mutagenesis and partial agonists. *J. Biol. Chem.* 279:9043–55

150. Cao L, Young MT, Broomhead HE, Fountain SJ, North RA. 2007. A Thr[339]-to-serine substitution in rat P2X$_2$ receptor second transmembrane domain causes constitutive opening and indiates a gating role for Lys[308]. *J. Neurosci.* 27:12916–23

151. Wilkinson WJ, Jiang LH, Surprenant A, North RA. 2006. Role of ectodomain lysines in the subunits of the heteromeric P2X$_{2/3}$ receptor. *Mol. Pharmacol.* 70:1159–63

152. Clyne JD, Wang LF, Hume RI. 2002. Mutational analysis of the conserved cysteines of the rat P2X$_2$ purinoceptor. *J. Neurosci.* 22:3873–80

153. Ennion SJ, Evans RJ. 2002. Conserved cysteine residues in the extracellular loop of the human P2X$_1$ receptor form disulfide bonds and are involved in receptor trafficking to the cell surface. *Mol. Pharmacol.* 61:303–11

154. Nagaya N, Tittle RK, Saar N, Dellal SS, Hume RI. 2005. An intersubunit zinc binding site in rat P2X$_2$ receptors. *J. Biol. Chem.* 280:25982–93

155. Marquez-Klaka B, Rettinger J, Bhargava Y, Eisele T, Nicke A. 2007. Identification of an intersubunit cross-link between substituted cysteine residues located in the putative ATP binding site of the P2X$_1$ receptor. *J. Neurosci.* 27:1456–66

156. Roberts JA, Digby HR, Kara M, El Ajouz S, Sutcliffe MJ, et al. 2008. Cysteine substitution mutagenesis and the effects of methanethiosulfonate reagents at P2X$_2$ and P2X$_4$ receptors support a core common mode of ATP action at P2X receptors. *J. Biol. Chem.* 283:20126–36

157. Yan Z, Liang Z, Obsil T, Stojilkovic SS. 2006. Participation of the Lys[313]-Ile[333] sequence of the purinergic P2X$_4$ receptor in agonist binding and transduction of signals to the channel gate. *J. Biol. Chem.* 281:32649–59

137. A P2X receptor homolog in a simple eukaryote has an important function as intracellular vacuolar protein.

158. Wilkinson WJ, Jiang LH, Surprenant A, North RA. 2006. Role of ectodomain lysines in the subunits of the heteromeric $P2X_{2/3}$ receptor. *Mol. Pharmacol.* 70:1159–63

159. Roberts JA, Evans RJ. 2007. Cysteine substitution mutants give structural insight and identify ATP binding and activation sites at P2X receptors. *J. Neurosci.* 27:4072–82

160. Rassendren F, Buell G, Newbolt A, North RA, Surprenant A. 1997. Identification of amino acid residues contributing to the pore of a P2X receptor. *EMBO J.* 16:3446–54

161. Egan TM, Haines WR, Voigt MM. 1998. A domain contributing to the ion channel of ATP-gated $P2X_2$ receptors identified by the substituted cysteine accessibility method. *J. Neurosci.* 18:2350–59

162. Migita K, Haines WR, Voigt MM, Egan TM. 2001. Polar residues of the second transmembrane domain influence cation permeability of the ATP-gated $P2X_2$ receptor. *J. Biol. Chem.* 276:30934–41

163. Egan TM, Khakh BS. 2004. Contribution of calcium ions to P2X channel responses. *J. Neurosci.* 24:3413–20

164. Samways DS, Egan TM. 2007. Acidic amino acids impart enhanced Ca^{2+} permeability and flux in two members of the ATP-gated P2X receptor family. *J. Gen. Physiol.* 129:245–56

165. Duckwitz W, Hausmann R, Aschrafi A, Schmalzing G. 2006. $P2X_5$ subunit assembly requires scaffolding by the second transmembrane domain and a conserved aspartate. *J. Biol. Chem.* 281:39561–72

166. Chaumont S, Jiang LH, Penna A, North RA, Rassendren F. 2004. Identification of a trafficking motif involved in the stabilization and polarization of P2X receptors. *J. Biol. Chem.* 279:29628–38

167. Jiang LH, Rassendren F, Mackenzie A, Zhang YH, Surprenant A, et al. 2005. *N*-Methyl-ᴅ-glucamine and propidium dyes utilize different permeation pathways at rat $P2X_7$ receptors. *Am. J. Physiol. Cell Physiol.* 289:C1295–302

168. Masin M, Kerschensteiner D, Dümke K, Rubio ME, Soto F. 2006. Fe65 interacts with $P2X_2$ subunits at excitatory synapses and modulates receptor function. *J. Biol. Chem.* 281:4100–8

169. Wilson HL, Wilson SA, Surprenant A, North RA. 2002. Epithelial membrane proteins induce membrane blebbing and interact with the $P2X_7$ receptor C terminus. *J. Biol. Chem.* 277:34017–23

170. Denlinger LC, Fisette PL, Sommer JA, Watters JW, Prabhu U, et al. 2001. The nucleotide receptor $P2X_7$ contains multiple protein- and lipid-interaction motifs including a potential binding site for bacterial lipopolysaccharide. *J. Immunol.* 167:1871–76

171. Denlinger LC, Sommer JA, Parker K, Gudipaty L, Fisette PL, et al. 2003. Mutation of a dibasic amino acid motif within the C terminus of the $P2X_7$ nucleotide receptor results in trafficking defects and impaired function. *J. Immunol.* 171:1304–11

172. Pelegrin P, Surprenant A. 2006. Pannexin-1 mediates large pore formation and interleukin-1β release by the ATP-gated $P2X_7$ receptor. *EMBO J.* 25:5071–82

173. Petrilli V, Dotert C, Muruve DA, Tschopp J. 2007. The inflammasome: a danger sensing complex triggering innate immunity. *Curr. Opin. Immunol.* 19:615–22

Activation of the Epithelial Sodium Channel (ENaC) by Serine Proteases

Bernard C. Rossier[1] and M. Jackson Stutts[2]

[1] Department of Pharmacology and Toxicology, University of Lausanne, CH-1005 Lausanne, Switzerland; email: Bernard.Rossier@unil.ch

[2] Cystic Fibrosis/Pulmonary Research and Treatment Center, University of North Carolina, Chapel Hill, North Carolina 27599-7248

Annu. Rev. Physiol. 2009. 71:361–79

First published online as a Review in Advance on October 17, 2008

The *Annual Review of Physiology* is online at physiol.annualreviews.org

This article's doi: 10.1146/annurev.physiol.010908.163108

Copyright © 2009 by Annual Reviews. All rights reserved

0066-4278/09/0315-0361$20.00

Key Words

sodium transport, channel-activating proteases, prostasin, furin, serpin

Abstract

The study of human monogenic diseases [pseudohypoaldosteronism type 1 (PHA-1) and Liddle's syndrome] as well as mouse models mimicking the salt-losing syndrome (PHA-1) or salt-sensitive hypertension (Liddle's syndrome) have established the epithelial sodium channel ENaC as a limiting factor in vivo in the control of ionic composition of the extracellular fluid, regulation of blood volume and blood pressure, lung alveolar clearance, and airway mucociliary clearance. In this review, we discuss more specifically the activation of ENaC by serine proteases. Recent in vitro and in vivo experiments indicate that membrane-bound serine proteases are of critical importance in the activation of ENaC in different organs, such as the kidney, the lung, or the cochlea. Progress in understanding the basic mechanism of proteolytic activation of ENaC is accelerating, but uncertainty about the most fundamental aspects persists, leaving numerous still-unanswered questions.

ENaC IS A LIMITING FACTOR IN THE CONTROL OF IONIC COMPOSITION OF THE EXTRACELLULAR FLUID

In aldosterone-responsive epithelial cells of kidney and colon, the epithelial sodium channel (ENaC) plays a critical role in the control of sodium balance, blood volume, and blood pressure (1). In glucocorticoid-responsive epithelia, as in distal lung airways, ENaC has a distinct role in controlling fluid reabsorption at the air-liquid interface, thereby determining the rate of mucociliary transport (2). In the inner ear, ENaC may play another important role in controlling the unusual ionic composition of the endolymph ($Na^+ < 1$ mM, $K^+ > 140$ mM), which is required for mechanotransduction in hair cells (3). ENaC is made of three subunits (α, β, γ) that are 30% homologous at the protein level (4). The membrane topology of each subunit predicts the presence of two transmembrane domains with short cytoplasmic N and C termini and a large extracellular loop. ENaC is characterized by high Na^+ selectivity ($P_{Na}/P_K > 20$), a low unitary conductance (4–5 pS in the presence of Na^+), gating kinetics characterized by long closing and opening times, and a high sensitivity to amiloride (K_i: 0.1 μM). Recent reviews address the questions of the structure-function relationship of ENaC and of ENaC regulation (5, 6).

Tissue-specific expression of ENaC occurs in Na^+-transporting epithelia sensitive to aldosterone. ENaC subunits are found in the aldosterone-sensitive distal nephron (ASDN), in the surface epithelia of the colon, and in the duct cells of exocrine glands. ENaC is expressed in epithelial cells of nontransporting epithelia (keratinocytes and hair follicles), which are mineralocorticoid responsive but apparently not involved in overall Na^+ balance (7). Aldosterone-dependent ENaC expression in taste buds suggests an important role of ENaC in salt tasting (8). ENaC is expressed in distal and proximal airways as well as in nasal mucosa (9). However, expression here is not under mineralocorticoid control but rather un-

der glucocorticoid control (10). In these tissues, ENaC participates in the control of the extracellular fluid at the air-cell interface by regulating the volume of the airway surface liquid (ASL) and mucociliary clearance (11). Lung epithelial fluid transport may be important in the resolution of pulmonary edema (12). In humans, imbalance of ENaC activity leads to a large variety of pathologies. Abnormally increased ENaC activity may lead to hypertension (Liddle's syndrome) in ASDN or contribute to decreased mucociliary transport in lung, as observed in cystic fibrosis (13). An abnormal decrease in ENaC activity may lead to severe renal salt-losing syndromes with a hypotensive phenotype [pseudohypoaldosteronism type 1 (PHA-1)] (14) and/or to respiratory distress syndrome in premature newborns (15).

Despite large changes in water and salt intake, the kidney is able to maintain extracellular osmolarity and volume within narrow margins (1). Such fine control requires specific factors or hormones: Among them, aldosterone and vasopressin play key roles. Hormones or various regulatory factors regulate ENaC activity by modifying the number of channels expressed at the cell surface (N) and/or channel open probability (P_o) and/or channel unitary ionic conductance (16). No experimental evidence supports the idea that ENaC activity is regulated through a change in the single-channel conductance. We are therefore left with the possibility of changing N and/or P_o. The experimental evidence for the specific control of N by aldosterone (17–20) and vasopressin (21) is now quite strong. A number of cytoplasmic factors [e.g., Na^+, pH, Ca^{2+}, protein kinase A (PKA), actin filaments] or external factors (e.g., self-inhibition by Na^+) affect P_o in patch-clamp experiments or in bilayer experiments (see review in References 5 and 6). The relevance of these effects to the regulation of ENaC in intact native cells remains, however, to be established. Naturally occurring mutations and structure-function relationship studies in the *Xenopus* oocyte system indicate that at least two protein domains are involved in ENaC gating (6). Mutation of an absolutely conserved glycine in the N terminus of the β subunit causes a severe pseudohypoaldosteronism (14). The mutation does not change the number of channels expressed at the plasma membrane but decreases the whole-cell P_o by roughly 50% (22). A second gating domain was identified through genetic screening in *Caenorhabditis elegans* of the ENaC/degenerin gene family. Mutations in *mec* genes at the degenerin site, located in the large extracellular loop preceding the second transmembrane domain M2, lead to a permanent opening of this mechanosensitive channel, causing cell swelling and neuronal death (23). A corresponding mutation in ENaC causes a threefold increase in P_o, an effect that is due to a shortening of channel closed times and an increase in the number of long openings [see review (6)].

ENaC AND ITS ACTIVATION BY SERINE PROTEASES

Serine proteases (SPs) (24) compose a very large gene family (~2% of identified genes) (25). The functions of SPs are extraordinarily diverse and important: proteolysis of food proteins in the intestine, blood clotting and fibrinolysis, humoral and cellular immunity, embryonic development and neuronal plasticity, extracellular matrix remodeling, hormone maturation, apoptosis, and fertilization, to mention a few. In the past decade, it has become clear that partial proteolysis of ENaC is associated with dramatically stimulated channel activity.

Placing this proteolysis in a physiological context is challenging because a variety of SPs affect ENaC. So far, ENaC-activating proteases belong to the PA or SB clans (26) of evolutionally related SPs. ENaC is regulated by some type II transmembrane serine proteases (TTSPs), a rapidly expanding PA clan S1 subfamily of cell surface–associated SPs including enterokinase, corin, human airway trypsin-like proteases, the human TMPRSS subfamily, and matriptase (MT-SP1) (27).

Furin-like protein convertases that may play a critical role in ENaC activity belong to the SB clan, which also includes subtilisin and kexin (28). Soluble proteases that stimulate ENaC are from the PA clan, family S1, and include trypsin, chymotrypsin, and elastase.

Through use of the same functional complementation assay (29) of the α subunit that had been used to clone the β and γ subunits of ENaC (4), an upregulator of αβγ ENaC was detected (30), subsequently identified as a SP, and named channel-activating protease-1 (CAP-1) (30, 31). Indirect evidence showed that CAP-1 was a glycosyl-phosphatidylinositol (GPI)-anchored protein and, as such, was presumably targeted to the apical membrane of epithelial cells (32, 33). *Xenopus* CAP-1 and mouse CAP-1 are orthologs of human prostasin, purified from seminal fluid (34) and cloned from human prostate (35). It was assumed that prostasin was a secreted SP, but evidence for a GPI anchor of prostasin was subsequently provided (36). These findings define a class of membrane-bound SPs (anchored only in the external leaflet of the membrane bilayer) and a novel regulatory pathway, which is highly conserved throughout evolution from *Xenopus* (xCAP-1) (30) to mouse (31), rat (37), and human (34). Subsequently, two additional SPs, mCAP-2 [homolog of human transmembrane protease serine 4 (TMPRSS4)] and mCAP-3 (MT-SP1/ matriptase or epithin), were identified by homology cloning and were found to increase the activity of ENaC coexpressed in *Xenopus* oocytes (38). Each of these membrane-bound SPs can activate ENaC by four- to sevenfold, depending on the experimental conditions; the effect is always an increase in P_o, without any significant change in N (38). These CAPs are members of an emerging subgroup of S1 proteases that are attached to plasma membranes either by a C-terminal transmembrane domain (type I), by a GPI-anchored linkage, or by an N-terminal proximal transmembrane domain (type II, or TTSP) (39). TTSPs share common structural features: a short N-terminal cytoplasmic tail, a single transmembrane domain, and a C-terminal SP domain facing the extracellular milieu. Between the transmembrane domain and the catalytic domain lies a linker region consisting of modular protein interaction domains (e.g., CUB, LDLR, group A scavenger receptor domain), suggesting an association of TTSPs with other proteins at the cell surface (27). A TTSP is typically synthesized as a zymogen with an inhibitory domain that is removed at the cell surface (40). The cell biology of relevant TTSPs may provide clues for understanding the molecular events underlying the regulation of ENaC by proteases. In the *Xenopus* oocyte expression system, cell surface expression of CAP-1 is required for activation of ENaC (33). When the GPI-anchored consensus motif is mutated, CAP-1 is secreted, and ENaC is no longer activated. Interestingly, catalytic mutants of CAP-1 significantly decreased ENaC activation but did not fully abolish the effect of xCAP-1. The data indicated the critical role of the GPI anchor in ENaC activation and suggested that catalytic and noncatalytic mechanisms are involved (33). A noncatalytic mechanism was also observed for mouse CAP-1 because a triple HDS catalytic dead mutant was still able to fully activate ENaC in the oocyte expression system (41).

Investigators have been aware of protease effects on ENaC activity since the mid-1990s. However, Hughey, Kleyman, and colleagues (42) were the first to recognize canonical cleavage sequences of furin-like protein convertases in the putative extracellular domains of the α and γ ENaC subunits. These authors determined that cleavage at these sites generated ENaC fragments that they had earlier linked to ENaC processing (43). A total of seven mammalian furin-like enzymes have been recognized (44). Furin, PACE4, PC5/6B, and PC7 are distinguished from the others in containing a transmembrane domain and sorting to the constitutive secretory pathway. Work so far has demonstrated a role of furin in ENaC cleavage and activity, but other related convertases may participate redundantly or in a tissue-specific manner. The intracellular loca-

tion of furin and its own processing have been intensely studied (45). Furin becomes active in the trans-Golgi network (TGN) but cycles between the cell surface and endocytic compartments (46). Identifying the points of ENaC exposure to convertases may be informative for understanding the degree and purpose of ENaC proteolysis prior to delivery to the cell surface. Membrane association is not required for SPs to regulate ENaC. S1 family proteases of the PA clan, including trypsin and elastase, activate ENaC from the extracellular mileu (47, 48). The activation of ENaC caused by CAP-1 coexpression in *Xenopus* oocytes can be mimicked by the external addition of trypsin, suggesting that both CAP-1 and trypsin act via the same mechanism. Outside-out patch-clamp experiments show that trypsin and elastase activate near-silent channels by dramatically increasing their P_o (48, 49). Appropriate SP inhibitors block these increases in P_o (30, 31). ENaC activation by application of SPs to the external face of excised membrane patches implies that channels that have not undergone proteolytic processing can reside at the cell surface and can be activated by encountering the appropriate active protease. With diverse SPs targeting ENaC both in intracellular compartments and at the cell surface, proteolytic activation may be the default state for ENaC.

In many epithelia (50–57) ENaC appears to be constitutively activated through the presence of endogenously expressed SPs because basal transepithelial Na^+ transport cannot be further activated by the addition of trypsin. However, the addition of aprotinin on the apical side of the epithelium typically inhibits baseline Na^+ transport. Exposure to trypsin can reverse this inhibition. In *Xenopus* kidney cells, aprotinin can block up to 90% of amiloride-sensitive electrogenic Na^+ transport (30, 56), whereas the mouse mpk-CCD$_{C14}$ cell line, derived from the cortical collecting duct, appears to be only 50% sensitive to aprotinin (31). In human bronchial epithelial cells, aprotinin and the humanized version of aprotinin (BAY-39–9437) inhibited basal I_{Na} current by approximately 30% (53). These findings suggested that ENaC activation either is achieved by a constitutive SP-independent mechanism or, alternatively, depends on more than one SP with different sensitivities to aprotinin and acting in combination within the same cells. That ENaC in native epithelia does not respond to trypsin indicates that SPs are not prevented from activating ENaC by cognate serpins, at least at steady state. However, some studies have found important effects of serpin expression on proteolytic regulation of ENaC (58, 59), perhaps indicating model system or tissue variability in SP-serpin balance for regulating ENaC.

ENaC AS A SUBSTRATE FOR SERINE PROTEASES AND ITS FUNCTIONAL CONSEQUENCES: RECENT DEVELOPMENTS

Conceptual Framework

One of the most relevant questions in this field is to identify the substrate(s) for the various SPs that have been implicated in ENaC activation. Recent reviews have also addressed this question (60–62). **Figure 1** shows two models for how SPS may act on ENaC: the colocalization model and the sequential model.

In both models, ENaC is postulated to be a substrate for CAP. Because CAPs express their

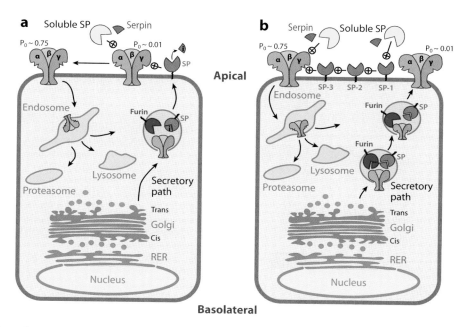

Figure 1

Schematic models of proteolytic regulation of ENaC. (*a*) The colocalization model emphasizes the presence of consensus recognition and cleavage sequences for furin and furin-like convertases in α and γ ENaC. Convertase(s) and substrate (ENaC) are expected to cotraffic along the same trans-Golgi path, potentially allowing partially proteolized ENaC heteromers to be delivered to the cell surface. At the cell surface, channel-activating serine proteases (SPs) become activated and, if appropriately colocalized, can complete proteolytic activation. Alternatively, certain soluble serine proteases present in the extracellular milieu can activate near-silent ENaC under the influence of soluble serine protease inhibitors (serpins). (*b*) The sequential model elaborates on the colocalization implicit in proteases directly cleaving ENaC to include the further possibility that a protease or proteases may be essential for activating ENaC but may act indirectly by participating in a protease cascade. As depicted, ENaC may be partially processed by convertases and then fully activated by a protease that requires activation by other membrane-bound or soluble proteases.

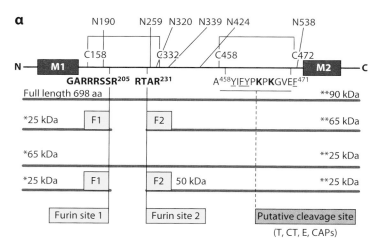

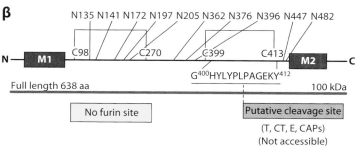

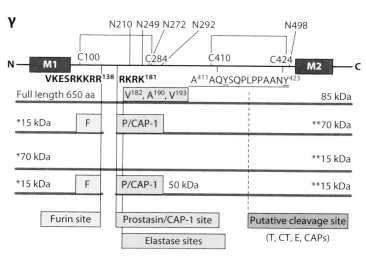

catalytic domain in the extracellular milieu, the large extracellular domain of ENaC (60 kDa) expresses cleavage sites for these proteases. However, proteolysis must be highly restricted because ENaC is activated but not degraded in the presence of external SPs such as trypsin, chymotrypsin (47), or elastase (49), whereas other SPs such as thrombin (47, 63) or kallikrein (47) are unable to activate the channel.

We therefore propose that the extracellular domain must be folded in such a way that the access of the enzyme to the cleavage site is limited and specific (**Figure 2**). Disulfide bridges [α: C158-C332 and C458-C472 (**Figure 2**, top);

Figure 2

Linear models for ENaC subunit cleavage. Each subunit has common features, including two transmembrane domains (M1 and M2) with short N and C termini, a large extracellular loop with 5–12 N-glycosylation sites, and multiple cysteines that participate in disulfides bridges critical for folding, trafficking, and/or activity. Known and suspected sites of ENaC cleavage are indicated. Molecular masses (kDa) of nonglycosylated fragments consistent with cleavage at the indicated sites, and that have been detected by N-terminal (*asterisks*) or C-terminal antibodies (*double asterisks*), are indicated. (*Top*) In the α subunit model, within the first cysteine disulfide bridge (C158-C332) two furin cleavage sites (F) are identified (*yellow boxes*). Putative trypsin (T), chymotrypsin (CT), elastase (E), or SPs within the second cysteine disulfide bridge (C458-C472) are indicated (*green box*). These sites may generate a 65–70-kDa fragment as detected by an N-terminal antibody and as observed in oocytes, kidney, and lung. (*Middle*) In the β subunit model, furin cleavage sites (F) are absent. Putative trypsin (T), chymotrypsin (CT), elastase (E), or CAP-induced cleavage sites are postulated to be protected by heavy glycosylation (ten potential N-glycosylation sites) and/or by the absence of conformational changes induced by furin. (*Bottom*) In the γ subunit model, only one furin cleavage site (F) is observed within the first cysteine disulfide bridge (C100-C284) (*yellow*). A CAP-1/prostasin cleavage site (*light blue box*) and elastase cleavage sites (*pink box*) are indicated. Putative trypsin (T), chymotrypsin (CT), elastase (E), or SPs within the second cysteine disulfide bridge (C410-C424) are indicated, but there are no experimental data confirming that they exist in vivo or in vitro.

γ: C100-C284 and C410-C424 (**Figure 2**, bottom)] have been identified in the extracellular domains of ENaC subunits (64) and may play in this respect a critical role, as shown in **Figure 2**. Recently, Sheng et al. (65) identified additional cysteines in the cysteine-rich domain of the extracellular loop that could form disulfide bridges. Cleavage within these loops theoretically can induce conformational changes, permitting proper ENaC subunit assembly (ER), trafficking (ER to plasma membrane), or activation (the P_o effect) at the apical membrane or degradation (lysosome or proteasome), compatible with the colocalization model (**Figure 1***a*) or, more likely, the sequential model (**Figure 1***b*). Cleavage of ENaC at furin sites on α and γ subunits may occur in intracellular compartments and result in the delivery of active ENaC onto the cell surface.

In Vitro Evidence of ENaC Proteolytic Cleavage and Activation

Hughey et al. (43) provided the first direct evidence of ENaC cleavage. They found that when tagged ENaC subunits were expressed alone in either Chinese hamster ovary or Madin-Darby canine kidney (MDCK) type 1 epithelia, the α subunit (95 kDa), β subunit (96 kDa), and γ subunit (93 kDa) each produced a single band on SDS gels by immunoblotting. However, coexpression of αβγ ENaC subunits revealed a second band for each subunit (65 kDa for α, 110 kDa for β, and 75 kDa for γ). The smaller size of the processed α and γ subunits was consistent with proteolytic cleavage. Hughey et al. (42) have subsequently implicated furin in ENaC proteolytic processing. They demonstrated that the α and γ ENaC subunits may be cleaved during maturation at consensus sites for furin cleavage. Using site-specific mutagenesis of the key P1 arginine in the R-X-X-R(P1) furin sequences in the α and γ subunits, ENaC expression in furin-deficient cells, and furin-specific inhibitors, the authors proposed that ENaC cleavage correlates with channel activity (42). Their analysis of ENaC subunits with furin sites mutated indicated a dominant con-

tribution of the two α subunit sites to ENaC basal current (I_{Na}), and a much smaller effect of the single γ subunit furin site. Harris et al. (67) used labeling of cell surface–expressed proteins (66) and differential expression of ENaC subunits to obtain a semiquantitative correlation of channel cleavage and amiloride-sensitive ENaC-mediated Na$^+$ transport (67). The α subunit alone (or in combination with the β and/or γ subunits) was efficiently transported to the cell surface; α ENaC contained in wild-type αβγ ENaC appeared at the cell surface as the predicted full-length (95-kDa) molecule and a 65-kDa fragment, whereas α(furin mutant) ENaC appeared only as full length. I_{Na} of α(furin mutant)βγ was decreased by 60%. The γ subunit expressed alone did not appear at the cell surface, γ coexpressed with α reached the surface but was not detectably cleaved, and γ in αβγ complexes appeared mainly as a 76-kDa species in the surface pool. When the γ(furin mutant) was expressed with α and β, γ ENaC appeared in the cell surface pool as the 93-kDa full-length protein. However, basal I_{Na} was similar to I_{Na} of αβγ. Thus, furin-mediated cleavage was not essential for participation of α or γ in αβγ heteromers, and basal I_{Na} was reduced by preventing furin-mediated cleavage of the α, but not the γ, subunit. Residual current in the absence of furin-mediated proteolysis may be due to nonfurin endogenous proteases, and a simple correlation between furin-mediated cleavage and ENaC activation could not be established (68).

Other approaches indirectly supported the importance of the α furin sites in proteolytic regulation of ENaC. For example, Carattino et al. (69) found that ENaC is inhibited by a peptide derived from proteolytic processing of the α subunit. Channels bearing an engineered α subunit lacking both furin consensus sites and the intervening 26-residue sequence exhibited normal activity in *Xenopus* oocytes. A synthetic version of the 26-residue tract inhibited wild-type ENaC. More recently, the same authors proposed that a eight-residue tract between the α furin sites is a key conserved inhibitory domain that provides epithelial cells with a reserve of inactive channels that can be activated as required by proteases (70). Sheng et al. (72) provided evidence of a different contribution of cleavage at the α subunit furin to normal activity of ENaC. They proposed that activation resulted partly by relieving the channel of Na$^+$ self-inhibition (72). Interestingly, approximately half of the extracellular cysteine residues in the α and γ ENaC subunits are required to establish the tertiary structure that ensures a proper Na$^+$ self-inhibition response, likely by the formation of multiple intrasubunit disulfide bonds (72). Chraibi & Horisberger (73) showed that Na$^+$ self-inhibition is an intrinsic property of Na$^+$ channels resulting from the coexpression of the α, β, and γ subunits in the *Xenopus* oocyte expression system. Extracellular Na$^+$-dependent inactivation had a large activation energy and could be abolished by treatment with extracellular proteases. Taken together, the data from Sheng et al. (72) and Chraibi & Horisberger (73) suggest that self-inhibition and protease activation have a common denominator and that the disulfide bridges play a critical role in this process, consistent with the general working hypothesis (**Figure 2**).

Some recent studies reveal a greater than previously expected importance of cleavage of the γ subunit in proteolytic regulation of ENaC. Bruns et al. (71) identified a CAP-1/prostasin cleavage site in the γ subunit at a tetrabasic site (RKRK) 39 amino acids distal to the γ subunit furin consensus site. Coexpression of ENaC and prostasin produced channels with high P$_o$. Protease activation was blocked by the mutation of either the furin cleavage site or the prostasin cleavage site in the γ subunit. A CAP-1/prostasin catalytic mutant (S238A) was unable to activate ENaC mutated at the prostasin tetrabasic site (the RKRK^{186} cleavage site) but was able to fully activate the wild-type channel [confirming Andreasen et al.'s (41) finding] and, surprisingly, induced γ subunit cleavage. Altogether, a simple enzyme-substrate reaction (involving CAP-1/prostasin) cannot explain the data of Andreasen et al. (41) and Bruns et al. (71). Other enzymes (e.g., a proteolytic cascade involving heterocatalytic activa-

tion of CAPs) and/or associated proteins (serpins or other factors) must be involved in ENaC activation, but they have not yet been identified (see **Figure 1**). These uncertainties aside, the same lab has continued with the model that proteases activate ENaC by excising an inhibitory sequence in the γ subunit (74). This group reports that full activation was obtained by dual cleavage in γ ENaC between the furin site and the polybasic tract (74). Surprisingly, given earlier results supporting α ENaC as the most important subunit for proteolytic activation (70), α ENaC cleavage was not required (74). Diadov et al. (75) recently implicated the same basic tract in γ ENaC (RKRK^{182} in rat γ ENaC) in proteolytic regulation of channel gating. These authors observed a gain-of-function effect from mutating the terminal lysine in the tract and speculated that the new sequence was targeted by endogenous furin-like convertases. In these researchers' studies, mutation of the α ENaC furin sites did not interfere with channel activation. Garcia-Caballero et al. (76) found that CAP-2/TMPRSS4 stimulates ENaC by direct cleavage of γ ENaC. Although CAP-2/TMPRSS4 cleaved the α, β, and γ subunits at multiple sites, the only cleavage that mediated increased I$_{Na}$ was at the γ ENaC furin site. In contrast to CAP-1/prostasin, which can fully activate ENaC with its catalytic activity disabled, CAP-2/TMPRSS4 catalytic activity was required. Interestingly, CAP-2/TMPRSS4 fully activated ENaC when the rat γ ENaC RKRK tract was replaced by QQQQ, leaving some question about the necessity of a second cleavage at this site. It is difficult to reconcile earlier studies demonstrating that preventing cleavage at the α ENaC furin sites substantially decreased ENaC basal currents (43, 67) with the recent studies showing that cleavage of γ ENaC alone is required for full activation by CAPs (74, 76), trypsin (75), or elastase (67). This difficulty suggests that a significant gap remains in our understanding of the mechanistic events by which proteases affect the activity of ENaC.

A significant aspect of proteolytic regulation of ENaC is how it interfaces with other recognized regulatory pathways. Knight et al. (77) showed that Liddle's syndrome mutations increased Na^+ transport through dual effects on epithelial Na^+ channel surface expression (increased N) and proteolytic cleavage (increased P_o), a dual effect already proposed by Firsov et al. (78) in 1996. Knight et al. (77) showed that the Liddle mutations increased ENaC expression at the cell surface and the fraction of ENaC at the cell surface that was cleaved (active). This disproportionate increase in cleavage was reproduced by expression of a dominant-negative Nedd4–2 or mutation of ENaC ubiquitination sites, interventions that disrupt ENaC endocytosis and lysosomal degradation. Conversely, overexpression of Nedd4–2 had the opposite effect, decreasing the fraction of cleaved ENaC at the cell surface. Thus, the data not only suggested that Nedd4–2 regulates epithelial Na^+ transport in part by controlling the relative expression of cleaved and uncleaved ENaC at the cell surface but also provided a mechanism by which Liddle's syndrome mutations alter ENaC activity (77). Recently Ruffieux-Daidié et al. (79) showed that reversal of Nedd4–2 ubiquitylation of ENaC occurs and involves the aldosterone-induced deubiquitylating enzyme Usp2–45. They showed in Hek293 cells stably transfected with ENaC that ENaC ubiquitylation and deubiquitylation are related to each other. Usp2–45 interacted with ENaC, deubiquitylated the channel, and caused a more-than-20-fold stimulation of ENaC activity. These events were accompanied by a modest increase in cell surface expression of ENaC and strong proteolytic cleavage of the α and γ ENaC subunits at their extracellular loops. Collectively, the data suggest a novel multistep mechanism for channel activation in which the aldosterone-induced Usp2–45 deubiquitylates ENaC, thereby activating ENaC via ENaC proteolysis and interfering with its internalization; this result in turn favors ENaC cleavage.

In the experiments described above, the enzyme-substrate interactions were observed at steady state, i.e., by expressing ENaC in various expression systems (*Xenopus* oocytes, MDCK,

HEK293, Fischer rat thyroid cells, etc.) that endogenously expressed SPs (e.g., furin) or, alternatively, by coexpressing ENaC and CAP-1/prostasin or ENaC regulatory proteins (e.g., Nedd4–2, Usp2–45). The phenotype is observed at steady state, i.e., 24 to 48 h after injection of the cRNAs (oocytes), a few days after transient infection, or a few weeks after stable cDNA transfection. Such protocols establish a correlation between cleaved substrate and ENaC activation (increased P_o). To establish a causal relationship requires acute exposure of ENaC (substrate) to the SP and a strict time- and dose-dependent correlation between ENaC cleavage at the cell surfaces and changes in P_o or I_{Na}. Caldwell et al. were the first to demonstrate a dramatic sixfold increase in P_o within a couple minutes of exposing outside-out patch membranes to either trypsin (48) or elastase (49). Performing cell surface biotinylation experiments, Harris et al. (67) showed that cell surface γ ENaC (but not α or β ENaC) exposed to human neutrophil elastase for 2 min was cleaved (as a 67-kDa fragment, distinct from the 75-kDa furin-mediated fragment) and correlated semiquantitatively with increased I_{Na}. Thus, the elastase-induced exogenous proteolysis pattern was distinct from the endogenous proteolysis pattern induced upon preferential assembly (68), suggesting that a causal relationship exists between γ ENaC cleavage and ENaC activation at the plasma membrane, both taking place within *2* min. Recently, Adebamiro et al. (63) confirmed and extended these findings by showing that the γ subunit is the specific substrate for neutrophil and pancreatic elastase (63). The authors demonstrated that γ subunit valines 182 and 193 are critical for channel activation by neutrophil elastase, whereas alanine 190 is critical for channel activation by pancreatic elastase. Strikingly, ENaC [which is resistant to the SP thrombin (47)] can be made thrombin sensitive by insertion of a novel thrombin consensus sequence within the γ subunit. This is compelling evidence for the importance of this segment in specific cleavage. Unfortunately, the cell surface cleaved fragments were not measured, preventing the establishment of a direct relationship between cleavage and activation.

The activation of SPs in the intracellular compartment or in the extracellular milieu must be tightly controlled. One level of control is achieved by transcriptional/translational regulation, by protease activation, and by degradation of mature enzyme. A second level of regulation is achieved by serine protease inhibitors (SPIs), which play a major role in this control (80). It is thus not surprising that severe genetic and acquired diseases can be caused by dysfunction of either SPs or SPIs. Serpins are endogenous SPIs. Protease nexin-1 (PN-1) associates with prostasin by binding irreversibly to the catalytic site. Wakida et al. (81) showed that the expression of PN-1 substantially decreased (by 70%) CAP-1/prostasin-induced I_{Na} in oocytes. The effect of PN-1 was also studied in M1 cells, a renal cell line derived from mouse cortical collecting duct: Aldosterone and transforming growth factor-β (TGF-β) reciprocally regulate the expression of PN-1, contributing to the upregulation of Na^+ transport by aldosterone and its antagonism by TGF-β, respectively. Wakida et al. (81) suggested that PN-1 may represent a new factor that contributes to the regulation of ENaC activity in the kidney.

In Vivo and In Vitro Evidence of ENaC Proteolytic Cleavage and Activation in Different Organs

The regulation of ENaC is notably tailored to its specialized functions in different epithelia. Differences in expression of proteases and cognate inhibitors appear to contribute to tissue-specific control of ENaC activity. The relative prominence of specific proteolytic activating pathways, such as convertases, CAPs, and soluble SPs, in different tissues may permit Na^+ conductance to be controlled through the P_o of ENaC already resident at the cell surface or by the redistribution of constitutively active ENaC between intracellular pools and the cell surface.

Kidney. Masilamani et al. (19) determined the abundance (semiquantitative western blot-

ting) and distribution (immunohistochemistry) of ENaC subunits in the principal cells of the rat renal collecting duct. Elevated circulating aldosterone (due to either dietary salt restriction or aldosterone infusion) markedly increased the abundance of α ENaC protein without increasing the abundance of the β and γ subunits. The authors were the first to show that aldosterone induced in vivo a shift in the molecular weight of γ ENaC from 85 kDa to $\approx$75 kDa, consistent with physiological proteolytic clipping of the extracellular loop. Frindt et al. (82) demonstrated that short-term (overnight) salt deprivation (with a twofold increase in plasma aldosterone) was sufficient to activate ENaC as measured by whole-cell patch clamp of ex vivo cortical collecting ducts and to induce γ ENaC cleavage, but did not change the abundance of any ENaC subunit. Ergonul et al. (83) have recently extended this work to investigate the processing of the proteins in rat kidney with changes in Na^+ or K^+ intake. When animals were maintained on a low-Na^+ diet for 7–9 days, the abundance of two forms of the α subunit, with apparent masses of 85 and 30 kDa, increased. Salt restriction also increased the abundance of the β subunit and produced an endoglycosidase H (Endo H)-resistant pool of this subunit. The abundance of the 90-kDa form of the γ subunit decreased, whereas that of the $\approx$75-kDa form increased and also exhibited Endo H–resistant glycosylation. These changes (measured by semiquantitative western blotting of whole kidney cortex) in the α and γ subunits were correlated with increases (measured by whole-cell patch clamp of ex vivo cortical collecting ducts) in Na^+ conductance elicited by a 4-h infusion with aldosterone. Changes in all three subunits correlated with decreases in Na^+ conductance when Na^+-deprived animals drank saline for 5 h. Ergonul et al. (83) concluded that ENaC subunits are mainly in an immature form in salt-replete rats. With Na^+ depletion, the subunits mature in a process that involves proteolytic cleavage and further glycosylation. Similar changes occurred in the α and γ subunits but not in the β subunit when animals were treated with exogenous aldosterone and in the β and γ subunits but not in the α subunit when animals were fed a high-K^+ diet. Changes in the processing and maturation of the channels occurred rapidly enough to affect the daily regulation of ENaC activity and Na^+ reabsorption by the kidney.

In the studies described above, western blots were performed on homogenates of renal cortex, preventing the identification of the cellular localization of the cleaved fragments and raising the question of whether these fragments were physiologically relevant to the observed changes in ENaC activity. To address this question, Frindt et al. (84) developed a method to study surface (apical membrane) expression of ENaC proteins in rat kidney by biotinylation of the extracellular surfaces of renal cells of rat kidneys perfused in situ with solutions containing cell-impermeant biotin. Membranes were solubilized and labeled proteins were isolated by the use of neutravidin beads. Most of the γ ENaC at the surface was smaller in molecular mass than was the full-length subunit, consistent with cleavage of this subunit in the extracellular moiety close to the first transmembrane domain. Insensitivity of the channels to trypsin, measured in principal cells of the cortical collecting duct by whole-cell patch-clamp recording, corroborated this finding. In salt-replete animals, a small but significant fraction of ENaC I_{Na} became trypsin sensitive, and ENaC subunits could be detected at the surface in all physiological conditions, suggesting the existence of near-silent channels at the apical membrane. However, increasing aldosterone levels in the animals by feeding them a low-Na^+ diet or infusing them directly with hormone via osmotic minipumps for 1 week before surface labeling increased the surface expression of the subunits by two- to fivefold. Salt repletion of Na^+-deprived animals for 5 h decreased surface expression. Changes in the surface density of ENaC subunits contributed significantly to the regulation of Na^+ transport in renal cells by mineralocorticoid hormone but did not fully account for increased channel activity (84). Collectively, the data suggest that aldosterone has a dual effect on N and P_o, as discussed above for the Liddle mutations (77, 78).

Lung. Tarran et al. (59) have studied, in human airway cells grown in an air-liquid interface, the mechanisms whereby airway epithelia regulate airway surface liquid (ASL) height to a fixed value of 7 μM. These researchers showed that the abundance of CAP-2 mRNA was high compared with CAP-1 mRNA. The cells also expressed a large number of serpins, among them hepatic growth factor (HA1). The abundance of CAP-1 and HA1 mRNA was increased in bronchial cells from cystic fibrosis patients. ASL volume regulation was sensitive to trypsin and aprotinin, a CAP inhibitor, which regulated Na^+ absorption via changes in ENaC activity in both normal and cystic fibrosis cultures. Myerburg et al. (52) studied Na^+ transport by determining I_{Na} in primary human airway epithelial cells grown on a porous substrate that was mounted in Ussing chambers. The authors showed that protease inhibitors are present in the ASL and prevent the activation of near-silent ENaC. When the ASL volume was increased, endogenous protease inhibitors became diluted, allowing for proteolytic activation of near-silent channels. In cystic fibrosis patients, the normally present near-silent pool of ENaC was constitutively active, and the α subunit underwent increased proteolytic processing. The proteolytic fragment of the γ subunit was not reported. Interestingly, a ≈75-kDa fragment rather than a furin-mediated α fragment (30 kDa) was observed at the cell surface, suggesting another cleavage site near the second transmembrane domain (see **Figure 2**). This fragment may be similar to those observed in oocyte cell surfaces (66, 68), in kidney homogenates (83), in A6 kidney cells (17), and at the surfaces of lung alveolar cells (85). The functional significance of α proteolytic fragments at the surface is not clear, but furin-dependent cleavage does not seem to participate in ENaC maturation, processing, or activation in airway cells.

Planès et al. (50) examined in vitro and in vivo regulation of transepithelial lung alveolar Na$^+$ transport by SPs. In vitro experiments showed that inhibition of endogenous SPs by apical aprotinin decreased ENaC-mediated currents in primary cultures of rat and mouse alveolar epithelial cells without affecting the abundance or the electrophoretic migration pattern of biotinylated α and β ENaC expressed at the cell surface. Aprotinin suppressed the increase in amiloride-sensitive short-circuit current induced by the β2-agonist terbutaline. RT-PCR experiments indicated that CAP-1, CAP-2, and CAP-3 mRNAs were expressed in mouse alveolar epithelial cells, whereas CAP-1 was also expressed in alveolar macrophages recovered by bronchoalveolar lavage. CAP-1 protein was detected by western blotting in rat and mouse alveolar epithelial cells, alveolar macrophages, and bronchoalveolar lavage fluid. Finally, in vivo experiments revealed that intra-alveolar treatment with aprotinin abolished the increase in Na$^+$-driven alveolar fluid clearance induced by terbutaline in an in situ mouse lung model, whereas trypsin potentiated it. Together, the data showed that endogenous membrane-bound and/or secreted SPs such as CAPs regulate alveolar Na$^+$ and fluid transport in vitro and in vivo in rodent lung (50).

Cochlea. Guipponi et al. (86) showed that the transmembrane SP (TMPRSS3) mutated in nonsyndromic autosomal recessive deafness (DFNB8/10) activates the ENaC in vitro. RT-PCR and RNA in situ hybridization on rat and mouse cochlea revealed that TMPRSS33 is expressed in the spiral ganglion, the cells supporting the organ of Corti, and the stria vascularis and is coexpressed with ENaC (3). The amiloride-sensitive ENaC, which is expressed in many Na$^+$-reabsorbing tissues such as the inner ear and is regulated by membrane-bound CAPs, is a potential substrate of TMPRSS3. In the *Xenopus* oocyte expression system, proteolytic processing of TMPRSS3 was associated with increased ENaC-mediated currents. In contrast, six TMPRSS3 mutants (D103G, R109W, C194F, W251C, P404L, and C407R) causing deafness and a mutant in the catalytic triad of TMPRSS3 (S401A) failed to undergo

proteolytic cleavage and activate ENaC. These data indicate that important signaling pathways in the inner ear are controlled by proteolytic cleavage and suggest the existence of an autocatalytic process by which TMPRSS3 would become active. ENaC may be a substrate of TMPRSS3 in the inner ear. To test this hypothesis will require a transgenic model with gene inactivation of *Tmprss3* (87).

Skin. Hummler and colleagues (88) have reviewed in detail the role of ENaC in the skin. Charles et al. (89) showed a postnatal requirement of ENaC for the maintenance of epidermal barrier function. Within hours after birth, mice deficient for the α subunit of the highly amiloride-sensitive ENaC (α ENaC/Scnn1a) suffered from significant increased dehydration. This was characterized by a loss of body weight (by 6% in 6 h) and increased transepidermal water loss, which was accompanied by a higher skin surface pH in one-day-old pups. Although early and late differentiation markers as well as tight-junction protein distribution and function seemed unaffected, deficiency of α ENaC severely disturbed SC lipid composition with decreased ceramide and cholesterol levels, and increased probarrier lipids, whereas covalently bound lipids are drastically reduced. All together, the data showed that ENaC deficiency results in progressive dehydration and, consequently, weight loss due to severe impairment of lipid formation and secretion. ENaC expression is required for the postnatal maintenance of the epidermal barrier function but not for its generation. Interestingly, this skin mouse phenotype of impaired barrier function bears some similarity to the phenotype of an engineered mouse deleted specifically for skin CAP-1. Leyvraz et al. (90) showed that the epidermal barrier function is dependent on the SP CAP-1/Prss8. They generated mice lacking the membrane-anchored CAP-1 [also termed protease serine S1 family member 8 (Prss8) and prostasin] in skin. These mice died within 60 h after birth. They presented a lower body weight and exhibited severe malformation of the SC. This aberrant skin development was accompanied by impaired skin barrier function, as evidenced by dehydration and skin permeability assays and transepidermal water loss measurements leading to rapid, fatal dehydration. Analysis of differentiation markers revealed no major alterations in CAP-1/Prss8-deficient skin even though the epidermal deficiency of CAP-1/Prss8 expression resulted in SC lipid composition, corneocyte morphogenesis, and the processing of profilaggrin. The examination of tight-junction proteins revealed an absence of occludin, which did not prevent the diffusion of a subcutaneously injected tracer (approximately 600 Da) toward the skin surface. This study shows that CAP-1/Prss8 expression in the epidermis is crucial for the epidermal permeability barrier and is thereby indispensable for postnatal survival. This skin phenotype is clearly more severe than the one observed by Charles et al. (89). It is not yet known whether ENaC is a substrate for CAP-1 in the skin. A genetic approach may address this question in mechanistic terms (see below).

PERSPECTIVE AND UNSOLVED QUESTIONS

Ten years after the identification of the first CAP, the field has finally taken a decisive turn. The mechanism of ENaC activation by SPs has begun to be understood in molecular terms.

ENaC Activation by Serine Proteases: To Be or Not To Be Cleaved?

Despite all the progress reviewed here, there is still doubt that any of the observed cleavages of the α or γ subunits precede the opening of ENaC. Recently, Fejes-Toth et al. (91) showed that gene inactivation of Sgk1 had no effect on ENaC activity yet decreased significantly the amount of cleaved γ subunit. Such a discrepancy is not easily reconciled with the proposed effect of aldosterone on Sgk1, Nedd-4–2, and ENaC. More worrisome for understanding proteolytic regulation of ENaC, the study of Fejes-Toth et al. (91) indicates that there is no strict correlation between γ sub-

unit cleavage and channel activation, as Harris et al. (68) also observe. A logical extension of this observation, that ENaC cleavage is coincident with some other process that regulates ENaC, is a warning that future research in the area requires an open mind.

Unsolved Questions

The previous decade contained many advances, but key areas of ENaC regulation by proteases remain to be fully explored. For example, the very first protease found to regulate ENaC was CAP-1, and although a fairly detailed narrative about CAP-1 regulation of ENaC has been built up, we still do not know the mechanism of regulation. Strong relationships exist between CAP-1 activation of ENaC and cleavage of a polybasic tract in γ ENaC (71), between CAP-1 expression and ENaC-mediated currents in an airway cell line (92), and between ENaC activity and expression levels of a serpin (PN-1) that tightly binds ENaC (81). Nonetheless, that ENaC activation can occur through catalytically dead CAP-1 indicates that we are missing other essential information about these observations (41).

What we know about CAP-1 regulation of ENaC evokes a more complicated potential explanation that involves serpins and other proteases. Three ENaC-activating CAPs have been identified (38). Are there others? Tissue kallikrein has been recently proposed as activating ENaC in kidney and colon (93). Are there ENaC-interacting proteins on the cell surface that may be involved? Among CAPs 1–3, a sequential relationship has been detected between CAP-3 and CAP-1 in skin (94). Knowing this fact does not immediately clarify CAP-1–ENaC regulation but does give substance to the notion that a cascade involving CAPs, serpins, and unknown surface proteins may mediate ENaC activation.

Similarly, the role of furin-like convertases in regulating ENaC has not been thoroughly examined. The identification of ENaC as a protein able to be cleaved by furin suggests that proteolytic processing of ENaC affecting its activity may occur, analogous to convertase processing of prohormones, zymogens, and other proteins whose activity is revealed by selective proteolysis along the protein secretory path. Indeed, some results are highly consistent with this notion. Mutation of the furin sequences in α ENaC has a dramatic reduction, in model systems, of ENaC basal current, which can then be recovered by external trypsin (42). Mutation of the single γ ENaC furin sequence has much less effect (68). These results imply that cleavage of α ENaC at the furin sites is the major event for ENaC activation. However, some data are difficult to reconcile with this conclusion. Principally owing to work with elastases, it is quite clear that γ ENaC is secondarily cleaved in a defined region 60–70 amino acids downstream of the γ subunit furin site (63, 67). The same two studies (63, 67) did not observe elastase cleavage of α ENaC yet found that trypsin cannot add to the stimulation by elastase (and vice versa) (49). Thus, in the context of ENaC residing at the surface, γ ENaC, and its actual cleavage, appears to be more of a contributor to maximally activated ENaC currents than is α ENaC. Support for a bigger relative contribution of γ ENaC to proteolytic activation can also be drawn from the fully stimulatory effect of CAP-1 coexpression on ENaC, which is solely attributed to γ ENaC (71). Reconciliation of these somewhat conflicting lines of research may come from a better understanding of the purpose of the susceptibility of the α and γ subunits to convertases. Perhaps cleavage of ENaC along its secretory route plays roles in ENaC trafficking, subcellular localization, and/or surface dwell time. For example, drawing on the available reagents for testing furin function in different intracellular compartments, we may learn where convertase processing of ENaC occurs, and such discoveries may yield some clues to its role. Careful studies of dwell time of different combinations of furin-sensitive subunits may help explain the relative contributions of the α and γ subunits to proteolytic regulation of ENaC.

Understanding the regulation of ENaC by CAPs and furin-like convertases requires knowledge of how the proteases encounter the channel. Presently, there is no compelling evidence for coexpression of ENaC and SPs in any relevant cellular compartment (ER, Golgi, TGN, exocytic vesicles, plama membrane, recycling vesicles, endocytic vesicles, lysosomes, and/or proteasomes). To address this cell biology and cell physiology question will require the establishment of a native, nontransfected, cellular model that will allow for the quantitative detection of native ENaC subunits and various CAPs and allow for the possibility to study the trafficking of these molecules by pulse chase experiments, cell surface labeling (apical versus basolateral), and immunohistochemistry and to correlate them to the function, i.e., amiloride-sensitive transepithelial Na^+ transport with a time resolution of less than 1 min. This is probably a realistic goal for in vitro studies, but the detection of a few hundred copies of ENaC channel protein per cell remains a challenge with the sensitivity of the polyclonal antibodies presently available. As discussed by Frindt et al. (84), the extension of in vivo (i.e., perfused kidney) experimental protocols may suffer real limitations in terms of time resolution and sensitivity for quantitation. Even if this problem is overcome, we will still not have a true in vivo situation.

The identification of SPs and serpins relevant to ENaC activation in vivo is a particularly challenging question but one that should illuminate the physiological importance of proteolytic regulation of ENaC. Indeed, the fact that we are dealing with membrane-bound SPs prevents an easy and formal identification of the substrate(s) involved. Unlike the secreted or soluble forms of SPs to which straightforward Michaelis-Menten kinetics may apply, allowing the determination of a minimum consensus sequence for specific cleavage, such conditions do not apply for membrane-bound SPs. The coexpression of ENaC with CAPs could generate within a membrane microdomain very close contact between the catalytic binding site and the cleavage site. Therefore, the affinity measured in solution may not apply within the membrane. We believe that, to overcome this major difficulty, the only reasonable and powerful technique is the genetic approach that takes advantage of human mendelian diseases (i.e., the case of TMPRSS3 in human deafness) and of the establishment of transgenic mouse models. For furin, the genetic approach has been successful in revealing its essential role during embryonic development, and it may be possible to study the involvement of furin postnatally with conditional and tissue-specific knockouts (95–97). The roles of SPs and SPIs have also been successfully studied in vivo in the skin by a genetic approach (94, 98). Tissue-specific, time-dependent, inducible gene expression requires the generation of the corresponding lox alleles of the gene of interest [i.e., CAP-1 (99)] and mice expressing the cre recombinase under a tissue-specific promoter. The methodology is available but time consuming.

Finally, a precise intermolecular sequence of events initiated by an identified cleavage within ENaC remains the essential goal and is a requisite both for ascertaining if cleavage of ENaC is the activating event in ENaC proteolytic stimulation as well as for fully understanding most of the secondary questions we speculate about above. No doubt, this goal will be reached more quickly now that structural information about ENaC can be derived by ASIC1-based modeling (100). Important aspects include the subunit(s) cleaved, the site(s) within the subunit, and a convincing mechanistic link between cleavage and a change in gating. Analogy models do have, however, important limitations and may be misleading. Ultimately, the 3-D crystal structure of ENaC and its cocrystallization with CAPs or associated proteins will be required to better understand cleavage-induced ENaC gating.

DISCLOSURE STATEMENT

The authors are not aware of any biases that might be perceived as affecting the objectivity of this review.

LITERATURE CITED

1. Verrey F, Hummler E, Schild L, Rossier BC. 2008. Mineralocorticoid action in the aldosterone-sensitive distal nephron. In *The Kidney: Physiology and Pathophysiology*, ed. RJ Alpern, SC Hebert, pp. 889–924. Burlington: Academic

2. Randell SH, Boucher RC. 2006. Effective mucus clearance is essential for respiratory health. *Am. J. Respir. Cell Mol. Biol.* 35:20–28

3. Grunder S, Muller A, Ruppersberg JP. 2001. Developmental and cellular expression pattern of epithelial sodium channel α, β and γ subunits in the inner ear of the rat. *Eur. J. Neurosci.* 13:641–48

4. Canessa CM, Schild L, Buell G, Thorens B, Gautschi I, et al. 1994. Amiloride-sensitive epithelial Na^+ channel is made of three homologous subunits. *Nature* 367:463–67

5. Palmer LG, Garty H. 2000. Epithelial Na channels. In *The Kidney: Physiology and Pathophysiology*, ed. G Giebisch, D Seldin, pp. 251–76. Philadelphia: Lippincott Williams & Wilkins

6. Kellenberger S, Schild L. 2002. Epithelial sodium channel/degenerin family of ion channels: a variety of functions for a shared structure. *Physiol. Rev.* 82:735–67

7. Brouard M, Casado M, Djelidi S, Barrandon Y, Farman N. 1999. Epithelial sodium channel in human epidermal keratinocytes: expression of its subunits and relation to sodium transport and differentiation. *J. Cell Sci.* 112(Pt. 19):3343–52

8. Lin W, Kinnamon SC. 1999. Co-localization of epithelial sodium channels and glutamate receptors in single taste cells. *Biol. Signals Recept.* 8:360–65

9. Rochelle LG, Li DC, Ye H, Lee E, Talbot CR, Boucher RC. 2000. Distribution of ion transport mRNAs throughout murine nose and lung. *Am. J. Physiol. Lung Cell Mol. Physiol.* 279:L14–24

10. Matalon S, O'Brodovich H. 1999. Sodium channels in alveolar epithelial cells: molecular characterization, biophysical properties, and physiological significance. *Annu. Rev. Physiol.* 61:627–61

11. Knowles MR, Boucher RC. 2002. Mucus clearance as a primary innate defense mechanism for mammalian airways. *J. Clin. Investig.* 109:571–77

12. Matthay MA, Folkesson HG, Clerici C. 2002. Lung epithelial fluid transport and the resolution of pulmonary edema. *Physiol. Rev.* 82:569–600

13. Boucher RC. 2007. Airway surface dehydration in cystic fibrosis: pathogenesis and therapy. *Annu. Rev. Med.* 58:157–70

14. Chang SS, Grunder S, Hanukoglu A, Rosler A, Mathew PM, et al. 1996. Mutations in subunits of the epithelial sodium channel cause salt wasting with hyperkalaemic acidosis, pseudohypoaldosteronism type 1. *Nat. Genet.* 12:248–53

15. Rossier BC, Pradervand S, Schild L, Hummler E. 2002. Epithelial sodium channel and the control of sodium balance: interaction between genetic and environmental factors. *Annu. Rev. Physiol.* 64:877–97

16. Rossier BC. 2002. Hormonal regulation of the epithelial sodium channel ENaC: N or P_o? *J. Gen. Physiol.* 120:67–70

17. Alvarez de la Rosa D, Li H, Canessa CM. 2002. Effects of aldosterone on biosynthesis, traffic, and functional expression of epithelial sodium channels in A6 cells. *J. Gen. Physiol.* 119:427–42

18. Loffing J, Zecevic M, Feraille E, Kaissling B, Asher C, et al. 2001. Aldosterone induces rapid apical translocation of ENaC in early portion of renal collecting system: possible role of SGK. *Am. J. Physiol. Ren. Physiol.* 280:F675–82

19. Masilamani S, Kim GH, Mitchell C, Wade JB, Knepper MA. 1999. Aldosterone-mediated regulation of ENaC α, β, and γ subunit proteins in rat kidney. *J. Clin. Investig.* 104:R19–23

20. Dahlmann A, Pradervand S, Hummler E, Rossier BC, Frindt G, Palmer LG. 2003. Mineralocorticoid regulation of epithelial Na^+ channels is maintained in a mouse model of Liddle's syndrome. *Am. J. Physiol. Renal. Physiol.* 285:F310–18

21. Morris RG, Schafer JA. 2002. cAMP increases density of ENaC subunits in the apical membrane of MDCK cells in direct proportion to amiloride-sensitive Na^+ transport. *J. Gen. Physiol.* 120:71–85

22. Grunder S, Schild L, Rossier BC. 1997. Identification of a N-terminal gating domain in ENaC subunits. *J. Am. Soc. Nephrol.* 8:A0161 (Abstr.)

23. Mano I, Driscoll M. 1999. DEG ENaC channels: a touchy superfamily that watches its salt. *Bioessays* 21:568–78

24. Hedstrom L. 2002. Serine protease mechanism and specificity. *Chem. Rev.* 102:4501–24

25. Rawlings ND, Barrett AJ. 2000. MEROPS: the peptidase database. *Nucleic Acids Res.* 28:323–25

26. Rawlings ND, Barrett AJ. 1993. Evolutionary families of peptidases. *Biochem. J.* 290:205–18

27. Hooper JD, Clements JA, Quigley JP, Antalis TM. 2001. Type II transmembrane serine proteases. Insights into an emerging class of cell surface proteolytic enzymes. *J. Biol. Chem.* 276:857–60

28. Page MJ, Di Cera E. 2008. Serine peptidases: classification, structure and function. *Cell. Mol. Life Sci.* 65:1220–36

29. Vallet V, Horisberger JD, Rossier BC. 1998. Epithelial sodium channel regulatory proteins identified by functional expression cloning. *Kidney Int. Suppl.* 67:S109–14

30. Vallet V, Chraibi A, Gaeggeler HP, Horisberger JD, Rossier BC. 1997. An epithelial serine protease activates the amiloride-sensitive sodium channel. *Nature* 389:607–10

31. Vuagniaux G, Vallet V, Jaeger NF, Pfister C, Bens M, et al. 2000. Activation of the amiloride-sensitive epithelial sodium channel by the serine protease mCAP1 expressed in a mouse cortical collecting duct cell line. *J. Am. Soc. Nephrol.* 11:828–34

32. Lisanti MP, Sargiacomo M, Graeve L, Saltiel AR, Rodriguez-Boulan E. 1988. Polarized apical distribution of glycosyl-phosphatidylinositol-anchored proteins in a renal epithelial cell line. *Proc. Natl. Acad. Sci. USA* 85:9557–61

33. Vallet V, Pfister C, Loffing J, Rossier BC. 2002. Cell-surface expression of the channel activating protease xCAP-1 is required for activation of ENaC in the *Xenopus* oocyte. *J. Am. Soc. Nephrol.* 13:588–94

34. Yu JX, Chao L, Chao J. 1994. Prostasin is a novel human serine proteinase from seminal fluid. Purification, tissue distribution, and localization in prostate gland. *J. Biol. Chem.* 269:18843–48

35. Yu JX, Chao L, Chao J. 1995. Molecular cloning, tissue-specific expression, and cellular localization of human prostasin mRNA. *J. Biol. Chem.* 270:13483–89

36. Chen LM, Skinner ML, Kauffman SW, Chao J, Chao L, et al. 2001. Prostasin is a glycosylphosphatidylinositol-anchored active serine protease. *J. Biol. Chem.* 276:21434–42

37. Adachi M, Kitamura K, Miyoshi T, Narikiyo T, Iwashita K, et al. 2001. Activation of epithelial sodium channels by prostasin in *Xenopus* oocytes. *J. Am. Soc. Nephrol.* 12:1114–21

38. Vuagniaux G, Vallet V, Jaeger NF, Hummler E, Rossier BC. 2002. Synergistic activation of ENaC by three membrane-bound channel-activating serine proteases (mCAP1, mCAP2, and mCAP3) and serum- and glucocorticoid-regulated kinase (Sgk1) in *Xenopus* oocytes. *J. Gen. Physiol.* 120:191–201

39. Netzel-Arnett S, Hooper JD, Szabo R, Madison EL, Quigley JP, et al. 2003. Membrane anchored serine proteases: a rapidly expanding group of cell surface proteolytic enzymes with potential roles in cancer. *Cancer Metastasis Rev.* 22:237–58

40. Szabo R, Bugge TH. 2008. Type II transmembrane serine proteases in development and disease. *Int. J. Biochem. Cell Biol.* 40:1297–316

41. Andreasen D, Vuagniaux G, Fowler-Jaeger N, Hummler E, Rossier BC. 2006. Activation of epithelial sodium channels by mouse channel activating proteases (mCAP) expressed in *Xenopus* oocytes requires catalytic activity of mCAP3 and mCAP2 but not mCAP1. *J. Am. Soc. Nephrol.* 17:968–76

42. Hughey RP, Bruns JB, Kinlough CL, Harkleroad KL, Tong Q, et al. 2004. Epithelial sodium channels are activated by furin-dependent proteolysis. *J. Biol. Chem.* 279:18111–14

43. Hughey RP, Mueller GM, Bruns JB, Kinlough CL, Poland PA, et al. 2003. Maturation of the epithelial Na$^+$ channel involves proteolytic processing of the α- and γ-subunits. *J. Biol. Chem.* 278:37073–82

44. Thomas G. 2002. Furin at the cutting edge: from protein traffic to embryogenesis and disease. *Nat. Rev. Mol. Cell Biol.* 3:753–66

45. Molloy SS, Anderson ED, Jean F, Thomas G. 1999. Bi-cycling the furin pathway: from TGN localization to pathogen activation and embryogenesis. *Trends Cell Biol.* 9:28–35

46. Han J, Wang Y, Wang S, Chi C. 2008. Interaction of Mint3 with Furin regulates the localization of Furin in the trans-Golgi network. *J. Cell Sci.* 121:2217–23

47. Chraibi A, Vallet V, Firsov D, Hess SK, Horisberger JD. 1998. Protease modulation of the activity of the epithelial sodium channel expressed in *Xenopus* oocytes. *J. Gen. Physiol.* 111:127–38

48. Caldwell RA, Boucher RC, Stutts MJ. 2004. Serine protease activation of near-silent epithelial Na$^+$ channels. *Am. J. Physiol. Cell Physiol.* 286:C190–94

49. Caldwell RA, Boucher RC, Stutts MJ. 2005. Neutrophil elastase activates near-silent epithelial Na$^+$ channels and increases airway epithelial Na$^+$ transport. *Am. J. Physiol. Lung Cell Mol. Physiol.* 288:L813–19

50. Planès C, Leyvraz C, Uchida T, Angelova MA, Vuagniaux G, et al. 2005. In vitro and in vivo regulation of transepithelial lung alveolar sodium transport by serine proteases. *Am. J. Physiol. Lung Cell Mol. Physiol.* 288:L1099–109

51. Donaldson SH, Hirsh A, Li DC, Holloway G, Chao J, et al. 2002. Regulation of the epithelial sodium channel by serine proteases in human airways. *J. Biol. Chem.* 277:8338–45

52. Myerburg MM, Butterworth MB, McKenna EE, Peters KW, Frizzell RA, et al. 2006. Airway surface liquid volume regulates ENaC by altering the serine protease-protease inhibitor balance: a mechanism for sodium hyperabsorption in cystic fibrosis. *J. Biol. Chem.* 281:27942–49

53. Bridges RJ, Newton BB, Pilewski JM, Devor DC, Poll CT, Hall RL. 2001. Na$^+$ transport in normal and CF human bronchial epithelial cells is inhibited by BAY 39–9437. *Am. J. Physiol. Lung Cell Mol. Physiol* 281:L16–23

54. Liu L, Hering-Smith KS, Schiro FR, Hamm LL. 2002. Serine protease activity in m-1 cortical collecting duct cells. *Hypertension* 39:860–64

55. Nakhoul NL, Hering-Smith KS, Gambala CT, Hamm LL. 1998. Regulation of sodium transport in M-1 cells. *Am. J. Physiol.* 275:F998–1007

56. Adebamiro A, Cheng Y, Johnson JP, Bridges RJ. 2005. Endogenous protease activation of ENaC: effect of serine protease inhibition on ENaC single channel properties. *J. Gen. Physiol.* 126:339–52

57. Orce GG, Castillo GA, Margolius HS. 1981. Kallikrein inhibitors decrease short-circuit current by inhibiting sodium uptake. *Hypertension* 3:II-92–95

58. Tarran R, Grubb BR, Gatzy JT, Davis CW, Boucher RC. 2001. The relative roles of passive surface forces and active ion transport in the modulation of airway surface liquid volume and composition. *J. Gen. Physiol.* 118:223–36

59. Tarran R, Trout L, Donaldson SH, Boucher RC. 2006. Soluble mediators, not cilia, determine airway surface liquid volume in normal and cystic fibrosis superficial airway epithelia. *J. Gen. Physiol.* 127:591–604

60. Hughey RP, Carattino MD, Kleyman TR. 2007. Role of proteolysis in the activation of epithelial sodium channels. *Curr. Opin. Nephrol. Hypertens.* 16:444–50

61. Kleyman TR, Myerburg MM, Hughey RP. 2006. Regulation of ENaCs by proteases: an increasingly complex story. *Kidney Int.* 70:1391–92

62. Planès C, Caughey GH. 2007. Regulation of the epithelial Na$^+$ channel by peptidases. *Curr. Top. Dev. Biol.* 78:23–46

63. Adebamiro A, Cheng Y, Rao US, Danahay H, Bridges RJ. 2007. A segment of γENaC mediates elastase activation of Na$^+$ transport. *J. Gen. Physiol.* 130:611–29

64. Firsov D, Robert-Nicoud M, Gruender S, Schild L, Rossier BC. 1999. Mutational analysis of cysteine-rich domains of the epithelium sodium channel (ENaC). Identification of cysteines essential for channel expression at the cell surface. *J. Biol. Chem.* 274:2743–49

65. Sheng S, Maarouf AB, Bruns JB, Hughey RP, Kleyman TR. 2007. Functional role of extracellular loop cysteine residues of the epithelial Na$^+$ channel in Na$^+$ self-inhibition. *J. Biol. Chem.* 282:20180–90

66. Michlig S, Harris M, Loffing J, Rossier BC, Firsov D. 2005. Progesterone down-regulates the open probability of the amiloride-sensitive epithelial sodium channel via a Nedd4–2-dependent mechanism. *J. Biol. Chem.* 280:38264–70

67. Harris M, Firsov D, Vuagniaux G, Stutts MJ, Rossier BC. 2007. A novel neutrophil elastase inhibitor prevents elastase activation and surface cleavage of the epithelial sodium channel expressed in *Xenopus laevis* oocytes. *J. Biol. Chem.* 282:58–64

68. Harris M, Garcia-Caballero A, Stutts MJ, Firsov D, Rossier BC. 2008. Preferential assembly of epithelial sodium channel (ENaC) subunits in *Xenopus* oocytes: role of furin-mediated endogenous proteolysis. *J. Biol. Chem.* 283:7455–63

69. Carattino MD, Sheng S, Bruns JB, Pilewski JM, Hughey RP, Kleyman TR. 2006. The epithelial Na$^+$ channel is inhibited by a peptide derived from proteolytic processing of its α subunit. *J. Biol. Chem.* 281:18901–7

70. Carattino MD, Passero CJ, Steren CA, Maarouf AB, Pilewski JM, et al. 2008. Defining an inhibitory domain in the α-subunit of the epithelial sodium channel. *Am. J. Physiol. Ren. Physiol.* 294:F47–52

71. Bruns JB, Carattino MD, Sheng S, Maarouf AB, Weisz OA, et al. 2007. Epithelial Na$^+$ channels are fully activated by furin- and prostasin-dependent release of an inhibitory peptide from the γ-subunit. *J. Biol. Chem.* 282:6153–60

72. Sheng S, Carattino MD, Bruns JB, Hughey RP, Kleyman TR. 2006. Furin cleavage activates the epithelial Na$^+$ channel by relieving Na$^+$ self-inhibition. *Am. J. Physiol. Ren. Physiol.* 290:F1488–96

73. Chraibi A, Horisberger JD. 2002. Na self inhibition of human epithelial Na channel: temperature dependence and effect of extracellular proteases. *J. Gen. Physiol.* 120:133–45

74. Carattino MD, Hughey RP, Kleyman TR. 2008. Proteolytic processing of the epithelial sodium channel γ subunit has a dominant role in channel activation. *J. Biol. Chem.* 37:25290–95

75. Diakov A, Bera K, Mokrushina M, Krueger B, Korbmacher C. 2008. Cleavage in the γ-subunit of the epithelial sodium channel (ENaC) plays an important role in the proteolytic activation of near-silent channels. *J. Physiol.* 586:4587–608

76. Garcia-Caballero A, Dang Y, He H, Stutts MJ. 2008. ENaC proteolytic regulation by channel-activating protease 2. *J. Gen. Physiol.* 132:521–35

77. Knight KK, Olson DR, Zhou R, Snyder PM. 2006. Liddle's syndrome mutations increase Na$^+$ transport through dual effects on epithelial Na$^+$ channel surface expression and proteolytic cleavage. *Proc. Natl. Acad. Sci. USA* 103:2805–8

78. Firsov D, Schild L, Gautschi I, Merillat AM, Schneeberger E, Rossier BC. 1996. Cell surface expression of the epithelial Na channel and a mutant causing Liddle syndrome: a quantitative approach. *Proc. Natl. Acad. Sci. USA* 93:15370–75

79. Ruffieux-Daidié D, Poirot O, Boulkroun S, Verrey F, Kellenberger S, Staub O. 2008. Deubiquitylation regulates activation and proteolytic cleavage of ENaC. *J. Am. Soc. Nephrol.* 19:2170–80

80. Krowarsch D, Cierpicki T, Jelen F, Otlewski J. 2003. Canonical protein inhibitors of serine proteases. *Cell. Mol. Life Sci.* 60:2427–44

81. Wakida N, Kitamura K, Tuyen DG, Maekawa A, Miyoshi T, et al. 2006. Inhibition of prostasin-induced ENaC activities by PN-1 and regulation of PN-1 expression by TGF-β1 and aldosterone. *Kidney Int.* 70:1432–38

82. Frindt G, Masilamani S, Knepper MA, Palmer LG. 2001. Activation of epithelial Na channels during short-term Na deprivation. *Am. J. Physiol. Ren. Physiol.* 280:F112–18

83. Ergonul Z, Frindt G, Palmer LG. 2006. Regulation of maturation and processing of ENaC subunits in the rat kidney. *Am. J. Physiol. Ren. Physiol.* 291:F683–93

84. Frindt G, Ergonul Z, Palmer LG. 2008. Surface expression of epithelial Na channel protein in rat kidney. *J. Gen. Physiol.* 131:617–27

85. Planès C, Blot-Chabaud M, Matthay MA, Couette S, Uchida T, Clerici C. 2002. Hypoxia and β2-agonists regulate cell surface expression of the epithelial sodium channel in native alveolar epithelial cells. *J. Biol. Chem.* 277:47318–24

86. Guipponi M, Vuagniaux G, Wattenhofer M, Shibuya K, Vazquez M, et al. 2002. The transmembrane serine protease (TMPRSS3) mutated in deafness DFNB8/10 activates the epithelial sodium channel (ENaC) in vitro. *Hum. Mol. Genet.* 11:2829–36

87. Guipponi M, Antonarakis SE, Scott HS. 2008. TMPRSS3, a type II transmembrane serine protease mutated in nonsyndromic autosomal recessive deafness. *Front. Biosci.* 13:1557–67

88. Guitard M, Leyvraz C, Hummler E. 2004. A nonconventional look at ionic fluxes in the skin: lessons from genetically modified mice. *News Physiol. Sci.* 19:75–79

89. Charles RP, Guitard M, Leyvraz C, Breiden B, Haftek M, et al. 2008. Postnatal requirement of the epithelial sodium channel for maintenance of epidermal barrier function. *J. Biol. Chem.* 283:2622–30

90. Leyvraz C, Charles RP, Rubera I, Guitard M, Rotman S, et al. 2005. The epidermal barrier function is dependent on the serine protease CAP1/Prss8. *J. Cell Biol.* 170:487–96

91. Fejes-Toth G, Frindt G, Naray-Fejes-Toth A, Palmer LG. 2008. Epithelial Na$^+$ channel activation and processing in mice lacking SGK1. *Am. J. Physiol.* 294:F1298–305

92. Tong Z, Illek B, Bhagwandin VJ, Verghese GM, Caughey GH. 2004. Prostasin, a membrane-anchored serine peptidase, regulates sodium currents in JME/CF15 cells, a cystic fibrosis airway epithelial cell line. *Am. J. Physiol. Lung Cell Mol. Physiol.* 287:L928–35

93. Picard N, Eladari D, El Moghrabi S, Planès C, Bourgeois S, et al. 2008. Defective ENaC processing and function in tissue kallikrein-deficient mice. *J. Biol. Chem.* 283:4602–11

94. List K, Hobson JP, Molinolo A, Bugge TH. 2007. Co-localization of the channel activating protease prostasin/(CAP1/PRSS8) with its candidate activator, matriptase. *J. Cell Physiol.* 213:237–45

95. Roebroek AJ, Taylor NA, Louagie E, Pauli I, Smeijers L, et al. 2004. Limited redundancy of the pro-protein convertase furin in mouse liver. *J. Biol. Chem.* 279:53442–50

96. Grapin-Botton A, Constam D. 2007. Evolution of the mechanisms and molecular control of endoderm formation. *Mech. Dev.* 124:253–78

97. Constam DB, Robertson EJ. 2000. Tissue-specific requirements for the proprotein convertase furin/SPC1 during embryonic turning and heart looping. *Development* 127:245–54

98. Carney TJ, von der Hardt S, Sonntag C, Amsterdam A, Topczewski J, et al. 2007. Inactivation of serine protease Matriptase1a by its inhibitor Hai1 is required for epithelial integrity of the zebrafish epidermis. *Development* 134:3461–71

99. Rubera I, Meier E, Vuagniaux G, Merillat AM, Beermann F, et al. 2002. A conditional allele at the mouse channel activating protease 1 (*Prss8*) gene locus. *Genesis* 32:173–76

100. Jasti J, Furukawa H, Gonzales EB, Gouaux E. 2007. Structure of acid-sensing ion channel 1 at 1.9 Å resolution and low pH. *Nature* 449:316–23

Recent Advances in Understanding Integrative Control of Potassium Homeostasis

Jang H. Youn[1] and Alicia A. McDonough[2]

[1]Department of Physiology and Biophysics, [2]Department of Cell and Neurobiology, University of Southern California Keck School of Medicine, Los Angeles, California 90089-9142; email: youn@usc.edu, mcdonoug@usc.edu

Annu. Rev. Physiol. 2009. 71:381–401

First published online as a Review in Advance on August 29, 2008

The *Annual Review of Physiology* is online at physiol.annualreviews.org

This article's doi: 10.1146/annurev.physiol.010908.163241

Copyright © 2009 by Annual Reviews. All rights reserved

0066-4278/09/0315-0381$20.00

Key Words

dietary potassium, potassium adaptation, potassium excretion, kidney, muscle, insulin

Abstract

The potassium homeostatic system is very tightly regulated. Recent studies have shed light on the sensing and molecular mechanisms responsible for this tight control. In addition to classic feedback regulation mediated by a rise in extracellular fluid (ECF) [K^+], there is evidence for a feedforward mechanism: Dietary K^+ intake is sensed in the gut, and an unidentified gut factor is activated to stimulate renal K^+ excretion. This pathway may explain renal and extrarenal responses to altered K^+ intake that occur independently of changes in ECF [K^+]. Mechanisms for conserving ECF K^+ during fasting or K^+ deprivation have been described: Kidney NADPH oxidase activation initiates a cascade that provokes the retraction of K^+ channels from the cell membrane, and muscle becomes resistant to insulin stimulation of cellular K^+ uptake. How these mechanisms are triggered by K^+ deprivation remains unclear. Cellular AMP kinase–dependent protein kinase activity provokes the acute transfer of K^+ from the ECF to the ICF, which may be important in exercise or ischemia. These recent advances may shed light on the beneficial effects of a high-K^+ diet for the cardiovascular system.

ECF: extracellular fluid

[K+]: potassium concentration

ICF: intracellular fluid

INTRODUCTION

Homeostatic control of extracellular fluid (ECF) K^+ concentration is critical for normal function of nerve and muscle cells because ECF $[K^+]$ is a major determinant of membrane potential (1, 2). In addition, a diet rich in K^+ lowers blood pressure, blunts the negative effects of dietary NaCl, and reduces the risk of kidney stones and bone loss (3). In mammals, the ECF $[K^+]$ ranges between 3.8 and 5 mM, whereas the intracellular fluid (ICF) $[K^+]$ is 120–140 mM (4, 5). Among major electrolytes, K^+ has the highest ratio of daily intake to extracellular pool size (i.e., turnover), that is, the most significant homeostatic challenge. To meet this challenge, the K^+ homeostatic system is very efficient at clearing plasma K^+ after a K^+-containing meal and is very efficient at maintaining ECF $[K^+]$ during fasting or K^+-poor diet, as evidenced by minimal changes in plasma $[K^+]$ during these challenges. The components of the K^+ homeostatic system are not yet fully understood but include the sensing of K^+ intake, the regulation of K^+ distribution between the ECF and the ICF, and the regulation of K^+ excretion (**Figure 1**). This review focuses on the significant recent progress in understanding the individual components of the K^+ homeostasis system and how they are linked.

SENSING K+ INTAKE: FEEDFORWARD VERSUS FEEDBACK CONTROL OF EXTRACELLULAR FLUID [K+]

Extracellular K+ Homeostasis: General Considerations

Extracellular K^+ homeostasis relies on the maintenance of total body K^+ content as well as a regulated distribution of K^+ between a very large intracellular pool (>90% of total body K^+) and a much smaller extracellular pool (~2% total body K^+). Total body K^+ content is maintained by continuously matching K^+ excretion to dietary K^+ intake. This task is accomplished by the kidneys, which are responsible for approximately 90% of K^+ excretion (6,

7). Thus, the kidneys play a predominant role in the maintenance of day-to-day K^+ balance. In concert with renal excretion, rapid shifts in K^+ from the ECF to the large ICF pool of K^+ are necessary to buffer ECF K^+. The importance of the concerted actions of kidney and muscle is best illustrated by considering what happens during a K^+-containing meal and during fasting. Total ECF K^+ is only ~70 mEq K^+, and ingesting a meal containing K^+-rich foods such as fruit and meal may add another 70 mEq K^+ to the bloodstream within a short time of its ingestion, which would double the ECF $[K^+]$ if there were not rapid adjustments to either transfer the K^+ to the ICF compartment or excrete it. In fact, after a meal there is little change in ECF K^+ because postprandial insulin stimulates muscle and liver to actively take up both glucose and the K^+ not excreted in the short term by the kidneys (8). After the meal, muscle activity releases K^+ into the ECF K^+, and somewhat mysteriously, the kidneys excrete the amount consumed in the meal. In the animal kingdom, maintaining ECF K^+ between meals is critically important because fasting can be prolonged, e.g., nine weeks in gray whales and elephant seals (9). During fasting or consumption of a low-K^+ diet, the kidney reduces K^+ excretion to near zero by reabsorbing more and secreting less K^+ (6, 10). Yet, K^+ loss in the stools and sweat persists and challenges the small pool of ECF K^+. To balance the discrepancy between K^+ input and output over time, muscle exhibits an altruistic specialization to donate ICF $[K^+]$ to the ECF (4, 11, 12). In these settings, the sensing of K^+ status and appropriate adjustments are critical to maintain a ratio of ECF to ICF $[K^+]$ compatible with normal nerve and muscle cell excitability. Simply put, the adjustment may be regulated by feedback or feedforward control of ECF K^+.

Feedback Control of K+ Homeostasis

It has long been appreciated that an increase in ECF $[K^+]$ leads to an increase in K^+ secretion in the collecting ducts mediated by direct stimulation of renal Na,K-ATPase,

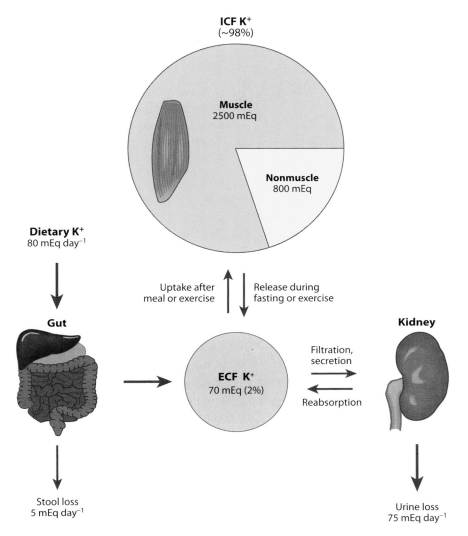

ICF K⁺
(~98%)

Muscle
2500 mEq

Nonmuscle
800 mEq

Dietary K⁺
80 mEq day⁻¹

Uptake after
meal or exercise

Release during
fasting or exercise

Gut

Kidney

Filtration,
secretion

ECF K⁺
70 mEq (2%)

Reabsorption

Stool loss
5 mEq day⁻¹

Urine loss
75 mEq day⁻¹

Figure 1

A schematic diagram illustrating daily K⁺ fluxes into and out of the extracellular fluid (ECF) pool in an average person. Approximately 98% of the body's K⁺ is located in the intracellular fluid (ICF), mainly in muscle, and only approximately 2% is located in the ECF. The ECF pool is regulated by input from the gut, output via the kidney and stools, and redistribution between the ECF and the ICF.

increased tubular flow, and an increase in aldosterone secretion. All three factors directly increase Na⁺ reabsorption via epithelial sodium channels (ENaCs), increasing the driving force for K⁺ secretion (13–15). The regulation of aldosterone synthesis by K⁺ has been recently reviewed (16). Conversely, a decrease in plasma [K⁺], mediated by either decreased intake or increased excretion, decreases K⁺ secretion by reducing ECF K⁺ stimulation of Na,K-ATPase activity, tubular flow, and aldosterone; these reductions decrease Na⁺ reabsorption via ENaC, which depresses the driving force for K⁺ secretion (7). In addition, decreasing plasma K⁺ increases expression of H,K-ATPases located in the distal nephron that effect active K⁺

ENaC: epithelial sodium channel

Feedback control

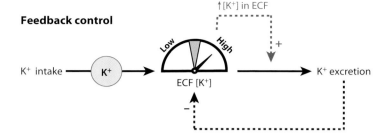

Feedforward control

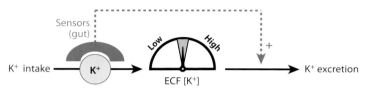

Figure 2

Schematic diagram illustrating control of extracellular fluid (ECF) [K⁺] via feedback versus feedforward mechanisms. (*Top*) In feedback control, a rise in ECF [K⁺] is the signal that initiates stimulation of K⁺ excretion by the kidney. The increased excretion brings ECF [K⁺] back toward the normal value. This process depends on the error signal of elevated ECF [K⁺] and stops when ECF [K⁺] is returned to the control range. (*Bottom*) In feedforward control, a local increase in [K⁺] in the gut is sensed during K⁺ intake and initiates stimulation of K⁺ excretion by the kidney independently of (i.e., before) a rise in ECF [K⁺], which helps to prevent a rise in ECF [K⁺].

reabsorption (7). Changes in renal K⁺ excretion help to normalize extracellular [K⁺], an example of classic feedback control, as illustrated in **Figure 2**.

Feedforward Control of K⁺ Homeostasis

Rabinowitz and colleagues challenged the traditional view of feedback control of ECF K⁺ two decades ago (17, 18). They confirmed that plasma K⁺ stimulates renal K⁺ excretion but that this response requires an elevation of plasma K⁺ above its normal range (19, 20). This group also provided evidence that aldosterone stimulates renal K⁺ excretion only at supraphysiological levels, with little effect within the physiological range of K⁺ (17, 20). In studies in sheep (21), meal intake over 1 h produced a pronounced kaliuresis in the ab-

sence of a change in plasma aldosterone concentration. Plasma [K⁺] was, indeed, increased after a K⁺-containing meal, but the magnitude of the increase (0.5 mEq L⁻¹) was too small, when reproduced by intravenous K⁺ infusion, to account for the meal-induced kaliuresis (20). Rabinowitz concluded that meal-induced increases in renal K⁺ excretion cannot be explained by changes in plasma K⁺ or aldosterone concentration. To explain the meal-induced kaliuresis, he proposed a kaliuretic reflex arising from sensors in the splanchnic bed (i.e., gut, portal circulation, and/or liver). According to this proposal, renal K⁺ excretion is increased, without (or before) increases in extracellular [K⁺], by a mechanism controlled by sensing of K⁺ intake (i.e., sensing of local increase in [K⁺] in splanchnic areas during K⁺ intake). Thus, a new concept of feedforward control of K⁺ homeostasis arose (**Figure 2**).

Feedback versus Feedforward Control: Teleological Considerations

Most homeostatic regulation in the body is under feedback control because it offers robust control (22). However, this type of control can be slow to respond to an external disturbance and inevitably mandates an error signal, a significant disturbance of the system. In contrast, feedforward control may allow a quick control of output function in anticipation of a rise of the signal. This type of control may add speed or accuracy to the control (at the expense of robustness). A combination of both feedback and feedforward control mechanisms may provide both robustness and accuracy (or speed) of the regulation.

Recent Evidence Supporting Feedforward Control of K⁺ Homeostasis

The studies of Morita and colleagues (23, 24) supported the idea of K⁺ sensing or feedforward control of K⁺ homeostasis. After intraportal delivery of KCl to anesthetized rats, these investigators observed increases in hepatic afferent nerve activity (HANA) and urinary K$^+$ excretion, the latter of which occurred in the absence of increases in plasma [K$^+$]. Bumetanide attenuated the effect of KCl on HANA, suggesting that the Na$^+$-K$^+$-2Cl$^-$ cotransporter is involved in the sensing of portal [K$^+$]. This research group suggested that the Na$^+$-K$^+$-2Cl$^-$ cotransporter serves as both a Na$^+$ and a K$^+$ sensor in the portal vein and is downregulated by high Na$^+$ and K$^+$ intake (25). One caveat in this group's design, however, is that the hepatoportal area may have been exposed to very high [K$^+$] and/or hypertonicity because KCl and/or NaCl solutions were injected into the portal vein as a bolus over a very short period of time in the experimental design. If so, membrane depolarization or liver shrinkage could have provoked the increase in HANA (26).

Additional evidence for feedforward regulation comes from our recent studies demonstrating that when dietary K$^+$ intake was reduced to 33% of the control level for two weeks, there was no change in plasma [K$^+$] or [aldosterone], yet there was a significant decrease in renal K$^+$ excretion along with a decrease in (insulin-stimulated) extrarenal cellular K$^+$ uptake (27, 28). We also routinely observed a decrease in (insulin-stimulated) extrarenal cellular K$^+$ uptake in rats fasted overnight before assay of K$^+$ uptake, compared with fed rats, with little if any decrease in plasma [K$^+$] (C.S. Choi, J. Youn & A. McDonough, unpublished). These studies demonstrate that, analogous to the response to increased K$^+$ intake, the body is able to sense reduced K$^+$ intake in the absence of even minor changes in extracellular [K$^+$] and, via unidentified signaling connection(s), both decreases renal K$^+$ excretion to match output to input as well as decreases insulin-stimulated cellular K$^+$ uptake to prevent a postprandial drop in plasma [K$^+$]. These results support the notion that K$^+$ intake is sensed independently of circulating [K$^+$] and support the operation of a feedforward control mechanism that is capable of close control of ECF [K$^+$].

Evidence that a Gut Factor Is Activated During Dietary K$^+$ Intake

Motivated by the evidence reviewed above, we directly tested the hypothesis that K$^+$ intake is sensed by putative K$^+$ sensors in the hepatic portal vein (29). We assayed renal K$^+$ excretion as a biomarker of sensitivity to this putative signal. In our first attempts, we infused K$^+$ for 2 h into the stomach, the hepatic portal vein, or a systemic vein of fasted rats. Plasma [K$^+$] and renal K$^+$ excretion profiles were indistinguishable between the three infusion groups, which seemed to provide evidence against portal or gut sensing of K$^+$ intake. Next, we fed the rats a K$^+$-free meal during the K$^+$ infusions to more closely mimic the reality of K$^+$ ingestion with a meal. In contrast to the fasting state, there was a dramatic effect of K$^+$ infusion route on the plasma [K$^+$] profile during K$^+$ infusion combined with a K$^+$-free meal. Intraportal and systemic infusions similarly increased plasma [K$^+$], whereas intragastric K$^+$ infusion did not significantly increase plasma [K$^+$]. However, renal K$^+$ excretion was similarly increased in all three K$^+$ infusion groups fed the K$^+$-deficient meal. That is, the ratio of change in K$^+$ excretion to change in plasma [K$^+$] (the renal efficiency) was markedly increased when the intragastric K$^+$ infusion was combined with a meal, the normal condition of dietary K$^+$ intake. Plasma [K$^+$] did not rise during the intragastric K$^+$ infusion because the meal either slowed the rate of K$^+$ absorption into the blood or increased K$^+$ cellular uptake or both. In any case, adding the meal to the K$^+$ infusion revealed an increase in renal K$^+$ secretion in the absence of the rise in plasma [K$^+$] that was seen when K$^+$ was infused in the unfed state. These data provide evidence for a gut, rather than a hepatoportal, factor that plays a role in K$^+$ homeostasis by a feedforward mechanism that senses K$^+$ intake during a meal and enhances renal clearance of plasma K$^+$. This response would, theoretically, minimize increases in plasma K$^+$ concentration during or after a meal.

GFR: glomerular filtration rate

Postprandial Changes in Hormones that Contribute to Feedforward Control of K+ Homeostasis

After a meal, insulin secretion by the pancreatic β-cells is stimulated by meal nutrients such as glucose and amino acids, and the rise in insulin then plays a critical role in the cellular disposal of the ingested nutrients. Because insulin also stimulates cellular K+ uptake, meal ingestion also leads to a rapid transfer of K+ from the ECF to the ICF of insulin-sensitive cells, buffering a postprandial increase in [K+]. In fact, intravenous K+ infusion results in a lower plasma [K+] profile in rats coincidentally fed a meal compared with fasted rats (29, 30). In addition, raising blood glucose alone with an oral load enhances extrarenal disposal of a K+ load by stimulating insulin secretion (31, 32). Oral glucose can even lower serum [K+] in patients with end-stage renal disease; such patients have negligible renal K+ excretion, demonstrating the importance of endogenous insulin secretion to prevent hyperkalemia in this group by transferring K+ from the ECF to the ICF (33). In both normal and pathological conditions, the consumption of K+ along with a meal containing nutrients that stimulate insulin release leads to the activation of plasma K+ clearance independent of plasma [K+]—evidence of feedforward control of extracellular K+ homeostasis by insulin.

Glucagon has also been implicated in feedforward control of ECF [K+] after a protein-rich (and thus also K+-rich) meal. Intraportal, but not intrarenal, infusion of a physiological dose of glucagon produces significant increases in renal blood flow and glomerular filtration rate (GFR), suggesting the existence of a hepatorenal axis (34). Earlier studies had determined that glucagon infusion also increases the excretion of inorganic ions, including K+ (35). Because cyclic AMP (cAMP), the intracellular second messenger of glucagon in the liver, is rapidly released from hepatocytes into the plasma after glucagon infusion (36) and because the kidney can take up and secrete cyclic nucleotides, Bankir and colleagues (37–39) pos-

tulated that extracellular circulating cAMP is the link between glucagon's action in the liver and the subsequent renal hemodynamic and ion transport effects. These researchers' results demonstrated that the renal increase in GFR after glucagon infusion depends on both glucagon and a rise in plasma cAMP and that cAMP alone can account for the increase in ion excretion. Specifically related to K+, the focus of this review, Bankir and colleagues' results demonstrated that a 100-min infusion of glucagon roughly doubled the excretion of K+ and lowered plasma [K+] (37). Importantly, they also calculated that the increased K+ excretion was not just a function of increased tubular flow rate (secondary to the increase in GFR). Rather, there was a parallel increase in the transtubular K+ gradient (the ratio of urine-to-plasma [K+] divided by the ratio of urine-to-plasma osmolality), with the increases in K+ excretion suggesting a specific effect on a K+ secretion pathway(s). The effect of glucagon on K+ excretion (but not those on urine output) was fully reversed 40 min after glucagon infusion was terminated (37). These effects of dietary protein to increase GFR and K+ secretion is evidence for another layer of feedforward control of renal K+ excretion to buffer a rise in plasma [K+] after a meal.

Putting It All Together: Feedforward Plus Feedback Control Mechanisms Control Plasma K+

The studies summarized above provide evidence for three distinct layers of feedforward regulation aimed to acutely buffer ECF [K+] after a K+-containing meal, all independent of a rise in ECF [K+]: (*a*) Insulin release rapidly stimulates cellular K+ uptake into extrarenal, insulin-responsive tissues; (*b*) glucagon, at least in part owing to cAMP released from the liver, rapidly increases renal K+ excretion after a protein-rich meal; and (*c*) an unidentified gut factor senses K+ ingestion, rapidly stimulates renal K+ excretion, and potentially stimulates extrarenal K+ uptake. If plasma [K+] increases despite these layers of control, feedback regulation is activated: Plasma K^+ can increase K^+ excretion, as discussed above, and aldosterone synthesis is stimulated, which increases K^+ secretion secondary to activation of ENaC (16). Aldosterone acts only after a finite lag period, not well suited for rapid control, but can chronically increase K^+ secretion until $[K^+]$ is normalized. The close physiological control of ECF $[K^+]$ is evidence of the efficacy of the combination of feedback with feedforward control.

Regulation of Renal Function by the Gastrointestinal Tract: New Directions

That the three layers of feedforward regulation (insulin, glucagon, gut factor) are activated by ingestion of K^+ with a meal raises the possibility of additional layers of regulation of K^+ homeostasis by the gut. The regulation of renal function by the gastrointestinal (GI) tract was very recently reviewed (26). Relevant to K^+ status, uroguanylin, produced in the small intestines, is stimulated by an oral Na^+ load or a high-salt diet and is both natriuretic and kaliuretic without changing GFR or renal blood flow (40). This suggests a fourth layer of K^+ regulation after a salt-containing meal. However, when the effects of uroguanylin were studied directly in principal cells of the isolated mouse collecting duct (the cells responsible for K^+ secretion in this region), researchers found that the signaling pathway is cGMP independent and that uroguanylin activates a G protein–coupled receptor that increases arachidonic acid concentration via phospholipase A_2 activation, a pathway known to inhibit apical ROMK (renal outer medulla potassium) channels in these cells (41, 42). Thus, inhibition of ROMK apparently does not explain the uroguanylin-mediated kaliuresis. Another potential layer of GI regulation of K^+ homeostasis is on the horizon now that it has been discovered that taste channels and receptors, analogous to those found in the tongue, are expressed in entero-endocrine cells of the intestine and respond to bitter, sweet, and umami by releasing signaling substances that can act on target tissues throughout the body (43, 44).

Whether dietary K^+ can alter gating of these receptors or channels in a manner that would alter renal K^+ excretion or the distribution of K^+ between ICF and ECF remains to be investigated.

MOLECULAR MECHANISMS OF RENAL K^+ ADAPTATION

When dietary K^+ intake is chronically increased or decreased, the kidneys respond by appropriately increasing or decreasing K^+ excretion, respectively. This so-called K^+ adaptation is critical for chronic K^+ balance and involves adaptations all along the nephron. Fine and final control of K^+ excretion is routinely assigned to the distal portion of the nephron, especially the collecting ducts. This region contains two cell types that play distinct roles in K^+ homeostasis and adaptation: (*a*) Principal cells can secrete K^+, and (*b*) intercalated cells (in the medullary collecting duct) can reabsorb K^+. It has been well established that high K^+ intake increases K^+ secretion in principal cells and that low K^+ intake suppresses K^+ secretion. The topic of chronic dietary K^+ regulation of renal K^+ transport has been reviewed recently (7, 45–47), and additional progress has been made more recently. In this section we review the signaling mechanisms known to connect a change in K^+ intake to a homeostatic change in K^+ excretion and the gaps in our knowledge that still remain.

Renal Responses to High K^+ Intake

To secrete K^+ from the ECF to the tubular lumen, the cation is pumped into the cell across the basolateral membrane via the Na^+ pump and is then transported from the cell to the tubular lumen via apical K^+ channels. The K^+ channel that secretes K^+ under normal flow conditions is a ROMK-like small-conductance K^+ (SK) channel originally cloned by Hebert and colleagues (48). In addition, both principal and intercalated cells express big-conductance K^+ (BK) channels that have a low open probability and density under normal conditions but increased activity under high-flow conditions

ROMK: renal outer medulla potassium channel

or high-K$^+$ diet [which increases flow to the cortical collecting duct (CCD)], recently reviewed by Sansom and colleagues (49, 50).

During chronic high K$^+$ intake or acute experimental infusion, the following adaptations can occur to increase K$^+$ secretion: (*a*) An increase in plasma K$^+$ stimulates Na,K-ATPase activity and cellular K$^+$ uptake, which increases the driving force for K$^+$ secretion in the collecting duct cells. (*b*) Proximal tubule and loop of Henle salt and water absorption is inhibited during an acute K$^+$ load (15, 51, 52), which increases tubule flow rate in the distal nephron. This increase in tubule flow rate can both activate K$^+$ secretion through BK channels (49) and increase the delivery and reabsorption of NaCl in the principal cells, which increases the electrochemical driving force for K$^+$ secretion through ROMK (7). (*c*) High plasma K$^+$ stimulates aldosterone synthesis and secretion, which, by increasing both Na$^+$ reabsorption via ENaC and Na,K-ATPase, increase the electrochemical gradient for K$^+$ secretion into the lumen and decrease K$^+$ reabsorption in the collecting ducts (7, 47). At the molecular level, there is evidence that ROMK and BK channel are expressed in both apical membranes and intracellular vesicles and that both redistribute to the apical membranes during a chronic high-K$^+$ diet (53, 54), without an apparent increase in total abundance (54, 55).

Many open questions remain in our understanding of what drives the renal response to a K$^+$ challenge (whether it is the normal amount consumed in a meal, a high-K$^+$ diet, or acute infusion). The effects of aldosterone and high plasma K$^+$ on the collecting ducts can directly increase K$^+$ secretion via ROMKs like SK, as described above. However, significant K$^+$ adaptation can occur independently of elevations in aldosterone (56) and with little or no change in plasma [K$^+$] (28, 57). Although we can postulate that a gut factor, glucagon, and uroguanylin are mediators, there is scant evidence that can explain regulation at the level of receptors and transporters. At the subcellular level, signaling studies in the thick ascending limb, a region in

which Na$^+$ and volume reabsorption is inhibited during high K$^+$ intake, show that a high-K$^+$ diet is associated with increases in both inducible nitric oxide synthase (iNOS) and nitric oxide (NO) production as well as heme oxygenase 2 and carbon monoxide (CO) production (58, 59). Both NO, via cGMP, and CO stimulate the 70-pS ROMK-related channel in the thick ascending limb apical membrane. The stimulation of the channel increases K$^+$ recycling in the medulla because of the parallel NKCC and is not thought to contribute to kaliuresis. Nonetheless, these studies reveal molecular mechanisms that may also be activated in the collecting duct to increase K$^+$ secretion during high K$^+$ intake.

Renal Responses to K$^+$ Restriction

During K$^+$ deprivation sufficient to lower plasma K$^+$, the following adaptations that decrease K$^+$ excretion occur. (*a*) GFR is reduced and fractional absorption in the proximal tubule increases, reducing delivery to the distal nephron (60). (*b*) The abundance of ROMKs is reduced by ∼50% (55). (*c*) The abundance of Na$^+$ transporters (NKCC2, NCC, and ENaC) is reduced by ∼50%, which can decrease the electrochemical gradient for K$^+$ secretion secondary to reduced Na$^+$ reabsorption. (*d*) The abundance of H,K-ATPase is increased, which is likely to facilitate K$^+$ reabsorption in the face of decreased ROMKs (61). (*e*) Finally, perhaps most important to reduce K$^+$ excretion, ROMK is phosphorylated and retracted from the apical membrane (7, 45, 46).

Wang and colleagues (7, 45, 46) have intensively investigated and recently reviewed the molecular mechanisms connecting intake of a nominally K$^+$-free diet to the redistribution of ROMK out of the apical membrane in the principal cells, thus reducing K$^+$ secretion and excretion. Wang and colleagues provided evidence that low K$^+$ intake for one week initiates the following chain of events: (*a*) Low K$^+$ stimulates the production of superoxide anions (62), at least in part by activating

NADPH oxidase II (63); (*b*) superoxide production increases expression of protein tyrosine kinase (PTK) such as c-Src by two to three days (62, 64) and decreases expression of protein phosphatase 2B (65); and (*c*) PTK phosphorylates ROMK (45), which provokes the redistribution of ROMK out of the apical membranes (53). In support of this chain of events, pretreatment with the reactive oxygen species (ROS) scavenger tempol prevents the increased expression of c-Src, blunts the phosphorylation of ROMK, and blunts the decrease in K^+ excretion (62, 63).

In addition, Wang and colleagues also explored the role of mitogen-activated protein kinase (MAPK) in mediating the effect of superoxide anions on ROMK channel activity and found that a one-day low K^+ intake increased the phosphorylation of p38 MAPK and extracellular signal–regulated kinase (ERK), both blunted by tempol, and that this was associated with decreased ROMK channel activity independent of and parallel to activation by PTK activity (66). This research group suggests that MAPK-induced inhibition of ROMK activity plays a predominant role in suppressing K^+ secretion in the early stage (<24 h) of K^+ depletion. Further study supports the hypothesis that ING4 (inhibitor of growth 4), which is stimulated by low K^+ intake dependent on superoxide anion production, may mediate the effect of a low-K^+ diet on MAPK to depress ROMK channel activity (67). In summary, a low-K^+ diet increases superoxide anion production, which, in parallel, increases the expression and activity of (*a*) PTK and (*b*) ING4 and MAPK, which in turn independently depress CCD K^+ channel activity and K^+ secretion (67).

Signals to the Kidney?

The signaling connection between a decrease in K^+ intake, a decrease in plasma $[K^+]$, and activation of NADPH oxidase has not been explored in detail. As discussed above, at the cellular level there is evidence that a high K^+ intake increases NO production (in the thick ascending limb at least) associated with increased K^+ channel activity, whereas low K^+ increases superoxide anion production, known to react with and lower the levels of NO (46, 68, 69) associated with decreased K^+ channel activity. Angiotensin II (AngII), an important renal hormone, stimulates Na^+ reabsorption via stimulating NADPH oxidase, increasing superoxide anion and depressing cellular NO levels (68, 69). AngII suppresses K^+ secretion (70), and Wei and colleagues (71) examined whether AngII also inhibits ROMK activity. Their studies confirmed their hypothesis and demonstrated that during a K^+-restricted diet (not during a normal-K^+ diet) AngII treatment inhibits ROMK channel activity, at least in part by increasing expression of NADPH oxidase and PTK. Whether renin secretion and/or local AngII production are increased during K^+ restriction in general remains to be determined.

Chronically reducing K^+ intake by two-thirds to 0.33% (normal rat chow is 1% K^+) activates renal mechanisms to retain K^+ but does not significantly depress plasma $[K^+]$ (28), evidence that a change in plasma K^+ is not a requisite error signal. Comparing the responses observed in rats fed 0.33% K^+ with normal K^+ versus responses in rats fed 0 K^+ is informative as to the signaling mechanisms driving the homeostatic responses. Rats fed a K^+-deficient diet long enough to cause a drop in plasma $[K^+]$ show a wide array of changes in transporter protein levels, including a decrease in muscle Na,K-ATPase $\alpha2$ subunit abundance (72), decreases in renal ROMK (55) and aquaporin 2 (AQP2) (73), increases in renal colonic H,K-ATPase (61), and a fall in muscle K^+ stores (4). These changes were not observed when rats were fed a 0.33% K^+ diet, even though urinary K^+ excretion was reduced 80%, evidence of significant K^+ conservation in the absence of an error signal, i.e., a change in plasma $[K^+]$ from normal levels. Even when rats were maintained on the 0.33% K^+ diet for 30 days, they did not manifest any changes in plasma $[K^+]$ or in the abundance of muscle Na,K-ATPase $\alpha2$

PTK: protein tyrosine kinase

ROS: reactive oxygen species

MAPK: mitogen-activated protein kinase

isoform, renal ROMK, H,K-ATPases, or AQP2. These results lead us to believe that the changes observed in response to the K^+-deficient diet may be provoked by an actual fall in plasma $[K^+]$. The K^+ adaptation in the 0.33%-K^+-diet-fed rats may be explained by a marked increase in the expression of the PTK cSrc along with the expected phosphorylation of ROMK, leading to the inactivation of ROMK via its retraction from the cell surface (28). This retraction of ROMK would increase net K^+ reabsorption in the CCDs mediated by existing pools of H,K-ATPases.

The findings after a "modest" reduction in K^+ intake demonstrate that there are two categories of responses to reduced K^+ intake: a first line of defense activated by an undefined signal that provokes renal K^+ conservation, and a second line of defense activated when plasma $[K^+]$ actually falls, in which ROMK and H,K-ATPase abundance changes occur, consistent with renal K^+ conservation. Very important to understanding K^+ homeostasis, these findings establish that a drop in plasma $[K^+]$ is not a requisite for setting off renal K^+ adaptation mediated by ROMK phosphorylation and the decreased distribution of ROMK in the apical membrane. Further study of models of "modest" reductions in K^+ intake will be very useful to determine the homeostatic signals that connect intake to output because such models eliminate the set of changes that occur secondarily to a fall in plasma $[K^+]$.

The proximal response to chronic K^+ deprivation described in the renal cells at this stage of our understanding involves increasing NADPH oxidase abundance and activity. That is, we do not understand the signaling upstream of this point. Whatever the signal, it can likely activate NADPH oxidase and superoxide anion production. Another important gap that should be addressed is to determine whether the acute increase in K^+ excretion seen after K^+ ingestion, mediated by the putative gut factor, involves a decrease in phosphorylation of ROMK and whether this decrease is mediated by the suppression of NADPH oxidase generation of superoxide anions. This information will

be critical to establish the nature and actions of the gut factor.

MOLECULAR MECHANISMS OF EXTRARENAL K^+ ADAPTATION

Critical Role of Insulin in Acute Regulation of Extracellular K^+

After a K^+-containing meal, plasma insulin prevents an excessive rise in plasma K^+ level by rapidly stimulating cellular K^+ uptake via Na^+ pumps (74). Without this action of insulin, ingested K^+ could produce life-threatening hyperkalemia. DeFronzo and coworkers (8) showed that insulin infusion caused a dose-dependent decline in plasma $[K^+]$ and that the majority of the decline in plasma $[K^+]$ was due to net K^+ uptake by the splanchnic bed (during the first hour) and peripheral tissues (during the second hour). Choi et al. (75) introduced the K-clamp technique for the quantification of insulin's action on cellular K^+ uptake in vivo at constant, basal plasma $[K^+]$. In the K-clamp technique, conscious rats are infused with insulin to stimulate K^+ (and glucose) uptake and then, on the basis of immediate determinations of plasma $[K^+]$ (and glucose), infused with K^+ (and glucose) appropriate to clamp plasma $[K^+]$ (and glucose) at basal levels. The amount of K^+ infused (K_{inf}) at steady state is equivalent to the sum of insulin's effects on cellular K^+ uptake and renal K^+ excretion. The initial studies showed that physiological concentrations of insulin do not significantly affect renal K^+ excretion (27, 75). Therefore, the K-clamp appears to be a good measure of insulin-stimulated cellular K^+ uptake (75). This technique has been used to determine that insulin's action on cellular K^+ uptake is subject to control by dietary K^+ intake.

Extrarenal Adaptation to Low K^+ Intake: Decreased Insulin Sensitivity

When K^+ intake is very low, K^+ output will exceed K^+ input despite renal K^+ conservation (via the colon, etc.). When this happens, the large stores of K^+ in the muscle

ICF are tapped to blunt the fall in ECF [K$^+$] (**Figure 2**). The K$^+$ shift from the ICF to the ECF has been attributed to a fall in muscle Na$^+$ pump levels (4). During dietary K$^+$ deprivation, Na$^+$ pump number decreases by more than 50%, estimated by ^{3}H-ouabain binding, Na,K-ATPase enzymatic activity, and ^{86}Rb uptake (12, 76). At the molecular level, K$^+$ deprivation provokes a specific decrease in the Na$^+$ pump α2 (not α1) catalytic isoform abundance in skeletal muscles (72, 77). This decrease in muscle Na$^+$ pumps theoretically protects against acute hypokalemia during insulin stimulation if an animal consumes a diet containing sugar but low K$^+$. We tested the hypothesis that insulin-stimulated cellular K$^+$ uptake decreases during K$^+$ deprivation in rats as a function of the decrease in muscle α2 abundance. After 10 days of K$^+$ deprivation, plasma [K$^+$] fell to 2.9 mM, Na,K-ATPase activity and α2 subunit levels fell by more than 50%, muscle K$^+$ stores decreased significantly (4), and insulin-mediated K$^+$ clearance to the ICF decreased to less than 10% of control. These findings provide evidence for a relationship between the Na,K-ATPase α2 pool size and insulin-stimulated K$^+$ uptake (75). However, when the same analyses were done after only two days of K$^+$ deprivation, plasma [K$^+$] fell slightly from 4.2 mM to 3.8 mM, there was not yet a significant fall in muscle Na,K-ATPase activity or expression, and insulin-mediated cellular K$^+$ uptake decreased to 20% of control—evidence against a direct relationship between the Na,K-ATPase α2 pool size and insulin-stimulated K$^+$ uptake (75). These data demonstrate that short-term K$^+$ deprivation leads to a significant insulin resistance of cellular K$^+$ uptake (with normal stimulation of glucose uptake) that precedes the decreases in muscle Na,K-ATPase expression, perhaps mediated by Na$^+$ pump inactivation or internalization. Longer K$^+$ deprivation does provoke a more profound decrease in insulin-sensitive cellular K$^+$ uptake that is likely secondary to the decrease in Na,K-ATPase α2 abundance (75). In addition, this study (75) demonstrated that insulin-stimulated cellular K$^+$ uptake, like renal K$^+$ excretion, is profoundly suppressed during K$^+$ deprivation. Thus, K$^+$ adaptation occurs in both the kidneys and extrarenal tissues in response to changes in K$^+$ intake. A correlation between K$^+$ intake and insulin-sensitive K$^+$ clearance is physiologically advantageous because it would, theoretically, blunt acute hypokalemia after a K$^+$-poor, carbohydrate-rich meal.

Differential Regulation of Insulin Action on K$^+$ Uptake versus Glucose Uptake

Of the many homeostatic systems of the body, the K$^+$ and glucose homeostatic systems are unique in that they share acute regulation by insulin. This feature suggests the potential for interactions or cross talk between the two systems. Insulin resistance with respect to glucose metabolism is usually associated with hyperinsulinemia because pancreatic β-cells increase insulin secretion to compensate for insulin resistance (78). The resulting hyperinsulinemic may provoke hypokalemia secondary to stimulating excessive cellular K$^+$ uptake unless insulin's action on K$^+$ uptake is similarly dampened. Impaired insulin action on K$^+$ fluxes has, indeed, been reported in obesity and diabetes (79, 80), which are both associated with insulin resistance with respect to glucose metabolism. However, insulin-mediated K$^+$ uptake is not altered in uremia, whereas insulin-mediated glucose uptake is markedly impaired (81). We determined the acute effect of insulin on K$^+$ cellular uptake in the high-fat-fed rat, a well-established model of insulin resistance with respect to glucose metabolism (82, 83). We discovered, inadvertently, that rats ate less of the typical 66% high-fat diet such that their dietary K$^+$ intake was reduced to one-third of that consumed by the rats eating a control diet of 1% K$^+$ (27). In this series, both insulin-stimulated K$^+$ uptake as well as insulin-stimulated glucose uptake were suppressed. Supplementing the high-fat diet with enough K$^+$ to match that consumed in the control diet group restored insulin action on K$^+$ uptake to control levels but did not correct insulin-stimulated glucose uptake (27).

These data indicate that insulin action on glucose uptake is selectively (i.e., without change in insulin action on K^+ uptake) impaired by high-fat feeding when K^+ intake is normal. These studies reveal two distinct varieties of insulin resistance: resistance to cellular glucose uptake observed in type 2 diabetes and resistance to cellular K^+ uptake observed during K^+ deprivation or fasting.

The molecular mechanisms responsible for the differential regulation of glucose versus K^+ cellular uptake by insulin have not been examined. Presumably, these pathways involve events that occur after insulin binds to its receptor, which is a common step in the regulation of K^+ and glucose uptake. The resistance of muscle cell glucose uptake during insulin resistance is attributed to an impairment of insulin signaling at the level of PI 3-kinase, which affects GLUT4 trafficking (recently reviewed in Reference 84). Insulin has been proposed to activate cellular K^+ uptake by activation of Na^+ pumps resident in the plasma membrane or by translocation of Na^+ pumps from intracellular pools to the plasma membrane, although this issue remains controversial (74, 85, 86). A rather new proposal is that the Na^+ pumps can be activated by the translocation of the Na^+ pump–associated protein phospholemman (PLM) to the plasma membrane; such translocation has been demonstrated in exercising muscles (87). Whether this PLM translocation also occurs during insulin stimulation remains to be determined. Establishing the connection between K^+ deprivation and decreased sensitivity of cell K^+ uptake to insulin is an important goal. Testing the hypothesis that the signaling cascade activated in the kidney during K^+ deprivation—namely increased activity of NADPH oxidase, superoxide anion production, and PTK activity—affects insulin stimulation of K^+ cellular uptake warrants consideration.

In a related line of investigation, we examined insulin-stimulated cellular K^+ uptake during chronic dexamethasone treatment. This was an important gap to explore because (a) it has long been known that dexamethasone treatment causes insulin resistance of cellular glucose clearance (88), (b) chronic treatment of patients or experimental animals with glucocorticoids leads to an increase in skeletal muscle Na^+ pumps by $\sim$50% (89, 90), and (c) this large increase in muscle Na^+ pumps could theoretically provoke an increase in insulin-driven cellular K^+ uptake after a meal and acute hypokalemia. Using the K-clamp, we assessed insulin-stimulated K^+ and glucose uptake in rats treated with dexamethasone for seven days, sufficient to raise Na,K-ATPase α2 levels by $\sim$50%. Interestingly, the measurements revealed depressed (rather than increased) cellular uptake of both K^+ and glucose in response to insulin, implicating a common step in insulin signaling that affects both glucose and K^+ uptake (91). These results provide novel evidence for insulin resistance of K^+ clearance as well as glucose clearance during glucocorticoid treatment. Although we were originally thinking that cellular clearance of K^+ may be higher in steroid-treated animals, the results suggest that the increase in Na,K-ATPase synthesis and abundance may be a compensation to counteract the dexamethasone-induced insulin resistance and that insulin-stimulated cell K^+ uptake may be even more suppressed if the Na,K-ATPase pool size is not increased.

Extrarenal Adaptation to High K^+ Intake

For more than 50 years, investigators have known that that rats maintained on a high-K^+ diet survive an acute K^+ load that is lethal to rats maintained on a control diet and that this adaptation is observed even in nephrectomized rats (92). This observation has led to the hypothesis that a major mechanism of the K^+ adaptation is extrarenal. In fact, it was later demonstrated that rats adapted to a 10% K^+ diet for 10 days exhibited (a) increased skeletal muscle (ouabain-binding) Na^+ pump activity and number as well as increased ^{86}Rb uptake during an acute K^+ load compared with control rats and (b) increased ^{86}Rb efflux upon termination of the K^+ infusion (93). These results demonstrate that skeletal muscle serves as a buffer compartment both during acute high K^+ intake, temporarily transferring K^+ from the ECF to the ICF until renal excretion can match K^+ intake, as well as during chronic K^+ deprivation by the net transfer of K^+ from the ICF to the ECF.

Rapid Reversal of K^+ Conservation State

In the wild it is common for large carnivorous mammals to experience long periods of fasting between large, protein-rich meals. The adaptations that occur to preserve ECF $[K^+]$ during prolonged fasts, e.g., reduced renal K^+ secretion and reduced insulin-stimulated K^+ uptake, can theoretically pose a risk of hyperkalemia when the fast is broken with a high-K^+ meal if the K^+ conservation mechanisms are not rapidly turned off. Sensing of K^+ intake in the gut and activation of a gut factor that communicates to key tissues in a feedforward manner may rapidly turn off the K^+ conservation mechanisms and prevent a rise in plasma $[K^+]$. Two relevant studies support this idea. First, Norgaard and colleagues (12) fed rats a K^+-deficient diet for three weeks, during which plasma K^+ fell to ~2 mM, muscle Na^+ pumps (ouabain binding) fell to 30% of control, and muscle K^+ stores fell by 40%. Nonetheless, a single day of a normal-K^+ diet corrected plasma and muscle K^+ stores to control levels. Sadre and colleagues deprived rats of K^+ for four weeks and then demonstrated that K^+ administration via a gastric tube normalized muscle K^+ stores within 24 h but provoked an increase in plasma $[K^+]$, which the researchers attributed to a decreased ability of the muscle to absorb the large K^+ dose (94). In both of these studies K^+ restoration was provided via the GI tract and may have stimulated a gut-derived factor to stimulate cellular K^+ uptake and renal K^+ excretion. An important question to answer is whether the gut is critical to the homeostatic adjustment to K^+ restoration in a K^+-depleted animal. This question could be assessed by infusing a K^+ load via another route.

Clearing Extracellular Fluid K^+ Back to the Intracellular Fluid during Exercise and Ischemia

K^+ homeostasis can also be significantly disrupted by exercise and/or ischemia independently of changes in K^+ intake. During ischemia, ECF $[K^+]$ bathing the muscle is significantly elevated because the Na^+ pumps are unable to actively clear K^+ from the ECF at a rate to match passive leaks. During exercise, $[K^+]$ is also elevated around contracting muscles because the rate of K^+ leaving the cell (via channels) rises more quickly than the Na^+ pump can actively pump K^+ back into the cell. Interstitial $[K^+]$ in these conditions can rise to 9–10 mM, bathing the exercising muscle (reviewed in detail recently in References 5 and 76). The high ECF $[K^+]$ can have life-threatening effects: Above 6.0 mM $[K^+]$ there are marked changes in the electrocardiogram, and at higher K^+ concentrations lethal arrhythmias may develop (95). When net K^+ influx rate rises above net K^+ efflux rate, homeostasis is restored. This can, theoretically, be accomplished by increasing active K^+ uptake (via Na,K-ATPase) or suppressing K^+ efflux (via channels or cotransporters). Clausen and colleagues (76) have provided evidence that the Na^+ pump rate is stimulated in exercising muscle by both the rise in ICF Na^+ as well as the local release of catecholamine and calcitonin gene-related peptide (CGRP). With chronic exercise training, the number of Na^+ pumps increases, which improves K^+ homeostasis by increasing Na^+ pump capacity (5, 76).

Increasing Na,K-ATPase activity to restore local K^+ homeostasis is not feasible in some cases of ischemia or exercise when the tissue becomes ATP depleted. Muscle ATP stores can become depleted and AMP levels increased owing to high ATPase activity during exercise or lack of O_2/substrate to make ATP in ischemia. We hypothesized that AMP-dependent protein kinase (AMPK) may play a role in normalizing local K^+ gradients (96). AMPK [reviewed recently (97)] acts as a cellular fuel gauge to maintain or restore energy balance by switching

AMPK: AMP-dependent protein kinase

on ATP-generating processes (e.g., cell glucose uptake, fatty acid oxidation) and switching off ATP-consuming processes (e.g., the synthesis of lipid, glycogen, and protein). Mice deficient in muscle and heart AMPK activity exhibit depressed exercise-induced glucose uptake, significant muscle weakness and impaired recovery from fatigue (98), and abnormalities in cardiac electrical conductance (99), all consistent with a problem with local K^+ homeostasis.

We specifically tested the hypothesis that infusing conscious rats with the AMPK activator 5-aminoimidazole-4-carboxamide-1-β-D-ribofuranoside (AICAR), to imitate the condition of an elevated AMP:ATP ratio in the ICF, would provoke an increased rate of clearance of K^+ from the ICF to the ECF (96). It was previously established that AICAR infusion increased cellular uptake of glucose (100). We found that a 3-h AICAR infusion decreased plasma $[K^+]$ from 4.0 mM to 3.3 mM in control rats, from 4.5 mM to 3.8 mM in rats fed a high-K^+ diet plus spironolactone, and from 3.2 mM to 2.5 mM in rats fed a K^+-deficient diet for one week, all evidence for a pronounced effect of this AMPK activator on K^+ homeostasis. There was no evidence for increased renal K^+ excretion, consistent with the idea of a net shift of K^+ from the ECF to the ICF. To test the hypothesis that the K^+ redistributed into muscle, we examined the response in mice lacking AMPK activity in heart and skeletal muscle (101) and found that the K^+-lowering response was very significantly blunted compared with control mice (96). We could detect no evidence for activation of Na,K-ATPase activity or translocation of Na,K-ATPase isoforms to the plasma membrane in samples in which translocation of GLUT4 to the membrane was evident (96), indicating that this response is not likely secondary to Na,K-ATPase activation. These results suggest a novel mechanism for redistributing K^+ from the ECF to the ICF owing to the activation of AMPK activity with AICAR. The role of AMPK as a cellular fuel gauge can be reconciled with the clearance of K^+ that accumulates in the ECF

during exercise or ischemia if one postulates that AMPK decreases a K^+ efflux pathway in muscle. This action would promote net cellular K^+ uptake (by decreasing the ratio of K^+ efflux via channels/leaks to K^+ influx via ATPases) without increasing ATP consumption. Establishing that a K^+ efflux pathway is regulated by AMPK and the molecular identity of the pathway(s) is an important goal because the answers could suggest novel therapeutic strategies to treat acute life-threatening hyperkalemia.

LINKING MECHANISMS OF K^+ HOMEOSTASIS TO THE CARDIOVASCULAR BENEFITS OF A K^+-RICH DIET

Beneficial effects of high dietary K^+ intake on blood pressure, cardiovascular disease, and stroke have been well documented and reviewed (102, 103). Numerous studies have addressed the beneficial effects of K^+, but fewer have addressed the molecular mechanisms by which such benefits occur. One hypothesis proposed by Young et al. (103) is that elevation in dietary K^+ increases plasma $[K^+]$, thereby inhibiting free radical formation, smooth muscle proliferation, and thrombus formation. However, although the relationship between high K^+ intake and cardiovascular benefits is very strong, the evidence that the benefit is mediated by elevated ECF K^+ is not. First, beneficial effects of K^+ supplementation have been observed without an increase in plasma $[K^+]$: Nearly doubling daily K^+ intake (by adding 60 mmol/70 kg body weight) for three days did not change serum $[K^+]$ in healthy men or women but did diminish platelet reactivity (104). Moreover, in another study of healthy volunteers, KCl or potassium citrate supplementation for six weeks (30 mmol day^{-1}) did not change plasma $[K^+]$ but did significantly lower blood pressure (105). Second, clinical studies investigating the relationship between serum $[K^+]$ and risk of cardiovascular events have shown a U-shaped association: Both significant hypokalemia and hyperkalemia are risk factors for cardiovascular events (106, 107). In addition, treating heart failure patients with a combination of spironolactone and ACE inhibitors, which elevates ECF [K$^+$] by decreasing K$^+$ excretion rather than increasing intake, significantly increases hyperkalemia-associated morbidity and mortality (108). These findings of increased risks of cardiovascular events with elevated serum [K$^+$] are at odds with the idea that high dietary K$^+$ intake brings about its beneficial effects by increasing plasma [K$^+$].

Taken together, these considerations suggest that a mechanism other than plasma [K$^+$] must be the mediator of the beneficial effects of dietary K$^+$ intake. This review supports the hypothesis that the process of dietary K$^+$ intake stimulates the mediator of the cardiovascular benefit of a high-K$^+$ diet. Alternatively, we cannot rule out the possibility that a high-K$^+$ diet inhibits the production of a K$^+$-conserving factor that has long-term detrimental effects on the cardiovascular system. Regarding this alternative, a strong candidate for a factor that both promotes K$^+$ conservation during K$^+$ deprivation and has long-term detrimental cardiovascular effects is ROS such as superoxide anions.

The studies of Wang and colleagues (7, 46, 63) have demonstrated that K$^+$ deprivation leads to an increase in NADPH oxidase and superoxide anion production. Such increases activate PTK, key to the retraction of renal K$^+$ channels out of the apical membrane, which reduces K$^+$ secretion and excretion (7, 46, 63). Work from our group has established that this response can occur in the absence of a fall in plasma [K$^+$]: Modest K$^+$ deprivation also increases PTK and ROMK phosphorylation (28). However, consuming a high-K$^+$ diet for one to three days depresses levels of renal PTK and increases ROMK channel activity fourfold (64). Determining the effect of a high-K$^+$ diet or acute K$^+$ intake on renal NADPH oxidase activity and superoxide anion production is an important gap to fill because K$^+$ intake can rapidly increase K$^+$ excretion (29, 57). In addition, determining whether ROS play a

role in extrarenal K$^+$ adaptation needs to be investigated.

Young et al. (103) have reviewed the effects of a diet rich in K$^+$ to suppress free radical production and the sequelae of free radicals on the cardiovascular system. More recent studies have provided specifics about the detrimental effects of a low-K$^+$ diet and the beneficial effects of a high-K$^+$ diet for the vasculature. Regarding detrimental effects, carotid arteries from rabbits fed a low-K$^+$ diet for one to three weeks have 100% more superoxide anion production along with an enhanced sensitivity to vasoconstrictors and reduced sensitivity to endothelium-dependent stimuli; these changes can be corrected by treatment with a superoxide dismutase mimetic (109). Two recent studies by Fujita and colleagues (110, 111) in Dahl salt–sensitive (DS) rats fed an 8% NaCl diet demonstrated that the cardiac dysfunction and vascular injury observed in this model were significantly reduced by supplementation with an 8% KCl diet. Specific to superoxide anion production, NADPH oxidase subunits' expression levels (p22^{phox}, p47^{phox}, and gp91^{phox}) were increased approximately tenfold in arteries from DS rats fed a high-salt diet, ROS generation was increased approximately fourfold, and these levels were returned to baseline in rats supplemented with a high-K$^+$ diet (111). Interestingly, treatment with a high-K$^+$ diet or the ROS scavenger tempol had a rather small effect on the salt-sensitive increase in blood pressure but still attenuated vascular neointimal hyperplasia (111) and improved left ventricular relaxation (110). Although there is emerging agreement that high K$^+$ intake can decrease the expression of NADPH oxidase, decrease superoxide anion production, and reduce cardiovascular injury and dysfunction, still unresolved is the initial connection between the high K$^+$ intake and the change in the NADPH expression. Further experiments will have to be aimed at determining whether K$^+$ ingestion through the gut is a critical link in the pathway. Accomplishing this aim could establish a natural and straightforward mechanism to improve cardiovascular health.

SUMMARY POINTS

1. Additional evidence has been collected for feedforward regulation of K^+ homeostasis: Dietary K^+ intake is sensed in the gut, and an unidentified gut factor is activated to stimulate renal K^+ excretion. This pathway may explain renal and extrarenal responses to altered K^+ intake that occur independently of changes in ECF $[K^+]$.

2. Molecular mechanisms responsible for conserving ECF K^+ during fasting or K^+ deprivation are emerging: (*a*) Kidney NADPH oxidase activation initiates a cascade that provokes phosphorylation and retraction of K^+ channels from the cell membrane, which decreases K^+ excretion, and (*b*) muscle becomes resistant to postprandial insulin stimulation of cellular K^+ uptake while maintaining normal insulin sensitivity of cellular glucose uptake. How these mechanisms are triggered by K^+ deprivation remains unclear.

3. A new link has been established between cellular energy metabolism and ion transport: Cellular AMPK activity provokes acute transfer of K^+ from the ECF to the ICF, which may be important during exercise or ischemia.

FUTURE ISSUES

1. Although it has been established that renal NADPH oxidase and ROS are increased during diets lacking K^+, leading to the phosphorylation and internalization of ROMK, the signaling connection between a decrease in K^+ intake and the activation of NADPH oxidase has not been determined. Future experiments could be aimed at determining whether K^+ ingestion through the gut is a critical link in the pathway.

2. Significant increases in K^+ excretion can occur independently of elevations in aldosterone and with little or no change in plasma $[K^+]$ that may be mediated by a gut factor. Future studies are warranted to establish the nature of the mediator and the underlying molecular mechanisms at the level of the nephrons, receptors, and transporters. This information is critical to determining the identity of the putative gut factor.

3. Determining the effect of a chronic high-K^+ diet or acute K^+ intake on renal NADPH oxidase activity and superoxide anion production is an important gap to fill, as are determinations of whether the kaliuresis involves a decrease in ROMK phosphorylation and whether this decrease is mediated by the suppression of NADPH oxidase generation of ROS.

4. Further study of models of "modest" reductions in K^+ intake will be very useful to determine the homeostatic signals that connect intake to output because such models eliminate the set of changes that occur secondarily to a fall in plasma $[K^+]$.

5. Studies in low-K^+-fed rats revealed a new variety of insulin resistance to cellular K^+ uptake. The molecular mechanisms for this selective insulin sensitivity are not understood. Testing the hypothesis that the signaling cascade activated in kidney during K^+ deprivation—namely increased activity of NADPH oxidase, superoxide anion production, and PTK activity—affects the insulin stimulation of K^+ cellular uptake warrants consideration.

6. A fasting animal rapidly turns off K$^+$ conservation mechanisms when breaking the fast with a high-K$^+$ meal, and the mechanisms are not understood. Perhaps the K$^+$ intake via the gut is critical to the homeostatic adjustment to K$^+$ restoration.

7. Establishing that a K$^+$ efflux pathway is regulated by AMPK and the molecular identity of the pathway(s) is an important goal because the answers could suggest novel therapeutic strategies to treat acute life-threatening hyperkalemia.

DISCLOSURE STATEMENT

The authors are not aware of any biases that might be perceived as affecting the objectivity of this review.

LITERATURE CITED

1. Kurtzman NA, Gonzalez J, DeFronzo R, Giebisch G. 1990. A patient with hyperkalemia and metabolic acidosis. *Am. J. Kidney Dis.* 15:333–56
2. Bia MJ, DeFronzo RA. 1981. Extrarenal potassium homeostasis. *Am. J. Physiol.* 240:F257–68
3. Institute of Medicine Panel on Dietary Reference Intakes. 2004. *Dietary Reference Intakes for Water, Potassium, Sodium, Chloride, and Sulfate.* Washington, D.C.: Natl. Acad.
4. Thompson CB, Choi C, Youn JH, McDonough AA. 1999. Temporal responses of oxidative vs glycolytic skeletal muscles to K$^+$ deprivation: Na$^+$ pumps and cell cations. *Am. J. Physiol.* 276:C1411–19
5. Sejersted OM, Sjogaard G. 2000. Dynamics and consequences of potassium shifts in skeletal muscle and heart during exercise. *Physiol. Rev.* 80:1411–81
6. Giebisch G, Wang W. 1996. Potassium transport: from clearance to channels and pumps. *Kidney Int.* 49:1624–31
7. Wang W. 2004. Regulation of renal K transport by dietary K intake. *Annu. Rev. Physiol.* 66:547–69
8. DeFronzo RA, Felig P, Ferrannini E, Wahren J. 1980. Effect of graded doses of insulin on splanchnic and peripheral potassium metabolism in man. *Am. J. Physiol.* 238:E421–27
9. Ortiz RM. 2001. Osmoregulation in marine mammals. *J. Exp. Biol.* 204:1831–44
10. Weiner ID, Wingo CS. 1997. Hypokalemia—consequences, causes, and correction. *J. Am. Soc. Nephrol.* 8:1179–88
11. Heppel LA. 1939. The electrolytes of muscle and liver in potassium depleted rats. *Am. J. Physiol.* 127:385–92
12. Norgaard A, Kjeldsen K, Clausen T. 1981. Potassium depletion decreases the number of ^{3}H-ouabain binding sites and the active Na-K transport in skeletal muscle. *Nature* 293:739–41
13. Giebisch G. 1998. Renal potassium transport: mechanisms and regulation. *Am. J. Physiol.* 274:F817–33
14. Woda CB, Bragin A, Kleyman TR, Satlin LM. 2001. Flow-dependent K$^+$ secretion in the cortical collecting duct is mediated by a maxi-K channel. *Am. J. Physiol. Renal Physiol.* 280:F786–93
15. Field MJ, Stanton BA, Giebisch GH. 1984. Differential acute effects of aldosterone, dexamethasone, and hyperkalemia on distal tubular potassium secretion in the rat kidney. *J. Clin. Investig.* 74:1792–802
16. Bassett MH, White PC, Rainey WE. 2004. The regulation of aldosterone synthase expression. *Mol. Cell. Endocrinol.* 217:67–74
17. Rabinowitz L. 1996. Aldosterone and potassium homeostasis. *Kidney Int.* 49:1738–42
18. Rabinowitz L. 1988. Model of homeostatic regulation of potassium excretion in sheep. *Am. J. Physiol.* 254:R381–88
19. Calo L, Borsatti A, Favaro S, Rabinowitz L. 1995. Kaliuresis in normal subjects following oral potassium citrate intake without increased plasma potassium concentration. *Nephron* 69:253–58
20. Rabinowitz L, Sarason RL, Yamauchi H. 1985. Effects of KCl infusion on potassium excretion in sheep. *Am. J. Physiol.* 249:F263–71

21. Rabinowitz L, Green DM, Sarason RL, Yamauchi H. 1988. Homeostatic potassium excretion in fed and fasted sheep. *Am. J. Physiol.* 254:R357–80

22. McDonough AA, Thompson CB, Youn JH. 2002. Skeletal muscle regulates extracellular potassium. *Am. J. Physiol. Renal Physiol.* 282:F967–74

23. Morita H, Fujiki N, Hagiike M, Yamaguchi O, Lee K. 1999. Functional evidence for involvement of bumetanide-sensitive $Na^+K^+2Cl^-$ cotransport in the hepatoportal Na^+ receptor of the Sprague-Dawley rat. *Neurosci. Lett.* 264:65–68

24. Morita H, Fujiki N, Miyahara T, Lee K, Tanaka K. 2000. Hepatoportal bumetanide-sensitive K^+-sensor mechanism controls urinary K^+ excretion. *Am. J. Physiol. Regul. Integr. Comp. Physiol.* 278:R1134–39

25. Tsuchiya Y, Nakashima S, Banno Y, Suzuki Y, Morita H. 2004. Effect of high-NaCl or high-KCl diet on hepatic Na^+- and K^+-receptor sensitivity and NKCC1 expression in rats. *Am. J. Physiol. Regul. Integr. Comp. Physiol.* 286:R591–96

26. Michell AR, Debnam ES, Unwin RJ. 2008. Regulation of renal function by the gastrointestinal tract: potential role of gut-derived peptides and hormones. *Annu. Rev. Physiol.* 70:379–403

27. Choi CS, Lee FN, McDonough AA, Youn JH. 2002. Independent regulation of in vivo insulin action on glucose versus K^+ uptake by dietary fat and K^+ content. *Diabetes* 51:915–20

28. Chen P, Guzman JP, Leong PK, Yang LE, Perianayagam A, et al. 2006. Modest dietary K^+ restriction provokes insulin resistance of cellular K^+ uptake and phosphorylation of renal outer medulla K^+ channel without fall in plasma K^+ concentration. *Am. J. Physiol. Cell Physiol.* 290:C1355–63

29. Lee FN, Oh G, McDonough AA, Youn JH. 2007. Evidence for gut factor in K^+ homeostasis. *Am. J. Physiol. Renal Physiol.* 293:F541–47

30. Jackson CA. 1992. Rapid renal potassium adaptation in rats. *Am. J. Physiol.* 263:F1098–104

31. Allon M, Dansby L, Shanklin N. 1993. Glucose modulation of the disposal of an acute potassium load in patients with end-stage renal disease. *Am. J. Med.* 94:475–82

32. Alvo M, Krsulovic P, Fernandez V, Espinoza AM, Escobar M, Marusic ET. 1989. Effect of a simultaneous potassium and carbohydrate load on extrarenal K homeostasis in end-stage renal failure. *Nephron* 53:133–37

33. Muto S, Sebata K, Watanabe H, Shoji F, Yamamoto Y, et al. 2005. Effect of oral glucose administration on serum potassium concentration in hemodialysis patients. *Am. J. Kidney Dis.* 46:697–705

34. Premen AJ. 1987. Splanchnic and renal hemodynamic responses to intraportal infusion of glucagon. *Am. J. Physiol.* 253:F1105–12

35. Pullman TN, Lavender AR, Aho I. 1967. Direct effects of glucagon on renal hemodynamics and excretion of inorganic ions. *Metabolism* 16:358–73

36. Strange RC, Mjos OD. 1975. The sources of plasma cyclic AMP: studies in the rat using isoprenaline, nicotinic acid and glucagon. *Eur. J. Clin. Investig.* 5:147–52

37. Ahloulay M, Dechaux M, Laborde K, Bankir L. 1995. Influence of glucagon on GFR and on urea and electrolyte excretion: direct and indirect effects. *Am. J. Physiol.* 269:F225–35

38. Ahloulay M, Dechaux M, Hassler C, Bouby N, Bankir L. 1996. Cyclic AMP is a hepatorenal link influencing natriuresis and contributing to glucagon-induced hyperfiltration in rats. *J. Clin. Investig.* 98:2251–58

39. Bankir L, Martin H, Dechaux M, Ahloulay M. 1997. Plasma cAMP: a hepatorenal link influencing proximal reabsorption and renal hemodynamics? *Kidney Int. Suppl.* 59:S50–56

40. Forte LR, London RM, Freeman RH, Krause WJ. 2000. Guanylin peptides: renal actions mediated by cyclic GMP. *Am. J. Physiol. Renal Physiol.* 278:F180–91

41. Sindic A, Schlatter E. 2007. Renal electrolyte effects of guanylin and uroguanylin. *Curr. Opin. Nephrol. Hypertens.* 16:10–15

42. Sindic A, Velic A, Basoglu C, Hirsch JR, Edemir B, et al. 2005. Uroguanylin and guanylin regulate transport of mouse cortical collecting duct independent of guanylate cyclase C. *Kidney Int.* 68:1008–17

43. Sternini C, Anselmi L, Rozengurt E. 2008. Enteroendocrine cells: a site of 'taste' in gastrointestinal chemosensing. *Curr. Opin. Endocrinol. Diabetes Obes.* 15:73–78

44. Egan JM, Margolskee RF. 2008. Taste cells of the gut and gastrointestinal chemosensation. *Mol. Interv.* 8:78–81

45. Lin DH, Sterling H, Wang WH. 2005. The protein tyrosine kinase-dependent pathway mediates the effect of K intake on renal K secretion. *Physiology (Bethesda)* 20:140–46

46. Wang WH. 2006. Regulation of ROMK (Kir1.1) channels: new mechanisms and aspects. *Am. J. Physiol. Renal Physiol.* 290:F14–19

47. Giebisch GH. 2002. A trail of research on potassium. *Kidney Int.* 62:1498–512

48. Hebert SC, Desir G, Giebisch G, Wang W. 2005. Molecular diversity and regulation of renal potassium channels. *Physiol. Rev.* 85:319–71

49. Pluznick JL, Sansom SC. 2006. BK channels in the kidney: role in K^+ secretion and localization of molecular components. *Am. J. Physiol. Renal Physiol.* 291:F517–29

50. Grimm PR, Sansom SC. 2007. BK channels in the kidney. *Curr. Opin. Nephrol. Hypertens.* 16:430–36

51. Brandis M, Keyes J, Windhager EE. 1972. Potassium-induced inhibition of proximal tubular fluid reabsorption in rats. *Am. J. Physiol.* 222:421–27

52. Sufit CR, Jamison RL. 1983. Effect of acute potassium load on reabsorption in Henle's loop in the rat. *Am. J. Physiol.* 245:F569–76

53. Lin DH, Sterling H, Yang B, Hebert SC, Giebisch G, Wang WH. 2004. Protein tyrosine kinase is expressed and regulates ROMK1 location in the cortical collecting duct. *Am. J. Physiol. Renal Physiol.* 286:F881–92

54. Najjar F, Zhou H, Morimoto T, Bruns JB, Li HS, et al. 2005. Dietary K^+ regulates apical membrane expression of maxi-K channels in rabbit cortical collecting duct. *Am. J. Physiol. Renal Physiol.* 289:F922–32

55. Mennitt PA, Frindt G, Silver RB, Palmer LG. 2000. Potassium restriction downregulates ROMK expression in rat kidney. *Am. J. Physiol. Renal Physiol.* 278:F916–24

56. Palmer LG, Antonian L, Frindt G. 1994. Regulation of apical K and Na channels and Na/K pumps in rat cortical collecting tubule by dietary K. *J. Gen. Physiol.* 104:693–710

57. Palmer LG, Frindt G. 1999. Regulation of apical K channels in rat cortical collecting tubule during changes in dietary K intake. *Am. J. Physiol.* 277:F805–12

58. Wang T, Sterling H, Shao WA, Yan Q, Bailey MA, et al. 2003. Inhibition of heme oxygenase decreases sodium and fluid absorption in the loop of Henle. *Am. J. Physiol. Renal Physiol.* 285:F484–90

59. Gu RM, Wei Y, Jiang HL, Lin DH, Sterling H, et al. 2002. K depletion enhances the extracellular Ca^{2+}-induced inhibition of the apical K channels in the mTAL of rat kidney. *J. Gen. Physiol.* 119:33–44

60. Walter SJ, Shore AC, Shirley DG. 1988. Effect of potassium depletion on renal tubular function in the rat. *Clin. Sci. (London)* 75:621–28

61. Silver RB, Soleimani M. 1999. H^+-K^+-ATPases: regulation and role in pathophysiological states. *Am. J. Physiol.* 276:F799–811

62. Babilonia E, Wei Y, Sterling H, Kaminski P, Wolin M, Wang WH. 2005. Superoxide anions are involved in mediating the effect of low K intake on c-Src expression and renal K secretion in the cortical collecting duct. *J. Biol. Chem.* 280:10790–96

63. Babilonia E, Lin D, Zhang Y, Wei Y, Yue P, Wang WH. 2007. Role of gp91[phox]-containing NADPH oxidase in mediating the effect of K restriction on ROMK channels and renal K excretion. *J. Am. Soc. Nephrol.* 18:2037–45

64. Wei Y, Bloom P, Lin D, Gu R, Wang WH. 2001. Effect of dietary K intake on apical small-conductance K channel in CCD: role of protein tyrosine kinase. *Am. J. Physiol. Renal Physiol.* 281:F206–12

65. Zhang Y, Lin DH, Wang ZJ, Jin Y, Yang B, Wang WH. 2008. K restriction inhibits protein phosphatase 2B (PP2B) and suppression of PP2B decreases ROMK channel activity in the CCD. *Am. J. Physiol. Cell Physiol.* 294:C765–73

66. Babilonia E, Li D, Wang Z, Sun P, Lin DH, et al. 2006. Mitogen-activated protein kinases inhibit the ROMK (Kir 1.1)-like small conductance K channels in the cortical collecting duct. *J. Am. Soc. Nephrol.* 17:2687–96

67. Zhang X, Lin DH, Jin Y, Wang KS, Zhang Y, et al. 2007. Inhibitor of growth 4 (ING4) is up-regulated by a low K intake and suppresses renal outer medullary K channels (ROMK) by MAPK stimulation. *Proc. Natl. Acad. Sci. USA* 104:9517–22

68. Wilcox CS. 2005. Oxidative stress and nitric oxide deficiency in the kidney: a critical link to hypertension? *Am. J. Physiol. Regul. Integr. Comp. Physiol.* 289:R913–35

69. Sachse A, Wolf G. 2007. Angiotensin II-induced reactive oxygen species and the kidney. *J. Am. Soc. Nephrol.* 18:2439–46

70. Wang T, Giebisch G. 1996. Effects of angiotensin II on electrolyte transport in the early and late distal tubule in rat kidney. *Am. J. Physiol.* 271:F143–49

71. Wei Y, Zavilowitz B, Satlin LM, Wang WH. 2007. Angiotensin II inhibits the ROMK-like small conductance K channel in renal cortical collecting duct during dietary potassium restriction. *J. Biol. Chem.* 282:6455–62

72. Thompson CB, McDonough AA. 1996. Skeletal muscle Na,K-ATPase α and β subunit protein levels respond to hypokalemic challenge with isoform and muscle type specificity. *J. Biol. Chem.* 271:32653–58

73. Amlal H, Krane CM, Chen Q, Soleimani M. 2000. Early polyuria and urinary concentrating defect in potassium deprivation. *Am. J. Physiol. Renal Physiol.* 279:F655–63

74. McKenna MJ, Gissel H, Clausen T. 2003. Effects of electrical stimulation and insulin on Na$^+$-K$^+$-ATPase ([^{3}H]ouabain binding) in rat skeletal muscle. *J. Physiol.* 547:567–80

75. Choi CS, Thompson CB, Leong PK, McDonough AA, Youn JH. 2001. Short-term K$^+$ deprivation provokes insulin resistance of cellular K$^+$ uptake revealed with the K$^+$ clamp. *Am. J. Physiol. Renal Physiol.* 280:F95–102

76. Clausen T. 2003. Na$^+$-K$^+$ pump regulation and skeletal muscle contractility. *Physiol. Rev.* 83:1269–324

77. Azuma KK, Hensley CB, Putnam DS, McDonough AA. 1991. Hypokalemia decreases Na$^+$-K$^+$-ATPase α2- but not α1-isoform abundance in heart, muscle, and brain. *Am. J. Physiol.* 260:C958–64

78. Bergman RN, Ader M, Huecking K, Van Citters G. 2002. Accurate assessment of β-cell function: the hyperbolic correction. *Diabetes* 51(Suppl. 1):S212–20

79. Arslanian S, Austin A. 1991. Impaired insulin mediated potassium uptake in adolescents with IDDM. *Biochem. Med. Metab. Biol.* 46:364–72

80. DeFronzo RA. 1988. Obesity is associated with impaired insulin-mediated potassium uptake. *Metabolism* 37:105–8

81. Alvestrand A, Wahren J, Smith D, DeFronzo RA. 1984. Insulin-mediated potassium uptake is normal in uremic and healthy subjects. *Am. J. Physiol.* 246:E174–80

82. Storlien LH, James DE, Burleigh KM, Chisholm DJ, Kraegen EW. 1986. Fat feeding causes widespread in vivo insulin resistance, decreased energy expenditure, and obesity in rats. *Am. J. Physiol.* 251:E576–83

83. Kim JK, Wi JK, Youn JH. 1996. Metabolic impairment precedes insulin resistance in skeletal muscle during high-fat feeding in rats. *Diabetes* 45:651–58

84. Graham TE, Kahn BB. 2007. Tissue-specific alterations of glucose transport and molecular mechanisms of intertissue communication in obesity and type 2 diabetes. *Horm. Metab. Res.* 39:717–21

85. Benziane B, Chibalin AV. 2008. Skeletal muscle sodium pump regulation: a translocation paradigm. *Am. J. Physiol. Endocrinol. Metab.* 295:E553–58

86. Kristensen M, Rasmussen MK, Juel C. 2008. Na$^+$-K$^+$ pump location and translocation during muscle contraction in rat skeletal muscle. *Pflüg. Arch.* 456:979–89

87. Rasmussen MK, Kristensen M, Juel C. 2008. Exercise-induced regulation of phospholemman (FXYD1) in rat skeletal muscle: implications for Na$^+$/K$^+$-ATPase activity. *Acta Physiol. (Oxf.)* 194:67–79

88. Weinstein SP, Paquin T, Pritsker A, Haber RS. 1995. Glucocorticoid-induced insulin resistance: Dexamethasone inhibits the activation of glucose transport in rat skeletal muscle by both insulin- and noninsulin-related stimuli. *Diabetes* 44:441–45

89. Dorup I, Clausen T. 1997. Effects of adrenal steroids on the concentration of Na$^+$-K$^+$ pumps in rat skeletal muscle. *J. Endocrinol.* 152:49–57

90. Thompson CB, Dorup I, Ahn J, Leong PK, McDonough AA. 2001. Glucocorticoids increase sodium pump α2- and β1-subunit abundance and mRNA in rat skeletal muscle. *Am. J. Physiol. Cell Physiol.* 280:C509–16

91. Rhee MS, Perianayagam A, Chen P, Youn JH, McDonough AA. 2004. Dexamethasone treatment causes resistance to insulin-stimulated cellular potassium uptake in the rat. *Am. J. Physiol. Cell Physiol.* 287:C1229–37

92. Alexander EA, Levinsky NG. 1968. An extrarenal mechanism of potassium adaptation. *J. Clin. Investig.* 47:740–48

93. Blachley JD, Crider BP, Johnson JH. 1986. Extrarenal potassium adaptation: role of skeletal muscle. *Am. J. Physiol.* 251:F313–18

94. Sadre M, Sheng HP, Fiorotto M, Nichols BL. 1987. Electrolyte composition changes of chronically K-depleted rats after K loading. *J. Appl. Physiol.* 63:765–69

95. Terkildsen JR CE, Smith NP. 2007. The balance between inactivation and activation of the Na$^+$-K$^+$ pump underlies the triphasic accumulation of extracellular K$^+$ ions during myocardial ischemia. *Am. J. Physiol. Heart Circ. Physiol.* 293:H3036–45

96. Zheng D, Perianayagam A, Lee DH, Brannan MD, Yang LE, et al. 2008. AMPK activation with AICAR provokes an acute fall in plasma [K$^+$]. *Am. J. Physiol. Cell Physiol.* 294:C126–35

97. Towler MC, Hardie DG. 2007. AMP-activated protein kinase in metabolic control and insulin signaling. *Circ. Res.* 100:328–41

98. Mu J, Barton ER, Birnbaum MJ. 2003. Selective suppression of AMP-activated protein kinase in skeletal muscle: update on 'lazy mice'. *Biochem. Soc. Trans.* 31:236–41

99. Gollob MH, Seger JJ, Gollob TN, Tapscott T, Gonzales O, et al. 2001. Novel PRKAG2 mutation responsible for the genetic syndrome of ventricular preexcitation and conduction system disease with childhood onset and absence of cardiac hypertrophy. *Circulation* 104:3030–33

100. Bergeron R, Previs SF, Cline GW, Perret P, Russell RR 3rd, et al. 2001. Effect of 5-aminoimidazole-4-carboxamide-1-β-D-ribofuranoside infusion on in vivo glucose and lipid metabolism in lean and obese Zucker rats. *Diabetes* 50:1076–82

101. Mu J, Brozinick JT Jr, Valladares O, Bucan M, Birnbaum MJ. 2001. A role for AMP-activated protein kinase in contraction- and hypoxia-regulated glucose transport in skeletal muscle. *Mol. Cell* 7:1085–94

102. Appel LJ, Brands MW, Daniels SR, Karanja N, Elmer PJ, Sacks FM. 2006. Dietary approaches to prevent and treat hypertension: a scientific statement from the American Heart Association. *Hypertension* 47:296–308

103. Young DB, Lin H, McCabe RD. 1995. Potassium's cardiovascular protective mechanisms. *Am. J. Physiol.* 268:R825–37

104. Kimura M, Lu X, Skurnick J, Awad G, Bogden J, et al. 2004. Potassium chloride supplementation diminishes platelet reactivity in humans. *Hypertension* 44:969–73

105. Braschi A, Naismith DJ. 2008. The effect of a dietary supplement of potassium chloride or potassium citrate on blood pressure in predominantly normotensive volunteers. *Br. J. Nutr.* 99:1284–92

106. Macdonald JE, Struthers AD. 2004. What is the optimal serum potassium level in cardiovascular patients? *J. Am. Coll. Cardiol.* 43:155–61

107. Cohen HW, Madhavan S, Alderman MH. 2001. High and low serum potassium associated with cardiovascular events in diuretic-treated patients. *J. Hypertens.* 19:1315–23

108. Juurlink DN, Mamdani MM, Lee DS, Kopp A, Austin PC, et al. 2004. Rates of hyperkalemia after publication of the Randomized Aldactone Evaluation Study. *N. Engl. J. Med.* 351:543–51

109. Yang BC, Li DY, Weng YF, Lynch J, Wingo CS, Mehta JL. 1998. Increased superoxide anion generation and altered vasoreactivity in rabbits on low-potassium diet. *Am. J. Physiol.* 274:H1955–61

110. Matsui H, Shimosawa T, Uetake Y, Wang H, Ogura S, et al. 2006. Protective effect of potassium against the hypertensive cardiac dysfunction: association with reactive oxygen species reduction. *Hypertension* 48:225–31

111. Kido M, Ando K, Onozato ML, Tojo A, Yoshikawa M, et al. 2008. Protective effect of dietary potassium against vascular injury in salt-sensitive hypertension. *Hypertension* 51:225–31

The Contribution of Epithelial Sodium Channels to Alveolar Function in Health and Disease

Douglas C. Eaton,[1,2,4] My N. Helms,[1,4] Michael Koval,[3] Hui Fang Bao,[1,4] and Lucky Jain[1,2,4]

Departments of [1]Physiology, [2]Pediatrics, and [3]Medicine and [4]The Center for Cell and Molecular Signaling, Emory University School of Medicine, Atlanta, Georgia 30322; email: deaton@emory.edu

Annu. Rev. Physiol. 2009. 71:403–23

First published online as a Review in Advance on October 2, 2008

The *Annual Review of Physiology* is online at physiol.annualreviews.org

This article's doi: 10.1146/annurev.physiol.010908.163250

Copyright © 2009 by Annual Reviews. All rights reserved

0066-4278/09/0315-0403$20.00

Key Words

ENaC, alveolar type 1 cells, alveolar type 2 cells, lung slice, CFTR, nonselective cation channels, single-channel recording

Abstract

Amiloride-sensitive epithelial sodium channels (ENaC) play an important role in lung sodium transport. Sodium transport is closely regulated to maintain an appropriate fluid layer on the alveolar surface. Both alveolar type I and II cells have several different sodium-permeable channels in their apical membranes that play a role in normal lung physiology and pathophysiology. In many epithelial tissues, ENaC is formed from three subunit proteins: α, β, and γ ENaC. Part of the diversity of sodium-permeable channels in lung arises from assembling different combinations of these subunits to form channels with different biophysical properties and different mechanisms for regulation. Thus, lung epithelium has enormous flexibility to alter the magnitude of salt and water transport. In lung, ENaC is regulated by many transmitter and hormonal agents. Regulation depends upon the type of sodium channel but involves controlling the number of apical channels and/or the activity of individual channels.

INTRODUCTION

Epithelial tissues, in general, are designed to form a barrier between two body compartments and to facilitate transport from one compartment to another, thereby regulating the composition of the compartments. Regulation of salt and water balance involves primary active transport of sodium (Na^+) and secondary active transport of chloride (Cl^-) followed by, in eukaryotes, osmotically driven water movement. That is, water movement depends upon the prior transport of osmotic equivalents, usually salt. Therefore, regulation of water balance requires the ability to regulate both salt reabsorption and salt secretion.

The alveolar surface of the lungs is a unique epithelium because it separates an internal vascular compartment (like most epithelia) from an air-filled compartment (unlike other epithelia) (**Figure 1**). As such, the primary physiological purpose of the alveolus is to promote the exchange of oxygen from the air space into the blood for CO_2 out of the blood. However, efficient gas exchange in the lungs depends upon having a thin liquid layer on the air-facing side of the alveolar epithelium. Proper gas exchange and alveolar function requires precise regulation of the amount of this luminal fluid. The amount of fluid on the airway surface represents a balance between the rate at which fluid is passively secreted from the vascular space through the paracellular space and tight junctions and the rate at which fluid is actively reabsorbed. Fluid appears in the airways because of the passive movement of fluid driven by hydrostatic pressure (mean value of approximately

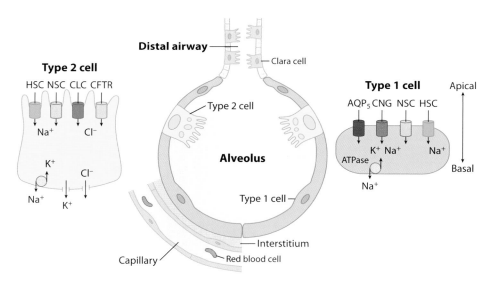

Figure 1

Schematic diagram of an alveolus and part of the distal airway. The alveolar epithelium is composed of two distinct cell types, alveolar type I (AT1) and type II (AT2) cells. (*Left*) AT2 cells, which cover only 2–5% of the internal surface area of the lung, are cuboidal cells that secrete pulmonary surfactant. AT2 cells contain ion channels, including two forms of amiloride-sensitive epithelial sodium channel (ENaC): a highly Na^+-selective form and a cation-nonselective form. In addition, AT2 cells have apical anion pathways, including the cystic fibrosis transmembrane regulator (CFTR) and Cl^- channels that are members of the CLC family of proteins. (*Right*) AT1 cells are large, squamous cells whose thin cytoplasmic extensions cover more than 95% of the internal surface area of the lung. Although less is known about the AT1 cells, they also appear to have a full complement of ion channels, including two forms of ENaC [highly selective cation (HSC) and nonselective cation (NSC) channels] and cyclic nucleotide–gated channels (CNG), and contribute to alveolar Na^+ absorption. They also have high concentrations of aquaporin 5 (AQP_5), which may contribute to either water or gas exchange. Adapted and used with permission from Reference 124.

15 mm Hg) out of the pulmonary capillaries across the airway epithelium. The amount secreted is relatively constant because, under normal physiological circumstances, the pulmonary capillary pressure and the permeability of the epithelium are relatively constant. However, under pathological conditions, when either the pulmonary blood pressure is abnormally elevated or the permeability of the epithelium is increased by inflammation or disease, passive secretion may increase dramatically. Regardless, even under normal levels of secretion, the alveoli would rapidly fill with fluid were it not for fluid reabsorption by the alveolar epithelium. Because fluid absorption depends upon the rate of ion transport, regulation of ion transport provides the mechanism by which the amount of airway fluid is regulated.

In airway epithelium, fluid is transported from the airway lumen into the interstitial spaces, and this process can be substantially inhibited by the addition of amiloride, an epithelial sodium channel (ENaC) inhibitor, to the alveolar space. Although all parts of the airway have ENaC and absorb Na^+, the alveolar epithelium covers more than 99% of the large internal surface area of the lung ($\sim$100–150 m^2 in humans), suggesting that alveolar epithelial cells are the major sites of Na^+ transport and fluid absorption in the adult lung. The alveolar epithelium is composed of two distinct cell types, alveolar type I (AT1) and type II (AT2) cells. AT2 cells, which cover only 2–5% of the internal surface area of the lung, are cuboidal cells that secrete pulmonary surfactant. AT2 cells contain ion channels, including the amiloride-sensitive ENaC and the cystic fibrosis transmembrane regulator (CFTR). AT1 cells are large, squamous cells whose thin cytoplasmic extensions cover more than 95% of the internal surface area of the lung. Although less is known about the AT1 cells, they also appear to have a full complement of ion channels, including ENaC, and contribute to alveolar Na^+ absorption (1, 2).

METHODS FOR EXAMINING FLUID BALANCE AND Na^+ REABSORPTION IN LUNG ALVEOLAR EPITHELIAL CELLS

If control of lung fluid balance depends upon regulation of the reabsorptive processes, then having methods to examine these processes is critical to understanding both normal fluid balance and the pathological fluid balance associated with lung edema and other diseases. In general, studying fluid balance involves either methods that examine the whole lung or methods that examine cellular components of the lung. There are both positive and negative reasons to use either approach.

Whole-Lung Approaches to Evaluate Lung Fluid Balance

Probably the simplest method for evaluating changes in lung fluid balance is to measure the wet-to-dry-weight ratio. This ratio is a measure of the fraction of total lung weight associated with lung fluid. As such, it represents interstitial, vascular, and airway surface fluid. If the interstitial and vascular volume remains relatively constant, then changes in the ratio represent changes in the amount of airway fluid. Because airway fluid is the balance between absorption and secretion, changes in the balance can be inferred from changes in the wet:dry ratio. Some information about mechanism can also be inferred if pharmacological agents that block specific reabsorptive pathways (e.g., amiloride) produce changes in the ratio. The method benefits from being simple but suffers from being often hard to interpret and being able to detect only relatively large changes in fluid balance.

More direct measurements of the ability of whole lung to clear instilled fluid provide more mechanistic information (3–6). In general, these methods involve instilling solutions of known composition that contain some volume marker into the distal airways and examining the rate at which the instillate is absorbed. Besides directly measuring instilled fluid uptake, there are many other variations on this theme, including the addition of isotopic markers to measure the properties of specific transport mechanisms, the addition of markers to the blood as a measure of secretion, and the addition of agents to gauge epithelial barrier function. The positive features of these measurements are that they provide information about the entire lung in situ and they have provided substantial mechanistic information about the mechanisms responsible for fluid reabsorption. The difficulty with such measurements is that it is often difficult to assign the specific mechanisms to particular cellular components of the lung.

Tissue Culture Models of Lung Epithelium

To understand the transport properties of individual types of lung epithelial cells, many investigators have developed tissue culture models to study lung cell transport. There are two categories of culture models: lung cell lines obtained from neoplastic tissue and primary cultures of lung epithelial cells. There are many tumor-derived lung cell lines that have characteristics of cells from different parts of the lung. A summary of these lines and their properties, all available from American Type Culture Collection, can be obtained by linking to **http://www.biotech.ist.unige.it/cldb/tis42. html**. These cell lines have provided an invaluable tool for examining the properties of presumptive lung ion channels and transporters and the signaling mechanisms that regulate them. Transport in many of these cell lines is amiloride-sensitive Na^+ transport, which suggests that they express ENaC (for reviews, see References 7 and 8). Single-channel recordings from some of these cells do reveal ion channels with all the properties of ENaC in renal cells or ENaC expressed in oocytes. Thus, these cultured cells can be a useful model in which to study ENaC properties and the mechanisms by which ENaC is regulated. However, these transformed cells may or may not reflect the properties of native cells. When

using these cells, investigators generally try to establish that the cells have biomarkers and other properties that correspond to specific types of lung cells. Some appear to be excellent models for AT1 or AT2 cells. Others appear to have a mixed phenotype, but regardless of the fidelity of phenotypic markers, all the transformed cells appear to form tight junctions poorly and, therefore, tend to form only low-resistance monolayers. This limits their usefulness because transepithelial current as a measure of ion transport is limited and difficult to interpret.

The alternative to a continuous cell line is to place specific types of lung cells into primary culture. The mechanical methods for isolating upper airway cells are well established, and such cells have been used extensively to investigate the pathology of cystic fibrosis (9). Isolation of alveolar cells requires more elaborate methods. Approaches for isolation of AT2 cells use enzymatic digestion and isolation by differential adherence (10). The method can produce AT2 cells of more than 95% purity. When placed in primary culture under appropriate conditions (see below), the cells have the expected biomarkers for AT2 cells and express ENaC. ENaC in these cells has been extensively studied by electrophysiological methods (7, 8, 11, 12). In culture, these cells spontaneously change phenotype, losing their AT2 surface markers and their characteristic AT2 columnar epithelial structure and transforming into a more squamous epithelium that expresses surface markers for AT1 cells. The time required to make this transformation depends upon culture conditions but varies from ~2 days to ~1 week (13). Interestingly, this is the same transition that AT2 cells undergo in vivo, although the transition in vivo takes somewhat longer. The AT2-AT1 transition can be viewed either positively or negatively. On the negative side, the transition limits the useful lifetime of AT2 cells in culture; however, the transition provides AT1-like cells in culture. The only question is how closely these cells reflect the properties of AT1 cells in vivo (14). Because of this question, it is important to compare the properties of these cells with those of AT1 cells in the lung. One approach to this problem has been to isolate AT1 cells with positive and negative immunoselection techniques that use antibodies specific for AT1 and AT2 surface membrane proteins. The separation methods are laborious and yield AT1 cells of varying purity; the trade-off is between total yield and purity (15, 16). Nonetheless, these isolated cells have been used to determine whether Na^+-transporting proteins and other transporters are present in AT1 cells (2, 14–16). However, the difficulty and length of the isolation bring into question whether these cells actually maintain a faithful phenotype of the AT1 cells in vivo.

Lung Slices as a Model for the In Vivo Properties of AT1 and AT2 Cells

Whole-lung models reflect the properties of alveolar cells in situ, but the interpretation of mechanistic studies can be difficult. In contrast, it is relatively easy to study transport mechanisms in preparations consisting of continuous lung cell lines or even in lung cells in primary culture. However, questions about the relationship between tissue culture models or cells in primary culture and alveolar cells in whole lungs always cloud our interpretation. Therefore, some alternative preparation that would reflect the properties of alveolar cells in situ is necessary. This need has been met by the development of a lung slice preparation (1, 17, 18). In this preparation, rat lungs are either sliced directly or first intratracheally instilled with warm low-melting-point agarose to expand the air spaces and provide support for the tissue during the slicing process. Excised lungs are removed en bloc and, if instilled with agarose, chilled in cold PBS (4°C) to solidify the agarose, after which a small block of tissue is separated from the largest lobe of the lung, and 100–250-μm-thick sections are prepared by use of a tissue-slicing unit (e.g., a Leica Vibratome). These slices have hemisections of alveoli on the surface of the slices with alveolar cells accessible to patch clamp electrodes, but to distinguish the cell types and the alveoli it

is useful to use fluorescein-labeled lectin from the cry-baby tree, *Erythrina cristagalli*, as a vital stain for ATI cells and fluorescently labeled LysoTracker (Invitrogen, Carlsbad, California) for AT2 cells (see **Figure 2**). Individual cells can be visualized, and patch clamp recordings can be obtained from both cell types. Our research group was the first to successfully use the lung slice model to study the contribution of epithelial Na^+ transport in AT1 and AT2 cells that make up the alveolar epithelium (1, 18), and this preparation is likely to be the method of choice for verifying the properties of AT1 and AT2 cells in native lung.

SINGLE-CHANNEL MEASUREMENTS FROM ALVEOLAR CELLS: A FAMILY OF AMILORIDE-SENSITIVE CHANNELS

In some epithelial tissues, Na^+ transport appears to be a relatively simple process, beginning with tissue reabsorption of Na^+ mediated, at a single-channel level, by Na^+ channels that are highly selective for Na^+ over K^+ and that are blocked by low concentrations of the drug amiloride (19–21). Channels with such properties are present in several epithelial preparations, but examination of single channels in several other Na^+-transporting epithelial cells, including lung cells, implies a more complicated picture of amiloride-sensitive channels (1, 7, 8, 22–25). Single-channel studies in lung alveolar epithelial cells have identified at least two different amiloride-sensitive channels with either high selectivity or no selectivity for Na^+ over K^+ in the apical membranes of a variety of cultured and native epithelial cells, including lung epithelial cells (**Table 1**). These channels differ not only in their ion selectivity but also in their unitary conductance and other biophysical characteristics; nevertheless, they both have been proposed to play some role in epithelial Na^+ reabsorption.

In lung AT2 cells, several investigators have described cation channels [as reviewed recently by Matalon et al. (7)]. Orser and colleagues

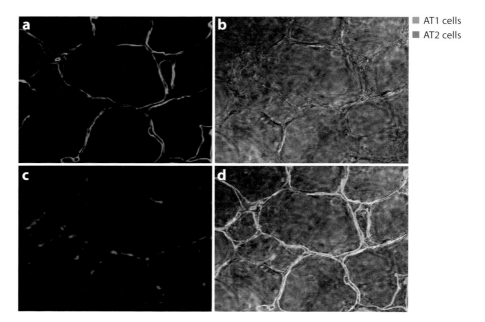

■ AT1 cells
■ AT2 cells

Figure 2

A lung slice showing AT1 and AT2 cells. Lung tissue slices 250 μm thick were prepared with a vibratome and labeled with fluorescein-labeled *Erythrina cristagalli* lectin (F-ECL). Panel *b* shows a DIC (differential interference contrast) image collected simultaneously with a confocal image at original 40 × magnification. Lung tissue slices were labeled with F-ECL, specific for AT1 cells in panel *a*, and LysoTracker Red (Invitrogen), specific for the lamellar bodies of AT2 in panel *c*. The confocal fluorescent and DIC images are merged in panel *d*. The characteristic squamous epithelial morphology of AT1 cells is apparent in panels *a*, *b*, and *d*. The position of the AT2 cells at the "corners" of the alveolar space can be seen in panels *c* and *d*. Used with permission from Reference 1.

(26) first described nonselective cation (NSC) channels in fetal lung cells. Subsequently, Feng and colleagues (27) described NSC channels in apical cell-attached and inside-out patches from rat adult AT2 cells. The channels have a Na^+-to-K^+ permeability ratio ($P_{Na}:P_K$) of close to 1, are voltage independent, and are inhibited by amiloride but remain in the closed state unless the intracellular calcium concentration is higher than 10 μM. Yue and colleagues (28) have reported the existence of amiloride-sensitive 25–27 pS channels in both cell-attached and inside-out patches of rat AT2 cells, which are activated by cyclic AMP (cAMP). Ion substitution studies showed that these channels have a Na:K selectivity of 6:1. Jain and coworkers (22, 24, 29) have also reported a NSC channel in the apical membrane patches of AT2 cells. This amiloride-sensitive channel has a unitary conductance of ~20 pS and a Na:K permeability ratio slightly greater than 1 and is to a certain extent dependent on intracellular calcium. Because of the diversity of amiloride-sensitive cation channels in airway epithelial cells, their properties are best differentiated by use of patch clamp methods to examine the properties of single channels.

Table 1 Comparison of amiloride-blockable channels in lung cells[a]

Channel property	Channel type	
	Highly selective	**Nonselective**
Na/K selectivity	>40	1.1
Unit conductance (pS)	6	21
Mean open time (s)	2	0.1
Mean closed time	3	0.3
Amiloride K_i	38	1800

[a]Data from References 22, 25, 29, and 30.

In the lung, the predominant cation channels appear to be of two types: (*a*) highly Na^+-selective cation (HSC) channels with a single-channel conductance of 4–6 pS and an Na^+/K^+ selectivity >40 (22, 30) and (*b*) NSC channels with a single-channel conductance of 19–24 pS and an Na^+/K^+ selectivity of ~1.4 (1, 18, 22, 29). The HSC channel in lung has biophysical properties identical to those of an epithelial Na^+ channel consisting of three homologous subunits—α, β, and γ—from rat colon (31). Additional expression studies of various combinations of the different ENaC subunits in heterologous expression systems and antisense knockdown experiments in lung cells confirmed that HSC channels are composed of α, β, and γ ENaC subunits and showed that NSC channels are composed of α subunits alone (25, 30). Culturing alveolar cells on permeable supports with dexamethasone, exposing cells to air interface, or elevating O_2 exposure to 95% was each effective in enhancing the expression of HSC channels and decreasing the expression of NSC channels (30). Both AT1 and AT2 cells contain both types of ENaC channels. The channel number per unit area of membrane of both types of ENaC channels, HSC and NSC, is approximately equal in AT1 and AT2 cells.

REGULATION OF ENaC IN THE LUNG

Excessive accumulation of fluid in the alveolar spaces frequently accompanies acute lung injury, and the failure of the lungs to rapidly clear this edema fluid leads to higher morbidity and mortality. Because alveolar fluid clearance is driven by active transport of Na^+, Cl^-, and water across the epithelial lining of air spaces, significant effort has focused on strategies to enhance this process during pathological fluid accumulation. This has often involved attempts to enhance the activity of ENaC at the apical membrane of alveolar epithelial cells. ENaC in the airways is regulated by a large variety of agents; these include transmitters interacting with G protein–coupled receptors (e.g., puriner-

gic, adrenergic, and dopaminergic agents), circulating hormones (e.g., glucocorticoids, angiotensin, and eicosanoids), chemokines (e.g., TNF-α, TGF-β, interleukins), and reactive oxygen and nitrogen species (ROS and RNS, respectively) [e.g., superoxide (O_2^-) and nitric oxide (NO)]. The large number of regulatory agents underscores the fact that regulation of lung Na^+ absorption is critical for normal lung fluid balance and function. Several agents have a particularly profound effect on Na^+ reabsorption (described below and summarized in **Figure 3**).

Regulation of ENaC by β-Adrenergic Agents

Because β-agonists can upregulate Na^+ transport across the alveolar epithelium in vivo and ex vivo, as well as across AT2 cells in primary culture (32–38), these agents may be useful in limiting alveolar edema and decreasing morbidity and mortality in patients with acute lung injury. Activation of β_2 receptors on AT2 cells stimulates adenylyl cyclase, which, in turn, increases intracellular cAMP levels, the number of HSC channels, and the activity of NSC channels in the apical membrane (22). The effects of increased cAMP are totally blocked by the β-antagonist propranolol and by the protein kinase A (PKA) blocker H89.

Regulation of ENaC by Purinergic Agonists

The luminal surface of airway epithelial cells contains several different types of purinergic receptors (39–42). These include adenosine type 1 and 2a (A1 and A2a) receptors. Adenosine at low concentrations (<100 nM) stimulates ENaC by increasing the open probability of single channels, presumably by activating A1 receptors. At higher concentrations (>1 μM), adenosine inhibits amiloride-sensitive Na^+ transport through A2a receptors (43). Besides adenosine receptors, there are several P2 purinergic receptors sensitive to ATP and UTP. The receptors are of both the

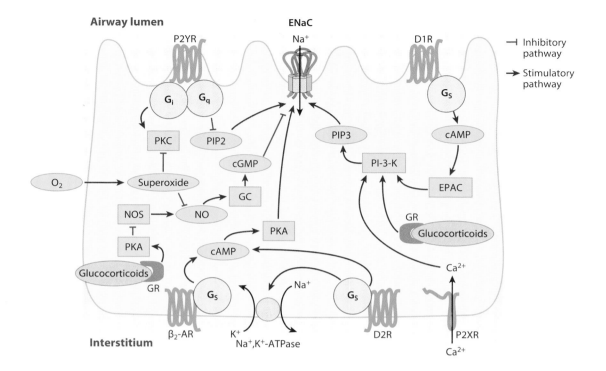

Figure 3

Major signaling pathways that regulate ENaC. For clarity, not all elements of the pathways are shown. A typical airway cell is shown with, starting from the top left, purinergic signaling mediated by P2Y receptors (P2YR). Purinergic signaling decreases ENaC activity by reducing the amount of membrane phosphatidylinositol-4,5-bis-phosphate (PIP2) by phospholipase C hydrolysis and by activating protein kinase C (PKC), which also inhibits ENaC. At the top right is the type 1 dopamine receptor (D1R), which promotes apical production of cyclic adenosine monophosphate (cAMP). This activates a complicated signaling pathway that involves EPAC activation of the small G protein Rap1 (not shown), which stimulates phosphatidylinositol-3-kinase (PI-3-K) to form phosphatidylinositol-3,4,5-trisphosphate (PIP3), which strongly stimulates ENaC. At the basolateral membrane is the type 2 dopamine receptor (D2R), which activates ENaC in a manner similar to the apical type 1 receptor but also activates the basolateral Na^+,K^+-ATPase. The basolateral membrane also contains β_2-adrenergic receptors (β_2-AR), which stimulate the production of an alternative pool of cAMP that activates protein kinase A (PKA) to promote ENaC insertion in the apical membrane. Glucocorticoids bind to their receptor (GR), which when bound activates a complicated signaling pathway that requires gene expression but finally leads to the activation of PI-3-K to form PIP3, which strongly stimulates ENaC. Also in the basolateral membrane are P2X receptors (P2XR) that respond to purines and allow calcium entry. The calcium also activates PI-3-K. G_s, G_i, and G_q are G proteins that stimulate or inhibit adenylyl cyclase or stimulate phospholipase, respectively. Nitric oxide (NO) is produced from nitric oxide synthase (NOS). NO activates guanylyl cyclase (GC) to produce cGMP, which inhibits ENaC. Glucocorticoids activate SGK1 (not shown), which inhibits NOS by phosphorylation and thereby results in constitutive increases in ENaC activity. By the same token, production of superoxide reduces NO to produce peroxynitrite (not shown) and leads to increased ENaC activity.

P2Y metabotropic and P2X ligand-gated ion channel types (44–46). P2Y2 receptors are activated by ATP or UTP and, when activated, strongly inhibit ENaC through a protein kinase C (PKC)-dependent mechanism. The Cl⁻ channel CFTR is often associated with ATP release. Some researchers have suggested that ATP, acting through P2Y2 receptors, allows for coordinated control of Na⁺ and Cl⁻ absorption across the apical membrane of AT2 cells. P2X receptors (probably P2X5 in the basolateral membrane) are, when activated by purines, ion channels themselves; however, they can also alter intracellular calcium and therefore would be expected to activate NSC channels and increase Na^+ transport in alveolar epithelial cells.

Regulation of ENaC by Dopamine

Dopamine increases lung liquid clearance under basal conditions and in situations in which edema accompanies lung injury (47–49). Alveolar epithelial cells appear to contain both type 1 and type 2 dopamine (D1 and D2) receptors. Stimulation of D1 or D2 receptors on the basolateral surface of AT2 cells activates Na^+,K^+-ATPase (50–52), but D1 activation of apical ENaC mediates some of the increase in lung fluid clearance (24). Apical D1 receptors stimulate the production of cAMP but do not activate PKA. Rather, the cAMP, through a complicated signaling pathway that involves activation of the small G protein Rap1 by cAMP activation of a GDP exchange protein, EPAC, finally increases the activity of individual ENaC proteins (24). The effect of dopamine on both ENaC and Na^+,K^+-ATPase suggests a coordinated response of both transporters to promote maximal increases in transport and alveolar fluid clearance. The action of dopamine also demonstrates that signaling within alveolar cells is compartmentalized. Basolateral production of cAMP by β-adrenergic receptors produces a PKA-dependent increase in the number of ENaC in the apical membrane, with little or no change in the activity of individual channels (22). In contrast, the apical production of cAMP by D1 receptors produces a PKA-independent increase in the activity of individual channels, with little or no change in the number of channels (24).

Regulation of ENaC by Steroids

Lung epithelial cells contain both mineralocorticoid and glucocorticoid receptors, but because circulating levels of aldosterone are usually low compared with glucocorticoids, it is likely that the major steroids regulating lung ENaC are glucocorticoids. Glucocorticoids regulate Na^+ absorption by short- and long-term processes: an initial phase that increases transport four- to sixfold in the first 2 to 6 h and a late phase that requires 12–48 h and increases transport another three- to fourfold (reviewed in Reference 53). A glucocorticoid, like other steroid hormones, enters target cells and binds to cytosolic glucocorticoid receptor complexes. After some rearrangement, the glucocorticoid-bound receptor acts as a DNA binding protein that targets steroid response elements on genetic DNA. Binding to the response elements alters gene expression. Increases in Na^+ transport can be measured within 1 h of exposure to glucocorticoids. This increase is dependent on gene transcription and translation; the gene products are generically referred to as steroid-induced proteins (SIPs). In the short term, glucocorticoids regulate Na^+ transport by inducing expression of the small G protein K-Ras2A and by subsequent K-Ras2A-induced activation of phosphatidylinositol phosphate-5-kinase (PIP-5-K) and phosphatidylinositol-3-kinase (PI-3-K) to produce phosphatidylinositol-3,4,5-phosphate, which ultimately increases the activity of individual ENaC channels (54–57). In the long term, glucocorticoids regulate Na^+ transport by inducing SIPs that alter the trafficking, assembly, and degradation of ENaC and thereby change the number of ENaC in the surface membrane. Glucocorticoids also apparently increase the number of HSC channels at the expense of NSC channels, ensuring that the epithelium becomes highly selective for Na^+ so that Na^+ is preferentially absorbed by the lung epithelium through a mechanism that involves altering rates of channel insertion and degradation (53).

Regulation of ENaC by Inflammatory Chemokines

Many inflammatory cytokines reduce ENaC activity and thus exacerbate the pulmonary edema associated with sepsis. TNF-α, IL-1β, and TGF-β all reduce ENaC activity via MAP kinase signaling pathways (58–61). Besides direct inhibition, IL-1β and TGF-β also interfere with glucocorticoid gene activation (62–65). Therefore, they not only reduce ENaC activity but also prevent glucocorticoid-mediated increases in ENaC activity and hamper the effectiveness of exogenously administered steroids that are often used to treat pulmonary edema.

Regulation of Lung ENaC by Reactive Oxygen Species

All cells produce ROS as part of their normal mitochondrial metabolism. Lung cells are no exception, but they are also exposed to relatively high levels of free oxygen in inspired air that can lead to the additional production of ROS. Conventional wisdom suggests that ROS are deleterious to cells and that, therefore, lung cells should be at higher risk for damage from ROS. It is true that exposure of cells to high concentrations of ROS for extended periods of time does result in programmed cell death. However, high concentrations of ROS, rather than producing adverse events, appear to play a significant role as signaling molecules that act as cellular detectors of the oxidation/reduction (redox) state of cells. Again, lung cells do not appear to be an exception to this general scheme.

The potential importance of oxygen-dependent signaling can be appreciated by remembering that, shortly after birth, the fetal lung must change from a fluid-secreting organ to a fluid-absorbing organ for independent breathing to occur. Perinatal exposure of lung cells to oxygen may be one signal for turning on the active transport of Na^+ necessary to promote fluid clearance from the newborn lung. Indeed, cell culture models of the alveolar epithelium show that increases in oxygen tension enhance ENaC activity. The oxygen sensitivity of ENaC may be due to a direct effect of oxygen; however, changes in oxygen tension also produce changes in ROS such as O_2^-. There does appear to be ROS regulation of ENaC because the addi-

tion of a cell-permeable O_2^- scavenger [2,2,6,6-tetramethylpiperidine 1-oxyl (TEMPO)] significantly decreases ENaC activity (66, 67). In fetal distal lung epithelial (FDLE) cells, Na^+ transport increases 6 h after an increase in oxygen tension (up to 100 mm Hg) (68). Besides the direct effect on Na^+ transport, ROS can influence transport properties of the alveolar epithelium in the long term by altering ENaC expression. Long-term exposure (48 h) to oxygen or ROS increases ENaC-subunit promoter activity (68), and following extended high-O_2 tension treatment, total ENaC protein increases (67, 68, 69). The promoter regions of ENaC subunits have NF-κB and AP-1 response elements (67, 70), and both transcription factors respond to the redox state of the cell. Redox-sensitive transcription factor regulation of ENaC gene expression may account for up-regulation of the ENaC protein following high O_2 exposure.

Cross Talk Between Reactive Oxygen Species Signaling and Steroid Regulation of ENaC

Steroid hormone and oxygen-radical signaling pathways are closely related because corticosteroids significantly increase O_2^- production in Na^+-transporting cells (66, 71). This makes O_2^- a likely second messenger in steroid hormone signaling pathways that alter ENaC function in the lung. Under physiological conditions that require the removal of alveolar fluid, steroid hormone receptor and O_2^- signaling may work in unison to increase net Na^+ transport and lung fluid clearance.

Treatment of fetal lung cells for two days with either corticosteroid or increased oxygen tension enhanced the Na^+ transport properties of FDLE cells to the same extent (72, 73). These results do not imply that ROS signaling requires the presence of steroids, because O'Brodovich and colleagues (74) have shown that increasing oxygen tension alone can increase ENaC activity in fetal lung cells in the absence of steroids.

Low Oxygen Tension and Lung Pathophysiology

Free-radical production is a natural consequence of cellular metabolism. Under normal conditions, the production of O_2^- appears to play an important role in regulating salt and water transport to maintain normal lung fluid balance. A lack of O_2^- production, under hypoxic conditions, decreases Na^+ transport and ENaC protein levels in rat lung studies (75, 76). These findings may be important because hypoxia-induced lung injury is likely to occur at high altitude. Patients with high-altitude pulmonary edema (HAPE) are unable to clear lung fluids and have inhibited epithelial Na^+ transport in the alveoli (reviewed in Reference 77). Effective treatment for HAPE includes classical stimulators of Na^+ reabsorption such as β-adrenergics and glucocorticoids and, possibly most important, oxygen therapy that will restore O_2^- (78, 79).

Regulation of Lung ENaC by Reactive Nitrogen Species

RNS such as NO and peroxynitrite ($ONOO^-$, formed from the reaction of O_2^- and NO) are also present in the alveolar epithelium. NO is well known for decreasing pulmonary vascular tone and lung capillary pressure, which are beneficial effects of NO in terms of enhancing oxygen delivery to tissue. However, in terms of lung fluid balance, NO is better known for its deleterious effects: Elevated levels of NO metabolites have been detected in pulmonary edema fluid (80), and several studies report NO inhibition of ENaC activity (66, 81–84). Direct NO inhibition of ENaC is certainly possible, given that there are several tyrosine and cysteine residues on the transmembrane domains and extracellular loops of ENaC subunits, respectively, that could be nitrated or nitrosylated directly by NO. However, there is currently no strong evidence for NO-mediated modification of ENaC. In contrast, the more traditional NO signaling pathway involving soluble guanylyl cyclase (sGC) activation and the production of cyclic guanosine 3′,5′-monophosphate (cGMP) does inhibit ENaC function (29).

Cross Talk Between Reactive Nitrogen Species Signaling and Reactive Oxygen Species Signaling in ENaC Regulation

NO is synthesized from the amino acid L-arginine in a reaction catalyzed by nitric oxide synthase (NOS). Because NO is an inhibitor of ENaC activity, it may be favorable for epithelial cells to decrease the production or activity of NO under edematous conditions. New evidence suggests that the bioavailability of NO may be limited directly by ROS or even regulatory proteins involved in steroid hormone signaling pathways. Increased O_2^- prevents the immediate NO inhibition of ENaC that is normally observed in Na^+-transporting epithelia (66). This effect is presumably due to the interaction of elevated O_2^- with NO to form $ONOO^-$. This may be an important mechanism regulating lung fluid balance because AT1 cells appear to generate very high levels of O_2^- compared with AT2 cells. This explains why exogenous NO given to reduce pulmonary vascular resistance and increase lung perfusion does not cause edema: NO presumably reduces ENaC activity in AT2 cells, but ENaC activity is unaffected in AT1 cells because in AT1 cells NO is scavenged by the high levels of O_2^-.

Cross Talk Between Reactive Nitrogen Species Signaling and Steroid Hormone Signaling in ENaC Regulation

The protein kinase SGK1 is a well-known positive modulator of ENaC function that is induced by steroid hormones (85, 86). However, it may also inhibit NO production via phosphorylation of inducible NO synthase. In vitro kinase assays indicate that SGK1 phosphorylates Ser733 and Ser903 residues of inducible nitric oxide synthase (iNOS) oligopeptides, and immunohistochemical studies show that iNOS and SGK1 proteins are closely located in the cytoplasm of lung cells (82). Interestingly, besides being present in the cytosol, SGK1 has also recently been detected in the mitochondria of epithelial cell lines (87). On the basis of these findings, and the characterization of a mitochondrial nitric oxide synthase (mtNOS) (88, 89), it is possible that SGK1 in the mitochondria regulates mtNOS production of NO, thereby permitting normal ENaC function in the cell.

THE ROLE OF TIGHT JUNCTIONS AND ALVEOLAR PARACELLULAR PERMEABILITY IN DETERMINING ENaC-MEDIATED FLUID REABSORPTION

As represented in **Figure 1**, the volume of alveolar fluid represents the balance between the secretion of fluid from the interstitium to the alveolar lumen and the ENaC-mediated reabsorption of fluid from the lumen back into the interstitium. This balance in general requires an effective barrier that restricts the passive movement of salt and water. Therefore, it is unreasonable to discuss the regulation of transcellular reabsorption without also recognizing the importance of tight junctions in determining fluid balance. Tight junctions are sites of cell-cell contact composed of several transmembrane and peripheral proteins assembled into a complex tethered to the cytoskeleton (90–92).

Central to determining the paracellular permeability of tight junctions are proteins in the claudin family (93–97). Claudins are transmembrane proteins that span the bilayer four times, contain two extracellular loop domains, and have the N and C termini oriented toward the cytoplasm. To date, researchers have identified 23 human claudins, including two claudins that exhibit splice variants (95, 98). Rodent and human lung epithelia express several claudins, including claudin-1, -3, -4, -5, -7, -8, and -18 (99–102). All these claudins are expressed in the alveolus. However, there are cell-specific differences in expression because AT2 cells are enriched for claudin-1, -3, and -18, in contrast to

AT1 cells, which are enriched for claudin-4, -7, and -8.

Claudins have two functions to regulate epithelial tight junctions because they provide a barrier to bulk flow of fluid and also produce the equivalent of paracellular ion channels. The effectiveness of tight junctions as a barrier to bulk fluid flow depends upon the claudin composition of the tight junctions. For example, in cultured alveolar epithelial cells, epidermal growth factor specifically increases claudin-4 and -7 and enhances barrier function (99), and in several other tissues, upregulation of claudins, such as claudin-2, -6, -10, and -15, increases tight junction permeability. However, increased expression of some claudins can result in impaired barrier function. In the lung, increased claudin-5 expression by airway or alveolar epithelia cells is associated with defective barrier function (100, 102, 103). The mechanism by which claudin-5 decreases lung barrier function is not clear at present and seems paradoxical in light of the absolute requirement for claudin-5 for endothelial barrier integrity (104). The observations suggest that claudin function differs, depending on the expression of other tight junction proteins in a given tissue (105–107).

Modulation of paracellular ion permeability by claudins is also complex. The selectivity of tight junction ion permeability in epithelial monolayers can be assessed by comparing the net transepithelial conductance and dilution potentials for specific anions and cations (typically Na^+ and Cl^-), with the caveat that these measurements also include some contribution from transcellular ion transport (108–111).

Measuring the effect of claudins on ion permeability following manipulation of claudin expression provides clues to the contribution of individual claudins to tight junction ion permeability and selectivity (94, 95). Several studies using mutated claudins and claudin-10 splice variants show that charged amino acid residues determine whether a particular claudin is preferentially anion or cation selective (95, 112–114); however, the contribution of an individual claudin to net ion permeability is also affected by the association with other claudins

(115, 116). For instance, in Madin Darby canine kidney (MDCK) cells that have cation-selective tight junctions, siRNA knockdown of claudin-4 or claudin-7 increased Na^+ permeability and decreased Cl^-. However, in porcine kidney (LLC-PK1) cells, which have anion-selective tight junctions, claudin-4 or claudin-7 knockdown decreased both Na^+ and Cl^- permeability (116). In contrast, overexpression of claudin-7 by LLC-PK1 cells increased Na^+ permeability and decreased Cl^- permeability (117). Clearly, claudin composition and stoichiometry will be key determinants of paracellular ion flux and underscore the necessity of making direct measurements in the tissue of interest. Because AT1 and AT2 cells express different claudins and are in direct contact (118), the permeability characteristics of AT2-AT1 cell junctions are likely to differ from those of AT1-AT1 cell junctions.

The notion that the claudins are sensitive to the presence of other types of claudins in tight junction strands suggests specificity of claudin-claudin interactions. Claudins can potentially interact in two different ways: laterally in the plane of the membrane (heteromeric interactions) and between adjacent cells by head-to-head binding (heterotypic interactions). Examining whether different claudins interact is difficult, particularly because most epithelial cells express four or more claudins. This has led to the use of fibroblasts as a claudin-null background to examine claudin compatibility (119, 120). However, fibroblasts lack several tight junction proteins, such as occludin, which may influence interactions between other tight junction proteins. HeLa cells are claudin null and so are another useful alternative system because they express occludin as well as other tight junction proteins (121). Using these systems, investigators have shown that several claudins, including claudin-1, claudin-2, and claudin-5, heterotypically interact with claudin-3 (119–121). Interestingly, claudin-3 and claudin-4 are not heterotypically compatible, despite the facts that they heteromerically interact and that their extracellular loop (EL) domains are highly conserved at the amino acid level (121). However, mutation of a single amino acid in EL 1 of claudin-3 to the corresponding amino acid in claudin-4 ($Asn_{44} \rightarrow Arg$) enables heterotypic binding to claudin-4, suggesting a structural basis for this specificity. But interpreting this is complicated by the lack of a detailed structural model for claudins. Such a model would also facilitate our understanding of the molecular basis for claudins in paracellular ion permeability and barrier function.

A NEW PARADIGM FOR REGULATION OF LUNG FLUID BALANCE

The broadly accepted paradigm for Na^+ transport in many epithelia, including the alveoli, involves a two-step process: first, movement of Na^+ from the alveolar space into the cells through cation channels in the apical cell membrane followed by active transport of the Na^+ across the basolateral membrane into the pulmonary interstitium via basolateral Na^+,K^+-ATPase. Despite almost universal agreement on this paradigm, for alveolar epithelium the devil is apparently in the details (20). In particular, there are several unresolved questions that, when resolved, could significantly affect our view of alveolar fluid clearance. The questions revolve around the contributions of the different cell types, the pathways for specific ions and water, and how these pathways are regulated to control fluid balance. The conventional wisdom for many years suggested that AT2 cells were the primary pathway for the movement of Na^+, that Cl^- moved through a paracellular pathway, and that AT1 cells were the pathway for water. More recently, it has become clear that AT1 cells have all the transport machinery necessary for mediating Na^+ reabsorption. Because AT1 cells make up 95% of the alveolar surface area, their potential for contributing to fluid balance is enormous. Another question involves the pathways for movement of anions. The character of these pathways is important because anion channels could form either co-ion pathways for the concomitant movement of Na^+ and Cl^- or a secretory pathway to increase alveolar fluid. Finally, alveolar fluid clearance appears to consist of a basal and a stimulated component of reabsorption; thus, it is important to understand the relative contributions of different ion channels and different cell types to these two components of clearance. These questions are all underscored by the differences between the robust increases in transport that can be produced by some treatments in in vitro models of alveolar transport compared with the often weak clearance response of the in vivo lung epithelium to the same treatments under conditions of pathologically abnormal fluid accumulation (1, 22).

As mentioned above, there has been long-standing controversy about the transport pathways involved in basal lung Na^+, Cl^-, and water movement and the precise role that ENaC plays in the absence of agents known to enhance ENaC activity (19). One view holds that ENaC contributes to basal lung fluid clearance, whereas the other camp suggests that ENaC only plays a role under extraordinary circumstances (at birth, in pulmonary edema, or after stimulation by β-agonists, etc.). The best evidence seems to suggest that the answer lies somewhere in between. Li & Folkesson (122) have shown—with an elegant approach using plasmid delivery of sequences that encode short hairpin RNA into the air spaces of the lungs to decrease α ENaC expression in intact epithelia—that amiloride-sensitive ENaC channels are responsible for most, if not all, of the β-agonist-induced increase in alveolar fluid clearance but that ENaC contributes less to baseline fluid clearance. Pretreatment with α ENaC siRNA blocked most of the terbutaline-associated increase in fluid transport. There is considerable evidence to show that, in animal models challenged with excess alveolar fluid, increasing the activity of ENaC and Na^+,K^+-ATPase can enhance edema resolution, raising the hopes for therapeutic interventions (4, 5, 22). However, pretreatment with α ENaC siRNA under unstimulated conditions blocked baseline lung fluid absorption by only 30%. This treatment also attenuated the amiloride sensitivity of lung fluid absorption. The investigators (122) did not explore the pathways re-

sponsible for the remaining 70% of alveolar fluid clearance, and there is no consensus about the ion transport mechanisms responsible for this large fraction of baseline fluid transport, although some investigators have attributed this to cyclic nucleotide–gated channels (17). The major caveat to this work (122) is that the presence of the target ENaC subunit was not evaluated in AT1 cells, and therefore it is not clear if the strategy worked equally well in these cells. This is important because, as pointed out above, there is a large body of evidence to show that AT1 cells also express ENaC and Na^+,K^+-ATPase and may have a significant role to play in alveolar salt and fluid transport (14, 23, 123).

All the recent data taken together suggest a revised paradigm of ion transport in the lung, in which ion transport occurs across the entire alveolar surface, rather than being limited to AT2 cells. From the ion channel characteristics of AT1 and AT2 cells and from the large differences in surface area of the two cell types, it appears that AT1 cells may be more instrumental in driving basal Na^+ transport than are the AT2 cells, whereas AT2 cells may play a role not only in Na^+ transport but also in regulating transcellular Cl^- and anion transport under conditions when salt and water transport is stimulated by various hormones or transmitter agents. AT2 cells may also mediate some aspects of Na^+ flux by modulating ENaC via CFTR, because CFTR can inhibit ENaC function and decrease Na^+ transport (40). The tight junctions also play a role in this new model. The fact that tight junctions in the healthy lung are anion selective implies that under basal conditions, most anion movement from lumen to interstitium is through the tight junctions around the AT1 cells and that cation movement from interstitium to lumen is restricted. However, under pathological conditions when the tight junctions become less selective, Na^+ can be actively transported transcellularly from lumen to interstitium only to leak back across the tight junctions and reduce effective alveolar fluid clearance. This new paradigm is also predicated on concerted changes in both apical Na^+ uptake and basolateral extrusion. However, it

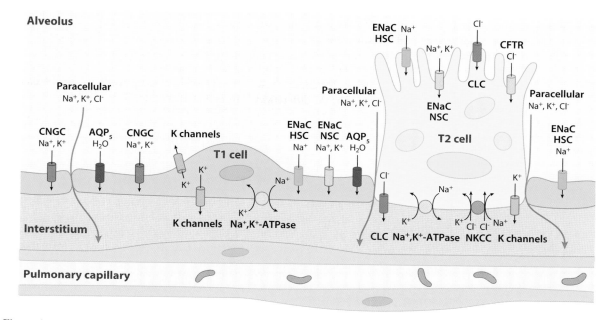

Alveolus

Figure 4

Schematic diagram of alveolar epithelial ion transport. Na^+ is absorbed from the apical surfaces of both AT1 and AT2 cells via ENaC (HSC and NSC channels). Electroneutrality is conserved with Cl^- movement through CFTR or CLC channels in AT2 cells and/or paracellularly through tight junctions. Na^+ is transported from the basal surface of both cell types into the interstitial space by Na^+,K^+-ATPase. K^+ may be transported from alveolar epithelial cells via K^+ channels located on the apical surface of AT1 cells or through potential basolateral K^+ channels in AT2 cells. Cyclic nucleotide–gated channels (CNGC) are an alternative pathway for the movement of Na^+ that would be amiloride insensitive. CNGC may also be a pathway for calcium entry into cells. If net ion transport is from the apical surface to the interstitium, an osmotic gradient is created. This gradient in turn directs water transport in the same direction, either through aquaporins or by diffusion. Used with permission from Reference 125.

should come as no surprise that the regulation of apical and basolateral Na^+ transport in the lung epithelium is likely coordinated (25). Many agents, such as steroids, β-agonists, and dopamine, that increase the number and/or activity of apical Na^+ channels (10) also have a similar effect on Na^+,K^+-ATPase (24). In this regard, lung epithelium differs from other epithelia responsible for Na^+ transport (like the colon and kidney) in that its primary goal is not net absorption of salt but the maintenance of a precisely regulated fluid layer through a combination of fluid absorption and secretion. Having control over the apical entry as well as the basolateral extrusion step would be useful under different physiological and pathological circumstances. In the final analysis, the ability of the lung epithelium to respond to sudden changes in fluid flux would be contingent on co-

ordinated changes in the activity of Na^+ channels and the pump (6) and the ability of the epithelium to move appropriate amounts of Cl^- (7, 13). A schematic diagram of this paradigm is shown in **Figure 4**.

CONCLUSIONS

Ion channels that are members of the ENaC family of proteins play an important role in lung fluid balance. Because the ENaC channel density/patch are similar in AT1 and AT2 cells and because AT1 cells cover >95% of the alveolar surface area, AT1 cells likely are responsible for the bulk of Na^+ transport in the lung. In alveolar AT1 and AT2 cells, there are a variety of different amiloride-sensitive, Na^+-permeable channels. This significant diversity appears to play a role in both normal lung physiology and pathological states. At least part of the diversity of Na^+-permeable channels in lung arises from assembling different combinations of α, β, and γ ENaC subunits to form channels with different biophysical properties and different mechanisms for regulation. In particular, when only α ENaC subunits are assembled together to form channels, the result is a 21–28 pS NSC channel that is sensitive to intracellular calcium and whose open probability is increased by cAMP. In marked contrast, when α, β, and γ ENaC subunits are assembled together, the result is a 4–6 pS channel that is highly selective for Na^+ over K^+ and is insensitive to intracellular calcium, and cAMP increases channel number with no effect on open probability.

In addition, the regulation of different ENaC channels in AT1 and AT2 cells means that each cell has the capacity to respond differently to hormonal and transmitter conditions within the lung. The diversity of channel expression and functional properties of the channels coupled with differential regulation of AT1 and AT2 cells lead to epithelial tissue in the lung that has enormous flexibility to alter the magnitude and regulation of salt and water transport. Better understanding of these channels and regulatory pathways will help delineate the mechanisms that regulate alveolar fluid balance and may provide the basis for developing strategies to prevent or treat the respiratory compromise associated with alveolar flooding.

DISCLOSURE STATEMENT

The authors are not aware of any biases that might be perceived as affecting the objectivity of this review.

ACKNOWLEDGMENTS

This work was supported in part by grants R24DK064399 and R37DK037963 to D.C.E., K99/R00HL092226 to M.N.H., R01HL071621 and R01HL063306 to L.J. and D.C.E., T32DK07656 support to H.F.B., and R01HL083120 to M.K.

LITERATURE CITED

1. Helms MN, Self J, Bao HF, Job LC, Jain L, Eaton DC. 2006. Dopamine activates amiloride-sensitive sodium channels in alveolar type I cells in lung slice preparations. *Am. J. Physiol. Lung Cell. Mol. Physiol.* 291:L610–18

2. Johnson MD, Bao HF, Helms MN, Chen XJ, Tigue Z, et al. 2006. Functional ion channels in pulmonary alveolar type I cells support a role for type I cells in lung ion transport. *Proc. Natl. Acad. Sci. USA* 103:4964–69

3. Matthay MA, Folkesson HG, Verkman AS. 1996. Salt and water transport across alveolar and distal airway epithelia in the adult lung. *Am. J. Physiol.* 270:L487–503

4. Matthay MA, Flori HR, Conner ER, Ware LB. 1998. Alveolar epithelial fluid transport: basic mechanisms and clinical relevance. *Proc. Assoc. Am. Physicians* 110:496–505

5. Sartori C, Matthay MA, Scherrer U. 2001. Transepithelial sodium and water transport in the lung. Major player and novel therapeutic target in pulmonary edema. *Adv. Exp. Med. Biol.* 502:315–38

6. Berthiaume Y, Folkesson HG, Matthay MA. 2002. Lung edema clearance: 20 years of progress. Invited review: alveolar edema fluid clearance in the injured lung. *J. Appl. Physiol.* 93:2207–13

7. Matalon S, Lazrak A, Jain L, Eaton DC. 2002. Lung edema clearance: 20 years of progress. Invited review: biophysical properties of sodium channels in lung alveolar epithelial cells. *J. Appl. Physiol.* 93:1852–59

8. Matalon S, O'Brodovich HM. 1999. Sodium channels in alveolar epithelial cells: molecular characterization, biophysical properties, and physiological significance. *Annu. Rev. Physiol.* 61:627–61

9. Zeitlin PL, Crawford I, Lu L, Woel S, Cohen ME, et al. 1992. CFTR protein expression in primary and cultured epithelia. *Proc. Natl. Acad. Sci. USA* 89:344–47

10. Dobbs LG. 1990. Isolation and culture of alveolar type II cells. *Am. J. Physiol.* 258:L134–47

11. Eaton DC, Chen J, Ramosevac S, Matalon S, Jain L. 2004. Regulation of Na$^+$ channels in lung alveolar type II epithelial cells. *Proc. Am. Thorac. Soc.* 1:10–16

12. Matalon S, Benos DJ, Jackson RM. 1996. Biophysical and molecular properties of amiloride-inhibitable Na$^+$ channels in alveolar epithelial cells. *Am. J. Physiol.* 271:L1–22

13. Dobbs LG, Pian MS, Maglio M, Dumars S, Allen L. 1997. Maintenance of the differentiated type II cell phenotype by culture with an apical air surface. *Am. J. Physiol.* 273:347–54

14. Gonzalez R, Yang YH, Griffin C, Allen L, Tigue Z, Dobbs L. 2005. Freshly isolated rat alveolar type I cells, type II cells, and cultured type II cells have distinct molecular phenotypes. *Am. J. Physiol. Lung Cell. Mol. Physiol.* 288:L179–89

15. Borok Z, Liebler JM, Lubman RL, Foster MJ, Zhou B, et al. 2002. Na transport proteins are expressed by rat alveolar epithelial type I cells. *Am. J. Physiol. Lung Cell. Mol. Physiol.* 282:L599–608

16. Johnson MD, Widdicombe JH, Allen L, Barbry P, Dobbs LG. 2002. Alveolar epithelial type I cells contain transport proteins and transport sodium, supporting an active role for type I cells in regulation of lung liquid homeostasis. *Proc. Natl. Acad. Sci. USA* 99:1966–71

17. Bourke S, Mason H, Borok Z, Kim KJ, Crandall ED, Kemp PJ. 2004. Development of a lung slice preparation for recording ion channel activity in alveolar epithelial type I cells. *Respir. Res.* 6:4–50

18. Helms MN, Jain L, Self JL, Eaton DC. 2008. Redox regulation of epithelial sodium channels (ENaC) examined in alveolar type 1 and 2 cells patch clamped in lung slice tissue. *J. Biol. Chem.* In press

19. Benos DJ. 1989. The biology of amiloride-sensitive sodium channels. *Hosp. Pract.* 24:149–55, 159–64

20. Smith PR, Benos DJ. 1991. Epithelial Na$^+$ channels. *Annu. Rev. Physiol.* 53:509–30

21. Garty H. 1994. Molecular properties of epithelial, amiloride-blockable Na$^+$ channels. *FASEB J.* 8:522–28

22. Chen XJ, Eaton DC, Jain L. 2002. β-Adrenergic regulation of amiloride-sensitive lung sodium channels. *Am. J. Physiol. Lung Cell. Mol. Physiol.* 282:609–20

23. Chen XJ, Jain L, Eaton DC. 2006. Purinergic regulation of amiloride-sensitive lung sodium channels. *Am. J. Physiol. Lung Cell. Mol. Physiol.* Under review

24. Helms MN, Chen XJ, Ramosevac S, Eaton DC, Jain L. 2005. Dopamine regulation of amiloride-sensitive sodium channels in lung cells. *Am. J. Physiol. Lung Cell. Mol. Physiol.* 291:L710–22

25. Jain L, Chen XJ, Malik B, Al-Khalili OK, Eaton DC. 1999. Antisense oligonucleotides against the α-subunit of ENaC decrease lung epithelial cation-channel activity. *Am. J. Physiol.* 276:L1046–51

26. Orser BA, Bertlik H, Fedorko L, O'Brodovich HM. 1990. Cation channels in apical membrane of fetal alveolar epithelium. *Can. J. Anaesth.* 37:S3

27. Feng ZP, Clark RB, Berthiaume Y. 1993. Identification of nonselective cation channels in cultured adult rat alveolar type II cells. *Am. J. Respir. Cell Mol. Biol.* 9:248–54

28. Yue G, Shoemaker RL, Matalon S. 1994. Regulation of low-amiloride-affinity sodium channels in alveolar type II cells. *Am. J. Physiol.* 267:L94–100

29. Jain L, Chen XJ, Brown LA, Eaton DC. 1998. Nitric oxide inhibits lung sodium transport through a cGMP-mediated inhibition of epithelial cation channels. *Am. J. Physiol.* 274:L475–84

30. Jain L, Chen XJ, Ramosevac S, Brown LA, Eaton DC. 2001. Expression of highly selective sodium channels in alveolar type II cells is determined by culture conditions. *Am. J. Physiol. Lung Cell. Mol. Physiol.* 280:L646–58

31. Canessa CM, Schild L, Buell G, Thorens B, Gautschi I, et al. 1994. Amiloride-sensitive epithelial Na$^+$ channel is made of three homologous subunits. *Nature* 367:463–67

32. Mutlu GM, Adir Y, Jameel M, Akhmedov AT, Welch L, et al. 2005. Interdependency of β-adrenergic receptors and CFTR in regulation of alveolar active Na$^+$ transport. *Circ. Res.* 96:999–1005

33. Folkesson HG, Matthay MA, Chapin CJ, Porta NF, Kitterman JA. 2002. Distal air space epithelial fluid clearance in near-term rat fetuses is fast and requires endogenous catecholamines. *Am. J. Physiol. Lung Cell. Mol. Physiol.* 282:L508–15

34. Planes C, Blot-Chabaud M, Matthay MA, Couette S, Uchida T, Clerici C. 2002. Hypoxia and β2-agonists regulate cell surface expression of the epithelial sodium channel in native alveolar epithelial cells. *J. Biol. Chem.* 277:47318–24

35. Marunaka Y, Niisato N, Ito Y. 2000. β agonist regulation of sodium transport in fetal lung epithelium: roles of cell volume, cytosolic chloride and protein tyrosine kinase. *J. Korean Med. Sci.* 15(Suppl.):S42–43

36. Saldias FJ, Lecuona E, Comellas AP, Ridge KM, Rutschman DH, Sznajder JI. 2000. β-Adrenergic stimulation restores rat lung ability to clear edema in ventilator-associated lung injury. *Am. J. Respir. Crit. Care Med.* 162:282–87

37. Minakata Y, Suzuki S, Grygorczyk C, Dagenais A, Berthiaume Y. 1998. Impact of β-adrenergic agonist on Na⁺ channel and Na⁺-K⁺-ATPase expression in alveolar type II cells. *Am. J. Physiol.* 275:414–22

38. Sakuma T, Folkesson HG, Suzuki S, Okaniwa G, Fujimura S, Matthay MA. 1997. β-Adrenergic agonist stimulated alveolar fluid clearance in ex vivo human and rat lungs. *Am. J. Respir. Crit. Care Med.* 155:506–12

39. Kishore BK, Ginns SM, Krane CM, Nielsen S, Knepper MA. 2000. Cellular localization of P2Y₂ purinoceptor in rat renal inner medulla and lung. *Am. J. Physiol. Ren. Fluid Electrolyte Physiol.* 278:F43–51

40. Homolya L, Watt WC, Lazarowski ER, Koller BH, Boucher RC. 1999. Nucleotide-regulated calcium signaling in lung fibroblasts and epithelial cells from normal and P2Y₂ receptor (–/–) mice. *J. Biol. Chem.* 274:26454–60

41. Taylor AL, Schwiebert LM, Smith JJ, King C, Jones JR, et al. 1999. Epithelial P2X purinergic receptor channel expression and function. *J. Clin. Investig.* 104:875–84

42. Gilfillan AM, Rooney SA. 1988. Functional evidence for involvement of P2 purinoceptors in the ATP stimulation of phosphatidylcholine secretion in type II alveolar epithelial cells. *Biochim. Biophys. Acta* 959:31–37

43. Ma HP, Ling BN. 1996. Luminal adenosine receptors regulate amiloride-sensitive, Na⁺ channels in A6 distal nephron cells. *Am. J. Physiol.* 270:F798–805

44. Barth K, Weinhold K, Guenther A, Linge A, Gereke M, Kasper M. 2008. Characterization of the molecular interaction between caveolin-1 and the P2X receptors 4 and 7 in E10 mouse lung alveolar epithelial cells. *Int. J. Biochem. Cell Biol.* In press

45. Communi D, Paindavoine P, Place GA, Parmentier M, Boeynaems JM. 1999. Expression of P2Y receptors in cell lines derived from the human lung. *Br. J. Pharmacol.* 127:562–68

46. Leipziger J. 2003. Control of epithelial transport via luminal P2 receptors. *Am. J. Physiol. Ren. Physiol.* 284:419–32

47. Saldias FJ, Comellas AP, Pesce L, Lecuona E, Sznajder JI. 2002. Dopamine increases lung liquid clearance during mechanical ventilation. *Am. J. Physiol. Lung Cell. Mol. Physiol.* 283:136–43

48. Barnard ML, Ridge KM, Saldias F, Friedman E, Gare M, et al. 1999. Stimulation of the dopamine 1 receptor increases lung edema clearance. *Am. J. Respir. Crit. Care Med.* 160:982–86

49. Barnard ML, Olivera WG, Rutschman DM, Bertorello AM, Katz AI, Sznajder JI. 1997. Dopamine stimulates sodium transport and liquid clearance in rat lung epithelium. *Am. J. Respir. Crit. Care Med.* 156:709–14

50. Dada LA, Sznajder JI. 2003. Mechanisms of pulmonary edema clearance during acute hypoxemic respiratory failure: role of the Na,K-ATPase. *Crit. Care Med.* 31:248–52

51. Guerrero C, Lecuona E, Pesce L, Ridge KM, Sznajder JI. 2001. Dopamine regulates Na-K-ATPase in alveolar epithelial cells via MAPK-ERK-dependent mechanisms. *Am. J. Physiol. Lung Cell. Mol. Physiol.* 281:L79–85

52. Lecuona E, Garcia A, Sznajder JI. 2000. A novel role for protein phosphatase 2A in the dopaminergic regulation of Na,K-ATPase. *FEBS Lett.* 481:217–20

53. Eaton DC, Malik B, Saxena NC, Al Khalili OK, Yue G. 2001. Mechanisms of aldosterone's action on epithelial Na⁺ transport. *J. Membr. Biol.* 184:313–19

54. Pochynyuk O, Tong Q, Staruschenko A, Ma HP, Stockand JD. 2006. Regulation of the epithelial Na⁺ channel (ENaC) by phosphatidylinositides. *Am. J. Physiol. Ren. Physiol.* 290:F949–57

55. Helms MN, Liu L, Liang YY, Al-Khalili O, Vandewalle A, et al. 2005. Phosphatidylinositol 3,4,5-trisphosphate mediates aldosterone stimulation of epithelial sodium channel (ENaC) and interacts with γ-ENaC. *J. Biol. Chem.* 280:40885–91

56. Ma HP, Eaton DC. 2005. Acute regulation of epithelial sodium channel by anionic phospholipids. *J. Am. Soc. Nephrol.* 16:3182–87

57. Ma HP, Saxena S, Warnock DG. 2002. Anionic phospholipids regulate native and expressed ENaC. *J. Biol. Chem.* 277:7641–44

58. Dagenais A, Frechette R, Yamagata Y, Yamagata T, Carmel JF, et al. 2004. Downregulation of ENaC activity and expression by TNF-α in alveolar epithelial cells. *Am. J. Physiol. Lung Cell. Mol. Physiol.* 286:301–11

59. Frank J, Roux J, Kawakatsu H, Su G, Dagenais A, et al. 2003. TGF-β1 decreases expression of the epithelial sodium channel αENaC and alveolar epithelial vectorial sodium and fluid transport via an ERK 1/2-dependent mechanism. *J. Biol. Chem.* 278:43939–50

60. Roux J, Kawakatsu H, Gartland B, Pespeni M, Sheppard D, et al. 2005. Interleukin-1β decreases expression of the epithelial sodium channel α-subunit in alveolar epithelial cells via a p38 MAPK-dependent signaling pathway. *J. Biol. Chem.* 280:18579–89

61. Willis BC, Kim KJ, Li X, Liebler J, Crandall ED, Borok Z. 2003. Modulation of ion conductance and active transport by TGF-β1 in alveolar epithelial cell monolayers. *Am. J. Physiol. Lung Cell. Mol. Physiol.* 285:L1190–91

62. Fillon S, Warntges S, Matskevitch J, Moschen I, Setiawan I, et al. 2001. Serum- and glucocorticoid-dependent kinase, cell volume, and the regulation of epithelial transport. *Comp. Biochem. Physiol. A* 130:367–76

63. Wagner CA, Ott M, Klingel K, Beck S, Melzig J, et al. 2001. Effects of the serine/threonine kinase SGK1 on the epithelial Na$^+$ channel (ENaC) and CFTR: implications for cystic fibrosis. *Cell Physiol. Biochem.* 11:209–18

64. Husted RF, Sigmund RD, Stokes JB. 2000. Mechanisms of inactivation of the action of aldosterone on collecting duct by TGF-β. *Am. J. Physiol. Ren. Fluid Electrolyte Physiol.* 278:F425–33

65. Husted RF, Zhang C, Stokes JB. 1998. Concerted actions of IL-1β inhibit Na$^+$ absorption and stimulate anion secretion by IMCD cells. *Am. J. Physiol.* 275:F946–54

66. Yu L, Bao HF, Self JL, Eaton DC, Helms MN. 2007. Aldosterone-induced increases in superoxide production counters nitric oxide inhibition of epithelial Na channel activity in A6 distal nephron cells. *Am. J. Physiol. Ren. Physiol.* 293:F1666–77

67. Rafii B, Tanswell AK, Otulakowski G, Pitkanen O, Belcastro-Taylor R, O'Brodovich H. 1998. O$_2$-induced ENaC expression is associated with NF-κB activation and blocked by superoxide scavenger. *Am. J. Physiol.* 275:L764–70

68. Baines DL, Ramminger SJ, Collett A, Haddad JJ, Best OG, et al. 2001. Oxygen-evoked Na$^+$ transport in rat fetal distal lung epithelial cells. *J. Physiol.* 532:105–13

69. Thome UH, Davis IC, Nguyen SV, Shelton BJ, Matalon S. 2003. Modulation of sodium transport in fetal alveolar epithelial cells by oxygen and corticosterone. *Am. J. Physiol. Lung Cell. Mol. Physiol.* 284:L376–85

70. Bremner HR, Freywald T, O'Brodovich HM, Otulakowski G. 2002. Promoter analysis of the gene encoding the β-subunit of the rat amiloride-sensitive epithelial sodium channel. *Am. J. Physiol. Lung Cell. Mol. Physiol.* 282:L124–34

71. Miyata K, Rahman M, Shokoji T, Nagai Y, Zhang GX, et al. 2005. Aldosterone stimulates reactive oxygen species production through activation of NADPH oxidase in rat mesangial cells. *J. Am. Soc. Nephrol.* 16:2906–12

72. Otulakowski G, Rafii B, Harris M, O'brodovich H. 2006. Oxygen and glucocorticoids modulate αENaC mRNA translation in fetal distal lung epithelium. *Am. J. Respir. Cell Mol. Biol.* 34:204–12

73. Thome UH, Davis IC, Nguyen SV, Shelton BJ, Matalon S. 2003. Modulation of sodium transport in fetal alveolar epithelial cells by oxygen and corticosterone. *Am. J. Physiol. Lung Cell. Mol. Physiol.* 284:L376–85

74. Otulakowski G, Rafii B, Harris M, O'Brodovich H. 2006. Oxygen and glucocorticoids modulate αENaC mRNA translation in fetal distal lung epithelium. *Am. J. Respir. Cell Mol. Biol.* 34:204–12

75. Planes C, Escoubet B, Blot-Chabaud M, Friedlander G, Farman N, Clerici C. 1997. Hypoxia downregulates expression and activity of epithelial sodium channels in rat alveolar epithelial cells. *Am. J. Respir. Cell Mol. Biol.* 17:508–18

76. Vivona ML, Matthay M, Chabaud MB, Friedlander G, Clerici C. 2001. Hypoxia reduces alveolar epithelial sodium and fluid transport in rats: reversal by β-adrenergic agonist treatment. *Am. J. Respir. Cell Mol. Biol.* 25:554–61

77. Mairbaurl H. 2006. Role of alveolar epithelial sodium transport in high altitude pulmonary edema (HAPE). *Respir. Physiol. Neurobiol.* 151:178–91

78. Gregorius DD, Dawood R, Ruh K, Nguyen HB. 2008. Severe high altitude pulmonary oedema: a patient managed successfully with noninvasive positive pressure ventilation in the Emergency Department. *Emerg. Med. J.* 25:243–44

79. Maggiorini M, Brunner-La Rocca HP, Peth S, Fischler M, Bohm T, et al. 2006. Both tadalafil and dexamethasone may reduce the incidence of high-altitude pulmonary edema: a randomized trial. *Ann. Int. Med.* 145:497–506

80. Zhu S, Ware LB, Geiser T, Matthay MA, Matalon S. 2001. Increased levels of nitrate and surfactant protein A nitration in the pulmonary edema fluid of patients with acute lung injury. *Am. J. Respir. Crit. Care Med.* 163:166–72

81. Guo Y, DuVall MD, Crow JP, Matalon S. 1998. Nitric oxide inhibits Na^+ absorption across cultured alveolar type II monolayers. *Am. J. Physiol.* 274:L369–77

82. Helms MN, Yu L, Malik B, Kleinhenz DJ, Hart CM, Eaton DC. 2005. Role of SGK1 in nitric oxide inhibition of ENaC in Na^+-transporting epithelia. *Am. J. Physiol. Cell Physiol.* 289:C717–26

83. Nielsen VG, Baird MS, Chen L, Matalon S. 2000. DETANONOate, a nitric oxide donor, decreases amiloride-sensitive alveolar fluid clearance in rabbits. *Am. J. Respir. Crit. Care Med.* 161:1154–60

84. Ruckes-Nilges C, Lindemann H, Klimek T, Glanz H, Weber WM. 2000. Nitric oxide has no beneficial effects on ion transport defects in cystic fibrosis human nasal epithelium. *Pflüg. Arch.* 441:133–37

85. Chen SY, Bhargava A, Mastroberardino L, Meijer OC, Wang J, et al. 1999. Epithelial sodium channel regulated by aldosterone-induced protein sgk. *Proc. Natl. Acad. Sci. USA* 96:2514–19

86. Naray-Fejes-Toth A, Canessa C, Cleaveland ES, Aldrich G, Fejes-Toth G. 1999. *sgk* is an aldosterone-induced kinase in the renal collecting duct. Effects on epithelial Na^+ channels. *J. Biol. Chem.* 274:16973–78

87. Cordas E, Naray-Fejes-Toth A, Fejes-Toth G. 2007. Subcellular location of serum- and glucocorticoid-induced kinase-1 in renal and mammary epithelial cells. *Am. J. Physiol. Cell Physiol.* 292:C1971–81

88. Elfering SL, Sarkela TM, Giulivi C. 2002. Biochemistry of mitochondrial nitric-oxide synthase. *J. Biol. Chem.* 277:38079–86

89. Ghafourifar P, Schenk U, Klein SD, Richter C. 1999. Mitochondrial nitric-oxide synthase stimulation causes cytochrome *c* release from isolated mitochondria. Evidence for intramitochondrial peroxynitrite formation. *J. Biol. Chem.* 274:31185–88

90. Matter K, Balda MS. 2003. Functional analysis of tight junctions. *Methods* 30:228–34

91. Schneeberger EE, Lynch RD. 2004. The tight junction: a multifunctional complex. *Am. J. Physiol. Cell Physiol.* 286:C1213–28

92. Tsukita S, Furuse M, Itoh M. 2001. Multifunctional strands in tight junctions. *Nat. Rev. Mol. Cell Biol.* 2:285–93

93. Angelow S, Kim KJ, Yu AS. 2006. Claudin-8 modulates paracellular permeability to acidic and basic ions in MDCK II cells. *J. Physiol.* 571:15–26

94. Angelow S, Yu AS. 2007. Claudins and paracellular transport: an update. *Curr. Opin. Nephrol. Hypertens.* 16:459–64

95. Van Itallie CM, Anderson JM. 2006. Claudins and epithelial paracellular transport. *Annu. Rev. Physiol.* 68:403–29

96. Angelow S, Kim KJ, Yu AS. 2005. Claudin-8 modulates paracellular permeability to acidic and basic ions in MDCK II cells. *J. Physiol.* 571:15–26

97. Koval M. 2006. Claudins: key pieces in the tight junction puzzle. *Cell Commun. Adhes.* 13:127–38

98. Niimi T, Nagashima K, Ward JM, Minoo P, Zimonjic DB, et al. 2001. *claudin-18*, a novel downstream target gene for the T/EBP/NKX2.1 homeodomain transcription factor, encodes lung- and stomach-specific isoforms through alternative splicing. *Mol. Cell. Biol.* 21:7380–90

99. Chen Y, Deng Y, Bao X, Reuss L, Altenberg GA. 2005. Mechanism of the defect in gap-junctional communication by expression of a connexin 26 mutant associated with dominant deafness. *FASEB J.* 19:1516–18

100. Coyne CB, Gambling TM, Boucher RC, Carson JL, Johnson LG. 2003. Role of claudin interactions in airway tight junctional permeability. *Am. J. Physiol. Lung Cell. Mol. Physiol.* 285:1166–78

101. Daugherty BL, Mateescu M, Patel AS, Wade K, Kimura S et al. 2004. Developmental regulation of claudin localization by fetal alveolar epithelial cells. *Am. J. Physiol. Lung Cell. Mol. Physiol.* 287:1266–73

102. Wang WJ, Mulugeta S, Russo SJ, Beers MF. 2003. Deletion of exon 4 from human surfactant protein C results in aggresome formation and generation of a dominant negative. *J. Cell Sci.* 116:683–92

103. Fernandez AL, Koval M, Fan X, Guidot DM. 2007. Chronic alcohol ingestion alters claudin expression in the alveolar epithelium of rats. *Alcohol* 41:371–79

104. Nitta T, Hata M, Gotoh S, Seo Y, Sasaki H, et al. 2003. Size-selective loosening of the blood-brain barrier in claudin-5-deficient mice. *J. Cell Biol.* 161:653–60

105. Furuse M, Furuse K, Sasaki H, Tsukita S. 2001. Conversion of zonulae occludentes from tight to leaky strand type by introducing claudin-2 into Madin-Darby canine kidney I cells. *J. Cell Biol.* 153:263–72

106. Turksen K, Troy TC. 2002. Permeability barrier dysfunction in transgenic mice overexpressing claudin 6. *Development* 129:1775–84

107. Laukoetter MG, Nava P, Lee WY, Severson EA, Capaldo CT, et al. 2007. JAM-A regulates permeability and inflammation in the intestine in vivo. *J. Exp. Med.* 204:3067–76

108. Van Driessche W, Kreindler JL, Malik AB, Margulies S, Lewis SA, Kim KJ. 2007. Interrelations/cross talk between transcellular transport function and paracellular tight junctional properties in lung epithelial and endothelial barriers. *Am. J. Physiol. Lung Cell. Mol. Physiol.* 293:L520–24

109. Pittet JF, Griffiths MJ, Geiser T, Kaminski N, Dalton SL, et al. 2001. TGF-β is a critical mediator of acute lung injury. *J. Clin. Investig.* 107:1537–44

110. Kim KJ, Borok Z, Ehrhardt C, Willis BC, Lehr CM, Crandall ED. 2005. Estimation of paracellular conductance of primary rat alveolar epithelial cell monolayers. *J. Appl. Physiol.* 98:138–43

111. Frank J, Roux J, Kawakatsu H, Su G, Dagenais A, et al. 2003. Transforming growth factor-β1 decreases expression of the epithelial sodium channel αENaC and alveolar epithelial vectorial sodium and fluid transport via an ERK1/2-dependent mechanism. *J. Biol. Chem.* 278:43939–50

112. Wen H, Watry DD, Marcondes MC, Fox HS. 2004. Selective decrease in paracellular conductance of tight junctions: role of the first extracellular domain of claudin-5. *Mol. Cell. Biol.* 24:8408–17

113. Colegio OR, Van Itallie C, Rahner C, Anderson JM. 2003. Claudin extracellular domains determine paracellular charge selectivity and resistance but not tight junction fibril architecture. *Am. J. Physiol. Cell Physiol.* 284:C1346–54

114. Alexandre MD, Jeansonne BG, Renegar RH, Tatum R, Chen YH. 2007. The first extracellular domain of claudin-7 affects paracellular Cl⁻ permeability. *Biochem. Biophys. Res. Commun.* 357:87–91

115. Van Itallie CM, Fanning AS, Anderson JM. 2003. Reversal of charge selectivity in cation or anion-selective epithelial lines by expression of different claudins. *Am. J. Physiol. Ren. Physiol.* 285:F1078–84

116. Hou J, Gomes AS, Paul DL, Goodenough DA. 2006. Study of claudin function by RNA interference. *J. Biol. Chem.* 281:36117–23

117. Alexandre MD, Lu Q, Chen YH. 2005. Overexpression of claudin-7 decreases the paracellular Cl⁻ conductance and increases the paracellular Na⁺ conductance in LLC-PK1 cells. *J. Cell Sci.* 118:2683–93

118. Wang F, Daugherty B, Keise LL, Wei Z, Foley JP, et al. 2003. Heterogeneity of claudin expression by alveolar epithelial cells. *Am. J. Respir. Cell Mol. Biol.* 29:62–70

119. Furuse M, Sasaki H, Tsukita S. 1999. Manner of interaction of heterogeneous claudin species within and between tight junction strands. *J. Cell Biol.* 147:891–903

120. Coyne CB, Gambling TM, Boucher RC, Carson JL, Johnson LG. 2003. Role of claudin interactions in airway tight junctional permeability. *Am. J. Physiol. Lung Cell. Mol. Physiol.* 285:L1166–78

121. Daugherty BL, Ward C, Smith T, Ritzenthaler JD, Koval M. 2007. Regulation of heterotypic claudin compatibility. *J. Biol. Chem.* 282:30005–13

122. Li T, Folkesson HG. 2006. RNA interference for α-ENaC inhibits rat lung fluid absorption in vivo. *Am. J. Physiol. Lung Cell. Mol. Physiol.* 290:L649–660

123. Angelow S, Kim KJ, Yu AS. 2006. Claudin-8 modulates paracellular permeability to acidic and basic ions in MDCK II cells. *J. Physiol.* 571:15–26

124. Matthay MA, Folkesson HG, Clerici C. 2002. Lung epithelial fluid transport and the resolution of pulmonary edema. *Physiol. Rev.* 82:569–600

125. Eaton DC, Jain L. 2006. Ion channels: an overview. In *Encyclopedia of Respiratory Medicine*, ed. SD Shapiro, GJ Laurent, 1:1–4. Oxford, UK: Elsevier

The Role of CLCA Proteins in Inflammatory Airway Disease

Anand C. Patel,[1] Tom J. Brett,[2,3,4]
and Michael J. Holtzman[2,4]

[1]Department of Pediatrics, [2]Department of Medicine, [3]Department of Biochemistry, and
[4]Department of Cell Biology and Physiology, Washington University School of Medicine,
St. Louis, Missouri 63110; email: holtzmanm@wustl.edu

Annu. Rev. Physiol. 2009. 71:425–49

First published online as a Review in Advance on
October 27, 2008

The *Annual Review of Physiology* is online at
physiol.annualreviews.org

This article's doi:
10.1146/annurev.physiol.010908.163253

Copyright © 2009 by Annual Reviews.
All rights reserved

0066-4278/09/0315-0425$20.00

Key Words

airway epithelial cell, asthma, calcium-activated chloride channel,
mucous cell metaplasia, chronic obstructive pulmonary disease
(COPD)

Abstract

Inflammatory airway diseases such as asthma and chronic obstructive
pulmonary disease (COPD) exhibit stereotyped traits that are variably
expressed in each person. In experimental mouse models of chronic lung
disease, these individual disease traits can be genetically segregated and
thereby linked to distinct determinants. Functional genomic analysis
indicates that at least one of these traits, mucous cell metaplasia, depends on members of the calcium-activated chloride channel (*CLCA*)
gene family. Here we review advances in the biochemistry of the CLCA
family and the evidence of a role for CLCA family members in the development of mucous cell metaplasia and possibly airway hyperreactivity
in experimental models and in humans. On the basis of this information,
we develop the model that CLCA proteins are not integral membrane
proteins with ion channel function but instead are secreted signaling
molecules that specifically regulate airway target cells in healthy and
disease conditions.

CLCA: chloride channel calcium activated

COPD: chronic obstructive pulmonary disease

INTRODUCTION

The calcium-activated chloride channel (CLCA) proteins were first isolated in 1991 and have since grown to a complex family that is preserved throughout the animal, plant, and microbial kingdoms. Despite uncertainty over the biological and pathological function of the CLCA family of proteins, they remain an intriguing target for therapeutic intervention. This practical interest in CLCA proteins was largely confined to the cancer field until new lines of research recognized a connection between *CLCA* gene expression and the development of inflammatory airway disease both in animal models and in humans with asthma, chronic obstructive pulmonary disease (COPD), and cystic fibrosis. This link between *CLCA* gene expression and airway disease is particularly interesting because each of these diseases is associated with excess production of airway mucus, and at least some members of the CLCA family are expressed selectively in mucous cells (1–3). Thus, the development of mucus hypersecretion in complex airway diseases or even after acute respiratory infection may be marked or driven by CLCA proteins. The possibility that CLCA controls mucus production is especially relevant to clinical aspects of airway diseases, because mucus hypersecretion is responsible for much of the morbidity and mortality associated with these conditions (4–11). Furthermore, there is limited understanding of the pathobiology underlying the development of hypersecretory diseases, and there are no effective and specific therapeutic strategies to inhibit mucus production or secretion in acute or chronic respiratory disease.

In this review, we provide the background biochemistry that defines the CLCA family, focusing on the genomic organization of the family and the structure-function relationships of the CLCA proteins. Previous reviews are also available on this topic (12–16), but we update this information to include more recent genome sequencing and identification of additional CLCA family members. We then sum-marize our knowledge of the functional biology of CLCA proteins, focusing on the role of specific CLCA proteins in causing disease. Significant research effort has aimed at the possible activity of CLCA proteins in the control of cell death and the development and spread of cancer, and we summarize these issues. However, we concentrate mainly on the role of CLCA proteins in the development of inflammatory airway disease. We devote specific attention to the association of CLCA family members with the inflammatory process that leads to overproduction of airway mucus and the often concomitant development of airway hyperreactivity, because these two disease traits are characteristic of airway disease and are likely responsible for much of the morbidity associated with this type of disease. We review in separate sections the evidence for CLCA function in animal models of airway disease and in humans with chronic obstructive lung disease. We conclude with a perspective on CLCA biology and a set of questions that need to be addressed for future research on CLCA proteins, because this field is just now poised for definitive understanding and critical therapeutic development.

BACKGROUND BIOCHEMISTRY

In this section, we review the genomic organization of the *CLCA* gene locus and provide a nomenclature to define all members of the *CLCA* gene family in humans and related animal species. We also review how this genomic information serves as a basis to define the structure of CLCA proteins, and how this structure may relate to function. The information is aimed at providing a background for the subsequent sections on the biological role of CLCA proteins in healthy and diseased tissue.

Genomic Organization and Gene Expression

The initial members of the CLCA family were identified independently in two laboratories on the basis of protein purification and were designated as bovine tracheal calcium-activated chloride channel (CaCC) and bovine lung endothelial cell adhesion molecule-1 (Lu-ECAM-1) (17–19). Subsequent molecular cloning indicated that the two gene products were homologous (13, 20). New nomenclature was developed, and CaCC was named bClca1 and Lu-ECAM-1 was designated bClca2. Since that time, additional *Clca* gene homologs have been discovered in multiple species. In addition, putative orthologs are being discovered in other species as genome sequences are completed. For example, sequencing of the genome for the sea squirt (*Ciona intestinalis*) led to the description of seven putative *CLCA* homologs in that species (14). At present, homologs or putative homologs of CLCA family members exist in at least 30 species (*Aedes aegypti, Anopheles gambiae, Bos taurus, Canis familiaris, Cavia porcellus, Ciona intestinalis, Ciona savignyi, Danio rerio, Dasypus novemcinctus, Echinops telfairi, Equus caballus, Erinaceus europaeus, Gallus gallus, Loxodonta africana, Macaca mulatta, Microcebus murinus, Mus musculus, Myotis lucifugus, Ochotona princeps, Ornithorhynchus anatinus, Oryctolagus cuniculus, Otolemur garnettii, Pan troglodytes, Pongo pygmaeus, Rattus norvegicus, Sorex araneus, Sper-*

mophilus tridecemlineatus, Strongylocentrotus purpuratus, Tupaia belangeri, and Xenopus tropicalis) on the basis of Ensembl GeneTreeView or NCBI Gene database query (21, 22). In many species, the genomic organization of the *CLCA* locus is highly conserved, with the *CLCA* genes contained in a single block that preserves the same ordering of CLCA family members (**Figure 1**). However, it is not yet certain that exon homologies are routinely preserved for noncoding promoter sequences that regulate gene expression.

A consequence of sequencing multiple species during the discovery of the *CLCA* gene family is an inconsistent nomenclature for the *CLCA* genes. A more unified nomenclature scheme was proposed to resolve some of these inconsistencies (12). However, since that time additional homologs have been discovered, and additional renaming has occurred. Two comprehensive reviews of CLCA cloning and sequencing have been recently published (14, 16). Here we update the nomenclature for selected species (human, mouse, bovine, rat, and horse) that are especially relevant to airway disease (**Table 1**). This nomenclature is based on structural homology for coding sequences

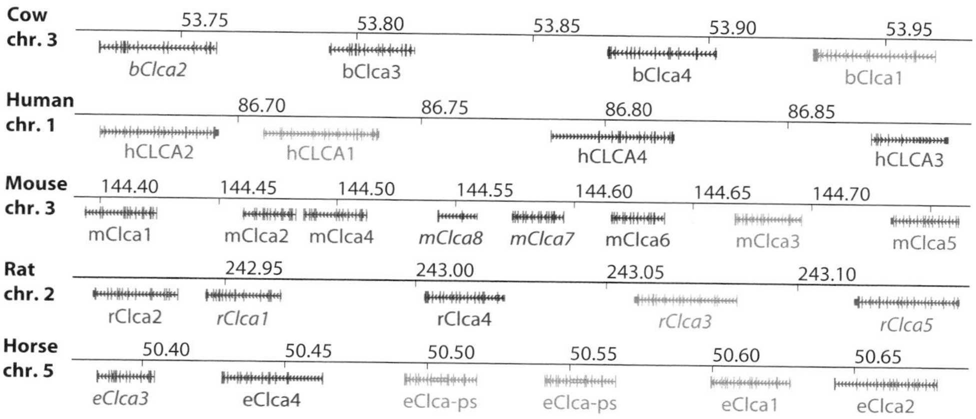

Figure 1

Map of syntenic CLCA loci in cow, human, mouse, rat, and horse genomes. Each color represents a homologous group of genes based on the alignment of coding sequences generated through the use of Clustal version 2.0 software (117). Position and gene structure diagrams are derived from UCSC Genome Browser (118). BLAT software was used to align transcripts to genome sequence for genes that were not annotated in the UCSC Genome Browser database (119). Italics denote genes with proposed changes in nomenclature.

Table 1 Unified nomenclature for CLCA family members. Gene assignments were based on analysis of human, mouse, cow, rat, and horse coding sequences with respect to the closest mouse homolog

CLCA name	Original name(s)	Reference(s)	GenBank accession (22)
hCLCA1	hCaCC-1	28, 129	NM_0001285
hCLCA2	hCaCC-3, CaCC	38, 129	NM_006536
hCLCA3		38	NM_004921
hCLCA4	hCaCC-2	129	NM_012128
mClca1		34, 130	NM_009899
mClca2		131	NM_030601
mClca3	gob-5	132	NM_017474
mClca4		35	NM_139148
mClca5		88	NM_178697
mClca6		88	NM_207208
mClca7	AI747448	23	NM_0001033199
mClca8	EG622139, A730041H10Rik	23	NM_001039222
bClca1	LOC507643	22, 133	XM_865947
bClca2	LOC784768	20	XM_001252288, U36445
bClca3	CaCC, bClca2	134	AF001261-AF001264, NM_181018
bClca4		22, 133	NM_001034300
rClca1	Rbclca2	135	AB256513, XM_001063517
rClca2	Rbclca1	26, 136	NM_001077356
rClca3	rClca3_Pred, rClca1[a]	22, 127	NM_001107449
rClca4	rClca6, Prp3	22, 127	NM_201419
rClca5	rClca2_Pred, rClca2[a]	22, 127	NM_001107450
eClca1		25	NM_001081799
eClca2		22, 137	XM_001496218
eClca3	LOC100052500	22, 118, 119, 137	XM_001495998
eClca4		22, 137	XM_001494684

[a]The rat genome is undergoing reannotation (127), but at present, the RatMap RGST naming resource has proposed different names assigned on the basis of homology to human sequences (128).

(**Figure 2**). We preserve the CLCA designation for this gene family, although this approach may have to be revised in the future if CLCA proteins prove to function as signaling molecules instead of ion channels.

The human CLCA (*hCLCA*) locus consists of four genes located on the short arm of chromosome 1 (1p31-p22), a region containing no other known genes (14). These *hCLCA* genes were identified by homology screens for *bClca1* and *bClca2* sequences (14). Similarly, there are four *Clca* genes in the cow and horse genomes and five *Clca* genes in rats. In contrast to these species, the mouse Clca (*mClca*) locus consists

of eight genes located on chromosome 3 (23). The basis for the extra *Clca* genes in the mouse is uncertain, but the extra genes appear to contribute to at least some degree of functional redundancy (24). One member each from the human and mouse *CLCA* gene families (*hCLCA3* and *mClca8*, respectively) may represent pseudogenes because these two genes contain premature stop codons and may not be expressed at detectable levels. However, there are conflicting views on the level of expression for these two genes, so it is still possible that they exert some type of biological function under some conditions.

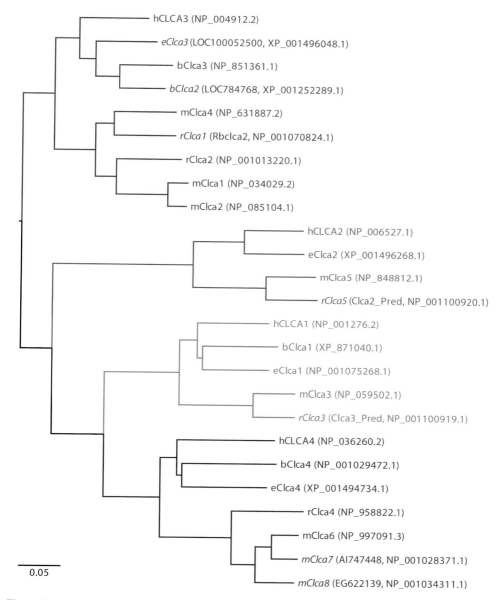

Figure 2

Sequence homology tree for CLCA family members in cow, human, mouse, rat, and horse. The homology tree was generated by alignment of protein sequences with Clustal (117). Initial visualization was performed by use of FigTree 1.1.2 software (120). The bar represents 5% sequence diversity. As described in **Figure 1**, each color represents a homologous group of genes based on the alignment of coding sequences, and italics denote genes with proposed changes in nomenclature.

The pattern of expression of the *CLCA* genes in different tissues was determined soon after gene identification and was perhaps the first insight into the functional bi-ology of CLCA proteins. Each *hCLCA* gene has a distinct pattern of expression based on mRNA detection in healthy human tissue (**Table 2**). Similar to the case for human *CLCA*

Table 2 Tissue distribution of *hCLCA* gene expression. Levels of expression were determined by detection of mRNA species for the corresponding *hCLCA* gene

Gene	Tissue	Reference(s)
hCLCA1	Intestinal goblet cells and basal crypt epithelial cells	34
	Airway mucous cells	58
	Uterus, stomach, testis, kidney, fetal spleen	129
	Brain	138
hCLCA2	Trachea, breast	47
	Trachea, uterus, prostate, testis, kidney	129
	Conjunctival and corneal epithelium	139
	Nasal epithelium	140
	Cornea, skin, vagina, esophagus, larynx	141, 142
hCLCA3	Lung, trachea, breast, spleen, thymus	38
hCLCA4	Brain, colon, bladder, uterus, prostate, stomach, testis, salivary gland, breast, small intestine, appendix, trachea	129
	Nasal epithelium	140

genes, the mouse Clca family members are expressed differentially across tissues (**Table 3**). Of the four horse Clca (*eClca*) genes, only *eClca1* has been characterized to any extent, and it is expressed in the nasal epithelium, trachea, subtracheal glands, mucous (goblet) cells of the airway epithelium, sweat glands, kidney, small intestine, and colon (25). The

Table 3 Tissue distribution of mClca gene expression. Levels of expression were determined by detection of mRNA species for the corresponding *mClca* gene

Gene	Tissue	Reference(s)
mClca1	Lung, aorta, spleen, bone marrow, lymph nodes, brain	143
	Kidney, skin, liver, spleen	130
	Intestine, cecum, brain, dorsal root ganglion	23
	Breast	48
mClca2	Breast	48, 131
	Thymus, colon, small intestine, bladder, epididymis, vesicular gland, skin, breast	143
	Intestine, cecum, dorsal root ganglion	23
mClca3	Stomach, small intestine, colon, uterus, trachea; limited to mucous cells	132
	Lung, stomach, small intestine, colon, uterus, trachea; limited to mucous cells	59
	Airway mucous cells	52
mClca4	Colon, small intestine, bladder, stomach, esophagus, uterus, skeletal muscle, heart, aorta, and lung; endothelial, smooth muscle, and epithelial cells	35
mClca5	Eye, spleen, heart, intestine, lung, skeletal muscle, stomach, testis	88
	Dorsal root ganglion	23
mClca6	Intestine, stomach, eye, spleen	88
	Small and large intestine; nongoblet cell enterocytes	33
mClca7	Intestine and cecum	23
mClca8	No transcripts detected	23

pattern of CLCA expression, particularly at the protein level, still needs to be fully determined. Similarly, there is limited information on the types of cells that express CLCA proteins. However, as further developed in the next section on inflammatory airway disease, the initial analysis indicates that specific CLCA family members (e.g., hCLCA1 in humans and mClca3 in mice) exhibit increased expression in airway mucous cells and are therefore associated with conditions of mucus production.

Protein Structure and Function

The proposed structural organization of CLCA proteins has undergone considerable revision in recent years. The initial structural models were based on the observation that the bClca1 protein appeared to regulate calcium-dependent chloride conductance in tracheal epithelial cells (20). The proposal that CLCA proteins may therefore function as ion channels was further supported by experiments in which transfection of 293T cells with several different CLCA isoforms from various species (human, mouse, rat, pig, and cow) all produced a chloride current in response to calcium ionophores or calcium release from the endoplasmic reticulum (16, 26), a process involving calcium/calmodulin-dependent protein kinase activity (27). Subsequent theoretical and experimental studies using hydropathy analysis and epitope insertion in concert with immunohistochemistry produced several different models of transmembrane topology with as many as five transmembrane-spanning segments (16). However, these models predicted a transmembrane pass within the common von Willebrand factor type A (VWA) domain, a structurally conserved soluble domain. In addition, some of the epitope insertion experiments likely resulted in misfolding of the protein because the hallmark proteolytic cleavage was blocked (28). Thus, even at this early stage of analysis, there were serious inconsistencies in the proposed model for CLCA proteins as transmembrane ion channels.

Recent experimental results and improved sequence analysis tools (including hidden Markov models for transmembrane segments and protein fold prediction) have led to a new model for CLCA structure. A revised scheme for the domain structure for mouse and human CLCA proteins is provided in **Figure 3**, and an annotated analysis of amino acid sequence for these proteins is provided in **Figure 4**. This updated analysis implies that the CLCA proteins are soluble secreted molecules (29–31) with the exception of a subset of CLCA proteins that contain a C-terminal membrane-anchoring region (32, 33). This anchoring region takes the form of a transmembrane α-helix or GPI (glycosylphosphatidylinositol) anchor that lacks the structural requirements to function as an ion channel (15, 29). A feature common to the CLCA family of proteins is the presence of a proteolytic cleavage site that is located approximately 240 amino acids from the C terminus. To date, all the CLCA isoforms that have been tested in mammalian cell culture appear to be processed similarly (29–37). In each case, a precursor 120-kDa glycoprotein is cleaved to produce ~85-kDa N-terminal and ~35-kDa C-terminal products, both of which contain numerous N-linked glycosylation sites (30, 33, 38). For hCLCA1, mClca3, and mClca4 (which have no identifiable transmembrane region), both the N- and C-terminal products are secreted into the media when expressed in human cell lines (29–31). In similar experiments with hCLCA2 (which has a transmembrane region), the N-terminal product is released into the media, whereas the C-terminal product remains cell surface associated (32). A recent report indicates that mClca4 contains luminal motifs that control its trafficking out of the endoplasmic reticulum (31). Mutation of the forward-trafficking motifs led to trapping of the full-length protein in the endoplasmic reticulum, implying that cleavage occurs after exit from this site. This sequence of events is consistent with the observation that full-length hCLCA1 (as well as the N- and C-terminal cleavage products) is found secreted into the media of hCLCA1-expressing HEK293 cells

VWA domain: von Willebrand factor type A domain

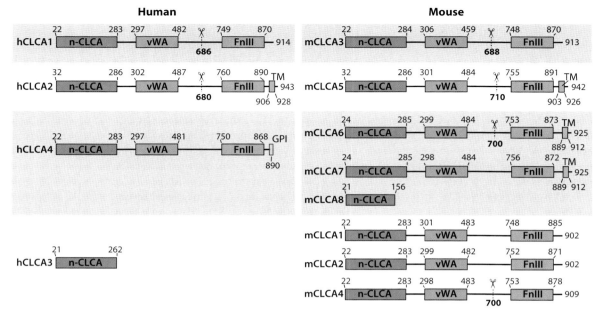

Figure 3

Structural scheme for human and mouse CLCA proteins. Each row contains the corresponding human and mouse homologs. Scissors denote the estimated location of proteolytic cleavage sites (shown in **bold**). The structures of hCLCA3 and mClca8 are abbreviated because the corresponding genes contain premature stop codons. Labels denote the following: n-CLCA, N-terminal CLCA domain; VWA, von Willebrand factor type A domain; FnIII, fibronectin type III domain; TM, transmembrane α-helix; GPI, glycosylphosphatidylinositol anchor.

(29), implying that cleavage may occur outside the cell.

One proposal for the nature of CLCA cleavage rests on an analysis of protein sequence. The N-terminal 280 amino acids comprise a cysteine-rich domain conserved only in the CLCA family. The function of this domain is unknown. However, a recent bioinformatics analysis suggested that this region contains a zinc metalloprotease domain (39). This finding raised the possibility that the CLCA proteins may self-cleave as part of normal processing. Furthermore, expression of an hCLCA1 mutant in which a predicted critical active-site glutamate was changed to glutamine totally abolished cleavage. However, in these experiments only the cell lysates (and not supernatants) were analyzed, so the mutant protein may have been misfolded. Therefore, the precise nature of the process for cleavage of CLCA proteins still needs to be defined. The function of the CLCA

protein cleavage remains uncertain, particularly because a specific cleavage site has not yet been precisely identified and characterized. Defining the site and mechanism for CLCA cleavage is critical to defining the function of the mature CLCA peptides.

Directly adjacent to the N-terminal CLCA (n-CLCA) domain is a VWA domain, a widely distributed protein-protein interaction domain (40). The VWA domains commonly engage proteins by coordinating a divalent cation at the binding interface, using a highly conserved set of five residues known as a metal ion–dependent adhesion site (MIDAS). All the CLCA VWA domains contain an intact MIDAS, with the exception of hCLCA2 and the homologous mClca5 (**Figure 4**). Ironically, the VWA domain of hCLCA2 was the first CLCA domain for which a binding partner was identified, and it was found to be a β4-integrin (41). This binding does not depend on MIDAS

MIDAS: metal ion–dependent adhesion site

but instead involves the consensus sequence F(S/N)R(I/L/V)(S/T)S found in some CLCA proteins and the specificity-determining loop (SDL) of β4-integrin (41–43). The hCLCA2, mCLCA1, and mCLCA2 proteins contain two of these β4-binding motifs (**Figure 4**). This interaction promotes the adhesion of hCLCA2-expressing cell lines to tumorigenic cell lines with upregulated β4-integrin expression and may be of importance in certain cancer metastases and colonizations. Furthermore, the binding of hCLCA2 to β4-integrin may initiate a novel signaling cascade that promotes tumor invasion.

In addition to cleavage sites and VWA interaction domains, the remaining region C-terminal to the VWA domain of CLCA proteins is rich in β-sheet structure as determined by a secondary structure prediction from JPRED 3 software (44). This region includes an easily detectable fibronectin type III (FnIII) domain. FnIII domains are protein-protein interaction modules of the immunoglobulin (Ig) superfamily. No function or binding partners have been identified for this portion of any CLCA family member. Taken together, this updated sequence analysis and recent experimental data indicate that the CLCA proteins cannot act in solo as ion channels and instead imply the revised view that they interact with other proteins in a possible signaling capacity.

MODELS OF INFLAMMATORY AIRWAY DISEASE

In this section, we review cell and animal models that were studied to define the role of CLCA proteins in the development of inflammatory airway diseases. However, before work on the function of CLCA in the lung even began, several reports suggested that CLCA proteins may have a role in regulating the development of cancer. For example, studies of melanoma in mice indicated that Clca proteins may promote metastasis to the lung (19, 45, 46). Related work indicated that interaction between hCLCA2 on endothelial cells and its β4-integrin ligand on breast cancer cells may mediate more aggressive tumor behavior (41, 43, 47). Similarly, CLCA family members with functional β4-integrin-binding domains may use this mechanism to activate focal adhesion kinase signaling to extracellular signal–regulated kinase (ERK) with the potential to regulate cell survival (42, 43). In contrast, other studies showed that human breast cancer cells lacked the usual expression of hCLCA2 and that overexpression of hCLCA2 decreased the capacity of these cells for invasion and tumor generation in nude mice (48). Similarly, mouse breast cancer cells expressed decreased amounts of mClca1, mClca2, and mClca5, and reexpression or overexpression of these proteins decreased cell growth and increased sensitivity to apoptosis (36, 49).

FnIII domain: fibronectin type III domain

ERK: extracellular signal–regulated kinase

Figure 4

(*a,b*) Annotated sequence alignment for the human and mouse CLCA proteins. Numbering is based on the sequence for hCLCA1. Sequences for full-length human and mouse proteins are shown for hCLCA1 (GI 110611231), mClca3 (GI 8567336), hCLCA2 (GI 5729769), mClca5 (GI 30520101), hCLCA4 (GI 6912310), mClca6 (GI 110556625), mClca7 (GI 85701714), mClca1 (GI 32964827), mClca2 (GI 13447394), and mClca4 (GI 20982843). N-terminal signal peptides were removed after identification by SignalP software (121). Protein folding was predicted by the use of PHYRE software (122). Sequence conservation is color coded as follows: magenta, invariant residues (with the exception of cysteine); blue, invariant cysteines; yellow, residues with a conservation score of 5 or greater as determined by ALSCRIPT software (123). Green boxes indicate the most strongly predicted sites of *N*-linked glycosylation by NetNGlyc software (124). Blue boxes denote β4-integrin-binding motifs with the consensus sequence F(S/N)R(I/L/V)(S/T)S. The hCLCA1 motif (amino acids 728–733) does not contain this strict consensus but still binds weakly to β4-integrin (43). Domains are defined and shaded in the same colors as depicted in **Figure 3**. The location of MIDAS (metal ion–dependent adhesion site) residues of the von Willebrand factor type A (VWA) domain and the site of the transmembrane (TM) α-helices were predicted by TMHMM software (125). The sequence of hCLCA4 is truncated after the position of the GPI (glycosylphosphatidylinositol) anchor as predicted by the Big-PI Predictor software (126).

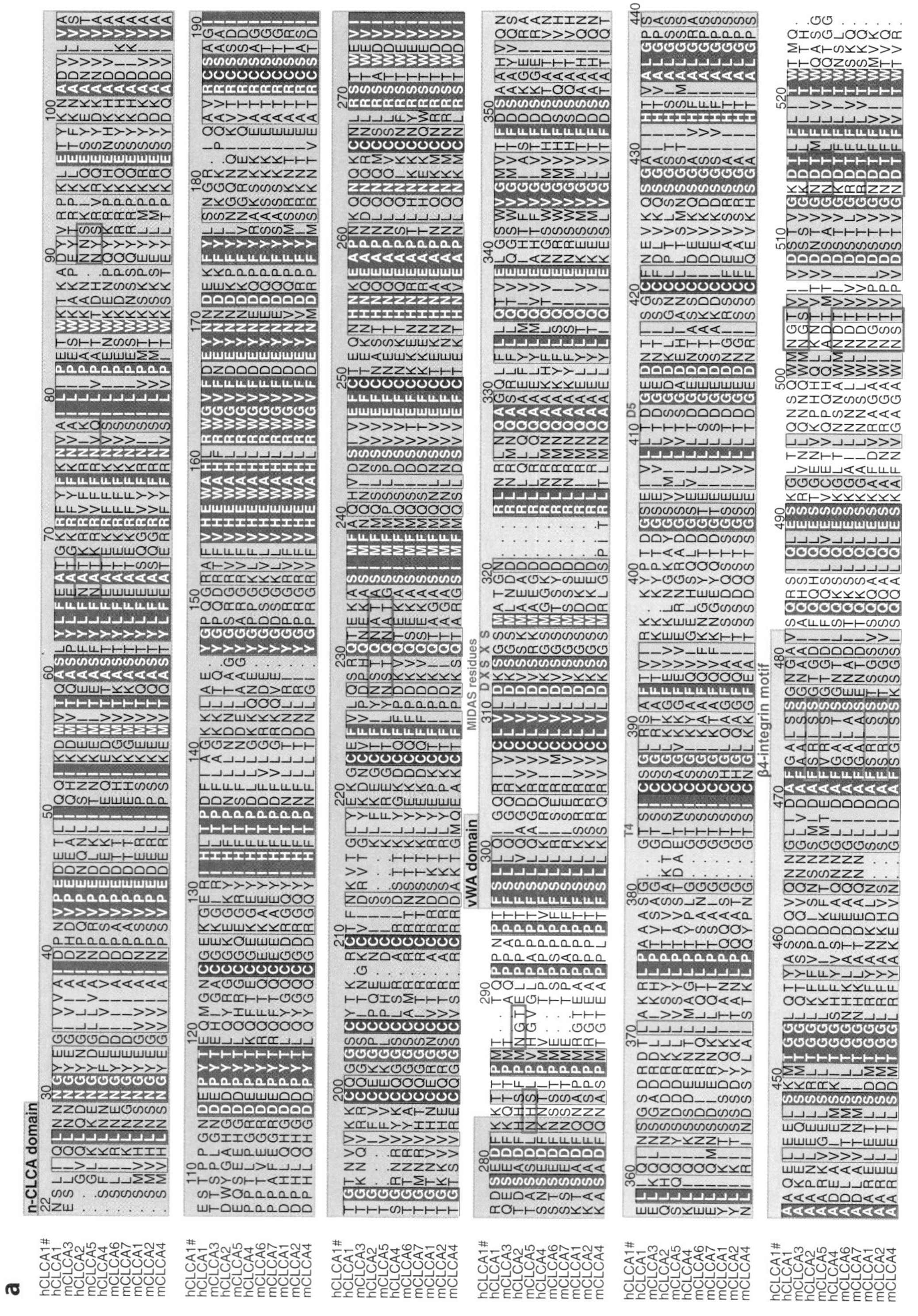

Figure 4
(Continued)

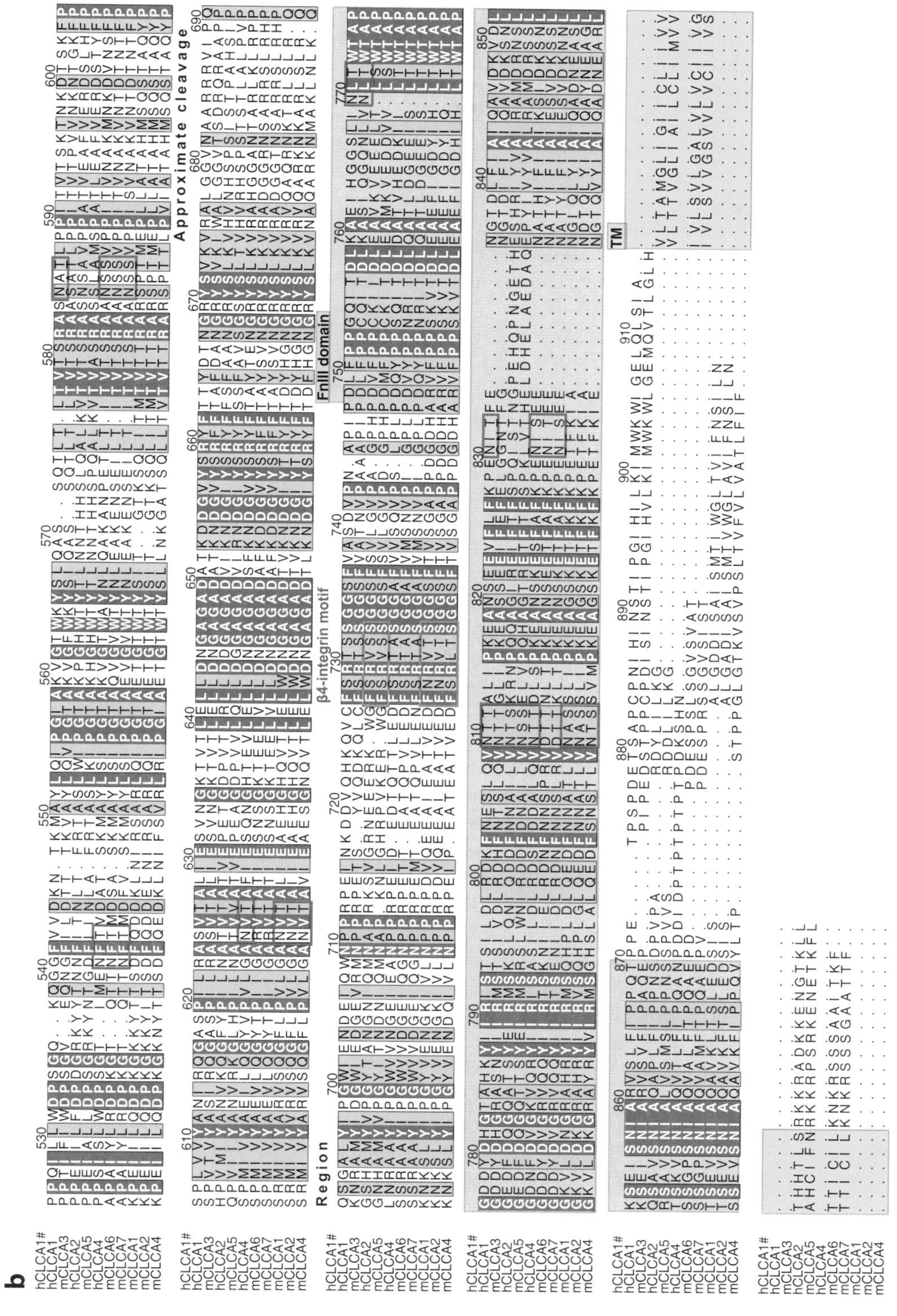

b

Figure 4

(Continued)

Decreased hCLCA2 expression is also observed in breast cancer specimens and appears to be due to DNA methylation (50). Recent work also shows that the *hCLCA2* gene may be deleted in mantle cell lymphoma specimens (51). These studies highlight the possible role of hCLCA2 as a tumor suppressor and set the stage for studies of hCLCA1 in inflammatory airway disease. A connection between the activities of CLCA proteins as tumor promoters or suppressors and their role as immune modifiers has not been examined.

Isolated Cell Models

Studies of isolated cells were perhaps the first to develop a role for CLCA proteins in immunity. In particular, overexpression of mClca3 or hCLCA1 in a human mucoepidermoid cell line NCI-H292 caused a significant increase in mucin production (52, 53). These findings imply that CLCA proteins may regulate the level of mucus production and the consequent efficiency of mucociliary clearance of particulate and microbial pathogens. Using this cell model, investigators explored the possibility that CLCA proteins worked as ion channels to somehow influence mucus production. For example, investigators used niflumic acid (a drug that was purported to be a specific inhibitor of chloride channel activity) to block the increase in mucin production found in NCI-H292 cells that expressed hCLCA1 (53). In related work, another group showed that niflumic acid suppressed ATP-mediated exocytosis of mucin in HT29-Cl.16E cells (54). Similarly, McNamara and colleagues (55) found that adenosine induction of mucin biosynthesis depended on transactivation of EGF receptor and that niflumic acid inhibited this type of transactivation. Other work with the niflumic acid derivative MSI-2216 also demonstrated inhibition of mucus production (56, 57). These studies inferred a role for CLCA proteins in the signaling pathway leading to mucin synthesis and secretion but provided no direct evidence for how CLCA proteins might be involved in this process. Because structural analysis now indicates that CLCA proteins cannot be membrane ion channels, there is a need for more detailed study for how CLCA proteins might control the development of mucous cells.

Although some progress has been made in establishing a role of hCLCA1 in mucous cell metaplasia in isolated cell models, there is little information on CLCA function in airway smooth muscle cultures. CLCA proteins may influence airway smooth muscle behavior because antisense oligonucleotides that knock down mClca expression appeared to inhibit airway hyperreactivity in ovalbumin-challenged mice (52). The available reports indicate that hCLCA1 (and the homologous mClca3) is localized to mucin granules of mucous cells in the lung, gastrointestinal tract, and uterus and is not expressed in airway smooth muscle (28, 34, 58, 59). However, mClca4 is expressed at significant levels in airway smooth muscle in mice (35). In addition, calcium-activated chloride flux is found in smooth muscle of the vasculature, urethra, and lymphatics (60–62). Because at least some CLCA/Clca proteins may depolarize cellular membranes (15), CLCA expression in airway smooth muscle may lead to bronchoconstriction. Similar to the case for mucous cells, CLCA function in smooth muscle cells still needs to be defined.

Animal Models

Perhaps the best evidence for a biological role of Clca proteins has been developed in animal models of inflammatory airway disease. In this case, a series of studies has established a link between the expression of Clca proteins in the lung and the development of inflammatory airway disease. One of the initial reports came from a study of a mouse model of allergic asthma that depends on sensitization and subsequent inhalation challenge with ovalbumin. Using this model, researchers reported that mClca3 appeared to regulate mucous cell metaplasia and airway hyperreactivity (52). Under these conditions, the administration of a full-length *mClca3* antisense oligonucleotide delivered with an adenoviral vector system inhibited the development of mucous cell metaplasia and airway hyperreactivity. These investigators did not recognize that this full-length construct might also interact with shared sequence homologies in other mClca family members. Nonetheless, expression of *mClca3* in the airway epithelium was sufficient to cause mucous cell metaplasia and airway hyperreactivity in these animals. This finding was consistent with the evidence that mClca3 expression was sufficient to drive mucin production in NCI-H292 cells (52, 53).

In addition to these studies of allergic asthma in mice, another research group studied the role of Clca proteins in the development of an allergic asthma-like illness in horses. This animal species develops a disease of episodic recurrent airway obstruction (also known as heaves) that also features mucous cell metaplasia and airway hyperreactivity (63–65). The pathogenesis of this equine disease also involves T helper type 2 (Th2) cytokine production (especially IL-4 and IL-13) (66). Furthermore, as in the airway disease that develops in mice and humans, Muc5ac is the predominant airway mucin that is expressed during the development of mucous cell metaplasia (67). Thus, the disease in horses manifests the characteristic features of human asthma and COPD as well as the allergen challenge and viral infection mouse models of chronic airway disease (68). Therefore, it appeared likely that equine Clca proteins would also be involved in the pathobiology of mucous cell metaplasia of the asthma-like disease that develops in horses. Indeed, investigators report a significant increase in eClca1 expression in the airway epithelium of horses with equine recurrent airway obstruction (25, 69).

In a distinct approach to understanding the basis for inflammatory airway disease, our lab also came to evaluate the role of the *Clca* gene family (24). Our group aimed to genetically segregate mucous cell metaplasia from airway hyperreactivity in a mouse model and thereby identify a molecular pathway that might be associated with a single disease trait. Others have also used inbred mouse strains to define a genetic influence on airway disease, but this work has generally concentrated on the development of airway hyperreactivity (70–78). In the usual case, these previous studies were directed at defining a genetic locus for a single quantitative trait rather than segregating one trait from another in a complex phenotype. Moreover, the usual allergen challenge model of airway disease did not often account for the chronic nature of the phenotype. In view of the well-described role of respiratory syncytial virus (RSV) as well as other viruses in chronic airway disease (79–84), we developed a mouse model of viral bronchiolitis. The infectious agent used in this model is Sendai virus (SeV), which is a mouse parainfluenza virus similar to other paramyxoviruses that more commonly infect humans. However, mice are relatively resistant to infection with human pathogens such as RSV. By contrast, SeV replicates at high efficiency in the mouse lung, and SeV infection causes injury and inflammation of the small airways (i.e., bronchiolitis) that is indistinguishable from the comparable condition in humans. This acute response is followed by a delayed but permanent switch to chronic airway disease that is characterized by mucus production (mucous cell metaplasia) and increased airway reactivity to inhaled methacholine (airway hyperreactivity) (24, 85, 86). We came to recognize that both of these traits are inducible on a long-term basis after viral bronchiolitis in the C57BL/6J strain of inbred mice (86). In contrast, we found that the Balb/cJ strain responded similarly during the acute infection but then failed to develop any chronic airway disease (24).

We took advantage of this difference in genetic susceptibility to develop an F2 intercross population with phenotypic extremes that exhibit one or the other disease trait, and we analyzed these extremes for gene expression, using oligonucleotide microarrays (24). This combined genetic and genomic strategy provided evidence of a selective association between expression of a member of the *mClca* gene

Th2: T helper type 2

SeV: Sendai virus

Stat: signal transducer and activator of transcription

family, i.e., *mClca3*, with the development of mucous cell metaplasia but not airway hyperreactivity. In support of the relationship between mClca3 expression and mucous cell metaplasia, we also found that transfer of the *mClca3* gene to the mouse airway (using an adeno-associated viral vector) was sufficient to produce mucous cell metaplasia but did not cause airway hyperreactivity (24). We next generated a mouse that was homozygous for a targeted null mutation of the *mClca3* gene (*mClca3*$^{-/-}$). However, *mClca3*$^{-/-}$ mice developed the same degree of mucous cell metaplasia (and airway hyperreactivity) as wild-type mice, in both viral and allergen-challenged models of chronic airway disease (24).

Because *mClca3* gene expression was sufficient but not necessary for the development of mucous cell metaplasia, we reasoned that another mechanism must also be capable of mediating the chronic change in airway behavior. We specifically questioned whether another *mClca* gene might be responsible for such an effect. Accordingly, we used the BLAST (87) program to search for homologs of the *mClca3* gene and consequently uncovered four additional *mClca* genes flanking *mClca3* on mouse chromosome 3. We tentatively named these genes *mClca5*, *mClca6*, *mClca7*, and *mClca8*; subsequently, these genes were given the same designations (23, 88), with homology to human CLCA family members as shown in **Figure 2**. Because other mClca family members are also expressed in airway tissue (88, 89), they may compensate for the loss of mClca3 and thereby allow for the development of mucous cell metaplasia in the setting of mClca3 deficiency (24). Indeed, we recently showed that mClca family members may exhibit functional redundancy in the development of inflammatory airway disease. In particular, mClca5 expression was also increased in concert with mucous cell metaplasia and airway hyperreactivity after viral infection (24). In addition, *mClca5* gene transfer to the airway epithelium was sufficient to cause mucous cell metaplasia but not airway hyperreactivity (24). Preliminary studies indicate that mClca6 expression in the air-

way epithelium is also sufficient for mucous cell metaplasia but not airway hyperreactivity (90).

Another group also recognized that mClca3 deficiency alone does not influence the development of mucous cell metaplasia after allergen challenge, but their analysis of genetic compensation was limited to expression levels of mClca1, mClca2, and mClca4 (91). In fact, our work indicates that mClca5, mClca6, and mClca7 are more suitable candidates for compensatory function. Given the extensive sequence homology for the mClca family members, it is likely that *mClca3* antisense oligonucleotides may bind and inhibit several family members. Under these conditions, *mClca* gene knockdown may inhibit the induction of mucous cell metaplasia and airway hyperreactivity after viral infection, but this possibility still needs to be formally tested.

The precise signals that regulate *mClca* gene expression in mouse models of inflammatory airway disease are also under study. Most studies show that Th2 cytokines (IL-4, IL-9, and IL-13) will drive increased expression of CLCA proteins in cell culture (53, 92–94). However, at least one group does not find this relationship (94, 95), suggesting that the state of the cultured cells may influence the responsiveness to cytokines. For example, the degree of cell differentiation may influence the expression of cytokine receptor or receptor signaling function. Indeed, IL-13 receptor expression and signaling function may be critical determinants of mucous cell metaplasia after viral infection (96). In that regard, Stat6 (signal transducer and activator of transcription 6), which transmits the signal from IL-4 and IL-13 receptors, is also required to upregulate *CLCA* gene expression (93). Similarly, increased Clca expression develops in concert with increased IL-13 expression in virus- and allergen-induced mouse models of airway diseases (24, 52, 91, 96, 97). Moreover, increased Clca expression develops after IL-13 instillation into the lung and in IL-13 transgenic mice (93, 94, 98, 99). Thus, Th2 cytokine signals regulate *CLCA* gene expression under a variety of conditions. This stimulatory effect of IL-13 may depend on Stat6-binding sites in the *CLCA* gene promoter, but the precise biochemical mechanism still needs to be defined.

INFLAMMATORY AIRWAY DISEASE IN HUMANS

In this section, we review the evidence of a role for CLCA proteins in inflammatory airway disease in humans. We concentrate on the two most common types of chronic airway disease, asthma and COPD, but we include reports of CLCA action in other diseases that also manifest mucus hypersecretion and airway hyperreactivity.

CLCA Polymorphism as Genetic Susceptibility Factor

Several lines of evidence have linked *CLCA* gene variation to the development of inflammatory airway disease in humans. In particular, two separate studies have used quantitative trait locus (QTL) linkage analyses of large cohorts of COPD patients to track the influence of *CLCA* genes on airway disease. Each of these studies found that lung function (marked by the degree of abnormality in FEV_1:FVC ratio, a measure of airflow obstruction) is linked to genetic change in a region of human chromosome 1 containing the *CLCA* family locus (100, 101). In addition, specific variations in the *hCLCA1* gene have been linked to susceptibility to airway disease. Thus, a haplotype analysis of single-nucleotide polymorphisms (SNPs) in the *hCLCA1* gene observed linkage to childhood and adult asthma as well as COPD in a Japanese population (102, 103). Analysis of patients with mild forms of cystic fibrosis (in which a residual gastrointestinal chloride current is maintained but increased gastrointestinal and airway mucus accumulations still occur) also detected linkage with genetic variations in the *CLCA* gene locus (104). Other investigators have shown that intestinal mucous secretion is improved when mClca3 expression is restored in mouse models of cystic fibrosis (37, 105), suggesting that a physiological level of Clca function is necessary for normal intestinal as well as airway mucous cell function.

CLCA Expression as a Biomarker

In studies of human tissue, several studies have shown that the levels of hCLCA1 mRNA and protein are significantly increased in the airway epithelium of asthmatic patients (58, 106–108). Preliminary studies suggest that increased levels of hCLCA1 protein can also be detected in bronchoalveolar lavage (BAL) fluid as cleaved N-terminal and C-terminal polypeptides in these types of patients (109). Upregulation of hCLCA1 has also been observed in airways of cystic fibrosis patients with high levels of mucus production (110, 111). Furthermore, in patients with COPD, lung samples may contain increased levels of hCLCA1 as well as hCLCA2 and hCLCA4 (108, 112, 113). Thus, hCLCA1 expression and perhaps hCLCA2 and hCLCA4 expression appear to be useful biomarkers of inflammatory airway disease. Because of the difficulties in accurately monitoring mucus production or mucous cell metaplasia in vivo, measurements of CLCA mRNA or protein may serve as a useful biomarker for quantifying this disease trait in the setting of airway inflammation. Indeed, one study of asthma patients already suggests that *hCLCA1* mRNA levels may decrease after treatment with inhaled glucocorticoids (107), and a preliminary report suggests that hCLCA1 levels may track with mucous cell metaplasia following withdrawal of glucocorticoid treatment (109). However, additional studies are needed to validate the use of CLCA mRNA or protein levels as a diagnostic or prognostic marker of activity of inflammatory airway disease.

CLCA Proteins as Therapeutic Targets

In addition to use as a biomarker, CLCA proteins (and particularly hCLCA1) may also be useful as specific therapeutic targets in the treatment of inflammatory airway disease, or even in general for controlling excessive mucous secretions. As noted above, one group has supported this approach by showing that knockdown of *mClca* gene expression using antisense directed against *mClca* transcripts blocked the

FEV_1: forced expiratory volume in 1 s

FVC: forced vital capacity

development of mucous cell metaplasia and airway hyperreactivity after ovalbumin allergen challenge in mice (52). However, this report has not yet been confirmed in this mouse model or other models, and there are no ongoing efforts to apply this strategy to humans with airway disease.

In addition to genetic knockout and knockdown strategies, other groups have pursued the potential for small molecules to inhibit CLCA function. In particular, niflumic acid and its derivatives were proposed initially as CLCA inhibitors. In support of this possibility, one group has shown that niflumic acid can inhibit mucous cell metaplasia induced by hCLCA1 expression or IL-13 administration (53). These findings led to the pursuit of talniflumate (a niflumic acid derivative) as a potential therapy for the overproduction of mucus in airway disease. Preclinical data in a mouse model of cystic fibrosis indicated that talniflumate improved survival by decreasing gastrointestinal obstruction caused by mucus (114). In addition, a phase II study of the drug found no toxicity in patients with cystic fibrosis (115). Talniflumate was also being considered for treatment of asthma and COPD, but drug development was eventually halted (115, 116). Another niflumic acid derivative, MSI-2216, also inhibited mucous secretion in both airway cell lines as well as primary cultures of upper airway epithelial cells, but further development of this drug was also stopped (56, 57).

Some of the likely problems for drug development of CLCA inhibitors have been the issues of drug specificity and mechanism of action. In the case of niflumic acid, the drug may block multiple members of the CLCA family rather than a single member (13, 88). This effect was attributed to inhibition of a common ion channel function among family members. However, this mechanism became less likely when it was recognized that CLCA proteins were unlikely to function as calcium-dependent chloride channels. Moreover, subsequent studies suggested that niflumic acid and its derivatives may act by directly or indirectly blocking activation of the Stat6 transcription factor (93). The niflumic acid class may thereby block the actions of IL-13 on mucus production because at least part of the IL-13 receptor signaling pathway travels through Stat6 to drive downstream expression of mucin genes. In any case, it is reasonable to conclude that more precise definition of the molecular mechanism for CLCA action will be needed to rationally develop specific inhibitors.

CONCLUSION

In sum, there is a critical clinical need for better understanding and treatment of inflammatory airway disease in general and mucous cell metaplasia in particular. Although work is at an early stage, several lines of evidence already suggest that CLCA proteins (particularly hCLCA1) may be master regulators of mucin gene expression in humans. We have gradually developed the proposal that the airway maintains a special program for host defense and that changes in this same program may lead to airway disease. The same argument can be made for mucus secretion. Thus, under normal circumstances a physiological level of mucus production allows for mucociliary clearance of particulate matter from the airway and aids in defense against respiratory pathogens. However, if mucus secretion is excessive, the condition leads to airway obstruction and consequent dysfunction. In both normal and abnormal conditions, there appears to be an immune axis that drives cytokine production and, in turn, a downstream mechanism that translates the cytokine signal into mucin gene expression. In humans, this downstream mechanism features hCLCA1, the first member of a newly defined family of CLCA proteins in humans. Additional CLCA family members may mediate airway hyperreactivity, but this issue requires further study. In either case, the mechanism for how hCLCA1 might signal to cause mucin gene expression and mucous cell metaplasia is a goal of ongoing studies. Initial analysis of CLCA structure and function suggests that CLCA proteins are secreted and cleaved into fragment polypeptides. We are particularly focused on defining the possibility that hCLCA1 is a secreted protein that is cleaved into an active polypeptide that in turn binds to a receptor on the cell surface and signals to activate mucin gene expression (**Figure 5**). This scheme for hCLCA1 (and orthologous mClca3, -5, and -6) control over epithelial behavior can be tested in isolated airway epithelial cells as well as in mouse models and then in patients with mucous cell metaplasia due to asthma, COPD, or other inflammatory airway diseases. A full understanding of this pathway will eventually provide multiple targets for therapeutic intervention, including modifications of hCLCA1 expression, processing, and signaling. A particular challenge for therapeutic strategies will be to block excessive mucus production without compromising host defense and thereby restore proper balance to the mucociliary system.

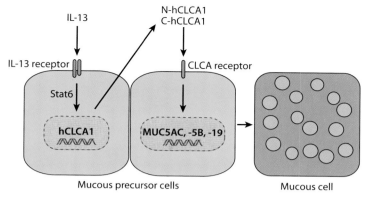

Figure 5

Proposed scheme for hCLCA1 regulation of mucous cell metaplasia. IL-13 generated by the innate and adaptive immune systems binds to IL-13 receptor, activates Stat6, and causes increased expression of hCLCA1. Following processing, the N-terminal hCLCA (N-hCLCA) and C-terminal hCLCA (C-hCLCA) peptides may be secreted and bind to a receptor, activate downstream signaling molecules, and cause mucin gene expression. Although depicted on neighboring cells, all these events may occur on the same cell as well.

SUMMARY POINTS

1. Genomic analysis and gene sequencing have defined a family of CLCA proteins in humans and animals named for the calcium-activated chloride flux associated with purified and recombinant CLCA proteins.

2. Bioinformatic analysis and experimental data indicate that CLCA proteins are not integral membrane channels but are instead secreted signaling proteins that are cleaved into polypeptide fragments.

3. Bioinformatic analysis also indicates that the CLCA proteins exhibit a stereotyped domain architecture that includes VWA and FnIII domains equipped for protein-protein interaction and possibly signaling function.

4. Gene expression analysis indicates that CLCA proteins, especially hCLCA1 in humans, are linked to the development of mucous cells in the airway as well as other tissues.

5. Functional studies combined with genetic and genomic screens indicate that mClca3 (as well as mClca5 and mClca6) is sufficient for the development of mucous cell metaplasia after allergen challenge and viral infection in mouse models.

6. Clinical studies indicate that CLCA levels (particularly hCLCA1 levels) are linked to the development of inflammatory airway disease, especially mucous cell metaplasia and possibly airway hyperreactivity.

7. Inflammatory airway diseases such as asthma, COPD, and cystic fibrosis are characterized by airway obstruction, at least in part because of mucin overproduction.

8. CLCA proteins may serve as useful biomarkers as well as therapeutic targets for the diagnosis and treatment of patients with inflammatory airway disease.

FUTURE ISSUES

1. What is the basis for redundancy of *CLCA/Clca* genes? Humans and other species express several *CLCA/Clca* genes with significant sequence homology. The variable number of *CLCA/Clca* genes among species suggests the occurrence of gene duplication, but the functional basis for this evolutionary approach still needs to be defined.

2. What is the pattern of tissue expression for the various *CLCA/Clca* genes under healthy and disease conditions, and how is this pattern determined? Although we have preliminary analysis of sequence homologies among CLCA family members, we still need to understand what determines cell and tissue specificity for expression and how those patterns translate into function.

3. How is *CLCA/Clca* gene expression regulated? In addition to tissue-specific patterns of expression, it appears that *CLCA* genes are upregulated in disease. For example, CLCA expression is likely induced by Th2 cytokines that are produced during inflammatory airway disease. However, the biochemical basis for how CLCA proteins translate the cytokine signal into mucin gene expression still needs to be defined.

4. What are the site, mechanism, and functional role of CLCA protein cleavage? CLCA family members appear to uniformly undergo cleavage, but the precise amino acid site for cleavage as well as the protease(s) involved still need to be determined. This information is critical to next explore the structure and function of the mature CLCA peptides.

5. What is the three-dimensional structure of CLCA proteins? The understanding of CLCA function as well as the rational design of CLCA inhibitors will be greatly aided by definition of CLCA structure. At present, we have no information on crystal structure for any portion of the CLCA proteins and little understanding of CLCA protein interactions.

6. Do CLCA proteins function as signaling molecules, and if so, what is the signaling pathway for CLCA/Clca proteins to regulate mucin gene expression? These questions are critical to understanding how CLCA proteins regulate cell function. The development of properly folded recombinant CLCA proteins and the concomitant assay of their functional effects is still needed.

7. What is the role of CLCA/Clca proteins in airway smooth muscle function? Specific CLCA proteins (notably hCLCA1 and mClca3) are under study for their capacity to regulate mucin gene expression and mucous cell metaplasia. However, study of other CLCA/Clca family members is needed to address their role in controlling smooth muscle cell function in airways and other tissues.

DISCLOSURE STATEMENT

The authors are not aware of any biases that might be perceived as affecting the objectivity of this review.

ACKNOWLEDGMENTS

The authors gratefully acknowledge their laboratory colleagues for valuable assistance and advice in the course of this work as well as support by grants from the National Institutes of Health

(National Heart, Lung, and Blood Institute and National Institute of Allergic and Infectious Diseases), Martin Schaeffer Fund, and Alan A. and Edith L. Wolff Charitable Trust.

LITERATURE CITED

1. Groneberg DA, Eynott PR, Lim S, Oates T, Wu R, et al. 2002. Expression of respiratory mucins in fatal status asthmaticus and mild asthma. *Histopathology* 40:367–73

2. Kuyper LM, Pare PD, Hogg JC, Lambert RK, Ionescu D, et al. 2003. Characterization of airway plugging in fatal asthma. *Am. J. Med.* 115:6–11

3. Hogg JC, Chu F, Utokaparch S, Woods R, Elliott WM, et al. 2004. The nature of small-airway obstruction in chronic obstructive pulmonary disease. *N. Engl. J. Med.* 350:2645–53

4. Vestbo J, Prescott E, Lange P. 1996. Association of chronic mucus hypersecretion with FEV1 decline and chronic obstructive pulmonary disease morbidity. Copenhagen City Heart Study Group. *Am. J. Respir. Crit. Care Med.* 153:1530–35

5. Ekberg-Aronsson M, Pehrsson K, Nilsson JA, Nilsson PM, Lofdahl CG. 2005. Mortality in GOLD stages of COPD and its dependence on symptoms of chronic bronchitis. *Respir. Res.* 6:98

6. Ekberg-Aronsson M, Lofdahl K, Nilsson JA, Lofdahl CG, Nilsson PM. 2008. Hospital admission rates among men and women with symptoms of chronic bronchitis and airflow limitation corresponding to the GOLD stages of chronic obstructive pulmonary disease—a population-based study. *Respir. Med.* 102:109–20

7. Lange P, Nyboe J, Appleyard M, Jensen G, Schnohr P. 1990. Relation of ventilatory impairment and of chronic mucus hypersecretion to mortality from obstructive lung disease and from all causes. *Thorax* 45:579–85

8. Prescott E, Lange P, Vestbo J. 1995. Chronic mucus hypersecretion in COPD and death from pulmonary infection. *Eur. Respir. J.* 8:1333–38

9. Pistelli R, Lange P, Miller DL. 2003. Determinants of prognosis of COPD in the elderly: mucus hypersecretion, infections, cardiovascular comorbidity. *Eur. Respir. J. Suppl.* 40:10s–14s

10. Jensen HH, Godtfredsen NS, Lange P, Vestbo J. 2006. Potential misclassification of causes of death from COPD. *Eur. Respir. J.* 28:781–85

11. Hogg JC, Chu FS, Tan WC, Sin DD, Patel SA, et al. 2007. Survival after lung volume reduction in chronic obstructive pulmonary disease: insights from small airway pathology. *Am. J. Respir. Crit. Care Med.* 176:454–59

12. Gruber AD, Fuller CM, Elble RC, Benos DJ, Pauli BU. 2000. The CLCA gene family: a novel family of putative chloride channels. *Curr. Genomics* 1:201

13. Fuller CM, Ji HL, Tousson A, Elble RC, Pauli BU, Benos DJ. 2001. Ca^{2+}-activated Cl^- channels: a newly emerging anion transport family. *Pflüg. Arch.* 443(Suppl. 1):S107–10

14. Gruber AD, Elble RC, Pauli BU, Fuller CM. 2002. Discovery and cloning of the CLCA gene family. In *Current Topics in Membranes*, ed. CM Fuller, pp. 367–87. London: Acad. Press

15. Hartzell C, Putzier I, Arreola J. 2004. Calcium-activated chloride channels. *Annu. Rev. Physiol.* 67:719–58

16. Loewen ME, Forsyth GW. 2005. Structure and function of CLCA proteins. *Physiol. Rev.* 85:1061–92

17. **Ran S, Benos DJ. 1991. Isolation and functional reconstitution of a 38-kDa chloride channel protein from bovine tracheal membranes. *J. Biol. Chem.* 266:4782–88**

18. **Ran S, Benos DJ. 1992. Immunopurification and structural analysis of a putative epithelial Cl^- channel protein isolated from bovine trachea. *J. Biol. Chem.* 267:3618–25**

19. Zhu D, Cheng CF, Pauli BU. 1991. Mediation of lung metastasis of murine melanomas by a lung-specific endothelial cell adhesion molecule. *Proc. Natl. Acad. Sci. USA* 88:9568–72

20. Cunningham SA, Awayda MS, Bubien JK, Ismailov II, Arrate MP, et al. 1995. Cloning of an epithelial chloride channel from bovine trachea. *J. Biol. Chem.* 270:31016–26

21. Flicek P, Aken BL, Beal K, Ballester B, Caccamo M, et al. 2008. Ensembl 2008. *Nucleic Acids Res.* 36:D707–14

22. Wheeler DL, Barrett T, Benson DA, Bryant SH, Canese K, et al. 2008. Database resources of the National Center for Biotechnology Information. *Nucleic Acids Res.* 36:D13–21

17, 18. These contemporaneous articles provided for the initial purification and characterization of Clca proteins (bClca1 and bClca2).

24. Demonstrated the activity of Clca family members to regulate mucous cell metaplasia (but not airway hyperreactivity) and to exhibit functional compensation in $mClca3^{-/-}$ mice.

27. Provided the first molecular cloning of a *Clca* gene (*bClca1*) and the first evidence that expression of recombinant Clca protein influences calcium-dependent chloride flux.

29, 30. The first articles to definitively demonstrate that hCLCA1 and mCLCA3 are not ion channels but instead are secreted soluble proteins.

41, 42. Identified and characterized the interaction between hCLCLA2 and β4-integrin and highlighted the role of this interaction in cancer metastases.

23. Al-Jumaily M, Kozlenkov A, Mechaly I, Fichard A, Matha V, et al. 2007. Expression of three distinct families of calcium-activated chloride channel genes in the mouse dorsal root ganglion. *Neurosci. Bull.* 23:293–99

24. Patel AC, Morton JD, Kim EY, Alevy Y, Swanson S, et al. 2006. Genetic segregation of airway disease traits despite redundancy of calcium-activated chloride channel family members. *Physiol. Genomics* 25:502–13

25. Anton F, Leverkoehne I, Mundhenk L, Thoreson WB, Gruber AD. 2005. Overexpression of eCLCA1 in small airways of horses with recurrent airway obstruction. *J. Histochem. Cytochem.* 53:1011–21

26. Yamazaki J, Okamura K, Ishibashi K, Kitamura K. 2005. Characterization of CLCA protein expressed in ductal cells of rat salivary glands. *Biochim. Biophys. Acta* 1715:132–44

27. Fuller CM, Ismailov II, Keeton DA, Benos DJ. 1994. Phosphorylation and activation of a bovine tracheal anion channel by Ca^{2+}/calmodulin-dependent protein kinase II. *J. Biol. Chem.* 269:26642–50

28. Gruber AD, Elble RC, Ji HL, Schreur KD, Fuller CM, Pauli BU. 1998. Genomic cloning, molecular characterization, and functional analysis of human CLCA1, the first human member of the family of Ca^{2+}-activated Cl^- channel proteins. *Genomics* 54:200–14

29. Gibson A, Lewis AP, Affleck K, Aitken AJ, Meldrum E, Thompson N. 2005. hCLCA1 and mCLCA3 are secreted nonintegral membrane proteins and therefore are not ion channels. *J. Biol. Chem.* 280:27205–12

30. Mundhenk L, Alfalah M, Elble RC, Pauli BU, Naim HY, Gruber AD. 2006. Both cleavage products of the mCLCA3 protein are secreted soluble proteins. *J. Biol. Chem.* 281:30072–80

31. Huan C, Greene KS, Shi B, Spizz G, Sun H, et al. 2008. mCLCA4 processing and secretion requires luminal sorting motifs. *Am. J. Physiol.* 295:C279–87

32. Elble RC, Walia V, Cheng HC, Connon CJ, Mundhenk L, et al. 2006. The putative chloride channel hCLCA2 has a single C-terminal transmembrane segment. *J. Biol. Chem.* 281:29448–54

33. Bothe MK, Braun J, Mundhenk L, Gruber AD. 2008. Murine mCLCA6 is an integral apical membrane protein of nongoblet cell enterocytes and colocalizes with the cystic fibrosis transmembrane conductance regulator. *J. Histochem. Cytochem.* 56:495–509

34. Gandhi R, Elble RC, Gruber AD, Schreur KD, Ji HL, et al. 1998. Molecular and functional characterization of a calcium-sensitive chloride channel from mouse lung. *J. Biol. Chem.* 273:32096–101

35. Elble RC, Ji G, Nehrke K, DeBiasio J, Kingsley PD, et al. 2002. Molecular and functional characterization of a murine calcium-activated chloride channel expressed in smooth muscle. *J. Biol. Chem.* 277:18586–91

36. Beckley JR, Pauli BU, Elble RC. 2004. Re-expression of detachment-inducible chloride channel mCLCA5 suppresses growth of metastatic breast cancer cells. *J. Biol. Chem.* 279:41634–41

37. Brouillard F, Bensalem N, Hinzpeter A, Tondelier D, Trudel S, et al. 2005. Blue native/SDS-PAGE analysis reveals reduced expression of mClCA3 protein in cystic fibrosis knock-out mice. *Mol. Cell. Proteomics* 4:1762–75

38. Gruber AD, Pauli BU. 1999. Molecular cloning and biochemical characterization of a truncated, secreted member of the human family of Ca^{2+}-activated Cl^- channels. *Biochim. Biophys. Acta* 1444:418–23

39. Pawlowski K, Lepisto M, Meinander N, Sivars U, Varga M, Wieslander E. 2006. Novel conserved hydrolase domain in the CLCA family of alleged calcium-activated chloride channels. *Proteins* 63:424–39

40. Whittaker CA, Hynes RO. 2002. Distribution and evolution of von Willebrand/integrin A domains: widely dispersed domains with roles in cell adhesion and elsewhere. *Mol. Biol. Cell* 13:3369–87

41. Abdel-Ghany M, Cheng HC, Elble RC, Pauli BU. 2001. The breast cancer β4 integrin and endothelial human CLCA2 mediate lung metastasis. *J. Biol. Chem.* 276:25438–46

42. Abdel-Ghany M, Cheng HC, Elble RC, Pauli BU. 2002. Focal adhesion kinase activated by β4 integrin ligation to mCLCA1 mediates early metastatic growth. *J. Biol. Chem.* 277:34391–400

43. Abdel-Ghany M, Cheng HC, Elble RC, Lin H, DiBasio J, Pauli BU. 2003. The interacting binding domains of the β4 integrin and calcium-activated chloride channels (CLCAs) in metastasis. *J. Biol. Chem.* 278:49406–16

44. Cole C, Barber JD, Barton GJ. 2008. The Jpred 3 secondary structure prediction server. *Nucleic Acids Res.* 36:W197–201

45. Zhu D, Pauli BU. 1993. Correlation between the lung distribution patterns of Lu-ECAM-1 and melanoma experimental metastases. *Int. J. Cancer* 53:628–33

46. Zhu D, Cheng CF, Pauli BU. 1992. Blocking of lung endothelial cell adhesion molecule-1 (Lu-ECAM-1) inhibits murine melanoma lung metastasis. *J. Clin. Investig.* 89:1718–24

47. Giancotti FG. 2007. Targeting integrin β4 for cancer and antiangiogenic therapy. *Trends Pharmacol. Sci.* 28:506–11

48. Gruber AD, Pauli BU. 1999. Tumorigenicity of human breast cancer is associated with loss of the Ca^{2+}-activated Cl^- channel CLCA2. *Cancer Res.* 59:5488–91

49. Elble RC, Pauli BU. 2001. Tumor suppression by a proapoptotic calcium-activated chloride channel in mammary epithelium. *J. Biol. Chem.* 276:40510–17

50. Li X, Cowell JK, Sossey-Alaoui K. 2004. CLCA2 tumour suppressor gene in 1p31 is epigenetically regulated in breast cancer. *Oncogene* 23:1474–80

51. Balakrishnan A, von Neuhoff N, Rudolph C, Kamphues K, Schraders M, et al. 2006. Quantitative microsatellite analysis to delineate the commonly deleted region 1p22.3 in mantle cell lymphomas. *Genes Chromosomes Cancer* 45:883–92

52. Nakanishi A, Morita S, Iwashita H, Sagiya Y, Ashida Y, et al. 2001. Role of gob-5 in mucus overproduction and airway hyperresponsiveness in asthma. *Proc. Natl. Acad. Sci. USA* 98:5175–80

53. Zhou Y, Shapiro M, Dong Q, Louahed J, Weiss C, et al. 2002. *A calcium-activated chloride channel blocker inhibits goblet cell metaplasia and mucus overproduction*. Novartis Found. Symp., 248th: Mucus Hypersecretion in Respiratory Disease

54. Bertrand CA, Danahay H, Poll CT, Laboisse C, Hopfer U, Bridges RJ. 2004. Niflumic acid inhibits ATP-stimulated exocytosis in a mucin secreting epithelial cell line. *Am. J. Physiol. Cell Physiol.* 286:C247–55

55. McNamara N, Gallup M, Khong A, Sucher A, Maltseva I, et al. 2004. Adenosine up-regulation of the mucin gene, *MUC2*, in asthma. *FASEB J.* 18:1770–72

56. Hauber HP, Daigneault P, Frenkiel S, Lavigne F, Hung HL, et al. 2005. Niflumic acid and MSI-2216 reduce TNF-α-induced mucin expression in human airway mucosa. *J. Allergy Clin. Immunol.* 115:266–71

57. Hauber HP, Goldmann T, Vollmer E, Wollenberg B, Hung HL, et al. 2007. LPS-induced mucin expression in human sinus mucosa can be attenuated by hCLCA inhibitors. *J. Endotoxin Res.* 13:109–16

58. Hoshino M, Morita S, Iwashita H, Sagiya Y, Nagi T, et al. 2002. Increased expression of the human Ca^{2+}-activated Cl^- channel 1 (CaCC1) gene in the asthmatic airway. *Am. J. Respir. Crit. Care Med.* 165:1132–36

59. Leverkoehne I, Gruber AD. 2002. The murine mCLCA3 (alias gob-5) protein is located in the mucin granule membranes of intestinal, respiratory, and uterine goblet cells. *J. Histochem. Cytochem.* 50:829–38

60. Kirkup AJ, Edwards G, Green ME, Miller M, Walker SD, Weston AH. 1996. Modulation of membrane currents and mechanical activity by niflumic acid in rat vascular smooth muscle. *Eur. J. Pharmacol.* 317:165–74

61. Sergeant GP, Hollywood MA, McHale NG, Thornbury KD. 2001. Spontaneous Ca^{2+} activated Cl^- currents in isolated urethral smooth muscle cells. *J. Urol.* 166:1161–66

62. Ledoux J, Greenwood IA, Leblanc N. 2005. Dynamics of Ca^{2+}-dependent Cl^- channel modulation by niflumic acid in rabbit coronary arterial myocytes. *Mol. Pharmacol.* 67:163–73

63. Lowell FC. 1964. Observations on heaves. An asthma-like syndrome in the horse. *J. Allergy Clin. Immunol.* 35:322–30

64. Davis E, Rush BR. 2002. Equine recurrent airway obstruction: pathogenesis, diagnosis, and patient management. *Vet. Clin. N. Am. Equine Pract.* 18:453–67

65. Leguillette R. 2003. Recurrent airway obstruction–heaves. *Vet. Clin. N. Am. Equine Pract.* 19:63–86

66. Horohov DW, Beadle RE, Mouch S, Pourciau SS. 2005. Temporal regulation of cytokine mRNA expression in equine recurrent airway obstruction. *Vet. Immunol. Immunopathol.* 108:237–45

67. Gerber V, Robinson NE, Venta RJ, Rawson J, Jefcoat AM, Hotchkiss JA. 2003. Mucin genes in horse airways: MUC5AC, but not MUC2, may play a role in recurrent airway obstruction. *Equine Vet. J.* 35:252–57

68. Holtzman MJ, Kim EY, Morton JD. 2005. Genetic and genomic approaches to complex lung diseases using mouse models. In *Computational Genetics and Genomics: Tools for Understanding Disease*, ed. G Peltz, pp. 99–141. Totawa, NJ: Humana

69. Range F, Mundhenk L, Gruber AD. 2007. A soluble secreted glycoprotein (eCLCA1) is overexpressed due to goblet cell hyperplasia and metaplasia in horses with recurrent airway obstruction. *Vet. Pathol.* 44:901–11

70. Levitt RC, Mitzner W. 1989. Autosomal recessive inheritance of airway hyperreactivity to 5-hydroxytryptamine. *J. Appl. Physiol.* 67:1125–32

71. Konno S, Adachi M, Matsuura T, Sunouchi K, Hoshino H, et al. 1993. Bronchial reactivity to methacholine and serotonin in six inbred mouse strains. *Arerugi* 42:42–47

72. Chiba Y, Yanagisawa R, Sagai M. 1995. Strain and route differences in airway responsiveness to acetylcholine in mice. *Res. Commun. Mol. Pathol. Pharmacol.* 90:169–72

73. Zhang LY, Levitt RC, Kleeberger SR. 1995. Differential susceptibility to ozone-induced airways hyperreactivity in inbred strains of mice. *Exp. Lung Res.* 21:503–18

74. Miyabara Y, Yanagisawa R, Shimojo N, Takano H, Lim HB, et al. 1998. Murine strain differences in airway inflammation caused by diesel exhaust particles. *Eur. Respir. J.* 11:291–98

75. Brewer JP, Kisselgof AB, Martin TR. 1999. Genetic variability in pulmonary physiological, cellular, and antibody responses to antigen in mice. *Am. J. Respir. Crit. Care Med.* 160:1150–56

76. De Sanctis GT, Singer JB, Jiao A, Yandava CN, Lee YH, et al. 1999. Quantitative trait locus mapping of airway responsiveness to chromosomes 6 and 7 in inbred mice. *Am. J. Physiol.* 277:L1118–23

77. Ewart SL, Kuperman D, Schadt E, Tankersley C, Grupe A, et al. 2000. Quantitative trait loci controlling allergen-induced airway hyperresponsiveness in inbred mice. *Am. J. Respir. Cell Mol. Biol.* 23:537–45

78. Van Oosterhout AJ, Jeurink PV, Groot PC, Hofman GA, Nijkamp FP, Demant P. 2002. Genetic analysis of antigen-induced airway manifestations of asthma using recombinant congenic mouse strains. *Chest* 121:13S

79. Graham BS, Perkins MD, Wright PF, Karzon DT. 1988. Primary respiratory syncytial virus infection in mice. *J. Med. Virol.* 26:153–62

80. Seemungal T, Harper-Owen R, Bhowmik A, Moric I, Sanderson G, et al. 2001. Respiratory viruses, symptoms, and inflammatory markers in acute exacerbations and stable chronic obstructive pulmonary disease. *Am. J. Respir. Crit. Care Med.* 164:1618–23

81. Rhode G, Wiethege A, Borg I, Kauth M, Bauer TT, et al. 2003. Respiratory viruses in exacerbations of chronic obstructive pulmonary disease requiring hospitalisation: a case-control study. *Thorax* 58:37–42

82. Borg I, Rohde G, Loseke S, Bittscheidt J, Schultze-Werninghaus G, et al. 2003. Evaluation of a quantitative real-time PCR for the detection of respiratory syncytial virus in pulmonary diseases. *Eur. Respir. J.* 21:944–51

83. Falsey AR, Hennessey PA, Formica MA, Cox C, Walsh EE. 2005. Respiratory syncytial virus infection in elderly and high-risk adults. *N. Engl. J. Med.* 352:1749–59

84. Beckham JD, Cadena A, Lin J, Piedra PA, Glezen WP, et al. 2005. Respiratory viral infections in patients with chronic obstructive pulmonary disease. *J. Infect.* 50:322–30

85. Tyner JW, Kim EY, Ide K, Pelletier MR, Roswit WT, et al. 2006. Blocking airway mucous cell metaplasia by inhibiting EGFR antiapoptosis and IL-13 transdifferentiation signals. *J. Clin. Investig.* 116:309–21

86. Walter MJ, Morton JD, Kajiwara N, Agapov E, Holtzman MJ. 2002. Viral induction of a chronic asthma phenotype and genetic segregation from the acute response. *J. Clin. Investig.* 110:165–75

87. Altschul SF, Gish W, Miller W, Myers EW, Lipman DJ. 1990. Basic local alignment search tool. *J. Mol. Biol.* 215:403–10

88. Evans CM, Williams OW, Tuvim MJ, Nigam R, Mixides GP, et al. 2004. Mucin is produced by Clara cells in the proximal airway of antigen-challenged mice. *Am. J. Respir. Cell Mol. Biol.* 31:382–94

89. Gruber AD, Schreur KD, Ji HL, Fuller CM, Pauli BU. 2004. Molecular cloning and transmembrane structure of hCLCA2 from human lung, trachea, and mammary gland. *Am. J. Physiol.* 276:C1261–70

90. Patel AC, Battaile JT, Alevy Y, Patterson GA, Swanson S, et al. 2007. Homologous mouse *Clca3,5,6* and human *hCLCA1,2,4* gene clusters control mucous cell metaplasia. *Am. J. Respir. Crit. Care Med.* 175:A499 (Abstr.)

91. Robichaud A, Tuck SA, Kargman S, Tam J, Wong E, et al. 2005. Gob-5 is not essential for mucus overproduction in preclinical murine models of allergic asthma. *Am. J. Respir. Cell Mol. Biol.* 33:303–14

92. Zhou Y, Dong Q, Louahed J, Dragwa C, Savio D, et al. 2001. Characterization of a calcium-activated chloride channel as a shared target of Th2 cytokine pathways and its potential involvement in asthma. *Am. J. Respir. Cell Mol. Biol.* 25:486–91

93. Nakano T, Inoue H, Fukuyama S, Matsumoto K, Matsumura M, et al. 2006. Niflumic acid suppresses interleukin-13-induced asthma phenotypes. *Am. J. Respir. Crit. Care Med.* 173:1216–21

94. Yasuo M, Fujimoto K, Tanabe T, Yaegashi H, Tsushima K, et al. 2006. Relationship between calcium-activated chloride channel 1 and MUC5AC in goblet cell hyperplasia induced by interleukin-13 in human bronchial epithelial cells. *Respiration* 73:347–59

95. Thai P, Chen Y, Dolganov G, Wu R. 2005. Differential regulation of MUC5AC/Muc5ac and hCLCA-1/mGob-5 expression in airway epithelium. *Am. J. Respir. Cell Mol. Biol.* 33:523–30

96. Kim EY, Battaile JT, Patel AC, You Y, Agapov E, et al. 2008. Persistent activation of an innate immune response translates respiratory viral infection into chronic lung disease. *Nat. Med.* 14:633–40

97. Holtzman MJ, Battaile JT, Patel AC. 2006. Immunogenetic programs for viral induction of mucous cell metaplasia. *Am. J. Respir. Cell Mol. Biol.* 35:29–39

98. Kuperman DA, Huang X, Koth LL, Chang GH, Dolganov GM, et al. 2002. Direct effects of interleukin-13 on epithelial cells cause airway hyperreactivity and mucus overproduction in asthma. *Nat. Med.* 8:885–89

99. Nath P, Leung SY, Williams AS, Noble A, Xie S, et al. 2007. Complete inhibition of allergic airway inflammation and remodelling in quadruple IL-4/5/9/13$^{-/-}$ mice. *Clin. Exp. Allergy* 37:1427–35

100. Silverman EK, Palmer LJ, Mosley JD, Barth M, Senter JM, et al. 2002. Genomewide linkage analysis of quantitative spirometric phenotypes in severe early-onset chronic obstructive pulmonary disease. *Am. J. Hum. Genet.* 70:1229–39

101. Palmer LJ, Celedon JC, Chapman HA, Speizer FE, Weiss ST, Silverman EK. 2003. Genome-wide linkage analysis of bronchodilator responsiveness and postbronchodilator spirometric phenotypes in chronic obstructive pulmonary disease. *Hum. Mol. Genet.* 12:1199–210

102. Kamada F, Suzuki Y, Shao C, Tamari M, Hasegawa K, et al. 2004. Association of the *hCLCA1* gene with childhood and adult asthma. *Genes Immun.* 5:540–47

103. Hegab AE, Sakamoto T, Uchida Y, Nomura A, Ishii Y, et al. 2004. *CLCA1* gene polymorphisms in chronic obstructive pulmonary disease. *J. Med. Genet.* 41:1–7

104. Ritzka M, Stanke F, Jansen S, Gruber AD, Pusch L, et al. 2004. The CLCA gene locus as a modulator of the gastrointestinal basic defect in cystic fibrosis. *Hum. Genet.* 115:483–91

105. Young FD, Newbigging S, Choi C, Keet M, Kent G, Rozmahel RF. 2007. Amelioration of cystic fibrosis intestinal mucous disease in mice by restoration of mCLCA3. *Gastroenterology* 133:1928–37

106. Toda M, Tulic MK, Levitt RC, Hamid Q. 2002. A calcium-activated chloride channel (HCLCA1) is strongly related to IL-9 expression and mucus production in bronchial epithelium of patients with asthma. *J. Allergy Clin. Immunol.* 109:246–50

107. Woodruff PG, Boushey HA, Dolganov GM, Barker CS, Yang YH, et al. 2007. Genome-wide profiling identifies epithelial cell genes associated with asthma and with treatment response to corticosteroids. *Proc. Natl. Acad. Sci. USA* 104:15858–63

108. Wang K, Wen FQ, Feng YL, Ou XM, Xu D, et al. 2007. Increased expression of human calcium-activated chloride channel 1 gene is correlated with mucus overproduction in Chinese asthmatic airway. *Cell Biol. Int.* 31:1388–95

109. Patel AC, Kim EY, Roswit WT, Swanson S, Christie C, et al. 2008. Potential utility of CLCA1 as a biomarker: response of hCLCA1/mClca3 to corticosteroid in natural and experimental asthma. *Am. J. Respir. Crit. Care Med.* 177:A75 (Abstr.)

110. Hauber HP, Manoukian JJ, Nguyen LH, Sobol SE, Levitt RC, et al. 2003. Increased expression of interleukin-9, interleukin-9 receptor, and the calcium-activated chloride channel hCLCA1 in the upper airways of patients with cystic fibrosis. *Laryngoscope* 113:1037–42

111. Hauber HP, Tsicopoulos A, Wallaert B, Griffin S, McElvaney NG, et al. 2004. Expression of HCLCA1 in cystic fibrosis lungs is associated with mucus overproduction. *Eur. Respir. J.* 23:846–50

112. Nakanishi A, Shigeru M. 2004. *U.S. Patent Appl. No. 2,004,185,500*

113. Szymkowski DE. 2006. *U.S. Patent No. 7,141,365 B2*

114. Walker NM, Simpson JE, Levitt RC, Boyle KT, Clarke LL. 2006. Talniflumate increases survival in a cystic fibrosis mouse model of distal intestinal obstructive syndrome. *J. Pharmacol. Exp. Ther.* 317:275–83

115. Knight D. 2004. Talniflumate (Genaera). *Curr. Opin. Investig. Drugs* 5:557–62

116. Donnelly LE, Rogers DF. 2003. Therapy for chronic obstructive pulmonary disease in the 21st century. *Drugs* 63:1973–98

117. Larkin MA, Blackshields G, Brown NP, Chenna R, McGettigan PA, et al. 2007. Clustal W and Clustal X version 2.0. *Bioinformatics* 23:2947–48

118. Karolchik D, Kuhn RM, Baertsch R, Barber GP, Clawson H, et al. 2008. The UCSC Genome Browser Database: 2008 update. *Nucleic Acids Res.* 36:D773–79

119. Kent WJ. 2002. BLAT—the BLAST-like alignment tool. *Genome Res.* 12:656–64

120. Rimbaut A. 2008. FigTree 1.1.2. **http://tree.bio.ed.ac.uk/software/figtree/**

121. Bendtsen JD, Nielsen H, von Heijne G, Brunak S. 2004. Improved prediction of signal peptides: SignalP 3.0. *J. Mol. Biol.* 340:783–95

122. Bennett-Lovsey RM, Herbert AD, Sternberg MJ, Kelley LA. 2008. Exploring the extremes of sequence/structure space with ensemble fold recognition in the program Phyre. *Proteins* 70:611–25

123. Barton GJ. 1993. ALSCRIPT: a tool to format multiple sequence alignments. *Protein Eng.* 6:37–40

124. Blom N, Sicheritz-Ponten T, Gupta R, Gammeltoft S, Brunak S. 2004. Prediction of post-translational glycosylation and phosphorylation of proteins from the amino acid sequence. *Proteomics* 4:1633–49

125. Krogh A, Larsson B, von Heijne G, Sonnhammer EL. 2001. Predicting transmembrane protein topology with a hidden Markov model: application to complete genomes. *J. Mol. Biol.* 305:567–80

126. Eisenhaber B, Bork P, Eisenhaber F. 1999. Prediction of potential GPI-modification sites in proprotein sequences. *J. Mol. Biol.* 292:741–58

127. Twigger SN, Pruitt KD, Fernandez-Suarez XM, Karolchik D, Worley KC, et al. 2008. What everybody should know about the rat genome and its online resources. *Nat. Genet.* 40:523–27

128. Petersen G, Ståhl F. 2008. RGST—Rat Gene Symbol Tracker, a database for defining official rat gene symbols. *BMC Genomics* 9:29

129. Agnel M, Vermat T, Culouscou JM. 1999. Identification of three novel members of the calcium-dependent chloride channel (CaCC) family predominantly expressed in the digestive tract and trachea. *FEBS Lett.* 455:295–301

130. Romio L, Musante L, Cinti R, Seri M, Moran O, et al. 1999. Characterization of a murine gene homologous to the bovine CaCC chloride channel. *Gene* 228:181–88

131. Lee D, Ha S, Kho Y, Kim J, Cho K, et al. 1999. Induction of mouse Ca^{2+}-sensitive chloride channel 2 gene during involution of mammary gland. *Biochem. Biophys. Res. Commun.* 264:933–37

132. Komiya T, Tanigawa Y, Hirohashi S. 1999. Cloning and identification of the gene *gob-5*, which is expressed in intestinal goblet cells in mice. *Biochem. Biophys. Res. Commun.* 255:347–51

133. Snelling WM, Chiu R, Schein JE, Hobbs M, Abbey CA, et al. 2007. A physical map of the bovine genome. *Genome Biol.* 8:R165

134. Elble RC, Widom J, Gruber AD, Abdel-Ghany M, Levine R, et al. 1997. Cloning and characterization of lung-endothelial cell adhesion molecule-1 suggest it is an endothelial chloride channel. *J. Biol. Chem.* 272:27853–61

135. Jeong SM, Park HK, Yoon IS, Lee JH, Kim JH, et al. 2005. Cloning and expression of Ca^{2+}-activated chloride channel from rat brain. *Biochem. Biophys. Res. Commun.* 334:569–76

136. Yoon IS, Jeong SM, Lee SN, Lee JH, Kim JH, et al. 2006. Cloning and heterologous expression of a Ca^{2+}-activated chloride channel isoform from rat brain. *Biol. Pharm. Bull.* 29:2168–73

137. Chowdhary BP, Raudsepp T. 2008. The horse genome derby: racing from map to whole genome sequence. *Chromosome Res.* 16:109–27

138. Zhang SJ, Steijaert MN, Lau D, Schutz G, Delucinge-Vivier C, et al. 2007. Decoding NMDA receptor signaling: identification of genomic programs specifying neuronal survival and death. *Neuron* 53:549–62

139. Itoh R, Kawamoto S, Miyamoto Y, Kinoshita S, Okubo K. 2000. Isolation and characterization of a Ca^{2+}-activated chloride channel from human corneal epithelium. *Curr. Eye Res.* 21:918–25

140. Mall M, Gonska T, Thomas J, Schreiber R, Seydewitz HH, et al. 2003. Modulation of Ca^{2+}-activated Cl^- secretion by basolateral K^+ channels in human normal and cystic fibrosis airway epithelia. *Pediatr. Res.* 53:608–18

141. Connon CJ, Kawasaki S, Yamasaki K, Quantock AJ, Kinoshita S. 2005. The quantification of hCLCA2 and colocalisation with integrin β4 in stratified human epithelia. *Acta Histochem.* 106:421–25

142. Connon CJ, Yamasaki K, Kawasaki S, Quantock AJ, Koizumi N, Kinoshita S. 2004. Calcium-activated chloride channel-2 in human epithelia. *J. Histochem. Cytochem.* 52:415–18

143. Leverkoehne I, Horstmeier BA, von Samson-Himmelstjerna G, Scholte BJ, Gruber AD. 2002. Real-time RT-PCR quantitation of mCLCA1 and mCLCA2 reveals differentially regulated expression in pre- and postnatal murine tissues. *Histochem. Cell Biol.* 118:11–17

Role of HDAC2 in the Pathophysiology of COPD

Peter J. Barnes

National Heart and Lung Institute, Imperial College, London SW3 6LY, United Kingdom; email: p.j.barnes@imperial.ac.uk

Annu. Rev. Physiol. 2009. 71:451–64

First published online as a Review in Advance on September 25, 2008

The *Annual Review of Physiology* is online at physiol.annualreviews.org

This article's doi:
10.1146/annurev.physiol.010908.163257

Copyright © 2009 by Annual Reviews.
All rights reserved

0066-4278/09/0315-0451$20.00

Key Words

histone acetylation, histone deacetylase, macrophage, theophylline, oxidative stress, nuclear factor-κB

Abstract

Chronic obstructive pulmonary disease (COPD), characterized by progressive inflammation in the small airways and lung parenchyma, is mediated by the increased expression of multiple inflammatory genes. The increased expression of these genes is regulated by acetylation of core histones, whereas histone deacetylase 2 (HDAC2) suppresses inflammatory gene expression. In COPD, HDAC2 activity and expression are reduced in peripheral lung and in alveolar macrophages, resulting in amplification of the inflammatory response. Corticosteroid resistance in COPD occurs because corticosteroids use HDAC2 to switch off activated inflammatory genes. The reduction in HDAC2 appears to be secondary to the increased oxidative and nitrative stress in COPD lungs. Antioxidants and inhibitors of nitric oxide synthesis may therefore restore corticosteroid sensitivity in COPD, but this can also be achieved by low concentrations of theophylline and curcumin, which act as HDAC activators.

INTRODUCTION

Chronic obstructive pulmonary disease (COPD) is a common disease of progressive airway obstruction as a result of emphysema and small-airway disease (chronic obstructive bronchiolitis). COPD results in gradually increasing shortness of breath and limitation of exercise (1, 2). COPD is a global health problem that is already the fourth most common cause of death in developed countries, with increasing mortality in developing countries (3). COPD is also a common cause of morbidity, and COPD exacerbations are among the most common reasons for acute hospital admissions, imposing a large burden on health resources. COPD now has a world-wide prevalence of more than 10% in men over the age of 40 years and is increasing toward this figure in women (4). Because of the enormous burden of disease and escalating health care costs, there is now renewed interest in the underlying cellular and molecular mechanisms of COPD (5) and a search for new therapies (6, 7). The definition of COPD adopted by the Global Initiative on Obstructive Lung Disease (GOLD) encompasses the idea that COPD is a chronic inflammatory disease, and much of the recent research on disease mechanisms has focused on the nature of this inflammatory response (2). COPD slowly progresses over many decades, leading to death from respiratory failure, but most patients die of comorbidities such as heart disease (heart failure and myocardial infarction) and lung cancer before this stage. Although the most common cause of COPD is chronic cigarette smoking, some patients, particularly in developing countries, develop the disease from inhalation of wood smoke from biomass fuels or other inhaled irritants (4). However, only ~25% of smokers develop COPD (8), suggesting that there may be genetic or host factors that predispose to its development, although these have not yet been identified. The airway obstruction is relentlessly progressive, and only smoking cessation reduces the rate of decline in lung function. As the disease becomes more severe, however, there is less benefit of smoking cessation, and lung inflammation persists in ex-smokers (9, 10).

In sharp contrast to asthma, COPD responds poorly to currently available therapies, particularly corticosteroids, and there is a pressing need for the development of effective anti-inflammation treatments (11). A major barrier to therapy of COPD is resistance to the anti-inflammatory effects of corticosteroids. The molecular mechanisms for this corticosteroid resistance are now being elucidated, particularly as the molecular basis for the anti-inflammatory effects of corticosteroids is better understood (12). An important mechanism of corticosteroid resistance in COPD, which is also linked to amplification of the inflammatory process, is a reduction in the critical nuclear enzyme histone deacetylase (HDAC)2 (13).

INFLAMMATION IN COPD

The progressive airflow limitation in COPD is due to two major pathological processes: remodeling and narrowing of small airways and destruction of the lung parenchyma with consequent destruction of the alveolar attachments of these airways as a result of emphysema. This results in diminished lung recoil, higher resistance to flow, and closure of small airways at higher lung volumes during expiration, trapping air in the lung. This leads to the characteristic hyperinflation of the lungs, which gives rise to the sensation of dyspnea and limits exercise capacity (14). Both the small-airway remodeling and narrowing and the emphysema are likely to be the results of chronic inflammation in the lung periphery (15). Quantitative studies have shown that the inflammatory response in small airways increases as the disease progresses, leading to peribronchiolar fibrosis (16). There is a specific pattern of inflammation in COPD airways and lung parenchyma, with increased numbers of macrophages. Macrophages play an important role in orchestrating the inflammation in COPD lungs through the release of multiple proinflammatory mediators (17):

Chronic obstructive pulmonary disease (COPD): a chronic lung disease with progressive airflow obstruction as a result of destruction of the lung parenchyma (emphysema) and fibrosis of small airways (chronic obstructive bronchiolitis)

Corticosteroids: anti-inflammatory drugs that suppress inflammatory genes after binding to a cytoplasmic glucocorticoid receptor

Corticosteroid resistance: a reduced or absent anti-inflammatory response to corticosteroids that is characteristic of some inflammatory diseases, such as COPD

Histone deacetylases (HDACs): nuclear enzymes that remove acetyl groups from acetylated histones and therefore suppress gene transcription

T lymphocytes with a predominance of CD8+ (cytotoxic) T cells and, in more severe diseases, B lymphocytes with increased numbers of neutrophils in the lumen (5).

The inflammatory response in COPD involves both innate and adaptive immune responses (18, 19). Multiple inflammatory mediators are increased in COPD and are derived from inflammatory cells and structural cells of the airways and lungs (20). Many cytokines and chemokines are involved in orchestrating the chronic inflammatory process (21). A similar pattern of inflammation is seen in smokers without airflow limitation, but in COPD this inflammation is amplified, with more inflammatory cells and higher concentrations of mediators. This inflammation is even further amplified during acute exacerbations of the disease, which are usually precipitated by bacterial and viral infections (22, 23) (**Figure 1**). The molecular basis of this amplification of inflammation is not yet completely understood, but recently the role of reduced HDAC activity in COPD has been highlighted as a potential mechanism for amplifying inflammatory gene expression. The amplification of inflammation and susceptibility to develop airflow limitation may be genetically determined, but the specific genes that are involved in these mechanisms have not yet been identified (24).

Cigarette smoke and other irritants in the respiratory tract may activate surface macrophages and airway epithelial cells to release chemotactic factors that then attract circulating leukocytes into the lungs. Among chemotactic factors, chemokines predominate and therefore play a key role in orchestrating the chronic inflammation in COPD lungs and its further amplification during acute exacerbations (25). These may be the initial inflammatory events occurring in all smokers. However, in smokers who develop COPD, this inflammation progresses into a more complicated inflammatory pattern of adaptive immunity and involves T and B lymphocyte infiltration and possibly dendritic cells, along with a complicated interacting array of cytokines and other mediators (19, 21).

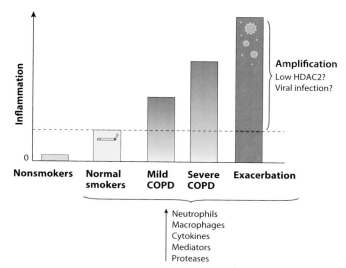

Figure 1

Amplification of lung inflammation in chronic obstructive pulmonary disorder (COPD). Normal smokers have a mild inflammatory response, which represents the normal (probably protective) reaction of the respiratory mucosa to chronic inhaled irritants. In COPD this same inflammatory response is markedly amplified, and this amplification increases as the disease progresses. The inflammation is further increased during exacerbations triggered by infective organisms. The molecular mechanisms of this amplification are currently unknown but may be determined by genetic factors or possibly latent viral infection. Oxidative stress is an important amplifying mechanism and may increase the expression of inflammatory genes through impairing the activity of histone deacetylase 2 (HDAC2), which is needed to switch off inflammatory genes.

The chronic inflammation in COPD is associated with increased expression of multiple inflammatory proteins, including cytokines, chemokines, inflammatory enzymes, receptors, and adhesion molecules. Increased expression of these inflammatory proteins is mainly at the level of gene expression and is regulated by proinflammatory transcription factors, such as nuclear factor-κB (NF-κB) (26). NF-κB is activated in COPD lungs and inflammatory cells, particularly in alveolar macrophages and airway epithelial cells (27), with further activation during exacerbations (28).

HISTONE ACETYLATION AND DEACETYLATION

Gene expression is regulated by acetylation of core histones that open up the chromatin structure (chromatin remodeling) to allow

Nuclear factor-κB (NF-κB): a proinflammatory transcription factor that switches on the transcription of multiple inflammatory genes, such as those encoding for cytokines and chemokines

Histone acetyltransferase (HAT): HAT activity in coactivator molecules acetylates lysine residues on core histones, which opens up the chromatin structure to make DNA available for gene transcription

Glucocorticoid receptors (GRs): bind corticosteroids and translocate to the nucleus, where they seek out and switch off activated inflammatory genes

transcription factors and RNA polymerase to bind to DNA, thus initiating gene transcription (29, 30). Gene expression is regulated by various coactivator molecules, such as CREB-binding protein and p300, all of which have intrinsic histone acetyltransferase (HAT) activity. Expression of inflammatory genes is regulated by increased acetylation of histone 4 (31, 32). Thus, epigenetic factors play a critical role in chronic inflammation (33). In COPD peripheral lung, airway biopsies, and alveolar macrophages, there is an increase in the acetylation of histones associated with the promoter region of inflammatory genes, such as CXCL8 (interleukin-8), that are regulated by NF-κB, and the degree of acetylation increases with disease severity (34). This increased acetylation of histones associated with inflammatory genes in COPD is not due to an increase in HAT activity in lungs or macrophages, however. By contrast, in asthma there is an increase in airway HAT activity that is correlated with increased expression of inflammatory genes (35).

Histone acetylation is reversed by HDACs. There are 11 HDAC isoenzymes that deacetylate histones and other proteins within the nucleus, and specific HDACs appear to be differentially regulated and to regulate different groups of genes (36). HDACs play a critical role in the suppression of gene expression by reversing the hyperacetylation of core histones. We have found that for the regulation of inflammatory genes, HDAC2 is of critical importance (31, 37). The expression of inflammatory genes is determined by a balance between histone acetylation (which activates transcription) and deacetylation (which switches off transcription) (**Figure 2**).

It has been increasingly recognized that many regulatory proteins, particularly transcription factors and nuclear receptors, are also regulated by acetylation that is controlled by HATs and HDACs (38). Acetylation plays a key role in the regulation of androgen and estrogen receptors, and we have now shown that this is also the case for glucocorticoid receptors (GRs) (37). GR is acetylated within the nucleus at specific lysine residues close to the hinge region and binds to its DNA binding site only in its acetylated form. However, to inhibit NF-κB-activated genes, HDAC2 must deacetylate the receptor (**Figure 3**).

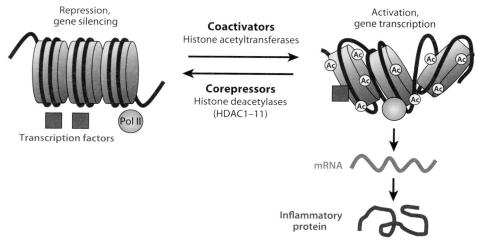

Figure 2

Chromatin remodeling and gene expression. Acetylation of core histones regulates gene activation and repression. Histone acetylation is mediated by coactivators that have intrinsic histone acetyltransferase activity, opening up the chromatin structure to allow the binding of RNA polymerase II (Pol II) and transcription factors. Gene repression is induced by histone deacetylases (HDACs), which reverse this acetylation.

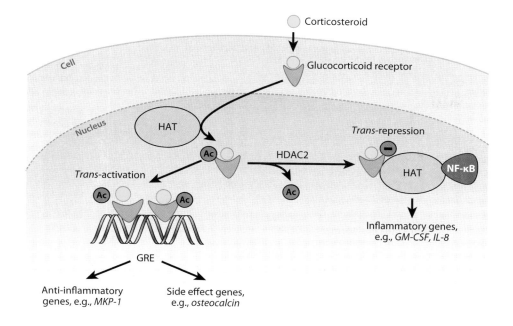

Figure 3

Acetylation of the glucocorticoid receptor (GR). After corticosteroid binding to GR, the receptor translocates to the nucleus, where it is acetylated by a histone acetyltransferase (HAT). This acetylation is necessary for GR to bind to its glucocorticoid receptor recognition element (GRE) in the promoter region of steroid-sensitive genes, which include the genes that encode proteins that mediate the side effects of corticosteroids, such as osteocalcin. The acetylated GR must be deacetylated by histone deacetylase 2 (HDAC2) to inhibit activated nuclear factor-κB (NF-κB), which in turn suppresses activated inflammatory genes.

HISTONE DEACETYLASE ACTIVITY AND EXPRESSION IN COPD

HDAC activity is reduced in alveolar macrophages of cigarette smokers compared with nonsmokers, and this reduction is correlated with increased expression of inflammatory genes in these cells (39). There is also a reduction in total HDAC activity in peripheral lung, bronchial biopsies, and alveolar macrophages from COPD patients, and this reduction is correlated with disease severity and with increased gene expression of CXCL8 and increased acetylation of histones associated with the NF-κB binding site on the CXCL8 promoter (34, 40) (**Figure 4**). There is a selective reduction in the expression of HDAC2, with lesser reductions in HDAC3

and HDAC5. In patients with very severe COPD (GOLD stage 4), the expression of HDAC2 was less than 5% of that seen in normal lung. In parallel with the reduced protein expression of HDAC2, there was also a reduction in its messenger RNA, suggesting that there may be a reduction in transcription of the HDAC2 gene or a reduced stability of its mRNA. However, almost nothing is known about the transcriptional regulation of HDAC genes, and this is an area of current research activity.

HDAC2 is reduced in alveolar macrophages from COPD patients to a greater extent than in macrophages from individuals with normal lung function, and this reduction is correlated with corticosteroid insensitivity. Restoration of HDAC2 expression to normal by transfection

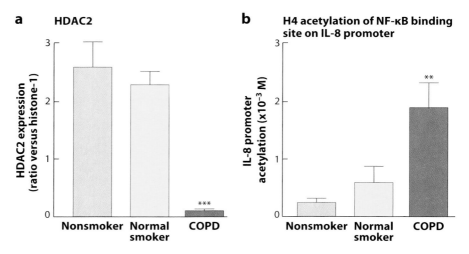

a HDAC2

b H4 acetylation of NF-κB binding site on IL-8 promoter

Figure 4

(*a*) Reduced histone deacetylase 2 (HDAC2) in peripheral lung of patients with severe chronic obstructive pulmonary disorder (COPD) compared with smokers with normal lung function and age-matched nonsmokers. (*b*) The increased acetylation of histones at the nuclear factor-κB (NF-κB) binding site on the promoter region of the CXCL8 (IL-8) gene. Double asterisks indicate $p < 0.01$; triple asterisks indicate $p < 0.001$. Adapted from Reference 34.

with a plasmid vector of HDAC2 reverses corticosteroid resistance in these cells, whereas transfection with an HDAC1 vector is without effect (37). This provides compelling evidence that the reduction in HDAC2 seen in COPD is linked to reduced corticosteroid responsiveness.

As discussed above, HDAC2 is required for the deacetylation-activated nuclear GR for GR to inhibit NF-κB activity and therefore the expression of inflammatory genes. The reduced activity of HDAC2 in COPD patients is associated with increased acetylation of GR, which may be a major mechanism accounting for corticosteroid resistance in COPD (37). In addition, the increased acetylated GR may promote gene activation and gene suppression by binding to GR recognition elements (GREs) in steroid-sensitive genes, such as genes involved in side effects of corticosteroids, including the osteocalcin gene, which is involved in osteoporosis, and pro-opiomelanocortin, which regulates the hypothalamo-pituitary-adrenal axis (**Figure 3**). Therefore, the reduced HDAC2 in COPD not only may lead to amplified inflam-

mation and corticosteroid resistance but also may increase the risk of corticosteroid-induced side effects.

CORTICOSTEROID RESISTANCE IN COPD

Corticosteroids are very effective in suppressing inflammation in asthmatic airways. An important molecular mechanism for the anti-inflammatory action of corticosteroids is the recruitment by activated GR of HDAC2 to activated inflammatory genes, thus reversing the acetylation of these inflammatory genes and silencing their transcription (12, 31, 41). In patients with COPD, the reduction in HDAC2 expression may thus account for the corticosteroid insensitivity that is seen in this disease (42). Corticosteroids provide little clinical benefit in COPD patients and fail to significantly reduce progression of the disease or to reduce its mortality (43, 44). This may reflect the fact that corticosteroids, even in high systemic doses, do not suppress inflammation in the lungs, in marked contrast to their high level

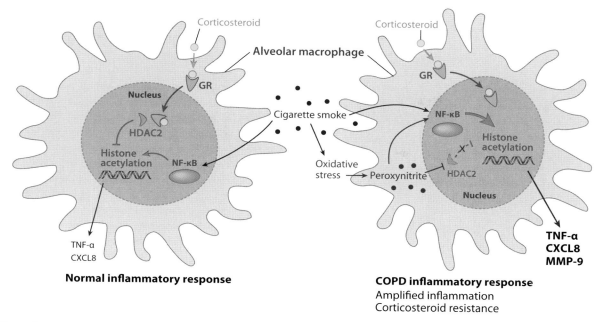

Figure 5

Proposed mechanism of corticosteroid resistance in chronic obstructive pulmonary disorder (COPD) patients. (*Left*) Stimulation of normal alveolar macrophages activates nuclear factor-κB (NF-κB) and other transcription factors to switch on histone acetyltransferase, leading to histone acetylation and subsequently to transcription of genes encoding inflammatory proteins, such as tumor necrosis factor-α (TNF-α) and CXCL8 (IL-8). Corticosteroids reverse this process by binding to glucocorticoid receptors (GR) and recruiting histone deacetylase 2 (HDAC2). This reverses the histone acetylation induced by NF-κB and switches off the activated inflammatory genes. (*Right*) In COPD patients, cigarette smoke activates macrophages, as in normal subjects, but oxidative stress (acting through the formation of peroxynitrite) impairs HDAC2 activity. This amplifies the inflammatory response to NF-κB activation but also reduces the anti-inflammatory effect of corticosteroids because HDAC2 is now unable to reverse histone acetylation. MMP-9, matrix metalloproteinase-9.

of efficacy in asthma (45–47), and fail to suppress the secretion of inflammatory proteins, such as CXCL8 and matrix metalloproteinase-9, from macrophages (48, 49). The reduction in HDAC2 expression in COPD cells may therefore account for not only the amplification of inflammation but also the insensitivity to the anti-inflammatory effects of corticosteroids (13) (**Figure 5**).

MECHANISMS OF HISTONE DEACETYLASE REDUCTION

The reasons for the reduction in HDAC activity, particularly HDAC2, in COPD are not yet completely understood. However, there is increasing evidence that the reduction may be due

to inactivation of the enzyme by oxidative and nitrative stress (42, 50) (**Figure 6**). Oxidative stress is increased in COPD and increases with disease severity (51–53). Nitrative stress is also increased in the peripheral lungs of COPD patients (54), and there is increased expression of inducible nitric oxide synthase (iNOS) in small airways and the lung parenchyma (55). Oxidative and nitrative stress leads to the rapid formation of peroxynitrite, which nitrates selected tyrosine residues on certain proteins. HDAC2, but not other isoforms of HDAC, shows increased tyrosine nitration in macrophages and peripheral lungs of COPD patients, which is correlated with increased expression of CXCL8 (56). Oxidative and nitrative stress induces corticosteroid resistance in macrophage-like cells

Oxidative stress: an excessive number of reactive oxygen species compared with the number of antioxidant molecules

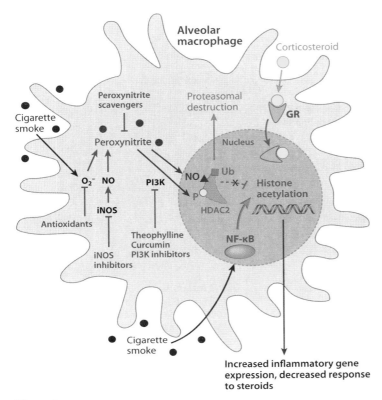

Figure 6

Possible mechanisms for decreased histone deacetylase (HDAC) activity and expression in chronic obstructive pulmonary disorder (COPD). Superoxide anions (O_2^-) and nitric oxide (NO) generated by cigarette smoke and inflammatory cells combine to form peroxynitrite. NO production from inflammatory cells is derived from inducible NO synthase (iNOS) in response to inflammatory stimuli. Peroxynitrite nitrates histone deacetylase 2 (HDAC2), possibly at a tyrosine (Tyr) residue within the catalytic site. This may inactivate HDAC2 and also lead to ubiquitination (Ub) of the enzyme that labels HDAC2 for degradation by the proteasome, resulting in its reduced expression. Oxidative stress also activates a phosphoinositide-3-kinase (PI3K) pathway that phosphorylates (P) and inactivates HDAC2. Loss of HDAC function then results in enhanced inflammatory gene expression and blocks the anti-inflammatory action of corticosteroids. HDAC function may be restored by antioxidants, iNOS inhibitors, or peroxynitrite scavengers, which reduce tyrosine nitration, or by theophylline, curcumin, or PI3K inhibitors, which restore HDAC function to normal (names of inhibitors are in *gray text*).

PI3K:
phosphoinositide-3-kinase

tivity of this enzyme but also leads to its ubiquitination, which marks it for degradation by the proteasome, resulting in the very low levels of HDAC2 protein in the lungs of patients with severe COPD (59). Oxidative stress also activates the phosphoinositide-3-kinase (PI3K) pathway, which results in phosphorylation of serine residues and inactivation of HDAC2 (60). The high level of oxidative/nitrative stress in COPD lungs, particularly as the disease progresses, may therefore result in increased tyrosine nitration, serine phosphorylation, and ubiquitination, leading to impaired HDAC2 catalytic activity and a reduction in protein levels, which thereby results in increased expression of inflammatory genes and impaired responses to corticosteroids. As discussed above, there is also a reduction in HDAC2 mRNA, indicating an additional reduction in transcriptional activity.

THERAPEUTIC IMPLICATIONS

Corticosteroid resistance in COPD is a major clinical barrier because steroids provide little clinical benefit and no alternative anti-inflammatory treatments are currently available (6, 7). Understanding the molecular basis for corticosteroid resistance in COPD provides several new therapeutic opportunities to reverse or bypass this resistance (**Figure 3**).

Antioxidants

Because oxidative/nitrative stress appears to be a mechanism that can lead to corticosteroid resistance, antioxidants and inhibitors of iNOS may be effective through inhibiting the generation of peroxynitrite. Currently available antioxidants, such as vitamins C and E and *N*-acetylcysteine, are not very potent and may not reduce oxidative stress in the lung sufficiently. Oral *N*-acetylcysteine failed to reduce exacerbations or the decline in lung function in COPD patients treated with or without inhaled corticosteroids (61). Selective iNOS inhibitors are now in clinical development and may be effective in reducing peroxynitrite

(U937 cells) in vitro, and this is mimicked by cigarette smoke extract and reversed by the antioxidant *N*-acetylcysteine. In vivo exposure of mice to cigarette smoke reduces HDAC activity in the lungs and induces a steroid-resistant neutrophilic inflammation (57, 58). Nitration of HDAC2 not only inactivates the catalytic ac-

formation (62) but may not prevent the generation of peroxynitrite from nitric oxide in cigarette smoke. More potent antioxidants and peroxynitrite scavenger drugs are now in development (63).

Theophylline

Theophylline, which has been used to treat airway diseases for more than 70 years, is an activator of HDACs. Low concentrations (10^{-6} M) increase HDAC activity and expression through a mechanism that is independent of phosphodiesterase inhibition or adenosine receptor antagonism, which together account for all the known side effects of theophylline (64). Low concentrations of theophylline are able to restore the activity and expression of HDAC2 to normal in alveolar macrophages in COPD patients and to restore the response of these cells to corticosteroids (65). This effect of theophylline is reversed by an HDAC inhibitor, confirming that the mechanism of action of theophylline is through HDAC activation. In a murine model of short-term cigarette smoke exposure to induce a steroid-resistant inflammation, low oral does of theophylline were effective in reducing lung inflammation only when combined with a corticosteroid (58). The clinical implications are that low doses of theophylline (that give a plasma concentration of 5–10 mg L^{-1}) may restore the responsiveness of COPD patients to corticosteroids, resulting in effective suppression of the inflammation and reduced progression of the disease. Clinical trials to test this idea are hampered by the facts that theophylline is inexpensive and research funding is difficult to obtain. The molecular mechanism whereby theophylline activates HDACs is currently unknown, but the activation appears to occur within the nucleus and through the inhibition of PI3K and is mimicked by a PI3K inhibitor, LY-294002. In the future, new drugs that activate HDAC2 selectively may be developed, especially as the molecular mechanisms of theophylline are further elucidated (66, 67).

Curcumin

Curcumin, a polyphenol found in curry powder, also reverses corticosteroid sensitivity–induced human monocytes by oxidative stress and cigarette smoke extract by restoring HDAC2 expression to normal (68). However, curcumin is very nonspecific and also inhibits HAT activity and NF-κB, making its effects difficult to interpret (69).

IMPLICATIONS FOR OTHER DISEASES

Oxidative/nitrative stress is increased in other inflammatory diseases, suggesting that reduction in HDAC activity may contribute to the amplification of inflammation and reduced responses to corticosteroids. Asthmatic patients who smoke have more severe asthma and show markedly reduced responses to corticosteroids (70). HDAC2 is markedly reduced in the airways of smoking asthmatic patients compared with nonsmoking asthmatics of a similar disease severity (71, 72). We have also shown that HDAC activity is also impaired in patients with severe asthma who have a high level of oxidative stress and who require high doses of corticosteroids for adequate control (73, 74). Oxidative stress is also increased in cystic fibrosis and interstitial lung disease, which are diseases that are resistant to corticosteroids (75–77), so it may be possible in the future to reverse corticosteroid insensitivity in these diseases by restoring HDAC2 activity. It is likely that in other severe inflammatory diseases, such as rheumatoid arthritis and inflammatory bowel disease, the increased oxidative stress may reduce HDAC2 activity, leading to increased inflammation and reduced responsiveness to corticosteroids.

CONCLUSIONS

In COPD patients, there is a reduction in HDAC activity in peripheral lung, airways, and alveolar macrophages, which worsens as the disease becomes more severe. This may account for the increased pulmonary inflammation and resistance to corticosteroids in COPD. There appears to be a selective reduction in HDAC2 expression, and this reduction may be due to oxidative and nitrative stress, which is increased in COPD lungs. Therapeutic options aimed at increasing HDAC activity, such as antioxidants, iNOS inhibitors, and theophylline, may be beneficial. Reduced HDAC activity may also occur in other inflammatory diseases, including asthmatic patients who smoke and those with severe disease, rheumatoid arthritis, or inflammatory bowel disease. This area of research may lead to the development of novel anti-inflammatory therapies aimed at increasing HDAC2 activity in the future.

SUMMARY POINTS

1. COPD is associated with a chronic inflammation of the lung parenchyma and small airways and a high level of oxidative stress.

2. Corticosteroids, which are effective in treating inflammatory lung diseases such as asthma, have no anti-inflammatory effects in COPD, making this disease difficult to treat with currently available therapies.

3. The anti-inflammatory effects of corticosteroids are mediated via the recruitment of the nuclear enzyme HDAC2, which deacetylates the hyperacetylated histone residues associated with inflammatory gene activation, thereby resulting in inflammatory gene suppression.

4. HDAC2 activity and expression are markedly reduced in COPD lungs, airways, and alveolar macrophages, which may account for the amplified inflammation and corticosteroid resistance seen in these patients.

5. The reduction in HDAC2 is secondary to oxidative and nitrative stress, which leads to tyrosine nitration, phosphorylation, and ubiquitination of HDAC2, resulting in loss of its activity and its degradation.

6. The reduction of HDAC2 by oxidative stress may be reversed by antioxidants, peroxynitrite scavengers, theophylline, curcumin, and PI3K inhibitors.

7. Theophylline reverses corticosteroid resistance on COPD cells and in smoking mice in vivo via inhibition of the PI3K pathway, which is activated by oxidative stress to restore HDAC2 activity.

8. These mechanisms may also be relevant in other inflammatory lung diseases, such as severe asthma, smoking asthma, cystic fibrosis, and interstitial lung disease, in which there is a high degree of oxidative stress and a poor response to corticosteroids.

FUTURE ISSUES

1. The mechanisms that cause a reduction in HDAC2 messenger RNA need to be elucidated by investigating the regulation of the HDAC2 gene.

2. The possibility of genetic polymorphisms in the oxidative stress–PI3K–HDAC2 pathways needs to be explored to determine whether they may contribute to the increased susceptibility of COPD patients to the effects of cigarette smoke compared with the majority of smokers who have normal lung function.

3. The effects of drugs other than theophylline that inhibit PI3K should be explored, and the specific isoenzyme of PI3K responsive for the downstream phosphorylation of HDAC2 needs to be identified.

4. It is important to establish whether low-dose theophylline is able to reverse corticosteroid resistance in patients with COPD in a small proof-of-concept study followed, if indicated, by a large prolonged clinical trial to determine whether there are any reductions in symptoms, exacerbations, and disease progression.

5. The role of HDAC2 needs to be explored in other steroid-insensitive chronic inflammatory diseases, including severe asthma, smoking asthma, cystic fibrosis, interstitial lung disease, as well as nonpulmonary inflammatory diseases, such as rheumatoid arthritis and inflammatory bowel disease.

6. The potential for theophylline and other drugs that may reverse corticosteroid resistance should be evaluated in these other inflammatory diseases.

DISCLOSURE STATEMENT

The author is not aware of any biases that might be perceived as affecting the objectivity of this review.

LITERATURE CITED

1. Barnes PJ. 2000. Chronic obstructive pulmonary disease. *New Engl. J. Med.* 343:269–80
2. Rabe KF, Hurd S, Anzueto A, Barnes PJ, Buist SA, et al. 2007. Global strategy for the diagnosis, management, and prevention of COPD—2006 update. *Am. J. Respir. Crit. Care Med.* 176:532–55
3. Barnes PJ. 2007. Chronic obstructive pulmonary disease: a growing but neglected epidemic. *PLoS Med.* 4:e112
4. Mannino DM, Buist AS. 2007. Global burden of COPD: risk factors, prevalence, and future trends. *Lancet* 370:765–73
5. Barnes PJ, Shapiro SD, Pauwels RA. 2003. Chronic obstructive pulmonary disease: molecular and cellular mechanisms. *Eur. Respir. J.* 22:672–88
6. Barnes PJ, Hansel TT. 2004. Prospects for new drugs for chronic obstructive pulmonary disease. *Lancet* 364:985–96
7. Barnes PJ. 2008. Frontrunners in novel pharmacotherapy of COPD. *Curr. Opin. Pharmacol.* 8:300–7
8. Lokke A, Lange P, Scharling H, Fabricius P, Vestbo J. 2006. Developing COPD: a 25 year follow up study of the general population. *Thorax* 61:935–39
9. Gamble E, Grootendorst DC, Hattotuwa K, O'Shaughnessy T, Ram FS, et al. 2007. Airway mucosal inflammation in COPD is similar in smokers and ex-smokers: a pooled analysis. *Eur. Respir. J.* 30:467–71
10. Willemse BW, ten Hacken NH, Rutgers B, Lesman-Leegte IG, Postma DS, Timens W. 2005. Effect of 1-year smoking cessation on airway inflammation in COPD and asymptomatic smokers. *Eur. Respir. J.* 26:835–45
11. Barnes PJ, Stockley RA. 2005. COPD: current therapeutic interventions and future approaches. *Eur. Respir. J.* 25:1084–106
12. Barnes PJ. 2006. How corticosteroids control inflammation. *Br. J. Pharmacol.* 148:245–54
13. Barnes PJ. 2006. Reduced histone deacetylase in COPD: clinical implications. *Chest* 129:151–55
14. O'Donnell DE, Laveneziana P. 2007. Dyspnea and activity limitation in COPD: mechanical factors. *COPD* 4:225–36
15. Hogg JC. 2004. Pathophysiology of airflow limitation in chronic obstructive pulmonary disease. *Lancet* 364:709–21

16. Hogg JC, Chu F, Utokaparch S, Woods R, Elliott WM, et al. 2004. The nature of small-airway obstruction in chronic obstructive pulmonary disease. *New Engl. J. Med.* 350:2645–53

17. Barnes PJ. 2004. Macrophages as orchestrators of COPD. *COPD* 1:59–70

18. Curtis JL, Freeman CM, Hogg JC. 2007. The immunopathogenesis of chronic obstructive pulmonary disease: insights from recent research. *Proc. Am. Thorac. Soc.* 4:512–21

19. Barnes PJ. 2008. Immunology of asthma and chronic obstructive pulmonary disease. *Nat. Immunol. Rev.* 8:183–92

20. Barnes PJ. 2004. Mediators of chronic obstructive pulmonary disease. *Pharm. Rev.* 56:515–48

21. Barnes PJ. 2008. Cytokine networks in asthma and chronic obstructive pulmonary disease. *J. Clin. Investig.* In press

22. Celli BR, Barnes PJ. 2007. Exacerbations of chronic obstructive pulmonary disease. *Eur. Respir. J.* 29:1224–38

23. Wilkinson TM, Hurst JR, Perera WR, Wilks M, Donaldson GC, Wedzicha JA. 2006. Effect of interactions between lower airway bacterial and rhinoviral infection in exacerbations of COPD. *Chest* 129:317–24

24. Silverman EK. 2006. Progress in chronic obstructive pulmonary disease genetics. *Proc. Am. Thorac. Soc.* 3:405–8

25. Donnelly LE, Barnes PJ. 2006. Chemokine receptors as therapeutic targets in chronic obstructive pulmonary disease. *Trends Pharmacol. Sci.* 27:546–53

26. Barnes PJ. 2006. Transcription factors in airway diseases. *Lab. Investig.* 86:867–72

27. Di Stefano A, Caramori G, Capelli A, Lusuardi M, Gnemmi I, et al. 2002. Increased expression of NF-kB in bronchial biopsies from smokers and patients with COPD. *Eur. Resp. J.* 20:556–63

28. Caramori G, Romagnoli M, Casolari P, Bellettato C, Casoni G, et al. 2003. Nuclear localisation of p65 in sputum macrophages but not in sputum neutrophils during COPD exacerbations. *Thorax* 58:348–51

29. Kouzarides T. 2007. Chromatin modifications and their function. *Cell* 128:693–705

30. Shahbazian MD, Grunstein M. 2007. Functions of site-specific histone acetylation and deacetylation. *Annu. Rev. Biochem.* 76:75–100

31. Ito K, Barnes PJ, Adcock IM. 2000. Glucocorticoid receptor recruitment of histone deacetylase 2 inhibits IL-1b-induced histone H4 acetylation on lysines 8 and 12. *Mol. Cell. Biol.* 20:6891–903

32. Barnes PJ, Adcock IM, Ito K. 2005. Histone acetylation and deacetylation: importance in inflammatory lung diseases. *Eur. Respir. J.* 25:552–63

33. Adcock IM, Ford P, Barnes PJ, Ito K. 2006. Epigenetics and airways disease. *Respir. Res.* 7:21

34. Ito K, Ito M, Elliott WM, Cosio B, Caramori G, et al. 2005. Decreased histone deacetylase activity in chronic obstructive pulmonary disease. *New Engl. J. Med.* 352:1967–76

35. Ito K, Caramori G, Lim S, Oates T, Chung KF, et al. 2002. Expression and activity of histone deacetylases (HDACs) in human asthmatic airways. *Am. J. Respir. Crit. Care Med.* 166:392–96

36. Thiagalingam S, Cheng KH, Lee HJ, Mineva N, Thiagalingam A, Ponte JF. 2003. Histone deacetylases: unique players in shaping the epigenetic histone code. *Ann. N.Y. Acad. Sci.* 983:84–100

37. Ito K, Yamamura S, Essilfie-Quaye S, Cosio B, Ito M, et al. 2006. Histone deacetylase 2-mediated deacetylation of the glucocorticoid receptor enables NF-kB suppression. *J. Exp. Med.* 203:7–13

38. Popov VM, Wang C, Shirley LA, Rosenberg A, Li S, et al. 2007. The functional significance of nuclear receptor acetylation. *Steroids* 72:221–30

39. Ito K, Lim S, Caramori G, Chung KF, Barnes PJ, Adcock IM. 2001. Cigarette smoking reduces histone deacetylase 2 expression, enhances cytokine expression and inhibits glucocorticoid actions in alveolar macrophages. *FASEB J.* 15:1100–2

40. Szulakowski P, Crowther AJ, Jimenez LA, Donaldson K, Mayer R, et al. 2006. The effect of smoking on the transcriptional regulation of lung inflammation in patients with chronic obstructive pulmonary disease. *Am. J. Respir. Crit. Care Med.* 174:41–50

41. Barnes PJ, Adcock IM. 2003. How do corticosteroids work in asthma? *Ann. Intern. Med.* 139:359–70

42. Barnes PJ, Ito K, Adcock IM. 2004. A mechanism of corticosteroid resistance in COPD: inactivation of histone deacetylase. *Lancet* 363:731–33

43. Yang IA, Fong KM, Sim EH, Black PN, Lasserson TJ. 2007. Inhaled corticosteroids for stable chronic obstructive pulmonary disease. *Cochrane Database Syst. Rev.* CD002991

44. Calverley PM, Anderson JA, Celli B, Ferguson GT, Jenkins C, et al. 2007. Salmeterol and fluticasone propionate and survival in chronic obstructive pulmonary disease. *N. Engl. J. Med.* 356:775–89

45. Keatings VM, Jatakanon A, Worsdell YM, Barnes PJ. 1997. Effects of inhaled and oral glucocorticoids on inflammatory indices in asthma and COPD. *Am. J. Respir. Crit. Care Med.* 155:542–48

46. Culpitt SV, Nightingale JA, Barnes PJ. 1999. Effect of high dose inhaled steroid on cells, cytokines and proteases in induced sputum in chronic obstructive pulmonary disease. *Am. J. Respir. Crit. Care Med.* 160:1635–39

47. Loppow D, Schleiss MB, Kanniess F, Taube C, Jorres RA, Magnussen H. 2001. In patients with chronic bronchitis a four week trial with inhaled steroids does not attenuate airway inflammation. *Respir. Med.* 95:115–21

48. Culpitt SV, Rogers DF, Shah P, de Matos C, Russell RE, et al. 2003. Impaired inhibition by dexamethasone of cytokine release by alveolar macrophages from patients with chronic obstructive pulmonary disease. *Am. J. Respir. Crit. Care Med.* 167:24–31

49. Russell RE, Culpitt SV, DeMatos C, Donnelly L, Smith M, et al. 2002. Release and activity of matrix metalloproteinase-9 and tissue inhibitor of metalloproteinase-1 by alveolar macrophages from patients with chronic obstructive pulmonary disease. *Am. J. Respir. Cell Mol. Biol.* 26:602–9

50. Rahman I, Marwick J, Kirkham P. 2004. Redox modulation of chromatin remodeling: impact on histone acetylation and deacetylation, NF-κB and proinflammatory gene expression. *Biochem. Pharmacol.* 68:1255–67

51. Montuschi P, Collins JV, Ciabattoni G, Lazzeri N, Corradi M, et al. 2000. Exhaled 8-isoprostane as an in vivo biomarker of lung oxidative stress in patients with COPD and healthy smokers. *Am. J. Respir. Crit. Care Med.* 162:1175–77

52. Paredi P, Kharitonov SA, Leak D, Ward S, Cramer D, Barnes PJ. 2000. Exhaled ethane, a marker of lipid peroxidation, is elevated in chronic obstructive pulmonary disease. *Am. J. Respir. Crit. Care Med.* 162:369–73

53. Bowler RP, Barnes PJ, Crapo JD. 2004. The role of oxidative stress in chronic obstructive pulmonary disease. *COPD* 2:255–77

54. Brindicci C, Ito K, Resta O, Pride NB, Barnes PJ, Kharitonov SA. 2005. Exhaled nitric oxide from lung periphery is increased in COPD. *Eur. Respir. J.* 26:52–59

55. Ricciardolo FL, Caramori G, Ito K, Capelli A, Brun P, et al. 2005. Nitrosative stress in the bronchial mucosa of severe chronic obstructive pulmonary disease. *J. Allergy Clin. Immunol.* 116:1028–35

56. Ito K, Tomita T, Barnes PJ, Adcock IM. 2004. Oxidative stress reduces histone deacetylase (HDAC)2 activity and enhances IL-8 gene expression: role of tyrosine nitration. *Biochem. Biophys. Res. Commun.* 315:240–45

57. Marwick JA, Kirkham PA, Stevenson CS, Danahay H, Giddings J, et al. 2004. Cigarette smoke alters chromatin remodelling and induces proinflammatory genes in rat lungs. *Am. J. Respir. Cell Mol. Biol.* 31:633–42

58. Fox JC, Spicer D, Ito K, Barnes PJ, Fitzgerald MF. 2007. Oral or inhaled corticosteroid combination therapy with low dose theophylline reverses corticosteroid insensitivity in a smoking mouse model. *Proc. Am. Thorac. Soc.* 2:A637 (Abstr.)

59. Osoata G, Adcock IM, Barnes PJ, Ito K. 2005. Oxidative stress causes HDAC2 reduction by nitration, ubiquitinylation and proteasomal degradation. *Proc. Am. Thorac. Soc.* 2:A755 (Abstr.)

60. Failla M, To Y, Ito M, Adcock IM, Barnes PJ, Ito K. 2007. Oxidative stress-induced PI3-kinase activation reduces HDAC activity and is inhibited by theophylline. *Proc. Am. Thorac. Soc.* 2:A45 (Abstr.)

61. Decramer M, Rutten-van Molken M, Dekhuijzen PN, Troosters T, van Herwaarden C, et al. 2005. Effects of N-acetylcysteine on outcomes in chronic obstructive pulmonary disease: a randomised placebo-controlled trial. *Lancet* 365:1552–60

62. Hansel TT, Kharitonov SA, Donnelly LE, Erin EM, Currie MG, et al. 2003. A selective inhibitor of inducible nitric oxide synthase inhibits exhaled breath nitric oxide in healthy volunteers and asthmatics. *FASEB J.* 17:1298–300

63. Kirkham P, Rahman I. 2006. Oxidative stress in asthma and COPD: antioxidants as a therapeutic strategy. *Pharmacol. Ther.* 111:476–94

64. Ito K, Lim S, Caramori G, Cosio B, Chung KF, et al. 2002. A molecular mechanism of action of theophylline: induction of histone deacetylase activity to decrease inflammatory gene expression. *Proc. Natl. Acad. Sci. USA* 99:8921–26

65. Cosio BG, Tsaprouni L, Ito K, Jazrawi E, Adcock IM, Barnes PJ. 2004. Theophylline restores histone deacetylase activity and steroid responses in COPD macrophages. *J. Exp. Med.* 200:689–95

66. Barnes PJ. 2005. Targeting histone deacetylase 2 in chronic obstructive pulmonary disease treatment. *Expert Opin. Ther. Targets* 9:1111–21

67. Marwick JA, Ito K, Adcock IM, Kirkham PA. 2007. Oxidative stress and steroid resistance in asthma and COPD: pharmacological manipulation of HDAC-2 as a therapeutic strategy. *Expert Opin. Ther. Targets* 11:745–55

68. Meja KK, Rajendrasozhan S, Adenuga D, Biswas SK, Sundar IK, et al. 2008. Curcumin restores corticosteroid function in monocytes exposed to oxidants by maintaining HDAC2. *Am. J. Respir. Cell Mol. Biol.* 39:312–23

69. Chen Y, Shu W, Chen W, Wu Q, Liu H, Cui G. 2007. Curcumin, both histone deacetylase and p300/CBP-specific inhibitor, represses the activity of nuclear factor kappa B and Notch 1 in Raji cells. *Basic Clin. Pharmacol. Toxicol.* 101:427–33

70. Thomson NC, Chaudhuri R, Livingston E. 2004. Asthma and cigarette smoking. *Eur. Respir. J.* 24:822–33

71. Murahidy A, Ito M, Adcock IM, Barnes PJ, Ito K. 2005. Reduction is histone deacetylase expression and activity in smoking asthmatics: a mechanism of steroid resistance. *Proc. Am. Thorac. Soc.* 2:A889 (Abstr.)

72. Ahmad T, Barnes PJ, Adcock IM. 2008. Overcoming steroid insensitivity in smoking asthmatics. *Curr. Opin. Investig. Drugs* 9:470–77

73. Hew M, Bhavsar P, Torrego A, Meah S, Khorasani N, et al. 2006. Relative corticosteroid insensitivity of peripheral blood mononuclear cells in severe asthma. *Am. J. Respir. Crit. Care Med.* 174:134–41

74. Bhavsar P, Hew M, Khorasani N, Torrego A, Barnes PJ, et al. 2008. Relative corticosteroid insensitivity of alveolar macrophages in severe asthma compared to non-severe asthma. *Thorax* 63:784–90

75. Montuschi P, Toni GC, Paredi P, Pantelidis P, du Bois RM, et al. 1998. 8-Isoprostane as a biomarker of oxidative stress in interstitial lung diseases. *Am. J. Respir. Crit. Care Med.* 158:1524–27

76. Paredi P, Kharitonov SA, Leak D, Shah PL, Cramer D, et al. 2000. Exhaled ethane is elevated in cystic fibrosis and correlates with carbon monoxide levels and airway obstruction. *Am. J. Respir. Crit. Care Med.* 161:1247–51

77. Montuschi P, Kharitonov SA, Ciabattoni G, Corradi M, van Rensen L, et al. 2000. Exhaled 8-isoprostane as a new noninvasive biomarker of oxidative stress in cystic fibrosis. *Thorax* 55:205–9

Aspirin-Sensitive Respiratory Disease

Sophie P. Farooque and Tak H. Lee

King's College London, MRC & Asthma UK Centre in Allergic Mechanisms of Asthma, Guy's Hospital, London SE1 9RT, England; email: tak.lee@kcl.ac.uk

Annu. Rev. Physiol. 2009. 71:465–87

The *Annual Review of Physiology* is online at physiol.annualreviews.org

This article's doi:
10.1146/annurev.physiol.010908.163114

Copyright © 2009 by Annual Reviews.
All rights reserved

0066-4278/09/0315-0465$20.00

Key Words

asthma, COX-1 and COX-2 inhibitors, cysteinyl leukotrienes, nasal polyps, PGE_2

Abstract

Aspirin-sensitive respiratory disease (ASRD) is a condition character-ized by persistent and often severe inflammation of the upper and lower respiratory tracts. Patients develop chronic eosinophilic rhinosinusi-tis, nasal polyposis, and asthma. The ingestion of aspirin and other cyclooxygenase-1 (COX-1) inhibitors induces exacerbations of airway disease that may be life-threatening. Thus, aspirin sensitivity is a phe-notypic marker for the syndrome, yet nearly all affected individuals can be desensitized by the administration of graded doses of aspirin, leading to long-term clinical benefits. Patients with aspirin sensitivity are often able to tolerate selective COX-2 inhibitors. The pathogenesis of ASRD is underpinned by abnormalities in eicosanoid biosynthesis and eicosanoid receptor expression coupled with intense mast cell and eosinophilic infiltration of the entire respiratory tract. This review fo-cuses on the molecular, cellular, and biochemical abnormalities char-acterizing ASRD and highlights unanswered questions in the literature and potential future areas of investigation.

ASRD: aspirin-sensitive respiratory disease

NSAIDS: nonsteroidal anti-inflammatory drugs

FESS: functional endoscopic sinus surgery

INTRODUCTION

On the tenth of August of 1897, Felix Hoffmann, a pharmacist working at the Bayer factory in Germany, succeeded in synthesizing acetylsalicylic acid. By March 1899 this compound had been named aspirin and was being marketed by Bayer. Aspirin has since become the most widely used medicine of all time (1, 2). In 1922, Widal described the association of aspirin sensitivity, asthma, and nasal polyps. He also pioneered the first aspirin challenges and desensitization (3). However, aspirin-sensitive respiratory disease (ASRD) became widely recognized only after Samter & Beer published two articles in the late 1960s (e.g., Reference 4), and the combination of aspirin sensitivity, asthma, and nasal polyposis became known as Samter's triad.

Today ASRD is recognized to be an aggressive phenotype of airway disease that often runs a protracted course. Compared with aspirin-tolerant individuals, patients with ASRD are more likely to experience irreversible airflow obstruction (5), to suffer frequent exacerbations (6), to be diagnosed with severe asthma and be prescribed high-dose oral steroids by their physicians, and more often to require intubation for their severe asthma (5).

DEFINITION

The clinical syndrome of ASRD is characterized by a combination of chronic rhinosinusitis progressing to chronic hyperplastic eosinophilic sinusitis, moderate-to-severe asthma, and nasal polyposis. Previous exposure to aspirin is not a risk factor for the development of ASRD, but once established, ingestion of aspirin or other nonsteroidal anti-inflammatory drugs (NSAIDS) induces an acute worsening of rhinosinusitis and asthma. Aspirin sensitivity therefore is a valuable phenotypic marker for a disease that often responds poorly to steroids and that is characterized by intractable inflammation of the upper and lower respiratory tracts.

CLINICAL FEATURES AND NATURAL HISTORY

ASRD has high morbidity, and symptoms are almost always lifelong once ASRD develops. There is no discernable relationship to an atopic diathesis (5), and ASRD may develop in patients with no prior history of IgE-mediated respiratory disease or manifest in individuals who already have allergic rhinitis and/or allergic asthma. However, unlike atopic asthma it is rare in childhood, with peak ages of onset between 29 and 34 years observed in two large studies documenting the natural history of the disease (7, 8). Typically the condition progresses from the upper to the lower respiratory tracts (4, 8), and approximately half of all patients relate a viral upper respiratory tract infection as the inciting event (8).

Like all adult asthma, ASRD affects women more frequently than men; in women the onset of symptoms is often earlier, and the disease is more aggressive (8). Persistent rhinorrhea and nasal congestion are usually the earliest symptoms. In an average patient, within two years of onset the disease progresses to the lower respiratory tract with asthma, and within four years aspirin sensitivity and nasal polyposis develop (8). Approximately 20% of aspirin-sensitive patients have evidence of generalized mast cell degranulation and experience urticaria (20%) or angioedema (8%) following aspirin challenge (8). In some series the percentage of patients with systemic reactions following aspirin challenge has been reported to be greater than 50%, suggesting heterogeneity within the phenotype (9).

Aspirin-sensitive individuals average one episode of sinusitis every ten weeks (7) and have marked thickening of the basement membrane suggestive of sinusoidal mucosal remodeling (10). At the time of surgery, subjects typically have a larger volume of polyp tissue than do aspirin-tolerant patients, and surgery often results in only a transient relief of symptoms. Six months following functional endoscopic sinus surgery (FESS) and polypectomy, patients with ASRD have significantly higher rates of symptom recurrence (nasal blockage, facial pain, postnasal drip, and anosmia) in addition to regrowth of nasal polyps (11). On average patients with ASRD will require ten times as many resections of their nasal polyps compared with their non-aspirin-sensitive counterparts (11), and the frequency of FESS polypectomy in ASRD correlates with the density of mast cells and eosinophils in polyp tissue (12). FESS leads to a reduction in lower respiratory tract symptoms with decreased inhaled corticosteroid use and improved asthma symptom scores at six months and one year follow-up (13).

PREVALENCE

The exact prevalence of ASRD in adults is unclear because many asthmatics routinely avoid NSAIDS and patients who experience a mild worsening of their asthma after ingesting NSAIDS may not correlate this with aspirin sensitivity. The prevalence of ASRD also varies depending on the population studied (severe asthmatics, patients with nasal polyps and chronic rhinosinusitis, patients with rhinitis alone) and the technique used to make the diagnosis (population-based questionnaire, physician assessment, or aspirin challenge). A higher prevalence of ASRD is invariably found when asthmatic patients are challenged with aspirin rather than when patient questionnaires are relied upon to make the diagnosis. If patient questionnaires are used as the sole means of assessing for the presence of aspirin sensitivity, the prevalence of ASRD ranges from 0.6% (14) to 2.5% (15) in the general population and from 3.8% (16) to 11% (15) in asthmatics. On the basis of oral aspirin challenges in selected asthmatic populations, the prevalence rises to approximately 20% (17). Conversely, in a study of more than 6000 patients with nasal polyps, it was estimated that 14% of individuals were also aspirin sensitive (18).

It is important to recognize ASRD clinically because it is associated with near-fatal asthma (8, 19), although it remains uncertain if this is because of accidental/unknown NSAID inges-

tion or because of the severity of underlying disease. In a survey of 500 patients with ASRD, 15% of patients who were diagnosed were unaware of their aspirin-sensitive status prior to challenge (8). Estimates vary, but an incidence of ASRD in 14% of rapid-onset, near-fatal/fatal asthma attacks has been reported (20), and it is estimated that between 11% and 15% of all intensive-care admissions with asthma involve aspirin-sensitive patients (21, 22). Lack of patient and doctor awareness, aggravated by limited access to aspirin challenges, therefore may have potentially grave consequences.

THE DIAGNOSIS OF ASRD

The mechanism of aspirin sensitivity remains unclear, but most commentators agree that reaction to aspirin is not immunologically mediated. There is no evidence of an IgE-mediated response to aspirin, and patients do not have positive skin prick tests to aspirin or other NSAIDS. The fact that on first exposure the effect of aspirin is often mimicked by a wide variety of other NSAIDS precludes drug hapten-antibody recognition and makes the scenario of immunological reactivity highly unlikely. Furthermore, despite avoidance of aspirin, chronic inflammation of both the upper and lower respiratory tracts is ongoing and progressive, confirming that the ingestion of NSAIDS is not responsible for inducing the disease. Szczeklik et al. (23) first proposed a nonallergic mechanism of ASRD more than 30 years ago.

Aspirin causes time-dependent covalent modification and irreversible inhibition of cyclooxygenase-1 (COX-1) and competes directly with arachidonic acid for binding to the cyclooxygenase site (24). Serine 530 is the site of acetylation of ovine COX-1 by aspirin (24). X-ray crystallographic studies have established that when serine 530 is acetylated by aspirin, the acetyl group protrudes into the cyclooxygenase active site and interferes with arachidonate binding (24). There is a correlation between in vitro potency of COX-1 inhibition by a drug and the risk of precipitation of aspirin-induced asthma (23). Aspirin also

COX: cyclooxygenase

PG: prostaglandin

5-LO:
5-lipooxygenase

CysLT: cysteinyl
leukotriene

CysLTR: cysteinyl
leukotriene receptor

EP: E-prostanoid

LX: lipoxin

acetylates cyclooxygenase-2 (COX-2) but does not block substrate binding. Hence the resulting enzyme maintains its cyclooxygenase activity and is still able to oxygenate arachidonic acid to exclusively form 15-hydroxy-5,8,11,13-eicosatetraenoic acid (15-HETE), which then undergoes transcellular biosynthesis to produce 15-epi-LXA$_4$ (lipoxin A$_4$) (25).

Presently, a diagnosis of aspirin sensitivity can be definitively established only by aspirin challenge in which patients receive increasing doses of oral aspirin, or inhaled or nasal lysine aspirin. As well as being time consuming, aspirin challenges are not without risk, and greater than a third of patients have been reported to experience late reactions following bronchial aspirin challenge (26). Flow-cytometric determination of basophil activation induced by aspirin (27) and aspirin-triggered 15-HETE generation in peripheral blood leukocytes (PBLs) may be highly specific and sensitive in vitro diagnostic tests (28), but this possibility has to be fully validated.

THE PATHOPHYSIOLOGY OF ASRD

Mediators and Mechanisms

Unlike preformed mediators such as histamine that are released upon mast cell or basophil stimulation and cytokines that typically require gene transcription and mRNA translation for production, the leukotrienes (LTs) and the prostaglandins (PGs) are synthesized de novo upon cellular activation.

Once released from the cell membrane by cytosolic phospholipase A$_2$ (cPLA$_2$), unesterified arachidonic acid may be metabolized through two main pathways: the COX pathway, which yields the prostanoids, or the 5-lipoxygenase (5-LO) pathway to form the unstable intermediary LTA$_4$. LTA$_4$ is metabolized to either LTB$_4$ or the cysteinyl leukotrienes (CysLTs), depending on distal enzymes differentially expressed in leukocytes (29) (**Figure 1**).

The pathophysiology of ASRD is characterized by abnormalities in the biosynthesis of eicosanoid mediators and eicosanoid receptor expression. The biochemical hallmark of ASRD is enhanced CysLT production both at baseline (30, 31) and following aspirin challenge (32, 33). The nasal and bronchial mucosa of aspirin-sensitive patients is rich with eosinophils and mast cells (34, 35), both abundant sources of CysLTs (36). Furthermore, there is upregulation of the cysteinyl leukotriene 1 receptor (CysLT$_1$R) on target cells in ASRD (37), and one consequence of this upregulation may be enhanced end-organ responsiveness to CysLTs.

Conversely, on inflammatory cells, there is downregulation of the E-prostanoid 2 (EP2) receptor, which binds the anti-inflammatory prostanoid prostaglandin E$_2$ (PGE$_2$) (38). There is evidence that PGE$_2$ ameliorates aspirin-induced disease, at least partly by inhibiting excessive CysLT synthesis (39, 40). The effect of reduced EP2 receptor expression in ASRD on downstream signaling pathways is as yet unclear. It is possible that the halting of the inflammatory process by PGE$_2$ is critically impaired owing either to deficient PGE$_2$ production and/or to abnormal PGE$_2$ receptor expression in aspirin-sensitive patients. Reduced synthesis of anti-inflammatory lipoxins (LXs) also correlates with ASRD (41–43). LXs are functional antagonists of the CysLTs, and therefore diminished capacity to generate LXs may contribute to the uncontrolled, chronic inflammation that characterizes this phenotype of airway disease.

Inflammatory Pathways

Biopsy studies from the respiratory tract in aspirin-sensitive patients show elevated numbers of eosinophils and mast cells as compared with their aspirin-tolerant counterparts (34, 35). Both mast cells and eosinophils are critical effector cells, important cellular sources of CysLTs, and are responsible for perpetuating the devastating inflammation seen in ASRD (36).

Extensive infiltration of eosinophils and mast cells was noted in bronchial biopsies from aspirin-sensitive patients ten years ago (34).

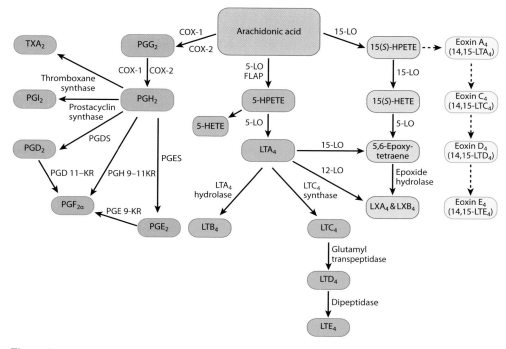

Figure 1

Schematic representation of the metabolism of arachidonic acid by the cyclooxygenase and the 5- and 15-lipoxygenase (LO) pathways. Once released from the cell membrane by cytosolic phospholipase A_2 (cPLA$_2$), arachidonic acid may be metabolized through two main pathways: the cyclooxygenase (COX) pathway to yield the prostanoids or the 5-lipoxygenase (5-LO) pathway to form the unstable intermediary LTA$_4$, which, then depending on distal enzymes differentially expressed in leukocytes, is metabolized to either LTB$_4$ or the cysteinyl leukotrienes (CysLTs). COX-1 and COX-2 catalyze a bifold reaction in which first arachidonic acid is oxidized to prostaglandin G_2 (PGG$_2$) and then PGG$_2$ is reduced to prostaglandin H_2 (PGH$_2$). PGH$_2$ serves as a substrate for the prostaglandin synthase (PGS) or prostaglandin ketoreductase (PGKR) enzymes, which are responsible for the production of the five principal bioactive prostanoids generated in vivo: PGE$_2$, PGF$_{2\alpha}$, PGD$_2$, PGI$_2$ (prostacyclin), and TXA$_2$ (thromboxane). 15-Lipoxygenase (15-LO) catalyzes the hydroperoxidation of arachidonic acid, with the insertion of one molecule of oxygen at position 15 to form 15-HPETE. The products of arachidonic acid also include lipoxins (LXs) and eoxins (EXs). The dotted arrows refer to an as-yet-unconfirmed pathway for EX synthesis (124).

Similar changes have since been observed in the upper airways with inferior turbinate biopsies from patients with ASRD demonstrating intense inflammation and elevated numbers of total and activated eosinophils, mast cells, and T lymphocytes (35) and nasal polyps from aspirin-sensitive patients revealing substantial eosinophilic and mast cell infiltration compared with aspirin-tolerant patients (44). Aspirin-sensitive patients show a significant increase in the mobilization of leukocyte and eosinophil progenitor cells from the bone mar-

row for up to 20 h after bronchial aspirin challenge (9). This increase, coupled with decreased apoptosis of inflammatory cells in the airway, is a likely contributory factor to the persistent inflammation characterizing the syndrome (44).

Mast cells are found in three distinct microsites in the airways of asthmatic patients: the bronchial smooth muscle, the bronchial epithelium, and the airway mucus glands (45). In ASRD, mast cells are seen in the epithelial cell layer but not in the mucosa of nasal polyps (46). In aspirin-sensitive patients,

ICAM-1: intercellular adhesion molecule-1

RV: rhinovirus

AHR: airway hyperresponsiveness

BAL: bronchoalveolar lavage

significantly higher levels of the mast cell chemoattractant stem cell factor (SCF) have been detected in both mRNA and protein levels from cultured epithelial cells taken from nasal polyp explants (12). The mechanism responsible for an increased expression of SCF in nasal polyp epithelial cells from aspirin-sensitive patients is not known. But because SCF promotes mast cell survival, acts as a major growth and differentiation factor (48, 49), and is chemotactic for human mast cells (50), its upregulation may partially explain the increased numbers of mast cells seen in ASRD and their ongoing activation.

The epithelium in asthma is particularly susceptible to infection with respiratory viruses. Rhinovirus (RV) is able to preferentially invade the bronchial epithelium in asthmatics, using intercellular adhesion molecule-1 (ICAM-1) and very low-density lipoprotein receptors to enter the cell (51). The release of tryptase by mast cells upregulates ICAM-1 expression in airway epithelium (52). Enhanced expression of ICAM-1 as well as of the adhesion molecule vascular cell adhesion molecule-1 (VCAM-1) and very late-activation antigen-4 (VLA-4) has been demonstrated in nasal polyps from aspirin-sensitive patients (53). In individuals who are clinically free of symptoms of a respiratory tract infection, human RV is found in the bronchial mucosa in significantly higher numbers of asthmatics (73%) versus normal controls (22%) (54). Furthermore, there is profound impairment of virus-induced interferon-β (IFN-β) mRNA expression in asthmatic cultures, and this leads to poor clearance of virus and apoptosis by epithelial cells (55). Investigators have observed a correlation between sequestering RV and diminished pulmonary function, eosinophilic inflammation of the bronchial mucosa, and serum eosinophilia (54).

Hence infective exacerbations in asthma arise on a background of impaired epithelial host defense (55), and chronic infection with RV is associated with clinically worse disease (54). ASRD is an acquired condition, and patients will often describe "a cold that never cleared up" as heralding the onset of the syndrome. This observation may be highly significant and has led commentators to propose that a chronic viral respiratory tract infection may be the underlying event that both incites and perpetuates the inflammatory cascade in aspirin sensitivity (56). Presently the ability of the airways in patients with ASRD to clear viral infections is little explored, and whether there is a substantial difference in IFN-β expression between aspirin-sensitive and aspirin-tolerant asthmatics is unknown.

Mast cells are a major source of the proinflammatory cytokine tumor necrosis factor-α (TNF-α) in the airways and also release the autacoid mediators histamine, prostaglandin D_2 (PGD$_2$), and LTC$_4$ (45), all of which are potent contractile agonists of airway smooth muscle and are capable of inducing airway hyperresponsiveness (AHR), excessive mucus secretion, and edema of the bronchial mucosa. There is conflicting evidence whether mast cells from aspirin-sensitive individuals following exposure to aspirin are directly activated to release these mediators (57–60).

Mast cells cultured from peripheral blood progenitors in aspirin-sensitive patients and activated by anti-IgE overproduced CysLTs by almost 200% compared with aspirin-tolerant patients (57). However, neither basal or IgE-induced release of histamine nor CysLTs were upregulated when the cells were incubated with aspirin. PGD$_2$ release from mast cells was also suppressed following incubation with aspirin in the same study (57). Conversely, in vivo studies suggest that mast cells are directly activated by aspirin in ASRD. Aspirin-sensitive patients have raised levels of the PGD$_2$ metabolite (9α,11β-PGF$_2$) and tryptase in their blood prior to exposure to aspirin, and following challenge in most patients, these levels rise further still (58). After segmental bronchial challenge of asthmatics with lysine aspirin, an aspirin cogener that can be made into an aerosol, a significant reduction in the numbers of tryptase-positive mast cells in the bronchial mucosa and an increase in mucosal infiltration of eosinophils are seen (59). Bronchoalveolar lavage (BAL) from aspirin-sensitive patients after segmental lysine aspirin challenge has shown a trend toward increased histamine release with no inhibition of PGD_2 production, although the same study found no increase in tryptase or eosinophil cationic protein (ECP) levels (60). Overall these findings suggest that chronic ongoing mast cell activation is a feature of ASRD at baseline, and this may be aggravated further as a result of aspirin provocation. However, a mechanism for direct mast cell activation by aspirin is unclear.

Mast cells secrete the proinflammatory cytokines interleukin (IL)-4, IL-5, and IL-13, and raised IL-4 and IL-6 levels have been measured in exhaled breath condensates from aspirin-sensitive patients (61). IL-5 promotes eosinophil development, priming, differentiation, and survival (62). It also stimulates translocation of 5-LO to the nucleus and upregulates 5-lipoxygenase-activating protein (FLAP) in human eosinophils, thereby enhancing the ability of eosinophils to produce CysLTs (63). In bronchial biopsies from aspirin-sensitive patients, IL-5 expression in mast cells and eosinophils is significantly increased (64). In vivo studies measuring IL-5, both in BAL and serum from aspirin-sensitive patients, have recorded no baseline increases in IL-5 production compared with levels in aspirin-tolerant asthmatics (60, 65). However, within 15 min of lysine aspirin challenge, an increase in numbers of eosinophils with a corresponding increase in IL-5 has been measured in BAL fluid in individuals with ASRD (60).

IL-13 is predominantly produced by Th2-polarized $CD4^+$ T cells, but it is also synthesized by non–T cells, including mast cells, basophils, and eosinophils (66). Its production is upregulated in human eosinophils in response to IL-5 and GM-CSF stimulation (67). IL-13 both enhances leukocyte production of LTD_4 (36) and increases expression of the $CysLT_1R$ on bronchial smooth muscle (68). Both IL-4 and IL-13 also enhance bronchial smooth-muscle proliferation in fibroblast growth factor 2–primed cells (69). IL-13 alone may mediate some of the main physiological consequences of asthma, namely AHR, mucus hypersecretion,

and subepithelial fibrosis (66). The importance of IL-13 in the pathogenesis of airway inflammation has been underscored by the correlation between IL-13 overexpression in sputum and bronchial biopsies and severe asthma (70). Furthermore, IL-13 selectively downregulates PGE_2 biosynthesis in human airway epithelial cells from both asthmatic and normal subjects (71). IL-13 inhibits COX-2 and mPGES-1 (a membrane-bound form of prostaglandin E synthase) while upregulating the PGE_2-metabolizing enzyme 15-prostaglandin dehydrogenase (15-PGDH), and these enzymatic changes correlate with decreased supernatant PGE_2 levels (71). These findings underline the likely importance of transcellular interactions between epithelial cells and leukocytes in vivo. IL-13 also induces apoptosis in lung epithelial cells and is implicated in airway remodeling by upregulating profibrotic genes in primary lung fibroblasts (72). We are not aware of any published studies examining IL-13 production in ASRD or the functional consequences of enhanced or diminished cytokine production.

Finally, a dysfunctional respiratory epithelium that is increasingly vulnerable to damage is a well-recognized feature of asthma. Researchers have extensively described the concept of a disordered epithelium orchestrating an inflammatory response that repairs incompletely and interacts with multiple environmental factors to produce a chronic wound scenario in chronic asthma (73). An elevation of epithelial EP1 and EP2 receptor expression has been observed in the inflamed nasal mucosa of aspirin-sensitive patients (38). This is significant because PGE_2 administration protects against airway obstruction, i.e., it is bronchoprotective; PGE_2 also promotes wound healing and repair in the respiratory epithelium (74).

The 5-Lipoxygenase Pathway, Cysteinyl Leukotrienes, and Their Receptors

The CysLTs are critical proinflammatory mediators in asthma. They have considerable

IL: interleukin

15-PGDH: 15-prostaglandin dehydrogenase

uLTE$_4$: urinary leukotriene E$_4$

LTC$_4$S: leukotriene C$_4$ synthase

importance in the pathogenesis of ASRD. As well as being potent activators of inflammatory cells, they are powerful bronchoconstrictor agonists. LTC$_4$ and LTD$_4$ are approximately 1000-fold more potent than histamine in contracting human bronchi in vitro (75). In mild asthmatics neither bronchial responsiveness to LTD$_4$ nor the amount of urinary LTE$_4$ (uLTE$_4$) that can be recovered without specific antigen challenge is attenuated by inhaled corticosteroids (76). CysLTs induce mucus hypersecretion from goblet cells and augment leukocyte adhesion to endothelial cells (36). They also act directly on endothelial cells to induce microvascular permeability (36). Moreover, they are implicated in airway remodeling and enhance the capacity of lung fibroblasts to proliferate to known mitogens and to lay down collagen (36). Finally, the CysLTs, while being generated by eosinophils, monocytes, and mast cells, can activate these cells in both an autocrine and a paracrine manner (36). The CysLTs act as chemoattractants for inflammatory leukocytes, and inhalation of LTE$_4$ increases infiltration of neutrophils and eosinophils into asthmatic airways (77).

A distinguishing feature of ASRD is upregulation of the 5-LO pathway at baseline, which is enhanced further with the addition of NSAIDS. Christie et al. (78) quantitated uLTE$_4$ levels as a measure of global CysLT production in aspirin-sensitive patients. A mean sixfold-higher resting uLTE$_4$ level was detected in aspirin-sensitive compared with normal or non-aspirin-sensitive patients prior to oral aspirin challenge, with a further fourfold increase following aspirin challenge (78) (**Figure 2**). Inhalation challenge with lysine aspirin also induced a similar rise in uLTE$_4$ (79). Long-term desensitization with oral aspirin resulted in significant blunting of uLTE$_4$ production, but levels nonetheless remained elevated compared with those of aspirin-tolerant individuals (80).

There is broad overlap in uLTE$_4$ concentrations between aspirin-tolerant and aspirin-sensitive patients, thereby precluding the clinical use of uLTE$_4$ as a diagnostic test for ASRD. Nevertheless, the increase in CysLT levels in all bodily fluids studied following aspirin challenge is a unique feature of ASRD (65, 78, 79, 81, 82). To date, whether enhanced CysLT production in ASRD is the result of a predetermined abnormality in the production of CysLTs or is the consequence of greater numbers and activation of eosinophils and mast cells remains unanswered—although these possibilities are not mutually exclusive.

LTC$_4$ synthase (LTC$_4$S) is a critical enzyme in CysLT production and conjugates LTA$_4$ with reduced glutathione in mast cells, basophils, eosinophils, and other inflammatory cells to yield LTC$_4$ (36). Circulating blood eosinophils from aspirin-sensitive patients have more mRNA for this enzyme (83), as do nasal polyps from aspirin-sensitive individuals compared with aspirin-tolerant subjects (43). Immunostaining in nasal polyps resected from aspirin-sensitive individuals demonstrates a fourfold-higher expression of LTC$_4$S in the polyp mucosa compared with expression

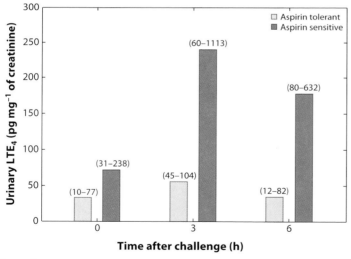

Figure 2

The time-dependent changes in urinary LTE$_4$ concentrations after lysine aspirin challenge in six aspirin-sensitive patients. After lysine aspirin challenge in patients with ASRD, urinary LTE$_4$ concentrations increased from 72 pg mg^{-1} (geometric mean) to 240 pg mg^{-1} at 3 h and to 178 pg mg^{-1} at 6 h. The increase in urinary LTE$_4$ concentrations over time was significant ($p < 0.025$). Aspirin-tolerant individuals displayed no significant increase in urinary LTE$_4$ concentrations. (Adapted with permission from Reference 79. Copyright American Thoracic Society.)

in aspirin-insensitive subjects (46). LTC_4S is also overexpressed in the bronchial mucosa in ASRD, with fivefold-higher expression compared with aspirin-tolerant subjects (84). This expression of LTC_4S in the bronchial mucosa appears to reflect enzymatic activity, with a correlation between CysLT levels measured in BAL fluid and the numbers of LTC_4S-immunoreactive cells. Furthermore, LTC_4S upregulation in aspirin-sensitive patients correlates with bronchial responsiveness to lysine aspirin challenge (84). The cells in the airways expressing elevated LTC_4S include eosinophils (46) and mast cells (85).

Increased expression of LTC_4S in ASRD may be genetically regulated. A transversion of adenine (A) to cytosine (C), 444 nucleotides upstream of the translational start site ($-444A/C$), in the promoter region of the LTC_4S gene correlates with an elevated risk of ASRD (86, 87). This single-nucleotide polymorphism (SNP) has been associated with an increased prevalence of ASRD in Japanese studies (86) and severe steroid-dependent ASRD in European populations (87). However, the SNP is only weakly associated with asthma, and not at all with ASRD, in Australian populations (88). It is not associated with asthma or ASRD in the United States (89). The Japanese study observed no relationship between this polymorphism and the LTC_4S activity found in eosinophils. The authors therefore hypothesized that, although this polymorphism might have some effect on the development of ASRD, it was probably acting in linkage disequilibrium with another causatively important and as-yet-unidentified mutation (86). The presence of the LTC_4 $-444C$ allele in aspirin-tolerant patients and the lack of its universal expression in ASRD confirm that this SNP does not represent a unique pathophysiological feature delineating aspirin-sensitive from aspirin-tolerant individuals. Future studies are required to understand why NSAIDS do not induce a rise in CysLT levels in non-aspirin-sensitive individuals.

CysLTs act on at least two rhodopsin-class G protein–coupled receptors: $CysLT_1R$ and $CysLT_2R$. The sequences of these two receptors are highly divergent, with a 38% homology (90). The evidence for separate cysteinyl leukotriene receptors (CysLTRs) was initially based on differences in biological activity of the individual LTs, the effects of LT receptor antagonists, and radioligand binding studies. In 1999 a previously cloned orphan G protein–coupled receptor was deorphanized and shown to have the pharmacological properties of the $CysLT_1R$ (91, 92). The screening of public databases for novel orphan GPCRs led to the identification of the $CysLT_2R$ in 2000 (93, 94). Both the $CysLT_1R$ and the $CysLT_2R$ are expressed on granulocytes, monocytes, and macrophages (36).

Whereas the $CysLT_1R$ mediates airway smooth-muscle contraction, mucus hypersecretion, microvascular leakage, and swelling of the airways, the $CysLT_2R$ mediates the inflammatory response, possibly through the modulation of chemokine gene transcription, and contributes to vascular permeability and tissue fibrosis (36). The expression of the $CysLT_1R$ is upregulated by proinflammatory cytokines, including IL-13, which augments $CysLT_1R$ expression on human lung fibroblasts, thereby permitting bronchial smooth muscle to respond to LTC_4 and release eotaxin (95). There is emerging evidence of pharmacological cross talk between both sets of receptors, and knockdown of the $CysLT_2R$ stimulates LTD_4-dependent proliferation of cord blood–derived human mast cells. The absence of the $CysLT_2R$ was associated with an upregulation of $CysLT_1R$ expression at the cell surface and enhanced LTD_4-induced ERK phosphorylation (96).

The discovery of the two CysLTRs rapidly opened the way for experiments to determine whether the expression of the $CysLT_1R$ or the $CysLT_2R$ differed in patients with aspirin sensitivity compared with their aspirin-tolerant counterparts. In 2002, Sousa et al. (37) reported that the percentage of inflammatory cells expressing the $CysLT_1R$ in nasal biopsies was fivefold higher in aspirin-sensitive patients compared with aspirin-tolerant individuals. There was no upregulation of the $CysLT_2R$ (97) or the LTB$_4$ receptor (37). Desensitization with aspirin led to downregulation of the CysLT$_1$R in nasal inflammatory cells (37), but substantial amounts of CysLTs continued to be produced (80). Sousa et al.'s (37) results may explain why subjects with ASRD demonstrate greater airway responsiveness to aerosolized LTE$_4$ than do non-aspirin-sensitive subjects (98, 99).

Several studies have reported biological properties of the CysLTs that are not readily explained on the basis of the recognized pharmacological properties of the CysLT$_1$R and CysLT$_2$R, suggesting that there may be further distinct, uncharacterized CysLT receptors or that postreceptor signaling may be differentially regulated in a ligand-specific fashion (100–103). Cumulative findings from a series of radioligand binding studies indicated that there are two separate receptors mediating the actions of LTC$_4$ and LTD$_4$ (100, 104, 105) and that the effects of LTE$_4$ may also be mediated through the LTD$_4$ receptor (100). In 1984, Lee and coworkers (100) established that prior exposure to LTE$_4$, but not to LTC$_4$ or LTD$_4$, enhanced contractility to histamine in guinea pig tracheal spirals in a time- and dose-dependent fashion. Furthermore, LTC$_4$, LTD$_4$, and LTE$_4$ all elicited the same magnitude of contraction of tracheal smooth muscle (100). Finally, the authors observed that the rank order of potency of CysLTs to induce contractility was different between guinea pig tracheal smooth muscle (LTE$_4$ > LTD$_4$ = LTC$_4$) and guinea pig parenchymal strips (LTD$_4$ > LTE$_4$ > LTC$_4$) (106). These observations lent further credence to the idea of three different receptors binding CysLTs.

Evidence for a third CysLT receptor subtype in human tissue was initially based on studies reporting that LTC$_4$-induced vasoconstriction of human and porcine isolated pulmonary arterial preparations was insensitive to CysLTR antagonists, including the nonselective CysLT$_1$/CysLT$_2$ receptor antagonist BAY u9773 (102, 103). At approximately the same time, Bandeira-Melo et al. (107) demonstrated that LTC$_4$ acts as an intracrine mediator of eosinophil cytokine secretion and that this is mediated by an intracellular G protein–coupled receptor distinct from CysLT$_1$ and CysLT$_2$.

Further indirect evidence for a third CysLTR in humans arose from the finding that, compared with nonasthmatic subjects, asthmatics show an extraordinary bronchial hyperresponsiveness to LTE$_4$ but that bronchial hyperreactivity to LTC$_4$ and LTD$_4$ is magnitudes smaller. Therefore, compared with the airways of normal subjects, the airways of asthmatics are approximately 14-, 15-, 6-, 9-, and 219-fold more responsive to histamine, methacholine, LTC$_4$, LTD$_4$, and LTE$_4$, respectively (108). Patients with ASRD demonstrate an even greater preferential bronchoconstrictor response to LTE$_4$ but not to LTC$_4$ (99) and are more-than-tenfold more vulnerable to the bronchoconstrictor effects of LTE$_4$ when compared with aspirin-tolerant individuals (98). Prior exposure to LTC$_4$, LTD$_4$, and LTE$_4$ also enhances subsequent airway responsiveness to histamine in asthmatic patients, but not in normal controls (109). LTE$_4$ is an end metabolite of CysLT biosynthesis and is therefore likely to persist for the longest time at the site of release. Taken together, these data suggest that LTE$_4$ is a critical mediator in asthma and in particular in ASRD.

CysLT$_1$R and CysLT$_2$R ligate the CysLTs with an affinity order of LTD4 > LTC4 ≫ LTE$_4$ (CysLT$_1$) (92) and LTC$_4$ = LTD$_4$ ≫ LTE$_4$ (CysLT$_2$), respectively (93). In view of the pharmacologically defined differential sensitivity of both CysLTRs to their ligands and the low affinity of both CysLT$_1$R and CysLT$_2$R to LTE$_4$, it is difficult to understand why LTE$_4$ but not LTD$_4$ or LTC$_4$ demonstrates such differential effects if the ability of the CysLTs to induce bronchoconstriction is mediated solely through CysLT$_1$R or CysLT$_2$R. Some commentators have suggested that chronic exposure to grossly raised CysLT concentrations may lead to dysregulation of the CysLT$_1$R, resulting in the receptor developing increased affinity for a ligand (LTE$_4$) that is normally several logs less potent (110).

More recently, indication of a putative third receptor in human bronchi has come from a study by Yoshisue et al. (111), who observed enhancement of mitogenesis of bronchial fibroblasts by LTD_4 in the presence of epidermal growth factor; neither two specific $CysLT_1R$ antagonists nor the dual antagonist (BAY u9773) prevented this enhancement. The absence of mRNA for both the $CysLT_1R$ and the $CysLT_2R$ in bronchial fibroblasts suggested that this mitogenic response may be mediated by a third, uncharacterized receptor (111). The orphan G protein–coupled receptor 17 is a dualistic receptor, intermediate in phylogeny between the CysLTRs and the P2Y pyrimidinergic receptors that bind both uracil nucleotides and CysLTs (112). It has been identified as a putative candidate for a third CysLT receptor and as yet has no assigned pathological or physiological role (36).

Aspirin Desensitization

Nearly all aspirin-sensitive individuals can be desensitized to aspirin (113, 114). Once successfully desensitized, most patients will experience improved symptomatic control, require fewer FESS polypectomies, have a reduced requirement for oral corticosteroids, and have fewer hospital admissions for asthma (115). Aspirin desensitization remains an add-on treatment in patients with moderate-to-severe ASRD. This is particularly true for those individuals in whom symptoms are poorly controlled despite optimal medical therapy, who require unacceptably frequent doses of oral corticosteroids, or who require multiple surgeries to resect nasal polyps.

In a series examining outcomes in 172 patients taking 650 mg of aspirin bd for more than a year, 67% of patients derived clinical benefit, whereas 9% did not improve and 14% stopped because of side effects, most often intractable urticaria or gastritis (115). Thus, aspirin desensitization produces definitive improvements in both upper and lower respiratory tract symptoms in the majority of patients with ASRD (115). In a typical desensitization regimen, patients receive escalating doses of oral aspirin over two to five days until 325–650 mg of as-

pirin twice daily is tolerated (115). Daily administration of up to 1300 mg of aspirin per a day may be required to maintain this state. The frequency of aspirin-induced side effects does not vary for maintenance doses of 325 mg or 625 mg of aspirin bd (115).

Despite its success as a therapeutic modality, the precise mechanism by which desensitization occurs is incompletely understood. Arm and colleagues demonstrated that prior to aspirin desensitization LTE_4 was 1870 times more potent than histamine in inducing a 35% fall in specific airway conductance. Within 24 h of successful desensitization with oral aspirin, there was a mean 33-fold decrease in airway responsiveness to LTE_4 but no change in airway responsiveness to histamine (98) (**Figure 3a,b**). These data suggested that successful aspirin desensitization may partly be explained by selective receptor downregulation. Investigators from the same laboratory noted that successful aspirin desensitization occurred despite substantial release of CysLTs after aspirin challenge (80), supporting the view that receptor tachyphylaxis rather than reduced CysLT production was responsible for the desensitized state. In 2002, Lee and colleagues (37) confirmed that aspirin desensitization was associated with significant downregulation in $CysLT_1R$ in inflammatory cells.

The 15-Lipoxygenase Pathway

Products from the 15-lipoxygenase (15-LO) pathway include 15-HPETE, LXs, and eoxins (EXs) (**Figure 1**). Comparatively little is known about the biological function of 15-LO, but its expression is upregulated by the Th2 cytokines IL-4 and IL-13 in the upper and lower respiratory tracts (116, 117), and its expression is significantly increased in the bronchial submucosa of patients with asthma or chronic bronchitis compared with control subjects (118).

Nicknamed the "good" lipids for asthma (119), LXs were discovered by Serhan and colleagues (120) more than 20 years ago. These compounds are trihydroxytetraene eicosanoids that are generated via the sequential actions

EX: eoxin

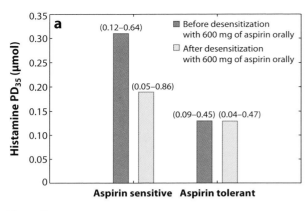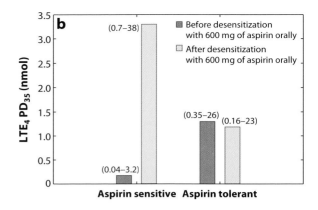

Figure 3

(*a*) The PD$_{35}$ [geometric mean (range)] to histamine before and after desensitization with 600 mg of oral aspirin in five aspirin-sensitive and five aspirin-tolerant subjects. Oral desensitization with 600 mg of aspirin was not associated with a significant change in the geometric mean histamine PD$_{35}$ in aspirin-sensitive or aspirin-tolerant patients. (*b*) The PD$_{35}$ [geometric mean (range)] to LTE$_4$ before and after desensitization with 600 mg of oral aspirin in the aspirin-sensitive and aspirin-tolerant subjects shown in panel *a*. Oral desensitization with 600 mg of aspirin led to the geometric mean PD$_{35}$ to LTE$_4$ increasing from 0.17 nmol to 3.3 nmol after desensitization. (Data from Reference 98.)

of two or more lipoxygenases. There are three principal biosynthetic pathways for their formation. The first involves cooperation between 15- and 5-LO, the second involves acetylation of COX-2 by aspirin to trigger 15-epi-LXA$_4$ biosynthesis, and the third involves interactions between 12-LO and 5-LO. Interaction of leukocytes with epithelium, endothelium, or platelets results in the formation of LXA$_4$ (5*S*,6*R*,15*S*-trihydroxy-7,9,13-*trans*-11-*cis*-ecosatetraenoic acid) and its positional isomer lipoxin B$_4$ (LXB$_4$; 5*S*,14*R*,15*S*-trihydroxy-6,10,12-*trans*-8-*cis*-ecosatetraenoic acid). The LXs modulate leukocyte trafficking and vascular tone, and in contrast to CysLTs they have potent anti-inflammatory effects (121). A reduction in their levels may be a feature of ASRD (41, 43, 122).

In patients with ASRD, LX synthesis from stimulated whole blood is reduced compared with aspirin-tolerant subjects (41, 122), although researchers have more recently suggested that diminished LXA$_4$ synthesis is a feature of severe asthma regardless of aspirin sensitivity (123). The only in vitro study to measure LXs from the respiratory tract in patients with aspirin sensitivity found significant downregu-

lation of 15-LO expression with a corresponding decrease in LXA$_4$ production (43). Patients with ASRD upregulate CysLT$_1$R expression on inflammatory leukocytes (37) but may have decreased levels of LXs to compete with CysLTs for binding at their cognate receptor. Therefore, diminished capacity to generate LXs may contribute to the uncontrolled, chronic inflammation that characterizes this phenotype of asthma.

In 2008, details of EXs, a novel group of compounds that are metabolites of the 15-LO pathway, were published. Feltenmark et al. (124) isolated human eosinophils from whole blood and stimulated them with arachidonic acid. This led to the formation of 15(*S*)-HETE, 12(*S*)-HETE, and a new product that had an UV absorbance maximum at 282 nm and a shorter retention time than did LTC$_4$ on RP-HPLC. Positive-ion electrospray tandem mass spectrometry identified the product as 14,15-LTC$_4$, which in eosinophils could be further metabolized to 14,15-LTD$_4$ and 14,15-LTE$_4$; these three products were named EXC$_4$, EXD$_4$, and EXE$_4$, respectively. Mast cells derived from cord blood and nasal polyps could also produce EXC$_4$ when stimulated with arachidonic acid.

Incubation of eosinophils with arachidonic acid favored EXC_4 production, whereas stimulation with calcium ionophore led almost exclusively to LTC_4 production. Receptor-mediated stimulation of eosinophils with LTC_4, PGD_2, and IL-5 also triggered the release of EXs. An endothelial cell monolayer was used to demonstrate that EXs induced vascular leakage and were almost as potent as CysLTs in this regard. However, EXs did not appear to induce smooth-muscle contractility (124). Clearly these potentially important proinflammatory mediators may induce many of the pathological features characteristic of ASRD, and studies will be required to elucidate any role they may have in the pathogenesis of ASRD.

PGE₂, E-Prostanoid Receptors, and ASRD

PGE_2 is the most abundant PG in the human body, and the lung has been described as a privileged site for its beneficial anti-inflammatory effects (125). In the airways PGE_2 is widely regarded as the prostanoid with the most prominent immunomodulatory, antifibrotic, and bronchodilating properties. It is the predominant PG synthesized by the bronchial epithelium, and injury to the airway epithelium decreases the production of PGE_2 and leads to an overproduction of CysLTs by leukocytes such as bronchial macrophages (126). As well as being generated by airway smooth muscle (127), PGE_2 also influences the bronchial remodeling process in the human lung by inhibiting the transition of fibroblasts into myofibroblasts via its actions on the EP2 receptor (128). PGE_2 therefore has the potential to reverse at least three of the cardinal features of ASRD: enhanced CysLT production, smooth-muscle hyperplasia, and airway epithelial damage.

It has been more than a decade since pretreatment with aerosolized PGE_2 was shown to attenuate lysine aspirin–induced bronchoconstriction in ASRD (40, 129) and to completely abolish any rise in $uLTE_4$ excretion (40). In IgE-activated mast cells cultured from peripheral blood cell progenitors in aspirin-sensitive patients, 10^{-6} M PGE_2 significantly reduced CysLT production while not inhibiting PGD_2 or histamine production (57). In a second in vitro study, stimulating peripheral blood leukocytes (PBLs) with aspirin prior to incubation with PGE_2 induced modest but significant decreases in CysLT production in aspirin-sensitive patients, and this inhibitory effect on CysLT production was also observed in aspirin-tolerant and control subjects (39). In cytokine-primed human neutrophils, PGE_2 inhibits CysLT biosynthesis by inhibiting 5-LO translocation to the nucleus (130), a step at the beginning of CysLT production.

PGE_2 is generated from COX-derived PGH_2 by prostaglandin E synthase (PGES), which occurs in multiple forms, two of which are membrane bound (mPGES-1 and mPGES-2) and one of which is cytosolic (cPGES). Of these enzymes, COX-1 along with mPGES-2 and cPGES are constitutively expressed, whereas COX-2 and mPGES-1 are induced by inflammatory stimuli. In vitro, mPGES-1 functionally couples to COX-2 and localizes to the nuclear membrane (131), whereas cPGES couples to COX-1 (131). The cellular function of mPGES-2 is less clear and appears to be constitutive; mPGES-2 is functionally coupled with both COX-1 and COX-2 (131). PGE_2 is rapidly degraded by the PGE_2-metabolizing enzyme 15-PGDH (132). There is evidence of a reciprocal relationship between COX-2 expression and 15-PGDH expression in the lung and in human airway epithelial cells in both asthmatic and normal subjects: IL-13 significantly downregulates COX-2 and mPGES-1 while upregulating 15-PGDH (71). The relative expression of mPGES and 15-PGDH in aspirin-sensitive patients is not yet defined.

The close relationship between CysLT production and its inhibition by PGE_2 in ASRD has led to suggestions that aspirin-sensitive patients are particularly dependent on the anti-inflammatory effects of PGE_2 and that the anti-inflammatory effects of PGE_2 are impaired because of reduced production, or abnormal EP receptor expression and signaling, or both. A number of studies in the literature address PGE_2 production in aspirin-sensitive patients. However, most of these have involved the stimulation of PBLs, which are remote from the site of the disease or prolonged culture of tissues in vitro. Kowalski et al. (42) examined eicosanoid production in PBLs from aspirin-sensitive patients. At baseline PBLs generated similar amounts of PGE_2 as compared with aspirin-tolerant patients. Stimulation with calcium ionophore significantly increased PGE_2 production, and incubation with aspirin inhibited PGE_2 biosynthesis, but there was no difference between both groups (42). Kowalski et al. (133) also measured PGE_2 production in cell culture supernatants from nasal polyp epithelial cell explants. This group found diminished PGE_2 production at baseline in aspirin-sensitive patients but observed, on treatment with aspirin, a similar relative inhibition of PGE_2 production in aspirin-sensitive and control subjects. Bronchial fibroblasts derived from aspirin-sensitive patients stimulated for 18 h with cytomix (a cell-stimulating mixture of lipopolysaccharide, IL-1α, and TNF-α) also produced less PGE_2 than did fibroblasts from aspirin-tolerant patients, although spontaneous production was equivalent in both groups (134). Although all these experiments have yielded valuable data, the use of individual cell types excluded the possibility of transcellular eicosanoid biosynthesis. In vivo studies have measured PGE_2 release in nasal lavage fluid, exhaled breath condensate, and BAL fluid in aspirin-sensitive and aspirin-tolerant patients (135–139). The findings from these studies are inconsistent, and it is difficult to draw any firm conclusions from the results. However, no in vivo study has found diminished levels of PGE_2 at baseline in aspirin-sensitive patients, and some studies have reported a relative overproduction of PGE_2 in aspirin-sensitive patients (135, 138).

PGE_2 ligates four G protein–coupled receptors designated EP1–4 (139). Expression of the EP2 receptor in nasal mucosal inflammatory cells in aspirin-sensitive patients is reduced (38). Furthermore, a functional SNP of the EP2 gene, believed to be associated with a decrease in the transcription level of the receptor, is associated with an increased risk of aspirin-sensitive asthma and rhinosinusitis (140). Together these findings suggest that diminished EP2 receptor expression may be a cause of the excessive CysLT production in ASRD.

The Cyclooxygenase Pathway and ASRD

Since Szczeklik first proposed that the precipitation of acute attacks of asthma and rhinosinusitis in susceptible patients was a consequence of aspirin inhibiting COX-1, a second isoform of COX (COX-2) has been discovered. Both isoforms manufacture prostanoids (**Figure 1**) and are products of distinct genes expressed in normal human respiratory epithelium. COX-1 is constitutively expressed and is concerned with basal prostanoid production or housekeeping functions, whereas COX-2 is upregulated in response to inflammatory stimuli. Immunostaining in human and porcine endothelial cells suggests that COX-1 is expressed and confined to both the perinuclear zone and the cytoplasm. COX-2 localizes to the nucleus alone in a resting-state cell (141), but with cell activation it relocates to the nuclear envelope and the cytoplasm (141).

The majority of aspirin-sensitive patients can tolerate selective COX-2 inhibitors (142). This has led to suggestions that there is a functional COX-2 deficiency in ASRD, COX-2 makes only a trivial contribution to local PGE_2 production in these individuals (143), and therefore aspirin-sensitive subjects do not experience an exacerbation of their symptoms when ingesting a selective COX-2 inhibitor, because there is little or no reduction in airway PGE_2 production. This idea is attractive, but the literature does not unequivocally support the two arms of this hypothesis (diminished COX-2 expression and decreased PGE_2 synthesis).

In 1999, using reverse transcriptase PCR, Picado et al. (144) reported that nasal polyps from aspirin-sensitive and aspirin-tolerant patients with ASRD expressed equivalent levels of COX-1 mRNA but significantly less COX-2 mRNA compared with levels in nasal mucosa from healthy patients. In 2004, a second study from the same research group used real-time PCR to assess relative expression of COX-1 and COX-2 mRNA in three groups of patients: aspirin-sensitive patients undergoing nasal polypectomy, aspirin-tolerant patients undergoing nasal polypectomy, and patients undergoing corrective nasal surgery (145). The investigators also sought to determine whether in the hour following surgery there was a time-dependent upregulation in COX mRNA. The expression of COX-2 mRNA at baseline was lower in both groups of patients with nasal polyps than levels in the nasal mucosa. Over a 1-h time course, the expression of COX-2 mRNA rose significantly in nasal polyps from aspirin-tolerant patients (reaching similar levels as in nasal mucosa), but this upregulation of COX-2 was not seen in the nasal polyps of aspirin-sensitive patients. A comparison between the three cohorts revealed that, although none of the patients undergoing corrective nasal surgery and only 6% of the aspirin-tolerant group were receiving oral corticosteroids at the time of surgery, almost 40% of the aspirin-sensitive group were receiving oral corticosteroids at the time of surgery. The lack of parity between the three groups of patients complicates the interpretation of this study's findings (145).

In a third study, PGE_2 and COX-2 mRNA was measured in homogenized sinonasal tissue of patients with chronic rhinosinusitis with and without nasal polyposis (CRS group), in sinonasal tissue of patients with nasal polyposis and aspirin sensitivity, and in normal nasal mucosa from healthy subjects (NM group) (43). The study groups were comparable with all patients undergoing polypectomy receiving topical nasal corticosteroids and none requiring oral corticosteroids at the time of the proce-

dure. Diminished levels of COX-2 were quantified in patients with nasal polyposis compared with individuals with chronic rhinosinusitis alone. This decreased expression of COX-2 was particularly enhanced in aspirin-sensitive patients and correlated positively with LT production and inversely with PGE_2 measurements. The authors concluded that changes in tissue eicosanoid metabolism appear to be related to the severity of eosinophilic inflammation, rather than aspirin sensitivity per se (43). Nonetheless, this study provides direct evidence supporting the hypothesis that patients with ASRD produce less PGE_2 owing to diminished COX-2 expression. In contrast, when the expression of COX-2 was examined in the bronchial mucosa of aspirin-sensitive and aspirin-tolerant patients with asthma, enhanced COX-2 expression was observed in aspirin-sensitive subjects, with a mean fourfold increase in the percentage of COX-2-expressing cells that were mast cells and a 2.5-fold increase in the number of eosinophils expressing COX-2 (146). Further studies are now clearly needed to elucidate the relative expression of COX-1 and COX-2, particularly in the bronchial mucosa, of patients with ASRD. The functional consequences of COX inhibition also warrant further investigation.

In this respect Mastalerz and colleagues (147) measured PGE_2 urinary metabolites and used these as a biomarker for systemic PGE_2 production in aspirin-tolerant and aspirin-sensitive patients. No differences in urinary PGE_2 metabolites were noted at baseline. However, after aspirin challenge, although non-aspirin-sensitive patients showed diminished systemic production of PGE_2, as demonstrated by a fall in measurable urinary PGE_2 metabolites, no parallel fall was noted in aspirin-sensitive patients. The administration of the selective COX-2 inhibitor celecoxib led to a decrease in the production of PGE_2 metabolites in both groups of patients and challenged the concept that aspirin-sensitive patients can tolerate COX-2 inhibitors because COX-2 induces relatively trivial quantities of

PGE$_2$. Although it is not possible to directly extrapolate these observations locally to the nasal and bronchial mucosa, the data indicate that COX-2 contributes considerably to overall PGE$_2$ production and support previous in vivo work that reported no significant decreases in PGE$_2$ levels after aspirin challenge in patients with ASRD, although levels decreased in all other groups (135).

CONCLUDING REMARKS

ASRD is a chronic inflammatory disorder of the upper and lower respiratory tracts that is characterized by enhanced CysLT production at baseline and subsequent to aspirin exposure (78). It is heterogeneous in its clinical presentation and is probably underrecognized. The inflammatory pathways underlying ASRD are complex, but despite the absence of any animal models, substantial progress has been made in the past ten years in understanding of the pathophysiology of this condition. Gradually a picture has emerged of a phenotype of airway disease in which multiple defects in the inflammatory pathways described above either enhance inflammation or lead to an inadequate anti-inflammatory response.

Approximately half of all patients with ASRD describe a flu-like illness triggering the syndrome (8). This observation, combined with the findings from a study in which pretreatment of aspirin-sensitive individuals with acyclovir abrogated both aspirin-induced bronchoconstriction (148) and the expected increase in uLTE$_4$ following aspirin challenge, has led some commentators to suggest an infectious component to the disease (56, 149). A latent viral infection could induce many of the features of ASRD, including enhanced CysLT biosynthesis, modification of transcription events in the COX pathway, and treatment-refractory inflammation (149). Furthermore, the reduced ability of the epithelium in asthma to clear RV is now well recognized (54, 55). The hypothesis suggesting that a viral insult in genetically susceptible individuals leads to a chronic viral infection that then drives ASRD is therefore attractive. But experiments to test this provocative hypothesis are lacking, and studies are required to determine the relevance of chronic infection in perpetuating the inflammatory mechanisms that underpin ASRD.

In recent years, there has been significant interest in the anti-inflammatory prostanoid PGE$_2$, its biosynthesis relative to COX-2 expression, and its capability to ameliorate ASRD at least partly by reducing excessive CysLT production (40). It remains to be determined whether PGE$_2$ production is depressed in ASRD. Decreased EP2 receptor expression has been noted in the nasal mucosa of aspirin-sensitive patients, and this may adversely affect local end-organ responses (38). This reduced expression may be genetically determined, as previously reported (140), and these important observations require urgent confirmation. Alternatively, or in addition, *Staphylococcus aureus* enterotoxin B (SEB) may downregulate PGE$_2$, EP2 receptor expression, and COX-2, so the role of SEB and other superantigens in contributing to ASRD should be tested (150).

Most aspirin-sensitive patients have moderate-to-severe disease, and medications currently available fail to control symptoms in a large proportion of this cohort. Novel therapeutic approaches targeting aspirin-sensitive patients are required. To date there are few specific treatments aimed specifically at managing patients with ASRD. Desensitization with aspirin remains one therapeutic modality specific to ASRD, but its mechanism of action is largely unknown; elucidation of the mechanism(s) will likely yield insights into not only desensitization but also the pathogenesis of ASRD.

In summary, ASRD is a complex phenotype of asthma characterized by severe inflammation of the airways that responds poorly to current medications. Increased understanding of the molecular, cellular, and biochemical basis of ASRD will likely open the way for new diagnostic techniques and therapeutic interventions.

DISCLOSURE STATEMENT

The authors are not aware of any biases that might be perceived as affecting the objectivity of this review.

ACKNOWLEDGMENTS

The research was supported by the Medical Research Council (UK) and Asthma UK.

LITERATURE CITED

1. Vane JR, Botting RM. 2003. The mechanism of action of aspirin. *Thromb. Res.* 110:255–58
2. Rinsema TJ. 1999. One hundred years of aspirin. *Med. Hist.* 43:502–7
3. Widal MF, Abrami P, Lenmoyez J. 1922. Anaphylaxie et idiosyncrasie. *Presse Med.* 30:189–92
4. Samter M, Beers RF Jr. 1968. Intolerance to aspirin. Clinical studies and consideration of its pathogenesis. *Ann. Int. Med.* 68:975–83
5. Mascia K, Haselkorn T, Deniz YM, Miller DP, Bleecker ER, Borish L. TENOR Study Group. 2005. Aspirin sensitivity and severity of asthma: evidence for irreversible airway obstruction in patients with severe or difficult-to-treat asthma. *J. Allergy Clin. Immunol.* 116:970–75
6. Koga T, Oshita Y, Kamimura T, Koga H, Aizawa H. 2006. Characterisation of patients with frequent exacerbation of asthma. *Respir. Med.* 100:273–78
7. Berges-Gimeno MP, Simon RA, Stevenson DD. 2002. The natural history and clinical characteristics of aspirin-exacerbated respiratory disease. *Ann. Allergy Asthma Immunol.* 89:474–78
8. Szczeklik A, Nizankowska E, Duplaga M. 2000. Natural history of aspirin-induced asthma. AIANE Investigators. European Network on Aspirin-Induced Asthma. *Eur. Respir. J.* 16:432–36
9. Makowska JS, Grzegorczyk J, Bienkiewicz B, Wozniak M, Kowalski ML. 2008. Systemic responses after bronchial aspirin challenge in sensitive patients with asthma. *J. Allergy Clin. Immunol.* 121:348–54
10. Rehl RM, Balla AA, Cabay RJ, Hearp ML, Pytynia KB, Joe SA. 2007. Mucosal remodeling in chronic rhinosinusitis. *Am. J. Rhinol.* 21:651–57
11. Kim JE, Kountakis SE. 2007. The prevalence of Samter's triad in patients undergoing functional endoscopic sinus surgery. *Ear Nose Throat J.* 86:396–99
12. Kowalski ML, Lewandowska-Polak A, Woźniak J, Ptasińska A, Jankowski A, et al. 2005. Association of stem cell factor expression in nasal polyp epithelial cells with aspirin sensitivity and asthma. *Allergy* 60:631–37
13. Awad OG, Fasano MB, Lee JH, Graham SM. 2008. Asthma outcomes after endoscopic sinus surgery in aspirin-tolerant versus aspirin-induced asthmatic patients. *Am. J. Rhinol.* 22:197–203
14. Kasper L, Sladek K, Duplaga M, Bochenek G, Liebhart L, Gladysz U. 2003. Prevalence of asthma with aspirin hypersensitivity in the adult population of Poland. *Allergy* 58:1064–66
15. Vally H, Taylor M, Thompson PJ. 2002. The prevalence of aspirin-intolerant asthma in Australian asthmatic patients. *Thorax* 57:569–74
16. Settipane G, Chafee F, Klein D. 1974. Aspirin intolerance. II. A prospective study in an atopic and normal population. *J. Allergy Clin. Immunol.* 53:200–4
17. McDonald J, Mathison DA, Stevenson DD. 1972. Aspirin intolerance in asthma—detection by challenge. *J. Allergy Clin. Immunol.* 50:198–207
18. Settipane GA, Chafee FH. 1977. Nasal polyps in asthma and rhinitis. A review of 6,037 patients. *J. Allergy Clin. Immunol.* 59:17–21
19. Yoshimine F, Hasegawa T, Suzuki E, Terada M, Koya T, et al. 2005. Contribution of aspirin-intolerant asthma to near fatal asthma based on a questionnaire survey in Niigata Prefecture, Japan. *Respirology* 10:477–84
20. Plaza V, Serrano J, Picado C, Sanchis J. High Risk Asthma Research Group. 2002. Frequency and clinical characteristics of rapid-onset fatal and near-fatal asthma. *Eur. Respir. J.* 19:846–52
21. Picado C, Montserrat JM, Roca J, Rodríguez-Roisín R, Estopá R, et al. 1983. Mechanical ventilation in severe exacerbation of asthma. Study of 26 cases with six deaths. *Eur. J. Respir. Dis.* 64:102–7

22. Bellomo R, McLaughlin P, Tai E, Parkin G. 1994. Asthma requiring mechanical ventilation. A low morbidity approach. *Chest* 105:891–96

23. Szczeklik A, Gryglewski RJ, Czerniawska-Mysik G. 1975. Relationship of inhibition of prostaglandin biosynthesis by analgesics to asthma attacks in aspirin-sensitive patients. *Br. Med. J.* 1:67–69

24. DeWitt DL, el-Harith EA, Kraemer SA, Andrews MJ, Yao EF, et al. 1990. The aspirin and heme-binding sites of ovine and murine prostaglandin endoperoxide synthases. *J. Biol. Chem.* 265:5192–98

25. Chiang N, Bermudez EA, Ridker PM, Hurwitz S, Serhan CN. 2004. Aspirin triggers antiinflammatory 15-epi-lipoxin A4 and inhibits thromboxane in a randomized human trial. *Proc. Natl. Acad. Sci. USA* 101:15178–83

26. Makowska JS, Grzegorczyk J, Bienkiewicz B, Wozniak M, Kowalski ML. 2008. Systemic responses after bronchial aspirin challenge in sensitive patients with asthma. *J. Allergy Clin. Immunol.* 121:348–54

27. Gamboa P, Sanz ML, Caballero MR, Urrutia I, Antépara I, et al. 2004. The flow-cytometric determination of basophil activation induced by aspirin and other nonsteroidal anti-inflammatory drugs (NSAIDs) is useful for in vitro diagnosis of the NSAID hypersensitivity syndrome. *Clin. Exp. Allergy* 34:1448–57

28. Kowalski ML, Ptasinska A, Jedrzejczak M, Bienkiewicz B, Cieslak M, et al. 2005. Aspirin-triggered 15-HETE generation in peripheral blood leukocytes is a specific and sensitive Aspirin-Sensitive Patients Identification Test (ASPITest). *Allergy* 60:1139–45

29. Folco G, Murphy RC. 2006. Eicosanoid transcellular biosynthesis: from cell-cell interactions to in vivo tissue responses. *Pharmacol. Rev.* 58:375–88

30. Higashi N, Taniguchi M, Mita H, Kawagishi Y, Ishii T, et al. 2004. Clinical features of asthmatic patients with increased urinary leukotriene E4 excretion (hyperleukotrienuria): involvement of chronic hyperplastic rhinosinusitis with nasal polyposis. *J. Allergy Clin. Immunol.* 113:277–83

31. Antczak A, Montuschi P, Kharitonov S, Gorski P, Barnes PJ. 2002. Increased exhaled cysteinyl-leukotrienes and 8-isoprostane in aspirin-induced asthma. *Am. J. Respir. Crit. Care Med.* 166:301–6

32. Szczeklik A, Sladek K, Dworski R, Nizankowska E, Soja J, et al. 1996. Bronchial aspirin challenge causes specific eicosanoid response in aspirin-sensitive asthmatics. *Am. J. Respir. Crit. Care Med.* 154:1608–14

33. Fischer AR, Rosenberg MA, Lilly CM, Callery JC, Rubin P, et al. 1994. Direct evidence for a role of the mast cell in the nasal response to aspirin in aspirin-sensitive asthma. *J. Allergy Clin. Immunol.* 94:1046–56

34. Sousa A, Pfister R, Christie PE, Lane SJ, Nasser SM, et al. 1997. Enhanced expression of cyclo-oxygenase isoenzyme 2 (COX-2) in asthmatic airways and its cellular distribution in aspirin-sensitive asthma. *Thorax* 52:940–45

35. Varga EM, Jacobson MR, Masuyama K, Rak S, Till SJ, et al. 1999. Inflammatory cell populations and cytokine mRNA expression in the nasal mucosa in aspirin-sensitive rhinitis. *Eur. Respir. J.* 14:610–15

36. Peters-Golden M, Henderson WR Jr. 2007. Leukotrienes. *N. Engl. J. Med.* 357:1841–54

37. Sousa AR, Parikh A, Scadding G, Corrigan CJ, Lee TH. 2002. Leukotriene-receptor expression on nasal mucosal inflammatory cells in aspirin-sensitive rhinosinusitis. *N. Engl. J. Med.* 347:1493–99

38. Ying S, Meng Q, Scadding G, Parikh A, Corrigan CJ, Lee TH. 2006. Aspirin-sensitive rhinosinusitis is associated with reduced E-prostanoid 2 receptor expression on nasal mucosal inflammatory cells. *J. Allergy Clin. Immunol.* 117:312–18

39. Celik G, Bavbek S, Misirligil Z, Melli M. 2001. Release of cysteinyl leukotrienes with aspirin stimulation and the effect of PGE$_2$ on this release from peripheral blood leucocytes in aspirin-induced asthmatic patients. *Clin. Exp. Allergy* 31:1615–22

40. Sestini P, Armetti L, Gambaro G, Pieroni MG, Refini RM, et al. 1996. Inhaled PGE$_2$ prevents aspirin-induced bronchoconstriction and urinary LTE4 excretion in aspirin-sensitive asthma. *Am. J. Respir. Crit. Care Med.* 153:572–75

41. Sanak M, Levy BD, Clish CB, Chiang N, Gronert K, et al. 2000. Aspirin-tolerant asthmatics generate more lipoxins than aspirin-intolerant asthmatics. *Eur. Respir. J.* 16:44–49

42. Kowalski ML, Ptasinska A, Bienkiewicz B, Pawliczak R, DuBuske L. 2003. Differential effects of aspirin and misoprostol on 15-hydroxyeicosatetraenoic acid generation by leukocytes from aspirin-sensitive asthmatic patients. *J. Allergy Clin. Immunol.* 112:505–12

43. Pérez-Novo CA, Watelet JB, Claeys C, Van Cauwenberge P, Bachert C. 2005. Prostaglandin, leukotriene, and lipoxin balance in chronic rhinosinusitis with and without nasal polyposis. *J. Allergy Clin. Immunol.* 115:1189–96

44. Kowalski ML, Grzegorczyk J, Pawliczak R, Kornatowski T, Wagrowska-Danilewicz M, Danilewicz M. 2002. Decreased apoptosis and distinct profile of infiltrating cells in the nasal polyps of patients with aspirin hypersensitivity. *Allergy* 57:493–500

45. Bradding P, Walls AF, Holgate ST. 2006. The role of the mast cell in the pathophysiology of asthma. *J. Allergy Clin. Immunol.* 117:1277–84

46. Adamjee J, Suh YJ, Park HS, Choi JH, Penrose JF, et al. 2006. Expression of 5-lipoxygenase and cyclooxygenase pathway enzymes in nasal polyps of patients with aspirin-intolerant asthma. *J. Pathol.* 209:392–99

47. Deleted in proof

48. Bischoff SC, Sellge G, Schwengberg S, Lorentz A, Manns MP. 1999. Stem cell factor-dependent survival, proliferation and enhanced releasability of purified mature mast cells isolated from human intestinal tissue. *Int. Arch. Allergy Immunol.* 118:104–7

49. Undem BJ, Lichtenstein LM, Hubbard WC, Meeker S, Ellis JL. 1994. Recombinant stem cell factor-induced mast cell activation and smooth muscle contraction in human bronchi. *Am. J. Respir. Cell Mol. Biol.* 11:646–50

50. Nilsson G, Butterfield JH, Nilsson K, Siegbahn A. 1994. Stem cell factor is a chemotactic factor for human mast cells. *J. Immunol.* 15:3717–23

51. Dreschers S, Dumitru CA, Adams C, Gulbins E. 2007. The cold case: Are rhinoviruses perfectly adapted pathogens? *Cell Mol. Life Sci.* 64:181–91

52. Cairns JA, Walls AF. 1996. Mast cell tryptase is a mitogen for epithelial cells: stimulation of IL-8 production and intercellular adhesion molecule-1 expression. *J. Immunol.* 156:275–83

53. Kupczyk M, Kupryś I, Danilewicz M, Bocheńska-Marciniak M, Murlewska A, et al. 2006. Adhesion molecules and their ligands in nasal polyps of aspirin-hypersensitive patients. *Ann. Allergy Asthma Immunol.* 96:105–11

54. Wos M, Sanak M, Soja J, Olechnowicz H, Busse WW, Szczeklik A. 2008. The presence of rhinovirus in lower airways of patients with bronchial asthma. *Am. J. Respir. Crit. Care Med.* 177:1082–89

55. Wark PA, Johnston SL, Bucchieri F, Powell R, Puddicombe S, Laza-Stanca V. 2005. Asthmatic bronchial epithelial cells have a deficient innate immune response to infection with rhinovirus. *J. Exp. Med.* 201:937–47

56. Szczeklik A. 1988. Aspirin-induced asthma as a viral disease. *Clin. Allergy* 18:15–20

57. Wang XS, Wu AY, Leung PS, Lau HY. 2007. PGE suppresses excessive anti-IgE induced cysteinyl leucotrienes production in mast cells of patients with aspirin exacerbated respiratory disease. *Allergy* 62:620–27

58. Bochenek G, Nagraba K, Nizankowska E, Szczeklik A. 2003. A controlled study of 9α,11β-PGF2 (a prostaglandin D2 metabolite) in plasma and urine of patients with bronchial asthma and healthy controls after aspirin challenge. *J. Allergy Clin. Immunol.* 111:743–49

59. Nasser S, Christie PE, Pfister R, Sousa AR, Walls A, et al. 1996. Effect of endobronchial aspirin challenge on inflammatory cells in bronchial biopsy samples from aspirin-sensitive asthmatic subjects. *Thorax* 51:64–70

60. Szczeklik A, Sladek K, Dworski R, Nizankowska E, Soja J, Sheller J. 1996. Bronchial aspirin challenge causes specific eicosanoid response in aspirin-sensitive asthmatics. *Am. J. Respir. Crit. Care Med.* 154:1608–14

61. Carpagnano GE, Resta O, Gelardi M, Spanevello A, Di Gioia G, et al. 2007. Exhaled inflammatory markers in aspirin-induced asthma syndrome. *Am. J. Rhinol.* 21:542–47

62. Rosenberg HF, Phipps S, Foster PS. 2007. Eosinophil trafficking in allergy and asthma. *J. Allergy Clin. Immunol.* 119:1303–10

63. Cowburn AS, Holgate ST, Sampson AP. 1999. IL-5 increases expression of 5-lipoxygenase-activating protein and translocates 5-lipoxygenase to the nucleus in human blood eosinophils. *J. Immunol.* 163:456–65

64. Sousa AR, Lams BE, Pfister R, Christie PE, Schmitz M, Lee TH. 1997. Expression of interleukin-5 and granulocyte-macrophage colony-stimulating factor in aspirin-sensitive and nonaspirin-sensitive asthmatic airways. *Am. J. Respir. Crit. Care Med.* 156:1384–89

65. Mastalerz L, Sanak M, Szczeklik A. 2001. Serum interleukin-5 in aspirin-induced asthma. *Clin. Exp. Allergy* 31:1036–40

66. Wills-Karp M. 2004. Interleukin-13 in asthma pathogenesis. *Immunol. Rev.* 202:175–90

67. Schmid-Grendelmeier P, Altznauer F, Fischer B, Bizer C, Straumann A, et al. 2002. Eosinophils express functional IL-13 in eosinophilic inflammatory diseases. *J. Immunol.* 169:1021–27

68. Espinosa K, Bossé Y, Stankova J, Rola-Pleszczynski M. 2003. CysLT1 receptor upregulation by TGF-β and IL-13 is associated with bronchial smooth muscle cell proliferation in response to LTD$_4$. *J. Allergy Clin. Immunol.* 111:1032–40

69. Bossé Y, Thompson C, Audette K, Stankova J, Rola-Pleszczynski M. 2008. Interleukin-4 and interleukin-13 enhance human bronchial smooth muscle cell proliferation. *Int. Arch. Allergy Immunol.* 146:138–48

70. Saha SK, Berry MA, Parker D, Siddiqui S, Morgan A, et al. 2008. Increased sputum and bronchial biopsy IL-13 expression in severe asthma. *J. Allergy Clin. Immunol.* 121:685–91

71. Trudeau J, Hu H, Chibana K, Chu HW, Westcott JY, Wenzel SE. 2006. Selective downregulation of prostaglandin E$_2$-related pathways by the Th2 cytokine IL-13. *J. Allergy Clin. Immunol.* 117:1446–54

72. Borowski A, Kuepper M, Horn U, K.üpfer U, Zissel G, et al. 2008. Interleukin-13 acts as an apoptotic effector on lung epithelial cells and induces profibrotic gene expression in lung fibroblasts. *Clin. Exp. Allergy* 38:619–28

73. Holgate ST, Holloway J, Wilson S, Bucchieri F, Puddicombe S, Davies DE. 2004. Epithelial-mesenchymal communication in the pathogenesis of chronic asthma. *Proc. Am. Thorac. Soc.* 1:93–98

74. Savla U, Appel HJ, Sporn PH, Waters CM. 2001. Prostaglandin E$_2$ regulates wound closure in airway epithelium. *Am. J. Physiol. Lung Cell Mol. Physiol.* 280:L421–31

75. Dahlén SE, Hedqvist P, Hammarstrom S, Samuelsson B. 1980. Leukotrienes are potent constrictors of human bronchi. *Nature* 288:484–86

76. Gyllfors P, Dahlén SE, Kumlin M, Larsson K, Dahlén B. 2006. Bronchial responsiveness to leukotriene D4 is resistant to inhaled fluticasone propionate. *J. Allergy Clin. Immunol.* 118:78–83

77. Laitinen LA, Laitinen A, Haahtela T, Vilkka V, Spur BW, Lee TH. 1993. Leukotriene E4 and granulocytic infiltration into asthmatic airways. *Lancet* 341:989–90

78. Christie PE, Tagari P, Ford-Hutchinson AW, Charlesson S, Chee P, Arm JA. 1991. Urinary leukotriene E4 concentrations increase after aspirin challenge in asthmatic subjects. *Am. Rev. Respir. Dis.* 102:5–29

79. Christie PE, Tagari P, Ford-Hutchinson AW, Black C, Markendorf A, et al. 1992. Urinary leukotriene E4 after lysine-aspirin inhalation in asthmatic subjects. *Am. Rev. Respir. Dis.* 146:1531–34

80. Nasser SM, Patel M, Bell GS, Lee TH. 1995. The effect of aspirin desensitization on urinary leukotriene E4 concentrations in aspirin-sensitive asthma. *Am. J. Respir. Crit. Care Med.* 151:1326–30

81. Picado C, Ramis I, Rosellò J, Prat J, Bulbena O, et al. 1992. Release of peptide leukotriene into nasal secretions after local instillation of aspirin in aspirin-sensitive asthmatic patients. *Am. Rev. Respir. Dis.* 145:65–69

82. Sanak M, Kielbasa B, Bochenek G, Szczeklik A. 2004. Exhaled eicosanoids following oral aspirin challenge in asthmatic patients. *Clin. Exp. Allergy* 34:1899–904

83. Sanak M, Pierzchalska M, Bazan-Socha S, Szczeklik A. 2000. Enhanced expression of the leukotriene C$_4$ synthase due to overactive transcription of an allelic variant associated with aspirin-intolerant asthma. *Am. J. Respir. Cell Mol. Biol.* 23:290–96

84. Cowburn AS, Sladek K, Soja J, Adamek L, Nizankowska E, Szczeklik A. 1998. Overexpression of leukotriene C4 synthase in bronchial biopsies from patients with aspirin-intolerant asthma. *J. Clin. Investig.* 101:834–46

85. Cai Y, Bjermer L, Halstensen TS. 2003. Bronchial mast cells are the dominating LTC$_4$S-expressing cells in aspirin-tolerant asthma. *Am. J. Respir. Cell Mol. Biol.* 29:683–93

86. Kawagishi Y, Mita H, Taniguchi M, Maruyama M, Oosaki R, et al. 2002. Leukotriene C4 synthase promoter polymorphism in Japanese patients with aspirin-induced asthma. *J. Allergy Clin. Immunol.* 109:936–42

87. Sanak M, Simon HU, Szczeklik A. 1997. Leukotriene C4 synthase promoter polymorphism and risk of aspirin-induced asthma. *Lancet* 350:1599–600

88. Kedda MA, Shi J, Duffy D, Phelps S, Yang I, et al. 2004. Characterization of two polymorphisms in the leukotriene C4 synthase gene in an Australian population of subjects with mild, moderate, and severe asthma. *J. Allergy Clin. Immunol.* 113:889–95

89. Van Sambeek R, Stevenson DD, Baldasaro M, Lam BK, Zhao J, et al. 2000. 5′ flanking region polymorphism of the gene encoding leukotriene C4 synthase does not correlate with the aspirin-intolerant asthma phenotype in the United States. *J. Allergy Clin. Immunol.* 106:72–76

90. Kanaoka Y, Boyce Y. 2004. Cysteinyl leukotrienes and their receptors: cellular distribution and function in immune and inflammatory responses. *J. Immunol.* 173:1503–10

91. Lynch KR, O'Neill GP, Liu Q, Im DS, Sawyer N, et al. 1999. Characterization of the human cysteinyl leukotriene CysLT1 receptor. *Nature* 399:789–93

92. Sarau HM, Ames RS, Chambers J, Ellis C, Elshourbagy N, et al. 1999. Identification, molecular cloning, expression, and characterization of a cysteinyl leukotriene receptor. *Mol. Pharmacol.* 56:657–63

93. Heise CE, O'Dowd BF, Figueroa DJ, Sawyer N, Nguyen T, et al. 2000. Characterization of the human cysteinyl leukotriene 2 receptor. *J. Biol. Chem.* 275:30531–36

94. Nothacker HP, Wang Z, Zhu Y, Reinscheid RK, Lin SH, Civelli O. 2000. Molecular cloning and characterization of a second human cysteinyl leukotriene receptor: discovery of a subtype selective agonist. *Mol. Pharmacol.* 58:1601–8

95. Chibana K, Ishii Y, Asakura T, Fukuda T. 2003. Up-regulation of cysteinyl leukotriene 1 receptor by IL-13 enables human lung fibroblasts to respond to leukotriene C4 and produce eotaxin. *J. Immunol.* 170:4290–95

96. Jiang Y, Borrelli LA, Kanaoka Y, Bacskai BJ, Boyce JA. 2007. CysLT2 receptors interact with CysLT1 receptors and down-modulate cysteinyl leukotriene dependent mitogenic responses of mast cells. *Blood* 110:3263–70

97. Corrigan C, Mallett K, Ying S, Roberts D, Parikh A, et al. 2005. Expression of the cysteinyl leukotriene receptors cysLT$_1$ and cysLT$_2$ in aspirin-sensitive and aspirin-tolerant chronic rhinosinusitis. *J. Allergy Clin. Immunol.* 115:316–22

98. Arm JP, O'Hickey SP, Spur BW, Lee TH. 1989. Airway responsiveness to histamine and leukotriene E4 in subjects with aspirin-induced asthma. *Am. Rev. Respir. Dis.* 140:148–53

99. Christie PE, Schmitz-Schumann M, Spur BW, Lee TH. 1993. Airway responsiveness to leukotriene C4 (LTC4), leukotriene E4 (LTE4) and histamine in aspirin-sensitive asthmatic subjects. *Eur. Respir. J.* 6:1468–73

100. Lee TH, Austen KF, Corey EJ, Drazen JM. 1984. Leukotriene E4-induced airway hyperresponsiveness of guinea pig tracheal smooth muscle to histamine and evidence for three separate sulfidopeptide leukotriene receptors. *Proc. Natl. Acad. Sci. USA* 81:4922–25

101. Pong SS, DeHaven RN. 1983. Characterization of a leukotriene D4 receptor in guinea pig lung. *Proc. Natl. Acad. Sci. USA* 80:7415–19

102. Bäck M, Norel X, Walch L, Gascard J, de Montpreville V, et al. 2000. Prostacyclin modulation of contractions of the human pulmonary artery by cysteinyl-leukotrienes. *Eur. J. Pharmacol.* 401:389–95

103. Walch L, Norel X, Bäck M, Gascard JP, Dahlén SE, Brink C. 2002. Pharmacological evidence for a novel cysteinyl-leukotriene receptor subtype in human pulmonary artery smooth muscle. *Br. J. Pharmacol.* 137:1339–45

104. Krilis S, Lewis RA, Corey EJ, Austen KF. 1983. Specific receptors for leukotriene C4 on a smooth muscle cell line. *J. Clin. Investig.* 72:1516–19

105. Pong SS, DeHaven RN, Kuehl FA Jr, Egan RW. 1983. Leukotriene C4 binding to rat lung membranes. *J. Biol. Chem.* 258:9616–19

106. Drazen JM, Lewis RA, Austen KF, Corey EJ. 1983. Pulmonary phamacology of the SRS-A leukotrienes. In *Leukotrienes and Prostacyclin*, ed. F Berti, G Folco, GP Velo, pp. 125–34. New York: Plenum

107. Bandeira-Melo C, Woods LJ, Phoofolo M, Weller PF. 2002. Intracrine cysteinyl leukotriene receptor-mediated signaling of eosinophil vesicular transport-mediated interleukin-4 secretion. *J. Exp. Med.* 196:841–50

108. Arm JP, O'Hickey SP, Hawksworth RJ, Fong CY, Crea AE, et al. 1990. Asthmatic airways have a disproportionate hyperresponsiveness to LTE4, as compared with normal airways, but not to LTC4, LTD4, methacholine, and histamine. *Am. Rev. Respir. Dis.* 142:1112–18

109. O'Hickey SP, Hawksworth RJ, Fong CY, Arm JP, Spur BW, Lee TH. 1991. Leukotrienes C4, D4, and E4 enhance histamine responsiveness in asthmatic airways. *Am. Rev. Respir. Dis.* 144:1053–57

110. Arm JP, Austen KF. 2002. Leukotriene receptors and aspirin sensitivity. *N. Engl. J. Med.* 347:1524–26

111. Yoshisue H, Kirkham-Brown J, Healy E, Holgate ST, Sampson AP, Davies DE. 2007. Cysteinyl leukotrienes synergize with growth factors to induce proliferation of human bronchial fibroblasts. *J. Allergy Clin. Immunol.* 119:132–40

112. Ciana P, Fumagalli M, Trincavelli ML, Verderio C, Rosa P, et al. 2006. The orphan receptor GPR17 identified as a new dual uracil nucleotides/cysteinyl-leukotrienes receptor. *EMBO J.* 25:4615–27

113. Pleskow WW, Stevenson DD, Mathison DA, Simon RA, Schatz M, Zeiger RS. 1982. Aspirin desensitization in aspirin-sensitive asthmatic patients: clinical manifestations and characterization of the refractory period. *J. Allergy Clin. Immunol.* 69:11–19

114. Stevenson DD. 2003. Aspirin desensitization in patients with AERD. *Clin. Rev. Allergy Immunol.* 24:159–68

115. Lee JY, Simon RA, Stevenson DD. 2007. Selection of aspirin dosages for aspirin desensitization treatment in patients with aspirin-exacerbated respiratory disease. *J. Allergy Clin. Immunol.* 119:157–64

116. Jayawickreme SP, Gray T, Nettesheim P, Eling T. 1999. Regulation of 15-lipoxygenase expression and mucus secretion by IL-4 in human bronchial epithelial cells. *Am. J. Physiol.* 276:L596–603

117. Nassar GM, Morrow JD, Roberts LJ 2nd, Lakkis FG, Badr KF. 1994. Induction of 15-lipoxygenase by interleukin-13 in human blood monocytes. *J. Biol. Chem.* 269:27631–34

118. Bradding P, Redington AE, Djukanovic R, Conrad DJ, Holgate ST. 1995. 15-Lipoxygenase immunoreactivity in normal and in asthmatic airways. *Am. J. Respir. Crit. Care Med.* 151:1201–4

119. Peters-Golden M. 2002. 'Good' lipids for asthma. *Nat. Med.* 8:931–32

120. Serhan CN, Hamberg M, Samuelsson B. 1984. Lipoxins: novel series of biologically active compounds formed from arachidonic acid in human leukocytes. *Proc. Natl. Acad. Sci. USA* 81:5335–39

121. Stevenson DD, Szczeklik A. 2006. Clinical and pathologic perspectives on aspirin sensitivity and asthma. *J. Allergy Clin. Immunol.* 118:773–86

122. Kowalski ML, Ptasinska A, Bienkiewicz B, Pawliczak R, DuBuske L. 2003. Differential effects of aspirin and misoprostol on 15 hydroxyeicosatetraenoic acid generation by leukocytes from aspirin-sensitive asthmatic patients. *J. Allergy Clin. Immunol.* 112:505–12

123. Celik GE, Erkekol FO, Misirligil Z, Melli M. 2007. Lipoxin A4 levels in asthma: relation with disease severity and aspirin sensitivity. *Clin. Exp. Allergy* 37:1494–501

124. Feltenmark S, Gautam N, Brunnström A, Griffiths W, Backman L, et al. 2008. Eoxins are proinflammatory arachidonic acid metabolites produced via the 15-lipoxygenase-1 pathway in human eosinophils and mast cells. *Proc. Natl. Acad. Sci. USA* 105:680–85

125. Vancheri C, Mastruzzo C, Sortino MA, Crimi N. 2004. The lung as a privileged site for the beneficial actions of PGE2. *Trends Immunol.* 25:40–46

126. Holgate ST, Peters-Golden M, Panettieri RA, Henderson WR Jr. 2003. Roles of cysteinyl leukotrienes in airway inflammation, smooth muscle function, and remodeling. *J. Allergy Clin. Immunol.* 111:S18–34

127. Charbeneau RP, Peters-Golden M. 2005. Eicosanoids: mediators and therapeutic targets in fibrotic lung disease. *Clin. Sci. (London)* 108:479–91

128. Kolodsick JE, Peters-Golden M, Larios J, Toews GB, Thannickal VJ, Moore BB. 2003. Prostaglandin E_2 inhibits fibroblast to myofibroblast transition via E prostanoid receptor 2 signaling and cyclic adenosine monophosphate elevation. *Am. J. Respir. Cell Mol. Biol.* 29:537–44

129. Szczeklik A, Mastalerz L, Nizankowska E, Cmiel A. 1996. Protective and bronchodilator effects of prostaglandin E and salbutamol in aspirin-induced asthma. *Am. J. Respir. Crit. Care Med.* 153:567–71

130. Flamand N, Surette ME, Picard S, Bourgoin S. 2002. Cyclic AMP-mediated inhibition of 5-lipoxygenase translocation and leukotriene biosynthesis in human neutrophils. *Mol. Pharmacol.* 62:250–56

131. Murakami M, Kudo I. 2006. Prostaglandin E synthase: a novel drug target for inflammation and cancer. *Curr. Pharm. Des.* 12:943–54

132. Bakhle YS. 1983. Synthesis and catabolism of cyclo-oxygenase products. *Br. Med. Bull.* 39:214–18

133. Kowalski ML, Pawliczak R, Wozniak J, Siuda K, Poniatowska M, et al. 2000. Differential metabolism of arachidonic acid in nasal polyp epithelial cells cultured from aspirin-sensitive and aspirin-tolerant patients. *Am. J. Respir. Crit. Care Med.* 161:391–98

134. Pierzchalska M, Szabó Z, Sanak M, Soja J, Szczeklik A. 2003. Deficient prostaglandin E2 production by bronchial fibroblasts of asthmatic patients, with special reference to aspirin-induced asthma. *J. Allergy Clin. Immunol.* 111:1041–48

135. Ferreri NR, Howland WC, Stevenson DD, Spiegelberg HL. 1988. Release of leukotrienes, prostaglandins, and histamine into nasal secretions of aspirin-sensitive asthmatics during reaction to aspirin. *Am. Rev. Respir. Dis.* 137:847–54

136. Picado C, Ramis I, Rosellò J, Prat J, Bulbena O, Plaza V. 1992. Release of peptide leukotriene into nasal secretions after local instillation of aspirin in aspirin-sensitive asthmatic patients. *Am. Rev. Respir.* 145:65–69

137. Sanak M, Kielbasa B, Bochenek G, Szczeklik A. 2004. Exhaled eicosanoids following oral aspirin challenge in asthmatic patients. *Clin. Exp. Allergy* 34:1899–904

138. Sladek K, Dworski R, Soja J, Sheller JR, Nizankowska E, et al. 1994. Eicosanoids in bronchoalveolar lavage fluid of aspirin-intolerant patients with asthma after aspirin challenge. *Am. J. Respir. Crit. Care Med.* 149:940–46

139. Narumiya S, Sugimoto Y, Ushikubi F. 1999. Prostanoid receptors: structures, properties, and functions. *Physiol Rev.* 79:1193–226

140. Jinnai N, Sakagami T, Sekigawa T, Kakihara M, Nakajima T, et al. 2004. Polymorphisms in the prostaglandin E2 receptor subtype 2 gene confer susceptibility to aspirin-intolerant asthma: a candidate gene approach. *Hum. Mol. Genet.* 13:3203–17

141. Parfenova H, Parfenov VN, Shlopov BV, Levine V, Falkos S, et al. 2001. Dynamics of nuclear localization sites for COX-2 in vascular endothelial cells. *Am. J. Physiol. Cell Physiol.* 281:C166–78

142. Gyllfors P, Bochenek G, Overholt J, Drupka D, Kumlin M, et al. 2003. Biochemical and clinical evidence that aspirin-intolerant asthmatic subjects tolerate the cyclooxygenase 2-selective analgetic drug celecoxib. *J. Allergy Clin. Immunol.* 111:1116–21

143. Szczeklik A, Sanak M. 2006. The broken balance in aspirin hypersensitivity. *Eur. J. Pharmacol.* 533:145–55

144. Picado C, Fernandez-Morata JC, Juan M, Roca-Ferrer J, Fuentes M, Xaubet A. 1999. Cyclooxygenase-2 mRNA is downexpressed in nasal polyps from aspirin-sensitive asthmatics. *Am. J. Respir. Crit. Care Med.* 160:291–96

145. Pujols L, Mullol J, Alobid I, Roca-Ferrer J, Xaubet A, Picado C. 2004. Dynamics of COX-2 in nasal mucosa and nasal polyps from aspirin-tolerant and aspirin-intolerant patients with asthma. *J. Allergy Clin. Immunol.* 114:814–19

146. Sousa A, Pfister R, Christie PE, Lane SJ, Nasser SM, et al. 1997. Enhanced expression of cyclo-oxygenase isoenzyme 2 (COX-2) in asthmatic airways and its cellular distribution in aspirin-sensitive asthma. *Thorax* 52:940–45

147. Mastalerz L, Sanak M, Gawlewicz-Mroczka A, Gielicz A, Cmiel A, Szczeklik A. 2008. Prostaglandin E2 systemic production in patients with asthma with and without aspirin hypersensitivity. *Thorax* 63:27–34

148. Szczeklik A, Stevenson DD. 2003. Aspirin-induced asthma: advances in pathogenesis, diagnosis, and management. *J Allergy Clin. Immunol.* 111:913–21

149. Yoshida S, Sakamoto H, Yamawaki Y, Shoji T, Akahori K, et al. 1998. Effect of acyclovir on bronchoconstriction and urinary leukotriene E4 excretion in aspirin-induced asthma. *J. Allergy Clin. Immunol.* 102:909–14

150. Pérez-Novo CA, Waeytens A, Claeys C, Cauwenberge PV, Bachert C. 2008. *Staphylococcus aureus* enterotoxin B regulates prostaglandin E2 synthesis, growth, and migration in nasal tissue fibroblasts. *J. Infect. Dis.* 197:1036–43

Immunobiology of Asthma

Qutayba Hamid and Meri Tulic

Meakins-Christie Labs, McGill University, Montreal, Québec H2X 2P2, Canada;
email: qutayba.hamid@mcgill.ca

Annu. Rev. Physiol. 2009. 71:489–507

First published online as a Review in Advance on
November 13, 2008

The *Annual Review of Physiology* is online at
physiol.annualreviews.org

This article's doi:
10.1146/annurev.physiol.010908.163200

Copyright © 2009 by Annual Reviews.
All rights reserved

0066-4278/09/0315-0489$20.00

Key Words

pathology, cytokines, chemokines, remodeling, immunology

Abstract

Asthma is characterized by chronic inflammation of the airways in which
there is an overabundance of eosinophils, mast cells, and activated T
helper lymphocytes. These inflammatory cells release mediators that
then trigger bronchoconstriction, mucus secretion, and remodeling.
The inflammatory mediators that drive this process include cytokines,
chemokines, growth factors, lipid mediators, immunoglobulins, and histamine. The inflammation in allergic asthma can be difficult to control.
This is mainly due to the development of an adaptive immunity to an allergen, leading to immunological memory. This leads to recall reactions
to the allergen, causing persistent inflammation and damage to the airways. Generally, in asthma inflammation is directed by Th2 cytokines,
which can act by positive feedback mechanisms to promote the production of more inflammatory mediators including other cytokines and
chemokines. This review discusses the role of cytokines and chemokines
in the immunobiology of asthma and attempts to relate their expression
to morphological and functional abnormalities in the lungs of asthmatic
subjects. We also discuss new concepts in asthma immunology, in particular the role of cytokines in airway remodeling and the interaction
between cytokines and infection.

INTRODUCTION

Histopathological changes in the bronchial and bronchiolar walls in asthma involve the mucosa (i.e., the epithelium and lamina propria), submucosa [with included airway smooth muscle (ASM) and mucus-secreting glands], and adventitia (the interface between airway and surrounding lung parenchyma) (1, 2). Information on the pathological changes in asthma has been obtained from studying airway sections from asthma deaths; patients who had asthma but who died from non-asthma-related causes; surgically resected lung; and endoscopic biopsies of mild, moderate, and severe asthmatic subjects. Tissue from transbronchial biopsies of severe asthmatics has also been studied for inflammation and remodeling

It is now accepted that asthma is a chronic inflammatory condition, and evidence of inflammation can be observed in mild, moderate, and severe disease. However, the relative magnitude, type of inflammatory cells, and site of the inflammatory infiltrate may differ among patients. Many cells are involved in the immune and inflammatory responses to allergens in asthma; these include T cells, eosinophils, mast cells, and neutrophils.

The role of the activated T lymphocyte in controlling and perpetuating chronic inflammation in asthma has received much attention (3). In some studies, T cell activation can be related to measures of asthma severity, such as the degree of airway narrowing or airway hyperresponsiveness (AHR), as well as the bronchial eosinophil response (4). The association between tissue eosinophilia and asthma is strong, but the degree of tissue eosinophilia varies with each case and with the duration of the terminal episode (5, 6).

Mast cells are usually found adjacent to blood vessels in the lamina propria in normal human airways, but in asthma they are observed in the airway epithelium (7), the airway mucous glands, and the ASM (8–10). Neutrophilic inflammation is found in severe persistent asthma (11), asthma exacerbations (12–14), sudden-onset fatal asthma (15), occu-pational asthma (16), nocturnal asthma (17), and even childhood asthma (18). Neutrophils may play a role in the pathophysiology of airway disease through their release of reactive oxygen species, cytokines, lipid mediators, and enzymes including elastase, cathepsin G, myeloperoxidases, and nonenzymatic defensins (19).

Dendritic cells and their subtypes are key antigen-presenting cells that respond rapidly to antigenic challenge with kinetics similar to those of neutrophils (20). They form an interface between innate and adaptive immunity and orchestrate both primary and secondary immune responses (21). They are present throughout the respiratory tree and number approximately 500 cells per mm^2 within the epithelium (22). Inflammatory cells and immune responses are regulated by a number of immune mediators that are secreted from inflammatory and structural cells. In this review, we focus on cytokines and chemokines and their possible role in the immunobiology of asthma.

CYTOKINES AND CHEMOKINES

Cytokines are a family of small glycosylated proteins that are involved in cell-to-cell signaling, cellular growth, differentiation, proliferation, chemotaxis, immunomodulation, immunoglobulin isotype switching, and apoptosis. The actions of cytokines are mediated through specific cytokine receptors on the surfaces of target cells. Although cytokines usually have effects on adjacent cells, they can act at a distance and can have effects on the cells producing the cytokines themselves. Many of these cytokines exhibit pleiotropy and have overlapping functions, making their individual roles in the pathogenesis of asthma and allergic disease difficult to differentiate.

Until recently, T lymphocytes and eosinophils were considered to be the major source of cytokines in asthma (**Figure 1**), but researchers now recognize that cytokines are produced not only by other inflammatory cells, but also by structural cells including epithelial, endothelial and ASM cells, and fibroblasts. To date, more than 30 different cytokines involved in asthma pathology have been identified, and this number continues to grow. Among these cytokines are T cell–derived molecules such as the so-called T helper–1 (Th1) cells [interleukin (IL)-2, interferon (IFN)-γ, and IL-12], Th2 cells (IL-4, -5, -9, -13, and -25), Th3 or T regulatory cytokines [IL-10 and transforming growth factor beta (TGF-β)], and Th-17 cells (IL-17A and -17F), all of which are involved in the regulation of cell-mediated and humoral immunities. Although the distinction between Th1 and Th2 cells in humans is not as clear as in mice, there is overwhelming evidence in the literature supporting the notion that allergic inflammation is driven by an imbalance between Th1 and Th2 cytokines, favoring the Th2 arm of the immune response (**Figure 2**). Recently, Th-17-associated cytokines were also implicated in asthma immunobiology (W. Ramli, J. Martin & Q. Hamid, unpublished data). Other cytokines include proinflammatory cytokines [IL-1β, -6, and -11; tumor necrosis factor (TNF)-α; and granulocyte/macrophage colony–stimulating factor (GM-CSF)] that are involved in innate host defense, anti-inflammatory cytokines (IL-10, IFN-γ, IL-12, and IL-18), growth factors [platelet-derived growth factor (PDGF), TGF-β, fibroblast growth factor (FGF), and epidermal growth factor (EGF)], and chemotactic cytokines or chemokines [RANTES, monocyte chemoattractant protein (MCP)-1–MCP-5, eotaxin, and IL-8]. In this review, which focuses on the role of cytokines and chemokines in asthma, cytokines and chemokines are broadly categorized on the basis of their functional activities and are subdivided into (*a*) eosinophil-associated cytokines, (*b*) immunoglobin E (IgE)-mediated cytokines, (*c*) remodeling-associated cytokines, and (*d*) immunomodulatory cytokines.

IMMUNOGLOBIN E–ASSOCIATED CYTOKINES

Asthma is clinically categorized into occupational, nonatopic, and atopic (allergic) forms;

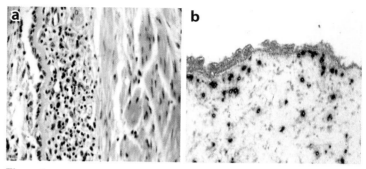

Figure 1

T cells are the predominant cells that infiltrate the airways in asthma and are the major source of cytokines. (*a*) A hematoxylin- and eosin-stained section showing T cell infiltration. (*b*) An in situ hybridization of section from a bronchial biopsy from a patient with asthma showing messenger RNA expression of interleukin-4 in T cells.

the vast majority of asthmatics have the atopic form. A central mediator in atopic asthma is the IgE antibody, which is produced by sensitized allergen-specific B cells (**Figure 2**). Allergens are antigens that can (*a*) elicit hypersensitivity or allergic reactions and (*b*) increase IgE levels in the serum in susceptible subjects subsequent to stimulation. By presenting the allergen fragments in conjunction with the major histocompatibility complex (MHC), B cells can activate specific Th2 cells to produce numerous cytokines, leading to further B cell activation and antibody release. IgE antibodies bind to the high-affinity IgE receptor Fc epsilon receptor I (FcεRI) that is present on mast cells, eosinophils, and basophils, thereby sensitizing these cells to antigen exposure. Subsequently, cross-linking of adjacent IgE–FcεRI complexes by allergens triggers (*a*) the degranulation of cytoplasmic vesicles containing histamine and (*b*) the de novo formation of eicosanoids and reactive oxygen species. This results in smooth muscle contraction, mucous secretion, and vasodilatation, all of which are hallmarks of asthma. IgE-producing B cells play a critical role in allergic inflammation; consequently, factors responsible for their activation—namely IgE-associated cytokines such as IL-4, IL-9, and IL-13—are of considerable interest.

IL-4 is vital for the regulation of growth, differentiation, activation, and function of B cells

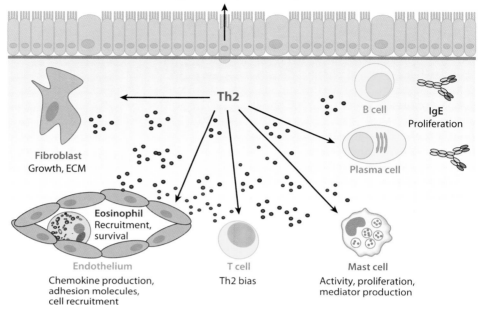

Epithelium
Mucus secretion, proliferation, chemokine production

Th2

Fibroblast
Growth, ECM

B cell

IgE
Proliferation

Plasma cell

Eosinophil
Recruitment, survival

Endothelium
Chemokine production, adhesion molecules, cell recruitment

T cell
Th2 bias

Mast cell
Activity, proliferation, mediator production

Figure 2

The possible effect of T helper–2 (Th2) cytokines on various cells in the lung and the cytokines' potential role in asthma. Abbreviations: ECM, extracellular matrix; IgE, immunoglobin E.

(23) (**Figure 2**). IL-4 exerts its actions by a specific cell-surface receptor composed of the IL-4Rα chain and the γ common chain. IL-4 increases expression of the antigen-presenting proteins, MHC class II molecules, on B cells, resulting in increased allergen-presentation capacity to Th2 cells (24). On the vasculature, IL-4 promotes expression of VCAM-1 on endothelium, thereby allowing for recruitment of eosinophils and other inflammatory cells such as T cells, monocytes, and basophils from the blood into the sites of inflammation (25). IL-4 also induces isotype switching, a process that leads to the production of IgE by B cells. After switching occurs, IL-4 potentiates IgE production. Furthermore, IL-4 enhances the IgE-mediated response by upregulating IgE receptors on inflammatory cells within the airway (26). Conversely, activation of IgE by IL-4 can be diminished by cross-regulation from Th1 cytokines (**Figure 1**). IFN-γ, a Th1 cytokine, can suppress isotype switch recombination to the IgE isotype in B cells activated by IL-4 (27). Additionally, IFN-γ also inhibits IL-4-induced expression of the low-affinity IgE receptor (28). IL-4Rα-deficient mice are more resistant to the development of features of asthma than are the IL-4-deficient mice (29). However, the latter mice still develop AHR, which suggests that there exist other cytokines that can initiate signal transduction via the IL-4 receptor.

IL-13 has a 70% sequence homology with IL-4, and it binds a heterodimer composed of the IL-4Rα chain and an IL-13Rα chain (30). Like IL-4, IL-13 is produced by Th2 cells and is found in high concentrations within allergic tissues (31, 32). Owing to the redundancy in IL-4Rα binding, IL-4 and IL-13 exhibit some degree of functional overlap. Similarly to IL-4, overexpression of IL-13 within the lungs results in IgE production, inflammation, mucus hypersecretion, eosinophilia, and upregulation of VCAM-1. However, the unique nature of IL-13 can be observed in its effects on airway sensitivity to contractile agonists, whereby the blocking of IL-13 prevents AHR in mice following antigen challenge (33). Accordingly, researchers hypothesize that IL-13 is the primary factor involved in the expression and induction of allergen-induced AHR.

Both IL-4 and IL-13 are critical in the induction and regulation of allergic asthma through their production of IgE. Interestingly, the effect of exogenous IL-13 depends on when it is given in relationship to allergen exposure: Its administration after initial allergen sensitization in mice has no effect on serum IgE levels (33). The emerging paradigm is that IL-13 induces features of the allergic response via its actions on epithelial and smooth muscle cells rather than through traditional effector pathways involving eosinophils and IgE-mediated events (34).

IL-9 is also relevant to IgE-dependent host responses. IL-9 is a Th2 cytokine, and its expression is regulated by a variety of mediators, in particular IL-2, which stimulates its production. Although IL-9 is produced by a variety of cell types including mast cells, eosinophils, and neutrophils, the major sources of this cytokine are Th lymphocytes. Transgenic mice overexpressing IL-9 have increased serum levels of all Ig isotypes, including IgE, and an associated accumulation of B cells in the lungs (35, 36). In vitro, IL-9 enhances IL-4-mediated IgE production by both human and murine B cells (37). IL-9 also stimulates protease production from mast cells and induces their expression of $Fc\epsilon RI$. This suggests that in addition to potentiating IgE production, IL-9 primes mast cells to respond to allergen challenge through increased cell-surface expression of $Fc\epsilon RI$ and the production of proinflammatory mediators.

In addition to influencing IgE-mediated immunity, IL-9 can coordinate a multitude of responses associated with asthma through direct, indirect, and synergistic actions. IL-9 has been identified as a T cell growth factor and is capable of stimulating the proliferation of activated T cells (38–40). IL-9-transgenic mice

demonstrate in vivo increased AHR, marked eosinophilia, mucous overproduction (41), and increased expression of eotaxin and MCP-1, -3, and -5 in airway epithelial cells (42).

The newly described cytokine IL-25 (also known as IL-17E) also seems to play a role in the regulation of IgE-mediated responses. IL-25 stimulates IgE synthesis and eosinophilia in mice models of allergic inflammation by stimulating the release of IL-4 and IL-5 cytokines (43). As the roles of novel cytokines such as IL-25 are clarified, especially with regard to the regulation of IgE-mediated inflammation, it becomes evident that the cytokine network regulating inflammation is broad and complex. Nonetheless, insofar as they prove to be critical mediators of the inflammatory process, the stimulatory effects of these cytokines make them obvious targets in the treatment of allergic diseases (**Figure 3**).

EOSINOPHIL-ASSOCIATED CYTOKINES

Eosinophils are prominent in allergic airway disease and are still considered by many to be the hallmark of asthma. Increased numbers of eosinophils in the bronchial mucosa as well as in the bronchoalveolar lavage (BAL) and sputum are consistent features of asthma, and BAL eosinophilia has been linked to development of the late airway response and asthma severity. Increased eosinophilia in asthmatics is observed not only in the large or the central airways of these patients but also in the peripheral parts of the lungs (44). Although terminal differentiation of eosinophils occurs within the bone marrow, recent evidence indicates that eosinophils can also differentiate locally at the site of inflammation and that the presence of eosinophils within allergic mucosal tissue is not solely due to infiltration of mature cells (45).

Teleologically, eosinophils form part of the host defense against parasitic infestation. The biological activity exerted by these cells is largely attributable to their release of prestored granular proteins, including eosinophil cationic protein, eosinophil peroxidase, and major basic

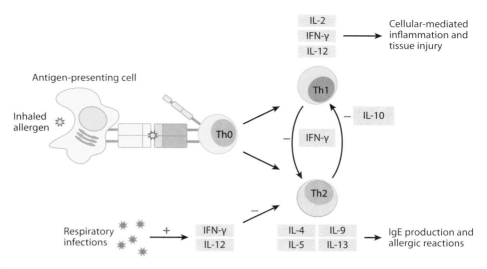

Figure 3

The regulation of T cells, cytokines, and factors that may bias cytokines' gene expression toward T helper–1 (Th1) or Th2 patterns. Abbreviations: IFN, interferon; IgE, immunoglobin E; IL, interleukin.

protein. These potent cytotoxic proteins have been found in high concentrations in the sputum of asthmatic patients and are thought to play an important role in the epithelial damage observed in asthmatic patients. In addition to cytotoxic proteins, eosinophils can synthesize and release oxygen radicals and lipid mediators (e.g., LTB$_4$, LTC$_4$, and PAF), as well as numerous cytokines (e.g., IL-1, -2, -3, -4, -5, -6, -10, -12, and -13; TNF-α; and GM-CSF) and chemokines (e.g., IL-8, RANTES, and MIP-1α). Recently, it was suggested that eosinophils may play a role in airway remodeling because they have the ability to synthesize and release fibrogenic cytokines including TGF-β, IL-11, IL-17, and IL-25 (46). Together, these eosinophil-associated cytokines are responsible for the tissue-destructive potency of this proinflammatory cell.

IL-5 is the most important Th2 cytokine associated with eosinophils, and it can regulate most aspects of eosinophil behavior including eosinophil growth, maturation, differentiation, survival, and activation (**Figure 2**). Although IL-5 is produced by T helper cells, cytotoxic T lymphocytes, and mast cells, eosinophils are the predominant sources of this cytokine. Hu-

man eosinophils express IL-5 messenger RNA (mRNA) and release IL-5 protein in vitro. Allergen challenge in the bronchial segments results in increased IL-5 mRNA expression in eosinophils in BAL fluid, with a 300-fold increase in IL-5 protein concentrations (47). IL-5 plays a central role in accumulation and activation of eosinophils in the lungs, an effect readily seen in IL-5-overexpressing transgenic mice that have lifelong eosinophilia (48). Moreover, IL-5 is a potent eosinophil chemoattractant (3), and it upregulates integrin receptor expression on eosinophils, thereby promoting adherence of eosinophils to VCAM-expressing endothelial cells and eosinophil accumulation.

Studies with IL-5 monoclonal antibodies in animal models of allergic inflammation clearly support a role for IL-5 in allergic disease; however, similar studies in humans have been disappointing. Although blocking IL-5 was effective in abolishing blood and sputum eosinophilia, it did not protect against the allergen-induced late airway response, nor did it have any effect on baseline AHR in mild asthmatics (49). These results question the role of eosinophils in airway responsiveness in humans and suggest that alternate mechanisms and/or factors are responsible for airway narrowing in asthmatic patients.

Although secondary in importance to IL-5, IL-3 and GM-CSF are also typically viewed as eosinophil-associated cytokines. IL-3 and GM-CSF are pluripotent hematopoietic growth factors that stimulate the formation of not only eosinophil lineages but also neutrophil, erythroid, and monocytic lineages. Increased expression of GM-CSF has been documented in the bronchial epithelium and in the eosinophils of asthmatics following endobronchial allergen challenge. GM-CSF is involved in the priming of eosinophils and accounts for increased eosinophil survival in the BAL fluid of asthmatic patients. In addition, GM-CSF may also be involved in development of chronic eosinophilia and airway remodeling of asthma (see Remodeling-Associated Cytokines section, below), as insertion of the *GM-CSF* gene in the epithelium of rats caused eosinophil accumulation in their lungs and irreversible fibrosis (50, 51).

REMODELING-ASSOCIATED CYTOKINES

It has long been known that architectural and structural changes occur in the airways of asthmatic patients. These changes include collagen (type III and IV) and fibronectin deposition, increased thickness of subepithelial basement membrane, goblet cell hyperplasia, increased ASM mass and size, angiogenesis, and fibrosis, all of which collectively contribute to the phenomenon known as airway remodeling. Some of these changes were first described in postmortem airway sections from status asthmaticus victims in the 1960s. More recently, airway remodeling was reported in patients with mild asthma and in children with difficult-to-treat asthma (52). The functional consequences of airway remodeling include persistent AHR and mucous hypersecretion, which contribute to increased susceptibility to asthma exacerbations. The mechanisms involved in airway remodeling are poorly understood, but research in the past three to five years suggests that the balance between matrix metalloproteinases and tissue inhibitors of metalloproteinases may play a role in this process. Moreover, the increase in ASM content, along with a change in the phenotype of fibroblasts to contractile myofibroblasts, may explain the permanent reduction in airway caliber, which is steroid insensitive and typical in patients with severe forms of the disease. The predominant remodeling-associated cytokines include TGF-β, PDGF, IL-6, IL-11, IL-13, IL-17, and IL-25 (**Figure 2**).

TGF-β is a potent profibrotic cytokine. Major sources of TGF-β include fibroblasts, eosinophils, and epithelial cells. However, macrophages, monocytes, neutrophils, ASM cells, and lymphocytes also produce this cytokine. TGF-β is detected in increased concentrations in baseline asthmatic BAL fluid before allergen challenge, and its levels increase even more after allergen challenge. Furthermore, TGF-β exerts an important influence on the turnover of extracellular matrix proteins. In in vitro tissue culture systems, TGF-β exhibits a pleiotropic nature depending on the cell type, culture conditions, and presence of other cytokines. TGF-β induces the proliferation and release of profibrotic and proinflammatory cytokines in fibroblasts and ASM cells, whereas in monocytes, lymphocytes, and epithelial cells, TGF-β inhibits their proliferation and cytokine release (53, 54). TGF-β is also a potent chemoattractant for many cell types, including monocytes, fibroblasts, and mast cells.

Recently, eosinophils were recognized as one of the most abundant sources of TGF-β not only in the asthmatic airways but also in (*a*) the nasal tissues of patients with nasal polyposis and (*b*) hypereosinophilic patients. Using in situ hybridization and immunocytochemistry, our laboratory has demonstrated TGF-β to be significantly elevated in both mild and severe asthmatics, compared with normal subjects, and we have shown that levels of TGF-β expression correlate with basement membrane thickness and disease severity in these patients (55). Approximately 65% of all TGF-β-positive cells are activated eosinophils, which are localized within the reticular lamina. The local production of TGF-β by eosinophils may be responsible for the subepithelial fibrosis observed in asthmatics. However, TGF-β can also inhibit eosinophil survival and function and may be involved in the repair process of airway epithelial cells. These effects of TGF-β illustrate the complex actions of this cytokine in asthma.

PDGF is not only a major mitogen but also a remodeling-associated cytokine. Its ability to stimulate proliferation of tissue-structural cells, including fibroblasts, epithelial cells, and vascular smooth muscle cells, is well known, and this cytokine has been implicated in alterations of lung function in several chronic lung diseases. Fibroblasts from asthmatic patients show enhanced responsiveness to PDGF, and it is known to activate fibroblasts to proliferate, secrete collagen, and contract collagen lattices (56). Once again, eosinophils are the predominant cellular sources of this cytokine. However, platelets, macrophages, airway epithelial and endothelial cells, vascular smooth muscle cells, and fibroblasts are also known to secrete PDGF. PDGF can be induced by both mechanical and oxidative stress as well as by exposure of cells to various cytokines including IFN-γ, TNF-α, IL-1, and TGF-β. Although PDGF plays an important role in airway remodeling, it is thought that this growth factor is likely to act in concert with other remodeling cytokines, in particular TGF-β, to alter the structural makeup of the airway wall in asthmatic airways.

Another remodeling-associated cytokine is IL-6. IL-6 was first noted for its antiviral activity and for its growth-promoting effects on B cells. IL-6 is produced by macrophages, monocytes, T and B cells, fibroblasts, epithelial cells, and endothelial cells, as well as ASM cells and eosinophils. This cytokine is consistently found in high concentrations in biological fluids and tissues from both animal models of allergic disease and asthmatic patients, but its exact role in asthma remains unclear. IL-6 can stimulate T and B cells' production of Th2 cytokines (57), thereby contributing to the generation and/or perpetuation of Th2-driven inflammation. Also, IL-6 is a potent stimulant of the acute-phase allergic response, and it was recently shown to be a potent smooth muscle mitogen. Mice overexpressing IL-6 in their airways have subepithelial fibrosis, collagen deposition, and increased accumulation of α smooth muscle actin–containing cells, but they do not have eosinophilia, mucous cell metaplasia, or AHR (58).

IL-11 and IL-13 have recently received much attention as key remodeling-associated cytokines because they are thought not only to cause fibrosis and collagen deposition but also to induce myofibroblast hyperplasia, airway obstruction, and AHR. Much of what is known about the role of these cytokines in airway remodeling comes from studies using transgenic mice. Histological analysis of mice in whose lungs IL-11 or IL-13 was constitutively overexpressed resulted in airway wall thickening, enlarged alveoli, subepithelial and adventitial tissue fibrosis, collagen I and III deposition, and increased numbers of contractile and proinflammatory cells when compared with littermate controls (59–61). In addition, IL-11 and IL-13 transgenic mice had baseline airway obstruction and were more responsive to methacholine challenge (62). Sources of IL-11 and IL-13 include epithelial cells, fibroblasts, eosinophils, and smooth muscle cells. Recent evidence suggests that one mechanism by which IL-13 induces tissue fibrosis is the selective stimulation and activation of TGF-β production (59).

We recently confirmed that the results obtained in IL-11-transgenic mice also hold true for human asthma. Using immunocytochemistry and in situ hybridization, we have demonstrated increased expression of IL-11 mRNA and protein in the epithelial and subepithelial layers of the airway walls in patients with severe asthma, but not in mild asthmatics or healthy controls (63). Furthermore, IL-11 expression was inversely correlated with the forced expiratory volume in the first second (FEV$_1$) in severe asthmatics, and the IL-11 mRNA–positive cells were localized to epithelial cells and major basic protein–positive eosinophils (64). A proposed mechanism by which IL-11 can induce these structural changes in the airways may be promotion of the synthesis of TIMP-1 (tissue inhibitor of metalloproteinases-1), which is elevated in sputum and biopsy samples from asthmatic patients and correlates with asthmatic airway obstruction (65). Although these studies clearly support a role for IL-11 in airway remodeling, other studies (61) suggest that levels of IL-11 may in fact increase as a result of normal airway repair. IL-11-transgenic mice exhibited selective inhibition of antigen-induced airway and parenchymal eosinophilia, Th2 inflammation, Th2-cytokine production, and VCAM-1 gene expression (61). These conflicting data point to the dual nature of IL-11, which acts both as a cytokine that promotes healing and as a cytokine that is capable of inducing local tissue fibrosis.

Other remodeling-associated cytokines recently described in the literature include IL-17 and IL-25, which are potent proinflammatory cytokines. IL-17 and IL-17A are produced exclusively by activated Th lymphocytes, whereas Th2 cells and mast cells secrete IL-25. Expression of IL-17 is markedly increased in asthmatic subjects (66). In mice, systemic overexpression of IL-17 induces neutrophilia via direct in vivo stimulation of IL-6 and IL-8, whereas overexpression of IL-25 results in increased Th2-cytokine gene expression (particularly IL-4, IL-5, IL-10, and IL-13), increased mucous production, elevated serum IgE and IgG1, and tissue eosinophilia (24). These pathological changes are not restricted to the lungs.

IMMUNOMODULATORY CYTOKINES

It is now generally accepted that adult atopic disease is characterized by the expression of T cell immunity to common airborne environmental allergens, which is polarized toward the Th2-cytokine profile; in nonatopics, Th1-skewed immunity is observed. As a result, shifting the balance from Th2 to Th1 immunomodulatory cytokines (**Figure 3**) such as IL-10, IL-12, and IFN-γ may be important for the treatment of allergic inflammation.

IL-10 is primarily known as an inhibitory cytokine; however, it can exert both immunosuppressive and immunostimulatory effects. IL-10 was originally identified as a product of murine Th2 cells, but in humans IL-10 is produced by Th0, Th1, and Th2 cells and also by activated monocytes, mast cells, and macrophages (43). In normal lungs, alveolar macrophages are the major source of IL-10, but IL-10's expression is significantly reduced in asthmatic individuals (67, 68). IL-10 curtails the effects of proinflammatory cytokines (e.g., TNF-α, IL-1β, IL-6, IL-8, and MIP-1α) released during an allergic reaction. In addition, IL-10 can inhibit eosinophil survival and migration by preventing the release of chemoattractants such as RANTES and IL-8 from human ASM cells (68). Moreover, IL-10 inhibits allergen-induced eosinophilia in sensitized mice and dampens their late-phase response to allergen challenge (47, 69). Its other actions include downregulation of the IL-4-induced isotype switching of activated B cells (25), which prevents IgE synthesis. IL-10 may inhibit Th2-driven inflammation, and it is also known to inhibit the differentiation of Th1 cells, thereby preventing the release of IFN-γ and IL-2. Moreover, IL-10 can interfere with the functions of monocytes and macrophages. IL-10 can inhibit MHC class II expression on the surface of antigen-presenting cells and can prevent superoxide and nitric oxide (NO) release from inflammatory cells (70).

Other members of the IL-10 family of cytokines are IL-19, IL-20, IL-22, and IL-24. Like IL-10, they are considered to be anti-inflammatory cytokines and are produced by a variety of cell types, including monocytes, keratinocytes, mast cells, and lymphocytes. However, unlike IL-10, these cytokines cannot inhibit the effects of proinflammatory mediators involved in allergic response. IL-19, IL-20, and IL-24 have not been extensively studied; however, studies of IL-22 have shown that this cytokine is involved in the induction of IgE-independent acute-phase response signals (71).

IFN-γ is the most important cytokine in cell-mediated immunity; it controls the balance of Th1/Th2 development (**Figure 1**). IFN-γ is produced by Th1 cells and has an inhibitory effect on Th2 cells. IFN-γ inhibits allergic responses by preventing isotype switching of IgE and IgE production in B cells (72). The main sources of IFN-γ are Th cells. However, IFN-γ can also be produced by cytotoxic T cells and natural killer (NK) cells. In addition to its potent inhibitory effect on Th2 cells, IFN-γ stimulates de novo expression of MHC class II molecules on epithelial and endothelial cells and upregulates their expression on macrophages/monocytes and on dendritic cells. Importantly, IFN-γ stimulates monocytes, NK cells, and neutrophils to increase their cytokine production, phagocytosis, adherence, respiratory burst, and NO production, thereby promoting cell-mediated cytotoxic responses at the site of inflammation.

In sensitized and allergen-challenged mice, nebulized IFN-γ prevents allergen-induced increase in Th2 cytokine production, AHR, and lung eosinophilia (73). This has been proposed to occur via upregulation of IL-10. T cell production of IFN-γ is reduced in the BAL of asthmatic patients, and the levels of IFN-γ correlate closely with disease severity (74). Clinical trials with IFN-γ in humans have proved disappointing, as no significant improvement in lung function was observed in steroid-dependent asthmatics despite reduced numbers of eosinophils in the blood (75).

IL-12 is produced by antigen-presenting cells including B cells, monocytes, macrophages, Langerhans cells, and dendritic cells, as well as neutrophils and mast cells. IL-12 promotes T cell differentiation toward a Th1-mediated response by stimulating NK and T cells to produce IFN-γ while suppressing the expansion and differentiation of IL-4-secreting Th2 cells (76). The biologically active form of IL-12 is a heterodimer consisting of the p40 subunit and the p35 subunit, which are expressed by different genes. The effects of IL-12 have been extensively studied in small animal models of allergic inflammation and consistently demonstrate that this cytokine is involved in reduction of allergen-specific IgE,

abolition of AHR, and airway eosinophilia (77, 78). However, this effect of IL-12 is critically dependent on the timing of its administration. The most effective protection against allergen-induced inflammation is observed when IL-12 is administered early in the active sensitization process and thus when it can act in synergy with IL-18 (79). IL-18, also known as IFN-γ-inducing factor, is a potent inducer of IFN-γ production by T cells, NK cells, and B cells (80). In asthmatic patients, IL-12 is significantly reduced in peripheral blood and in airway biopsy specimens, compared with healthy controls. IL-12 mRNA levels increase in biopsies of asthmatic patients following treatment with corticosteroids (31), and although the administration of IL-12 has failed to show any effects on AHR or the late-phase asthmatic responses in mild asthmatics, IL-12 is effective in suppressing blood and sputum eosinophilia in these patients (81).

CHEMOKINES

Chemokines are small cytokines (8–10 kDa) involved primarily in a process called chemotaxis, whereby they attract and regulate leukocyte trafficking into the tissues by binding specific seven-transmembrane-spanning G protein–coupled receptors (**Figure 4**). To date, more than 40 chemokines have been described and have been classified into four subclasses according to their structure: CXC, CC, C, and CX3C. The two main groups are CXC (α chemokines) and CC (β chemokines). CXC chemokines include IL-8 and IP-10, which primarily target neutrophils. Eotaxin, RANTES, MCP-1–MCP-4, macrophage inhibitory protein (MIP)-1α, and MIP-1β are typical CC chemokines, which target monocytes, T cells, and eosinophils. For this reason, CC chemokines are thought to have the greatest relevance in the pathogenesis of asthma. Levels of chemokines in both BAL and biopsy samples of asthmatics are higher than in control patients (43).

Eotaxin and RANTES, acting in synergy with IL-5, are the most important eosinophil chemoattractants in allergic inflammation. These chemokines are produced by the majority of inflammatory cells, and recently their expression was described in ASM cells and fibroblasts. Unlike RANTES, which binds many chemokine receptors (CCRs) including CCR1, CCR3, and CCR5, eotaxin binds specifically to CCR3, which is highly expressed on and has selective chemoattractant activity for eosinophils (82). In addition, eotaxin induces α4- and β1-integrin expression on eosinophils, allowing for firm adhesion of eosinophils to endothelium and transmigration into the site of inflammation (**Figure 4**). More importantly, eotaxin and RANTES are produced in high concentrations in the lungs of asthmatic patients.

MCPs and MIPs are monocyte/macrophage chemoattractants and activating factors. To date, four MCPs (MCP-1–MCP–4) and two MIPs (MIP-1α and -1β) have been described. Increased levels of MCP-1 and MCP-3 have been detected in BAL of asthmatic patients (83), and increased expression of MCP-4 has been reported in the sputum (84), BAL (85), bronchial mucosa (61), and small airways (64) of asthmatics as well as in the noses of patients with allergic rhinitis (65). MCP-1 binds CCR2, MCP-2 binds CCR3, MCP-3 binds CCR1 and CCR3, and MCP-4 binds CCR2, CCR3, and CCR5. MCP-1 immunoreactivity has been demonstrated in human eosinophils; furthermore, MCP-2, MCP-3, and, in particular, MCP-4 are also thought to be potent eosinophil chemoattractants (85). Moreover, MCP-4 attracts not only eosinophils and monocytes but also lymphocytes and basophils. MIP-1α binds CCR1 and CCR5, and MIP-1β binds CCR5 exclusively.

Although the primary role of chemokines is chemotaxis, chemokines have a variety of other functions, including direct effects on T cell differentiation. MIP-1α, MIP-1β, and RANTES can promote the development of IFN-γ-producing Th1 cells by stimulating IL-12 production from antigen-producing cells. However, MCP-1, MCP-2, MCP-3, and MCP-4 can increase T cell production of IL-4 and can

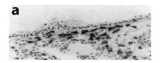

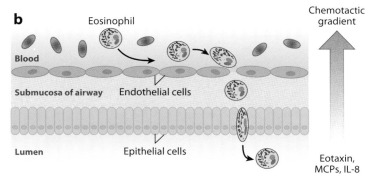

Figure 4

(*a*) Histology showing real eosinophils accumulating in the lung. (*b*) Illustration of the role of chemokines in the pathogenesis of asthma. MCP, monocyte chemoattractant protein.

decrease antigen-producing cells' production of IL-12, resulting in a Th2 phenotype.

CYTOKINES, ASTHMA, AND INFECTION

The reasons for the increasing prevalence of allergic respiratory diseases in developed countries and in undeveloped countries that develop a Western lifestyle remain unclear. For many years, lower respiratory tract infections in early life have been recognized as primary triggers of asthma exacerbations in young children. Using both epidemiological and virology data, prospective studies have convincingly shown that viral and not bacterial respiratory infections precipitate reactive airway symptoms (86). It is now believed that the development of bacteria-induced, nonwheezing lower respiratory tract infection in childhood may protect against the development of atopy and asthma in later life. This line of thought comes from experimental evidence that suggests that the principal trigger for normal postnatal maturation of the immune system is the commensal microbial flora, particularly those of the gastrointestinal tract. Microbial exposure helps to skew the

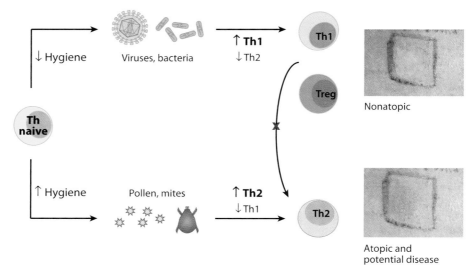

Figure 5

The possible mechanism of the hygiene hypothesis and the effect of early infections on cytokine gene expression. Abbreviations: Th, T helper cell; Treg, T regulatory cell.

immune response away from allergic phenotype and toward the normal adult nonatopic immune response (87). The longer the immune system takes to adapt postnatally to its functionally mature state, the greater is the risk for allergic sensitization (88).

The hygiene hypothesis suggests that decreasing levels of exposure to infections and/or commensal microbial stimuli in developed countries, particularly during the induction of primary Th1/Th2 responses to allergens in early life, may be responsible for the increased prevalence of asthma (**Figure 5**). There is ample epidemiological evidence to support this hypothesis (89). The reported increase in atopy inversely correlates with a steady decline in the extent to which people in Western societies are exposed to infectious diseases such as whooping cough, measles, tuberculosis, and influenza (90). Allergic diseases appear to increase with advancing socioeconomic development and occur more frequently in industrialized countries than in developing areas (91), and spending one's early years resident on a farm protects against the development of allergies (92, 93).

Bacterial lipopolysaccharide (LPS) or endotoxin has been suggested as a potential mediator of these effects. LPS is a major component of the outer membrane of ubiquitous gram-negative bacteria. Gram-negative infections make up a significant proportion of clinical respiratory tract infections among children in early life; thus, there is ongoing and chronic environmental exposure to gram-negative organisms and their components/products. In common house dust, LPS makes up a significant proportion of dust weight, and a significant correlation has been reported between domestic LPS exposure and clinical severity of asthma in adults (94). The first direct in vivo evidence that environmental exposure to LPS early in life (before polarized Th cell responses are established) protects against allergen sensitization was reported by Gereda and colleagues (95), who have demonstrated that (*a*) the homes of allergen-sensitive infants (9–24 months of age) contain lower concentrations of LPS in house dust than do the homes of nonsensitive infants and (*b*) the lower concentrations are associated with reduced proportions of IFN-γ-producing Th cells. Consistent with this hypothesis, three distinct LPS phenotypes in humans have been described—sensitive, intermediate, and hyporesponsive—on the basis of reduction in FEV_1 following inhalation of increasing doses of LPS and in vitro production of IL-6 and IL-8 from peripheral blood monocytes and alveolar macrophages (95).

Experimental results in animal models of asthma have supported the hygiene hypothesis. They have shown that treatment with microbes [e.g., BCG (96) and *Lactobacillus* (97)] or microbial products [LPS (98) and CpG DNA (99, 100)] inhibits allergic sensitization, eosinophilic inflammation, and AHR in these animals. In a similar animal model of allergic disease, we have shown that both the timing of exposure and the dose are critical to the in vivo effect of LPS (98). That study demonstrated that, although LPS exposure during early sensitization protects against the development of IgE and consequent allergic inflammation, in marked contrast, exposure after allergen challenge further exacerbates the allergic response.

Toll is a *Drosophila* receptor that is involved in antifungal immune responses. Toll-like receptors (TLRs) are a large family of evolutionarily conserved receptors from Toll that sense invasion by microorganisms through the recognition of specific pathogen-associated molecule patterns and that produce immediate innate responses. To date, 10 TLRs (TLR1–10) have been identified in humans and in mice (88) TLRs are single-transmembrane-domain receptors that have a cytoplasmic signaling portion homologous to the IL-1 receptor. Although the TLRs differ in their extracellular domain structure, similar cytoplasmic domains allow TLRs to use the same signaling molecules. All TLRs signal through an adaptor protein known as myeloid differentiation factor 88 (MyD88). Following activation, MyD88 recruits the IL-1R/IL-1R-associated protein kinase (IRAK) complex to the TLR; IRAK becomes phosphorylated and then activates the TNF receptor–associated factor 6. This process leads to activation of JNK (c-Jun N-terminal kinase), MAPK (mitogen-activated protein ki-

nase), and NF-κB (nuclear factor–kappa B) pathways, leading to a cascade of events including the release of cytokines and the activation of antigen-presenting cells.

Different TLRs recognize different ligands. TLR3 is a cell-surface receptor for double-stranded RNA; hence, it may be implicated in viral recognition. TLR5 is specific for bacterial flagellin, whereas TLR9 is a receptor for unmethylated CpG motifs, which are abundant in bacterial DNA. In mammals, TLR4 is the principal receptor responsible for LPS-induced signal transduction (101). Recognition of LPS by TLR4 is aided by two accessory proteins: CD14 and MD-2. TLR4 is expressed at particularly high levels by cells of the innate immune system, such as monocytes, dendritic cells, macrophages, and endothelial cells (102–105). TLR4 expression is thought to be related to LPS sensitivity (106), and it was recently demonstrated in murine macrophages that TLR expression and function decline with age (107, 108).

We have shown that bacterial LPS can prevent local allergen-induced inflammation in the nasal mucosa of atopic children. This occurs by downregulation of local Th2 and by upregulation of Th1 cytokines in the tissue, proliferation, and activation of TLR4+IL10+ and CD25+ Th cells, as well as by increased expression of anti-inflammatory cytokine IL-10 (109). These events occur locally without systemic recruitment of inflammatory cells and are orchestrated through the TLR4-dependent pathway. TLR4 is an important bridge between innate and adaptive immunity that potentially drives the molecular mechanisms governing the hygiene hypothesis and that may help explain why reduced exposure to bacterial products may lead to delayed or skewed development of the immune system and more atopic disease.

CONCLUSIONS

Bronchial asthma is a complex, chronic disease of the airways that is characterized by reversible AHR, airway remodeling, and inflammation. These pathological and physiological changes occur even in mild asthmatics and can be detected in asthmatic children. Within the past decade, one of the most striking advances in study of asthma has been the recognition that cytokines and chemokines play an integral role in orchestrating, perpetuating, and amplifying the underlying processes of this disease. Therapy for asthma may lie in specific targeting of cytokine and chemokine receptors rather than in global immunosuppression. Additionally, it has become clear that bacterial products play a role in the maturation of early immune responses and that they can modulate allergic inflammation in young children. Only by understanding the immunobiology of asthma can we begin to provide a rational basis of novel drug design and make progress in identifying those individuals at risk for this disease.

DISCLOSURE STATEMENT

The authors are not aware of any biases that might be perceived as affecting the objectivity of this review.

ACKNOWLEDGMENTS

We would like to thank the Strauss Foundation, Fonds de la Recherche en Santé Québec (FRSQ), and Canadian Institutes of Health Research (CIHR) for supporting Q. Hamid's laboratory. We would also like to thank our colleagues who helped by reviewing the text and providing ideas for illustrations.

LITERATURE CITED

1. Huber H, Koessler KK. 1922. The pathology of bronchial asthma. *Arch. Intern. Med.* 30:689–760
2. Hogg JC. 1993. The pathology of asthma. In *Asthma*, Vol. IV. *Physiology, Immunopharmacology and Treatment*, ed. KF Austen, L Lichtenstein, AB Kay, ST Holgate, pp. 17–25. Oxford: Blackwell Sci.
3. Larché M, Robinson DS, Kay AB. 2003. The role of T lymphocytes in the pathogenesis of asthma. *J. Allergy Clin. Immunol.* 111:449–61
4. Robinson DS, Bentley AM, Hartnell A, Kay AB, Durham SR. 1993. Activated memory T helper cells in bronchoalveolar lavage from atopic asthmatics. Relationship to asthma symptoms, lung function and bronchial responsiveness. *Thorax* 48:26–32
5. Gleich GJ, Motojima S, Frigas E, Kephart GM, Fujisawa T, Kravis LP. 1980. The eosinophilic leucocyte and the pathology of fatal bronchial asthma: evidence for pathologic heterogeneity. *J. Allergy Clin. Immunol.* 80:412–15
6. Azzawi M, Johnston PW, Majumdar S, Kay AB, Jeffery PK. 1992. T lymphocytes and activated eosinophils in asthma and cystic fibrosis. *Am. Rev. Resp. Dis.* 145:1477–82
7. Bradding P, Roberts JA, Britten KM, Montefort S, Djukanovic R, et al. 1994. Interleukin-4, -5, and -6 and tumour necrosis factor–α in normal and asthmatic airways: evidence for the human mast cell as a source of these cytokines. *Am. J. Respir. Cell Mol. Biol.* 10:471–80
8. Carroll NG, Mutavdzic S, James AL. 2002. Increased mast cells and neutrophils in submucosal mucous glands and mucus plugging in patients with asthma. *Thorax* 57:677–82
9. Brightling CE, Bradding P, Symon FA, Holgate ST, Wardlaw AJ, Pavord ID. 2002. Mast cell infiltration of airway smooth muscle in asthma. *N. Engl. J. Med.* 346:1699–705
10. Bradding P, Walls AF, Holgate ST. 2006. The role of the mast cell in the pathophysiology of asthma. *J. Allergy Clin. Immunol.* 117:1277–84
11. Jatakanon A, Uasuf C, Maziak W, Lim S, Chung KF, Barnes PJ. 1999. Neutrophilic inflammation in severe persistent asthma. *Am. J. Respir. Crit. Care Med.* 160:1532–39
12. Ordonez CL, Shaughnessy TE, Matthay MA, Fahy JV. 2000. Increased neutrophil numbers and IL-8 levels in airway secretions in acute severe asthma: clinical and biologic significance. *Am. J. Respir. Crit. Care Med.* 161:1185–90

13. Fahy JV, Kim KW, Liu J, Boushey HA. 1995. Prominent neutrophilic inflammation in sputum from subjects with asthma exacerbation. *J. Allergy Clin. Immunol.* 95:843–52

14. Qiu Y, Zhu J, Bandi V, Guntupalli KK, Jeffery PK. 2007. Bronchial mucosal inflammation and up-regulation of CXC chemoattractants and receptors in severe exacerbations of asthma. *Thorax* 63:475–82

15. Sur S, Crotty TB, Kephart GM, Hyma BA, Colby TV, et al. 1993. Sudden-onset fatal asthma. A distinct entity with few eosinophils and relatively more neutrophils in the airway submucosa? *Am. Rev. Respir. Dis.* 148:713–19

16. Anees W, Huggins V, Pavord ID, Robertson AS, Burge PS. 2002. Occupational asthma due to low molecular weight agents: eosinophilic and noneosinophilic variants. *Thorax* 57:231–36

17. Martin RJ, Cicutto LC, Smith HR, Ballard RD, Szefler SJ. 1991. Airways inflammation in nocturnal asthma. *Am. Rev. Respir. Dis.* 143:351–57

18. McDougall CM, Helms PJ. 2006. Neutrophil airway inflammation in childhood asthma. *Thorax* 61:739–41

19. Borregaard N, Cowland JB. 1997. Granules of the human neutrophilic polymorphonuclear leukocyte. *Blood* 89:3503–21

20. Moller GM, Overbeek SE, Van Helden–Meeuwsen CG, Van Haarst JM, Prens EP, et al. 1996. Increased numbers of dendritic cells in the bronchial mucosa of atopic asthmatic patients: downregulation by inhaled corticosteroids. *Clin. Exp. Allergy* 26:517–24

21. Masten BJ, Olson GK, Tarleton CA, Rund C, Schuyler M, et al. 2006. Characterization of myeloid and plasmacytoid dendritic cells in human lung. *J. Immunol.* 177:7784–93

22. Holt PG, Schon-Hegrad MA, Oliver J, Holt BJ, McMenamin PG. 1990. A contiguous network of dendritic antigen-presenting cells within the respiratory epithelium. *Int. Arch. Allergy Appl. Immunol.* 91:155–59

23. Tangye SG, Ferguson A, Avery DT, Ma CS, Hodgkin PD. 2002. Isotype switching by human B cells is division-associated and regulated by cytokines. *J. Immunol.* 169:4298–306

24. Seder RA, Paul WE. 1994. Acquisition of lymphokine-producing phenotype by CD4+ T cells. *Annu. Rev. Immunol.* 12:635–73

25. Chung KF, Barnes PJ. 1999. Cytokines in asthma. *Thorax* 54:825–57

26. Vercelli D, Jabara HH, Lee BW, Woodland N, Geha RS, Leung DY. 1988. Human recombinant interleukin 4 induces Fc ε R2/CD23 on normal human monocytes. *J. Exp. Med.* 167:1406–16

27. Xu L, Rothman P. 1994. IFN-γ represses ε germline transcription and subsequently down-regulates switch recombination to ε. *Int. Immunol.* 6:515–21

28. Denoroy MC, Yodoi J, Banchereau J. 1990. Interleukin 4 and interferons α and γ regulate Fc ε R2/CD23 mRNA expression on normal human B cells. *Mol. Immunol.* 27:129–34

29. Gavett SH, O'Hearn DJ, Karp CL, Patel EA, Schofield BH, et al. 1997. Interleukin-4 receptor blockade prevents airway responses induced by antigen challenge in mice. *Am. J. Physiol.* 272:L253–61

30. Schleimer RP, Sterbinsky SA, Kaiser J, Bickel CA, Klunk DA, et al. 1992. IL-4 induces adherence of human eosinophils and basophils but not neutrophils to endothelium. Association with expression of VCAM-1. *J. Immunol.* 148:1086–92

31. Naseer T, Minshall EM, Leung DY, Laberge S, Ernst P, et al. 1997. Expression of IL-12 and IL-13 mRNA in asthma and their modulation in response to steroid therapy. *Am. J. Respir. Crit. Care Med.* 155:845–51

32. Humbert M, Durham SR, Kimmitt P, Powell N, Assoufi B, et al. 1997. Elevated expression of messenger ribonucleic acid encoding IL-13 in the bronchial mucosa of atopic and nonatopic subjects with asthma. *J. Allergy Clin. Immunol.* 99:657–65

33. Wills-Karp M, Luyimbazi J, Xu X, Schofield B, Neben TY, et al. 1998. Interleukin-13: central mediator of allergic asthma. *Science* 282:2258–61

34. Wills-Karp M, Chiaramonte M. 2003. Interleukin-13 in asthma. *Curr. Opin. Pulm. Med.* 9:21–27

35. McLane MP, Haczku A, van de Rijn M, Weiss C, Ferrante V, et al. 1998. Interleukin-9 promotes allergen-induced eosinophilic inflammation and airway hyperresponsiveness in transgenic mice. *Am. J. Respir. Cell Mol. Biol.* 19:713–20

36. Vink A, Warnier G, Brombacher F, Renauld JC. 1999. Interleukin 9-induced in vivo expansion of the B-1 lymphocyte population. *J. Exp. Med.* 189:1413–23

37. Dugas B, Renauld JC, Pene J, Bonnefoy JY, Peti-Frere C, et al. 1993. Interleukin-9 potentiates the interleukin-4-induced immunoglobulin (IgG, IgM and IgE) production by normal human B lymphocytes. *Eur. J. Immunol.* 23:1687–92

38. Uyttenhove C, Simpson RJ, Van Snick J. 1988. Functional and structural characterization of P40, a mouse glycoprotein with T cell growth factor activity. *Proc. Natl. Acad. Sci. USA* 85:6934–38

39. Houssiau FA, Renauld JC, Stevens M, Lehmann F, Lethe B, et al. 1993. Human T cell lines and clones respond to IL-9. *J. Immunol.* 150:2634–40

40. Schmitt E, Van Brandwijk R, Van Snick J, Siebold B, Rude E. 1989. TCGF III/P40 is produced by naïve murine CD4+ T cells but is not a general T cell growth factor. *Eur. J. Immunol.* 19:2167–70

41. Louahed J, Toda M, Jen J, Hamid Q, Renauld JC, et al. 2000. Interleukin-9 upregulates mucus expression in the airways. *Am. J. Respir. Cell Mol. Biol.* 22:649–56

42. Dong Q, Louahed J, Vink A, Sullivan CD, Messler CJ, et al. 1999. IL-9 induces chemokine expression in lung epithelial cells and baseline airway eosinophilia in transgenic mice. *Eur. J. Immunol.* 29:2130–39

43. Borish LC, Steinke JW. 2003. Cytokines and chemokines. *J. Allergy Clin. Immunol.* 111:S460–75

44. Minshall EM, Hogg JC, Hamid QA. 1998. Cytokine mRNA expression in asthma is not restricted to the large airways. *J. Allergy Clin. Immunol.* 101:386–90

45. Cameron L, Christodoulopoulos P, Lavigne F, Nakamura Y, Eidelman D, et al. 2000. Evidence for local eosinophil differentiation within allergic nasal mucosa: inhibition with soluble IL-5 receptor. *J. Immunol.* 164:1538–45

46. Hurst SD, Muchamuel T, Gorman DM, Gilbert JM, Clifford T, et al. 2002. New IL-17 family members promote Th1 or Th2 responses in the lung: in vivo function of the novel cytokine IL-25. *J. Immunol.* 169:443–53

47. Ohnishi T, Kita H, Weiler D, Sur S, Sedgwick JB, et al. 1993. IL-5 is the predominant eosinophil-active cytokine in the antigen-induced pulmonary late-phase reaction. *Am. Rev. Respir. Dis.* 147:901–7

48. Mould AW, Ramsay AJ, Matthaei KI, Young IG, Rothenberg ME, Foster PS. 2000. The effect of IL-5 and eotaxin expression in the lung on eosinophil trafficking and degranulation and the induction of bronchial hyperreactivity. *J. Immunol.* 164:2142–50

49. Leckie MJ, ten Brinke A, Khan J, Diamant Z, O'Connor BJ, et al. 2000. Effects of an interleukin-5 blocking monoclonal antibody on eosinophils, airway hyper-responsiveness, and the late asthmatic response. *Lancet* 356:2144–48

50. Xing Z, Ohkawara Y, Jordana M, Graham F, Gauldie J. 1996. Transfer of granulocyte-macrophage colony-stimulating factor gene to rat lung induces eosinophilia, monocytosis, and fibrotic reactions. *J. Clin. Investig.* 97:1102–10

51. Adach K, Suzuki M, Sugimoto T, Suzuki S, Niki R, et al. 2002. Granulocyte colony–stimulating factor exacerbates the acute lung injury and pulmonary fibrosis induced by intratracheal administration of bleomycin in rats. *Exp. Toxicol. Pathol.* 53:501–10

52. Payne DN, Rogers AV, Adelroth E, Bandi V, Guntupalli KK, et al. 2003. Early thickening of the reticular basement membrane in children with difficult asthma. *Am. J. Respir. Crit. Care Med.* 167:78–82

53. Wahl SM, Hunt DA, Wakefield LM, McCartney-Francis N, Wahl LM, et al. 1987. Transforming growth factor type β induces monocyte chemotaxis and growth factor production. *Proc. Natl. Acad. Sci. USA* 84:5788–92

54. Kehrl JH, Wakefield LM, Roberts AB, Jakowlew S, Alvarez-Mon M, et al. 1986. Production of transforming growth factor β by human T lymphocytes and its potential role in the regulation of T cell growth. *J. Exp. Med.* 163:1037–50

55. Minshall EM, Leung DY, Martin RJ, Song YL, Cameron L, et al. 1997. Eosinophil-associated TGF-β1 mRNA expression and airways fibrosis in bronchial asthma. *Am. J. Respir. Cell Mol. Biol.* 17:326–33

56. Clark RA, Folkvord JM, Hart CE, Murray MJ, McPherson JM. 1989. Platelet isoforms of platelet-derived growth factor stimulate fibroblasts to contract collagen matrices. *J. Clin. Investig.* 84:1036–40

57. Akira S, Taga T, Kishimoto T. 1993. Interleukin-6 in biology and medicine. *Adv. Immunol.* 54:1–78

58. DiCosmo BF, Geba GP, Picarella D, Elias JA, Rankin JA, et al. 1994. Airway epithelial cell expression of interleukin-6 in transgenic mice. Uncoupling of airway inflammation and bronchial hyperreactivity. *J. Clin. Investig.* 94:2028–35

59. Lee CG, Homer RJ, Zhu Z, Lanone S, Wang X, et al. 2001. Interleukin-13 induces tissue fibrosis by selectively stimulating and activating transforming growth factor beta(1). *J. Exp. Med.* 194:809–21

60. Chen LC, Zhang Z, Myers AC, Huang SK. 2001. Cutting edge: altered pulmonary eosinophilic inflammation in mice deficient for Clara cell secretory 10-kDa protein. *J. Immunol.* 167:3025–28

61. Wang J, Homer RJ, Hong L, Cohn L, Lee CG, et al. 2000. IL-11 selectively inhibits aeroallergen-induced pulmonary eosinophilia and Th2 cytokine production. *J. Immunol.* 165:2222–31

62. Zhu Z, Homer RJ, Wang Z, Chen Q, Geba GP, et al. 1999. Pulmonary expression of interleukin-13 causes inflammation, mucus hypersecretion, subepithelial fibrosis, physiologic abnormalities, and eotaxin production. *J. Clin. Investig.* 103:779–88

63. Zhu Z, Lee CG, Zheng T, Chupp G, Wang J, et al. 2001. Airway inflammation and remodeling in asthma. Lessons from interleukin-11 and interleukin-13 transgenic mice. *Am. J. Resp. Crit. Care Med.* 164:S67–S70

64. Minshall E, Chakir J, Laviolette M, Molet S, Zhu Z, et al. 2000. IL-11 expression is increased in severe asthma: association with epithelial cells and eosinophils. *J. Allergy Clin. Immunol.* 105:232–38

65. Hermann JA, Hall MA, Maini RN, Feldmann M, Brennan FM. 1998. Important immunoregulatory role of interleukin-11 in the inflammatory process in rheumatoid arthritis. *Arthritis Rheum.* 41:1388–97

66. Molet S, Hamid Q, Davoine F, Nutku E, Taha R, et al. 2001. IL-17 is increased in asthmatic airways and induces human bronchial fibroblasts to produce cytokines. *J. Allergy Clin. Immunol.* 108:430–38

67. Borish L, Aarons A, Rumbyrt J, Cvietusa P, Negri J, Wenzel S. 1996. Interleukin-10 regulation in normal subjects and patients with asthma. *J. Allergy Clin. Immunol.* 97:1288–96

68. John M, Lim S, Seybold J, Jose P, Robichaud A, et al. 1998. Inhaled corticosteroids increase interleukin-10 but reduce macrophage inflammatory protein-1α, granulocyte-macrophage colony–stimulating factor, and interferon-γ release from alveolar macrophages in asthma. *Am. J. Respir. Crit. Care Med.* 157:256–62

69. Tulic MK, Knight DA, Holt PG, Sly PD. 2001. Lipopolysaccharide inhibits the late-phase response to allergen by altering nitric oxide synthase activity and interleukin-10. *Am. J. Respir. Cell Mol. Biol.* 24:640–46

70. Cunha FQ, Moncada S, Liew FY. 1992. Interleukin-10 (IL-10) inhibits the induction of nitric oxide synthase by interferon-γ in murine macrophages. *Biochem. Biophys. Res. Comm.* 182:1155–59

71. Lecart S, Morel F, Noraz N, Pene J, Garcia M, et al. 2002. IL-22, in contrast to IL-10, does not induce Ig production, due to absence of a functional IL-22 receptor on activated human B cells. *Int. Immunol.* 14:1351–56

72. Stirling RG, Chung KF. 2000. New immunological approaches and cytokine targets in asthma and allergy. *Eur. Respir. J.* 16:1158–74

73. Lack G, Bradley KL, Hamelmann E, Renz H, Loader J, et al. 1996. Nebulized IFN-γ inhibits the development of secondary allergic responses in mice. *J. Immunol.* 157:1432–39

74. Koning H, Neijens HJ, Baert MR, Oranje AP, Savelkoul HF. 1997. T cell subsets and cytokines in allergic and nonallergic children. I. Analysis of IL-4, IFN-γ and IL-13 mRNA expression and protein production. *Cytokine* 9:416–26

75. Boguniewicz M, Schneider LC, Milgrom H, Newell D, Kelly N, et al. 1993. Treatment of steroid-dependent asthma with recombinant interferon-γ. *Clin. Exp. Allergy* 23:785–90

76. Manetti R, Parronchi P, Giudizi MG, Piccinni MP, Maggi E, et al. 1993. Natural killer cell stimulatory factor (interleukin 12 [IL-12]) induces T helper type 1 (Th1)-specific immune responses and inhibits the development of IL-4-producing Th cells. *J. Exp. Med.* 177:1199–204

77. Lee YL, Fu CL, Ye YL, Chiang BL. 1999. Administration of interleukin-12 prevents mite Der p 1 allergen–IgE antibody production and airway eosinophil infiltration in an animal model of airway inflammation. *Scand. J. Immunol.* 49:229–36

78. Kips JC, Brusselle GJ, Joos GF, Peleman RA, Tavernier JH, et al. 1996. Interleukin-12 inhibits antigen-induced airway hyperresponsiveness in mice. *Am. J. Respir. Crit. Care Med.* 153:535–39

79. Hofstra CL, Van Ark I, Hofman G, Kool M, Nijkamp FP, Van Oosterhout AJ. 1998. Prevention of Th2-like cell responses by coadministration of IL-12 and IL-18 is associated with inhibition of antigen-induced airway hyperresponsiveness, eosinophilia, and serum IgE levels. *J. Immunol.* 161:5054–60

80. Nakanishi K, Yoshimoto T, Tsutsui H, Okamura H. 2001. Interleukin-18 is a unique cytokine that stimulates both Th1 and Th2 responses depending on its cytokine milieu. *Cytokine Growth Factor Rev.* 12:53–72

81. Bryan SA, O'Connor BJ, Matti S, Leckie MJ, Kanabar V, et al. 2000. Effects of recombinant human interleukin-12 on eosinophils, airway hyper-responsiveness, and the late asthmatic response. *Lancet* 356:2149–53

82. Ponath PD, Qin S, Post TW, Wang J, Wu L, et al. 1996. Molecular cloning and characterization of a human eotaxin receptor expressed selectively on eosinophils. *J. Exp. Med.* 183:2437–48

83. Miotto D, Christodoulopoulos P, Olivenstein R, Taha R, Cameron L, et al. 2001. Expression of IFN-γ-inducible protein; monocyte chemotactic proteins 1, 3, and 4; and eotaxin in TH1- and TH2-mediated lung diseases. *J. Allergy Clin. Immunol.* 107:664–70

84. Taha RA, Laberge S, Hamid Q, Olivenstein R. 2001. Increased expression of the chemoattractant cytokines eotaxin, monocyte chemotactic protein-4, and interleukin-16 in induced sputum in asthmatic patients. *Chest* 120:595–601

85. Lamkhioued B, Garcia-Zepeda EA, Abi-Younes S, Nakamura H, Jedrzkiewicz S, et al. 2000. Monocyte chemoattractant protein (MCP)-4 expression in the airways of patients with asthma. Induction in epithelial cells and mononuclear cells by proinflammatory cytokines. *Am. J. Respir. Crit. Care Med.* 162:723–32

86. Johnston SL, Pattemore PK, Sanderson G, Smith S, Lampe F, et al. 1995. Community study of role of viral infections in exacerbations of asthma in 9-11-year-old children. *Brit. Med. J.* 310:1225–29

87. Sudo N, Sawamura S, Tanaka K, Aiba Y, Kubo C, Koga Y. 1997. The requirement of intestinal bacterial flora for the development of an IgE production system fully susceptible to oral tolerance induction. *J. Immunol.* 159:1739–45

88. Holt PG. 1995. Environmental factors and primary T cell sensitisation to inhalant allergens in infancy: reappraisal of the role of infections and air pollution [review]. *Pediatr. Allergy Immunol.* 6:1–10

89. Liu AH, Redmon AH Jr. 2001. Endotoxin: friend or foe? *Allergy Asthma Proc.* 22:337–40

90. Cookson WOCM, Moffatt MF. 1997. Asthma: an epidemic in the absence of infection? *Science* 275:41–42

91. Yemaneberhan H, Bekele Z, Venn A, Lewis S, Parry E, Britton J. 1997. Prevalence of wheeze and asthma and relation to atopy in urban and rural Ethiopia. *Lancet* 350:85–90

92. Von Ehrenstein OS, Von Mutius E, Illi S, Baumann L, Bohm O, von Kries R. 2000. Reduced risk of hay fever and asthma among children of farmers. *Clin. Exp. Allergy* 30:187–93

93. Riedler J, Eder W, Oberfeld G, Schreuer M. 2000. Austrian children living on a farm have less hay fever, asthma and allergic sensitization. *Clin. Exp. Allergy* 30:194–200

94. Michel O, Kips J, Duchateau J, Vertongen F, Robert L, et al. 1996. Severity of asthma is related to endotoxin in house dust. *Am. J. Resp. Crit. Care Med.* 154:1641–46

95. Gereda JE, Leung DYM, Thatayatikom A, Streib JE, Price MR, et al. 2000. Relation between house-dust endotoxin exposure, type 1 T cell development, and allergen sensitisation in infants at high risk of asthma. *Lancet* 355:1680–83

96. Kline JN, Cowden JD, Hunninghake GW, Schutte BC, Watt JL, et al. 1999. Variable airway responsiveness to inhaled lipopolysaccharide. *Am. J. Respir. Crit. Care Med.* 160:297–303

97. Erb KJ, Holloway JW, Sobeck A, Moll H, Le Gros G. 1998. Infection of mice with *Mycobacterium bovis*–Bacillus Calmette-Guérin (BCG) suppresses allergen-induced airway eosinophilia. *J. Exp. Med.* 187:561–69

98. Murosaki S, Yamamoto Y, Ito K, Inokuchi T, Kusaka H, et al. 1998. Heat-killed *Lactobacillus* plantarum L-137 suppresses naturally fed antigen-specific IgE production by stimulation of IL-12 production in mice. *J. Allergy Clin. Immunol.* 102:57–64

99. Tulic MK, Wale JL, Holt PG, Sly PD. 2000. Modification of the inflammatory response to allergen challenge after exposure to bacterial lipopolysaccharide. *Am. J. Respir. Cell Mol. Biol.* 22:604–12

100. Broide D, Schwarze J, Tighe H, Gifford T, Nguyen MD, et al. 1998. Immunostimulatory DNA sequences inhibit IL-5, eosinophilic inflammation, and airway hyperresponsiveness in mice. *J. Immunol.* 161:7054–62

101. Kline JN, Waldschmidt TJ, Businga TR, Lemish JE, Weinstock JV, et al. 1998. Modulation of airway inflammation by CpG oligodeoxynucleotides in a murine model of asthma. *J. Immunol.* 160:2555–59

102. Chow JC, Young DW, Golenbock DT, Christ WJ, Gusovsky F. 1999. Toll-like receptor–4 mediates lipopolysaccharide-induced signal transduction. *J. Biol. Chem.* 274:10689–92

103. Medzhitov R, Preston-Hurlburt P, Janeway CA Jr. 1997. A human homologue of the *Drosophila* Toll protein signals activation of adaptive immunity. *Nature* 388:394–97

104. Rock FL, Hardiman G, Timans JC, Kastelein RA, Bazan JF. 1998. A family of human receptors structurally related to *Drosophila* Toll. *Proc. Natl. Acad. Sci. USA* 95:588–93

105. Muzio M, Natoli G, Saccani S, Levrero M, Mantovani A. 1998. The human toll signaling pathway: divergence of nuclear factor κB and JNK/SAPK activation upstream of tumor necrosis factor receptor–associated factor 6 (TRAF6). *J. Exp. Med.* 187:2097–101

106. Zhang FX, Kirschning CJ, Mancinelli R, Xu XP, Jin Y, et al. 1999. Bacterial lipopolysaccharide activates nuclear factor–κB through interleukin-1 signaling mediators in cultured human dermal endothelial cells and mononuclear phagocytes. *J. Biol. Chem.* 274:7611–14

107. Nomura F, Akashi S, Sakao Y, Sato S, Kawai T, et al. 2000. Cutting edge: Endotoxin tolerance in mouse peritoneal macrophages correlates with down-regulation of surface Toll-like receptor 4 expression. *J. Immunol.* 164:3476–79

108. Renshaw M, Rockwell J, Engleman C, Gewirtz A, Katz J, Sambhara S. 2002. Cutting edge: impaired Toll-like receptor expression and function in aging. *J. Immunol.* 169:4697–701

109. Tulic MK, Fiset PO, Manoukian JJ, Frenkiel S, Lavigne F, et al. 2004. Role of toll-like receptor 4 in protection by bacterial lipopolysaccharide in the nasal mucosa of atopic children but not adults. *Lancet* 363:1689–97

Noncontractile Functions of Airway Smooth Muscle Cells in Asthma

Omar Tliba and Reynold A. Panettieri, Jr.

Pulmonary, Allergy and Critical Care Division, Airways Biology Initiative, University of Pennsylvania, Philadelphia, Pennsylvania 19104; email: rap@mail.med.upenn.edu

Annu. Rev. Physiol. 2009. 71:509–35

First published online as a Review in Advance on October 13, 2008

The *Annual Review of Physiology* is online at physiol.annualreviews.org

This article's doi: 10.1146/annurev.physiol.010908.163227

Copyright © 2009 by Annual Reviews. All rights reserved

0066-4278/09/0315-0509$20.00

Key Words

synthetic function, airway remodeling, mesenchymal cells, airway hyperresponsiveness, hyperplasia, hypertrophy

Abstract

Although pivotal in regulating bronchomotor tone in asthma, airway smooth muscle (ASM) also modulates airway inflammation and undergoes hypertrophy and hyperplasia, contributing to airway remodeling in asthma. ASM myocytes secrete or express a wide array of immunomodulatory mediators in response to extracellular stimuli, and in chronic severe asthma, increases in ASM mass may render the airway irreversibly obstructed. Although the mechanisms by which ASM secretes cytokines and chemokines are the same as those regulating immune cells, there exist unique ASM signaling pathways that may provide novel therapeutic targets. This review provides an overview of our current understanding of the proliferative as well as the synthetic properties of ASM.

INTRODUCTION

Hypertrophy: a state of increased cell size

Hyperplasia: a state of increased airway smooth muscle cell number

Airway remodeling: structural changes in the airways manifested by increases in lamina reticularis, ASM, mucus glands, angiogenesis, and vasculogenesis

Asthma, which is characterized by airway hyperresponsiveness and inflammation, remains a common and potentially devastating respiratory disease. Currently, approximately 1 in 20 Americans have asthma; for children, recent estimates suggest an incidence as high as 10%. Although asthma typically induces reversible airway obstruction, in some patients airflow obstruction can become fixed. Such obstruction may be a consequence of persistent structural changes in the airway wall caused by the frequent stimulation of airway smooth muscle (ASM) by contractile agonists, inflammatory mediators, and growth factors. Importantly, the bronchoconstriction evoked by smooth muscle shortening promotes airway obstruction and constitutes the hallmark of asthma, as shown in

Figure 1. Although ASM functions as the primary effector cell that regulates bronchomotor tone, evidence now suggests that ASM may undergo hypertrophy and/or hyperplasia and modulate inflammatory responses by secreting chemokines and cytokines. This article reviews current studies focusing on noncontractile functions of ASM cells such as signaling pathways modulating ASM growth and inflammatory mediator–induced effects on ASM synthetic responses.

SIGNALING PATHWAYS REGULATING ASM GROWTH

Airway remodeling that occurs in asthma has been attributed in part to increased smooth muscle mass. Although increased ASM mass is due to ASM cell growth, the mechanisms

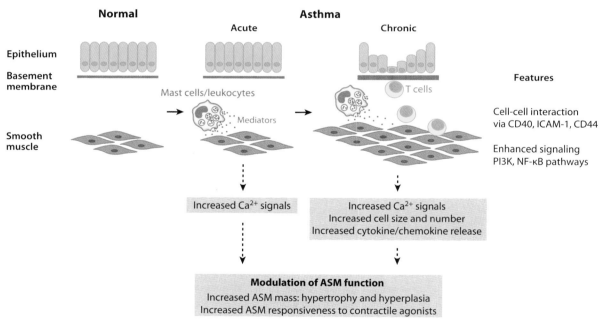

Figure 1

Factors affecting airway smooth muscle (ASM) function in acute exacerbations of asthma and in chronic disease. During acute inflammation, a variety of mediators, such as cytokines, can modulate ASM contractile function by enhancing calcium signaling to agonists. These mediators regulate the recruitment and activation of eosinophils and T lymphocytes in the airway mucosa, a characteristic histopathological feature of chronic disease. The persistence of airway inflammation via the production of cytokines and chemokines by both inflammatory as well as structural cells has the potential to directly stimulate ASM proliferation. ASM proliferation can occur as a result of T cell–ASM interaction mediated by cell surface expression of various CAM proteins such as intracellular adhesion molecule-1 (ICAM-1), CD40, and CD44. Enhanced activation of phosphatidylinositol 3-kinase (PI3K) or the transcription factor nuclear factor-κB (NF-κB) in ASM may also stimulate mitogenic and synthetic functions. From Reference 190 with permission.

that regulate ASM proliferation remain unclear. Mitogenic agents such as polypeptide growth factors, inflammatory mediators, and cytokines stimulate ASM growth. The observation that contractile agonists induce or augment ASM proliferation may be a critical link between the chronic stimulation of muscle contraction and myocyte proliferation (1). Although the specific cellular and molecular mechanisms by which agonists induce cell growth remain to be elucidated, similarities exist between signal transduction processes activated by these agents and those of known growth factors. The identification of the critical regulatory sites that mediate ASM growth may improve our understanding of the mechanisms that regulate airway inflammation and may provide new therapeutic targets to alter airway remodeling seen in patients with chronic irreversible airflow obstruction.

An increase in ASM mass constitutes a common histopathological manifestation reported in fatal asthma (2, 3). Subsequent studies reveal that the increase in ASM mass was due primarily to hyperplasia rather than hypertrophy. Studies have suggested that hypertrophy and hyperplasia of ASM occur in specific airway generations rather than globally throughout the lung (4). Although early studies characterized increases in ASM mass in tissue derived from patients who died of asthma, new evidence using quantitative morphometric techniques demonstrates that increases in ASM mass occur even in patients with mild asthma (5–7).

Given the substantial airway inflammation that occurs in asthma, there exist a plethora of mediators that stimulate the growth of human ASM in vivo. Although there has been substantial progress in examining cell signaling pathways that regulate cultured ASM cell proliferation, far fewer studies have characterized the growth factors and the cognate receptors that mediate ASM mitogenesis in vivo. Despite the paucity of evidence, platelet-derived growth factor (PDGF) and basic fibroblast growth factor (FGF), among others, are potent and effective smooth muscle mitogens that contribute to the asthma diathesis as described in **Table 1**. Although disparate inflammatory me-

Table 1 Putative smooth muscle growth factors found in human bronchoalveolar lavage fluid or in airway smooth muscle in vivo[a]

Growth factor	Ligand	Receptor	Reference(s)
Fibroblast growth factor	√	?	181
Epidermal growth factor	√	√	182
Platelet-derived growth factor (PDGF)-β	√	√	183, 184
PDGF-AA isoform	√	√	185
Thrombin	√	√	186
Sphingosine-1-phosphate	√	?	115
Insulin-like growth factor-1	√	√	187
Transforming growth factor-β	√	?	188
Tryptase	√	?	189
Lipopolysaccharide	?	?	N/A

[a]A check mark indicates that the role of the molecules is identified; a question mark indicates that the role of the molecules remains unknown.

diators may modulate cell growth, the downstream signaling pathways by which mitogens induce proliferation likely converge on several critical pathways.

Signaling Mechanism Regulating ASM Growth in Response to Growth Factors and Contractile Agonists

Smooth muscle cell proliferation is stimulated by mitogens that fall into two broad categories: (a) those that activate receptors with intrinsic receptor tyrosine kinase (RTK) activity and (b) those that mediate their effects through receptors coupled to heterotrimeric GTP-binding proteins (G proteins) and activate non-receptor-linked tyrosine kinases found in the cytoplasm, as shown in **Figure 2**. Although both pathways increase cytosolic calcium through the activation of phospholipase C (PLC), different PLC isoenzymes appear to be involved. Activated PLC hydrolyzes phosphatidylinositol bisphosphate (PIP_2) to inositol trisphosphate (IP_3) and diacylglycerol (DAG). These second messengers activate other cytosolic tyrosine kinases as well as serine and threonine kinases (protein kinases C, G, and N) that have pleiotropic effects including the activation of proto-oncogenes, which are a family of cellular genes (c-*onc*) that control normal

RTK: receptor tyrosine kinase

PLC: phospholipase C

PI3K: phosphatidylinositol 3-kinase

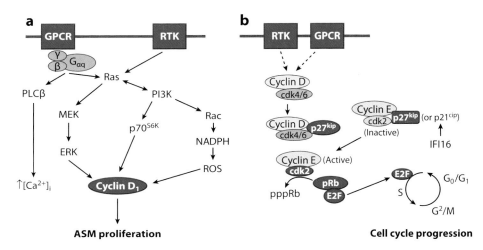

ASM proliferation

Cell cycle progression

Figure 2

Mitogenic signaling pathways in airway smooth muscle (ASM). (*a*) Mitogens initiate human ASM cell proliferation by activating pathways that involve receptor tyrosine kinase (RTK) or receptors coupled to heterotrimeric G proteins (GPCR). Upon activation of the cognate receptor, sequential downstream signaling events are mediated through Ras, phosphatidylinositol 3-kinase (PI3K), p70^{S6} kinase (p70^{S6K}), Rac, MAPK/ERK kinase (MEK), and extracellular signal–regulated kinase (ERK), which may induce cyclin D$_1$ expression. In human ASM, PI3K appears to be necessary and sufficient to modulate growth factor–induced ASM cell proliferation, whereas other pathways appear to be necessary but not sufficient. (*b*) The upstream signaling events described in panel *a* lead to a complicated interaction of molecules that promote progression through the cell cycle. Ultimately, the hyperphosphorylation of retinoblastoma protein in conjunction with elongation factor 2F (E2F) promotes the progression of the cell through a phase of the cell cycle and modulates the expression of critical proteins necessary for cell division.

cellular growth and differentiation. This section focuses on the role of PLC activation and the phosphatidylinositol 3-kinase (PI3K) signaling pathway in ASM mitogenesis.

Phospholipase C activation. Receptors with intrinsic tyrosine kinase activity and those coupled to G proteins activate specific PLC isoforms, the critical regulatory enzymes in the activation of the phosphoinositide (PI) pathway. The γ family of PLC contains Src-homology SH2 and SH3 domains and is regulated by tyrosine phosphorylation. In ASM cells, growth factors that activate receptors with intrinsic tyrosine kinases include PDGF and epidermal growth factor (EGF) in human ASM cells (8–10) and insulin-like growth factor-1 (IGF-1) in bovine and rabbit ASM cells (11–13). However, the role of PLCγ_1 activation in modulating ASM cell growth remains unknown.

G proteins are composed of three distinct subunits, α, β, and γ, the latter two existing as a tightly associated complex, as shown in **Figure 2** (14). Although the α subunit was considered the functional component important in downstream signaling events, the $\beta\gamma$ subunits also play a critical role in modulating cell function (15). Advances in single-cell microinjection techniques in combination with the development of neutralizing antibodies to specific G$_\alpha$ subunits enabled investigators to characterize the role of G protein activation in cell proliferation. These techniques reveal that both thrombin and bradykinin require G$_q$ activation to mobilize cytosolic calcium, to generate IP$_3$, and to induce mitogenesis; thrombin, but not bradykinin, induces cell growth by stimulating G$_{i2}$ (16). These studies determined that a single mitogen may require functional coupling to distinct subtypes of G proteins to stimulate cell growth and provided a mechanism to explain

SH: Src homology

why some, but not all, agonists induce cell proliferation while mobilizing comparable levels of cytosolic calcium.

Recently, researchers have explored the role of PLC activation and IP$_3$ in mediating contractile agonist–induced ASM cell growth. Several agonists, which mediate their effects through G protein–coupled receptors (GPCRs), induce ASM cell proliferation. Histamine (17) and serotonin induce canine and porcine ASM cell proliferation. In contrast, endothelin-1, leukotriene D$_4$ (LTD$_4$), and U-46619, a thromboxane A$_2$ mimetic, induce rabbit ASM cell growth (18), and thrombin, lysophosphatidic acid, and sphingosine-1-phosphate (S-1-P) induce mitogenesis in human ASM cells (9). Although the mechanisms that mediate these effects are unknown, agonist-induced mitogenesis is probably modulated by activation of G proteins in a manner similar to that described in vascular smooth muscle (VSM). In human ASM cells, thrombin, but not bradykinin, stimulated ASM cell proliferation despite a fivefold-greater increase in [^{3}H]-inositol phosphate formation in cells treated with bradykinin as compared with those treated with thrombin. Inhibition of PLC activation had no effect on thrombin- or EGF-induced myocyte proliferation. In addition, pertussis toxin completely inhibited thrombin-induced ASM cell growth but had no effect on PI turnover induced by either thrombin or bradykinin (9). Taken together, these studies suggest that thrombin induces human ASM cell growth by activation of a pathway that is sensitive to pertussis toxin and independent of PLC activation or PI turnover.

Compared with RTK-dependent growth factors, contractile agonists, with the exception of thrombin and S-1-P, appear to be less effective human ASM mitogens (19, 20). In rabbit ASM cells, endothelin-1 induces cell proliferation by activating phospholipase A$_2$, which subsequently leads to the microenvironmental availability of thromboxane A$_2$ and LTD$_4$ (18, 21). In human ASM cells, however, endothelin-1, thromboxane A$_2$, and LTD$_4$ appear to have little effect on mitogenesis even though these

agonists induce increases in cytosolic calcium (9, 22, 23). Clearly, interspecies variability exists with regard to contractile agonist–induced cell proliferation. These models, however, may prove useful in dissecting downstream signaling events that modulate the differential effects of contractile agonists on ASM cell proliferation.

Phosphatidylinositol 3-kinase signaling pathway. The PI3K signaling pathway, a highly conserved signal transduction network, regulates a variety of cellular functions including cell proliferation, differentiation, transformation, cell motility, and apoptosis (24, 25). A subfamily of lipid kinases, PI3Ks catalyze the addition of phosphate molecules specifically to the 3′ position of the inositol ring of PIs (26). Tumor suppressor PTEN (phosphatases and tensin homolog) negatively regulates PI3K signaling by specifically dephosphorylating the -OH in the 3′ position of the inositol ring of the lipid second messenger phosphatidylinositol-3,4,5-trisphosphate (27–29). The 3′-phosphoinositides function as second messengers and activate downstream effector molecules such as phosphoinositide-dependent kinase 1 (PDK1), p70 S6 kinase (p70^{S6K}), protein kinase Cζ, and serine-threonine kinase Akt (26, 30). The ability of PI3K to regulate diverse functions may be due to the existence of multiple isoforms that have specific substrate specificities and that reside in unique cytoplasmic locations within the cell (31).

PI3K isoforms are divided into three classes on the basis of their structure and substrate specificity. Class IA PI3Ks exist as cytoplasmic heterodimers composed of a 110-kDa (p110α, β, or δ) catalytic subunit and an 85-kDa (p85, p55, or p50) adaptor protein. Although catalytic subunits p110α and p110β are ubiquitously expressed in mammalian cells, catalytic subunit p110δ expression occurs predominantly in lymphocytes and lymphoid tissues and, therefore, may play a role in PI3K-mediated signaling of immune responses (32). Receptor and nonreceptor tyrosine kinases activate class IA isoforms, whereas the $\beta\gamma$ subunits of GPCRs

PDK1: phosphoinositide-dependent kinase 1

Akt: serine/threonine protein kinase B

activate class IB p110γ (33). Class II isoforms mainly associate with phospholipid membranes in the endoplasmic reticulum and Golgi apparatus (34). Class III isoforms, structurally related to yeast vesicular sorting protein Vps34p (35), use only membrane phosphatidylinositol as a substrate and generate phosphatidylinositol-3-monophosphate.

Growth factors, insulin, cytokines, and cell-cell and cell-matrix adhesion activate PI3K by both receptor and nonreceptor tyrosine kinases that stimulate Akt. Downstream of PI3K, the tumor suppressor complex tuberous sclerosis complex (TSC)1/TSC2 directly inhibits phosphorylation of Akt (36–38). TSC2 forms a complex with tumor suppressor TSC1 and regulates mTOR/S6K1 signaling by directly modulating small GTPase Rheb activity via the GTPase-activating protein (GAP) domain of TSC2 (39). Rheb binds to Raptor (40, 41) and controls the activity of the mTOR/Raptor complex (mTORC1), which, in turn, phosphorylates and activates p70^{S6K}, as shown in **Figure 2**. mTOR forms components of the rapamycin-sensitive mTOR/Raptor (TORC1), which phosphorylates S6K1, and of the rapamycin-insensitive mTOR/Rictor (TORC2), which phosphorylates Akt (42) and activates Rac1 (43). The PI3K-TSC1/TSC2-Rheb-mTOR-S6K1 signaling pathway remains highly conserved across species, modulating cell growth and proliferation (24, 25, 39).

Studies demonstrate that the PI3K signaling cascade plays a critical role in regulating human airway and pulmonary arterial VSM cell proliferation (44, 45). Inhibitors of PI3K abrogate DNA synthesis in bovine ASM and porcine and rat VSM cells stimulated with PDGF, basic FGF, angiotensin II, and serum (46–49). Stimulation of α_1 adrenergic receptors with noradrenaline promoted mitogenesis in human VSM cells in a PI3K-dependent manner (50). In rat thoracic aorta VSM cells, PI3K inhibitors completely blocked angiotensin II–induced Ras activation but had little effect on mitogen-activated protein kinase (MAPK) activation or protein synthesis (51). Thrombin, which induces human ASM cell growth by ac-

tivating a receptor presumably coupled to both G_i and G_q proteins (9), requires PI3K activation to stimulate growth (52, 53). In bovine ASM, the mitogenic effects of PDGF or endothelin-1 are PI3K- or p70^{S6K}-dependent (48). These data suggest that in both airway and VSM cells, PI3K mediates mitogenic signaling induced by numerous agents.

In numerous cell types, PI3K activates p70^{S6K} in response to serum and growth factors (54). p70^{S6K} is a critical enzyme for mitogen-induced cell cycle progression through the G_1 phase and translational control of mRNA transcripts that contain a polypyrimidine tract at their transcriptional start site (55). EGF and thrombin significantly stimulate p70^{S6K}, and wortmannin, LY294002, and rapamycin completely block this activation in ASM cells (52). Moreover, transient expression of constitutively active p110* PI3K activates p70^{S6K} in the absence of stimulation with mitogens, whereas overexpression of a dominant-negative Δp85 PI3K abolished EGF- and thrombin-induced p70^{S6K} activation in human ASM (52). Thus, EGF and thrombin induce activation of p70^{S6K} in human ASM cells, and mitogen-induced activation of p70^{S6K} appears to be PI3K-dependent.

A recent study demonstrated that GPCR activation by inflammatory and contractile agents synergizes with RTK activation to augment human ASM growth. In EGF-stimulated cells, GPCR-mediated potentiation appears independent of increased EGFR or p42/p44 MAPK activation but requires sustained activation of p70^{S6K} for several hours after the initial early phase of activation. These findings not only provide insight into mechanisms by which inflammation may contribute to ASM hyperplasia/hypertrophy in diseases such as asthma and chronic obstructive pulmonary disorder (COPD), but also suggest a mechanism by which GPCRs and RTKs interact to promote cell growth (45, 52, 56–58).

Src protein tyrosine kinase integrates GPCR- and RTK-induced signaling pathways that regulate cell proliferation, migration, differentiation, and gene transcription, as shown in **Figure 2** (59, 60). Because more than one

MAPK: mitogen-activated protein kinase

pathway activates Src, Src may integrate signals originating through GPCRs and RTKs. The cellular functions regulated by Src were identified by the use of predominantly overexpressed Src and its mutants in established immortalized cell lines and in vitro assays. Selective cell-permeable Src inhibitors PP1 and PP2 were used to demonstrate Src inhibition of DNA synthesis in guinea pig ASM (61) and in rat aortic smooth muscle cells (62). Others demonstrate that Src is necessary and sufficient for human ASM DNA synthesis. Interestingly, PP2 had differential effects on EGF- versus PDGF- and thrombin-induced DNA synthesis: IC_{50} for EGF was ~5.5 μM, a magnitude higher than the IC_{50} for PDGF and thrombin, which was ~0.3 μM for both agonists. The sensitivity of thrombin-induced DNA synthesis to Src inhibition is potentially due to a critical role of Src in modulating mitogenic signaling from activated GPCRs to its downstream targets (63, 64). The differential Src regulation of cell proliferation correlated with differential activation of Src-associated PI3K. Thus, upon stimulation of human ASM cells with PDGF or thrombin, PI3K activity was associated with Src, but not in cells stimulated with EGF. Because cell proliferation requires PI3K activation, the lack of Src-associated PI3K activation may explain the insensitivity of EGF-induced ASM DNA synthesis to blocking Src. Furthermore, EGF-dependent PI3K activation in human ASM cells also requires activation of the ErbB2 receptor of EGF receptor family (65). Collectively, these data suggest a differential role of Src in growth-promoting signaling stimulated by EGF, PDGF, and thrombin.

Modulatory Role of Cytokines

Although levels of interleukin (IL)-1β, IL-6, and tumor necrosis factor (TNF) are increased in the bronchoalveolar lavage fluid derived from asthmatics (66), whether these cytokines stimulate ASM proliferation in vitro remains controversial. IL-1β and IL-6 stimulate hyperplasia and hypertrophy of cultured guinea pig ASM cells (67); however, other investigators have shown that IL-1β (68) and IL-6 (69) are not mitogenic for human ASM cells. Studies showed that TNF (30 pM) had little mitogenic effect on human ASM cells (69). However, others reported that the proliferative effect of TNF on human ASM cells was concentration-dependent and TNF (0.3–30 pM) was promitogenic, although at higher concentrations (300 pM), the mitogenic effect was abolished. Such conflicting reports may be due to cytokine-induced cyclo-oxygenase 2–dependent prostanoid production (68). Cyclo-oxygenase products, such as prostaglandin E_2 (PGE_2), inhibit DNA synthesis (68). Therefore, cytokine-induced proliferative responses in ASM may be greater under conditions of cyclo-oxygenase inhibition, in which the expression of growth-inhibitory prostanoids, such as PGE_2, is limited (67, 68, 70).

Modulatory Role of the Extracellular Matrix

Airway remodeling, a key feature of persistent asthma, is also characterized by the deposition of extracellular matrix (ECM) proteins in the airways (71, 72). ECM proteins (collagen I, III, and V; fibronectin; tenascin; hyaluronan; versican; and laminin 2/β2) are increased in profusion in asthmatic airways (71, 73–75). Components of the ECM also modulate mitogen-induced ASM growth. Fibronectin and collagen I increase human ASM cell mitogenesis in response to PDGF-BB or thrombin, whereas laminin inhibits proliferation (76). The increase in cell proliferation was accompanied by a decrease in expression of smooth muscle cell contractile proteins such as α-actin, calponin, and myosin heavy chain, suggesting that ECM may also modulate smooth muscle phenotype. Human ASM cells also secrete ECM proteins in response to asthmatic sera (77), suggesting a cellular source for ECM deposition in airways and implicating a novel mechanism in which ASM cells may modulate autocrine proliferative responses.

Molecular Mechanisms Modulating ASM Cell Cycle

Extracellular stimuli transduce proliferative responses that move the cell through the cell cycle, which comprises distinct phases termed G_1, S (DNA synthesis), G_2, and M (mitosis), as shown in **Figure 2**. ASM growth requires activating cell cycle events similar to those described in other cell types (17, 22). As cells enter the cycle from G_0/G_{1A}, one or more D-type cyclins (D_1, D_2, and D_3) are expressed as part of the delayed early response to mitogen. Progression through G_1 phase initially depends on holoenzymes composed of one or more of the D-type cyclins (D_1, D_2, and/or D_3) in association with cyclin-dependent kinases CDK4 or CDK6, followed by activation of cyclin E–CDK2 complex as cells approach the G_1-S transition. Together, cyclin E and CDK2 act to hyperphosphorylate retinoblastoma protein (pRb), which then releases the elongation factor E2F that activates DNA polymerase. This step, termed the restriction point, represents the point of no return; cell commitment to undergo DNA synthesis and mitosis is inevitable. In ASM cells, S phase is commonly detected by the incorporation of radiolabeled thymidine from culture medium in ASM cells (17, 22) or by immunofluorescent detection of the thymidine analog 5-bromo-2′-deoxyuridine (78). At each phase of the G_1-to-S transition, CDK inhibitors (CKIs) can also constrain CDK activities. CKIs are assigned to two families on the basis of their structures and CDK targets: (*a*) The INK4 family ($p16^{INK4a}$, $p15^{INK4b}$, $p18^{INK4c}$, and $p19^{INK4d}$) specifically inhibits the catalytic subunits of CDK4 and CDK6, and (*b*) the Cip/Kip family ($p21^{Cip1}$, $p27^{Kip1}$, and $p57^{Kip2}$) inhibits the activities of cyclin D–, cyclin E–, and cyclin A–dependent kinases (79).

Evidence suggests that extracellular signal–regulated kinase (ERK) activation induces expression of cyclin D_1 in ASM cells. Hence, several studies have focused on the transcriptional regulation of ERK-induced cyclin D_1 accumulation. The promoter region of cyclin D_1 (80) contains multiple *cis* elements potentially important for transcriptional activation, including binding sites for simian virus 40 protein 1 (Sp1), activator protein-1 (AP-1), signal transducers and activators of transcription (STAT), nuclear factor-κB (NF-κB), and cAMP response element binding protein/activating transcription factor-2 (CREB/ATF-2) (81, 82). Mitogen-induced ERK activation, thymidine incorporation, and Elk-1 and AP-1 reporter activity were abrogated by MAPK/ERK kinase 1 (MEK1) inhibition. Such studies suggest a linkage between ERK activation, transcription factor activation, cyclin D_1 expression, and ASM proliferation. Similarly, MEK1 inhibition also attenuated expression of c-Fos (83), suggesting that c-Fos may be one or both of the dimer pairs in the AP-1 transcription factor complex responsible for cyclin D_1 expression in ASM cells. Whether ERK-dependent transcriptional regulation of cyclin D_1 gene expression is via direct *cis* activation with AP-1 dimers (composed of c-Fos) or via Elk-1-mediated transactivation still requires further investigation. In addition, cyclin D_1 protein, but not cyclin D_1 mRNA levels, was affected by MEK1 inhibition (84), suggesting that posttranscriptional control of cyclin D_1 protein levels may also occur independently of the MEK1/ERK signaling pathways. Interestingly, in bovine ASM, inhibition of MEK1 and ERK activity attenuates PDGF-induced DNA synthesis, suggesting that activation of MEK1 and ERKs is required for proliferation (85). In human ASM (82), mitogens, including EGF, PDGF-BB, and thrombin, activated ERK1 and ERK2 and stimulated ASM growth responses. Studies such as these suggest that the ERK pathway plays a key signaling role in mediating mitogen-induced ASM proliferation. Additionally, NF-κB has been implicated in regulating smooth muscle proliferation in response to serum, PDGF, and transforming growth factor β1 (TGFβ1) (86–89). Together, these data support the role of NF-κB and ERK in regulating ASM proliferative responses, and hence these factors may be potential therapeutic targets in the treatment of ASM hyperplasia. However, because these are ubiquitous molecules, a strategy for targeting delivery to ASM would be required.

Antiproliferative Signaling Pathways in ASM

β_2-Agonists activate the β_2-adrenergic receptor/G_s/adenylyl cyclase pathway to elevate cAMP in ASM cells. Albuterol, fenoterol, and salmeterol inhibit mitogen-induced proliferation of human ASM cells (90–92). β_2-Agonists and other cAMP-elevating agents are thought to arrest the G_1 phase of the cell cycle by posttranscriptionally inhibiting cyclin D_1 protein levels by action on a proteasome-dependent degradation pathway (93). Another mechanism was suggested by Musa et al. (94), who examined the effects of forskolin, an activator of adenylate cyclase, on DNA synthesis, cyclin D_1 gene expression, CREB phosphorylation, and DNA binding in bovine ASM. This study showed that by increasing cAMP in ASM cells, forskolin suppressed cyclin D_1 gene expression through phosphorylation and transactivation of CREB. On the basis of this observation, it seems that the effect of cAMP on cyclin D_1 gene expression is exerted by way of *cis* repression of the cyclin D_1 promoter.

In human ASM, glucocorticoids (GCs) arrest ASM cells in the G_1 phase (95), reducing thrombin-stimulated increases in cyclin D_1 protein and messenger RNA levels and attenuating pRb phosphorylation through a pathway either downstream of or parallel to the ERK cascade (95). Dexamethasone has differential inhibitory effects on RTK- and GPCR-mediated ASM cell growth, suggesting that steroid effects on ASM mitogenesis are complex. In vitro studies showed that the antiproliferative effects of the corticosteroids were impaired by contact of ASM with collagen, but not by contact of ASM on laminin (96), suggesting that, in the collagen-rich microenvironment of the inflamed and fibrotic asthmatic airway, ASM may contribute to the lack of corticosteroid responsiveness in patients with severe asthma.

SIGNALING PATHWAYS REGULATING ASM SYNTHETIC FUNCTIONS

A wide variety of cells reside in or infiltrate through the inflamed submucosa in asthma and can potentially undergo cell-cell interactions, as shown in **Figures 1** and **3**. Eosinophils, macrophages, and in particular lymphocytes may initiate or perpetuate the asthma diathesis by secreting proinflammatory mediators or by expressing cell adhesion molecules (CAMs) that may act directly or indirectly on ASM. Although many cell-cell interactions likely contribute to airway hyperresponsiveness in asthma, T cells, mast cells, and ASM can directly interact via CAMs. The capacity for ASM cells to respond to and secrete a myriad of cytokines and growth factors potentially impugns the status of ASM as immunomodulatory cells, as detailed in **Table 2**. In response to cytokines such as IL-1β, TNFα, and interferon (IFN)γ, ASM cells express a host of CAMs that promote interactions between ASM and inflammatory cells. Further advances in understanding the immunoregulatory potential of ASM revealed that cytokines also upregulate the expression of Toll-like receptors (TLRs) in ASM cells, as shown in **Figure 3**. These receptors serve as pattern-recognition molecules that modulate innate and adaptive immune and inflammatory responses to microbial infection, tissue injury, and inflammation as described in **Tables 2** and **3**. In this section, we review the recent advances describing immunomodulatory functions of ASM cells.

Adhesion Molecules

The expression and activation of a cascade of CAMs that include selectins, integrins, and CD31, as well as the local production of chemoattractants, evoke leukocyte adhesion and transmigration to lymph nodes and sites of inflammation involving nonlymphoid tissues. The mechanisms responsible for extravasation of leukocytes from the circulation during the establishment of a local inflammatory response are rapidly being delineated. However, the

GC: glucocorticoid

CAM: cell adhesion molecule

TLR: Toll-like receptor

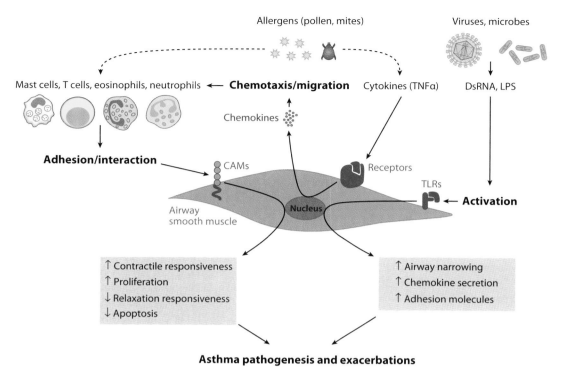

Figure 3

Environmental challenges induce asthma exacerbations that, in part, are mediated by alterations in airway smooth muscle (ASM) function. Allergens as well as viruses and bacterial infections are common stimuli for asthma exacerbations. Traditionally, these environmental challenges are thought to be mediated through airway inflammation and trafficking leukocytes. Contemporary thought suggests that structural cells, namely ASM, may modulate inflammatory responses by altering cell adhesion molecule (CAM) expression or by secreting chemokines and cytokines. The paracrine and autocrine secretion of chemokines and cytokines may then also alter the responsiveness of ASM to contractile agonists and agents that promote bronchodilation. Repeated asthma exacerbations may induce chronic alterations in ASM manifested by myocyte hypertrophy and hyperplasia. Other abbreviations used: DsRNA, double-stranded ribonucleic acid; LPS, lipopolysaccharide endotoxin; TLRs, Toll-like receptors; TNFα, tumor necrosis factor α. From Reference 191 with permission.

subsequent interactions of the infiltrating leukocytes with other cell types in the bronchial submucosa or with the ECM that may sustain the inflammatory response remain unclear. Infiltrating inflammatory cells bind to airway structural cells through specific CAMs, and such cell-cell interactions perpetuate airway inflammation (97). Early studies suggested that anti-ICAM-1 (intracellular adhesion molecule-1) antibodies inhibited allergen-induced airway hyperresponsiveness and inflammation in monkeys (98). In addition to mediating cell contact, some of the CAMs may also function as costimulatory molecules that contribute to the activation of structural cells (99).

More recently, a focus on CAMs on ASM suggests that specific CAMs are expressed in ASM both in vivo and in vitro and mediate cell-cell interactions. In situ hybridization and immunohistochemical analyses of lung tissue revealed the in vivo expression of CAMs in ASM (100, 101). Specifically, after lipopolysaccharide (LPS) stimulation of rat lungs, ASM markedly enhanced ICAM-1 expression at both the protein and mRNA levels (100). Using in vivo human bronchial tissue transplanted onto the flank of SCID mice, Lazaar and colleagues (101) demonstrated a marked increase in ICAM-1 and vascular cellular adhesion molecule-1 (VCAM-1) expression after the injection of

TNFα, a cytokine that is produced in considerable quantities in asthmatic airways (102). Further in vitro studies confirmed the expression of ICAM-1 and VCAM-1 on cultured ASM that was inducible by a wide range of inflammatory mediators such as TNFα, IL-1β, and IFNγ (101, 103). Although the function of CAMs on ASM remains incompletely defined, surface expression of CAMs on ASM may be pivotal in regulating ASM cell interactions with a variety of inflammatory cells relevant for the pathogenesis of asthma (101, 104–106). Other studies suggest that activated T cells avidly adhere to cultured ASM, an interaction that is mediated through ICAM-1, VCAM-1, and CD44 (101). The latter interaction enhances T cell binding, increases bronchoconstrictor responses to acetylcholine, and impairs relaxation responses to isoproterenol (104). More recently, investigators demonstrated that CD4$^+$ T cells interact with ASM in vivo. Adoptive transfer of CD4$^+$ T cells from sensitized rats revealed a marked increase in proliferation and an inhibition of apoptosis of airway myocytes in naive recipients upon repeated allergen challenge. The enhanced proliferation in ASM in vivo results in increases in ASM mass and airway hyperresponsiveness. Additionally, genetically modified CD4$^+$ T cells expressing enhanced green fluorescent protein (GFP) were localized by confocal microscopy to be juxtaposed to the ASM. These findings are clinically relevant and imply that CD4$^+$ T cells may directly modulate ASM function through cell-cell interactions in vivo (107). Furthermore, other inflammatory cells, including eosinophils (105) and neutrophils (106), also adhere to ASM in vitro. The attachment of such cells to ASM decreases in the presence of anti-ICAM-1 and anti-VCAM-1 antibodies. Furthermore, mast cell–ASM interactions in vivo in subjects with asthma also demonstrated that cell-cell attachment may modulate and alter stromal cell function (108). The identification of the critical regulatory sites that modulate CAM expression on airway myocytes and disrupt cell-cell adherence may provide new therapeutic approaches

Table 2 Immunomodulatory proteins expressed by human airway smooth muscle cells[a]

Cytokines	Chemokines	CAMs	Growth factors	Others
IL-1β	IL-8	ICAM-1	IGF-1	CD40
IL-5	RANTES	VCAM-1	PDGF	HLA-DR
IL-6	Eotaxin	CD44	SCF	FcγRII
IL-17	TARC	LFA-1		FcγRIII
IFNβ	Fractalkine			NO
VEGF	MCP-1, -2, -3			PGE$_2$
GM-CSF				TLRs
TGFβ				
LIF				
IP10				

[a]Abbreviations used: CAM, cell adhesion molecule; FcγR, lymphocyte Fc gamma receptor; GM-CSF, granulocyte macrophage colony–stimulating factor; ICAM-1, intracellular adhesion molecule-1; IGF-1, insulin-like growth factor-1; IFN, interferon; IL, interleukin; IP10, interferon gamma inducible 10kD protein; LFA, lymphocyte function–associated antigen; LIF, leukemia inhibitory factor; MCP, monocyte chemotactic protein; NO, nitric oxide; PDGF, platelet-derived growth factor; PGE$_2$, prostaglandin E$_2$; RANTES, regulated on activation, normal T cells expressed and secreted; SCF, stem cell factor; TARC, thymus- and activation-regulated chemokine; TGFβ, transforming growth factor β; VCAM-1, vascular cell adhesion molecule-1; VEGF, vascular endothelial growth factor; TLR, Toll-like receptor.

to alter the airway remodeling seen in patients with chronic airflow obstruction.

Cytokine and Chemokine Expression

Cytokines and chemokines play a central role in regulating inflammatory and immune responses in chronic lung diseases such as asthma and COPD. Indeed, in vivo studies using selective inhibitors as well as neutralizing antibodies against various cytokines and chemokines demonstrate their importance in antigen-induced airway inflammation (leukocyte infiltration) and hyperresponsiveness in animal models (109–111). Studies in knockout or transgenic mice also illustrate the importance of cytokines in the abnormal airway changes induced by allergen challenge in sensitized animals (112). ASM may provide a potential target for cytokines secreted by immunocytes. In human ASM cells, cytokines alter proinflammatory gene expression in an autocrine or a paracrine manner (113). Evidence

Table 3 Novel molecules regulating the immunomodulatory functions of airway smooth muscle[a]

Stimulus	Receptor	Effects
PGN, Pam$_3$CSK$_4$	TLR2	↑↑ IL-6, CXCL8, eotaxin secretion
LPS, pLPS	TLR4	↑↑ IL-6, CXCL8, eotaxin secretion
DsRNA, poly(I:C)	TLR3	↑↑ IL-6, CXCL8, CXCL10, eotaxin secretion
IL-17	IL-17R	↑↑ CXCL8 and eotaxin secretion, ↑↑ neutrophil chemotaxis
LL-37	Purinergic P2	↑↑ CXCL8 secretion
VIP	VIPR	↑↑ Mast cell chemotaxis, ↑↑ fractalkine function
Fibronectin, type I collagen	β1 integrin	↑↑ IL-1β-induced eotaxin and RANTES secretion

[a]Abbreviations used: dsRNA, double-stranded ribonucleic acid; IL, interleukin; LPS, lipopolysaccharide; PGN, peptidoglycan; RANTES, regulated on activation, normal T cells expressed and secreted; TLR, Toll-like receptor; VIP, vasoactive intestinal peptide.

convincingly demonstrates that ASM cells secrete a number of cytokines and chemoattractants as detailed in **Table 2**.

IL-6, a pleiotropic cytokine, induces smooth muscle cell hyperplasia (67) but also modulates B and T cell proliferation and immunoglobulin secretion. IL-6 secretion by ASM cells is inducible by multiple stimuli, including IL-1β, TNFα, TGFβ, and S-1-P (69, 114–117). Interestingly, transgenic expression of IL-6 in the murine lung evokes a peribronchiolar inflammatory infiltrate but promotes airway hyporesponsiveness. This intriguing dual role for IL-6 in controlling local inflammation and in regulating airway reactivity (118, 119) is consistent with the known ability of IL-6 to inhibit TNF and IL-1β secretion. ASM cells may also play a role in promoting both the recruitment and the survival of eosinophils by secretion of granulocyte macrophage colony–stimulating factor (GM-CSF) and IL-5 (120–122). Finally, additional cytokines that are secreted by human ASM cells include IL-1β, IFNβ, and other IL-6 family cytokines, such as leukemia inhibitory factor and IL-11, which are secreted following exposure of ASM cells to viral particles (116, 117, 123–125).

Autocrine IFNβ secretion regulates ASM inflammatory gene expression. In ASM cells, TNFα activates Janus kinase 1 (JAK1), tyrosine kinase2 (Tyk2), and STAT1- and STAT2-dependent gene expression via the autocrine action of IFNβ (126). Autocrine IFNβ differentially regulates TNFα-induced inflammatory gene expression by suppressing IL-6 expression and promoting RANTES (regulated on activation, normal T cells expressed and secreted) secretion. Although functional cross talk between IFNγ and TNFα occurs in other cell types (mostly hemopoietic cells), this study (126) was the first to demonstrate secretion of IFNβ by TNFα in airway structural cells. The autocrine secretion of IFNβ is a novel signaling component by which TNFα regulates ASM function in human ASM cells.

NF-κB activation modulates IFN signaling in ASM cells. IFNs interact with other inflammatory mediators such as TNFα and promote the synergistic release of inflammatory mediators from ASM cells (127). In some instances, IFNs may antagonize TNFα inflammatory responses by inhibiting the NF-κB pathway. IFNγ inhibits TNFα-induced NF-κB-dependent genes, including those encoding IL-6 and eotaxin in ASM cells (128), and IFNγ suppresses TNFα-inducible gene expression that includes vascular endothelial growth factor (129), IL-17 receptor (130), and TLR3 expression (131). Multiple mechanisms underlying the inhibitory effect of IFN on NF-κB pathways have been proposed; these include the inhibition of NF-κB DNA binding, the prevention of inhibitor of kappa B (IκB) degradation, and the regulation of TNFα receptor 1 via STAT interaction (127). The use of trichostatin A, a specific histone deacetylase inhibitor, reverses IFNγ-inhibitory effects on TNFα-inducible genes and NF-κB-dependent gene expression in ASM cells (128). These findings suggest that IFNγ negatively regulates expression of proinflammatory genes induced by TNFα. The mechanism for such negative regulation involves increased histone deacetylase activity, which results in transcriptional repression and thereby impaired NF-κB function. A better understanding of the inhibitory mechanisms exerted by IFNγ on TNFα-inducible inflammatory genes may offer new insight into the design of alternative approaches for the treatment of airway inflammation in asthma.

Chemokine expression in ASM cells. Chemokines play a central role in the recruitment and trafficking of inflammatory cells along diffusion gradients. After the initiation of injury or inflammation, chemokines provide a diffusion gradient for cell trafficking (113). Chemokines can be categorized by their molecular structure and by the degree of selectivity for distinct inflammatory cell populations (132). For example, eotaxin, RANTES, and IL-5 primarily recruit eosinophils, although eotaxin and RANTES affect other cell types; CXCL8 markedly recruits neutrophils; monocyte chemotactic proteins (MCPs) recruit monocytes; thymus- and activation-regulated chemokine (TARC) recruits lymphocytes; and stem cell factor recruits mast cells. Many of the aforementioned chemokines, which recruit and activate leukocytes, are found in bronchoalveolar lavage fluid and lung tissue of subjects with asthma. In murine models of allergen-induced airway hyperresponsiveness, neutralizing MCP-5, eotaxin, RANTES, and MCP-1 dramatically reduces airway hyperresponsiveness as well as leukocyte migration (109). Intranasal delivery of a recombinant poxvirus-derived viral CC-chemokine inhibitor protein also improves pulmonary function and decreases inflammation of the airway and lung parenchyma (133). In a chronic allergen exposure murine model, the administration of a CCR3 antagonist reduced eosinophil numbers in the airway wall tissue that was accompanied by a decrease in airway remodeling parameters (111). Together these

studies demonstrate that in vivo chemokines promote and perpetuate airway inflammation during allergen exposure.

Although many cells have the capacity to secrete chemokines, new evidence suggests that ASM may be a prominent source of chemokines in the submucosa. Immunohistochemical and in situ hybridization studies revealed that MCP-1, RANTES, and fractalkine (FKN) are expressed in ASM of bronchial biopsies in subjects with asthma (134–136). CXCL10, a potent chemokine for activated T cells, natural killer (NK) cells, and mast cells that bind to CXCR3, is also expressed in ASM in subjects with asthma or COPD (137, 138). Expression of CXCL10 in ASM cells and of CXCR3 (the CXCL10 receptor) in mast cells was seen in ASM in vivo (137). In murine models of allergen-induced airway hyperresponsiveness, eotaxin, an eosinophil-specific chemokine mediator, is markedly expressed in ASM tissue (139). The expression of chemokine receptors also exists in ASM, as demonstrated in subjects with asthma who express strong immunoreactivity for CCR3 (the eotaxin receptor) (140), a receptor that has been linked to the pathogenesis of asthma (141). To further understand the mechanisms by which chemokines are expressed, in vitro studies showed that in response to specific inflammatory mediators, cultured ASM cells also express and secrete a variety of chemokines such as eotaxin, RANTES, CXCL8, MCP-1, MCP-2, MCP-3, and TARC (142). Although the precise physiological relevance of chemokine receptor expression in ASM remains unclear, there is no doubt that the chemokine levels increase in bronchoalveolar lavage fluid in subjects with asthma, and the increased levels may be mediated in part by ASM. The recent identification of the infiltration of mast cells into ASM bundles may also suggest that mast cells, via diffusion gradient of chemokines, traffic to the submucosa (108). Activated ASM supernatant from subjects with asthma exhibits chemotactic activity for purified lung mast cells and subsequently elicits their migration toward ASM. The precise mechanisms by which this occurs remain unclear but can serve as a new therapeutic target in decreasing airway infiltration of immunocytes and inflammatory cells in asthma. Blocking CXCL10 decreased mast cell migration into the ASM bundles (108). In parallel studies, El-Shazly and colleagues (135) demonstrated that FKN also facilitated smooth muscle–induced mast cell chemotaxis.

Mediators that regulate chemokine secretion by ASM cells. IL-17, a recently described T lymphocyte–derived cytokine that is increased in the airways of patients with asthma (143), promotes neutrophil migration. In ASM cells, IL-17 induced CXCL8 release and was associated with concomitant activation of p38 MAPK, p42/p44 ERK, JNK, and NF-κB signaling pathways (144). Using promoter deletion constructs, investigators characterized the importance of AP-1 and NF-κB in mediating IL-17-induced IL-8 expression. Studies in ASM cells showed that IL-17A, either alone or with IL-1β, significantly increased levels of eotaxin-1/CCL11, an important chemoattractant for eosinophils in the lungs (145). In addition to the MAPK pathway (p38, JNK, and p42/44 ERK), STAT3 also played a role in mediating IL-17A-induced eotaxin release (145). Collectively, these findings suggest that ASM cells may amplify eosinophilic infiltration in response to IL-17A by enhancing eotaxin expression levels.

Several studies have identified molecules that stimulate chemokine secretion by ASM, as summarized in **Figure 4**. For example, the antimicrobial protein human cathelicidin antimicrobial peptide LL-37, produced mainly by mast cells and neutrophils, stimulates IL-8 secretion by ASM cells. The LL-37 effect

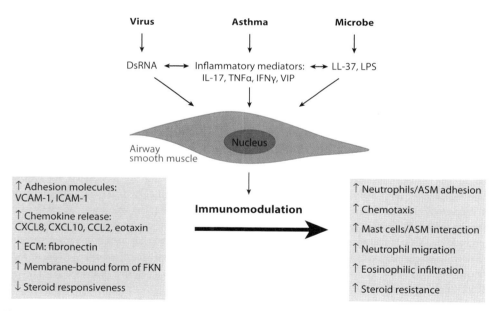

Figure 4

Potential mechanisms by which airway smooth muscle (ASM) may serve as an immunomodulatory tissue. Viruses, bacteria, and allergens may induce alterations in ASM through the secretion of soluble mediators or by the direct activation of cognate receptors on ASM. ASM secretion of chemokines, cytokines, and extracellular matrix can have a profound effect on trafficking leukocytes and steroid responsiveness. Abbreviations used: DsRNA, double-stranded ribonucleic acid; ECM, extracellular matrix; FKN, fractalkine; ICAM-1, intracellular adhesion molecule-1; IFNγ, interferon γ; IL-17, interleukin-17; LPS, lipopolysaccharide endotoxin; TNFα, tumor necrosis factor α; VCAM-1, vascular cellular adhesion molecule-1; VIP, vasoactive intestinal peptide. From Reference 192 with permission.

was dependent on activation of the ERK1/2, p38, and Src signaling pathways (146). Other studies investigated the role of ECM on ASM cells in modulating chemokine release (146, 147). Compared with cells obtained from normal volunteers, ASM cells from subjects with asthma express an increased amount of eotaxin, and an enhanced autocrine fibronectin secretion requires the engagement of $\alpha5\beta1$ integrin (146). Others showed that fibronectin and type I collagen enhanced IL-1β-dependent ASM secretion of eotaxin and RANTES release through a $\beta1$ integrin–dependent mechanism (147). These data suggest that the ECM environment surrounding the ASM cell amplifies chemokine release and further enhances cellular infiltration during inflammation and remodeling. Moreover, the neuropeptide vasoactive intestinal peptide can modulate FKN function in ASM cells (135). In several cell types, FKN, a CX3C chemokine, is expressed as a soluble or a membrane-bound moiety (148) and induces both migration and adhesion of leukocytes. Vasoactive intestinal peptide modulates the subcellular distribution of FKN, which may then favor the adhesion of ASM cells to mast cells through the FKN receptor, CX3CR1, which is highly expressed in mast cells (148). Collectively, these studies provide support for the potential role of ASM cells not only in the regulation of airway inflammation but also in airway leukocyte infiltration and retention.

Toll-Like Receptors

Mammalian TLRs are molecules for recognizing bacterial and viral components to evoke inflammatory responses, as summarized in **Figures 3** and **4**. Airway infections due to viruses exacerbate asthma and prompted investigators to study whether activation of TLRs in the airways promotes airway inflammatory responses. Accordingly, several TLR and TLR ligands have been associated with the asthma diathesis (149). Interest has focused on TLR function in ASM cells because microbial products such as LPS, a major component of the external membranes of gram-negative bacte-

ria, modulate ASM hyperresponsiveness to contractile agonists (150).

TLRs are recognized by different pathogen-associated molecular patterns and represent a family of 11 receptor subtypes. ASM cells express most TLR isoforms, as determined by total mRNA level (TLR1 to TLR10) and at the protein level. However, the pattern of TLR expression is more restricted; expression of TLR3 is far greater than that of other receptors (131). Inflammatory mediators such as TNFα, IL-1β, and IFNγ upregulate the expression of different TLRs in ASM cells (131). In addition, ASM cell stimulation with TLR2 agonists [synthetic bacterial lipopeptide (Pam$_3$CSK$_4$) or peptidoglycan (PGN)], with TLR3 agonists [polyinosinic polycytidylic acid, polyriboinosinic polyribocytidylic acid, poly(I:C), or viral double-stranded RNA (dsRNA)], or with a TLR2/4 agonist (LPS) increases the expression of several key immunomodulatory proteins that include IL-6, eotaxin, and the chemokines CXCL8 (IL-8) and CXCL10. These studies demonstrate potential physiologically relevant consequences of TLR activation in ASM cells. In vivo studies using immunohistochemical as well as RT-PCR analyses also demonstrated that TLR2, TLR3, and TLR4 are expressed in the smooth muscle layer of murine tracheal tissue (151). Treatment of murine tracheal tissues with the TLR2/4 agonist LPS or the TLR3 agonist poly(I:C) enhanced bradykinin-induced contractions. Simultaneous LPS and poly(I:C) treatment synergistically enhanced bradykinin-induced contraction (150), suggesting that, in addition to the inflammatory actions of LPS and poly(I:C), TLR expression on ASM also modulates airway hyperresponsiveness via altering ASM contractility.

The interaction of ASM cells with immune cells such as monocytes and mast cells dramatically amplifies TLR-mediated local inflammatory responses. Peripheral blood monocyte/ASM cell coculture enhanced TLR2- and TLR4-mediated IL-6, CCL2, and CXCL8 secretion (152). Monocytes may play a role in the initiation of inflammatory responses, and interaction with stromal cells may amplify such effects. Additionally, treatment of ASM cells with poly(I:C) stimulates the recruitment of mast cell lines to ASM cells (152). Oliver et al. (153) showed that rhinovirus infection enhanced IL-8 release from ASM derived from subjects with asthma, suggesting that viral infection and activation of ASM cells via the TLRs may contribute to asthma exacerbations through the regulation of mast cell activation and trafficking, as described in **Figure 4**. Together, these observations suggest that ASM cells may mediate inflammatory processes during viral and microbial infections.

Mechanisms Inhibiting ASM Synthetic Function

Mechanisms regulating synthetic function of ASM may provide novel therapeutic approaches. A variety of extracellular molecules modulate chemokine and/or cytokine expression in ASM.

Effects of intracellular cAMP–elevating agents on cytokine-induced synthetic responses. In asthma, β-agonist bronchodilators increase intracellular cAMP ([cAMP]$_i$) and stimulate cAMP-dependent protein kinase in ASM. In a similar manner, PGE$_2$, which is produced in large quantities at sites of inflammation, increases [cAMP]$_i$ in human ASM cells and is a potent and effective bronchodilator (154). [cAMP]$_i$-mobilizing agents in ASM cells also modulate cytokine-induced synthetic function (155). In TNFα-stimulated ASM cells, the expression of both eotaxin and RANTES is effectively inhibited by isoproterenol, PGE$_2$, dibutyl [cAMP]$_i$, or the phosphodiesterase inhibitors rolipram and cilomast (114, 156, 157). TNFα-induced interleukin (IL)-8 secretion is also inhibited by the combination of [cAMP]$_i$-mobilizing agents (158). Similarly, S-1-P, which activates a G$_s$ protein–coupled receptor and increases [cAMP]$_i$, abrogates TNFα-induced RANTES secretion in ASM cells (115). In contrast to the effects of [cAMP]$_i$ on chemokine secretion, pharmacologic agents that increase [cAMP]$_i$ stimulate secretion of IL-6 in hu-

man ASM cells (114) that modulates basal IL-6 promoter activity (159). More recently, researchers have shown that increases in cAMP limit secretion of GM-CSF by ASM cells, and cyclo-oxygenase inhibitors reduce PGE$_2$ and enhance cytokine-induced secretion of GM-CSF (160, 161). Accordingly, phosphodiesterase type IV inhibitors have been found in vitro to reduce GM-CSF secretion and in animal models to reduce antigen-induced airway hyperresponsiveness (161, 162). Activation of [cAMP]$_i$-dependent pathways inhibits, in part, TNFα-mediated induction of both ICAM-1 and VCAM-1 expression, as well as adhesion of activated T cells to ASM cells. The basal expression levels of ICAM-1 and VCAM-1, as well as the binding of activated T cells to unstimulated ASM, were resistant to increases in [cAMP]$_i$ (103). Thus, changes in [cAMP]$_i$ modulate cytokine-induced expression of CAMs and T cell adhesion to ASM cells. Taken together, current evidence suggests that some but not all proinflammatory functions in ASM cells are inhibited by [cAMP]$_i$-mobilizing agents. The efficacy of β-agonists in asthma may, in part, be due to inhibiting ASM secretion of proinflammatory molecules.

Effects of glucocorticoids on cytokine-induced synthetic responses. Although GCs are effective anti-inflammatory agents in asthma, the precise mechanisms by which GCs improve lung function in asthma remain unclear. Most anti-inflammatory effects of GCs are mediated via the GC receptor α isoform (GRα), which suppresses expression of inflammatory genes through transactivation or transrepression (163). Alternative splicing mechanisms induce transcription of another GC receptor isoform, GRβ (164). Cytokine-induced secretions of RANTES (114, 165, 166), monocyte chemoattractant protein (166), eotaxin (157), GM-CSF (122), and IL-6 (69) are abrogated by corticosteroids. In most studies, corticosteroid and [cAMP]$_i$-mobilizing agents act additively to inhibit chemokine and cytokine secretion, and the effects of dexamethasone on RANTES expression inhibit the transcription factor activator protein-1 (159). In contrast, dexamethasone has little effect on TNFα-induced or IL-1β-induced NF-κB activation in human ASM cells (167). Furthermore, cytokine-induced ICAM-1 expression in ASM cells, which is completely dependent on NF-κB activation, was also unaffected by dexamethasone, with IL-6 secretion being only modestly inhibited (159). In contrast, IL-1β-induced cyclo-oxygenase 2 expression was completely abrogated (167–169). Collectively, the anti-inflammatory effects of steroids in asthma may not solely modulate cytokine-induced NF-κB activation but are probably regulated by pathways such as AP-1 or other transcription factors.

Effect of cytokine combinations on ASM glucocorticoid responsiveness. The treatment of ASM cells with a combination of IFNs and TNFα impairs steroid inhibition of gene expression such as CD38, RANTES, and ICAM-1 by a mechanism involving the upregulation of the GRβ isoform (170). Although the mechanism of the synergy remains unknown, steroids augment IFNγ/TNFα-induced FKN and TLR2 expression in ASM (131, 171). Although the pathological role of the GRβ isoform is not well understood, reports demonstrate a correlation between steroid resistance in individuals with asthma and the expression levels of GRβ (172). More importantly, increased GRβ expression in the airways has been detected in patients who died of asthma (173). Because GRβ may act as a dominant-negative inhibitor of steroid action in other cell types (174), GRβ may mediate steroid insensitivity in inflammatory diseases (175). GRβ overexpression in ASM cells also prevents GC-induced transactivation and inhibits cytokine-induced proinflammatory gene expression (170).

In a GRβ-independent manner, short-term treatment of ASM cells with IFNs and TNFα partially inhibits steroid transactivation through the cellular accumulation of interferon regulatory factor-1 (IRF-1) (176). IRF-1 is an early response gene involved in diverse transcriptional regulatory processes (177), and an association exists between IRF-1 polymorphisms and childhood atopic asthma (178). Early steroid dysfunction seen after short incubation with IFNs and TNFα was recapitulated by the enhancement of IRF-1 cellular levels using constitutively active IRF-1 that dose-dependently inhibited glucocorticoid response element (GRE)-dependent gene transcription (176). Reducing IRF-1 cellular levels using siRNA approaches in TNF/IFN-treated ASM cells also significantly restored GC transactivation. These findings demonstrate that IRF-1 may serve as an alternative GRβ-independent mechanism mediating steroid insensitivity induced by cytokines. Because expression of IRF-1 is increased after viral infections (179) and because IRF-1 suppresses steroid signaling in ASM cells (176), IRF-1 may mediate the reduced steroid responsiveness seen in patients with asthma experiencing viral infections (180).

CONCLUSIONS

In summary, ASM contributes to the pathogenesis of asthma at multiple levels beyond its contractile functions. ASM, exposed to a variety of mediators and cytokines, can undergo phenotypic changes and secrete chemokines and cytokines, which may participate in or even perpetuate the mucosal inflammatory changes via the activation and recruitment of inflammatory cells. In chronic severe asthma, ASM hypertrophy and hyperplasia may render the asthmatic airway irreversibly obstructed. These new findings may provide unique therapeutic targets to decrease cell migration/infiltration and to disrupt cell-cell adherence and may ultimately reverse either airway remodeling or ongoing airway inflammation. Further elucidation of the cellular and molecular mechanisms that regulate noncontractile functions of ASM will offer new therapeutic targets in the treatment of asthma, chronic bronchitis, and emphysema.

Steroid insensitivity: incomplete response of a subject with asthma manifested by a lack of or a partial increase in pulmonary function and/or the elimination of symptoms in the presence of maximally dosed oral glucocorticoids

SUMMARY POINTS

1. Airway smooth muscle (ASM), the pivotal tissue regulating bronchomotor tone, undergoes hypertrophy and hyperplasia and secretes and expresses immunomodulatory molecules.

2. Contractile agonists and growth factors induce ASM cell proliferation, which increases ASM mass in asthma.

3. The signaling pathways by which ASM growth occurs converge on phosphatidylinositol 3-kinase and p70 S6 kinase activation.

4. Activation of protein kinase A inhibits ASM cell proliferation, whereas glucocorticoids abrogate some, but not all, growth responses.

5. ASM expresses cell adhesion molecules that promote binding of lymphocytes, eosinophils, and mast cells.

6. The consequences of cell-cell interactions between ASM and trafficking leukocytes include a decrease in ASM apoptosis and the induction of mitogenesis as well as the secretion of chemokines and cytokines.

7. Stimulation of ASM with combinations of cytokines typically seen after viral infections induces glucocorticoid insensitivity in myocytes.

8. Downstream signaling pathways modulating chemokine/cytokine secretion in ASM require, in part, NF-κB and p38 mitogen-activated protein kinase activation.

DISCLOSURE STATEMENT

The authors are not aware of any biases that might be perceived as affecting the objectivity of this review.

ACKNOWLEDGMENTS

The authors acknowledge grant support for Omar Tliba (K99 HL089409, National Heart, Lung, and Blood Institute, National Institutes of Health; RG-49324-N, American Lung Association; Parker B. Francis Fellowship Award) and for Reynold A. Panettieri, Jr. (R01 HL080676, R01 HL064063, R01 HL081824, and R01 HL077735, National Heart, Lung, and Blood Institute, National Institutes of Health; P30 ES013508, National Institute of Environmental Health Sciences).

LITERATURE CITED

1. Panettieri RA Jr, Kotlikoff MI. 1998. Cellular and molecular mechanisms regulating airway smooth muscle cell physiology and pharmacology. In *Pulmonary Diseases and Disorders*, ed. AP Fishman, JA Elias, JA Fishman, MA Grippi, LR Kaiser, RM Senior, pp. 107–17. New York: McGraw-Hill

2. Dunnill MS. 1960. The pathology of asthma, with special reference to the changes in the bronchial mucosa. *J. Clin. Pathol.* 13:27–33

3. Heard BE, Hossain S. 1973. Hyperplasia of bronchial muscle in asthma. *J. Pathol.* 110:319–31

4. Ebina M, Takahashi T, Chiba T, Motomiya M. 1993. Cellular hypertrophy and hyperplasia of airway smooth muscle underlying bronchial asthma. A 3-D morphometric study. *Am. Rev. Respir. Dis.* 148:720–26

5. Bush A. 2008. How early do airway inflammation and remodeling occur? *Allergol. Int.* 57:11–19

6. Tillie-Leblond I, de Blic J, Jaubert F, Wallaert B, Scheinmann P, Gosset P. 2008. Airway remodeling is correlated with obstruction in children with severe asthma. *Allergy* 63:533–41

7. Woodruff PG, Dolganov GM, Ferrando RE, Donnelly S, Hays SR, et al. 2004. Hyperplasia of smooth muscle in mild to moderate asthma without changes in cell size or gene expression. *Am. J. Respir. Crit. Care Med.* 169:1001–6

8. Hirst SJ, Barnes PJ, Twort CHC. 1992. Quantifying proliferation of cultured human and rabbit airway smooth muscle cells in response to serum and platelet-derived growth factor. *Am. J. Respir. Cell Mol. Biol.* 7:574–81

9. Panettieri RA Jr, Hall IP, Maki CS, Murray RK. 1995. α-Thrombin increases cytosolic calcium and induces human airway smooth muscle cell proliferation. *Am. J. Respir. Cell Mol. Biol.* 13:205–16

10. Stewart AG, Grigoriadis G, Harris T. 1994. Mitogenic actions of endothelin-1 and epidermal growth factor in cultured airway smooth muscle. *Clin. Exp. Pharmacol. Physiol.* 21:277–85

11. Cohen P, Noveral JP, Bhala A, Nunn SE, Herrick DJ, Grunstein MM. 1995. Leukotriene D$_4$ facilitates airway smooth muscle cell proliferation via modulation of the IGF axis. *Am. J. Physiol. Lung Cell. Mol. Physiol.* 269:L151–57

12. Kelleher MD, Abe MK, Chao T-SO, Jain M, Green JM, et al. 1995. Role of MAP kinase activation in bovine tracheal smooth muscle mitogenesis. *Am. J. Physiol. Lung Cell. Mol. Physiol.* 268:L894–901

13. Noveral JP, Bhala A, Hintz RL, Grunstein MM, Cohen P. 1994. The insulin-like growth factor axis in airway smooth muscle cells. *Am. J. Physiol. Lung Cell. Mol. Physiol.* 267:L761–65

14. Hepler JR, Gilman AG. 1992. G proteins. *Trends Biochem. Sci.* 17:383–87

15. Sternweis PC. 1994. The active role of βγ in signal transduction. *Curr. Opin. Cell Biol.* 6:198–203

16. Lamorte VJ, Harootunian AT, Spiegel AM, Tsien RY, Feransisco JR. 1993. Mediation of growth factor induced DNA synthesis and calcium mobilization by G$_q$ and G$_{i2}$. *J. Cell Biol.* 121:91–99

17. Panettieri RA Jr, Yadvish PA, Kelly AM, Rubinstein NA, Kotlikoff MI. 1990. Histamine stimulates proliferation of airway smooth muscle and induces c-fos expression. *Am. J. Physiol. Lung Cell. Mol. Physiol.* 259:L365–71

18. Noveral JP, Grunstein MM. 1992. Role and mechanism of thromboxane-induced proliferation of cultured airway smooth muscle cells. *Am. J. Physiol. Lung Cell. Mol. Physiol.* 263:L555–61

19. Panettieri RA Jr. 1994. Airways smooth muscle cell growth and proliferation. In *Airways Smooth Muscle: Development and Regulation of Contractility*, ed. D Raeburn, MA Giembycz, pp. 41–68. Basel, Switzerland: Birkhauser Verlag

20. Panettieri RA Jr, Cohen MD, Bilgen G. 1993. Airway smooth muscle cell proliferation is inhibited by microinjection of the catalytic subunit of cAMP dependent kinase. *Am. Rev. Respir. Dis.* 147:A252 (Abstr.)

21. Noveral JP, Rosenberg SM, Anbar RA, Pawlowski NA, Grunstein MM. 1992. Role of endothelin-1 in regulating proliferation of cultured rabbit airway smooth muscle cells. *Am. J. Physiol. Lung Cell. Mol. Physiol.* 263:L317–24

22. Panettieri RA Jr, Murray RK, DePalo LR, Yadvish PA, Kotlikoff MI. 1989. A human airway smooth muscle cell line that retains physiological responsiveness. *Am. J. Physiol. Cell Physiol.* 256:C329–35

23. Panettieri RA Jr, Tan EML, Ciocca V, Luttmann MA, Leonard TB, Hay DWP. 1998. Effects of LTD$_4$ on human airways smooth muscle cell proliferation, matrix expression, and contraction in vitro: differential sensitivity to cysteinyl leukotriene receptor antagonists. *Am. J. Respir. Cell Mol. Biol.* 19:453–61

24. Engelman JA, Ji L, Cantley LC. 2006. The evolution of phosphatidylinositol 3-kinases as regulators of growth and metabolism. *Nat. Rev. Genet.* 7:606–19

25. Krymskaya VP. 2007. Targeting phosphatidylinositol 3-kinase pathway in airway smooth muscle: rationale and promise. *BioDrugs* 21:85–95

26. Rameh LE, Cantley LC. 1999. The role of phosphoinositide 3-kinase lipid products in cell function. *J. Biol. Chem.* 274:8347–50

27. Li J, Yen C, Liaw D, Podsypanina K, Bose S, et al. 1997. PTEN, a putative protein tyrosine phosphatase gene mutated in human brain, breast, and prostate cancer. *Science* 275:1943–47

28. Maehama T, Taylor GS, Dixon JE. 2001. PTEN and myotubularin: novel phosphoinositide phosphatases. *Annu. Rev. Biochem.* 70:247–79

29. Steck PA, Pershouse MA, Jasser SA, Yung WKA, Lin HK, et al. 1997. Identification of a candidate tumour suppressor gene, *MMAC1*, at chromosome 10q23.3 that is mutated in multiple advanced cancers. *Nat. Genet.* 15:356–62

30. Toker A. 2000. Protein kinases as mediators of phosphoinositide 3-kinase signaling. *Mol. Pharmacol.* 57:652–58

31. Vanhaesebroeck B, Waterfield MD. 1999. Signaling by distinct classes of phosphoinositide 3-kinases. *Exp. Cell Res.* 253:239–54

32. Chantry D, Vojtek A, Kashishian A, Holtzman D, Wood C, et al. 1997. p110δ, a novel phosphatidyl-inositol 3-kinase catalytic subunit that associates with p85 and is expressed predominantly in leukocytes. *J. Biol. Chem.* 272:19236–41

33. Stoyanov B, Volonia S, Hanck T, Rubio I, Loubtchenkov M, et al. 1995. Cloning and characterization of a G protein-activated human phosphoinositide 3-kinase. *Science* 269:690–93

34. Domin J, Gaidarov I, Smith MEK, Keen JH, Waterfield MD. 2000. The class II phosphoinositide 3-kinase PI3K-C2α is concentrated in the trans-Golgi network and present in clathrin-coated vesicles. *J. Biol. Chem.* 275:11943–50

35. Volinia S, Dhand R, Vanhaesebroeck B, MacDougall L, Stein R, et al. 1995. A human phosphatidylinositol 3-kinase complex related to the yeast Vps34p-Vps15p protein sorting system. *EMBO J.* 14:3339–48

36. Dan HC, Sun M, Yang L, Feldman RI, Sui XM, et al. 2002. PI3K/AKT pathway regulates TSC tumor suppressor complex by phosphorylation of tuberin. *J. Biol. Chem.* 277:35364–70

37. Manning BD, Tee AR, Longdon MN, Blenis J, Cantley LC. 2002. Identification of the tuberous sclerosis complex-2 tumor suppressor gene product tuberin as a target of the phosphoinositide 3-kinase/Akt pathway. *Mol. Cell* 10:151–62

38. Potter CJ, Pedraza LG, Xu T. 2002. Akt regulates growth by directly phosphorylating Tsc2. *Nat. Cell Biol.* 4:658–65

39. Goncharova EA, Krymskaya VP. 2008. Pulmonary lymphangioleiomyomatosis (LAM): progress and current challenges. *J. Cell. Biochem.* 103:369–82

40. Long X, Ortiz-Vega S, Lin Y, Avruch J. 2005. Rheb binding to mammalian target of rapamycin (mTOR) is regulated by amino acid sufficiency. *J. Biol. Chem.* 280:23433–36

41. Long X, Lin Y, Ortiz-Vega S, Yonezawa K, Avruch J. 2005. Rheb binds and regulates the mTOR kinase. *Curr. Biol.* 15:702–13

42. Sarbassov DD, Guertin DA, Ali SM, Sabatini DM. 2005. Phosphorylation and regulation of Akt/PKB by the rictor-mTOR complex. *Science* 307:1098–101

43. Jacinto E, Loewith R, Schmidt A, Lin S, Ruegg MA, et al. 2004. Mammalian TOR complex 2 controls the actin cytoskeleton and is rapamycin insensitive. *Nat. Cell Biol.* 6:1122–28

44. Goncharova EA, Ammit AJ, Irani C, Carroll RG, Eszterhas AJ, et al. 2002. Phosphatidylinositol 3-kinase is required for proliferation and migration of human pulmonary vascular smooth muscle cells. *Am. J. Physiol. Lung Cell. Mol. Physiol.* 283:L354–63

45. Krymskaya VP, Ammit AJ, Hoffman RK, Eszterhas AJ, Panettieri RA Jr. 2001. Activation of class IA phosphatidylinositol 3-kinase stimulates DNA synthesis in human airway smooth muscle cells. *Am. J. Physiol. Lung Cell. Mol. Physiol.* 280:L1009–18

46. Bacqueville D, Casagrande F, Perret B, Chap H, Darbon JM, Breton-Douillon M. 1998. Phosphatidyl-inositol 3-kinase inhibitors block aortic smooth muscle cell proliferation in mid-late G1 phase: effect on cyclin-dependent kinase 2 and the inhibitory protein p27[KIP1]. *Biochem. Biophys. Res. Commun.* 244:630–36

47. Saward L, Zahradka P. 1997. Angiotensin II activates phosphatidylinositol 3-kinase in vascular smooth muscle cells. *Circ. Res.* 81:249–57

48. Scott PH, Belham CM, Al-Hafidh J, Chilvers ER, Peacock AJ, et al. 1996. A regulatory role for cAMP in phosphatidylinositol 3-kinase/p70 ribosomal S6 kinase-mediated DNA synthesis in platelet-derived-growth-factor-stimulated bovine airway smooth-muscle cells. *Biochem. J.* 318:965–71

49. Weiss RH, Apostolids A. 1995. Dissociation of phosphatidylinositol 3-kinase activity and mitogenic inhibition in vascular smooth muscle cells. *Cell Signal.* 7:113–22

50. Hu ZW, Shi XY, Hoffman BB. 1996. α_1 adrenergic receptors activate phosphatidylinositol 3-kinase in human vascular muscle cells. *J. Biol. Chem.* 271:8977–82

51. Takahashi T, Kawahara Y, Okuda M, Ueno H, Takeshita A, Yokoyama M. 1997. Angiotensin II stimulates mitogen-activated protein kinases and protein synthesis by a Ras-independent pathway in vascular smooth muscle cells. *J. Biol. Chem.* 272:16018–22

52. Krymskaya VP, Penn RB, Orsini MJ, Scott PH, Plevin RJ, et al. 1999. Phosphatidylinositol 3-kinase mediates mitogen-induced human airways smooth muscle cell proliferation. *Am. J. Physiol. Lung Cell. Mol. Physiol.* 277:L65–78

53. Hu Q, Klippel A, Muslin AJ, Fantl WJ, Williams LT. 1995. Ras-dependent induction of cellular response by constitutively active phosphatidylinositol-3 kinase. *Science* 268:100–2

54. Chung J, Grammer TC, Lemon KP, Kazlauskas A, Blenis J. 1994. PDGF- and insulin-dependent pp70^{S6k} activation mediated by phosphatidylinositol-3-OH kinase. *Nature* 370:71–75

55. Jefferies HBJ, Fumagalli S, Dennis PB, Reinhard C, Pearson RB, Thomas G. 1997. Rapamycin suppresses 5'TOP mRNA translation through inhibition of p70^{s6k}. *EMBO J.* 16:3693–704

56. Billington CK, Kong KC, Bhattacharyya R, Wedegaertner PB, Panettieri RA, et al. 2005. Cooperative regulation of p70^{S6} kinase by receptor tyrosine kinases and G protein-coupled receptors augments airway smooth muscle growth. *Biochemistry* 44:14595–605

57. Krymskaya VP, Orsini MJ, Eszterhas AJ, Brodbeck KC, Benovic JL, et al. 2000. Mechanisms of proliferation synergy by receptor tyrosine kinase and G protein-coupled receptor activation in human airway smooth muscle. *Am. J. Respir. Cell Mol. Biol.* 23:546–54

58. Krymskaya VP, Goncharova EA, Ammit AJ, Lim PN, Goncharov DA, et al. 2005. Src is necessary and sufficient for human airway smooth muscle cell proliferation and migration. *FASEB J.* 19:428–30

59. Brown MT, Cooper JA. 1996. Regulation, substrates and functions of src. *Biochim. Biophys. Acta* 1287:121–49

60. Thomas SM, Brugge JS. 1997. Cellular functions regulated by Src family kinases. *Annu. Rev. Cell Dev. Biol.* 13:513–609

61. Tsang F, Hwa Choo H, Dawe GS, Fred Wong WS. 2002. Inhibitors of the tyrosine kinase signaling cascade attenuated thrombin-induced guinea pig airway smooth muscle cell proliferation. *Biochem. Biophys. Res. Commun.* 293:72–78

62. Sayeski PP, Ali MS. 2003. The critical role of c-Src and the Shc/Grb2/ERK2 signaling pathway in angiotensin II-dependent VSMC proliferation. *Exp. Cell Res.* 287:339–49

63. Luttrell LM, Ferguson SSG, Daaka Y, Miller WE, Maudsley S, et al. 1999. β-Arrestin-dependent formation of β$_2$ adrenergic receptor-Src protein kinase complexes. *Science* 283:655–61

64. Marinissen MJ, Gutkind JS. 2001. G-protein-coupled receptors and signaling networks: emerging paradigms. *Trends Pharmacol. Sci.* 22:368–76

65. Krymskaya VP, Hoffman R, Eszterhas A, Kane S, Ciocca V, Panettieri RA Jr. 1999. EGF activates ErbB-2 and stimulates phosphatidylinositol 3-kinase in human airway smooth muscle cells. *Am. J. Physiol. Lung Cell. Mol. Physiol.* 276:246–55

66. Broide DH, Lotz M, Cuomo AJ, Coburn DA, Federman EC, Wasserman SI. 1992. Cytokines in symptomatic asthma airways. *J. Allergy Clin. Immunol.* 89:958–67

67. De S, Zelazny ET, Souhrada JF, Souhrada M. 1995. IL-1β and IL-6 induce hyperplasia and hypertrophy of cultured guinea pig airway smooth muscle cells. *J. Appl. Physiol.* 78:1555–63

68. Belvisi MG, Saunders M, Yacoub M, Mitchell JA. 1998. Expression of cyclo-oxygenase-2 in human airway smooth muscle is associated with profound reductions in cell growth. *Br. J. Pharmacol.* 125:1102–8

69. McKay S, Hirst SJ, Bertrand-de Haas M, de Jonste JC, Hoogsteden HC, et al. 2000. Tumor necrosis factor-α enhances mRNA expression and secretion of interleukin-6 in cultured human airway smooth muscle cells. *Am. J. Respir. Cell Mol. Biol.* 23:103–11

70. Stewart AG, Tomlinson PR, Fernandes DJ, Wilson JW, Harris T. 1995. Tumor necrosis factor α modulates mitogenic responses of human cultured airway smooth muscle. *Am. J. Respir. Cell Mol. Biol.* 12:110–19

71. Laitinen LA, Laitinen A. 1995. Inhaled corticosteroid treatment and extracellular matrix in the airways in asthma. *Int. Arch. Allergy Immunol.* 107:215–16

72. Roberts CR. 1995. Is asthma a fibrotic disease? *Chest* 107:111S–17S

73. Altraja A, Laitinen A, Virtanen I, Kampe M, Simonsson B, et al. 1996. Expression of laminins in the airways in various types of asthmatic patients: a morphometric study. *Am. J. Respir. Cell Mol. Biol.* 15:482–88

74. Bousquet J, Vignola AM, Chanez P, Campbell AM, Bonsignore G, Michel FB. 1995. Airways remodelling in asthma: no doubt, no more? *Int. Arch. Allergy Immunol.* 107:211–14

75. Roberts CR, Burke A. 1998. Remodelling of the extracellular matrix in asthma: proteoglycan synthesis and degradation. *Can. Respir. J.* 5:48–50

76. Hirst SJ, Twort CHC, Lee TH. 2000. Differential effects of extracellular matrix proteins on human airway smooth muscle cell proliferation and phenotype. *Am. J. Respir. Cell Mol. Biol.* 23:335–44

77. Johnson PRA, Black JL, Cralin S, Ge Q, Underwood PA. 2000. The production of extracellular matrix proteins by human passively sensitized airway smooth-muscle cells in culture: the effect of beclomethasone. *Am. J. Respir. Crit. Care Med.* 162:2145–51

78. Ammit AJ, Kane SA, Panettieri RA Jr. 1999. Activation of K-p21ras and N-p21ras, but not H-p21ras, is necessary for mitogen-induced human airway smooth muscle proliferation. *Am. J. Respir. Cell Mol. Biol.* 21:719–27

79. Sherr CJ, Roberts JM. 1995. Inhibitors of mammalian G1 cyclin-dependent kinases. *Genes Dev.* 9:1149–63

80. Herber B, Truss M, Beato M, Muller R. 1994. Inducible regulatory elements in the human cyclin D1 promoter. *Oncogene* 9:2105–7

81. Nagata D, Suzuki E, Nishimatsu H, Satonaka H, Goto A, et al. 2000. Transcriptional activation of the cyclin D1 gene is mediated by multiple *cis*-elements, including SP1 sites and a cAMP-responsive element in vascular endothelial cells. *J. Biol. Chem.* 276:662–69

82. Orsini MJ, Krymskaya VP, Eszterhas AJ, Benovic JL, Panettieri RA Jr, Penn RB. 1999. MAPK superfamily activation in human airway smooth muscle: Mitogenesis requires prolonged p42/p44 activation. *Am. J. Physiol. Lung Cell. Mol. Physiol.* 277:479–88

83. Lee J-H, Johnson PRA, Roth M, Hunt NH, Black JL. 2001. ERK activation and mitogenesis in human airway smooth muscle cells. *Am. J. Physiol. Lung Cell. Mol. Physiol.* 280:L1019–29

84. Ravenhall C, Guida E, Harris T, Koutsoubos V, Stewart A. 2000. The importance of ERK activity in the regulation of cyclin D1 levels and DNA synthesis in human cultured airway smooth muscle. *Br. J. Pharmacol.* 131:17–28

85. Karpova AK, Abe MK, Li J, Liu PT, Rhee JM, et al. 1997. MEK1 is required for PDGF-induced ERK activation and DNA synthesis in tracheal monocytes. *Am. J. Physiol. Lung Cell. Mol. Physiol.* 272:L558–65

86. Chiou YL, Shieh JJ, Lin CY. 2006. Blocking of Akt/NF-κB signaling by pentoxifylline inhibits platelet-derived growth factor-stimulated proliferation in Brown Norway rat airway smooth muscle cells. *Pediatr. Res.* 60:657–62

87. Ibe BO, Portugal AM, Raj JU. 2006. Levalbuterol inhibits human airway smooth muscle cell proliferation: therapeutic implications in the management of asthma. *Int. Arch. Allergy Immunol.* 139:225–36

88. Vignola AM, Merendino AM, Chiappara G, Chanez P, Pace E, et al. 1997. Markers of acute airway inflammation. *Monaldi Arch. Chest Dis.* 52:83–85

89. Xie S, Sukkar MB, Issa R, Khorasani NM, Chung KF. 2007. Mechanisms of induction of airway smooth muscle hyperplasia by transforming growth factor-β. *Am. J. Physiol. Lung Cell. Mol. Physiol.* 293:L245–53

90. Stewart AG, Tomlinson PR, Wilson JW. 1997. β$_2$-Adrenoceptor agonist-mediated inhibition of human airway smooth muscle cell proliferation: importance of the duration of β$_2$-adrenoceptor stimulation. *Br. J. Pharmacol.* 121:361–68

91. Tomlinson PR, Wilson JW, Stewart AG. 1994. Inhibition by salbutamol of the proliferation of human airway smooth muscle cells grown in culture. *Br. J. Pharmacol.* 111:641–47

92. Tomlinson PR, Wilson JW, Stewart AG. 1995. Salbutamol inhibits the proliferation of human airway smooth muscle cells grown in culture: relationship to elevated cAMP levels. *Biochem. Pharmacol.* 49:1809–19

93. Stewart AG, Harris T, Fernandes DJ, Schachte LC, Koutsoubos V, et al. 1999. β$_2$-Adrenergic receptor agonists and cAMP arrest human cultured airway smooth muscle cells in the G1 phase of the cell cycle: role of proteasome degradation of cyclin D1. *Mol. Pharmacol.* 56:1079–86

94. Musa NL, Ramakrishnan M, Li J, Kartha S, Liu P, et al. 1999. Forskolin inhibits cyclin D1 expression in cultured airway smooth-muscle cells. *Am. J. Respir. Cell Mol. Biol.* 20:352–58

95. Fernandes D, Guida E, Koutsoubos V, Harris T, Vadiveloo P, et al. 1999. Glucocorticoids inhibit proliferation, cyclin D1 expression, and retinoblastoma protein phosphorylation, but not activity of the extracellular-regulated kinases in human cultured airway smooth muscle. *Am. J. Respir. Cell Mol. Biol.* 21:77–88

96. Bonacci JV, Schuliga M, Harris T, Stewart AG. 2006. Collagen impairs glucocorticoid actions in airway smooth muscle through integrin signalling. *Br. J. Pharmacol.* 149:365–73

97. Tang ML, Fiscus LC. 2001. Important roles for L-selectin and ICAM-1 in the development of allergic airway inflammation in asthma. *Pulm. Pharmacol. Ther.* 14:203–10

98. Wegner CD, Gundel RH, Reilly P, Haynes N, Letts LG, Rothlein R. 1990. Intercellular adhesion molecule-1 (ICAM-1) in the pathogenesis of asthma. *Science* 247:456–59

99. van Seventer GA, Shimuzu Y, Shaw S. 1991. Roles of multiple accessory molecules in T-cell activation. *Curr. Opin. Immunol.* 3:294–303

100. Beck-Schimmer B, Schimmer RC, Warner RL, Schmal H, Nordblom G, et al. 1997. Expression of lung vascular and airway ICAM-1 after exposure to bacterial lipopolysaccharide. *Am. J. Respir. Cell Mol. Biol.* 17:344–52

101. Lazaar AL, Albelda SM, Pilewski JM, Brennan B, Puré E, Panettieri RA Jr. 1994. T lymphocytes adhere to airway smooth muscle cells via integrins and CD44 and induce smooth muscle cell DNA synthesis. *J. Exp. Med.* 180:807–16

102. Bradding P, Roberts JA, Britten KM, Montefort S, Djukanovic R, et al. 1994. Interleukin-4, -5, and -6 and tumor necrosis factor-α in normal and asthmatic airways: evidence for the human mast cell as a source of these cytokines. *Am. J. Respir. Cell Mol. Biol.* 10:471–80

103. Panettieri RA Jr, Lazaar AL, Puré E, Albelda SM. 1995. Activation of cAMP-dependent pathways in human airway smooth muscle cells inhibits TNF-α-induced ICAM-1 and VCAM-1 expression and T lymphocyte adhesion. *J. Immunol.* 154:2358–65

104. Hakonarson H, Kim C, Whelan R, Campbell D, Grunstein MM. 2001. Bi-directional activation between human airway smooth muscle cells and T lymphocytes: role in induction of altered airway responsiveness. *J. Immunol.* 166:293–303

105. Hughes JM, Arthur CA, Baracho S, Carlin SM, Hawker KM, et al. 2000. Human eosinophil-airway smooth muscle cell interactions. *Mediators Inflamm.* 9:93–99

106. Lee CW, Lin WN, Lin CC, Luo SF, Wang JS, et al. 2006. Transcriptional regulation of VCAM-1 expression by tumor necrosis factor-α in human tracheal smooth muscle cells: involvement of MAPKs, NF-κB, p300, and histone acetylation. *J. Cell. Physiol.* 207:174–86

107. Ramos-Barbon D, Presley JF, Hamid QA, Fixman ED, Martin JG. 2005. Antigen-specific CD4[+] T cells drive airway smooth muscle remodeling in experimental asthma. *J. Clin. Investig.* 115:1580–89

108. Brightling CE, Bradding P, Symon FA, Holgate ST, Wardlaw AJ, Pavord ID. 2002. Mast-cell infiltration of airway smooth muscle in asthma. *N. Engl. J. Med.* 346:1699–705

109. Gonzalo JA, Lloyd CM, Wen D, Albar JP, Wells TN, et al. 1998. The coordinated action of CC chemokines in the lung orchestrates allergic inflammation and airway hyperresponsiveness. *J. Exp. Med.* 188:157–67

110. Lukacs NW, Kunkel SL, Allen R, Evanoff HL, Shaklee CL, et al. 1995. Stimulus and cell-specific expression of C-X-C and C-C chemokines by pulmonary stromal cell populations. *Am. J. Physiol. Lung Cell. Mol. Physiol.* 268:L856–61

111. Wegmann M, Goggel R, Sel S, Sel S, Erb KJ, et al. 2007. Effects of a low-molecular-weight CCR-3 antagonist on chronic experimental asthma. *Am. J. Respir. Cell Mol. Biol.* 36:61–67

112. Kanehiro A, Lahn M, Makela MJ, Dakhama A, Joetham A, et al. 2002. Requirement for the p75 TNF-α receptor 2 in the regulation of airway hyperresponsiveness by γδ T cells. *J. Immunol.* 169:4190–97

113. Howarth PH, Knox AJ, Amrani Y, Tliba O, Panettieri RA Jr, Johnson M. 2004. Synthetic responses in airway smooth muscle. *J. Allergy Clin. Immunol.* 114:S32–50

114. Ammit AJ, Hoffman RK, Amrani Y, Lazaar AL, Hay DWP, et al. 2000. TNFα-induced secretion of RANTES and IL-6 from human airway smooth muscle cells: modulation by cAMP. *Am. J. Respir. Cell Mol. Biol.* 23:794–802

115. Ammit AJ, Hastie AT, Edsall LC, Hoffman RK, Amrani Y, et al. 2001. Sphingosine 1-phosphate modulates human airway smooth muscle cell functions that promote inflammation and airway remodeling in asthma. *FASEB J.* 15:1212–14

116. Elias JA, Wu Y, Zheng T, Panettieri RA Jr. 1997. Cytokine- and virus-stimulated airway smooth muscle cells produce IL-11 and other IL-6-type cytokines. *Am. J. Physiol. Lung Cell. Mol. Physiol.* 273:648–55

117. Hedges JC, Singer CA, Gerthoffer WT. 2000. Mitogen-activated protein kinases regulate cytokine gene expression in human airway myocytes. *Am. J. Respir. Cell Mol. Biol.* 23:86–94

118. DiCosmo BF, Geba GP, Picarella D, Elias JA, Rankin JA, et al. 1994. Airway epithelial cell expression of interleukin-6 in transgenic mice. Uncoupling of airway inflammation and bronchial hyperreactivity. *J. Clin. Investig.* 94:2028–35

119. Wang J, Homer RJ, Chen Q, Elias JA. 2000. Endogenous and exogenous IL-6 inhibit aeroallergen-induced Th2 inflammation. *J. Immunol.* 165:4051–61

120. Hakonarson H, Maskeri N, Carter C, Chuang S, Grunstein MM. 1999. Autocrine interaction between IL-5 and IL-1β mediates altered responsiveness of atopic asthmatic sensitized airway smooth muscle. *J. Clin. Investig.* 104:657–67

121. Hallsworth MP, Soh CPC, Twort CHC, Lee TH, Hirst SJ. 1998. Cultured human airway smooth muscle cells stimulated by interleukin-1β enhance eosinophil survival. *Am. J. Respir. Cell Mol. Biol.* 19:910–19

122. Saunders MA, Mitchell JA, Seldon PM, Yacoub MH, Barnes PJ, et al. 1997. Release of granulocyte-macrophage colony stimulating factor by human cultured airway smooth muscle cells: suppression by dexamethasone. *Br. J. Pharmacol.* 120:545–46

123. Hakonarson H, Carter C, Maskeri N, Hodinka R, Grunstein MM. 1999. Rhinovirus-mediated changes in airway smooth muscle responsiveness: induced autocrine role of interleukin-1β. *Am. J. Physiol. Lung Cell. Mol. Physiol.* 277:L13–21

124. Knight DA, Lydell CP, Zhou D, Weir TD, Schellenberg RR, Bai TR. 1999. Leukemia inhibitory factor (LIF) and LIF receptor in human lung: distribution and regulation of LIF release. *Am. J. Respir. Cell Mol. Biol.* 20:834–41

125. Rodel J, Assefa S, Prochnau D, Woytas M, Hartmann M, et al. 2001. Interferon-β induction by *Chlamydia pneumoniae* in human smooth muscle cells. *FEMS Immunol. Med. Microbiol.* 32:9–15

126. Tliba O, Tliba S, Huang CD, Hoffman RK, DeLong P, et al. 2003. TNFα modulates airway smooth muscle function via the autocrine action of IFNβ. *J. Biol. Chem.* 278:50615–23

127. Tliba O, Amrani Y. 2008. Airway smooth muscle cell as an inflammatory cell: lessons learned from interferon signaling pathways. *Proc. Am. Thorac. Soc.* 5:106–12

128. Keslacy S, Tliba O, Baidouri H, Amrani Y. 2007. Inhibition of TNFα-inducible inflammatory genes by IFNγ is associated with altered NF-κB transactivation and enhanced HDAC activity. *Mol. Pharmacol.* 71:609–18

129. Wen FQ, Liu X, Manda W, Terasaki Y, Kobayashi T, et al. 2003. TH2 cytokine-enhanced and TGF-β-enhanced vascular endothelial growth factor production by cultured human airway smooth muscle cells is attenuated by IFN-γ and corticosteroids. *J. Allergy Clin. Immunol.* 111:1307–18

130. Lajoie-Kadoch S, Joubert P, Letuve S, Halayko AJ, Martin JG, et al. 2006. TNF-α and IFN-γ inversely modulate expression of the IL-17E receptor in airway smooth muscle cells. *Am. J. Physiol. Lung Cell. Mol. Physiol.* 290:L1238–46

131. Sukkar MB, Xie S, Khorasani NM, Kon OM, Stanbridge R, et al. 2006. Toll-like receptor 2, 3, and 4 expression and function in human airway smooth muscle. *J. Allergy Clin. Immunol.* 118:641–48

132. Riffo-Vasquez Y, Spina D. 2002. Role of cytokines and chemokines in bronchial hyperresponsiveness and airway inflammation. *Pharmacol. Ther.* 94:185–211

133. Dabbagh K, Xiao Y, Smith C, Stepick-Biek P, Kim SG, et al. 2000. Local blockade of allergic airway hyperreactivity and inflammation by the poxvirus-derived pan-CC-chemokine inhibitor vCCI. *J. Immunol.* 165:3418–22

134. Berkman N, Krishnan VL, Gilbey T, Newton R, O'Connor B, et al. 1996. Expression of RANTES mRNA and protein in airways of patients with mild asthma. *Am. J. Respir. Crit. Care Med.* 154:1804–11

135. El-Shazly A, Berger P, Girodet PO, Ousova O, Fayon M, et al. 2006. Fraktalkine produced by airway smooth muscle cells contributes to mast cell recruitment in asthma. *J. Immunol.* 176:1860–68

136. Sousa AR, Lane SJ, Nakhosteen JA, Yoshimura T, Lee TH, Poston RN. 1994. Increased expression of the monocyte chemoattractant protein-1 in bronchial tissue from asthmatic subjects. *Am. J. Respir. Cell Mol. Biol.* 10:142–47

137. Brightling CE, Ammit AJ, Kaur D, Black JL, Wardlaw AJ, et al. 2005. The CXCL10/CXCR3 axis mediates human lung mast cell migration to asthmatic airway smooth muscle. *Am. J. Respir. Crit. Care Med.* 171:1103–8

138. Hardaker EL, Bacon AM, Carlson K, Roshak AK, Foley JJ, et al. 2004. Regulation of TNF-α- and IFN-γ-induced CXCL10 expression: participation of the airway smooth muscle in the pulmonary inflammatory response in chronic obstructive pulmonary disease. *FASEB J.* 18:191–93

139. Li D, Wang D, Griffiths-Johnson DA, Wells TNC, Williams TJ, et al. 1997. Eotaxin protein and gene expression in guinea-pig lungs: constitutive expression and upregulation after allergen challenge. *Eur. Respir. J.* 10:1946–54

140. Joubert P, Lajoie-Kadoch S, Labonte I, Gounni AS, Maghni K, et al. 2005. CCR3 expression and function in asthmatic airway smooth muscle cells. *J. Immunol.* 175:2702–8

141. Ying S, Robinson DS, Meng Q, Rottman J, Kennedy R, et al. 1997. Enhanced expression of eotaxin and CCR3 mRNA and protein in atopic asthma. Association with airway hyperresponsiveness and predominant colocalization of eotaxin mRNA to bronchial epithelial and endothelial cells. *Eur. J. Immunol.* 27:3507–16

142. Lazaar AL, Panettieri RA Jr. 2006. Airway smooth muscle as a regulator of immune responses and bronchomotor tone. *Clin. Chest Med.* 27:53–69

143. Molet S, Hamid Q, Davoine F, Nutku E, Taha R, et al. 2001. IL-17 is increased in asthmatic airways and induces human bronchial fibroblasts to produce cytokines. *J. Allergy Clin. Immunol.* 108:430–38

144. Wuyts WA, Vanaudenaerde BM, Dupont LJ, Van Raemdonck DE, Demedts MG, Verleden GM. 2005. Interleukin-17-induced interleukin-8 release in human airway smooth muscle cells: role for mitogen-activated kinases and nuclear factor-κB. *J. Heart Lung Transplant.* 24:875–81

145. Rahman MS, Yamasaki A, Yang J, Shan L, Halayko AJ, Gounni AS. 2006. IL-17A induces eotaxin-1/CC chemokine ligand 11 expression in human airway smooth muscle cells: role of MAPK (Erk1/2, JNK, and p38) pathways. *J. Immunol.* 177:4064–71

146. Chan V, Burgess JK, Ratoff JC, O'Connor BJ, Greenough A, et al. 2006. Extracellular matrix regulates enhanced eotaxin expression in asthmatic airway smooth muscle cells. *Am. J. Respir. Crit. Care Med.* 174:379–85

147. Peng Q, Lai D, Nguyen TT, Chan V, Matsuda T, Hirst SJ. 2005. Multiple β1 integrins mediate enhancement of human airway smooth muscle cytokine secretion by fibronectin and type I collagen. *J. Immunol.* 174:2258–64

148. Imai T, Hieshima K, Haskell C, Baba M, Nagira M, et al. 1997. Identification and molecular characterization of fractalkine receptor CX3CR1, which mediates both leukocyte migration and adhesion. *Cell* 91:521–30

149. Braun-Fahrlander C, Riedler J, Herz U, Eder W, Waser M, et al. 2002. Environmental exposure to endotoxin and its relation to asthma in school-age children. *N. Engl. J. Med.* 347:869–77

150. Luo SF, Wang CC, Chiu CT, Chien CS, Hsiao LD, et al. 2000. Lipopolysaccharide enhances bradykinin-induced signal transduction via activation of Ras/Raf/MEK/MAPK in canine tracheal smooth muscle cells. *Br. J. Pharmacol.* 130:1799–808

151. Bachar O, Adner M, Uddman R, Cardell LO. 2004. Toll-like receptor stimulation induces airway hyperresponsiveness to bradykinin, an effect mediated by JNK and NF-κB signaling pathways. *Eur. J. Immunol.* 34:1196–207

152. Morris GE, Whyte MK, Martin GF, Jose PJ, Dower SK, Sabroe I. 2005. Agonists of toll-like receptors 2 and 4 activate airway smooth muscle via mononuclear leukocytes. *Am. J. Respir. Crit. Care Med.* 171:814–22

153. Oliver BG, Johnston SL, Baraket M, Burgess JK, King NJ, et al. 2006. Increased proinflammatory responses from asthmatic human airway smooth muscle cells in response to rhinovirus infection. *Respir. Res.* 7:71

154. Hall IP, Widdop S, Townsend P, Daykin K. 1992. Control of cyclic AMP levels in primary cultures of human tracheal smooth muscle cells. *Br. J. Pharmacol.* 107:422–28

155. Lazaar AL, Panettieri RA Jr. 2001. Airway smooth muscle as an immunomodulatory cell: a new target for pharmacotherapy? *Curr. Op. Pharmacol.* 1:259–64

156. Hallsworth MP, Twort CH, Lee TH, Hirst SJ. 2001. β₂-Adrenoceptor agonists inhibit release of eosinophil-activating cytokines from human airway smooth muscle cells. *Br. J. Pharmacol.* 132:729–41

157. Pang L, Knox AJ. 2001. Regulation of TNF-α-induced eotaxin release from cultured human airway smooth muscle cells by β₂-agonists and corticosteroids. *FASEB J.* 115:261–69

158. Pang L, Knox AJ. 2000. Synergistic inhibition by β₂-agonists and corticosteroids on tumor necrosis factor-α-induced interleukin-8 release from cultured human airway smooth-muscle cells. *Am. J. Respir. Cell Mol. Biol.* 23:79–85

159. Ammit AJ, Lazaar AL, Irani C, O'Neill GM, Gordon ND, et al. 2002. Tumor necrosis factor-α-induced secretion of RANTES and interleukin-6 from human airway smooth muscle cells: modulation by glucocorticoids and β-agonists. *Am. J. Respir. Cell Mol. Biol.* 26:465–74

160. Bonazzi A, Bolla M, Buccellati C, Hernandez A, Zarini S, et al. 2000. Effect of endogenous and exogenous prostaglandin E₂ on interleukin-1β-induced cyclooxygenase-2 expression in human airway smooth-muscle cells. *Am. J. Respir. Crit. Care Med.* 162:2272–77

161. Lazzeri N, Belvisi MG, Patel HJ, Yacoub MH, Fan Chung K, Mitchell JA. 2001. Effects of prostaglandin E₂ and cAMP elevating drugs on GM-CSF release by cultured human airway smooth muscle cells. Relevance to asthma therapy. *Am. J. Respir. Cell Mol. Biol.* 24:44–48

162. Kanehiro A, Ikemura T, Makela MJ, Lahn M, Joetham A, et al. 2001. Inhibition of phosphodiesterase 4 attenuates airway hyperresponsiveness and airway inflammation in a model of secondary allergen challenge. *Am. J. Respir. Crit. Care Med.* 163:173–84

163. Leung DY, Bloom JW. 2003. Update on glucocorticoid action and resistance. *J. Allergy Clin. Immunol.* 111:3–22

164. Hollenberg SM, Weinberger C, Ong ES, Cerelli G, Oro A, et al. 1985. Primary structure and expression of a functional human glucocorticoid receptor cDNA. *Nature* 318:635–41

165. John M, Hirst SJ, Jose PJ, Robichaud A, Berkman N, et al. 1997. Human airway smooth muscle cells express and release RANTES in response to T helper 1 cytokines. Regulation by T helper 2 cytokines and corticosteroids. *J. Immunol.* 158:1841–47

166. Pype JL, Dupont LJ, Menten P, Van Coillie E, Opdenakker G, et al. 1999. Expression of monocyte chemotactic protein (MCP)-1, MCP-2, and MCP-3 by human airway smooth-muscle cells. Modulation by corticosteroids and T-helper 2 cytokines. *Am. J. Respir. Cell Mol. Biol.* 21:528–36

167. Amrani Y, Lazaar AL, Panettieri RA Jr. 1999. Up-regulation of ICAM-1 by cytokines in human tracheal smooth muscle cells involves an NF-κB-dependent signaling pathway that is only partially sensitive to dexamethasone. *J. Immunol.* 163:2128–34

168. Belvisi MG, Saunders MA, Haddad E-B, Hirst SJ, Yacoub MH, et al. 1997. Induction of cyclo-oxygenase-2 by cytokines in human cultured airway smooth muscle cells: novel inflammatory role of this cell type. *Br. J. Pharmacol.* 120:910–16

169. Pang L, Knox AJ. 1997. Effect of interleukin-1β, tumour necrosis factor-α and interferon-γ on the induction of cyclo-oxygenase-2 in cultured human airway smooth muscle cells. *Br. J. Pharmacol.* 121:579–87

170. Tliba O, Cidlowski JA, Amrani Y. 2006. CD38 expression is insensitive to steroid action in cells treated with tumor necrosis factor-α and interferon-γ by a mechanism involving the up-regulation of the glucocorticoid receptor β isoform. *Mol. Pharmacol.* 69:588–96

171. Sukkar MB, Issa R, Xie S, Oltmanns U, Newton R, Chung KF. 2004. Fractalkine/CX3CL1 production by human airway smooth muscle cells: induction by IFN-γ and TNF-α and regulation by TGF-β and corticosteroids. *Am. J. Physiol. Lung Cell. Mol. Physiol.* 287:L1230–40

172. Leung DY, Hamid Q, Vottero A, Szefler SJ, Surs W, et al. 1997. Association of glucocorticoid insensitivity with increased expression of glucocorticoid receptor β. *J. Exp. Med.* 186:1567–74

173. Christodoulopoulos P, Leung DY, Elliott MW, Hogg JC, Muro S, et al. 2000. Increased number of glucocorticoid receptor-β-expressing cells in the airways in fatal asthma. *J. Allergy Clin. Immunol.* 106:479–84

174. Oakley RH, Sar M, Cidlowski JA. 1996. The human glucocorticoid receptor β isoform. Expression, biochemical properties, and putative function. *J. Biol. Chem.* 271:9550–59

175. Pujols L, Mullol J, Torrego A, Picado C. 2004. Glucocorticoid receptors in human airways. *Allergy* 59:1042–52

176. Tliba O, Damera G, Banerjee A, Gu S, Baidouri H, et al. 2008. Cytokines induce an early steroid resistance in airway smooth muscle cells: novel role of interferon regulatory factor-1. *Am. J. Respir. Cell Mol. Biol.* 38:463–72

177. Kroger A, Koster M, Schroeder K, Hauser H, Mueller PP. 2002. Activities of IRF-1. *J. Interferon Cytokine Res.* 22:5–14

178. Nakao F, Ihara K, Kusuhara K, Sasaki Y, Kinukawa N, et al. 2001. Association of IFN-γ and IFN regulatory factor 1 polymorphisms with childhood atopic asthma. *J. Allergy Clin. Immunol.* 107:499–504

179. Mamane Y, Heylbroeck C, Genin P, Algarte M, Servant MJ, et al. 1999. Interferon regulatory factors: the next generation. *Gene* 237:1–14

180. Yamada K, Elliott WM, Hayashi S, Brattsand R, Roberts C, et al. 2000. Latent adenoviral infection modifies the steroid response in allergic lung inflammation. *J. Allergy Clin. Immunol.* 106:844–51

181. Redington AE, Roche WR, Madden J, Frew AJ, Djukanovic R, et al. 2001. Basic fibroblast growth factor in asthma: measurement in bronchoalveolar lavage fluid basally and following allergen challenge. *J. Allergy Clin. Immunol.* 107:384–87

182. Fedorov IA, Wilson SJ, Davies DE, Holgate ST. 2005. Epithelial stress and structural remodelling in childhood asthma. *Thorax* 60:389–94

183. Aubert JD, Hayashi S, Hards J, Bai TR, Pare PD, Hogg JC. 1994. Platelet-derived growth factor and its receptor in lungs from patients with asthma and chronic airflow obstruction. *Am. J. Physiol. Lung Cell. Mol. Physiol.* 266:L655–63

184. Taylor IK, Sorooshian M, Wangoo A, Haynes AR, Kotecha S, et al. 1994. Platelet-derived growth factor-β mRNA in human alveolar macrophages in vivo in asthma. *Eur. Respir. J.* 7:1966–72

185. Leung TF, Wong GW, Ko FW, Li CY, Yung E, et al. 2005. Analysis of growth factors and inflammatory cytokines in exhaled breath condensate from asthmatic children. *Int. Arch. Allergy Immunol.* 137:66–72

186. Terada M, Kelly EA, Jarjour NN. 2004. Increased thrombin activity after allergen challenge: a potential link to airway remodeling? *Am. J. Respir. Crit. Care Med.* 169:373–77

187. Chetty A, Andersson S, Lassus P, Nielsen HC. 2004. Insulin-like growth factor-1 (IGF-1) and IGF-1 receptor (IGF-1R) expression in human lung in RDS and BPD. *Pediatr. Pulmonol. Suppl.* 37:128–36

188. Pascual RM, Peters SP. 2005. Airway remodeling contributes to the progressive loss of lung function in asthma: an overview. *J. Allergy Clin. Immunol.* 116:477–86

189. von Ungern-Sternberg BS, Sly PD, Loh RK, Isidoro A, Habre W. 2006. Value of eosinophil cationic protein and tryptase levels in bronchoalveolar lavage fluid for predicting lung function impairment in anaesthetised, asthmatic children. *Anaesthesia* 61:1149–54

190. Ammit AJ, Panettieri RA Jr. 2003. Airway smooth muscle cell hyperplasia: a therapeutic target in airway remodeling in asthma? *Prog. Cell Cycle Res.* 5:49–57

191. Tliba O, Amrani Y, Panettieri RA Jr. 2008. Is airway smooth muscle the "missing link" modulating airway inflammation in asthma? *Chest* 133:236–42

192. Tliba O, Panettieri RA Jr. 2008. Regulation of inflammation by airway smooth muscle. *Curr. Allergy Asthma Rep.* 8:262–68

Cumulative Indexes

Contributing Authors, Volumes 67–71

G

Gainetdinov RR, 69:511–34
Gan W-B, 71:261–82
Garty H, 68:429–57
Geibel JP, 71:205–17
Gekle M, 67:573–94
Gendler SJ, 70:431–57
Gioeli D, 69:171–99
Goswami T, 69:87–112
Greene GL, 67:309–33

H

Hall RA, 68:489–503
Haller T, 67:595–621
Hamid Q, 71:489–507
Hamm LL, 69:317–40
Hammes SR, 69:171–99
Hao C-M, 70:357–77
Harrell JC, 67:285–308
Hartzell C, 67:719–58
Hathaway HJ, 70:165–90
Hattrup CL, 70:431–57
Hebert SC, 71:205–17
Heifets BD, 71:283–306
Helms MN, 71:403–23
Helmuth B, 67:177–201
Heneghan AF, 69:201–20
Hermann GE, 68:277–303
Hewitt SC, 67:285–308
Hock MB, 71:177–203
Hoffman JF, 70:1–22
Holtzman MJ, 71:425–49
Houck LD, 71:161–76
Hsieh PCH, 68:51–66
Hung CCY, 68:221–49
Hupfeld CJ, 69:561–77

I

Ishiguro H, 67:377–409

J

Jain L, 71:403–23
Jentsch TJ, 67:779–807

K

Kahle KT, 70:329–55
Kahn CR, 68:121–56

Kahn-Kirby AH, 68:717–34
Kalaany NY, 68:157–89
Kamm RD, 68:505–39
Karlish SJD, 68:429–57
Kashikar ND, 70:93–117
Kaupp UB, 70:93–117
Kelliher KR, 71:141–60
Kingsolver JG, 67:177–201
Kissler S, 67:147–73
Knowles MR, 69:423–50
Kobayashi T, 67:39–67
Koh SD, 68:305–41
Korach KS, 67:285–308
Koval M, 71:403–23
Kralli A, 71:177–203
Kültz D, 67:225–57
Kumar R, 69:341–59

L

Landowski CP, 70:257–71
Lange CA, 69:171–99
Lauder GV, 70:143–63
Lee RT, 68:51–66
Lee TH, 71:465–87
Lee W-K, 70:119–42
Lefkowitz RJ, 69:483–510
Lehnart SE, 67:69–98
Leinders-Zufall T, 71:115–40
Leinwand LA, 71:1–18
Lerch MM, 69:249–69
Lesser MP, 68:251–76
Lichtman AH, 68:67–95
Lifton RP, 70:329–55
Lisowski LK, 68:51–66
Liu L, 70:51–71
Liu-Chen L-Y, 68:489–503
Loffing J, 68:459–88
Loukoianov A, 70:405–29
Luczak ED, 71:1–18

M

Madden PGA, 70:143–63
Makrides V, 67:557–72
Mangelsdorf DJ, 68:157–89
Marder E, 69:291–316
Marker PC, 69:171–99
Markovich D, 69:361–75
Marks AR, 67:69–98
Mäser P, 71:59–82
McDonough AA, 71:381–401

McGuckin MA, 70:459–86
McNally EM, 71:37–57
Melvin JE, 67:445–69
Michell AR, 70:379–403
Milano SK, 69:451–82
Minke B, 68:647–82
Mitchell RN, 68:67–95
Miura MT, 69:201–20
Moore CAC, 69:451–82
Mulugeta S, 67:663–96
Munger SD, 71:115–40
Münzberg H, 70:537–56
Murphy E, 69:51–67
Murphy KG, 70:239–55
Murthy KS, 68:343–72
Myers MG, 70:537–56

N

Nencioni A, 67:147–73
Nettles KW, 67:309–33
Neuhofer W, 67:531–55
Nielsen S, 70:301–27
Nilius B, 68:683–715
North RA, 71:333–59

O

Olefsky JM, 69:561–77
Omran H, 69:423–50
Oprea TI, 70:165–90
O'Rourke B, 69:19–49
Owsianik G, 68:683–715

P

Panettieri RA Jr, 71:509–35
Parnas M, 68:647–82
Patel AC, 71:425–49
Petersen OH, 70:273–99
Phillips PA, 69:249–69
Poët M, 67:779–807
Popa SM, 70:213–38
Praetorius HA, 67:515–29
Praetorius J, 70:301–27
Premont RT, 69:511–34
Prossnitz ER, 70:165–90
Putzier I, 67:719–58

R

Ramadan T, 67:557–72
Ramirez J-M, 69:113–43

Chapter Titles, Volumes 67–71

Endocrinology

Gastrointestinal Physiology